de Gruyter Studies in Mathematics 29

Editors: Carlos Kenig · Andrew Ranicki · Michael Röckner

de Gruyter Studies in Mathematics

1 Riemannian Geometry, 2nd rev. ed., *Wilhelm P. A. Klingenberg*
2 Semimartingales, *Michel Métivier*
3 Holomorphic Functions of Several Variables, *Ludger Kaup and Burchard Kaup*
4 Spaces of Measures, *Corneliu Constantinescu*
5 Knots, *Gerhard Burde and Heiner Zieschang*
6 Ergodic Theorems, *Ulrich Krengel*
7 Mathematical Theory of Statistics, *Helmut Strasser*
8 Transformation Groups, *Tammo tom Dieck*
9 Gibbs Measures and Phase Transitions, *Hans-Otto Georgii*
10 Analyticity in Infinite Dimensional Spaces, *Michel Hervé*
11 Elementary Geometry in Hyperbolic Space, *Werner Fenchel*
12 Transcendental Numbers, *Andrei B. Shidlovskii*
13 Ordinary Differential Equations, *Herbert Amann*
14 Dirichlet Forms and Analysis on Wiener Space, *Nicolas Bouleau and Francis Hirsch*
15 Nevanlinna Theory and Complex Differential Equations, *Ilpo Laine*
16 Rational Iteration, *Norbert Steinmetz*
17 Korovkin-type Approximation Theory and its Applications, *Francesco Altomare and Michele Campiti*
18 Quantum Invariants of Knots and 3-Manifolds, *Vladimir G. Turaev*
19 Dirichlet Forms and Symmetric Markov Processes, *Masatoshi Fukushima, Yoichi Oshima and Masayoshi Takeda*
20 Harmonic Analysis of Probability Measures on Hypergroups, *Walter R. Bloom and Herbert Heyer*
21 Potential Theory on Infinite-Dimensional Abelian Groups, *Alexander Bendikov*
22 Methods of Noncommutative Analysis, *Vladimir E. Nazaikinskii, Victor E. Shatalov and Boris Yu. Sternin*
23 Probability Theory, *Heinz Bauer*
24 Variational Methods for Potential Operator Equations, *Jan Chabrowski*
25 The Structure of Compact Groups, *Karl H. Hofmann and Sidney A. Morris*
26 Measure and Integration Theory, *Heinz Bauer*
27 Stochastic Finance, *Hans Föllmer and Alexander Schied*
28 Painlevé Differential Equations in the Complex Plane, *Valerii I. Gromak, Ilpo Laine and Shun Shimomura*

Werner Fenchel · Jakob Nielsen

Discontinuous Groups of Isometries in the Hyperbolic Plane

Edited by
Asmus L. Schmidt

Walter de Gruyter
Berlin · New York 2003

Editor
Asmus L. Schmidt
University of Copenhagen
Institute for Mathematical Sciences
Universitetsparken 5
2100 Copenhagen
Denmark

Mathematics Subject Classification 2000: Primary: 51-02; secondary: 30F35, 30F60, 51M10, 57M60, 11Fxx

Keywords: Discontinuous group, discrete group, hyperbolic geometry, non-euclidean geometry, Riemann surface, low dimensional topology

Photo credit: p. xvii, p. xviii – courtesy of The Royal Danish Academy of Sciences and Letters (© The Royal Danish Academy of Sciences and Letters)

♾ Printed on acid-free paper which falls within the guidelines of the ANSI to ensure permanence and durability.

Library of Congress – Cataloging-in-Publication Data

Fenchel, W. (Werner), 1905–
Discontinuous groups of isometries in the hyperbolic plane / Werner Fenchel, Jakob Nielsen ; edited by Asmus L. Schmidt.
p. cm. – (De Gruyter studies in mathematics ; 29)
Includes bibliographical references and index.
ISBN 3 11 017526 6 (cloth : alk. paper)
1. Discontinuous groups. 2. Isometries (Mathematics). I. Nielsen, Jakob, 1890–1959. II. Schmidt, Asmus L. III. Title. IV. Series.
QA612.14 .F46 2002
514′.2–dc21 2002068997

ISBN 3-11-017526-6

Bibliographic information published by Die Deutsche Bibliothek

Die Deutsche Bibliothek lists this publication in the Deutsche Nationalbibliografie; detailed bibliographic data is available in the Internet at <http://dnb.ddb.de>.

Printed in Germany.
Cover design: Rudolf Hübler, Berlin.
Typeset using the authors' TEX files: I. Zimmermann, Freiburg.
Printing and binding: Hubert & Co. GmbH & Co. KG, Göttingen.

Preface

This book by Jakob Nielsen (1890–1959) and Werner Fenchel (1905–1988) has had a long and complicated history. In 1938–39, Nielsen gave a series of lectures on discontinuous groups of motions in the non-euclidean plane, and this led him – during World War II – to write the first two chapters of the book (in German). When Fenchel, who had to escape from Denmark to Sweden because of the German occupation, returned in 1945, Nielsen initiated a collaboration with him on what became known as the Fenchel–Nielsen manuscript. At that time they were both at the Technical University in Copenhagen. The first draft of the Fenchel–Nielsen manuscript (now in English) was finished in 1948 and it was planned to be published in the Princeton Mathematical Series. However, due to the rapid development of the subject, they felt that substantial changes had to be made before publication.

When Nielsen moved to Copenhagen University in 1951 (where he stayed until 1955), he was much involved with the international organization UNESCO, and the further writing of the manuscript was left to Fenchel. The archives of Fenchel now deposited and catalogued at the Department of Mathematics at Copenhagen University contain two original manuscripts: a partial manuscript (manuscript 0) in German containing Chapters I–II (§§1–15), and a complete manuscript (manuscript 1) in English containing Chapters I–V (§§1–27). The archives also contain part of a correspondence (first in German but later in Danish) between Nielsen and Fenchel, where Nielsen makes detailed comments to Fenchel's writings of Chapters III–V. Fenchel, who succeeded N. E. Nørlund at Copenhagen University in 1956 (and stayed there until 1974), was very much involved with a thorough revision of the curriculum in algebra and geometry, and concentrated his research in the theory of convexity, heading the International Colloquium on Convexity in Copenhagen 1965. For almost 20 years he also put much effort into his job as editor of the newly started journal Mathematica Scandinavica. Much to his dissatisfaction, this activity left him little time to finish the Fenchel–Nielsen project the way he wanted to.

After his retirement from the university, Fenchel – assisted by Christian Siebeneicher from Bielefeld and Mrs. Obershelp who typed the manuscript – found time to finish the book *Elementary Geometry in Hyperbolic Space*, which was published by Walter de Gruyter in 1989 shortly after his death. Simultaneously, and with the same collaborators, he supervised a typewritten version of the manuscript (manuscript 2) on discontinuous groups, removing many of the obscure points that were in the original manuscript. Fenchel told me that he contemplated removing parts of the introductory Chapter I in the manuscript, since this would be covered by the book mentioned above; but to make the Fenchel–Nielsen book self-contained he ultimately chose not to do so. He did decide to leave out §27, entitled *The fundamental group*.

As editor, I started in 1990, with the consent of the legal heirs of Fenchel and Nielsen, to produce a TEX-version from the newly typewritten version (manuscript 2). I am grateful to Dita Andersen and Lise Fuldby-Olsen in my department for having done a wonderful job of typing this manuscript in AMS-TEX. I have also had much help from my colleague Jørn Børling Olsson (himself a student of Käte Fenchel at Aarhus University) with the proof reading of the TEX-manuscript (manuscript 3) against manuscript 2 as well as with a general discussion of the adaptation to the style of TEX. In most respects we decided to follow Fenchel's intentions. However, turning the typewritten edition of the manuscript into TEX helped us to ensure that the notation, and the spelling of certain key-words, would be uniform throughout the book. Also, we have indicated the beginning and end of a proof in the usual style of TEX.

With this TEX-manuscript I approached Walter de Gruyter in Berlin in 1992, and to my great relief and satisfaction they agreed to publish the manuscript in their series Studies in Mathematics. I am most grateful for this positive and quick reaction. One particular problem with the publication turned out to be the reproduction of the many figures which are an integral part of the presentation. Christian Siebeneicher had at first agreed to deliver these in final electronic form, but by 1997 it became clear that he would not be able to find the time to do so. However, the publisher offered a solution whereby I should deliver precise drawings of the figures (Fenchel did not leave such for Chapters IV and V), and then they would organize the production of the figures in electronic form. I am very grateful to Marcin Adamski, Warsaw, Poland, for his fine collaboration concerning the actual production of the figures.

My colleague Bent Fuglede, who has personally known both authors, has kindly written a short biography of the two of them and their mathematical achievements, and which also places the Fenchel–Nielsen manuscript in its proper perspective. In this connection I would like to thank The Royal Danish Academy of Sciences and Letters for allowing us to include in this book reproductions of photographs of the two authors which are in the possession of the Academy.

Since the manuscript uses a number of special symbols, a list of notation with short explanations and reference to the actual definition in the book has been included. Also, a comprehensive index has been added. In both cases, all references are to sections, not pages.

We considered adding a complete list of references, but decided against it due to the overwhelming number of research papers in this area. Instead, a much shorter list of monographs and other comprehensive accounts relevant to the subject has been collected.

My final and most sincere thanks go to Dr. Manfred Karbe from Walter de Gruyter for his dedication and perseverance in bringing this publication into existence.

Copenhagen, October 2002 *Asmus L. Schmidt*

Contents

Life and work of the Authors

Jakob Nielsen[1] was born on October 15, 1890 in the village Mjels in Northern Schleswig (then under Germany), where his father owned a small farm. After attending the village school Jakob was taken to Rendsburg in 1900, where he went to the Realgymnasium. In 1908 he entered the University of Kiel and attended lectures in physics, chemistry, geology, biology, and literature. Only after some terms did mathematics take a prominent place. When Max Dehn was attached to the university at the end of 1911 he introduced Jakob Nielsen to topology and group theory at the level of current research. Their contact developed into a lifelong friendship.

In the summer of 1913 Nielsen graduated from the university with the doctor's thesis "Kurvennetze auf Flächen", which already points towards his later achievements. But shortly afterwards he was called up for service in the German navy, attached to the coast defence artillery. The war had broken out, and he was sent first to Belgium and then in April 1915 to Constantinople as one of the German officers functioning as advisers to the Turkish government on the defence of the Bosporus and the Dardanelles. He found time to write two short papers, published in 1917 and 1918, in continuation of his thesis and dealing with finitely generated free groups.

Back in Germany after the war had ended, Nielsen spent the summer term of 1919 in Göttingen, where he met Erich Hecke and later accompanied him to Hamburg as his assistent and "Privatdozent"; they too became close friends. From that period we have two papers of Nielsen both dealing with the fixed point problem for surface mappings.

Already in 1920 Jakob Nielsen was appointed professor at the Institute of Technology in Breslau, where he resumed contact with Dehn. In lectures here Nielsen formulated clearly the central problem he had set himself to solve: to determine and investigate the group of homotopy classes of homeomorphisms of a given surface. One link of this investigation, namely the proof that every automorphism of the fundamental group of a closed surface is induced by a homeomorphism, had been communicated to him by Dehn, who never published it. It is characteristic of Nielsen that whenever he needed this theorem, or merely some idea resembling its proof, he would stress his debt to Dehn.

The stay in Breslau became a brief one, for later in 1920 North Schleswig was reunited with Denmark after a referendum, and Jakob Nielsen opted for Denmark. He moved to Copenhagen the year after together with his wife Carola (née von Pieverling), and here he became a lecturer at the Royal Veterinary and Agricultural College. Quickly he became a treasured member of the Danish mathematical community. He met frequently with Harald Bohr and Tommy Bonnesen, and they followed each other's work with keen interest.

[1] What is written above about Jakob Nielsen and his work is largely an extract of Werner Fenchel's comprehensive memorial address at a meeting in the Danish Mathematical Society on 7 December 1959, printed in Acta Mathematica **103** (1960), vii–xix.

In a purely group theoretic paper by Nielsen, from 1921, a major result is that every subgroup of a finitely generated free group is likewise free. His proof is based on an ingenious method of reduction of systems of generators. In 1927 the theorem was extended by Otto Schreier to arbitrary free groups, and under the name of the Nielsen–Schreier theorem it contributes now one of the bases of the theory of infinite groups. Two other papers, from 1924, continue earlier investigations of the group of automorphisms of a given group.

Along with these and other investigations Jakob Nielsen took up the study of discontinuous groups of isometries of the non-euclidean plane and devoted several papers (1923, 1925, 1927) to this subject. His interest in it arose from the fact that the fundamental group of a surface of genus greater than 1 admits representations by such discontinuous groups.

These apparently somewhat desultory investigations turned out to be stones that went to the erection of an impressive building. Hints of this are to be found in some lectures given by Nielsen in Hamburg in 1924 and in Copenhagen in 1925, at the 6th Scandinavian Congress of Mathematicians. But in its final form it appeared in three long memoirs (300 pages in all) from the years 1927, 1929, and 1932 in Acta Mathematica under the common title "Untersuchungen zur Topologie der geschlossenen zweiseitigen Flächen". Here we find again the notions and methods he had previously used or developed: the universal covering surface interpreted as the non-euclidean plane, the latter represented by the conformal model in the interior of the unit circle; the fundamental group as a discontinuous group of isometries of the non-euclidean plane; the mappings of the latter onto itself which lie over a given surface mapping, and the automorphisms induced by them. As an essential new tool comes here the following theorem: Every mapping of the non-euclidean plane onto itself which lies over some surface mapping can be extended continuously to the points of the unit circle, representing the points at infinity of the non-euclidean plane, and the mapping of the circumference which arises in this way depends only on the homotopy class of the surface mapping. A two-dimensional problem is hereby reduced to a one-dimensional one. – With these memoirs Jakob Nielsen had broken new ground, and they gave him great international reputation.

In 1925 Nielsen became professor of theoretical mechanics at the Technical University in Copenhagen. Here he worked out his textbook on that subject in two volumes, published in 1933–34, in which he exploited more recent mathematical tools. A third volume about aerodynamics was added later. Nielsen's lectures demanded much of his students; he had an unusual power of expressing himself with great lucidity, but also with great terseness.

It is not possible here to mention the many papers, about 20, among them several comprehensive ones, which Jakob Nielsen published in the years after 1935, most of them carrying on his investigations on surface mappings. By means of the powerful tools developed in the previous papers, he succeeded in solving a series of related problems. In 1937 he gave a complete classification of the periodic mappings of a surface onto itself, and in 1942 a fourth great memoir, "Abbildungsklassen endlicher

Jakob Nielsen

Werner Fenchel

Ordnung", was published in Acta Mathematica. It deals with a problem to which he had been led in the third of the above mentioned Acta papers, and which he had solved there in some special cases: Does every homotopy class of surface mappings which is of finite order, in the sense that a certain power of it is the class of the identity mapping, contain a periodic mapping, that is, a mapping the same power of which is the identity? The proof that this is the case is extremely difficult and makes up all the 90 pages long paper. One cannot but admire the intellectual vigour with which this investigation is carried out. Finally I shall mention one more large paper: "Surface transformation classes of algebraically finite type" from 1944, in which more general classes of surface mappings are thoroughly investigated.

On several occasions Jakob Nielsen lectured at the Mathematical Institute of Copenhagen University to a small circle of young mathematicians on subjects that occupied him in connection with his research. Of special interest is a series of lectures on discontinuous groups of isometries of the non-euclidean plane, given in the year 1938–39; here he took up the theory for a certain class of these groups for its own sake. He realized the need for investigating the theory of discontinuous groups of motions in the non-euclidean plane in its full generality and from the bottom, in view of its many important fields of applications. Gradually it became clear, however, that this task, which Jakob Nielsen took up together with Werner Fenchel, was considerably more extensive than anticipated.

Although his heart was at this project, Jakob Nielsen could only devote to it a moderate part of his great working power, for since the end of the 1939–1945 war he was deeply engaged in international cooperation, especially the work of UNESCO, where he was a highly esteemed member of the Executive Board from 1952 to 1958.

In 1951 Jakob Nielsen was nominated Harald Bohr's successor at the University of Copenhagen. Here he lectured with delight and zeal to young mathematicians on subjects close to his heart. But the growing demands made upon him by his UNESCO work, with frequent journeys abroad, which interrupted his lectures, caused him in 1955 to resign his professorship; and after finishing his UNESCO work he could devote himself wholeheartedly to the work on the monograph with Fenchel. Jakob Nielsen succeeded in surmounting a difficulty which had long prevented a satisfactory conclusion. But already in January 1959 he was stricken with the disease which carried him off on the 3rd of August.

Werner Fenchel was born on the 3rd of May 1905 in Berlin, son of a representative. Already in highschool his deep interest in physics led him into mathematical studies far beyond the school curriculum. Aged eighteen he entered the University of Berlin, where he attended lectures by Einstein among others. With the growing demands of mathematical knowledge needed to understand the theory of relativity, Fenchel eventually concentrated foremost on mathematics. Towards the end of his study he succeeded in proving that the total curvature of a closed curve in space is at least 2π. He presented his result in the mathematics colloquium, and afterwards Erhard Schmidt decided right away that this would be suitable for a doctoral thesis.

Soon after graduating from the university in 1928 Fenchel was lucky to become assistent of Edmund Landau in Göttingen. At this leading centre of mathematics, counting Hilbert among its professors, Werner Fenchel met Harald Bohr, who was guest lecturing, and also briefly Jakob Nielsen for the first time. A Rockefeller stipend allowed Fenchel to spend a semester in Rome, studying differential geometry with Levi-Cività, and also to visit Bohr in Copenhagen in the spring of 1931. Here he also met Bonnesen, with whom he wrote in the following years the Ergebnisse tract "Theorie der konvexen Körper", published in 1934. Reprinted in 1976, it has become a classic in convexity theory.

In 1933 Werner Fenchel, like so many others, had to leave Germany. Invited by Harald Bohr he went to Copenhagen with his wife Käte (née Sperling), a group theorist. Here he continued assisting Otto Neugebauer in editing the Zentralblatt für Mathematik. He also translated and adapted Jakob Nielsen's textbook on theoretical mechanics to German. Inspired by Bohr's theory of almost periodic functions Fenchel wrote with him a paper on stable almost period motions (1936); and in a paper with Jessen (1935) he showed that every almost periodic motion on certain types of surfaces can be deformed continuously and almost periodically into a periodic motion. A paper by Fenchel from 1937 deals with motions in a euclidean space which are almost periodic modulo isometries. Retrospectively, these investigations of almost periodic motions may be seen as forerunners to the theory of dynamical systems.

The cooperation with Bonnesen led Fenchel to new contributions to the theory of convex bodies as developed by Brunn and Minkowski. He succeeded in solving a long standing problem about extension of Minkowski's inequalities for mixed volumes (1936). The Brunn-Minkowski theory had been developed in two extreme cases, the convex body being either smoothly bounded or a polytope. In a memoir from 1938 Fenchel and Jessen succeeded, independently of A. D. Alexandrov, in extending the theory to general convex bodies.

The German occupation of Denmark during the 1939–45 war forced in 1943 Werner and Käte to leave their new home country. Helped by Marcel Riesz they found refuge in Lund, together with their little son Tom. After the end of the war they returned to Denmark, where Fenchel in 1947 had his first tenure position, at the Technical University in Copenhagen, and here he succeeded in 1951 Jakob Nielsen as professor of theoretical mechanics.

Werner Fenchel visited the United States with his family in 1950–51, staying at U.S.C. in Los Angeles with his close friend Herbert Busemann, next at Stanford with Polýa and Szegö, and finally in Princeton at the Institute for Advanced Study and Princeton University. In a short paper from 1949 Fenchel had sketched ideas which were to lead to a far-reaching development in convexity theory. He associated with each convex function on a euclidean space a conjugate function, likewise convex, and established the basic properties of this concept of duality. This theory entered naturally in a series of lectures he gave at Princeton University, and mimeographed notes were written. These certainly ought to have been properly published, but copies soon began to circulate widely and had a great impact on research in this field.

Back in Denmark, Werner Fenchel seems to have put the duality theory aside, his publications from the 1950's dealing with other aspects of convexity and with geometrical and topological topics. In the light of the development in the theory of topological vector spaces it was, however, clear to Fenchel that it was desirable to extend his theory of conjugate convex functions to these very general spaces, and thereby widen its applicability. Thus encouraged, one of his students, Arne Brøndsted, carried out that project in a comprehensive paper published in 1964.

In a pioneering monograph "Convex Analysis" from 1970, R. T. Rockafellar applied the duality theory to create a theory of convex optimization based on the ideas of Kuhn and Tucker. This aspect of mathematical optimization has become an integral part of theoretical economics. Earlier, the author had spent a year in Copenhagen with Fenchel. In the preface to his book Rockafellar emphasizes the great impact Fenchel's lecture notes from Princeton had on his own perception of convexity theory, and he writes: "It is highly fitting, therefore, that this book be dedicated to Fenchel as honorary co-author".

In 1956 Fenchel had succeeded N.E. Nørlund as professor at the University of Copenhagen. He was an inspiring lecturer, with a delightful ability of vizualizing his subject. The newly started Journal Mathematica Scandinavica had Fenchel as a very dedicated editor during nearly twenty years. Likewise for many years he was secretary of the Danish Mathematical Society, and from 1958 to 1962 its chairman. In 1965 he organized a big international colloquium on convexity theory in Copenhagen.

After the war Werner Fenchel had joined Jakob Nielsen in pursuing the study of discontinuous groups of isometries of the hyperbolic plane. This led to a joint paper in 1948, and in the same year Fenchel published two more articles on that topic. As described in the above outline of Nielsen's work, their project of developing the theory from its basis with the aim of giving a comprehensive presentation of it turned out to be much bigger than foreseen. Provisional sketches of their work had circulated in a few copies among researchers in the field and excited keen interest.

After Jakob Nielsen's death in 1959 Werner Fenchel continued the project alone – no less so after his retirement from the university in 1974. In periods he was assisted by younger colleagues: Asmus Schmidt, Nils Andersen, Troels Jørgensen (then in Copenhagen), and Christian Siebeneicher in Bielefeld. And late in his life Fenchel succeeded in completing the body of the manuscript.

While working on the Nielsen project, Werner Fenchel had realized the need for a comprehensive exposition of the underlying hyperbolic geometry, also in higher dimensions and based on the conformal model. And shortly before his death on 24 January 1988 he had completed the manuscript to the monograph "Elementary Geometry in Hyperbolic Space", which was published the year after in the de Gruyter Studies.

Chapter I
Möbius transformations and non-euclidean geometry

§1 Pencils of circles – inversive geometry

1.1 Notations. The following considerations are based on the plane of all complex numbers, this plane being closed as usual by a point at infinity, in other words on Riemann's sphere of complex numbers. In general, the points of the plane as well as the corresponding complex numbers are denoted by small Latin letters, real numbers by Greek letters. The straight lines of the plane are considered as circles passing through the point at infinity; even single points will occasionally be included among the circles and are then spoken of as *zero-circles*. The circles of the plane in this general sense – as well as other subsets of the plane – will be denoted by calligraphic capitals. In this paragraph some definitions and theorems concerning *pencils* of circles are enumerated for subsequent use.

1.2 Three kinds of pencils. An *elliptic pencil*[1] consists of all circles passing through two different points u and v, the common points of the pencil. Each point of the plane

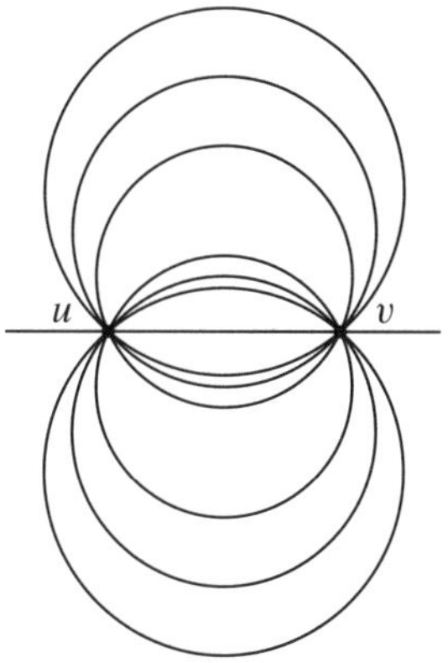

Figure 1.1

other than u and v lies on exactly one circle of the pencil. If one of the common points,

[1] Editor's note: the names elliptic and hyperbolic pencil have been switched as compared with the first edition of the Fenchel–Nielsen manuscript. It is now in accordance with common usage, cf. [15], [31], [55]. Earlier the expression coaxal circles were in use, cf. [25], [45]

in the sequel usually u, is termed the negative and the other the positive, the pencil is said to be *directed.* It is often appropriate to think of a directed elliptic pencil as made up of circular arcs joining u and v and directed from u towards v.

A *parabolic pencil* consists of all circles touching each other in some definite point u, the common point or zero-circle of the pencil. A direction of the circles in u is called the direction of the pencil. Each point of the plane other than u lies on exactly one circle of the pencil.

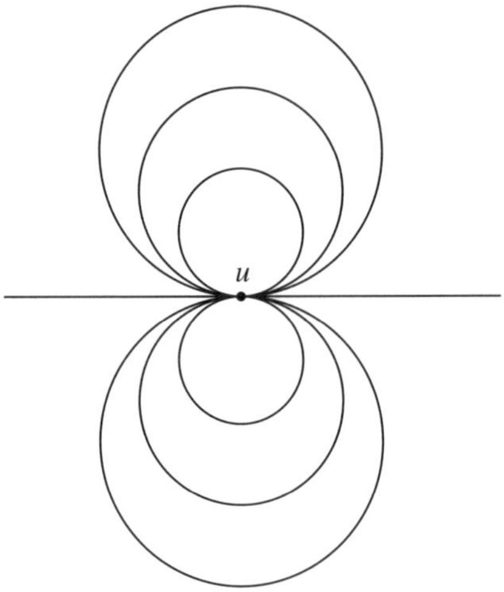

Figure 1.2

A *hyperbolic pencil* consists of all circles which are orthogonal to all circles of an elliptic pencil. The common points u and v of the elliptic pencil are included in the hyperbolic pencil as zero-circles. If none of these is at infinity, the hyperbolic pencil is made up of all apollonian circles for the points u and v, i.e. each circle of the pencil is the locus of all points whose distances from u and v are in a fixed ratio. If one of the zero-circles is at infinity, the pencil consists of all circles with the other zero-circle as their common centre. The two zero-circles are separated by every other circle of the pencil. Each point of the plane lies on exactly one circle of the pencil. If one of the zero-circles is termed the negative and the other the positive, the pencil is said to be directed. In that case the circles of the pencil are directed in accordance with the usual orientation of the complex plane when seen from the positive zero-circle.

1.3 Determination of pencils. The hyperbolic and elliptic pencil based on two different points u and v are called *conjugate*. The conjugate of a parabolic pencil is a parabolic pencil with the same common point and with a direction at right angles to the direction of the first pencil. Two different circles determine exactly one pencil of which they are members. This pencil is elliptic, parabolic, or hyperbolic according as

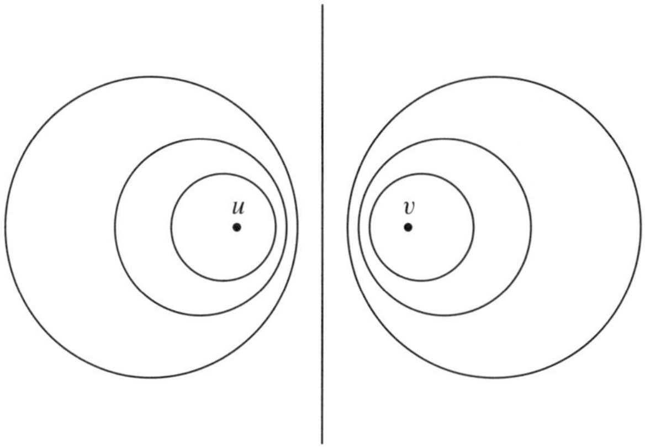

Figure 1.3

the two given circles intersect, touch, or have no point in common. (In this connection a zero-circle touches another circle if the point lies on the circle.)

Two different circles determine exactly one pencil of circles to which they are orthogonal. This pencil is hyperbolic, parabolic, or elliptic, according as the two given circles intersect, touch, or have no point in common. This pencil is conjugate to the pencil of which the two given circles are members. In this connection a zero-circle is orthogonal to another circle, if the point lies on the circle.

If three circles do not belong to one and the same pencil, the necessary and sufficient condition for the existence of exactly one circle orthogonal to all three is the following: If at least two of the circles intersect, the pair of intersection points of the first circle with the second are not separated on the first circle by the pair of its intersection points (if any) with the third. (In this connection a zero-circle has to be considered as orthogonal to itself.)

1.4 Inverse points. Two points are called *inverse with respect to a circle* $\mathcal{K}$, which is not a zero-circle, if they are the zero-circles of a hyperbolic pencil to which $\mathcal{K}$ belongs; in other words if they are the common points of an elliptic pencil orthogonal to $\mathcal{K}$. To every point x not on $\mathcal{K}$ there exists exactly one inverse, the second zero-circle of the hyperbolic pencil determined by the zero-circle x and the circle $\mathcal{K}$. Two points inverse with respect to $\mathcal{K}$ are separated by $\mathcal{K}$. The inverse of a point on $\mathcal{K}$ is, by definition, the point itself. The mapping which assigns to a point of the plane its inverse with respect to $\mathcal{K}$ is called the *inversion with respect to* $\mathcal{K}$.

§2 Cross-ratio

2.1 Definition and identities. The *cross-ratio* of two pairs of points x_1, y_1 and x_2, y_2 (thus of four points x_1, y_1, x_2, y_2 given in this order) is denoted by $(x_1 y_1 x_2 y_2)$ and defined as the complex number

$$(x_1 y_1 x_2 y_2) = \frac{x_2 - x_1}{x_2 - y_1} : \frac{y_2 - x_1}{y_2 - y_1} = \frac{(x_2 - x_1)(y_2 - y_1)}{(x_2 - y_1)(y_2 - x_1)}.$$

This definition has a meaning if no three among the four points coincide. The cross-ratio assumes the special values 0, ∞ and 1 in the following cases respectively, and in these cases only: If the two first or the two second points of the pairs coincide; if the first point of one pair coincides with the second of the other; if the two points of one pair coincide. Given any three different points x_1, y_1, x_2 there exists exactly one point y_2 such that the cross-ratio assumes a prescribed value. The following relations hold:

$$(x_2 y_2 x_1 y_1) = (y_1 x_1 y_2 x_2) = (x_1 y_1 x_2 y_2) \tag{1}$$

$$(y_1 x_1 x_2 y_2) = (x_1 y_1 y_2 x_2) = \frac{1}{(x_1 y_1 x_2 y_2)} \tag{2}$$

$$(x_1 x_2 y_1 y_2) = (y_2 y_1 x_2 x_1) = 1 - (x_1 y_1 x_2 y_2) \tag{3}$$

$$(x_1 y_1 x_2 z)(x_1 y_1 z\, y_2) = (x_1 y_1 x_2 y_2). \tag{4}$$

2.2 Amplitude and modulus. First, let the four points x_1, y_1, x_2, y_2 be different and none of them at infinity. Let x_1 and y_1 be joined by two circular arcs passing through x_2 and y_2 respectively, and let the half-tangents $y_1 s$ and $y_1 t$ of these circular arcs at y_1 by drawn (see Fig. 2.1). Counting the sign of angles in accordance with the

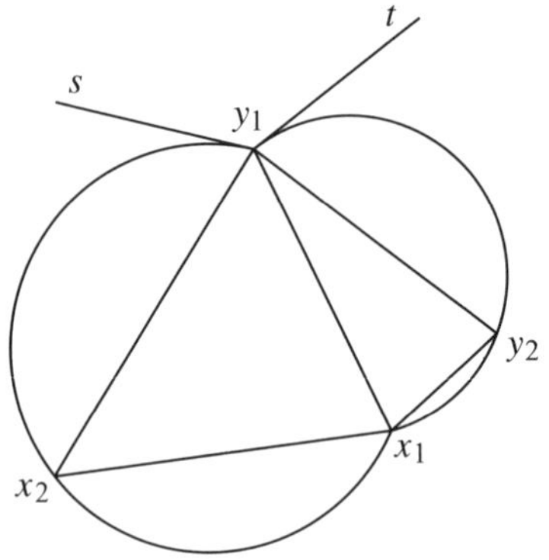

Figure 2.1

orientation of the complex plane, Fig. 2.1 illustrates the following relation:

$$\begin{aligned} sy_1t &= sy_1x_2 + x_2y_1y_2 + y_2y_1t \\ &= x_2y_1y_2 + y_1x_1x_2 + y_2x_1y_1 = x_2y_1y_2 + y_2x_1x_2 \\ &= \operatorname{amp}\frac{y_2 - y_1}{x_2 - y_1} + \operatorname{amp}\frac{x_2 - x_1}{y_2 - x_1} = \operatorname{amp}(x_1y_1x_2y_2). \end{aligned}$$

Hence the amplitude of the cross-ratio of two pairs of points equals the angle between the two circular arcs joining the points of one pair and passing each through one point of the other pair. In particular, the condition for the cross-ratio being real is that all four points are on one circle, the cross-ratio being negative or positive according as the pairs x_1, y_1 and x_2, y_2 separate or do not separate each other on that circle.

Moreover, $|x_2 - x_1|/|x_2 - y_1|$ and $|y_2 - x_1|/|y_2 - y_1|$ equals the ratio of distances of the points x_2 and y_2 respectively from the points x_1 and y_1. The first ratio remains unaltered if x_2 is displaced on the apollonian circle for x_1 and y_1 passing through x_2, thus on a circle of the hyperbolic pencil determined by x_1 and y_1 as zero-circles; and equally for y_2. In particular, x_2 and y_2 may be replaced by the intersection points x_2' and y_2' of these two circles with any circular arc joining x_1 and y_1. Hence

$$|(x_1y_1x_2y_2)| = (x_1y_1x_2'y_2')$$

this cross-ratio being positive. The condition for

$$|(x_1y_1x_2y_2)| = 1$$

is that x_2 and y_2 are on the same apollonian circle for x_1 and y_1.

The necessary and sufficient conditions for these special cases may be so formulated: The cross-ratio for two pairs of points is real (in particular: positive) if the points of one pair are situated on one circle (in particular: circular arc) of the elliptic pencil which is determined by the other pair as common points; in this case the two pairs are called *concyclical.* The cross-ratio for two pairs of points has modulus 1, if the points of one pair are situated on one circle of the hyperbolic pencil which is determined by the other pair as zero-circles.

It is easily seen that this holds even if the point at infinity or coincidences of points are admitted, with the restriction that coincident points of one pair cannot, of course, play the rôle of common points or zero-circles of the above pencils.

2.3 Harmonic pairs. Two pairs of points are called *harmonic*, if

$$(x_1y_1x_2y_2) = -1. \tag{5}$$

If x_1 and y_1 are chosen as common points of an elliptic and as zero-circles of a hyperbolic pencil the necessary and sufficient condition for the validity of (5) is that x_2 and y_2 are the intersection points of a circle of one pencil with a circle of the other.

Now, conjugate pencils are orthogonal. Hence, if $\mathcal{K}$ is the circle passing through two harmonic, and thus concyclical pairs x_1, y_1 and x_2, y_2 and $\mathcal{K}_1$ and $\mathcal{K}_2$ are circles orthogonal to $\mathcal{K}$ and passing through x_1 and y_1 and through x_2 and y_2 respectively, then $\mathcal{K}_1$ and $\mathcal{K}_2$ are mutually orthogonal. x_1 and y_1 are inverse with respect to $\mathcal{K}_2$, and so are x_2 and y_2 with respect to $\mathcal{K}_1$. Conversely, if three circles are mutually orthogonal, each of them cuts the two others in harmonic pairs.

If the points x and x' are inverse with respect to the circle $\mathcal{K}$, every circle through x and x' will cut $\mathcal{K}$ in a pair of points which is harmonic with the pair x, x'.

If none of the four points is at infinity, equation (5) may by written

$$(x_2 - x_1)(y_2 - y_1) + (x_2 - y_1)(y_2 - x_1) = 0$$

or

$$2(x_1 y_1 + x_2 y_2) - (x_1 + y_1)(x_2 + y_2) = 0.$$

This equation obviously holds in the case when three of the four points coincide, in which case no cross-ratio is defined. In the sequel it is appropriate to include this case in the term *harmonic pairs*.

§3 Möbius transformations, direct and reversed

3.1 Invariance of the cross-ratio. The set of linear fractional transformations

$$x \mapsto x' = \frac{ax + b}{cx + d}, \qquad ad - bc \neq 0 \tag{1}$$

with complex coefficients constitute a group of bijective mappings of the closed complex plane onto itself. Multiplication of the matrix $A = \left(\begin{smallmatrix} a & b \\ c & d \end{smallmatrix}\right)$ of coefficients of the transformation (1) by a factor $\neq 0$ does not affect the transformation, and so by a suitable choice of such a factor the determinant $ad - bc$ can be given the value 1.

It is easily seen that all angles are preserved under the mapping by the *Möbius transformation* (1): If $x = x(\tau)$ is a parametric representation of some differentiable curve, amp $\frac{dx}{d\tau}$ is equal to the angle from the real axis to the tangent of the curve directed in the sense of the increase of τ. For the image of the curve the corresponding angle is

$$\operatorname{amp} \frac{dx'}{dt} = \operatorname{amp} \frac{d}{dt} \frac{ax + b}{cx + d} = \operatorname{amp} \frac{dx}{dt} + \operatorname{amp} \frac{ad - bc}{(cx + d)^2}. \tag{2}$$

Thus amp $\frac{dx}{dt}$ is increased by an amount which depends on the point considered but which is independent of the direction of the curve at that point. Hence the angle subtended at the intersection point of two curves remains unaltered by the mapping both in magnitude and in sign. This remains valid for the point at infinity if angles are measured at the point $x = 0$ after performing the transformation $x \mapsto x' = \frac{1}{x}$.

Let x_1, y_1, x_2, y_2 be any four points, no three of which coincide, and x_1', y_1', x_2', y_2' their images under the Möbius transformation (1). If none of the four points is $-\frac{d}{c}$ or ∞, then

$$\frac{x_2' - x_1'}{x_2' - y_1'} = \frac{\frac{ax_2+b}{cx_2+d} - \frac{ax_1+b}{cx_1+d}}{\frac{ax_2+b}{cx_2+d} - \frac{ay_1+b}{cy_1+d}} = \frac{x_2 - x_1}{x_2 - y_1} \cdot \frac{cy_1 + d}{cx_1 + d},$$

and likewise, since the second factor of the right-hand member does not depend on x_2,

$$\frac{y_2' - x_1'}{y_2' - y_1'} = \frac{y_2 - x_1}{y_2 - y_1} \cdot \frac{cy_1 + d}{cx_1 + d}.$$

Hence

$$(x_1' y_1' x_2' y_2') = (x_1 y_1 x_2 y_2),$$

showing the invariance of the cross-ratio under the transformation (1). Continuity then shows this even holds in the special cases excluded above.

Since the reality of the cross-ratio characterizes the concyclical disposition of four points, any Möbius transformation maps circles onto circles. Combining this property with the property of isogonality, it follows that the circles of a pencil are mapped onto the circles of a pencil of the same kind. In particular, two points inverse with respect to some circle are mapped onto two points which are inverse with respect to the image of that circle.

3.2 Determination by three points. Let x_1, x_2, x_3 and x_1', x_2', x_3' be any two triples each made up of three different points. Then there is exactly one Möbius transformation (1) carrying x_1 into x_1', x_2 into x_2' and x_3 into x_3': Denoting by x' the image-point of an arbitrary point x, the invariance of the cross-ratio yields the equation

$$(x_1' x_2' x_3' x') = (x_1 x_2 x_3 x)$$

from which x' is calculated as a linear fractional function of x with the required property; the determinant of this transformation is

$$(x_1 - x_2)(x_1 - x_3)(x_2 - x_3)(x_1' - x_2')(x_1' - x_3')(x_2' - x_3')$$

and thus does not vanish in virtue of the conditions stated. Consequently, a Möbius transformation leaving three points fixed is the identical transformation.

3.3 Reversed transformations. Transformations like

$$x \mapsto x' = \frac{a\bar{x} + b}{c\bar{x} + d}, \qquad ad - bc \neq 0, \tag{3}$$

$\bar{x}$ denoting the conjugate of x, produce bijective mappings of the closed complex plane onto itself reversing orientation. They are the composition of an inversion with respect to the real axis and a Möbius transformation and may be called *reversed Möbius transformations*. Angles are left unaltered in magnitude but are reversed in sign. Cross-ratios are replaced by their conjugate values. Circles are mapped onto circles and pencils of circles onto pencils of the same kind. Since the product of two reversed Möbius transformations is a direct Möbius transformation, the set of all Möbius transformations, direct and reversed, constitute a group.

For any two prescribed triples x_1, x_2, x_3 and x_1', x_2', x_3' each consisting of three different points there exists exactly one reversed Möbius transformation carrying x_1 into x_1', x_2 into x_2' and x_3 into x_3', this transformation being calculated from the equation

$$(x_1'x_2'x_3'x') = (\overline{x}_1\overline{x}_2\overline{x}_3\overline{x}). \tag{4}$$

3.4 Inversions. If the transformation (4) leaves fixed the points x_1, x_2, x_3, i.e. if $x_1 = x_1'$, $x_2 = x_2'$, $x_3 = x_3'$, each point of the circle $\mathfrak{C}$ passing through these three points remains fixed; for if x is a point on this circle the cross-ratios in (4) are real and hence

$$(x_1x_2x_3x') = (x_1x_2x_3x).$$

This equation implies $x' = x$. In consequence of the properties of reversed transformations described above, every circle orthogonal to $\mathfrak{C}$ is mapped onto itself. The pair of common points u and v of the pencil must then be mapped onto itself. Now, u and v cannot be left fixed individually, since in that case every point of every circle of the pencil would be invariant, and the transformation cannot be identical since it is reversed. Hence u and v are interchanged. The transformation thus carries every point of the plane into its inverse with respect to $\mathfrak{C}$ and is called *inversion* with respect to $\mathfrak{C}$.

A reversed Möbius transformation leaving three points fixed is the inversion with respect to the circle passing through these three points.

§4 Invariant points and classification of Möbius transformations

4.1 The multiplier. The invariant points of a Möbius transformation (3.1) with matrix $A = \binom{a\ b}{c\ d}$ are determined by the equation

$$x = \frac{ax + b}{cx + d}$$

or

$$cx^2 + (d - a)x - b = 0.$$

In case $c = 0$ the point $x = \infty$ has to be included among its roots. In case $c = d - a = b = 0$, the transformation is the identity; this case needs no consideration.

The equation then has one or two roots according as

$$D = (d - a)^2 + 4bc = (\text{tr } A)^2 - 4 \det A$$

is equal to zero or different from zero. Let u and v denote the invariant points of the mapping (different or equal), x any other point and x' its image. In virtue of (3.3) and the fact that $u = u', \ v = v'$, a short calculation yields

$$k = (uvxx') = \frac{d + a - \sqrt{D}}{d + a + \sqrt{D}} \tag{1}$$

(or the reciprocal value dependent on the choice of u and v after a definite value for $\sqrt{D}$ has been fixed). Thus k is an invariant of the transformation. It is called the *multiplier* of the transformation (3.1). One has:

$$k + 2 + k^{-1} = \frac{(\text{tr } A)^2}{\det A}.$$

Moreover, from (3.2) one can calculate the increase of the amplitude in an invariant point:

$$\operatorname{amp} \frac{ad - bc}{(cu + d)^2} = -\operatorname{amp} k, \quad \operatorname{amp} \frac{ad - bc}{(cv + d)^2} = +\operatorname{amp} k. \tag{2}$$

4.2 Two invariant points. At first, let D be different from zero, thus u and v different. In virtue of the invariance of the cross-ratio one has

$$(uvx_0'x') = (uvx_0x).$$

On multiplying by $(uvxx_0')$ one gets from (4) in §2

$$(uvxx') = (uvx_0x_0'),$$

showing once more the invariance of k. This constant is neither 0 nor 1 nor ∞, since all four points are different, x not being invariant. Conversely, under the same conditions the equation

$$(uvxx') = k \tag{3}$$

determines a Möbius transformation with two different invariant points u and v. Obviously, the multiplier of the product of two such transformations with the same invariant points is the product of the corresponding multipliers.

The image of any circular arc joining u and v is a circular arc joining u and v; hence the circular arcs of the elliptic pencil with u and v as common points are interchanged. In consequence of the isogonality this also holds for the circles of the conjugate hyperbolic pencil with u and v as zero-circles. As stated in §2 and confirmed by calculation in Section 1, the amplitude of k measures the angle through which the circular arcs of the elliptic pencil are rotated about u or v; likewise the modulus of k characterizes the displacement of the circles of the hyperbolic pencil.

The necessary and sufficient condition for k being positive is that x and x' are on the same circular arc of the elliptic pencil; each of these circular arcs is then mapped onto itself. These transformations are called *hyperbolic*. The elliptic pencil with u and v as common points is called the *fundamental pencil* of the hyperbolic transformation.

The necessary and sufficient condition for the modulus of k being 1 is that x and x' are on the same circle of the hyperbolic pencil; each of these circles is then mapped onto itself. These transformations are called *elliptic*. The hyperbolic pencil with u and v as zero-circles is called the *fundamental pencil* of the elliptic transformation. Among these elliptic transformations is included the particular case $k = -1$, in which u, v and x, x' are harmonic; in the two conjugate pencils determined by u and v the two circles passing through x intersect again in x'. Thus this transformation is involutory since it interchanges the intersection points. It will be called the *involution with respect to the pair of points u, v*. A transformation (3.1) with two invariant points u and v, which interchanges two different points x and x' is the involution with respect to u, v. For the equation

$$k = (uvxx') = (uvx'x) = \frac{1}{k}$$

yields $k = -1$, since $k \neq 1$.

If neither $k > 0$ nor $|k| = 1$, no circular arc of the elliptic pencil and no circle of the hyperbolic pencil is mapped onto itself and the transformation is called *loxodromic*.

4.3 One invariant point. Secondly, let D be zero, thus u and v coincide. These transformations with only one invariant point, u, are called *parabolic*. Since in this case $D = (\operatorname{tr} A)^2 - 4 \det A = 0$ one gets $\operatorname{tr} A \neq 0$ and (1) yields $k = 1$, $\operatorname{amp} k = 0$. Thus the directions in u are left unaltered. Hence any circle through u is mapped onto a circle touching the former in u. Any parabolic pencil with u as common point is mapped onto itself.

Let x be any point other than u and x' its image, and draw the circle through u, x and x'. Its image must touch it in u and pass through x' and therefore coincides with the circle itself. Hence every point other than u lies on a circle which passes through u and coincides with its image. Two such circles have only u in common, since a second common point obviously would be invariant. These circles therefore form a parabolic pencil with u as common point. Thus there exists exactly one parabolic pencil with u as common point, whose circles are mapped onto themselves individually. It is called the *fundamental pencil* of the parabolic transformation. The direction of this pencil in u is called the *fundamental direction* of the parabolic transformation. Conversely, a Möbius transformation which reproduces the circles of a parabolic pencil individually, is parabolic, or the identity. For the common point u of the pencil is invariant, and if there is another invariant point v the circles of the elliptic pencil with u and v as common points are reproduced individually, since the directions in u either remain fixed or are reversed. Every other point of the plane is the intersection of a circle of the parabolic pencil and a circle of the elliptic pencil and thus remains fixed. The transformation, therefore, is the identity.

A parabolic transformation is uniquely determined by the invariant point u, another point x and its image x'. For the image y' of any other point y is situated both on the circle through x' which touches the circle through u, x and y in u and on the circle through y which touches the circle through u, x and x' in u; the latter belongs to the fundamental pencil. – If, in particular, y is on the circle through u, x and x', one may first construct the image z' of an arbitrary point z outside that circle, and then let z and z' play the role of x and x'.

No parabolic transformation can interchange two points. For if u is the invariant point, x any other point, and x' its image, the circle through u, x and x' is mapped onto itself in such a way that all its points are displaced in a definite direction without passing through u. Hence the image of x' is separated from x by x' and u and, therefore, cannot coincide with x. In reviewing the different types investigated it comes out that the involution with respect to a pair of points is the only type of transformation which interchanges two points.

4.4 Transformations with an invariant circle. Which are the Möbius transformations (other than the identity) which map a prescribed circle $\mathcal{K}$ onto itself and each of the two regions determined by $\mathcal{K}$ in the plane onto itself?

First, let it be assumed that $\mathcal{K}$ contains no invariant point of the transformation. Let u be an invariant point and denote by v its inverse with respect to $\mathcal{K}$. Since $\mathcal{K}$ is mapped onto itself, a pair of inverse points with respect to $\mathcal{K}$ are mapped onto a pair of inverse points with respect to $\mathcal{K}$. Since u is left fixed, v must be so too. So there is one invariant point in each of the two regions. $\mathcal{K}$ belongs to the hyperbolic pencil with u and v as zero-circles and, since $\mathcal{K}$ is mapped onto itself, the transformation is elliptic.

Secondly, let $\mathcal{K}$ contain two invariant points. $\mathcal{K}$ belongs to the elliptic pencil determined by these points as common points. Since $\mathcal{K}$ is mapped onto itself, so is every other circle of this pencil. Moreover, since the regions are reproduced individually, the same holds for every circular arc of the pencil. The transformation, therefore, is hyperbolic, $k > 0$.

Thirdly, let $\mathcal{K}$ contain one invariant point. If there were another outside $\mathcal{K}$, its inverse with respect to $\mathcal{K}$ would be invariant too and there would be more than two in all, which is impossible. The transformation, therefore, is parabolic, and $\mathcal{K}$ together with the invariant point as zero-circle determines the parabolic pencil whose circles are mapped onto themselves individually.

These three cases form a complete list of direct Möbius transformations of the required nature. In each case the circle $\mathcal{K}$ belongs to the fundamental pencil of the transformation. As far as reversed transformations are concerned, these can be characterized in the following way: Since each of the regions is mapped onto itself with orientation reversed, $\mathcal{K}$ must be so too. Hence there are exactly two invariant points u and v on $\mathcal{K}$. Let $\mathcal{C}$ denote the circle through u and v at right angles to $\mathcal{K}$. In virtue of the isogonality and of the invariance of u and v, $\mathcal{C}$ is mapped onto itself and, in particular, in consequence of the conservation of the regions, each of the two arcs into which it falls by u and v is mapped onto itself. Let x be a point on such an arc and x' its image. If one combines the hyperbolic transformation which has u and v as invariant points and carries x into x' with the inversion with respect to $\mathfrak{C}$, one gets a reversed Möbius transformation which maps the three points u, v, x in the same way as the transformation considered and therefore is that transformation itself. In the particular case $x = x'$ the hyperbolic transformation is the identity and the transformation considered is the inversion with respect to $\mathfrak{C}$.

4.5 Permutable transformations. In the sequel Möbius transformations, both direct and reversed, will by denoted by small *gothic* characters except for the identity for which the symbol 1 is generally used. For products of such symbols it is understood that the transformation indicated by the last symbol is the first performed, then the preceding one and so on. The inverse of a Möbius transformation $\mathfrak{f}$ is a Möbius transformation of the same kind; it is denoted by $\mathfrak{f}^{-1}$. Let $\mathfrak{f}$ and $\mathfrak{g}$ be two such transformations and u an invariant point for $\mathfrak{f}$. Then the transformation $\mathfrak{g}\mathfrak{f}\mathfrak{g}^{-1}$ evidently has $\mathfrak{g}u$ as invariant point, i.e. the image of u by $\mathfrak{g}$. From this is inferred:

If two transformations commute then each maps the set of invariant points of the other onto itself. □

There are several possibilities for two transformations to commute:

Let first both be direct Möbius transformations. If one is parabolic, the other must be so too and with the same invariant point; for a parabolic transformation can neither leave the invariant points of a non-parabolic fixed individually nor interchange them. If both are non-parabolic, each must leave the invariant points of the other fixed individually or interchange them. In the first case they have their invariant points in common; in the second case both must be involutions with respect to pairs of points, and these pairs must be harmonic.

These above necessary conditions prove also to be sufficient: For two parabolic transformations, $\mathfrak{f}$ and $\mathfrak{g}$, with the same invariant point u but with different fundamental directions this is inferred from the above mentioned invariance of the parabolic pencils with u as common point under the transformations $\mathfrak{f}$ and $\mathfrak{g}$ (cf. Fig. 4.1): The image $\mathfrak{C}_1$ of the circle through u, x and $\mathfrak{f}x$ by $\mathfrak{g}$ contains $\mathfrak{g}x$ and $\mathfrak{g}\mathfrak{f}x$; since $\mathfrak{C}_1$ belongs to the fundamental pencil for $\mathfrak{f}$, and thus is mapped onto itself by $\mathfrak{f}$, it also contains $\mathfrak{f}\mathfrak{g}x$.

The circle $\mathfrak{C}_2$ through u and $\mathfrak{f}x$ which belongs to the fundamental pencil of $\mathfrak{g}$, and thus is mapped onto itself by $\mathfrak{g}$, contains $\mathfrak{g}\mathfrak{f}x$. As it is the image of the circle through u, x and $\mathfrak{g}x$ by $\mathfrak{f}$, it also contains $\mathfrak{f}\mathfrak{g}x$. Hence $\mathfrak{g}\mathfrak{f}x$ and $\mathfrak{f}\mathfrak{g}x$ coincide with the intersection of $\mathfrak{C}_1$ and $\mathfrak{C}_2$. In addition it is seen that the product $\mathfrak{f}\mathfrak{g} = \mathfrak{g}\mathfrak{f}$ is again parabolic, since for every point $x \neq u$ the points x and $\mathfrak{f}\mathfrak{g}x$ lie on two different circles which touch at u and thus cannot coincide. If $\mathfrak{f}$ and $\mathfrak{g}$ have the same fundamental direction and thus the same fundamental pencil, it is evident that $\mathfrak{f}\mathfrak{g}$ and $\mathfrak{g}\mathfrak{f}$ are parabolic, since they reproduce individually the circles of that pencil. Now, let $\mathfrak{h}$ be a parabolic transformation with the same invariant point u but with another fundamental direction; $\mathfrak{h}$ is thus permutable with $\mathfrak{f}$ and $\mathfrak{g}$. Then $\mathfrak{f}\mathfrak{h}^{-1}$ and $\mathfrak{h}\mathfrak{g}$ are parabolic with u as invariant point, and their fundamental directions are different, since for $x \neq u$ (cf. Fig. 4.2) the points x, $\mathfrak{h}\mathfrak{g}x$ and $\mathfrak{f}\mathfrak{h}^{-1}(\mathfrak{h}\mathfrak{g}x) = \mathfrak{f}\mathfrak{g}x$ are not situated on a circle through u. They are, therefore,

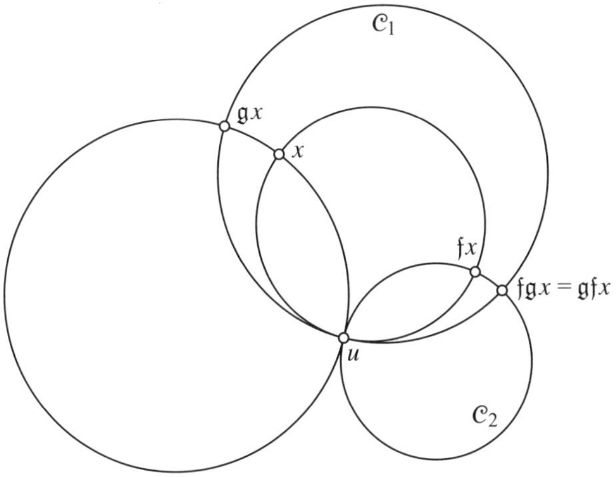

Figure 4.1

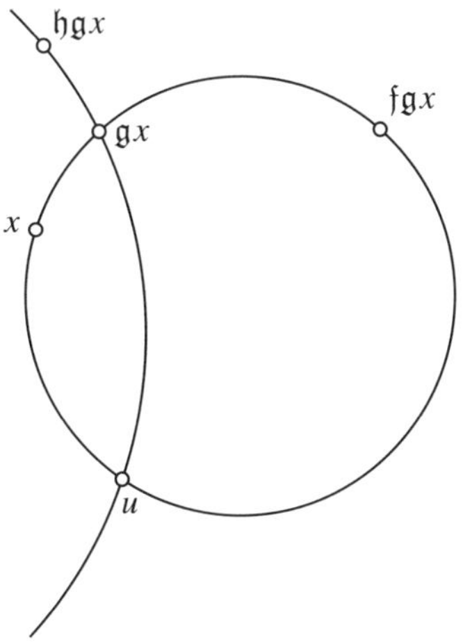

Figure 4.2

permutable in consequence of the case already dealt with and it follows that

$$\mathfrak{f}\mathfrak{g} = \mathfrak{f}\mathfrak{h}^{-1} \cdot \mathfrak{h}\mathfrak{g} = \mathfrak{h}\mathfrak{g}\mathfrak{f}\mathfrak{h}^{-1} = \mathfrak{g}\mathfrak{f}.$$

That two direct non-parabolic transformations $\mathfrak{f}$ and $\mathfrak{g}$ with the same invariant points are permutable, is inferred from the fact that in consequence of (3) the multipliers of $\mathfrak{fg}$ and $\mathfrak{gf}$ are both equal to the product of the multipliers of $\mathfrak{f}$ and $\mathfrak{g}$ and that a direct transformation is uniquely determined by its multiplier and its invariant points. – That two involutions $\mathfrak{f}$ and $\mathfrak{g}$ with respect to two pairs of harmonic points are permutable can be seen as follows: Since $\mathfrak{fg}$ interchanges the invariant points of $\mathfrak{g}$, $\mathfrak{fg}$ is involutory, hence $\mathfrak{fgfg} = 1$. In virtue of $\mathfrak{f} = \mathfrak{f}^{-1}$, $\mathfrak{g} = \mathfrak{g}^{-1}$ this can be written $\mathfrak{fgf}^{-1}\mathfrak{g}^{-1} = 1$, thus $\mathfrak{fg} = \mathfrak{gf}$.

As far as reversed transformations are concerned, only inversions with respect to circles are taken into account. If the inversion with respect to a circle $\mathfrak{C}$ is to be permutable with a direct transformation, the latter must map $\mathfrak{C}$ onto itself, since $\mathfrak{C}$ consists of invariant points of the inversion. This necessary condition is also sufficient for the permutableness, since the direct transformation then carries any two points which are inverse with respect to $\mathfrak{C}$ into two points which are also inverse with respect to $\mathfrak{C}$. – The inversions with respect to two different circles $\mathfrak{C}$ and $\mathfrak{C}'$ can only be permutable if the circle $\mathfrak{C}'$ which consists of the invariant points of the second inversion, is reproduced by the inversion with respect to $\mathfrak{C}$, i.e. if $\mathfrak{C}$ and $\mathfrak{C}'$ are orthogonal. On the other hand, this is sufficient; for the inversions with respect to $\mathfrak{C}$ and $\mathfrak{C}'$ are involutory, and their product is the involution with respect to the pair of intersection points of $\mathfrak{C}$ and $\mathfrak{C}'$, hence also involutory. – In all, the following result is obtained, the indicated conditions being necessary and sufficient:

Two direct Möbius transformations are permutable if they have their invariant points in common, or if they are involutions with respect to harmonic pairs of points. A direct transformation is permutable with the inversion with respect to a circle if it maps that circle onto itself. The inversions with respect to two circles are permutable if the circles are mutually orthogonal. □

§5 Complex distance of two pairs of points

5.1 Definition. Let any two pairs of points x, y and x', y' be given in this order, the points of the single pairs likewise being given in the indicated order. It is first assumed that the two pairs have no point in common, whereas coincidence of the points of the single pairs is not excluded. It will first be shown that there exists exactly one pair of points u, v which is harmonic with both of the given pairs; in the case of coincidences harmonicity is taken in the generalized sense indicated at the end of §2. If $x \neq y$, there exists exactly one direct Möbius transformation carrying x into y, y into x and x' into y'. Since that transformation interchanges two points, it is the involution with respect to a certain pair of points u, v (§4.3); therefore it also carries y' into x'. In case $x' = y'$, this point will at the same time be one of the points u and v. – If $x = y$ but $x' \neq y'$, one can start with x', y' in an analogous way. – If both $x = y$ and $x' = y'$, the required solution is found by putting $x = y = u$ and $x' = y' = v$, or conversely.

In all cases u and v are different. Since the two pairs x, y and x', y' are assumed without common point, in such expressions as e.g. $(uvxx')$ no three points coincide, and the cross-ratio therefore has a meaning.

First, consider the normal case of no coincidence, thus x, y, x', y' being four different points. Then

$$(uvxx') = (uvyx)(uvxx')(uvx'y') = (uvyy'),$$

since the two factors added in the intermediate term assume the value -1 in consequence of harmonicity. This expression is the multiplier (4.3) of the Möbius transformation with u, v as invariant points which carries x into x'; that it also carries y into y' is also evident from the fact that it must carry two points which are harmonic with u, v into two points which again are harmonic with u, v.

The logarithm, taken with reversed sign, of this multiplier is called *the complex distance of the pair of points x, y and x', y'*, given in this order:

$$a = \delta + \varphi i = -\log(uvxx') = -\log(uvyy'). \tag{1}$$

After a fixed choice of the notation u and v has been made, a is uniquely determined except for multiples of $2\pi i$. If the pairs or the points u and v are interchanged, the sign of a is reversed. From the relations

$$\begin{aligned}(uvxy') &= (uvxy)(uvyy') = -(uvyy')\\ (uvyx') &= (uvyx)(uvxx') = -(uvxx')\end{aligned} \tag{2}$$

it is seen that a is increased by πi if the points of one pair are interchanged.

The values of δ and φ may be deduced from the equation

$$\log(uvxx') = \log|uvxx'| + i\ \mathrm{amp}(uvxx')$$

using the results of §2.2:

Draw through x and x' the circles $\mathcal{K}$ and $\mathcal{K}'$ of the hyperbolic pencil with u and v as zero-circles and the circular arcs $\mathcal{B}$ and $\mathcal{B}'$ of the conjugate elliptic pencil, and draw an arbitrary arc $\mathcal{B}_0$ of the latter cutting $\mathcal{K}$ and $\mathcal{K}'$ at points x_0 and x_0' (Fig. 5.1). Then

$$\delta = -\log(uvx_0x_0'),$$

and φ is the angle from $\mathcal{B}$ to $\mathcal{B}'$ when measured in u (not in v as in §2.2 because of the reversed sign).

If coincidence takes place in at least one pair, $x = y$ say, this point is at the same time u or v, and there is no transformation with u and v as invariant points carrying the first pair into the second. But the cross-ratios in (1) exist and evidently take the values 0 or ∞. Accordingly one has to put $a = 0$ or $a = \infty$.

The definition of complex distance has to be extended to the case, hitherto excluded, where the two pairs have at least one point in common. Let for instance $x = x'$. If

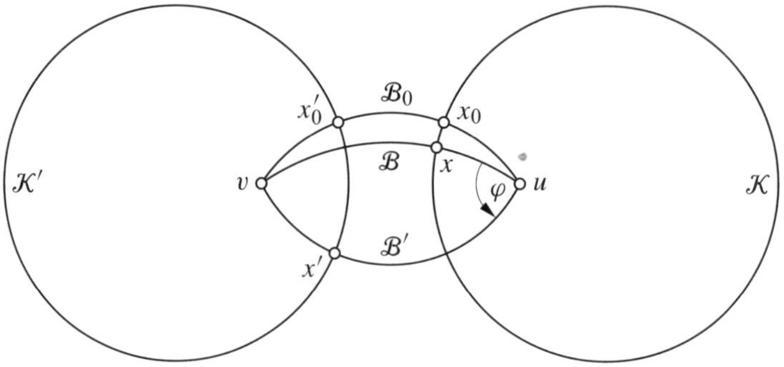

Figure 5.1

both u and v are chosen in this point the condition of common harmonicity (in the generalized sense of §2) is maintained; this is true whether y and y' coincide or not. In this case the symbol $(uvxx')$ is devoid of meaning, but another cross-ratio in (1), $(uvyy') = (uuyy')$, exists provided y and y' do not coincide with $x = x' = u = v$ and it takes the value 1. Thus $a = 0$: If the two first points or the two second points of the pairs coincide, the distance is zero. – If the first point of one pair coincides with the second of the other, $x = y'$ or $y = x'$, and if again both u and v are chosen in this common point in order to fulfill the condition of common harmonicity, both cross-ratios in (1) are devoid of meaning. This case may be treated by the remark that it reduces to the former case by the interchange of the points of one pair. According to a previous statement the distance then has to be $a = \pi i$. These are the only cases in which the distances 0 and πi occur. In both cases they may be justified by considerations of continuity.

5.2 Relations between distances. Let five pairs of points $x_1, y_1 : x_2, y_2 : x_3, y_3 : x_4, y_4 : x_5, y_5 :$ be given. The order of the two points in the single pairs is as indicated. As to the succession of the pairs, only their *cyclical* order matters. This order is indicated by the subscripts these being, in the sequel, only taken into account modulo 5. Let furthermore any two neighbouring pairs be harmonic; this is expressed by the equation

$$(x_\nu y_\nu x_{\nu+1} y_{\nu+1}) = -1 \tag{3}$$

or

$$2(x_\nu y_\nu + x_{\nu+1} y_{\nu+1}) - (x_\nu + y_\nu)(x_{\nu+1} + y_{\nu+1}) = 0, \tag{4}$$

ν ranging over all values modulo 5. Now, the complex distance of the two pairs next to the pair x_ν, y_ν is, according to Section 1

$$a_\nu = -\log (x_\nu y_\nu x_{\nu-1} x_{\nu+1}) = -\log (x_\nu y_\nu y_{\nu-1} y_{\nu+1}).$$

The relations governing these five distances a_ν will now be established. Since cross-ratios and, therefore, the relations looked for are invariant under Möbius transformations, it can be assumed that $x_3 = 0,\ \ y_3 = \infty$.

From (3), taken for $\nu = 2$ and $\nu = 3$, it is inferred that

$$x_2 + y_2 = 0, \qquad x_4 + y_4 = 0.$$

This together with (4), taken for $\nu = 1$ and $\nu = 4$ yields

$$x_1 y_1 = x_2^2, \qquad x_5 y_5 = x_4^2. \tag{5}$$

Now, from the definition of a_ν,

$$e^{-a_3} = (x_3 y_3 x_2 x_4) = (0\,\infty\, x_2 x_4) = \frac{x_2}{x_4},$$

hence from (4), taken for $\nu = 5$, and (5)

$$\cosh a_3 = \frac{e^{a_3} + e^{-a_3}}{2} = \frac{x_2^2 + x_4^2}{2x_2 x_4} = \frac{x_1 y_1 + x_5 y_5}{2x_2 x_4} = \frac{x_1 + y_1}{2x_2} \cdot \frac{x_5 + y_5}{2x_4}.$$

On the other hand

$$e^{-a_2} = (x_2 y_2 x_1 x_3) = (x_2\ -x_2\, x_1\, 0) = \frac{x_2 - x_1}{x_2 + x_1},$$

hence from (5)

$$\coth a_2 = \frac{e^{a_2} + e^{-a_2}}{e^{a_2} - e^{-a_2}} = \frac{x_1^2 + x_2^2}{2x_1 x_2} = \frac{x_1 + y_1}{2x_2}.$$

Likewise

$$e^{-a_4} = (x_4 y_4 x_3 x_5) = (x_4\ -x_4\, 0\, x_5) = \frac{x_4 + x_5}{x_4 - x_5},$$

hence from (5)

$$\coth a_4 = \frac{e^{a_4} + e^{-a_4}}{e^{a_4} - e^{-a_4}} = -\frac{x_4^2 + x_5^2}{2x_4 x_5} = -\frac{x_5 + y_5}{2x_4}.$$

From these formulae one gets

$$\cosh a_3 = -\coth a_2 \coth a_4.$$

Since all distances are defined by cross-ratios, the result is independent of the above special choice of $x_3,\ y_3$ hence one gets generally by permutation of subscripts

$$\cosh a_\nu = -\coth a_{\nu-1} \coth a_{\nu+1} \quad (\nu \bmod 5). \tag{6}$$

From the five relations (6) five others are deduced by elimination;

$$\begin{aligned}\cosh a_{\nu-2} \cosh a_{\nu+2} &= \coth a_{\nu-3} \coth a_{\nu-1} \coth a_{\nu+1} \coth a_{\nu+3} \\ &= -\coth a_{\nu-3} \cdot \cosh a_\nu \cdot \coth a_{\nu+3} \\ &= -\coth a_{\nu+2} \cosh a_\nu \coth a_{\nu-2},\end{aligned}$$

hence

$$\cosh a_\nu = -\sinh a_{\nu-2} \sinh a_{\nu+2} \quad (\nu \bmod 5). \tag{7}$$

By means of (6) and (7) all distances a_ν can be found whenever two of them are known. In concise form:

The cosh *of any of the distances, with sign reversed, equals the product of the* coth *of the neighbouring and also the product of the* sinh *of the non-neighbouring distances.* □

§6 Non-euclidean metric

6.1 Terminology of non-euclidean geometry. Let $\mathcal{E}$ be a fixed circle and $\mathcal{D}$ one of the two regions into which is decomposes the complex plane. In the sequel $\mathcal{E}$ is taken to be the unit circle $|x| = 1$ and $\mathcal{D}$ its interior $|x| < 1$. The exterior of $\mathcal{E}$ is denoted by $\mathcal{D}^*$. The region $\mathcal{D}$ is now looked upon as the non-euclidean plane.

The points of D will be called *proper points* of the non-euclidean geometry, the points of $\mathcal{E}$ *infinite points*, and those of $\mathcal{D}^*$ *improper points*. The concepts of *accumulation points, convergent sequences of points, closed and open sets* are taken in the ordinary sense in relation to the sphere of complex numbers. The *closure* of a set $\mathcal{M}$ will be denoted by $\overline{\mathcal{M}}$. Thus for instance $\overline{\mathcal{D}} = \mathcal{D} \cup \mathcal{E}$. On the other hand, the same concepts will be used in non-euclidean sense, properly speaking, i.e. in relation to $\mathcal{D}$. In that case such expressions as *sequences convergent on* $\mathcal{D}$, *sets closed on* $\mathcal{D}$ will be used. Thus a sequence of points of $\mathcal{D}$ convergent to a point of $\mathcal{E}$ is divergent on $\mathcal{D}$. The *closure on* $\mathcal{D}$ of a subset $\mathcal{M}$ of $\mathcal{D}$ is the intersection $\mathcal{D} \cap \overline{\mathcal{M}}$ of $\mathcal{D}$ and $\overline{\mathcal{M}}$. It will be denoted by $\tilde{\mathcal{M}}$.

The *non-euclidean straight lines* are, by definition, the circular arcs in $\mathcal{D}$ at right angles to $\mathcal{E}$. The extremities of such an arc on $\mathcal{E}$ are called *end-points* or *infinite points* of the non-euclidean line. If one of these is taken as the negative, the other as the positive, the line is said to be *directed*, the direction being from the negative towards the positive end-point. Occasionally even the complementary arc inside $\mathcal{D}^*$ of a non-euclidean line is taken into account and spoken of as its improper part.

If two circles which are orthogonal to $\mathcal{E}$ intersect, the intersection points are inverse with respect to $\mathcal{E}$. The one situated in $\mathcal{D}$ is the intersection point of the two non-euclidean lines. If two circles which are orthogonal to $\mathcal{E}$ touch each other, $\mathcal{E}$ is a member of the parabolic pencil which is orthogonal to both and thus passes through their point of contact.

Hence there are three possibilities for the mutual situation of two non-euclidean lines: They intersect in a proper point and are then called *concurrent*; or they have one end-point in common and are then called *parallel* or *asymptotic*; or they have no common point and are then called *divergent*. The first or second or third case occurs according as the two pairs of end-points separate each other on $\mathcal{E}$ or have one point in common or do not separate each other on $\mathcal{E}$.

For two non-concurrent, directed lines two cases can be distinguished: Their directions are called *opposite* if they are parallel with end-points of opposite sign coinciding, or if they are divergent and the pair of positive end-points is separated on $\mathcal{E}$ by the negative pair; otherwise their directions are said to be *concordant*.

From §1.3 it is seen that there is always exactly one non-euclidean straight line passing through two different points, whether proper or infinite. The part joining the two points is called the *join* of the two points. In particular the join is called a *segment* if both points are proper, and a *half-line* if one is proper, the other infinite; it is a full line, if both are infinite. A subset of $\overline{\mathcal{D}} = \mathcal{D} \cup \mathcal{E}$ is called *convex* if, given any two points of the subset, their join belongs to the subset. The intersection of any finite or infinite number of convex sets is convex. Segments, half-lines and complete lines are, themselves, examples of convex sets; the same is true of *half-planes*, i.e. the subsets into which $\mathcal{D} \cup \mathcal{E}$ or $\mathcal{D}$ are decomposed by some line, whether these half-planes are regarded as open, closed, or closed on $\mathcal{D}$. The intersection of all convex sets containing a given set $\mathcal{M}$ is termed the *convex hull* of $\mathcal{M}$.

The measurement of angles in non-euclidean geometry is taken to be the same as in euclidean geometry. Hence, in particular, orthogonality means the same in both geometries. From §1 is inferred:

Given any point proper or infinite, and any line which has not the given point as one of its end-points, there exists exactly one line passing through the point and perpendicular to the line. Two divergent lines have exactly one common perpendicular, whereas two parallel lines or two concurrent lines have none. The part of the perpendicular between the point and the line, resp. the part of the perpendicular between two divergent lines is called their join. □

The set of non-euclidean lines which are situated on the circles of a pencil in the sense of inversive geometry is called a *pencil of lines*. Three different cases occur: An elliptic pencil of circles determines a non-euclidean pencil of lines if (and only if) its common points are inverse with respect to $\mathcal{E}$. The pencil then is the set of all non-euclidean lines through the common point u situated in $\mathcal{D}$. The circles of the conjugate pencil are called *non-euclidean circles* with u as their centre. – A parabolic pencil of circles determine a non-euclidean pencil of lines if (and only if) its common point u lies on $\mathcal{E}$ and its direction is at right angles to $\mathcal{E}$. The pencil then is the set of all parallel lines with common infinite point at u. The circles of the conjugate pencil touch $\mathcal{E}$ at u and are called *horocycles* with u as their (infinite) centre.

– A hyperbolic pencil of circles determines a non-euclidean pencil of lines if (and only if) its zero-circles u and v are situated on $\mathcal{E}$. The pencil of lines then is the set of all perpendiculars to the non-euclidean line uv. The circular arcs in $\mathcal{D}$ of the conjugate pencil (which join u and v and include the line uv itself) are called *hypercycles* of uv.

6.2 Distance and angle. The *non-euclidean distance* is derived from the concept of complex distance of two pairs of points as defined in §5. To do this, each point x of $\overline{\mathcal{D}}$ is associated with its inverse x^* with respect to $\mathcal{E}$, the pair x^*, x representing the point x. An infinite point x is accordingly represented by the pair x, x. A directed line is represented by the pair of its end-points ordered in correspondence with the direction. So three types of pairs are involved: Two points inverse with respect to $\mathcal{E}$, two points coincident on $\mathcal{E}$ and two different points of $\mathcal{E}$. For reasons of brevity these are spoken of as *elliptic*, *parabolic* and *hyperbolic* pairs of points.

Let x^*, x and y^*, y be two elliptic pairs. The circle orthogonal to $\mathcal{E}$ and passing through x and y represents the non-euclidean line xy and cuts $\mathcal{E}$ in a pair of points u, v which is harmonic both with x^*, x and with y^*, y (Fig. 6.1).

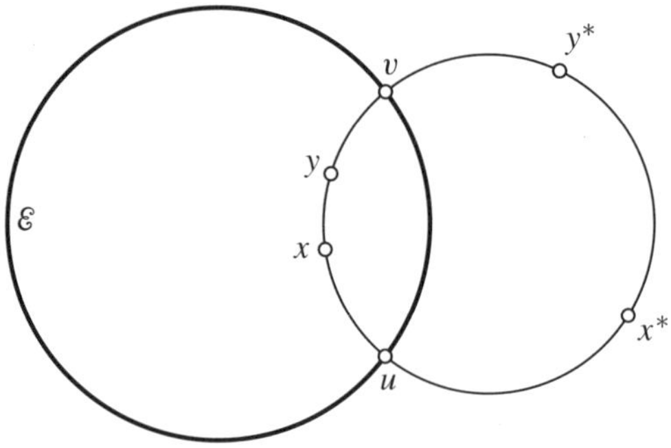

Figure 6.1

Since the pairs x, y and u, v are concyclical and do not separate each other, the cross-ratio $(uvxy)$ is positive and, therefore, the complex distance $-\log(uvxy)$ is real up to the addition of an arbitrary multiple of $2\pi i$. Its real value is called the *non-euclidean distance* of the points x and y and denoted by

$$[x, y] = -[y, x] \tag{1}$$

Its sign depends on the order in which u and v are taken, thus on the direction of the line on which x and y are situated. If x is fixed, the distance varies from $-\infty$ to ∞ as y describes the line from u to v. If z is another point of the line and z^* its inverse, then u, v is also harmonic with z^*, z and from

$$(uvxy)(uvyz) = (uvxz)$$

the relation

$$[x, y] + [y, z] = [x, z]$$

is deduced for the distance of three points on a directed line.

Let x^*, x be an elliptic and g_1, g_2 a hyperbolic pair and $\mathcal{G}$ the directed line $g_1 g_2$ (Fig. 6.2).

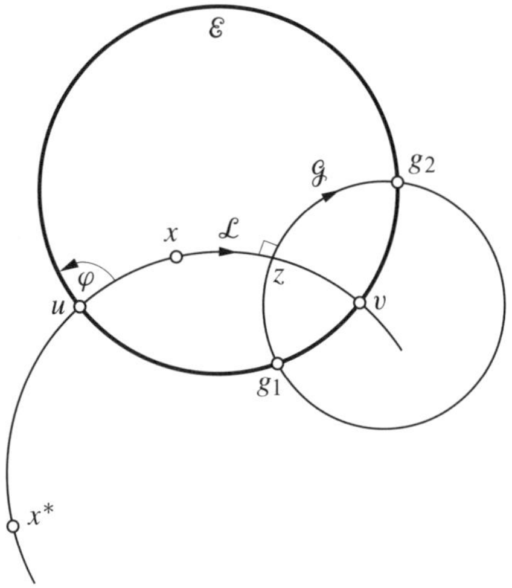

Figure 6.2

Let $\mathcal{L}$ be the perpendicular to $\mathcal{G}$ through x and u, v its end-points. Then u, v is harmonic with x^*, x and (in virtue of the theorem on three mutually orthogonal circles, §2.3) with g_1, g_2. If z denotes the proper intersection point of $\mathcal{L}$ and $\mathcal{G}$, the real part of the complex distance $-\log(uvxg_2)$ of x^*, x and g_1, g_2 equals the real value of $-\log(uvxz)$, thus the non-euclidean distance of x and z, or the distance

$$[x, \mathcal{G}] = -[\mathcal{G}, x] \tag{2}$$

of the point x from the line $\mathcal{G}$ measured on the perpendicular. According to §5 the imaginary part φ of the complex distance i.e. factor of i equals the angle subtended at u from the arc uxv to the arc ug_2v, which is also the angle from the directed line $\mathcal{L}$ to the directed line $\mathcal{G}$, thus $\varphi = \pm\frac{\pi}{2}$. If u and v are interchanged, both the distance and the angle φ are reversed in sign. The direction of $\mathcal{L}$ can be made dependent on the direction of $\mathcal{G}$ by taking $\mathcal{L}$ as the positive perpendicular of $\mathcal{G}$ according to the usual orientation of the complex plane; in that case one gets $\varphi = -\frac{\pi}{2}$. – The pairs x^*, x and g_1, g_2 are harmonic, if x lies on $\mathcal{G}$, and in that case only; the distance then is 0.

Let g_1, g_2 and h_1, h_2 be two hyperbolic pairs and $\mathcal{G}$ and $\mathcal{H}$ the directed lines determined by these pairs respectively. It is first assumed that the pairs do not separate each other on $\mathcal{E}$. The lines $\mathcal{G}$ and $\mathcal{H}$ are thus divergent and have a common perpendicular

$\mathcal{L}$. The end-points u and v of $\mathcal{L}$ are the common harmonic pair for g_1, g_2 and h_1, h_2 (Fig. 6.3).

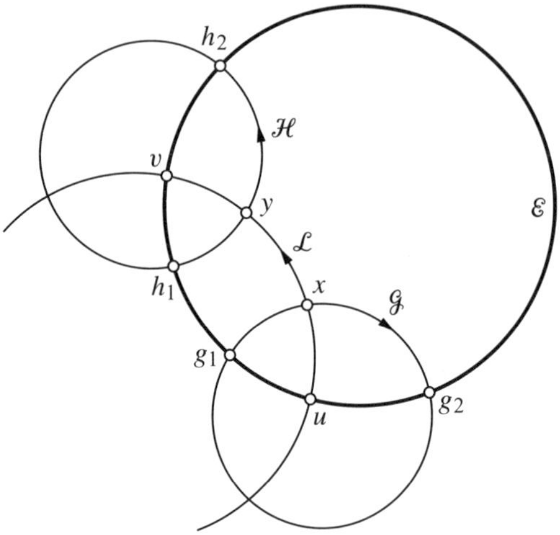

Figure 6.3

If $\mathcal{L}$ cuts $\mathcal{G}$ and $\mathcal{H}$ in x and y respectively, the real part of $-\log(uvg_1h_1)$ equals $-\log(uvxy)$ and thus measures the distance

$$[\mathcal{G}, \mathcal{H}] = -[\mathcal{H}, \mathcal{G}] \tag{3}$$

of $\mathcal{G}$ and $\mathcal{H}$ taken on their common perpendicular. The imaginary part φ is 0, if the directions of the lines $\mathcal{G}$ and $\mathcal{H}$ are concordant (as in Fig. 6.3), and π, if they are opposite.

The quantities (1), (2), (3) are the non-euclidean lengths of the join of the two objects considered.

If the pairs g_1, g_2 and h_1, h_2 separate each other on $\mathcal{E}$, the lines $\mathcal{G}$ and $\mathcal{H}$ intersect (Fig. 6.4).

Let u denote the proper and v the improper intersection point. These are then inverse with respect to $\mathcal{E}$ and constitute the common harmonic pair of the given pairs (§2.3). The cross-ratio (uvg_1h_1) has its modulus equal to 1; thus the complex distance $-\log(uvg_1h_1)$ is purely imaginary. Here φ equals the angle from $\mathcal{G}$ to $\mathcal{H}$

$$[\mathcal{G}, \mathcal{H}] = -[\mathcal{H}, \mathcal{G}] \tag{4}$$

subtended at u. The two given hyperbolic pairs are harmonic, if the corresponding lines are mutually orthogonal, and in that case only.

If the pairs g_1, g_2 and h_1, h_2 have one point in common, $\mathcal{G}$ and $\mathcal{H}$ are parallel and, in consequence of §5.1, the complex distance is 0 or πi, according as the directions

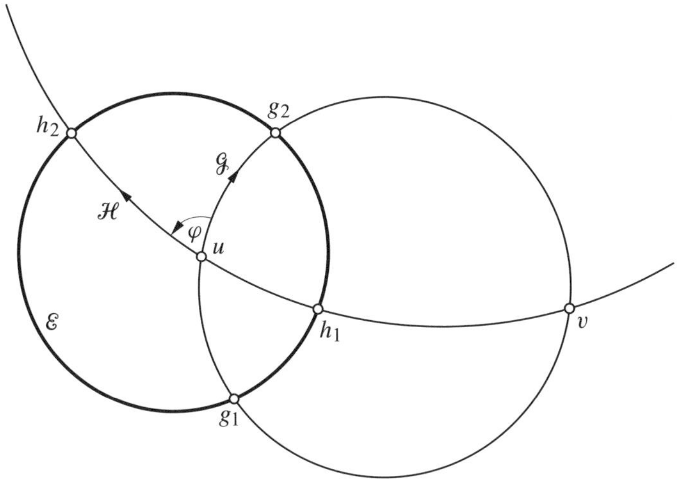

Figure 6.4

of the lines are concordant or opposite. Thus here too φ measures the angle between the two directed lines when taken in their common (infinite) point.

The distance of a parabolic pair x, x from any other pair with which it has no point in common is, by (5.1), equal to ∞. Its distance from a pair with which it has a point in common is not defined.

§7 Isometric transformations

7.1 Motions and reversions. All concepts of non-euclidean geometry introduced in §6 are invariant under direct Möbius transformations carrying $\mathcal{E}$ and $\mathcal{D}$ into themselves; the same is true in the case of reversed Möbius transformations apart from the fact that the sign of angles is reversed or more generally, complex distances are replaced by their conjugate values; such transformations therefore will be termed non-euclidean *motions* or *reversions* respectively. Configurations in $\mathcal{D} \cup \mathcal{E}$ which are carried into each other by a motion or reversion are called *congruent* or *symmetric* respectively.

Let u, v, w and u', v', w' be two sets of infinite points, each set consisting of three different points. Then, from §3.2 or §3.3, exactly one motion or reversion exists carrying u into u', v into v' and w into w', this being a motion or a reversion according as the senses of the sequences uvw and $u'v'w'$ on $\mathcal{E}$ are in accordance or not. Hence any two asymptotic triangles are congruent. – If x and x' are two proper points and $\mathcal{G}$ and $\mathcal{G}'$ two directed lines through x and x' respectively, exactly one motion and exactly one reversion exist carrying x into x' and $\mathcal{G}$ into $\mathcal{G}'$. For if u and u' are the negative and v and v' the positive end-points of $\mathcal{G}$ and $\mathcal{G}'$ respectively, u must pass into u' and v into v'. There is exactly one direct Möbius transformation carrying u into u', x into x', and v into v', and also one reversed. Now, since $\mathcal{E}$ passes through u and v and is orthogonal to $\mathcal{G}$, it is seen that $\mathcal{E}$ must pass into itself, and since x passes into x', the region $\mathcal{D}$ passes into itself. Hence the first transformation is a motion and the second a reversion.

From this and from the invariance of distances and angles is inferred: Any distance can be laid off from a prescribed point in a prescribed direction. Any angle can be laid off from a prescribed line in a prescribed point and half-plane.

7.2 Classification of motions and reversions. A characterization of those direct and reversed Möbius transformations which carry a circular region into itself has been given in §4.4. From this flows a classification of motions and reversions.

Motions:

1) Elliptic transformations, the invariant points being inverse with respect to $\mathcal{E}$; say u in $\mathcal{D}$ and $v = u^*$ in $\mathcal{D}^*$. In non-euclidean terms all lines through u are rotated about u through the angle $\varphi = -\operatorname{amp} k$, k being the multiplier of the transformation (§4.1); cf. Fig. 6.4. The non-euclidean circles with centre u are transformed into themselves individually. Consequently, each of them is the locus of the points at constant distance from u. This type of motion is called *rotation* with u as *centre* and φ as *angle of rotation*. If $\varphi = \pi$, the motion is a half-turn about u. If a rotation is not a half-turn, the angle of rotation can be normalized by taking $-\pi < \varphi < \pi$, and a definite sense of rotation can then be spoken of according to the sign of φ. In the following a rotation is represented by the symbol $\circlearrowleft$, a half-turn by a single dot or by $\circlearrowleft_2$).

2) Parabolic transformations, the invariant point u being on $\mathcal{E}$ and the fundamental direction tangent to $\mathcal{E}$. The pencil of parallel lines with u as their end-point is carried into itself as a whole. The horocycles with centre u are carried into themselves individually. Consequently, two of these horocycles cut out segments of equal length on all lines with the end-point u. This type of motion is called *limit-rotation* with u as *limit-centre* (often simply called *centre*). It is termed positive or negative according as it displaces the infinite points other than u in the positive or negative sense. A limit-rotation will be represented by $\hookrightarrow$, occasionally by $\circlearrowleft_\infty$ if it is treated together with rotations.

3) Hyperbolic transformations, the invariant points u and v being on $\mathcal{E}$. The pencil of perpendiculars of the line $\mathcal{A}$ joining u and v is carried into itself as a whole. Each of these perpendiculars is displaced through the amount $\lambda = -\log k$, measured on $\mathcal{A}$, in the direction from u towards v; cf. Fig. 6.1. In particular λ is the displacement of the points on $\mathcal{A}$. The hypercycles belonging to $\mathcal{A}$ are carried into themselves individually. Consequently each of them is the locus of the points at constant distance from $\mathcal{A}$. This

type of motion is called *translation* with $\mathcal{A}$ as *axis* and λ as *displacement.* The order of the points u and v determines a direction of $\mathcal{A}$, and the displacement is endowed with a sign by the above definition. If no direction of $\mathcal{A}$ is prescribed in advance, a definite direction is assigned to $\mathcal{A}$ by the motion itself by choosing the denotation u and v such that $\lambda > 0$. The points u and v are then spoken of as the *negative and positive fundamental point of the translation.* A translation will be represented by $\rightarrow$.

From §4 is inferred that a motion which carries a non-euclidean circle, or a horocycle or a hypercycle into itself is a rotation or a limit-rotation, or a translation respectively.

Reversions:

1) Inversion with respect to a circle $\mathcal{L}$ at right angles to $\mathcal{E}$. The points of the non-euclidean line $\mathcal{L}$ are left fixed individually. This type of reversion is called *reflection in the line* $\mathcal{L}$, and $\mathcal{L}$ is called *line of reflection.* The symbol of a reflection will be $-$.

2) Every other reversion can be obtained as the product of a translation and a reflection in the axis $\mathcal{A}$ of the translation. By §4.5 these two transformations are permutable. The points of $\mathcal{A}$ are displaced by the amount of the displacement λ of the translation. This type of reversion is called *reversed translation* (or *glide-reflection*) with $\mathcal{A}$ as *axis* and λ as *displacement.* The symbol of a reversed translation will be $\rightleftharpoons$.

7.3 Products of reflections and half-turns. Motions and reversions are denoted by small *Gothic* characters except that the symbol 1 is used for the identity. In particular, the letter $\mathfrak{s}$ is set apart for involutory transformations, i.e. reflections and half-turns. Thus the relation

$$\mathfrak{s}^2 = 1$$

is always valid. The product of two involutory transformations is replaced by its inverse if the factors are interchanged. As previously stated, the transformations of a product are thought of as being performed in the order from right to left. The axis of a translation, or reversed translation $\mathfrak{f}$ is denoted by $\mathcal{A}(\mathfrak{f})$ and the centre of a rotation $\mathfrak{f}$ by $c(\mathfrak{f})$. Accordingly displacement and angle of rotation are marked $\lambda(\mathfrak{f})$ and $\varphi(\mathfrak{f})$. The limit-centre of the limit-rotation $\mathfrak{f}$ is denoted by $u(\mathfrak{f})$. The line of a reflection $\mathfrak{s}$ is denoted by $\mathcal{L}(\mathfrak{s})$.

The product $\mathfrak{s}\mathfrak{s}'$ of two reflections $\mathfrak{s}$ and $\mathfrak{s}'$ is a motion and, in particular, a rotation, limit-rotation, or translation according as $\mathcal{L}(\mathfrak{s})$ and $\mathcal{L}(\mathfrak{s}')$ are concurrent, parallel or divergent. In the first case the intersection point is the invariant point of the product, and the angle from $\mathcal{L}(\mathfrak{s}')$ to $\mathcal{L}(\mathfrak{s})$ is half the angle of rotation. In the second case the common infinite point u of $\mathcal{L}(\mathfrak{s})$ and $\mathcal{L}(\mathfrak{s}')$ is the invariant point of the product and every horocycle with u as centre is reproduced by the reflections and thus by the product. In the third case the common perpendicular of $\mathcal{L}(\mathfrak{s})$ and $\mathcal{L}(\mathfrak{s}')$ is reproduced and the distance from $\mathcal{L}(\mathfrak{s}')$ to $\mathcal{L}(\mathfrak{s})$ is half the displacement of the translation.

Conversely, every motion can be conceived as the product of two reflections, the number of possible products forming a one-parameter family. In the case of a rotation $\mathfrak{f}$ any two lines through $c(\mathfrak{f})$ which subtend an angle $\frac{1}{2}\varphi(\mathfrak{f})$ can be chosen as lines of reflection. If a limit-rotation $\mathfrak{f}$ is given by its infinite centre $u(\mathfrak{f})$ together with a point x and its image $\mathfrak{f}x$, the points x and $\mathfrak{f}x$ being situated on a horocycle with centre $u(\mathfrak{f})$, one can choose the line joining $u(\mathfrak{f})$ and x and the line joining $u(\mathfrak{f})$ and the bisection point of the segment $x, \mathfrak{f}x$ as lines of reflection. A translation $\mathfrak{f}$ is obtained by the product of reflections at any two perpendiculars of $\mathcal{A}(\mathfrak{f})$ distant $\frac{1}{2}\lambda(\mathfrak{f})$ from each other.

If $\mathfrak{s}$ and $\mathfrak{s}'$ denote two half-turns with different centres, the product $\mathfrak{s}\mathfrak{s}'$ is a translation, since the line $c(\mathfrak{s})\ c(\mathfrak{s}')$ is reproduced. The displacement equals $2[c(\mathfrak{s}'), c(\mathfrak{s})]$. Conversely, a given translation $\mathfrak{f}$ is the product of two half-turns about any two points of $\mathcal{A}(\mathfrak{f})$ distant $\frac{1}{2}\lambda(\mathfrak{f})$ from each other.

If $\mathfrak{s}$ denotes a half-turn and $\mathfrak{s}'$ a reflection, and if $c(\mathfrak{s})$ is not on $\mathcal{L}(\mathfrak{s}')$ the product $\mathfrak{s}\mathfrak{s}'$ is a reversed translation. Evidently the axis $\mathcal{A}(\mathfrak{s}\mathfrak{s}')$ of the product is the perpendicular of $\mathcal{L}(\mathfrak{s}')$ through $c(\mathfrak{s})$, since this line is reproduced both by $\mathfrak{s}$ and by $\mathfrak{s}'$. The displacement of the product is $2[\mathcal{L}(\mathfrak{s}'),\ c(\mathfrak{s})]$. The product is a reflection if, and only if, $c(\mathfrak{s})$ is situated on $\mathcal{L}(\mathfrak{s}')$. Conversely, any reversed transformations $\mathfrak{f}$ can be obtained as the product of a half-turn and a reflection by choosing the centre of the half-turn on $\mathcal{A}(\mathfrak{f})$ and the line of reflection at right angles to $\mathcal{A}(\mathfrak{f})$ at the distance $\frac{1}{2}\lambda(\mathfrak{f})$ from the centre, which can be done in infinitely many ways. Any reflection $\mathfrak{f}$ can be replaced by the product of a reflection in a line at right angles to $\mathcal{A}(\mathfrak{f})$ and a half-turn about the intersection point of these two lines.

7.4 Transforms. If $\mathfrak{f}$ and $\mathfrak{g}$ are any motions or reversions, the motion or reversion

$$\mathfrak{f}' = \mathfrak{g}\mathfrak{f}\mathfrak{g}^{-1} \tag{1}$$

is said to be derived from $\mathfrak{f}$ through *transformation by* $\mathfrak{g}$ or to be the *transform of* $\mathfrak{f}$ *by* $\mathfrak{g}$. Then $\mathfrak{f}$ is the transform of $\mathfrak{f}'$ by $\mathfrak{g}^{-1}$. From

$$\mathfrak{f}_1' = \mathfrak{g}\mathfrak{f}_1\mathfrak{g}^{-1}, \quad \mathfrak{f}_2' = \mathfrak{g}\mathfrak{f}_2\mathfrak{g}^{-1}$$

one gets

$$\mathfrak{f}_1'\mathfrak{f}_2' = \mathfrak{g}\mathfrak{f}_1\mathfrak{f}_2\mathfrak{g}^{-1}.$$

Thus the transform of a product by $\mathfrak{g}$ is the product of the transforms of each factor by $\mathfrak{g}$. If $\mathfrak{f}$ and $\mathfrak{f}'$ are given in (1), the transformer $\mathfrak{g}$ is only determined up to a right-hand factor permutable with $\mathfrak{f}$. The addition of such a factor evidently has no effect in (1); on the other hand, from

$$\mathfrak{g}\mathfrak{f}\mathfrak{g}^{-1} = \mathfrak{h}\mathfrak{f}\mathfrak{h}^{-1}$$

it is inferred that $\mathfrak{g}^{-1}\mathfrak{h}$ is permutable with $\mathfrak{f}$.

Let u be an invariant point of $\mathfrak{f}$ in (1). Then the image $\mathfrak{g}u$ of u by $\mathfrak{g}$ is an invariant point of $\mathfrak{f}'$, as already mentioned in §4.5. Likewise any invariant point of $\mathfrak{f}'$ is carried into one of $\mathfrak{f}$ by $\mathfrak{g}^{-1}$. Thus $\mathfrak{g}$ maps the set of invariant points of $\mathfrak{f}$ upon the set of invariant points of $\mathfrak{f}'$.

Since the product of two reversions is a motion and the product of a reversion and a motion is a reversion, both of the elements $\mathfrak{f}$ and $\mathfrak{f}'$ are motions or both reversions. Let $\mathfrak{s}$ first be a reflection in the line $\mathcal{L}(\mathfrak{s})$. Then $\mathfrak{s}' = \mathfrak{g}\mathfrak{s}\mathfrak{g}^{-1}$ is a reversion and, since it leaves the points of the line $\mathfrak{g}\mathcal{L}(\mathfrak{s})$ fixed individually, it is the reflection in that line. Thus a reflection is transformed by $\mathfrak{g}$ by applying the motion or reversion $\mathfrak{g}$ to the line of reflection. If the motion $\mathfrak{f}$ is produced as the product $\mathfrak{s}_1\mathfrak{s}_2$ of two reflections, it is seen that $\mathfrak{f}' = \mathfrak{g}\mathfrak{f}\mathfrak{g}^{-1}$ is the product of the reflections in the lines $\mathfrak{g}\mathcal{L}(\mathfrak{s}_1)$ and $\mathfrak{g}\mathcal{L}(\mathfrak{s}_2)$. In virtue of

$$[\mathfrak{g}\mathcal{L}(\mathfrak{s}_1), \mathfrak{g}\mathcal{L}(\mathfrak{s}_2)] = [\mathcal{L}(\mathfrak{s}_1), \mathcal{L}(\mathfrak{s}_2)]$$

it can be concluded that $\mathfrak{f}$ and $\mathfrak{f}'$ are motions of equal type (rotation, limit-rotation, translation) with equal angles of rotation (except for sign, if $\mathfrak{g}$ is a reversion) or displacements. On the other hand these conditions are sufficient for $\mathfrak{f}$ and $\mathfrak{f}'$ being the transforms of each other. For if $\mathfrak{f} = \mathfrak{s}_1\mathfrak{s}_2$ and $\mathfrak{f}' = \mathfrak{s}_1'\mathfrak{s}_2'$ and

$$[\mathcal{L}(\mathfrak{s}_1'), \mathcal{L}(\mathfrak{s}_2')] = [\mathcal{L}(\mathfrak{s}_1), \mathcal{L}(\mathfrak{s}_2)]$$

then the two pairs of lines $\mathcal{L}(\mathfrak{s}_1)$, $\mathcal{L}(\mathfrak{s}_2)$ and $\mathcal{L}(\mathfrak{s}_1')$, $\mathcal{L}(\mathfrak{s}_2')$, each taken in the order indicated, are congruent or symmetric. In particular, any limit-rotation can be carried into any other by transformation. So there is no essential invariant of limit-rotations.

Two reversed translations are the transforms of each other if their displacements are equal, and in that case only; this is immediately seen from their decomposition into reflections and translations.

§8 Non-euclidean trigonometry

8.1 The special trigonometric formulae. The investigations of §5.2 dealt with a configuration consisting of five pairs of points in cyclical order, any two neighbouring pairs being harmonic. These investigations are now resumed on the assumption that all pairs are of the special nature which in §6 were defined as the elliptic, parabolic and hyperbolic types. At least one of any two neighbouring pairs must then be hyperbolic, since two elliptic, two parabolic or an elliptic and a parabolic pair cannot be harmonic. If x^*, x is an elliptic pair, any pair y, z harmonic with x^*, x is the intersection of a circle through x^* and x with a circle of the hyperbolic pencil determined by x^* and x (§2.3), and y and z can thus neither be inverse with respect to $\mathcal{E}$, nor coincide on $\mathcal{E}$; if x, x is a parabolic pair, then any pair harmonic with x, x must have one of its points coinciding with x and thus cannot be a parabolic pair different from x, x; cf. (7.2). In the sequel parabolic pairs play essentially the same rôle as elliptic pairs and, for reasons of brevity, they may be regarded as included among the elliptic pairs as a special case if not mentioned specifically. From what has been said it follows that at least three of the five pairs are hyperbolic. Two neighbouring hyperbolic pairs determine two mutually orthogonal lines, and two neighbouring pairs one of which is hyperbolic, the other elliptic, determine a line and a point on that line; thus different cases occur according as there are five, four, or three hyperbolic pairs: A pentagon, quadrangle or triangle with five, three or one right angle. These will in short be spoken of as *right-angled pentagon, right-angled quadrangle* and *right-angled triangle*. The elliptic pairs (if any) determine the vertices of those angles which are not right angles; in particular, for parabolic pairs these angles are zero. Any two non-adjacent sides of a pentagon or quadrangle are divergent, since they have a common perpendicular. Thus all polygons are without double points and, moreover, they are convex, since all vertices are in the same half-plane determined by the line of any side. Every side carries a direction depending on the order in which the corresponding hyperbolic pair is taken. It is assumed that the order of each hyperbolic pair is so chosen as to effect a description of the polygon in the positive (as usually counterclockwise) sense. For elliptic pairs the proper point is always the second point of the pair; cf. §6.

The three cases in question are illustrated by Fig. 8.1–8.3.

As a result of §5 the relations (5.6) and (5.7) are valid between the pertinent five complex distances a_ν, and these relations have now to be interpreted in terms of the positive real sides and angles of the configurations. This requires the consideration of four different types of elements in comparison with the results of §6:

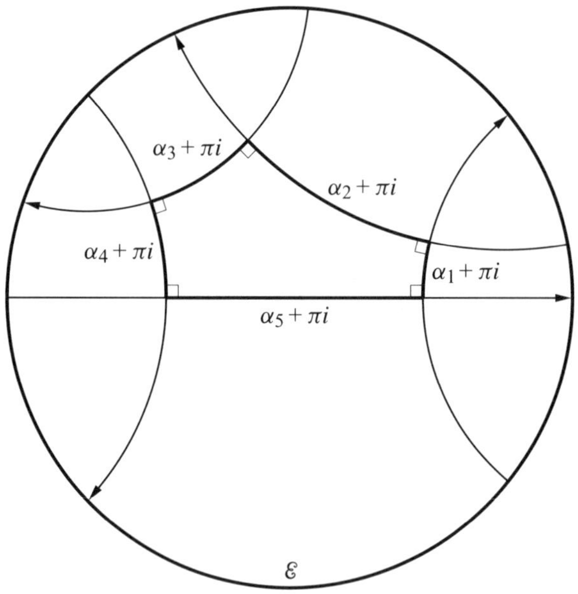

Figure 8.1

1) Sides which join the vertices of two right angles. A side of this type marks the distance between two lines. If the positive number α denotes its length, the complex

distance between the corresponding two hyperbolic pairs is (Fig. 6.3)

$$a = \alpha + \pi i,$$

the imaginary part arising from the fact that the directions of the two lines are opposite. This type covers the case of the right-angled pentagon completely and occurs in the case of the right-angled quadrangle.

2) Sides which join the vertices of a right and an oblique angle. (Quadrangle and triangle.) A side of this type marks the distance between a point and a line. If the side is directed from the point towards the line, the elliptic is the first and the hyperbolic the second pair and the side is the negative perpendicular of the line. Hence, if the positive number β denotes the length of the side, the complex distance from the elliptic to the hyperbolic pair is (Fig. 6.2)

$$a = \beta + \frac{\pi}{2} i.$$

If the side is directed from the line towards the point, it is the positive perpendicular of the line, but at the same time the pairs are interchanged, and the result is the same.

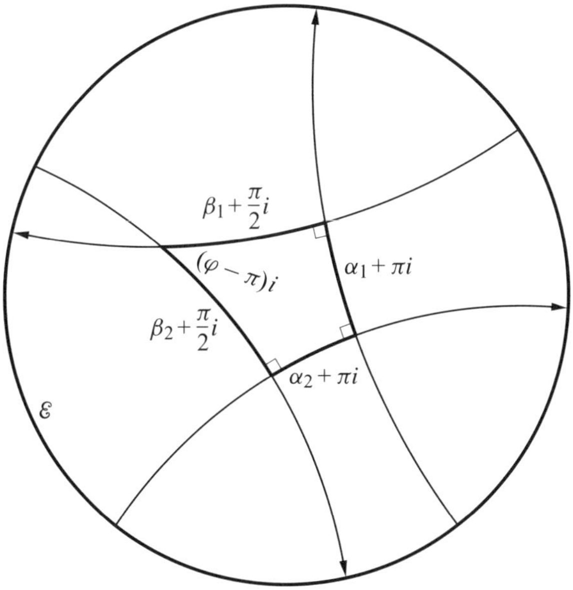

Figure 8.2

3) Sides which join the vertices of two oblique angles. (Triangle.) A side of this type marks the distance between two points. If γ denotes its length, even the complex distance between the two corresponding elliptic pairs is (Fig. 6.1)

$$a = \gamma.$$

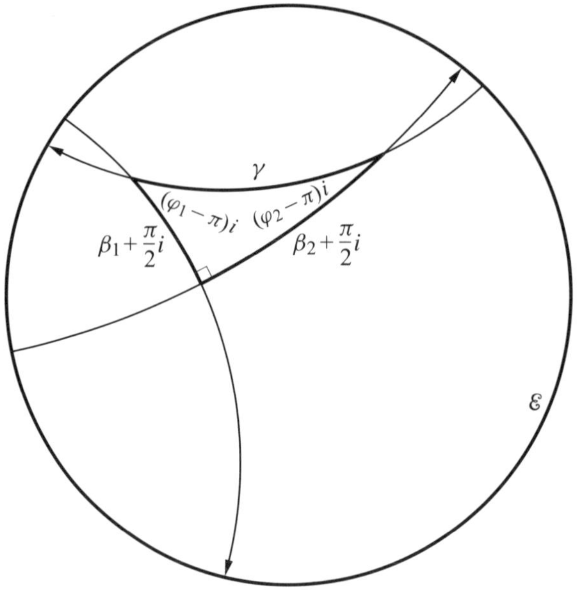

Figure 8.3

4) Oblique angles. (One in the quadrangle, two in the triangle.) Let φ denote the magnitude of such an angle inside the polygon. So far only the inequality $0 < \varphi < \pi$ is available (though it will later appear that actually $\varphi < \frac{\pi}{2}$). In order to determine the complex distance from the first to the second of the hyperbolic pairs pertaining to the sides which form the angle, it is observed that the first side is turned through an angle $\pi - \varphi$ about the intersection point into the second side. This would determine the distance in the case of Fig. 6.4, but since the intersection point is the second point of the elliptic pair harmonic with both hyperbolic pairs, the sign has to be reversed. Hence

$$a = (\varphi - \pi)i.$$

In each of the figures 8.1–8.3 the complex distances are inscribed on the corresponding side or angle. If these values are inserted in the formulae (6) and (7) of §5.2 and the transformation formulae

$$\begin{aligned}
\sinh(x \pm \pi i) &= -\sinh x \\
\cosh(x \pm \pi i) &= -\cosh x \\
\coth(x \pm \pi i) &= \coth x \\
\sinh(x + \tfrac{1}{2}\pi i) &= i\,\cosh x \\
\cosh(x + \tfrac{1}{2}\pi i) &= i\,\sinh x
\end{aligned}$$

$$\coth(x + \tfrac{1}{2}\pi i) = \tanh x$$
$$\sinh(\varphi - \pi)i = -i \sin\varphi$$
$$\cosh(\varphi - \pi)i = -\cos\varphi$$
$$\coth(\varphi - \pi)i = -i \cot\varphi$$

are taken into account, the following relations between the real elements in each of the polygons are easily deduced:

Right-angled Pentagon:

$$\cosh\alpha_\nu = \coth\alpha_{\nu-1}\cot\alpha_{\nu+1} \quad (\nu \bmod 5) \tag{V.1}$$

$$\cosh\alpha_\nu = \sinh\alpha_{\nu-2}\sinh\alpha_{\nu+2} \quad (\nu \bmod 5) \tag{V.2}$$

Right-angled Quadrangle:

$$\cos\varphi = \tanh\beta_1 \tanh\beta_2 \tag{IV.1}$$
$$\cos\varphi = \sinh\alpha_1 \sinh\alpha_2 \tag{IV.2}$$
$$\cosh\alpha_1 = \tanh\beta_1 \coth\alpha_2 \tag{IV.3}$$
$$\cosh\alpha_1 = \sin\varphi \cosh\beta_2 \tag{IV.4}$$
$$\sinh\beta_1 = \cot\varphi \coth\alpha_1 \tag{IV.5}$$
$$\sinh\beta_1 = \cosh\beta_2 \sinh\alpha_2 \tag{IV.6}$$

Right-angled Triangle:

$$\cosh\gamma = \cot\varphi_1 \cot\varphi_2 \tag{III.1}$$
$$\cosh\gamma = \cosh\beta_1 \cosh\beta_2 \tag{III.2}$$
$$\cos\varphi_1 = \coth\gamma \tanh\beta_1 \tag{III.3}$$
$$\cos\varphi_1 = \cosh\beta_2 \sin\varphi_2 \tag{III.4}$$
$$\sinh\beta_1 = \cot\varphi_1 \tanh\beta_2 \tag{III.5}$$
$$\sinh\beta_1 = \sin\varphi_2 \sinh\gamma. \tag{III.6}$$

In each of the cases triangle and quadrangle four equations can be added by interchanging the subscripts 1 and 2.

8.2 Properties of non-euclidean metric. From these formulae several fundamental properties of non-euclidean hyperbolic metric can be deduced. It has to be remembered that the hyperbolic functions are monotonous for an argument $\xi > 0$ and are all increasing apart from the coth, which is decreasing, and that the following inequalities hold:

$$\cosh\xi \geqq 1, \quad \sinh\xi \geqq 0, \quad 0 \leqq \tanh\xi < 1, \quad \coth\xi > 1.$$

From (III.2) is inferred that $\gamma > \beta_1,\ \gamma > \beta_2$: In a right-angled triangle the hypotenuse is greater than each of the other sides. Thus the shortest junction of a point and a line is the perpendicular. Furthermore, the inequality of triangles holds: In a triangle every side is smaller than the sum of the two others. For the projections of two sides on the third are smaller than the sides projected, and the sum of the projections is equal to or greater than the third side. It is a consequence of the inequality of triangles that the straight lines are the shortest paths in the non-euclidean metric.

From (IV.4) is inferred that $\alpha_1 < \beta_2$: In the right-angled quadrangle a side joining the vertices of two right angles is smaller than the opposite side. This implies that the common perpendicular of two divergent lines is the shortest joining segment. For a joining segment which is orthogonal to neither of the lines is greater than the perpendicular drawn from one of its end-points to the other line, and this perpendicular in turn, in virtue of the quality of right-angled quadrangles just quoted, is greater than the common perpendicular. Hence by a translation with displacement λ every point x not on the axis $\mathcal{A}$ of the translation is displaced more than λ. For, if perpendiculars to $\mathcal{A}$ are drawn through x and its image point, the shortest junction of these perpendiculars is the displacement λ of the translation.

From (III.3) is inferred that $\cos\varphi_1 > 0$, hence, since $0 < \varphi_1 < \pi$: In a right-angled triangle the two other angles are acute. Hence in any triangle at most one angle can be obtuse. For, let $\mathcal{P}$ be the join of the vertex v and the opposite side $\mathcal{S}$; if $\mathcal{P}$ is inside the triangle, then both angles adjacent to $\mathcal{S}$ are in right-angled triangles, and therefore are acute; if $\mathcal{P}$ is outside the triangle, then one of the angles adjacent to $\mathcal{S}$ is acute, and the angle at v, being the difference of two acute angles, is acute. From this can be concluded: At least one of the perpendiculars from the vertices of a triangle lies inside the triangle, namely the one starting from the vertex of the greatest angle.

From (III.1) is inferred that $\cot\varphi_1\ \cot\varphi_2 > 1$. In virtue of $0 < \varphi_1 < \frac{\pi}{2},\quad 0 < \varphi_2 < \frac{\pi}{2}$ this implies $\varphi_1 + \varphi_2 < \frac{\pi}{2}$: In a right-angled triangle the sum of the angles is inferior to π. Through decomposing an arbitrary triangle into two right-angled triangles by means of a perpendicular inside the triangle, this is seen to hold in the general case: In any triangle the sum of the angles is $< \pi$.

8.3 Area. In the sequel the notation polygonal region is taken to mean a region bounded by a finite number of closed polygons; among the vertices of these polygons there may be infinite points, and the region need not be connected. Any polygonal region $\mathcal{P}$ can be decomposed into a finite number of simply connected polygonal regions and, in particular, into triangles; for by the extensions of its sides the region is cut into convex polygons. Let ε_0 denote the number of vertices, ε_1 the number of sides, ε_2 the number of regions present in any such decomposition, and let ε_0' and ε_1' denote the numbers of those among these vertices and sides respectively which are situated on the boundary of $\mathcal{P}$. Then the number

$$\chi = \varepsilon_0 - \varepsilon_1 + \varepsilon_2$$

is called the *characteristic of the polygonal region* $\mathcal{P}$ and

$$\chi' = \varepsilon_0' - \varepsilon_1'$$

the boundary-characteristic of $\mathcal{P}$. These numbers are known to be independent of the decomposition applied on $\mathcal{P}$. Furthermore, let n be the number of vertices originally on the boundary of $\mathcal{P}$, and let φ_ν, $\nu = 1, 2, \ldots, n$ denote the sum of the interior angles of $\mathcal{P}$ at the vertex number ν, the sum φ_ν being in the interval from 0 to 2π. Then the *area* $\Phi(\mathcal{P})$ of $\mathcal{P}$ is defined by the equation

$$\Phi(\mathcal{P}) = \sum_{\nu=1}^{n} \left(1 - \frac{\varphi_\nu}{\pi}\right) - 2\chi + \chi'. \tag{1}$$

If $\mathcal{P}$ is simply connected, the characteristics are $\chi = 1$, $\chi' = 0$, and the formula for the area takes the form

$$\Phi(\mathcal{P}) = n - 2 - \frac{1}{\pi} \sum_{\nu=1}^{n} \varphi_\nu .$$

In particular, for a triangle one gets

$$\Phi(\mathcal{P}) = 1 - \frac{1}{\pi} \sum_{\nu=1}^{3} \varphi_\nu , \tag{2}$$

this being a positive number, since the sum of the angles is $< \pi$. It is seen from (2) that, on this definition, the unity of area is the area of a triangle with all three angles zero, thus with all three vertices infinite.

It is evident that this measure of area is invariant under motions and reversions. It will now be shown that it is additive, i.e. if $\mathcal{P}$ is decomposed into two polygonal regions $\mathcal{P}_1$ and $\mathcal{P}_2$, then

$$\Phi(\mathcal{P}) = \Phi(\mathcal{P}_1) + \Phi(\mathcal{P}_2). \tag{3}$$

To prove this, decompose $\mathcal{P}_1$ and $\mathcal{P}_2$ into simply connected polygonal regions. These two decompositions taken together constitute a decomposition of $\mathcal{P}$. Those sides and vertices which belong to the boundaries of $\mathcal{P}_1$ and $\mathcal{P}_2$ simultaneously are inner sides and vertices of the decomposition of $\mathcal{P}$ with the exception of such vertices (if any) which terminate polygonal chains used in cutting $\mathcal{P}$ into $\mathcal{P}_1$ and $\mathcal{P}_2$ and which, therefore, are on the boundary of $\mathcal{P}$. Let ε_0'' ($\geqq 0$) denote the number of these exceptional vertices. If symbols pertaining to $\mathcal{P}_1$ or $\mathcal{P}_2$ are marked with subscripts 1 and 2 respectively, one gets

$$-2\chi + \chi' = -2\chi_1 + \chi_1' - 2\chi_2 + \chi_2' + \varepsilon_0''.$$

To prove this equation, one only has to verify that every element not common to $\mathcal{P}_1$ and $\mathcal{P}_2$ is counted equally often in the two terms, that the same is true of those elements common to $\mathcal{P}_1$ and $\mathcal{P}_2$ which are interior to $\mathcal{P}$, and, finally, that the ε_0'' exceptional points are counted once in each of the six characteristics. – On the other hand the following relation is valid:

$$\sum_{\nu=1}^{n}\left(1-\frac{\varphi_\nu}{\pi}\right)=\sum_{\nu=1}^{n_1}\left(1-\frac{\varphi_{1,\nu}}{\pi}\right)+\sum_{\nu=1}^{n_2}\left(1-\frac{\varphi_{2,\nu}}{\pi}\right)-\varepsilon_0''.$$

For, if on the right such terms in both sums which pertain to the same vertex are taken together, they cancel, if the vertex is an inner point of $\mathcal{P}$, and they exceed the corresponding term on the left by 1, if the vertex is one of the exceptional vertices on the boundary of $\mathcal{P}$; thus in all the difference is ε_0''. The proof then is completed by addition of the two equations.

Since every polygonal region is decomposable into triangles, it is inferred from (2) and (3) that $\Phi(\mathcal{P})$ is positive for all $\mathcal{P}$. So $\Phi(\mathcal{P})$ has the fundamental properties of the measure of areas. In fact, such a measure is completely determined by the above properties up to a constant factor which can be fixed by the choice of a unit area. As this fact, however, will not be needed in the sequel, a more detailed discussion is omitted.

8.4 Sine amplitude. It is possible to associate with a non-euclidean triangle another quantity which presents an analogy with the euclidean measure of areas especially with respect to trigonometry but which does not possess the additive quality (3). Let $\gamma_1, \gamma_2, \gamma_3$ be the sides and $\varphi_1, \varphi_2, \varphi_3$ the opposite angles of some triangle and η_1, η_2, η_3 the perpendiculars from the vertices. Then by (III.6)

$$\sinh\gamma_1 \sinh\eta_1 = \sinh\gamma_1 \sinh\gamma_2 \sin\varphi_3 = \sinh\gamma_2 \sinh\eta_2.$$

Thus this magnitude, which will be denoted by σ is invariant for any permutation of the subscripts 1, 2, 3 and so belongs to the triangle as a whole; it is called the *sine amplitude* of the triangle:

$$\sigma = \sinh\gamma_\nu \sinh\eta_\nu = \sinh\gamma_{\nu-1} \sinh\gamma_\nu \sin\varphi_{\nu+1} \quad (\nu \bmod 3). \tag{4}$$

It is possible to express σ in terms of the sides only:

$$\sigma^2 = \begin{vmatrix} 1 & \cosh\gamma_1 & \cosh\gamma_2 \\ \cosh\gamma_1 & 1 & \cosh\gamma_3 \\ \cosh\gamma_2 & \cosh\gamma_3 & 1 \end{vmatrix}.$$

This formula is mentioned here without proof, as it will not be needed.

8.5 Projection of lines. Let $\mathcal{g}_0$ and $\mathcal{g}$ denote two lines, let δ denote their distance in case they are divergent, and φ their angle of intersection in case they are concurrent. The case of two parallel lines is included by putting $\delta = \varphi = 0$. Draw perpendiculars to $\mathcal{G}_0$ from all points of $\mathcal{G}$. The foot of all these perpendiculars constitute a segment or half-line of $\mathcal{G}_0$, which will be called the *projection of* $\mathcal{G}$ *on* $\mathcal{G}_0$; only if $\mathcal{G}$ and $\mathcal{G}_0$ are identical this projection is the whole of $\mathcal{G}_0$. The end-points of the said segment or half-line are the projections of the infinite points of $\mathcal{G}$. Let ρ denote the length of the projection. If $\mathcal{G}_0$ and $\mathcal{G}$ are divergent (Fig. 8.4), formula (IV.2) of Section 1 yields

$$\sinh\frac{\rho}{2}\sinh\delta = 1;$$

for the common perpendicular of $\mathcal{G}_0$ and $\mathcal{G}$ bisects the projection, since the figure is

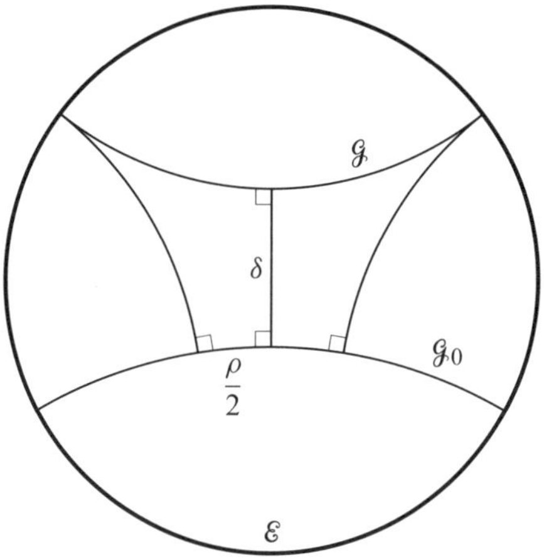

Figure 8.4

symmetric with respect to that perpendicular. If $\mathcal{G}_0$ and $\mathcal{G}$ are concurrent (Fig. 8.5), their intersection point bisects the projection, since the figure is reproduced by a half-turn around that point, and formula (III.4) of Section 1 yields

$$\cosh\frac{\rho}{2}\sin\varphi = 1.$$

These relations imply the following fact: Let $\mathcal{G}_0$ be a fixed line and κ a positive constant. All those lines $\mathcal{G}$ whose projection ρ on $\mathcal{G}_0$ is inferior to κ have a positive lower bound for their distance δ from $\mathcal{G}_0$, if divergent, or for their acute angle of intersection φ with $\mathcal{G}_0$, if concurrent.

8.6 Separation of collections of lines. The following elementary theorem of rather special nature will be useful in the sequel:

Let $\mathcal{G}_1$, $\mathcal{G}_2$, $\mathcal{G}_3$ *be three lines any two of which are divergent and which do not belong to one pencil, thus have no common perpendicular. Let* $\mathcal{G}_1'$, $\mathcal{G}_2'$, $\mathcal{G}_3'$ *be the*

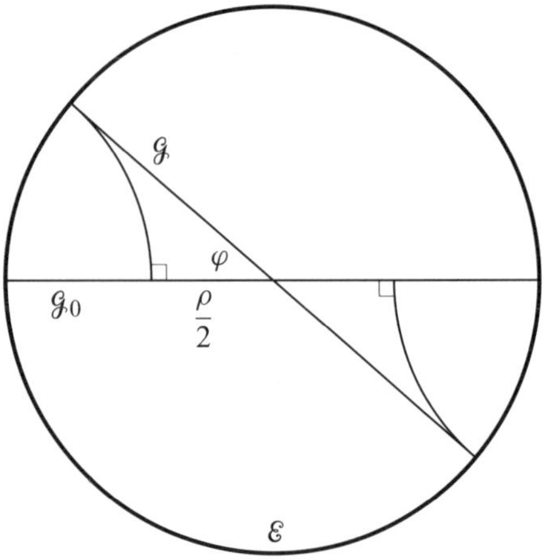

Figure 8.5

common perpendiculars of $\mathcal{g}_2$ and $\mathcal{g}_3$, of $\mathcal{g}_3$ and $\mathcal{g}_1$ and of $\mathcal{g}_1$ and $\mathcal{g}_2$. They are pairwise divergent, have no common perpendicular and separate each other in the same way as the lines $\mathcal{g}_1$, $\mathcal{g}_2$, $\mathcal{g}_3$, i.e. if the given lines do not separate each other, the same is true of the other set; and if one of the given lines separates the two others, the corresponding line of the other collection separates the two other lines.

Proof. To prove this, it is first assumed that $\mathcal{g}_1$, $\mathcal{g}_2$, $\mathcal{g}_3$ do not separate each other (Fig. 8.6). Then $\mathcal{g}_i$ and $\mathcal{g}_i'$ are divergent ($i = 1, 2, 3$); for otherwise $\mathcal{g}_i'$ would meet

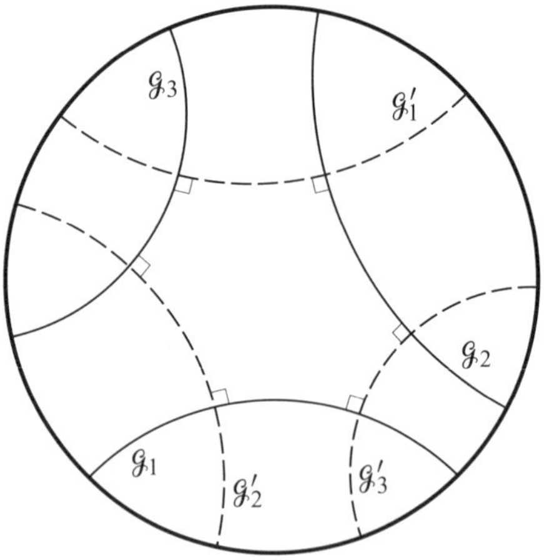

Figure 8.6

$\mathcal{G}_1$,$\mathcal{G}_2$ and $\mathcal{G}_3$, and these three lines then would separate each other. From this is inferred that the two other lines of the collection of $\mathcal{G}_i'$ are in the same half-plane determined by $\mathcal{G}_i'$, and thus the lines of the second collection do not separate each other.

Then, let $\mathcal{G}_2$ separate $\mathcal{G}_1$ and $\mathcal{G}_3$ (Fig. 8.7). From what has already been proved follows that one $\mathcal{G}_1'$, $\mathcal{G}_2'$, $\mathcal{G}_3'$ separates the two others (since the two collections play

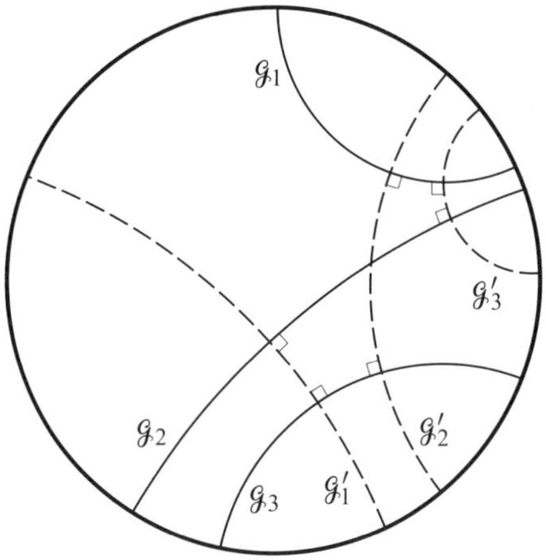

Figure 8.7

equivalent rôles). Let it be assumed that this were not $\mathcal{G}_2'$, but $\mathcal{G}_3'$, say. Then $\mathcal{G}_3$ and $\mathcal{G}_3'$ would intersect and the two collections of lines $\mathcal{G}_1, \mathcal{G}_3', \mathcal{G}_2, \mathcal{G}_2'$ and $\mathcal{G}_2, \mathcal{G}_1', \mathcal{G}_3, \mathcal{G}_2'$ would generate two right-angled quadrangles the oblique angles of which would have π as their sum. However, this is impossible, since the oblique angle of a right-angled quadrangle is acute as seen, e.g. from (IV.1) in Section 1. □

8.7 The general trigonometric formulae. Let six pairs of points x_ν, y_ν, $\nu = 1, 2, 3, 4, 5, 6$, be given in this order, the order of the points in each pair being as indicated. As in §5.2 only cyclical order of the pairs matters. Let furthermore any two neighbouring pairs be harmonic, and let a_ν denote the complex distance of the pairs $x_{\nu-1}, y_{\nu-1}$ and $x_{\nu+1}, y_{\nu+1}$ (ν mod 6). Further let x, y be the common harmonic pair of x_1, y_1 and x_4, y_4; then also y, x is. We put

$$-\log(xyx_4x_1) = -\log(yxx_1x_4) = b,$$

and for the complex distance of x_6, y_6 and x, y, which both are harmonic with x_1, y_1,

$$-\log(x_1y_1x_6x) = a_1',$$

and for the complex distance of x, y and x_2, y_2, which both are harmonic with x_1, y_1,

$$-\log(x_1 y_1 x x_2) = a_1''.$$

Then

$$a_1' + a_1'' = -\log(x_1 y_1 x_6 x_2) = a_1.$$

Similarly, we put for the complex distance of x_3, y_3 and x, y, which both are harmonic with x_4, y_4,

$$-\log(x_4 y_4 x_3 x) = a_4',$$

and for the complex distance of x, y and x_5, y_5, which both are harmonic with x_4, y_4,

$$-\log(x_4 y_4 x x_5) = a_4'',$$

so that

$$a_4' + a_4'' = -\log(x_4 y_4 x_3 x_5) = a_4.$$

Now the pairs $x, y; x_1, y_1; x_2, y_2; x_3, y_3; x_4, y_4$, taken in this cyclical order, satisfy the conditions of §5.2. The corresponding complex distances, taken in the same cyclical order, are

$$a_1'', \quad a_2, \quad a_3, \quad a_4', \quad b.$$

In the same way the pairs $y, x; x_4, y_4; x_5, y_5; x_6, y_6; x_1, y_1$, taken in this cyclical order, satisfy the conditions of §5.2, and the corresponding complex distances, taken in the same cyclical order, are

$$a_4'' + \pi i, \quad a_5, \quad a_6, \quad a_1' + \pi i, \quad b,$$

since the replacement of x by y produces the addition of πi in a_4'' and a_1'. Hence by applying equation (7) of §5.2 one gets

$$\begin{aligned} \cosh a_1'' &= -\sinh a_3 \sinh a_4', & \cosh(a_1' + \pi i) &= -\sinh a_5 \sinh(a_4'' + \pi i), \\ \sinh a_1'' &= -\cosh a_3 \sinh^{-1} b, & \sinh(a_1' + \pi i) &= -\cosh a_5 \sinh^{-1} b, \end{aligned}$$

whence by applying the transformation formulae of Section 1

$$\cosh a_1' = -\sinh a_5 \sinh a_4'', \quad \sinh a_1' = \cosh a_5 \sinh^{-1} b.$$

Hence, using $\sinh^{-2} b = \coth^2 b - 1$,

$$\begin{aligned} \cosh a_1 &= \cosh a_1'' \cosh a_1' + \sinh a_1'' \sinh a_1' \\ &= \sinh a_3 \sinh a_5 \sinh a_4' \sinh a_4'' - \cosh a_3 \cosh a_5 \sinh^{-2} b \\ &= \sinh a_3 \sinh a_5 \sinh a_4' \sinh a_4'' - \cosh a_3 \cosh a_5 \coth^2 b + \cosh a_3 \cosh a_5. \end{aligned}$$

By applying equation (6) of §5.2 one gets

$$-\coth a_3 \coth b = \cosh a_4' \tag{3}$$

$$-\coth a_5 \coth b = \cosh(a_4'' + \pi i) = -\cosh a_4'', \tag{4}$$

thus by inserting

$$\cosh a_1 = \sinh a_3 \sinh a_5 (\sinh a_4' \sinh a_4'' + \cosh a_4' \cosh a_4'') + \cosh a_3 \cosh a_5,$$

and finally

$$\cosh a_1 = \cosh a_3 \cosh a_5 + \sinh a_3 \sinh a_5 \cosh a_4.$$

Since indices only matter mod 6, the general formula is

$$\cosh a_\nu = \cosh a_{\nu-2} \cosh a_{\nu+2} + \sinh a_{\nu-2} \sinh a_{\nu+2} \cosh a_{\nu+3} \quad (\nu \bmod 6). \tag{7}$$

This formula (7) is now applied to six pairs of points each of which is either elliptic or hyperbolic with respect to the circle $\mathcal{E}$. Since two elliptic pairs cannot be harmonic, the two neighbours of an elliptic pair are hyperbolic, and the number of elliptic pairs cannot exceed three. There are thus 6, 5, 4, or 3 hyperbolic pairs. We denote these four cases by VI, V, IV, and III, respectively, and treat them in this order.

Case VI. Each hyperbolic pair corresponds to a directed line in $\mathcal{D}$. Let these lines in turn be denoted by $\mathcal{G}_1, \mathcal{G}_2, \mathcal{G}_3, \mathcal{G}_4, \mathcal{G}_5, \mathcal{G}_6$. Every two of the lines $\mathcal{G}_1, \mathcal{G}_3, \mathcal{G}_5$ are divergent, because they have a common normal. We further assume that none of them separates the two others, this being the only case we need in the sequel. The same then holds for $\mathcal{G}_2, \mathcal{G}_4, \mathcal{G}_6$ as seen in Section 6. Let $\mathcal{G}_{\nu+1}$ be the positive normal of $\mathcal{G}_\nu (\nu \bmod 6)$. The six lines form a convex hexagon with all its angles right.

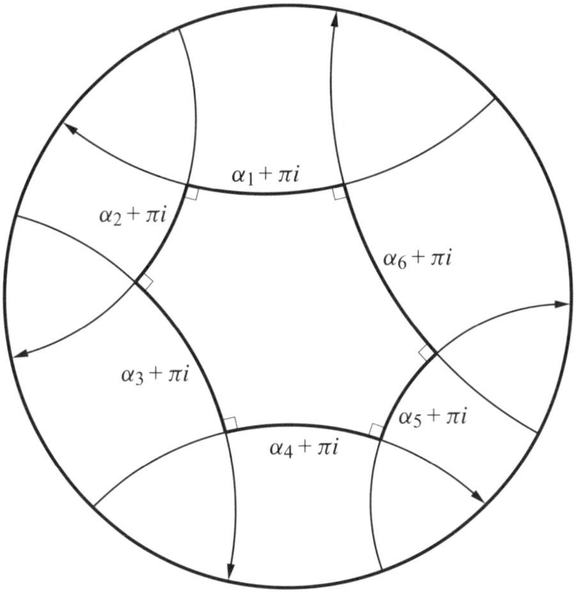

Figure 8.8

The complex distance of the pairs number $\nu - 1$ and $\nu + 1$ takes the value

$$a_\nu = \alpha_\nu + \pi i, \quad \alpha_\nu > 0,$$

for the same reason as in Fig. 8.1.

These distances are indicated in Fig. 8.8. Hence (7) yields, taking into account the transformation formulae of Section 1:

$$\begin{aligned}\cosh \alpha_\nu = &- \cosh \alpha_{\nu-2} \cosh \alpha_{\nu+2} \\ &+ \sinh \alpha_{\nu-2} \sinh \alpha_{\nu+2} \cosh \alpha_{\nu+3} \quad (\nu \bmod 6),\end{aligned} \tag{VI}$$

where the α are positive numbers indicating the lengths of the sides of the hexagon.

Case V. The five lines corresponding to the hyperbolic pairs form a pentagon in $\mathcal{D}$. The elliptic pair furnishes a vertex of that pentagon. Let φ be the angle of the pentagon at that vertex. The other four angles are right. If the direction of the five lines is as indicated in Fig. 8.9. and the elliptic vertex is to the left of the three lines not passing through it, then the pentagon is convex and $(\varphi - \pi)i$ is the complex distance of the two hyperbolic pairs neighbouring the elliptic pair. If β_1 and β_2 are the positive numbers indicating the lengths of these two sides, then $\beta_1 + \frac{\pi}{2}i$ and $\beta_2 + \frac{\pi}{2}i$ are the complex distances of the elliptic pair from its next neighbouring hyperbolic pairs. The three other distances are $\alpha_1 + \pi i$, $\alpha_3 + \pi i$, $\alpha_2 + \pi i$, where α_3 is the length of the side opposite to the angle φ. See Fig. 8.9. and compare Fig. 8.2.

Equation (7) yields in virtue of the transformation formulae of Section 1:

$$\cos \varphi = - \cosh \alpha_1 \cosh \alpha_2 + \sinh \alpha_1 \sinh \alpha_2 \cosh \alpha_3 \tag{V a}$$

$$\sinh \beta_1 = - \sinh \beta_2 \cosh \alpha_3 + \cosh \beta_2 \sinh \alpha_3 \cosh \alpha_2 \tag{V b}$$

$$\cosh \alpha_1 = - \cosh \alpha_2 \cos \varphi + \sinh \alpha_2 \sin \varphi \sinh \beta_2 \tag{V c}$$

$$\cosh \alpha_3 = \sinh \beta_1 \sinh \beta_2 - \cosh \beta_1 \cosh \beta_2 \cos \varphi. \tag{V d}$$

From (Vb) and (Vc) two other formulae result by the interchange of the subscripts 1 and 2.

Case IV. The four lines corresponding to the four hyperbolic pairs form a quadrangle in $\mathcal{D}$. One of these lines does not pass through any of the two vertices corresponding to the two elliptic pairs. We only need the case where these two vertices are on the same side of the line; then the quadrangle is convex. If the direction of the four lines is as indicated in Fig. 8.10, then the six complex distances are as indicated in that figure. One of them is real, namely γ. The two angles of the quadrangle at the side of length α are right.

Equation (7) yields in virtue of the transformation formulae of Section 1:

$$\cosh \gamma = - \sinh \beta_1 \sinh \beta_2 + \cosh \beta_1 \cosh \beta_2 \cosh \alpha \tag{IV a}$$

$$\cos \varphi_1 = - \cos \varphi_2 \cosh \alpha + \sinh \varphi_2 \sinh \alpha \sinh \beta_2 \tag{IV b}$$

$$\sinh \beta_1 = \sinh \beta_2 \cosh \gamma - \cosh \beta_2 \sinh \gamma \cos \varphi_2 \tag{IV c}$$

$$\cosh \alpha = \cos \varphi_1 \cos \varphi_2 + \sin \varphi_1 \sin \varphi_2 \cosh \gamma. \tag{IV d}$$

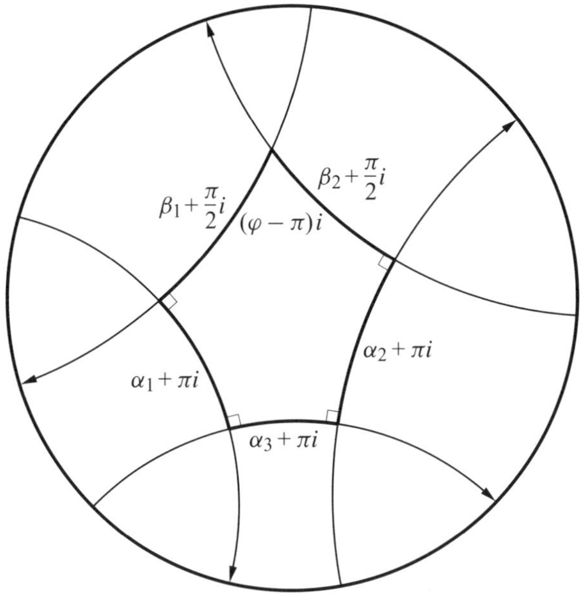

Figure 8.9

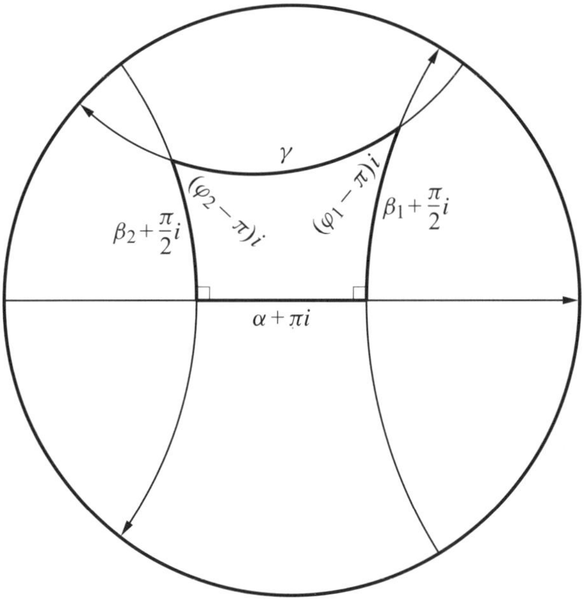

Figure 8.10

From (IVb) and (IVc) two other formulae result by the interchange of the subscripts 1 and 2.

Case III. The three lines corresponding to the three hyperbolic pairs form a triangle in $\mathcal{D}$. Each of the three elliptic pairs furnishes one vertex. If the direction of the three lines is as in Fig. 8.11, the six complex distances are those indicated in the figure. Three of them are real, the lengths of the sides of the triangle.

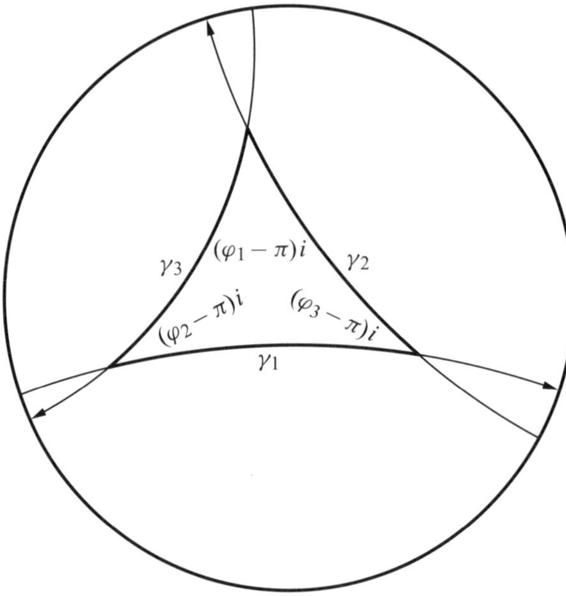

Figure 8.11

Equation (7) yields in virtue of the transformation formulae of Section 1:

$$\cosh \gamma_\nu = \cosh \gamma_{\nu-1} \cosh \gamma_{\nu+1} - \sinh \gamma_{\nu-1} \sinh \gamma_{\nu+1} \cos \varphi_\nu \quad (\nu \bmod 6) \qquad \text{(IIIa)}$$

$$\cos \varphi_\nu = -\cos \varphi_{\nu-1} \cos \varphi_{\nu+1} + \sin \varphi_{\nu-1} \sin \varphi_{\nu+1} \cosh \gamma_\nu \quad (\nu \bmod 6). \qquad \text{(IIIb)}$$

If parabolic pairs occur, the corresponding φ becomes zero, and the lengths of the sides including it become infinite. Equation (Va) retains its meaning for $\varphi = 0$, and so does (IVb), if $\varphi_1 = 0$ while φ_2 remains positive. Also (IIIb) retains its meaning if $\varphi_\nu = 0$ while $\varphi_{\nu-1}$ and $\varphi_{\nu+1}$ remain positive.

§9 Products and commutators of motions

9.1 Three motions with product 1. The investigations of the first four sections of this paragraph are intended to study in some detail the type of the product $\mathfrak{f}\mathfrak{g}$ of two given motions $\mathfrak{f}$ and $\mathfrak{g}$, where $\mathfrak{g} \neq \mathfrak{f}^{-1}$. If one puts

$$\mathfrak{h} = \mathfrak{g}^{-1}\mathfrak{f}^{-1},$$

the problem can be thus formulated: To determine the types of three motions $\mathfrak{f}$, $\mathfrak{g}$, $\mathfrak{h}$ for which

$$\mathfrak{f}\mathfrak{g}\mathfrak{h} = 1.$$

If the motions $\mathfrak{f}$ and $\mathfrak{g}$ are permutable, $\mathfrak{h}$ is permutable with both, and it is seen from §4.5 that all three motions are of the same type and have the same invariant points, these conditions being, on the other hand, sufficient for the permutableness. Two motions cannot interchange the invariant points of each other, thus the case mentioned in §4.5 of two involutions with respect to two harmonic pairs of points does not occur. If the three motions are rotations, they have their centre of rotation in common, and the sum of their angles of rotation with sign included equals $0 \pmod{2\pi}$. If they are translations, they have their axis in common, and the sum of their displacements equals 0 when taken with sign included according to an arbitrary direction of the common axis. If they are limit-rotations, they have their limit-centres in common. In this case there is no numerical invariant corresponding to the angle of a rotation or the displacement of a translation, but a substitute may be defined in the following way. Consider a directed horocycle $\mathcal{O}$ with limit-centre u, and let x, y, z be any three points of $\mathcal{O}$. Let $\mathfrak{f}$ denote the limit-rotation with limit-centre u for which $y = \mathfrak{f}x$; it exists because x and y are on the same horocycle with centre u. If we put $t = \mathfrak{f}z$, then $[x, z] = [\mathfrak{f}x, \mathfrak{f}z] = [y, t]$. Also the directed arc xz of $\mathcal{O}$ is congruent with the directed arc yt of $\mathcal{O}$, since the former is carried into the latter by $\mathfrak{f}$. Conversely, congruent arcs of $\mathcal{O}$ correspond to chords of $\mathcal{O}$ of equal length. Interchanging the rôle of y and z, let $\mathfrak{g}$ be that limit-rotation with limit-centre u for which $z = \mathfrak{g}x$. Then, since $\mathfrak{f}$ and $\mathfrak{g}$ are permutable,

$$\mathfrak{g}y = \mathfrak{g}\mathfrak{f}x = \mathfrak{f}\mathfrak{g}x = \mathfrak{f}z = t.$$

Hence $[x, y] = [z, t]$, and the corresponding directed arcs of $\mathcal{O}$ are congruent. Thus, without even introducing an analytic expression for the non-euclidean length of arcs of a horocycle (which could be done), one may measure the displacement of points of $\mathcal{O}$, brought about by a limit-rotation with limit-centre u, by directed arcs of $\mathcal{O}$, such arcs belonging to different initial points on $\mathcal{O}$ being congruent, and the product of two limit-rotations about u is then determined by the addition of directed arcs. Thus for the product $\mathfrak{f}\mathfrak{g}$

$$\operatorname{arc} xy + \operatorname{arc} xz = \operatorname{arc} xy + \operatorname{arc} yt = \operatorname{arc} xt$$

and by introducing $\mathfrak{h} = \mathfrak{g}^{-1}\mathfrak{f}^{-1}$ one gets for the product $\mathfrak{f}\mathfrak{g}\mathfrak{h}$

$$\operatorname{arc} xy + \operatorname{arc} yt + \operatorname{arc} tx,$$

which is the zero arc. – In the same way angles of rotations with centre c could be measured by arcs of a chosen directed circle with centre c, and displacements of translations with axis $\mathcal{A}$ by acts of a chosen directed hypercycle belonging to $\mathcal{A}$, and the analogy is then complete.

In the sequel it is assumed that $\mathfrak{f}$ and $\mathfrak{g}$ are not permutable. If it is possible to represent each of the motions $\mathfrak{f}$ and $\mathfrak{g}$ as product of two involutory transformations (reflections or half-turns) in such a way that the second factor of $\mathfrak{f}$ is the same as the first of $\mathfrak{g}$, thus

$$\mathfrak{f} = \mathfrak{s}\mathfrak{s}', \quad \mathfrak{g} = \mathfrak{s}'\mathfrak{s}''$$

then the third motion is

$$\mathfrak{h} = \mathfrak{s}''\mathfrak{s},$$

and the classification looked for will result from the mutual disposition of the lines of reflection $\mathcal{L}(\mathfrak{s})$, $\mathcal{L}(\mathfrak{s}')$, $\mathcal{L}(\mathfrak{s}'')$ or centres of half-turns $c(\mathfrak{s})$, $c(\mathfrak{s}')$, $c(\mathfrak{s}'')$. All three are reflections or all three half-turns, since $\mathfrak{f}$ and $\mathfrak{g}$ are motions. Now, such a representation is always possible with the only exception of the case in which $\mathfrak{f}$ and $\mathfrak{g}$ have an infinite invariant point in common, thus are either two translations with parallel axes or a translation and a limit-rotation the centre of which is end-point of the axis of translation. (If both were limit-rotations with common invariant point, they would be permutable.)

As this exceptional case plays a rather unessential rôle later on, it is briefly dealt with beforehand. Let $\mathfrak{f}$ and $\mathfrak{g}$ have the infinite point u as common invariant point. Then u is also invariant point of $\mathfrak{f}\mathfrak{g}$. So this product can only be a translation or limit-rotation. The latter requires that horocycles with centre u are reproduced by $\mathfrak{f}\mathfrak{g}$. Any two such horocycles cut out segments of equal length on all lines of the pencil with infinite point u. Therefore $\mathfrak{f}\mathfrak{g}$ is a limit-rotation if $\mathfrak{f}$ and $\mathfrak{g}$ are translations with displacements of equal amount but opposite sign, and in that case only.

9.2 Composition by half-turns. In the expositions to follow it is not necessary to keep limit-rotations apart from rotations, and they will therefore be thought of as included among the rotations unless mentioned specifically. The representation of $\mathfrak{f}$ and $\mathfrak{g}$ by means of three reflections or by means of three half-turns are referred to as cases I and II respectively.

As seen from §7.3, case II occurs if $\mathfrak{f}$ and $\mathfrak{g}$ are translations with concurrent axes, and in that case only. The intersection point of these axes then has to be chosen as the centre $c(\mathfrak{s}')$ of the half-turn $\mathfrak{s}'$, and hereupon $c(\mathfrak{s})$ and $c(\mathfrak{s}'')$ are uniquely determined by the fact that

$$[c(\mathfrak{s}'), c(\mathfrak{s})] = \tfrac{1}{2}\lambda(\mathfrak{f}), \quad [c(\mathfrak{s}''), c(\mathfrak{s}')] = \tfrac{1}{2}\lambda(\mathfrak{g}),$$

that $c(\mathfrak{s})$ is situated on $\mathcal{A}(\mathfrak{f})$ and $c(\mathfrak{s}'')$ on $\mathcal{A}(\mathfrak{g})$ and that the directions from $c(\mathfrak{s}')$ towards $c(\mathfrak{s})$ and from $c(\mathfrak{s}'')$ towards $c(\mathfrak{s}')$ agree with the directions of the translations $\mathfrak{f}$ and $\mathfrak{g}$. All three motions $\mathfrak{f}$, $\mathfrak{g}$, $\mathfrak{h}$ are translations, and their axes determine a triangle whose vertices are the centres of the half-turns $\mathfrak{s}$, $\mathfrak{s}'$, $\mathfrak{s}''$ and whose sides equal half the displacements of $\mathfrak{f}$, $\mathfrak{g}$ and $\mathfrak{h}$ (Fig. 9.1).

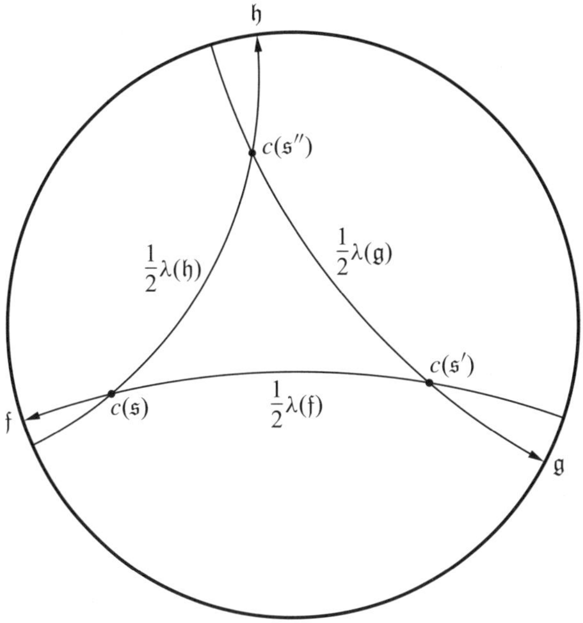

Figure 9.1

Conversely, if any three proper points are chosen as centres of three half-turns, any two of these will have a translation as their product, and the product (suitably arranged) of all three translations will be 1. If all three points are on the same line, the translations are permutable; if not, one is in the case II just treated.

9.3 Composition by reflections. The cases hitherto not treated are the following: $\mathfrak{f}$ and $\mathfrak{g}$ translations with non-concurrent axes, $\mathfrak{f}$ translation and $\mathfrak{g}$ rotation, the centre c not being end-point of $\mathcal{A}$, (or conversely), and both $\mathfrak{f}$ and $\mathfrak{g}$ rotations. These are all contained in case I, for there always exists a (uniquely determined) line at right angles to the axes of the translations (if any) and passing through the centres of the rotations (if any) pertaining to $\mathfrak{f}$ and $\mathfrak{g}$. This line is, according to the situation in question, the common perpendicular of $\mathcal{A}(\mathfrak{f})$ and $\mathcal{A}(\mathfrak{g})$, the perpendicular from $c(\mathfrak{f})$ on $\mathcal{A}(\mathfrak{g})$ or the line joining $c(\mathfrak{f})$ and $c(\mathfrak{g})$. This line has to be taken as line $\mathcal{L}(\mathfrak{s}')$ of the reflection $\mathfrak{s}'$. The lines $\mathcal{L}(\mathfrak{s})$ and $\mathcal{L}(\mathfrak{s}'')$ are then uniquely determined in the following way. If $\mathfrak{f}$ is a translation, $\mathcal{L}(\mathfrak{s})$ is at right angles to $\mathcal{A}(\mathfrak{f})$, its distance from $\mathcal{L}(\mathfrak{s}')$ equals $\frac{1}{2}\lambda(\mathfrak{f})$ and it is situated on that side of $\mathcal{L}(\mathfrak{s}')$ toward which $\mathfrak{f}$ points. If $\mathfrak{f}$ is a rotation, $\mathcal{L}(\mathfrak{s})$ passes through $c(\mathfrak{f})$ at an angle $\frac{1}{2}\varphi(\mathfrak{f})$ with $\mathcal{L}(\mathfrak{s}')$ and the shortest turn from $\mathcal{L}(\mathfrak{s}')$ to $\mathcal{L}(\mathfrak{s})$ coincides with the sense of rotation of $\mathfrak{f}$. If $\mathfrak{f}$ is a limit-rotation, in which case $\varphi(\mathfrak{f})$

has to be replaced by zero, $\mathcal{L}(\mathfrak{s})$ bisects the segment joining a point of $\mathcal{L}(\mathfrak{s}')$ with its image by $\mathfrak{f}$ just as in the case of a rotation. (This interpretation of the equivalent of the angle of rotation in the case of limit-rotations will not be repeated in similar cases hereafter.) In a similar way $\mathcal{L}(\mathfrak{s}'')$ is obtained by $\mathfrak{g}$, the direction from $\mathcal{L}(\mathfrak{s}'')$ to $\mathcal{L}(\mathfrak{s}')$ or the shortest turn from $\mathcal{L}(\mathfrak{s}'')$ to $\mathcal{L}(\mathfrak{s}')$ agreeing with the direction of translation or the sense of rotation of $\mathfrak{g}$. The motion $\mathfrak{h}$ is a translation or a rotation according as $\mathcal{L}(\mathfrak{s})$ and $\mathcal{L}(\mathfrak{s}'')$ are divergent or have a point in common. So the number of translations among the motions $\mathfrak{f}$, $\mathfrak{g}$, $\mathfrak{h}$ can be 3, 2, 1 or 0. Accordingly case I splits up into four sub-cases, which will be denoted by

$$\mathrm{I}(\to\to\to), \quad \mathrm{I}(\to\to \odot), \quad \mathrm{I}(\to \odot \odot), \quad \mathrm{I}(\odot \odot \odot)$$

the symbols in the parenthesis hinting to translations and rotations, and limit-rotations being, in this context, comprised among rotations.

To survey the different possibilities in question, it is convenient to take as starting point any three lines and to make them play the rôle of lines of reflection $\mathcal{L}(\mathfrak{s})$, $\mathcal{L}(\mathfrak{s}')$, $\mathcal{L}(\mathfrak{s}'')$. Then the products $\mathfrak{f} = \mathfrak{s}\mathfrak{s}'$, $\mathfrak{g} = \mathfrak{s}'\mathfrak{s}''$, $\mathfrak{h} = \mathfrak{s}''\mathfrak{s}$ are three motions with 1 as their product. The case of permutableness, i.e. the case of the three lines belonging to one pencil, having been excluded, it is always the matter of case I. Any two of the three lines have a common perpendicular or a common point (proper or infinite). These common perpendiculars or points are the axes or centres of the motions $\mathfrak{f}$, $\mathfrak{g}$ and $\mathfrak{h}$.

If every pair of the lines $\mathcal{L}(\mathfrak{s})$, $\mathcal{L}(\mathfrak{s}')$, $\mathcal{L}(\mathfrak{s}'')$ are divergent, $\mathfrak{f}$, $\mathfrak{g}$ and $\mathfrak{h}$ are translations; thus the case is $\mathrm{I}(\to\to\to)$. The three lines of reflection together with the three axes of translation determine a hexagon with six right angles. The sides opposite the lines of reflection are measured by half the displacements. According to §8.6 the lines $\mathcal{L}(\mathfrak{s})$, $\mathcal{L}(\mathfrak{s}')$, $\mathcal{L}(\mathfrak{s}'')$ separate each other in the same way as do the lines $\mathcal{A}(\mathfrak{g})$, $\mathcal{A}(\mathfrak{h})$, $\mathcal{A}(\mathfrak{f})$ in this order. If the lines of one collection do not separate each other (compare Fig. 8.6), the hexagon is convex (main case $\mathrm{I}(\to\to\to)$ Fig. 9.2), if they separate each other, the hexagon has a double point (by-case $\mathrm{I}(\to\to\to)$ – compare Fig. 8.7). If two of the lines of reflection have a point in common – let it be $\mathcal{L}(\mathfrak{s})$ and $\mathcal{L}(\mathfrak{s}')$ – while the third, $\mathcal{L}(\mathfrak{s}'')$, has no point in common with them, it is the case of a rotation $\mathfrak{f}$ and two translations $\mathfrak{g}$ and $\mathfrak{h}$, thus $\mathrm{I}(\to\to \odot)$. The five lines $\mathcal{L}(\mathfrak{s})$, $\mathcal{L}(\mathfrak{s}')$, $\mathcal{L}(\mathfrak{s}'')$, $\mathcal{A}(\mathfrak{g})$, $\mathcal{A}(\mathfrak{h})$ and the centre $c(\mathfrak{f})$ determine a pentagon with four right angles. The angle at $c(\mathfrak{f})$ is $\frac{1}{2}\varphi(\mathfrak{f})$ and the segments cut out on $\mathcal{A}(\mathfrak{g})$ and $\mathcal{A}(\mathfrak{h})$ are $\frac{1}{2}\lambda(\mathfrak{g})$ and $\frac{1}{2}\lambda(\mathfrak{h})$.

Fig. 9.3 illustrates the case in which the common point of $\mathcal{L}(\mathfrak{s})$ and $\mathcal{L}(\mathfrak{s}')$ is situated in the strip between $\mathcal{A}(\mathfrak{g})$ and $\mathcal{A}(\mathfrak{h})$; in this case the pentagon is convex. If the angle of the pentagon at $c(\mathfrak{f})$ is acute, or at most $\frac{1}{2}\pi$, we speak of the main case $\mathrm{I}(\to\to \odot)$; this will be of special interest in the sequel. If the common point of $\mathcal{L}(\mathfrak{s})$ and $\mathcal{L}(\mathfrak{s}')$ is situated outside the strip between $\mathcal{A}(\mathfrak{g})$ and $\mathcal{A}(\mathfrak{h})$, the pentagon has a double point. If the common point is situated on $\mathcal{A}(\mathfrak{g})$ or $\mathcal{A}(\mathfrak{h})$, the pentagon degenerates into a quadrangle. Examples of these by-cases of $\mathrm{I}(\to\to \odot)$ are easily constructed.

If two of the lines of reflection are divergent – let it be $\mathcal{L}(\mathfrak{s})$ and $\mathcal{L}(\mathfrak{s}'')$ – while each of them has a point in common with the third, $\mathcal{L}(\mathfrak{s}')$, it is the case of two rotations $\mathfrak{f}$ and $\mathfrak{g}$ and one translation $\mathfrak{h}$, thus $\mathrm{I}(\to \odot \odot)$. The four lines $\mathcal{L}(\mathfrak{s})$, $\mathcal{L}(\mathfrak{s}')$, $\mathcal{L}(\mathfrak{s}'')$, $\mathcal{A}(\mathfrak{h})$

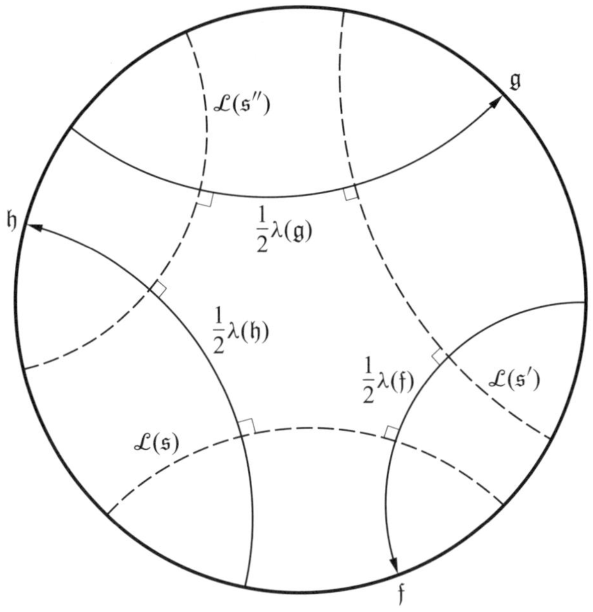

Figure 9.2

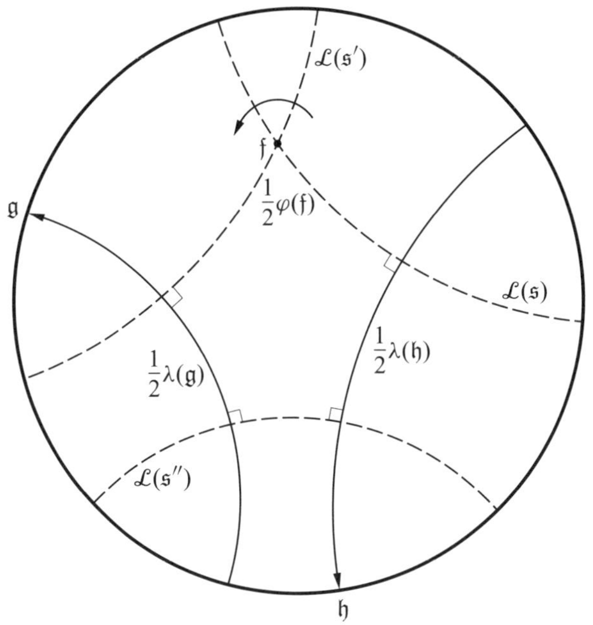

Figure 9.3

and the centres $c(\mathfrak{f})$ and $c(\mathfrak{g})$ determine a quadrangle with two adjacent right angles. The angles at $c(\mathfrak{f})$ and $c(\mathfrak{g})$ are $\frac{1}{2}\varphi(\mathfrak{f})$ and $\frac{1}{2}\varphi(\mathfrak{g})$, and the side opposite $\mathcal{L}(\mathfrak{s}')$ is $\frac{1}{2}\lambda(\mathfrak{h})$ (Fig. 9.4).

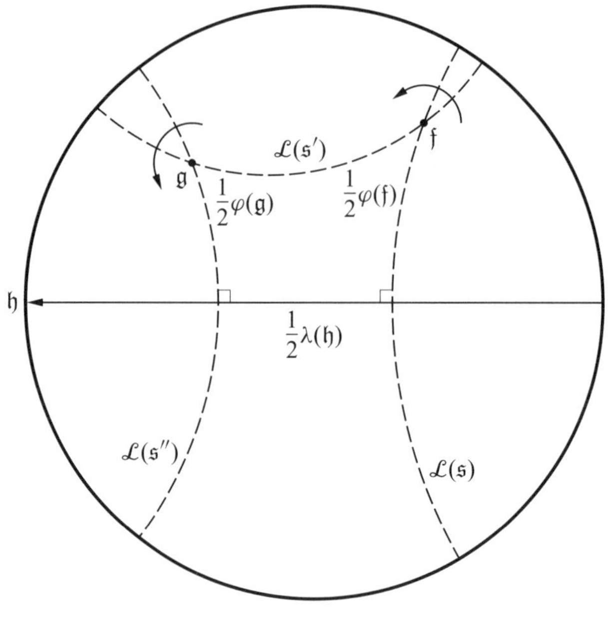

Figure 9.4

Here a special case has to be mentioned in which $\mathcal{L}(\mathfrak{s}')$ is the common perpendicular of $\mathcal{L}(\mathfrak{s})$ and $\mathcal{L}(\mathfrak{s}'')$, thus $\frac{1}{2}\varphi(\mathfrak{f}) = \frac{1}{2}\varphi(\mathfrak{g}) = \frac{\pi}{2}$. In that case the quadrangle degenerates into the segment $c(\mathfrak{f})c(\mathfrak{g})$ (Fig. 9.5).

This case is marked $\mathrm{I}(\rightarrow \circlearrowleft_2 \circlearrowleft_2)$, the subscript 2 indicating that both rotations are half-turns, thus of period 2. It is now assumed that not both of the angles $\frac{1}{2}\varphi(\mathfrak{f})$ and $\frac{1}{2}\varphi(\mathfrak{g})$ are right. The main case $\mathrm{I}(\rightarrow \circlearrowleft \circlearrowleft)$ is here taken to be the case in which there are adjacent angles at $c(\mathfrak{f})$ and $c(\mathfrak{g})$ on the same side of $\mathcal{L}(\mathfrak{s}')$ which are both smaller than or at most one equal to $\frac{\pi}{2}$ (Fig. 9.4). The quadrangle then is convex with interior angles at $c(\mathfrak{f})$ and $c(\mathfrak{g})$ smaller than or at most one equal to $\frac{\pi}{2}$. The possible by-cases are the following: A convex quadrangle with an obtuse angle; a quadrangle with a double point; a quadrangle degenerated into a triangle. The main case can be characterized by the fact that the sense of both rotations is the same; in the particular case one of them is a half-turn.

If every pair of the lines of reflection are non-divergent, $\mathfrak{f}$, $\mathfrak{g}$, and $\mathfrak{h}$ are rotations; thus the case is $\mathrm{I}(\circlearrowleft \circlearrowleft \circlearrowleft)$. The three lines $\mathcal{L}(\mathfrak{s})$, $\mathcal{L}(\mathfrak{s}')$, $\mathcal{L}(\mathfrak{s}'')$ and the three centres $c(\mathfrak{f}), c(\mathfrak{g}), c(\mathfrak{h})$ determine a triangle the angles of which are half the angles of rotation of $\mathfrak{f}$, $\mathfrak{g}$ and $\mathfrak{h}$ (Fig. 9.6). This case is spoken of as the main case or by-case $\mathrm{I}(\circlearrowleft \circlearrowleft \circlearrowleft)$

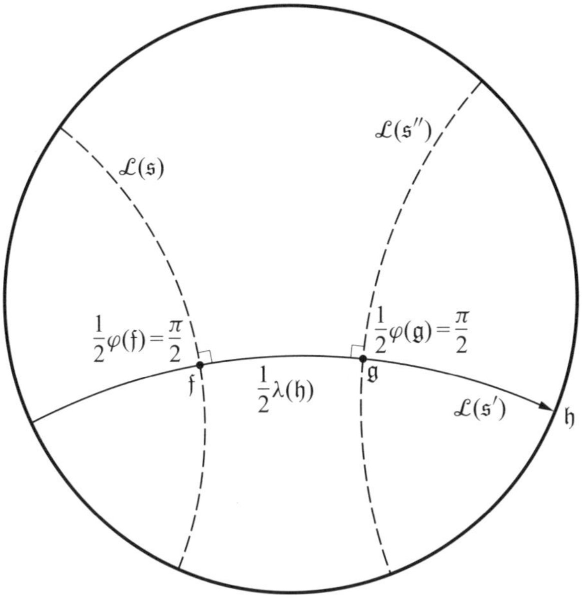

Figure 9.5

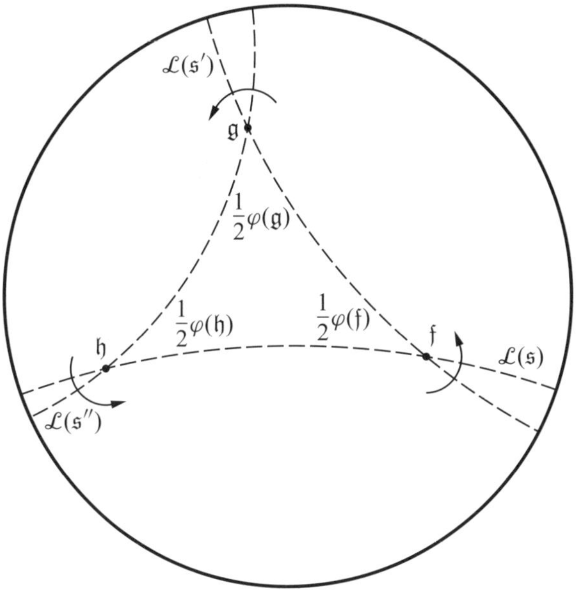

Figure 9.6

according as the triangle is acute or obtuse. The main case is characterized by the fact that the senses of all three rotations is the same; in particular, one of them can be a half-turn.

By a common definition, the main cases are those cases in which the polygons determined by the lines of reflection and axes of translation are convex and all their angles are acute.

Wherever it is necessary in the sequel to distinguish between rotations and limit-rotations, the latter will be marked $\odot_\infty$ or $\hookrightarrow$ instead of $\odot$ in the notation of types.

9.4 Calculation of invariants in the symmetric case. Let two translations $\mathfrak{g}$ and $\mathfrak{h}$ with equal displacements $\lambda(\mathfrak{g}) = \lambda(\mathfrak{f}) = \lambda$ and with divergent axes at the distances $[\mathcal{A}(\mathfrak{g}), \mathcal{A}(\mathfrak{h})] = \delta$ and opposite directions be given. If one puts $\mathfrak{f} = \mathfrak{h}^{-1}\mathfrak{g}^{-1}$, what is the value of the invariant of $\mathfrak{f}$ (i.e. the displacement $\lambda(\mathfrak{f})$ or angle of rotation $\varphi(\mathfrak{f})$, as the case may be)? The case $\mathrm{I}(\rightarrow\rightarrow\rightarrow)$ or $\mathrm{I}(\rightarrow\rightarrow \odot)$ occurs according as $\mathfrak{f}$ is a translation or rotation. In the case $\mathrm{I}(\rightarrow\rightarrow\rightarrow)$ the perpendicular bisecting the segment of $\mathcal{L}(\mathfrak{s}'')$ between $\mathcal{A}(\mathfrak{g})$ and $\mathcal{A}(\mathfrak{h})$ (cf. Fig. 9.2) is a line of symmetry of the figure, thus only the main case can arise. In the case $\mathrm{I}(\rightarrow\rightarrow \odot)$ the same line is line of symmetry, and the pentagon therefore is convex (cf. Fig. 9.3); the angle at $c(\mathfrak{f})$ may possibly be obtuse. In both cases the said perpendicular cuts the hexagon or pentagon into two congruent parts which are right-angled pentagons or right-angled quadrangles respectively. The formulae §8(V.2) and (IV.2) yield in the two cases

$$\left.\begin{aligned}\cosh\tfrac{1}{4}\lambda(\mathfrak{f})\\ \cos\tfrac{1}{4}\varphi(\mathfrak{f})\end{aligned}\right\} = \sinh\frac{\delta}{2}\sinh\frac{\lambda}{2}.$$

In particular, it is inferred that $\mathfrak{f}$ is a translation, limit-rotation, or rotation according as

$$\sinh\frac{\delta}{2}\sinh\frac{\lambda}{2} \gtreqless 1.$$

The rôles of δ and λ can be interchanged.

9.5 Relations for three commutators. The commutator of two transformations $\mathfrak{f}$ and $\mathfrak{g}$ is defined as the transformation

$$\mathfrak{k}_{\mathfrak{f}\mathfrak{g}} = \mathfrak{k}(\mathfrak{f}\mathfrak{g}) = \mathfrak{f}\mathfrak{g}\mathfrak{f}^{-1}\mathfrak{g}^{-1}.$$

$\mathfrak{f}$ and $\mathfrak{g}$ are permutable, if $\mathfrak{k}_{\mathfrak{f}\mathfrak{g}} = 1$, and conversely. One gets

$$\mathfrak{k}_{\mathfrak{f}\mathfrak{g}} = \mathfrak{k}_{\mathfrak{g}\mathfrak{f}}^{-1}.$$

Let three transformations be represented as products of three involutory transformations $\mathfrak{s}, \mathfrak{s}', \mathfrak{s}''$: $\mathfrak{f} = \mathfrak{s}\mathfrak{s}'$, $\mathfrak{g} = \mathfrak{s}'\mathfrak{s}''$, $\mathfrak{h} = \mathfrak{s}''\mathfrak{s}$, thus $\mathfrak{f}\mathfrak{g}\mathfrak{h} = 1$. Then

$$\begin{aligned} \mathfrak{k}_{\mathfrak{f}\mathfrak{g}} &= (\mathfrak{s}\mathfrak{s}''\mathfrak{s}')^2 = (\mathfrak{s}\mathfrak{g}^{-1})^2 = (\mathfrak{h}^{-1}\mathfrak{s}')^2 \\ \mathfrak{k}_{\mathfrak{g}\mathfrak{h}} &= (\mathfrak{s}'\mathfrak{s}\mathfrak{s}'')^2 = (\mathfrak{s}'\mathfrak{h}^{-1})^2 = (\mathfrak{f}^{-1}\mathfrak{s}'')^2 \\ \mathfrak{k}_{\mathfrak{h}\mathfrak{f}} &= (\mathfrak{s}''\mathfrak{s}'\mathfrak{s})^2 = (\mathfrak{s}''\mathfrak{f}^{-1})^2 = (\mathfrak{g}^{-1}\mathfrak{s})^2. \end{aligned} \tag{1}$$

From this is inferred that

$$\begin{aligned} \mathfrak{s}\mathfrak{k}_{\mathfrak{f}\mathfrak{g}}\mathfrak{s} &= \mathfrak{k}_{\mathfrak{h}\mathfrak{f}} \\ \mathfrak{s}'\mathfrak{k}_{\mathfrak{g}\mathfrak{h}}\mathfrak{s}' &= \mathfrak{k}_{\mathfrak{f}\mathfrak{g}} \\ \mathfrak{s}''\mathfrak{k}_{\mathfrak{h}\mathfrak{f}}\mathfrak{s}'' &= \mathfrak{k}_{\mathfrak{g}\mathfrak{h}}, \end{aligned} \tag{2}$$

and that

$$\begin{aligned} \mathfrak{h}\mathfrak{k}_{\mathfrak{f}\mathfrak{g}}\mathfrak{h}^{-1} &= \mathfrak{k}_{\mathfrak{g}\mathfrak{h}} \\ \mathfrak{f}\mathfrak{k}_{\mathfrak{g}\mathfrak{h}}\mathfrak{f}^{-1} &= \mathfrak{k}_{\mathfrak{h}\mathfrak{f}} \\ \mathfrak{g}\mathfrak{k}_{\mathfrak{h}\mathfrak{f}}\mathfrak{g}^{-1} &= \mathfrak{k}_{\mathfrak{f}\mathfrak{g}}. \end{aligned} \tag{3}$$

Thus the three commutators of $\mathfrak{f}$, $\mathfrak{g}$, $\mathfrak{h}$ are transformed into each other both by these transformations themselves and by the involutory transformations $\mathfrak{s}, \mathfrak{s}', \mathfrak{s}''$. The same is true of the inverse commutators $\mathfrak{k}_{\mathfrak{g}\mathfrak{f}}, \mathfrak{k}_{\mathfrak{h}\mathfrak{g}}, \mathfrak{k}_{\mathfrak{f}\mathfrak{h}}$.

9.6 Foot-triangle. In this section case I is considered, i.e. $\mathfrak{s}, \mathfrak{s}', \mathfrak{s}''$ are reflections in three lines $\mathcal{L}(\mathfrak{s})$, $\mathcal{L}(\mathfrak{s}')$, $\mathcal{L}(\mathfrak{s}'')$ which do not belong to one pencil of lines. Thus the commutators (1) are $\neq 1$. A product of three reflections, like $\mathfrak{s}\mathfrak{s}''\mathfrak{s}'$, is a reversion and cannot be a reflection, since by (1) its square is $\neq 1$. Hence it is a reversed translation, and the commutators (1) are translations.

In order to construct their axes, consider the configuration consisting of the three reflection lines $\mathcal{L}(\mathfrak{s})$, $\mathcal{L}(\mathfrak{s}')$, $\mathcal{L}(\mathfrak{s}'')$; this will be called a *trilateral*. By the perpendicular $\mathcal{T}$ of one of the sides of the trilateral, say $\mathcal{L}(\mathfrak{s})$, is meant the line orthogonal to $\mathcal{L}(\mathfrak{s})$ and also orthogonal to the common perpendicular $\mathcal{A}(\mathfrak{g})$ of the two other sides, $\mathcal{L}(\mathfrak{s}')$ and $\mathcal{L}(\mathfrak{s}'')$, in case these are divergent, or passing through their common point $c(\mathfrak{g})$ in case they are non-divergent. The intersection point p of $\mathcal{T}$ and $\mathcal{L}(\mathfrak{s})$ is called the *foot* of the perpendicular. $\mathcal{T}'$, p' and $\mathcal{T}''$, p'' belong in the same way to $\mathcal{L}(\mathfrak{s}')$ and $\mathcal{L}(\mathfrak{s}'')$ respectively. Fig. 9.7–9.10 illustrate the main cases.

In these the trilateral determines a convex hexagon, pentagon, quadrangle and triangle, respectively. The triangle $pp'p''$ is called the *foot-triangle*. The by-cases indicated in Section 3 can be treated by a certain generalization of these concepts. However, since the main cases are those needed in the sequel, we restrict ourselves to these cases. Then the following theorem holds:

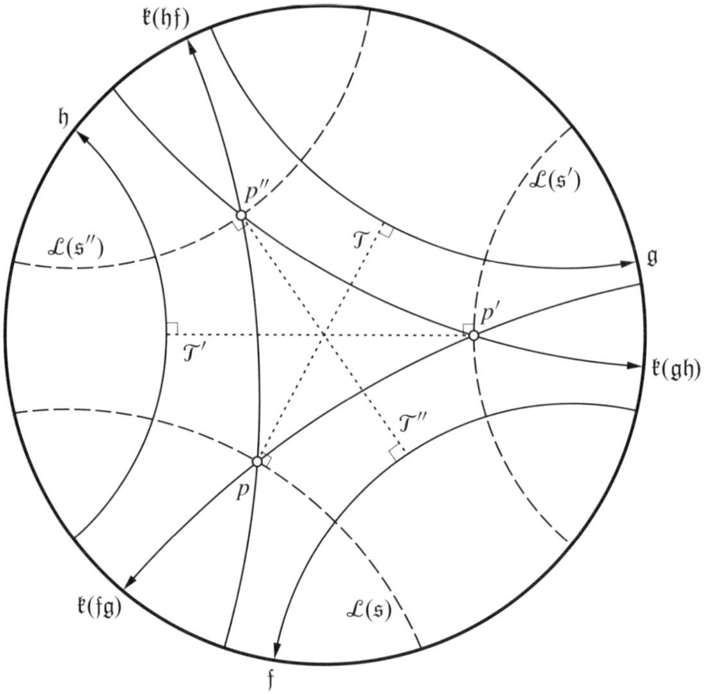

Figure 9.7

The axes of the three commutators coincide with the sides of the foot-triangle.

Proof. Owing to the cyclical symmetry, it is sufficient to prove that the axis of the reversed translation $\mathfrak{s}\mathfrak{g}^{-1} = \mathfrak{h}^{-1}\mathfrak{s}'$, the square of which is $\mathfrak{k}_{\mathfrak{f}\mathfrak{g}}$, passes through the foot-points p and p' situated on $\mathcal{L}(\mathfrak{s})$ and $\mathcal{L}(\mathfrak{s}')$, respectively. Let $\mathfrak{p}$, $\mathfrak{p}'$, $\mathfrak{p}''$ denote half-turns with centres p, p', p'', respectively and $\mathfrak{t}$, $\mathfrak{t}'$, $\mathfrak{t}''$ reflexions in $\mathcal{T}$, $\mathcal{T}'$, $\mathcal{T}''$, respectively. Then

$$\mathfrak{s} = \mathfrak{p}\mathfrak{t}, \quad \mathfrak{g}^{-1} = \mathfrak{t}\mathfrak{u},$$

where $\mathfrak{u}$ denotes the reflection at a suitable line at right angles to $\mathcal{A}(\mathfrak{g})$ or passing through $c(\mathfrak{g})$ according as $\mathfrak{g}$ is a translation or rotation. Hence

$$\mathfrak{s}\mathfrak{g}^{-1} = \mathfrak{p}\mathfrak{u},$$

which shows that the axis of $\mathfrak{s}\mathfrak{g}^{-1}$ passes through p. Similarly

$$\mathfrak{s}' = \mathfrak{t}'\mathfrak{p}', \quad \mathfrak{h}^{-1} = \mathfrak{v}\mathfrak{t}', \quad \mathfrak{h}^{-1}s' = \mathfrak{v}\mathfrak{p}'.$$

Here, $\mathfrak{p}'$ is a half-turn, $\mathfrak{t}'$ a reflection and $\mathfrak{v}$ a reflection in a line orthogonal to $\mathcal{A}(\mathfrak{h})$ or passing through $c(\mathfrak{h})$; it is inferred that the axis of $\mathfrak{h}^{-1}\mathfrak{s}'$ passes through p'. Thus $\mathfrak{k}_{\mathfrak{f}\mathfrak{g}}$ passes through p and p'.

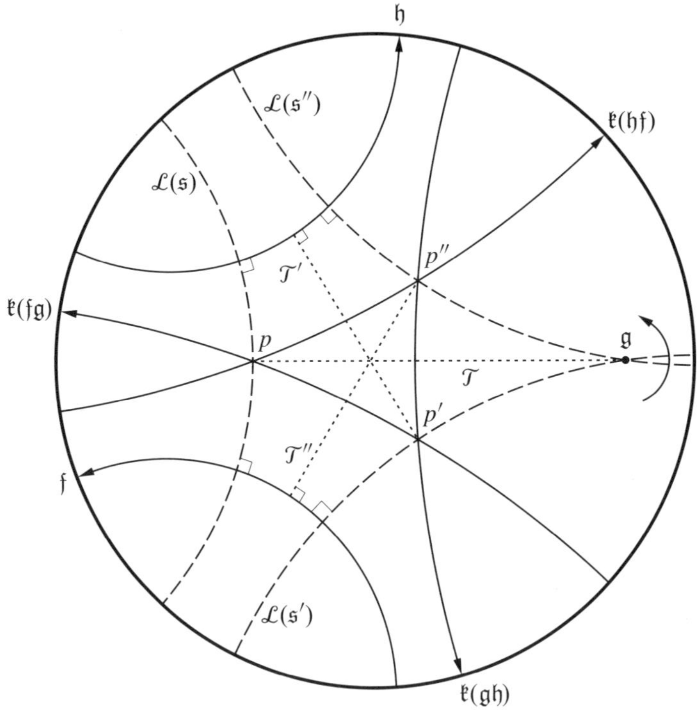

Figure 9.8

It is inferred from the relations (2) that two commutator axes which intersect at a point of a side of the trilateral are interchanged by the reflection in that side. This implies that the perpendiculars and the sides of the trilateral bisect the angles of the foot-triangle, and that the three perpendiculars of a trilateral intersect in one point.

In the cases I(→→ ⦿), I(→ ⦿ ⦿), I(⦿ ⦿ ⦿) one of the rotations can, in particular, be a half-turn. This implies that one more angle is right in the corresponding polygon, and two sides of the foot-triangle coincide while the third gets undetermined. The above quoted property of bisection carries with it the determination of that third side. Two half-turns can only occur in case I(→ ⦿ ⦿), namely the case previously marked I(→ $⦿_2$ $⦿_2$). In that case all three commutator axes coincide with the axis of translation ($\mathfrak{h}$ in Fig. 9.5). □

9.7 Circumscribed trilateral. In this section case II is considered. So let $\mathfrak{s}$, $\mathfrak{s}'$, $\mathfrak{s}''$ be half-turns whose centres $c(\mathfrak{s})$, $c(\mathfrak{s}')$, $c(\mathfrak{s}'')$ form a triangle. The commutators are found here in a dually corresponding way.

Let η, η', η'' denote the altitudes of the triangle drawn from $c(\mathfrak{s})$, $c(\mathfrak{s}')$, $c(\mathfrak{s}'')$, respectively. They pass through the same point in case the triangle is a trilateral. Let $\mathcal{P}$, $\mathcal{P}'$, $\mathcal{P}''$ denote lines at right angles to them through the same vertices. In the

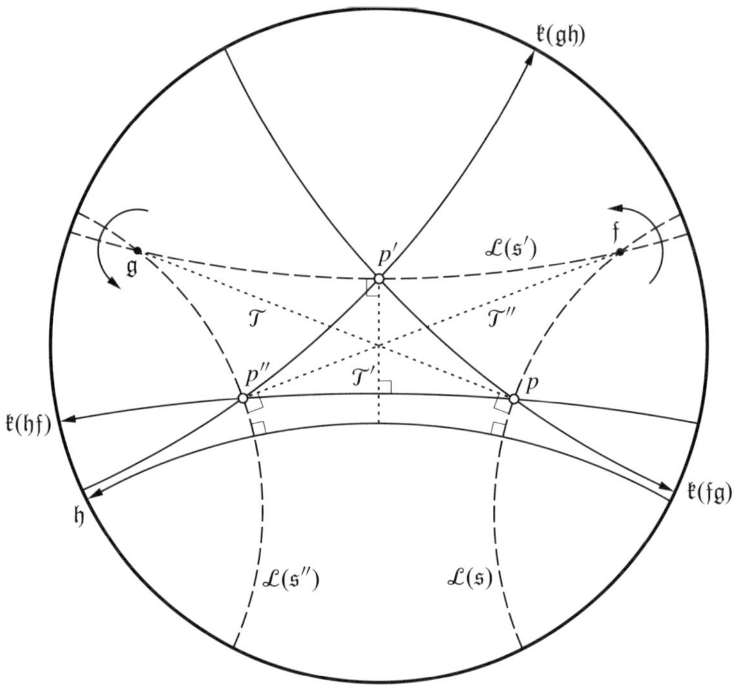

Figure 9.9

motion $\mathfrak{s}\mathfrak{g}^{-1} = \mathfrak{h}^{-1}\mathfrak{s}'$, the square of which is $\mathfrak{k}_{\mathfrak{f}\mathfrak{g}}$, let $\mathfrak{s}$ be replaced by the product of the reflections in $\mathcal{P}$ and η, and $\mathfrak{g}^{-1}$ by the product of the reflections in η and at another line orthogonal to $\mathcal{A}(\mathfrak{g})$. Then $\mathfrak{s}\mathfrak{g}^{-1}$ is the product of reflections in $\mathcal{P}$ and in this other line. Thus, if $\mathfrak{s}\mathfrak{g}^{-1}$ and hence $\mathfrak{k}_{\mathfrak{f}\mathfrak{g}}$ is a translation, its axis must be orthogonal to $\mathcal{P}$; if $\mathfrak{s}\mathfrak{g}^{-1}$ and hence $\mathfrak{k}_{\mathfrak{f}\mathfrak{g}}$ is a rotation, its centre must be on $\mathcal{P}$. From the expression $\mathfrak{h}^{-1}\mathfrak{s}'$ determining the same element, it is seen that the axis of $\mathfrak{k}_{\mathfrak{f}\mathfrak{g}}$ is orthogonal to $\mathcal{P}'$ or the centre of $\mathfrak{k}_{\mathfrak{f}\mathfrak{g}}$ is on $\mathcal{P}'$. Now, it is seen from (2) that the three commutators can be transformed into one another, and thus are all three translations, or all three rotations. Hence the following statement:

The lines $\mathcal{P}$, $\mathcal{P}'$, $\mathcal{P}''$ drawn through the vertices $c(\mathfrak{s})$, $c(\mathfrak{s}')$, $c(\mathfrak{s}'')$ at right angles to the altitudes of the triangle form a trilateral, the circumscribed trilateral. Two cases occur: Either any two of the sides are divergent; in that case each of the commutators is a translation with its axis orthogonal to two of these lines (case II(→), *Fig.* 9.11). *Or any two of the sides of the circumscribed trilateral have a point in common; in that case the commutators are rotations with these common points as centres (case* II(↻), *Fig.* 9.12). □

It is inferred from (2) that the centres of the half-turns s, s', s'' bisect the distances

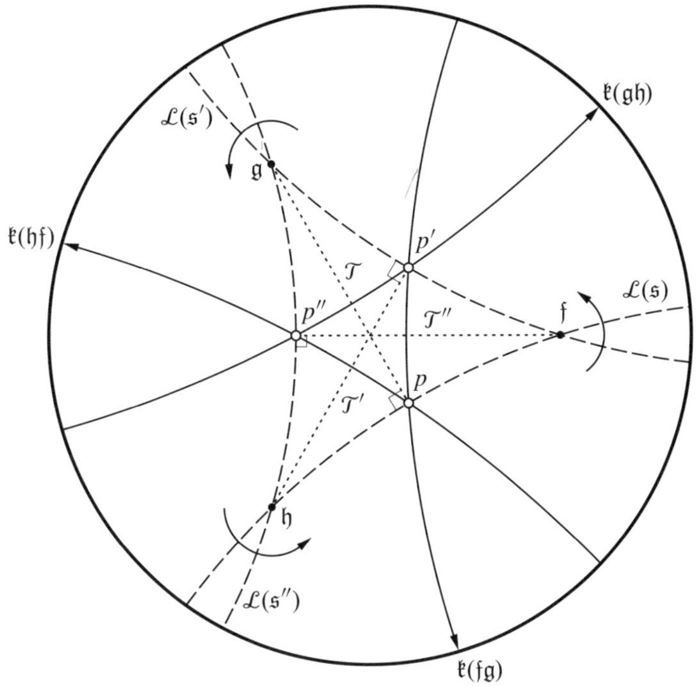

Figure 9.10

between the axes or the centres of the commutators. The triangle $c(\mathfrak{s})c(\mathfrak{s}')c(\mathfrak{s}'')$ is formed by the mid-points of the sides of the circumscribed trilateral $\mathcal{P}\mathcal{P}'\mathcal{P}''$.

9.8 Invariant of commutators. In concluding, let the invariant of the commutators be computed in case II. According to (1)

$$\mathfrak{k}_{\mathfrak{f}\mathfrak{g}} = \mathfrak{h}^{-1} \cdot \mathfrak{s}'\mathfrak{h}^{-1}\mathfrak{s}'.$$

The axis of $\mathfrak{s}'\mathfrak{h}^{-1}\mathfrak{s}'$ is the image of the axis of $\mathfrak{h}^{-1}$ by the half-turn $\mathfrak{s}'$. These two axes are, therefore, divergent, and their mutual distance is $2\eta'$. Moreover, the translations $\mathfrak{h}^{-1}$ and $\mathfrak{s}'\mathfrak{h}^{-1}\mathfrak{s}'$ have equal displacements $\lambda(\mathfrak{h}) = 2[c(\mathfrak{s}), c(\mathfrak{s}'')]$ with opposite directions. This corresponds to the case treated in §9.4. If $\lambda(\mathfrak{k})$ or $\varphi(\mathfrak{k})$ denotes the displacement or the angle of rotation of the commutator $\mathfrak{k}_{\mathfrak{f}\mathfrak{g}}$, then

$$\left.\begin{array}{r}\cosh\frac{1}{4}\lambda(\mathfrak{k})\\ \cos\frac{1}{4}\varphi(\mathfrak{k})\end{array}\right\} = \sinh\eta' \, \sinh[c(\mathfrak{s}), c(\mathfrak{s}'')] = \sigma,$$

σ denoting the sine amplitude (§8.4) of the triangle $c(\mathfrak{s})c(\mathfrak{s}')c(\mathfrak{s}'')$. Hence:

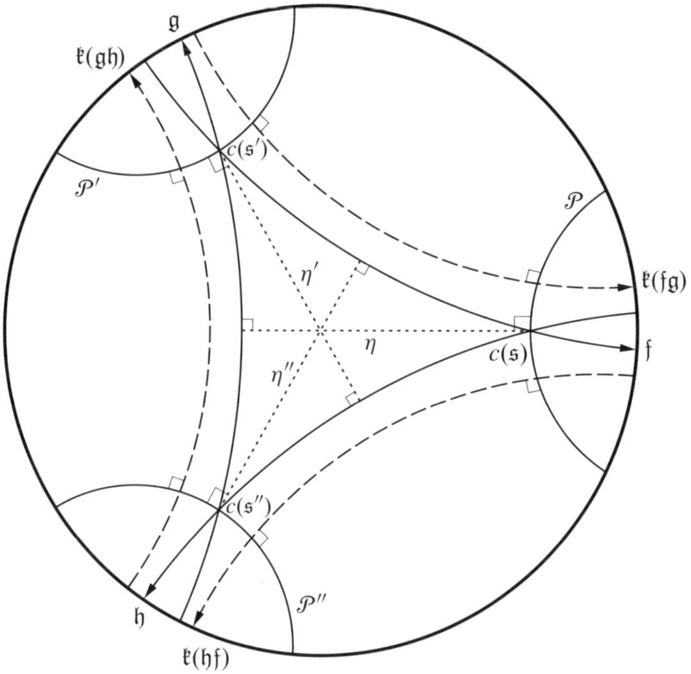

Figure 9.11

The commutators are translations, limit-rotations, or rotations according as the sine amplitude of the triangle $c(\mathfrak{s})c(\mathfrak{s}')c(\mathfrak{s}'')$ *is greater than, equal to, or smaller than* 1. □

9.9 Some auxiliary results. For later use we indicate the following auxiliary propositions:

The product $\mathfrak{r}\mathfrak{s}$ *of a rotation* $\mathfrak{r}$ *and a reflection* $\mathfrak{s}$ *in a line* $\mathcal{L}$ *not passing through the centre* c *of* $\mathfrak{r}$ *is a reversed translation.*

Proof. The rotation $\mathfrak{r}$ is representable as the product $\mathfrak{s}'\mathfrak{s}''$ of two reflections in lines $\mathcal{L}'$ and $\mathcal{L}''$ through c such that $\mathcal{L}''$ is the normal of $\mathcal{L}$ through c. Then $\mathfrak{r}\mathfrak{s} = \mathfrak{s}'\mathfrak{s}''\mathfrak{s} = \mathfrak{s}'\mathfrak{r}_1$, where $\mathfrak{r}_1$ is a half-turn; its centre c_1 is not on $\mathcal{L}'$. Hence $\mathfrak{s}'\mathfrak{r}_1$ is a reversed translation with the normal of $\mathcal{L}'$ through c_1 as axis. □

The product $\mathfrak{s}_1\mathfrak{s}_2\mathfrak{s}_3$ *of three reflections is a reflection if and only if the corresponding reflection lines* $\mathcal{L}_1$, $\mathcal{L}_2$, $\mathcal{L}_3$ *belong to one pencil.*

Proof. The condition is sufficient: If $\mathcal{L}_1$, $\mathcal{L}_2$, $\mathcal{L}_3$ belong to an elliptic pencil with the point p of $\mathcal{D}$ as common point, then the element $\mathfrak{s}_1\mathfrak{s}_2\mathfrak{s}_3$, which is a reversion and thus

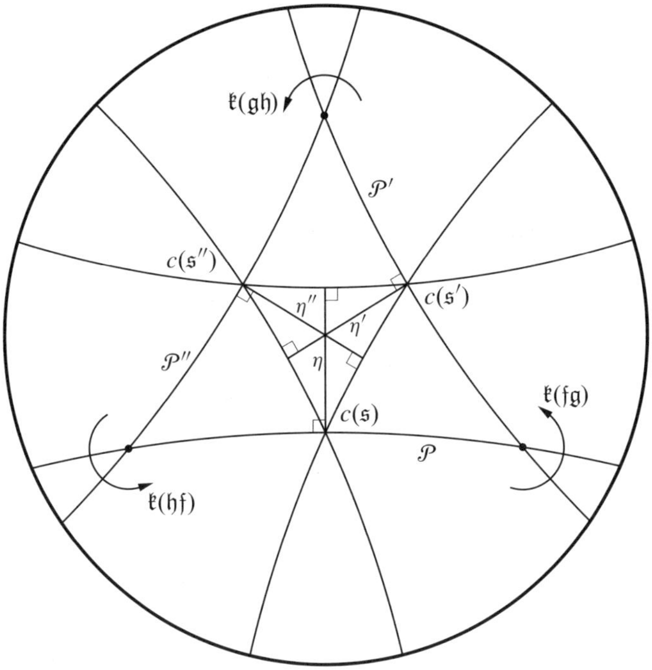

Figure 9.12

$\neq 1$, leaves p fixed and is thus a reflection. The same holds in the case of a parabolic pencil, since $\mathfrak{s}_1\mathfrak{s}_2\mathfrak{s}_3$ leaves an arbitrary horocycle with centre in the common infinite point invariant, and in the case of a hyperbolic pencil, since $\mathfrak{s}_1\mathfrak{s}_2\mathfrak{s}_3$ leaves the common normal of $\mathcal{L}_1$, $\mathcal{L}_2$, $\mathcal{L}_3$ invariant while interchanging its end-points.

The condition is necessary: $\mathcal{L}_2$ and $\mathcal{L}_3$ determine a pencil of lines. Assume that $\mathcal{L}_1$ does not belong to that pencil. Whatever the disposition of the three lines in $\mathcal{D}$, the product $\mathfrak{s}_2\mathfrak{s}_3$ can be replaced by the product $\mathfrak{s}_2'\mathfrak{s}_3'$ of reflections $\mathfrak{s}_2'$ and $\mathfrak{s}_3'$ in lines $\mathcal{L}_2'$ and $\mathcal{L}_3'$ belonging to the same pencil and chosen such that $\mathcal{L}_2'$ intersects $\mathcal{L}_1$, say in point c. Then $\mathfrak{s}_1\mathfrak{s}_2\mathfrak{s}_3 = \mathfrak{s}_1\mathfrak{s}_2'\mathfrak{s}_3' = \mathfrak{r}\mathfrak{s}_3'$, where $\mathfrak{r}$ is a rotation with centre c not on $\mathcal{L}_3$. Hence, by the preceding theorem, the product is a reversed translation. This proves the proposition. □

The bisecting normals of the sides of a triangle belong to one pencil.

Proof. Let x_1, x_2, x_3 be any three different points in $\mathcal{D}$, and $\mathcal{L}_1$, $\mathcal{L}_2$, $\mathcal{L}_3$ the bisecting normals of the segments x_2x_3, x_3x_1, x_1x_2 respectively, and let $\mathfrak{s}_1$, $\mathfrak{s}_2$, $\mathfrak{s}_3$ denote the reflections in $\mathcal{L}_1$, $\mathcal{L}_2$, $\mathcal{L}_3$. Then $\mathfrak{s}_2\mathfrak{s}_1\mathfrak{s}_3x_1 = \mathfrak{s}_2\mathfrak{s}_1x_2 = \mathfrak{s}_2x_3 = x_1$. The reversion $\mathfrak{s}_2\mathfrak{s}_1\mathfrak{s}_3$ leaves x_1 fixed and is thus a reflection. The preceding theorem then shows that $\mathcal{L}_1$, $\mathcal{L}_2$, $\mathcal{L}_3$ belong to one pencil. □

Chapter II
Discontinuous groups of motions and reversions

§10 The concept of discontinuity

10.1 Some notations and definitions. The circle (= circular disc) in the non-euclidean plane $\mathcal{D}$ with non-euclidean centre a and non-euclidean radius ρ will be denoted by $\mathfrak{C}(a;\rho)$. Let I indicate an infinite set and $\{\mathcal{M}_\nu\}_{\nu\in I}$ denote a family of subsets of $\mathcal{D}$. A point a of $\mathcal{D}$ is called an *accumulation point of the family* $\{\mathcal{M}_\nu\}$, if for an arbitrary $\rho > 0$ the intersection $\mathcal{M}_\mu \cap \mathfrak{C}(a,\rho)$ is non-empty for infinitely many $\mu \in I$. If no such accumulation point in $\mathcal{D}$ exists, it is said that *the sets $\mathcal{M}_\nu$ do not accumulate in $\mathcal{D}$.*

Let $\mathfrak{B}$ denote an arbitrary collection of motions and reversions. The point set $\mathcal{M}' \subset \mathcal{D}$ is called *equivalent to the point set $\mathcal{M} \subset \mathcal{D}$ with respect to* $\mathfrak{B}$, if there is an element $\mathfrak{b}$ in $\mathfrak{B}$ such that the image of $\mathcal{M}$ by $\mathfrak{b}$ is $\mathcal{M}'$; this is written $\mathcal{M}' = \mathfrak{b}\mathcal{M}$. The totality of all sets which are equivalent to $\mathcal{M}$ with respect to $\mathfrak{B}$ is denoted by $\mathfrak{B}\mathcal{M}$ and is called the *equivalence class* of $\mathcal{M}$ with respect to $\mathfrak{B}$. In this definition two sets $\mathfrak{b}_1\mathcal{M}$ and $\mathfrak{b}_2\mathcal{M}$ from $\mathfrak{B}\mathcal{M}$ are taken as different individuals of the equivalence class, if $\mathfrak{b}_1$ and $\mathfrak{b}_2$ are different elements from $\mathfrak{B}$ without regard to the identity or non-identity of $\mathfrak{b}_1\mathcal{M}$ and $\mathfrak{b}_2\mathcal{M}$ as point-sets. The sets of the equivalence class are thus in one-to-one correspondence with the elements of $\mathfrak{B}$. There will later-on be certain particular cases where it is convenient to depart from this rule, but they will then be indicated specifically.

If in particular $\mathfrak{B}$ is a group, the inverse of an element and the product of any two of its elements thus being defined, the concept of equivalence is a symmetric and transitive one, and any two sets equivalent with respect to $\mathfrak{B}$ determine one and the same equivalence class.

Two groups $\mathfrak{B}_1$ and $\mathfrak{B}_2$ of motions and reversions are said to be *of equal effect in a point set $\mathcal{B}$* of $\overline{\mathcal{D}}$, if any two points of $\mathcal{B}$ which are equivalent with respect to $\mathfrak{B}_1$ are also equivalent with respect to $\mathfrak{B}_2$, and conversely. If $\mathfrak{B}_2$ is a subgroup of $\mathfrak{B}_1$, then it has only to be verified that any element of $\mathfrak{B}_1$ carrying a point of $\mathcal{B}$ into a point of $\mathcal{B}$ belongs to $\mathfrak{B}_2$.

10.2 Discontinuity in $\mathcal{D}$. In the sequel $\mathfrak{G}$ denotes a group of motions and reversions. $\mathfrak{G}$ is said to be *discontinuous* in a point $x \in \overline{\mathcal{D}}$, if x is not an accumulation point of the equivalence class $\mathfrak{G}x$.

Assume that the point $a \in \mathcal{D}$ is an accumulation point for the equivalence class $\mathfrak{G}x$. Given any circle $\mathfrak{C}(a;\rho)$, there are infinitely many different elements $\mathfrak{g}_\nu \in \mathfrak{G}$ such that $\mathfrak{g}_\nu x$ is inside $\mathfrak{C}(a;\rho)$. If a itself belongs to $\mathfrak{G}x$, say $a = \mathfrak{h}x$, $\mathfrak{h} \in \mathfrak{G}$, then the points $\mathfrak{h}^{-1}\mathfrak{g}_\nu x$ are inside $\mathfrak{C}(x;\rho)$, and the $\mathfrak{h}^{-1}\mathfrak{g}_\nu$ are infinitely many different elements of $\mathfrak{G}$. If a does not belong to $\mathfrak{G}x$, then the $\mathfrak{g}_\nu x$ cannot coincide in one point. If $\mathfrak{g}_1 x$ and $\mathfrak{g}_2 x$, say, are different points inside $\mathfrak{C}(a;\rho)$, then $\mathfrak{g}_1^{-1}\mathfrak{g}_2 x$ is inside $\mathfrak{C}(x;2\rho)$ without coinciding with x. Since in both cases ρ is arbitrarily small, x is an accumulation point for $\mathfrak{G}x$. Hence the conclusion:

If $\mathfrak{G}$ is discontinuous in the point $x \in \mathfrak{D}$, then the equivalence class $\mathfrak{G}x$ has no accumulation point in $\mathfrak{D}$. □

Remark. If $\mathfrak{G}$ is a finite group, then $\mathfrak{G}x$ contains a finite number of points, and $\mathfrak{G}$ is thus discontinuous in x for every $x \in \overline{\mathfrak{D}}$. – The following statements therefore refer to infinite groups $\mathfrak{G}$. A group $\mathfrak{G}$ is called *discontinuous in a point set*, if it is discontinuous in every point of the set. Now the following proposition holds:

If a group $\mathfrak{G}$ is discontinuous at one point of $\mathfrak{D}$, then it is discontinuous in the whole of $\mathfrak{D}$.

Proof. This assertion is equivalent to the following: If the equivalence class of one point x_0 of $\mathfrak{D}$ has an accumulation point in $\mathfrak{D}$, then the same is true of the equivalence class of any other point x of $\mathfrak{D}$. Let p_0 be an accumulation point of $\mathfrak{G}x_0$ in $\mathfrak{D}$. Inside the circle $\mathfrak{C}(p_0;\rho)$ an infinite sequence $\mathfrak{g}_1 x_0,\ \mathfrak{g}_2 x_0,\dots$ of points of $\mathfrak{G}x_0$ is selected, $\mathfrak{g}_1,\mathfrak{g}_2,\dots$ being different elements of $\mathfrak{G}$. In virtue of

$$[\mathfrak{g}_\nu x,\ \mathfrak{g}_\nu x_0] = [x, x_0]$$

one gets

$$[\mathfrak{g}_\nu x,\ p_0] \leqq [\mathfrak{g}_\nu x,\ \mathfrak{g}_\nu x_0] + [\mathfrak{g}_\nu x_0, p_0] < \rho + [x, x_0].$$

Hence the points $\mathfrak{g}_1 x,\ \mathfrak{g}_2 x,\dots$ of $\mathfrak{G}x$ are inside a circle with centre p_0 and radius $\rho + [x, x_0]$ and thus accumulate in $\mathfrak{D}$. □

If a number $\rho > 0$, a point $x \in \mathfrak{D}$ and a group $\mathfrak{G}$ discontinuous in $\mathfrak{D}$ are given, all but a finite number of the elements of $\mathfrak{G}$ displace x more than ρ. For, inside and on $\mathfrak{C}(x;\rho)$ there can only be a finite number of points of $\mathfrak{G}x$. In particular, any rotation in $\mathfrak{G}$ must be of finite order. A group $\mathfrak{G}$ discontinuous in $\mathfrak{D}$ is *finite or enumerable.* For, there are only finitely many points of the equivalence class $\mathfrak{G}x$ of a point $x \in \mathfrak{D}$ in the circle $\mathfrak{C}(0;n)$, since it does not accumulate in $\mathfrak{D}$. Since $\bigcup_{n\in\mathbb{N}} \mathfrak{C}(0;n) = \mathfrak{D}$, an enumeration of $\mathfrak{G}$ results.

Let $\mathfrak{H}$ be a subgroup of $\mathfrak{G}$. If $\mathfrak{G}$ is discontinuous in $\mathfrak{D}$, the same is evidently true of $\mathfrak{H}$. On the other hand, if $\mathfrak{H}$ is discontinuous in $\mathfrak{D}$ and the index j of $\mathfrak{H}$ in $\mathfrak{G}$ is finite, then $\mathfrak{G}$ is discontinuous in $\mathfrak{D}$. For the equivalence class $\mathfrak{H}x$ of a point $x \in \mathfrak{D}$ does not accumulate in $\mathfrak{D}$, and $\mathfrak{G}x$ is composed of j point sets each of which is congruent or symmetric with $\mathfrak{H}x$, thus $\mathfrak{G}x$ does not accumulate in $\mathfrak{D}$.

If $\mathfrak{G}$ contains reversions, the motions contained in $\mathfrak{G}$ form a subgroup $\mathfrak{H}$ with index 2, thus a normal subgroup. (If $\mathfrak{n}$ is a reversion in $\mathfrak{G}$, all reversions of $\mathfrak{G}$ are contained in $\mathfrak{n}\mathfrak{H}$, and all in $\mathfrak{H}\mathfrak{n}$.) Hence the theorem:

A necessary and sufficient condition for a group of motions and reversions being discontinuous in $\mathfrak{D}$ is that the subgroup of motions contained is discontinuous in $\mathfrak{D}$. □

10.3 The distance function. Let $\mathfrak{G}$ be a group of motions and reversions which is discontinuous in $\mathfrak{D}$. Let $\varepsilon(\mathfrak{G}; x)$ – in short $\varepsilon(x)$ if $\mathfrak{G}$ is given – denote the minimum distance of a point $x \in \mathfrak{D}$ from all other points of its equivalence class $\mathfrak{G}x$. Then $\varepsilon(x) \geqq 0$; in particular, $\varepsilon(x) = 0$ requires that x is an invariant point of a transformation contained in $\mathfrak{G}$, in which case several points of the set $\mathfrak{G}x$ coincide in x. It is evident that $\varepsilon(x)$ takes the same value in equivalent points; thus one may write $\varepsilon(\mathfrak{G}x)$ instead of $\varepsilon(x)$.

For any two points x and x' of $\mathfrak{D}$ the inequality

$$\varepsilon(x) \leqq \varepsilon(x') + 2[x, x'] \tag{1}$$

holds. For, let $\mathfrak{g}x'$ be an equivalent point at minimum distance from x', thus $\varepsilon(x') = [x', \mathfrak{g}x']$. Then

$$\varepsilon(x) \leqq [x, \mathfrak{g}x] \leqq [x, x'] + [x', \mathfrak{g}x'] + [\mathfrak{g}x', \mathfrak{g}x] = \varepsilon(x') + 2[x, x'].$$

Since the rôles of x and x' can be interchanged in (1), this inequality can be written

$$|\varepsilon(x') - \varepsilon(x)\ | \leqq 2[x, x']. \tag{2}$$

This states that the function $\varepsilon(x)$ is uniformly continuous in $\mathfrak{D}$. It will be called the *distance function* of the group $\mathfrak{G}$.

10.4 Regular and singular points. In relation to a given group $\mathfrak{G}$ discontinuous in $\mathfrak{D}$, a point $x \in \mathfrak{D}$ will be called *regular*, if $\varepsilon(\mathfrak{G}; x) \neq 0$, and *singular*, if $\varepsilon(\mathfrak{G}; x) = 0$. Thus the set of singular points with respect to $\mathfrak{G}$ consists of the centres of rotation contained in $\mathfrak{G}$, if any, and the lines of reflection in $\mathfrak{G}$, if any. Since ε is a continuous function, the regular points form an open set, and the singular points a set closed on $\mathfrak{D}$.

If x is regular, then there are no points equivalent with respect to $\mathfrak{G}$ inside the circle $\mathfrak{C}(x; \frac{1}{2}\varepsilon(x))$.

Proof. For, assume that y and z are (different or coincident) points inside that circle, and that $z = \mathfrak{g}y,\ \mathfrak{g} \in \mathfrak{G},\ \mathfrak{g} \neq 1$. Then

$$\begin{aligned}[x, \mathfrak{g}x] \leqq [x, z] + [z, \mathfrak{g}x] &= [x, z] + [\mathfrak{g}y, \mathfrak{g}x] = [x, z] + [y, x] \\ &< \frac{1}{2}\varepsilon(x) + \frac{1}{2}\varepsilon(x) = \varepsilon(x).\end{aligned}$$

Since $\mathfrak{g} \neq 1$ and x is regular, $\mathfrak{g}x$ does not coincide with x, and the inequality thus contradicts the definition of $\varepsilon(x)$. □

The circle $\mathfrak{C}(x; \frac{1}{2}\varepsilon(x))$ *is the largest circle with that property.* For, if $\mathfrak{g} \in \mathfrak{G}$ is chosen such that $[x, \mathfrak{g}x] = \varepsilon(x)$, hence $\mathfrak{g} \neq 1$, and if $\rho > \frac{1}{2}\varepsilon(x)$, then the circle $\mathfrak{C}(x; \rho)$ intersects its image by $\mathfrak{g}$. If y is a point inside both circles, then both y and $\mathfrak{g}^{-1}y$ are inside $\mathfrak{C}(x; \rho)$.

The above proposition states in particular: *There are no singular points inside* $\mathfrak{C}(x; \frac{1}{2}\varepsilon(x))$.

Let $\mathfrak{C}(x; \rho)$ denote an arbitrary circle in $\mathfrak{D}$. If $\mathfrak{C}(x; \rho)$ contains the centres of infinitely many different rotations $\mathfrak{g}_\nu$, $1 \leqq \nu < \infty$, belonging to $\mathfrak{G}$, then all points $\mathfrak{g}_\nu x$ are inside the circle $\mathfrak{C}(x; 2\rho)$. Likewise, if $\mathfrak{C}(x; \rho)$ is cut by infinitely many different lines of reflections of $\mathfrak{G}$, say by the lines of the reflections $\mathfrak{s}_\nu$, $1 \leqq \nu < \infty$, then all points $\mathfrak{s}_\nu x$ are inside $\mathfrak{C}(x; 2\rho)$. In both cases the equivalence class $\mathfrak{G}x$ possesses accumulation points in $\mathfrak{D}$, and $\mathfrak{G}$ is thus not discontinuous in $\mathfrak{D}$. Hence the proposition:

For a group $\mathfrak{G}$ *discontinuous in* $\mathfrak{D}$ *neither the centres of rotations of* $\mathfrak{G}$ *nor the lines of reflections of* $\mathfrak{G}$ *accumulate in* $\mathfrak{D}$. *The rotations belonging to any centre of* $\mathfrak{G}$ *form a subgroup of finite order of* $\mathfrak{G}$. □

From this follows that the set of regular points is not empty. The set of singular points may be empty, namely if $\mathfrak{G}$ neither contains rotations nor reflections.

10.5 Fundamental domains and fundamental polygons. A *fundamental domain* for a group $\mathfrak{G}$ discontinuous in $\mathfrak{D}$ is any subset of $\mathfrak{D}$ in which every equivalence class of points of $\mathfrak{D}$ is represented by exactly one point. Since singular points are multiple points of their equivalence class, coincident points of an equivalence class have to be reckoned as one point in the above definition of a fundamental domain: It does not contain two *different* points which are equivalent with respect to $\mathfrak{G}$. The whole of $\mathfrak{D}$ is completely covered by the equivalence class of a fundamental domain, and apart from singular points it is simply covered.

A *fundamental polygon* $\mathfrak{P}$ for a group $\mathfrak{G}$ discontinuous in $\mathfrak{D}$ means a connected region closed on $\mathfrak{D}$ with the following three properties:

1) $\mathfrak{P}$ is in $\mathfrak{D}$ bounded by parts of straight lines; each of these parts is either a segment, a half-line, or a full line, and will be called a *side* of $\mathfrak{P}$. Boundary components of $\mathfrak{P}$ in $\mathfrak{D}$ are finite or infinite chains of sides; in particular, a chain may consist of one full line, or one or both of its ends may be half-lines. The collection of boundary chains of $\mathfrak{P}$ in $\mathfrak{D}$ is finite or enumerable, and it is required that the vertices of $\mathfrak{P}$ do not accumulate in $\mathfrak{D}$. The full boundary of $\mathfrak{P}$ on $\overline{\mathfrak{D}}$ may include arcs of $\mathfrak{E}$ and isolated points of $\mathfrak{E}$: they do not belong to $\mathfrak{P}$.

2) Any two members of the equivalence class $\mathfrak{G}\mathfrak{P}$ have no inner point in common.

3) Every bounded subset of $\mathcal{D}$ is covered by a finite number of members of the class $\mathfrak{G}\mathcal{P}$.

In consequence of 3) every point of $\mathcal{D}$ belongs to at least one member of the class $\mathfrak{G}\mathcal{P}$. Taken together with 2) this implies that a fundamental domain for $\mathfrak{G}$ may be obtained from $\mathcal{P}$ by the omission of certain boundary points.

$\mathcal{P}$ contains in its interior no singular point. For, if x were an inner point of $\mathcal{P}$ invariant by the rotation or reflection $\mathfrak{g} \in \mathfrak{G}$, then $\mathcal{P}$ and $\mathfrak{g}\mathcal{P}$ would have the inner point x in common. Centres of rotations of an order greater than two belonging to $\mathcal{P}$ are vertices of $\mathcal{P}$. Let c_1 be a centre of order 2 which is an inner point of a side $\mathcal{S}$ of $\mathcal{P}$, and let $\mathfrak{s}_1$ denote the half-turn about c_1. Then c_1 is also an inner point of a side of $\mathfrak{s}_1\mathcal{P}$, this side coinciding, at least in part, with $\mathcal{S}$. Let c_2 be another centre of order 2 in the interior of the same side $\mathcal{S}$ of $\mathcal{P}$, and let $\mathfrak{s}_2$ denote the half-turn about c_2. Then c_2 is an inner point of a side of $\mathfrak{s}_2\mathcal{P}$. Since $\mathfrak{s}_1\mathcal{P}$ and $\mathfrak{s}_2\mathcal{P}$ are different polygons of the class $\mathfrak{G}\mathcal{P}$, both of them must have a vertex on $\mathcal{S}$ between c_1 and c_2.

Let again $\mathcal{S}$ denote one of the sides of $\mathcal{P}$. If $\mathcal{S}$ is a segment, only a finite number of polygons of the class $\mathfrak{G}\mathcal{P}$ have points in common with $\mathcal{S}$. If $\mathcal{S}$ is a half-line or line, the same is true of every bounded segment of $\mathcal{S}$. Those points of $\mathcal{S}$ which belong to at least two of these polygons other than $\mathcal{P}$ are isolated; for they must be vertices of certain polygons $\mathfrak{g}\mathcal{P}$, $\mathfrak{g} \in \mathfrak{G}$, and the set of all vertices of all the polygons $\mathfrak{G}\mathcal{P}$ has no accumulation point in $\mathcal{D}$ in virtue of 1) and 3).

As an addition to the above definition of $\mathcal{P}$ we agree from now on to include among the vertices of $\mathcal{P}$ such inner points of the original sides of $\mathcal{P}$ which are common to at least three polygons of the collection $\mathfrak{G}\mathcal{P}$ or which are centres of order 2. The polygonal angle of $\mathcal{P}$ corresponding to vertices of this new type is equal to π.

This addition does not affect the assumptions made in the definition of the fundamental polygon. It should be noted that any two vertices of $\mathcal{P}$ which are centres of half-turns with π as corresponding polygonal angle are separated on the boundary of $\mathcal{P}$ by other vertices.

Let $\mathcal{S}_1, \mathcal{S}_2, \ldots$ denote the sides of $\mathcal{P}$ according to this extension of their definition. Then for each of these sides $\mathcal{S}_\nu$ there is exactly one polygon $\mathfrak{g}_\nu\mathcal{P}$ different from $\mathcal{P}$ which has this side in common with $\mathcal{P}$. Moreover, $\mathcal{S}_\nu$ is a complete side of $\mathfrak{g}_\nu\mathcal{P}$; for, in an end-point of $\mathcal{S}_\nu$ the angle of $\mathfrak{g}_\nu\mathcal{P}$ is either different from π, and the point is then a vertex of $\mathfrak{g}_\nu\mathcal{P}$ on that account, or that angle is equal to π, and the point is then a vertex of $\mathfrak{g}_\nu\mathcal{P}$ according to the extended definition of vertices. From this is inferred: To every side $\mathcal{S}_\nu$ of $\mathcal{P}$ corresponds exactly one side $\mathcal{S}'_\nu = \mathfrak{g}_\nu^{-1}\mathcal{S}_\nu$ of $\mathcal{P}$ equivalent to $\mathcal{S}_\nu$; they are not necessarily different, since $\mathcal{S}_\nu$ may be carried into itself by a reflection. Moreover it is possible that two different polygons of $\mathfrak{G}\mathcal{P}$ have two sides in common since they may have the centre of a half-turn on their common boundary.

According to 3) a vertex v of $\mathcal{P}$ belongs to a finite number of the polygons of $\mathfrak{G}\mathcal{P}$. If these polygons are denoted by $\mathcal{P}$, $\mathfrak{g}'\mathcal{P}$, $\mathfrak{g}''\mathcal{P}, \ldots, \mathfrak{g}^{(\sigma)}\mathcal{P}$, then v, ${\mathfrak{g}'}^{-1}v$, ${\mathfrak{g}''}^{-1}v$, ${\mathfrak{g}^{(\sigma)}}^{-1}v$ are (not necessarily different) vertices of $\mathcal{P}$, comprising exactly all vertices of $\mathcal{P}$ which are equivalent to v. This set of vertices is called a *cycle of vertices*. All the vertices of $\mathcal{P}$ are divided into complete cycles of vertices in a uniquely determined way.

The following statement holds for any fundamental polygon:

If certain sides and vertices are omitted from a fundamental polygon to the effect that only one side of every equivalent pair and only one vertex of every vertex cycle are retained, then a fundamental domain for the group is obtained.

10.6 Generation of $\mathfrak{G}$. We first state the following theorem:

If $\mathcal{S}_1, \mathcal{S}_2, \dots$ denote the sides of a fundamental polygon $\mathcal{P}$ (finite or enumerable in number) and $\mathfrak{g}_1\mathcal{P}, \mathfrak{g}_2\mathcal{P}, \dots$ are the polygons of the collection $\mathfrak{G}\mathcal{P}$ adjacent to these sides, then the elements $\mathfrak{g}_1, \mathfrak{g}_2, \dots$ generate the group $\mathfrak{G}$.

Proof. Let $\mathfrak{h}$ denote an arbitrary element of $\mathfrak{G}$, and let $\mathfrak{G}'$ denote the subgroup of $\mathfrak{G}$ generated by $\mathfrak{g}_1, \mathfrak{g}_2, \dots$. It has to be proved that $\mathfrak{h}$ belongs to $\mathfrak{G}'$. A chain of fundamental polygons $\mathcal{P} = \mathfrak{h}_0\mathcal{P}, \mathfrak{h}_1\mathcal{P}, \mathfrak{h}_2\mathcal{P}, \dots, \mathfrak{h}_s\mathcal{P} = \mathfrak{h}\mathcal{P}$ exists which joins $\mathcal{P}$ and $\mathfrak{h}\mathcal{P}$ in such a way that every two neighbouring members of the chain have a side in common. For, let $\mathcal{T}'$ be a straight segment joining an inner point of $\mathcal{P}$ with an inner point of $\mathfrak{h}\mathcal{P}$, and $\mathcal{C}$ an open circular disc containing $\mathcal{T}'$. Since $\mathcal{C}$ contains only a finite number of vertices of the collection $\mathfrak{G}\mathcal{P}$, one may, if necessary, by a slight variation of the end-points of $\mathcal{T}'$ replace $\mathcal{T}'$ by a segment $\mathcal{T}$ which also joins an inner point of $\mathcal{P}$ with an inner point of $\mathfrak{h}\mathcal{P}$ and which passes through no vertex. In virtue of 3) this segment $\mathcal{T}$ only passes through a finite number of fundamental polygons, and these polygons form in the order in which $\mathcal{T}$ cuts them, a chain of the required nature. Now the identity $\mathfrak{h}_0$ belongs to $\mathfrak{G}'$, and we proceed by induction: Let $\mathfrak{h}_\sigma$, $0 \leqq \sigma < s$, belong to $\mathfrak{G}'$. The polygons $\mathfrak{h}_\sigma\mathcal{P}$ and $\mathfrak{h}_{\sigma+1}\mathcal{P}$ have a side in common. The element $\mathfrak{h}_\sigma^{-1}$ carries these polygons into $\mathcal{P}$ and $\mathfrak{h}_\sigma^{-1}\mathfrak{h}_{\sigma+1}\mathcal{P}$ and these have, therefore, a side in common, say $\mathcal{S}_\tau$. Thus $\mathfrak{h}_\sigma^{-1}\mathfrak{h}_{\sigma+1}\mathcal{P}$ is the same as $\mathfrak{g}_\tau\mathcal{P}$, hence $\mathfrak{h}_\sigma^{-1}\mathfrak{h}_{\sigma+1} = \mathfrak{g}_\tau$, or

$$\mathfrak{h}_{\sigma+1} = \mathfrak{h}_\sigma\mathfrak{g}_\tau$$

which shows that $\mathfrak{h}_{\sigma+1}$ belongs to $\mathfrak{G}'$. This proves the assertion. □

The following special consequence of the theorem may be emphasized:

If a group possesses a fundamental polygon with a finite number of sides, then the group can be generated by a finite number of its elements. □

10.7 Relations for $\mathfrak{G}$. Returning to the general case of a finite or infinite collection $\mathfrak{g}_1, \mathfrak{g}_2, \dots$ of generators determined in the preceding section, we want to make a survey of all relations between these generators. It is noted that, if there is a half-turn among the generators and thus two consecutive sides of $\mathcal{P}$ correspond by that half-turn, two different of the symbols $\mathfrak{g}_\nu$ denote that half-turn according to the definition of sides and generators.

The polygons $\mathfrak{G}\mathcal{P}$ form a certain *network* W in $\mathcal{D}$. We now construct what one may call the *dual network* W^* in the following way: In the interior of $\mathcal{P}$ we choose an individual point p. Likewise, on every side $\mathcal{S}_\nu$ of $\mathcal{P}$ we choose an inner point p_ν of that side subject to the condition that, if $\mathfrak{g}_\nu$ carries $\mathcal{S}_\mu$ into $\mathcal{S}_\nu$, it also carries p_μ into p_ν; the point in question can thus only be freely chosen on one of each pair of equivalent sides of $\mathcal{P}$. We then join p and p_ν (for all values of ν) by a polygonal line inside $\mathcal{P}$ in such a way, that these lines only have p in common; if $\mathcal{P}$ is convex, one may for instance take each polygonal line to consist of one straight segment. The images of this set of polygonal lines by all elements of $\mathfrak{G}$ taken together form the dual network W^*.

Consider the side $\mathcal{S}_\nu$ of $\mathcal{P}$. Along this side $\mathfrak{g}_\nu\mathcal{P}$ is adjacent to $\mathcal{P}$. The polygonal line joining p and p_ν in $\mathcal{P}$ followed by the polygonal line joining p_ν and $\mathfrak{g}_\nu p$ in $\mathfrak{g}_\nu\mathcal{P}$, this being the image by $\mathfrak{g}_\nu$ of the polygonal line joining $\mathfrak{g}_\nu^{-1}p_\nu$ and p in $\mathcal{P}$, will be called the *oriented elementary path* $\mathfrak{g}_\nu$ joining p and $\mathfrak{g}_\nu p$. The image of this path by an arbitrary element $\mathfrak{h}$ of $\mathfrak{G}$ is called the *oriented elementary path* $\mathfrak{g}_\nu$ joining $\mathfrak{h}p$ and $\mathfrak{h}\mathfrak{g}_\nu p$; its starting point $\mathfrak{h}p$ is carried into its end-point $\mathfrak{h}\mathfrak{g}_\nu p$ by the element $\mathfrak{h}\mathfrak{g}_\nu\mathfrak{h}^{-1}$ of $\mathfrak{G}$, but we inscribe the (oriented) path with the signature $\mathfrak{g}_\nu$ only. Thus the whole equivalence class of oriented elementary paths which one gets by letting $\mathfrak{h}$ range over the whole of $\mathfrak{G}$ bears the same signature $\mathfrak{g}_\nu$. – Moreover, the following fact has to be remembered: Let $\mathcal{S}_\mu$ be the side of $\mathcal{P}$ equivalent to $\mathcal{S}_\nu$; (in the special case of $\mathfrak{g}_\nu$ being a reflection $\mathcal{S}_\mu$ and $\mathcal{S}_\nu$ are identical). The image by $\mathfrak{g}_\nu$ of the elementary path $\mathfrak{g}_\mu$ joining p and $\mathfrak{g}_\mu p$ is the elementary path joining $\mathfrak{g}_\nu p$ and p and thus bears the signature $\mathfrak{g}_\mu$ in this direction. The two signatures inscribed on each elementary path of W^* are both a generator with exponent $+1$, one for each direction. This yields the relation

$$\mathfrak{g}_\nu\mathfrak{g}_\mu = 1 \tag{3}$$

which in the case of a reflection $\mathfrak{g}_\nu$ becomes

$$\mathfrak{g}_\nu^2 = 1 \tag{3a}$$

We may speak of the points of the set $\mathfrak{G}p$ as the *vertices of* W^*. Then the elementary paths emanating from any vertex of W^* bear the signature of all generators $\mathfrak{g}_1, \mathfrak{g}_2, \dots$, each of them exactly once.

Now let

$$\mathfrak{g} = \mathfrak{g}_{\nu_1}\mathfrak{g}_{\nu_2}\cdots\mathfrak{g}_{\nu_s} \tag{4}$$

be an arbitrary product of generators each with exponent $+1$. Starting in an arbitrary polygon $\mathfrak{h}\mathcal{P}$ of the network W we associate with the expression $\mathfrak{g}$ a chain of polygons

$$\mathfrak{h}\mathcal{P},\quad \mathfrak{h}\mathfrak{g}_{\nu_1}\mathcal{P},\quad \mathfrak{h}\mathfrak{g}_{\nu_1}\mathfrak{g}_{\nu_2}\mathcal{P},\dots,\mathfrak{h}\mathfrak{g}_{\nu_1}\mathfrak{g}_{\nu_2}\cdots\mathfrak{g}_{\nu_s}\mathcal{P}. \tag{5}$$

This is in fact a chain, since any two neighbours such as $\mathfrak{h}\mathfrak{g}_{\nu_1}\dots\mathfrak{g}_{\nu_\sigma}\mathcal{P}$ and $\mathfrak{h}\mathfrak{g}_{\nu_1}\dots\mathfrak{g}_{\nu_\sigma}\mathfrak{g}_{\nu_{\sigma+1}}\mathcal{P}$ are the images of the two mutually adjacent polygons $\mathcal{P}$ and $\mathfrak{g}_{\nu_{\sigma+1}}\mathcal{P}$ by an element of $\mathfrak{G}$, here $\mathfrak{h}\mathfrak{g}_{\nu_1} \dots \mathfrak{g}_{\nu_\sigma}$, and therefore have a side in common. The first polygon of the chain is carried into the last one by $\mathfrak{h}\mathfrak{g}\mathfrak{h}^{-1}$. Since $\mathfrak{h}$ is arbitrary, any product (4) corresponds to an equivalence class of chains of polygons. – Conversely, if such an equivalence class is given, we consider in the class the particular chain which has $\mathcal{P}$ as its first polygon. If this chain is $\mathcal{P}\mathcal{P}_1\mathcal{P}_2 \dots \mathcal{P}_s$, a corresponding product (4) is formed by determining the factors in turn by

$$\mathfrak{g}_{\nu_{\sigma+1}}\mathcal{P} = \mathfrak{g}_{\nu_\sigma}^{-1} \cdots \mathfrak{g}_{\nu_1}^{-1}\mathcal{P}_{\sigma+1}.$$

This is illustrated very conveniently by the dual network W^*. If a product (4) is given, and if we start from the point $\mathfrak{h}p$ of the polygon $\mathfrak{h}\mathcal{P}$ and proceed from there along the elementary path marked $\mathfrak{g}_{\nu_1}$ to the next vertex of W^*, from there along the elementary path marked $\mathfrak{g}_{\nu_2}$ to the next one and so forth, we get a path on W^* which follows the chain of polygons (5). Conversely, any chain of polygons determines a path on W^* from which we read directly the corresponding sequence of factors in the product (4). If two adjacent polygons of the chain have two sides in common (case of a half-turn), one of these two sides has to be chosen as the connecting one in order that the product of generators be formally unique.

If two consecutive factors in (4) are generators $\mathfrak{g}_\nu$ and $\mathfrak{g}_\mu$ for which (3) holds, this means that a path on W^* corresponding to the product has a certain elementary path traversed forward and backward in immediate succession.

Any relation connecting the generators can be written in the form

$$\mathfrak{g}_{\nu_1}\mathfrak{g}_{\nu_2} \cdots \mathfrak{g}_{\nu_s} = 1. \tag{6}$$

It corresponds to a closed chain of polygons and hence to a closed path, a *cycle,* on W^*. Conversely, every cycle on W^* corresponds to a closed chain of polygons and to a relation between the generators with positive exponents, and this relation reads directly from the cycle in question. In particular, the closed chain of polygons surrounding a vertex of W corresponds to a cycle of W^* which constitutes a single mesh of the network W^*. Equivalent vertices of W yield equivalent cycles of W^*. It is therefore sufficient to examine the vertices of the polygon $\mathcal{P}$ itself. The number of corresponding relations between the generators is twice the number of cycles of vertices of $\mathcal{P}$, the factor 2 arising from the fact that every cycle can be described in both senses. This number is finite or infinite, according as $\mathcal{P}$ has a finite or infinite number of vertices.

In abbreviated form, let the system of equations

$$H_1 = 1, \quad H_2 = 1, \dots \tag{7}$$

express the relations derived in this way from the cycles of vertices of $\mathcal{P}$, each of the symbols H_i denoting a certain product of generators $\mathfrak{g}_\nu$ with positive exponents, the corresponding cycle on W^* being described in a definite sense from a definite vertex of W^*. Then also

$$\Phi_i H_i \Phi_i^{-1} = 1$$

is a valid relation, the symbol Φ_i denoting an arbitrary product of generators $\mathfrak{g}_\nu$. We now prove the following theorem:

In every relation (6) *the left-hand product can be obtained from a certain product of factors* $\Phi_i H_i \Phi_i^{-1}$ *by omitting inverse generators* $\mathfrak{g}_\nu \mathfrak{g}_\nu^{-1}$ *or* $\mathfrak{g}_\nu^{-1} \mathfrak{g}_\nu$ *in immediate succession and using relations* (3).

Remark. We may call the omission or addition of such an inverse pair an *identical alteration.*

Proof. It is immediately seen that the assertion is true for relations $\overline{H}_i = 1$, $\overline{H}_i$ denoting a cyclical permutation of the generators in the product H_i. In considering the general case, the left-hand product

$$\mathfrak{g} = \mathfrak{g}_{\nu_1} \cdots \mathfrak{g}_{\nu_s}$$

in (6) yields a cycle on the network W^*, if $\mathfrak{g}$ is described from a chosen vertex of W^*. Now the vertices of W are divided into two classes I and II by the cycle $\mathfrak{g}$ of W^*: Those from which one can go towards infinity on W without meeting the cycle $\mathfrak{g}$, and those from which one cannot. The class II only contains a finite number of vertices of W, because the cycle $\mathfrak{g}$ on W^* is bounded. We assume that this class is not empty. Let a path on W start from a vertex of class II and end in a vertex of class I, and let V_{II} be the last vertex of class II on it and V_{I} the following vertex, which then belongs to class I. It is inferred that the side $V_{\mathrm{II}} V_{\mathrm{I}}$ of W is crossed by a certain oriented elementary path π^* of W^* occurring at least once in the cycle $\mathfrak{g}$ of W^* in one of its directions. Let $\mathfrak{g}_{\nu_r}$ be the first of the factors of $\mathfrak{g}$ which corresponds to traversing π^*. We then write the relation pertaining to the vertex V_{II} of W in the form

$$\overline{H} = \mathfrak{g}_{\nu_r} \mathfrak{h} = 1$$

with $\mathfrak{h}$ as an abbreviation for the rest of the factors. Then

$$\begin{aligned} \mathfrak{g} &= \mathfrak{g}_{\nu_1} \cdots \mathfrak{g}_{\nu_{r-1}} \mathfrak{g}_{\nu_r} \mathfrak{g}_{\nu_{r+1}} \cdots \mathfrak{g}_{\nu_s} \\ &= \mathfrak{g}_{\nu_1} \cdots \mathfrak{g}_{\nu_{r-1}} (\mathfrak{g}_{\nu_r} \mathfrak{h}) \mathfrak{g}_{\nu_{r-1}}^{-1} \cdots \mathfrak{g}_{\nu_1}^{-1} \cdot \mathfrak{g}_{\nu_1} \cdots \mathfrak{g}_{\nu_{r-1}} \mathfrak{h}^{-1} \mathfrak{g}_{\nu_{r+1}} \cdots \mathfrak{g}_{\nu_s} \\ &= \Phi \overline{H} \Phi^{-1} \cdot \mathfrak{g}' \end{aligned}$$

where $\mathfrak{g}'$ is the product $\mathfrak{g}_{\nu_1} \ldots \mathfrak{g}_{\nu_{r-1}} \mathfrak{h}^{-1} \mathfrak{g}_{\nu_{r+1}} \cdots \mathfrak{g}_{\nu_s}$. In this expression $\mathfrak{h}^{-1}$ can be replaced by a product of generators with positive exponents by means of (3). The effect of the process then is that the number of times π^* is contained in the cycle $\mathfrak{g}$ has decreased by one, and no vertex of class I has changed to class II, because the process only affects the mesh of W^* which encloses V_{II}. After a finite number of repetitions of the process, equal to the number of times the cycle $\mathfrak{g}$ traverses π^* in one direction or the other, the vertex V_{II} has changed to the class I. This method of splitting off factors of the form $\Phi H \Phi^{-1}$ thus eventually reduces the product $\mathfrak{g}$ to a product for which the corresponding cycle on W^* yields an empty class II, and which therefore reduces to identity by the use of (3) only. – This proves the assertion. □

Hence the following theorem:

The elements $\mathfrak{g}_\nu$ *of the discontinuous group* $\mathfrak{G}$ *which carry a fundamental polygon* $\mathcal{P}$ *into all polygons adjacent to its sides constitute a set of generators of* $\mathfrak{G}$, *and any relation between them can be derived from relations* (3) *and* (7) *by identical alterations.* □

This system of generators and relations can be reduced by the usual convention that the introduction of a generator $\mathfrak{g}_\nu$ includes the applicability of its inverse $\mathfrak{g}_\nu^{-1}$. In all cases where $\mathfrak{g}_\nu$ is not a reflection the factors $\mathfrak{g}_\nu$ and $\mathfrak{g}_\mu$ in (3) are different in the sense that they correspond to different sides of $\mathcal{P}$. One of them can thus be eliminated, and at the same time the corresponding relation (3) cancels. Let

$$\mathfrak{k}_1,\ \mathfrak{k}_2,\dots \tag{8}$$

denote the generators retained in this process, and let

$$R_1 = 1,\ \ R_2 = 1,\dots \tag{9}$$

denote the relations obtained by insertion of the $\mathfrak{k}$'s partly from relation (3a), if reflections occur at all, and partly from relations (7), where it is now sufficient to write one of the two relations hitherto attached to an individual cycle of vertices. With these conventions we have the following theorem:

The elements (8), *one for each pair of equivalent sides whether coincident or not, constitute a system of generators of* $\mathfrak{G}$, *and the relations* (9), *one for each reflection among the generators and one for each cycle of vertices, constitute a complete system of defining relations with respect to these generators.* □

The completeness of the system (9) means that any product of the $\mathfrak{k}$'s which is equal to the identity can be brought into the form of a product of elements $\Phi_i R_i^{\pm 1} \Phi_i^{-1}$ by identical alterations.

We add the remark that the case of a half-turn with two consecutive sides in common with two neighbour polygons does not involve any exception. In this case $\mathcal{P}$ has two consecutive sides, say $\mathcal{S}_\nu$ and $\mathcal{S}_\rho$, in common with a neighbour polygon, which is both $\mathfrak{g}_\nu\mathcal{P}$ and $\mathfrak{g}_\rho\mathcal{P}$. These two sides form an equivalent pair, hence $\mathcal{S}_\rho$ plays the rôle of $\mathcal{S}_\mu$ above, and we get from (3)

$$\mathfrak{g}_\nu\mathfrak{g}_\rho = 1. \tag{10}$$

Thus the sides inscribed $\mathfrak{g}_\nu$ in one direction are inscribed $\mathfrak{g}_\rho$ in the opposite direction. The cycle of W corresponding to the common vertex is therefore according to (7)

$$\mathfrak{g}_\nu\mathfrak{g}_\nu = 1. \tag{11}$$

This then becomes one of the relations (9). Eliminating $\mathfrak{g}_\rho$ by (10), one gets $\mathfrak{g}_\rho = \mathfrak{g}_\nu^{-1} = \mathfrak{g}_\nu$ in virtue of (11).

10.8 Normal domains. In the Sections 5–7 we investigated the properties of fundamental polygons without raising the question whether such polygons exist for any given group $\mathfrak{G}$ discontinuous in $\mathcal{D}$. In the present section we construct certain polygons which prove to be special examples of fundamental polygons, and we indicate their additional properties.

Let x_0 be an arbitrary regular point for $\mathfrak{G}$ in $\mathcal{D}$ (Section 4). Then all points of the equivalence class $\mathfrak{G}x_0$ are different. The set of all those points x of $\mathcal{D}$ which satisfy the inequality

$$[x, x_0] \leqq [x, \mathfrak{g}x_0] \tag{12}$$

for every element $\mathfrak{g}$ of $\mathfrak{G}$ is called the *normal domain* $\mathcal{N}(\mathfrak{G}; x_0)$ with x_0 as *central point*. If there can be no doubt as to the group in question, the shorter notation $\mathcal{N}(x_0)$ will often be used. $\mathcal{N}(\mathfrak{G}; x_0)$ consists of all points of $\mathcal{D}$ whose distance from x_0 is at most equal to their distance from the rest of $\mathfrak{G}x_0$. For any fixed element $\mathfrak{g} \neq 1$ of $\mathfrak{G}$ those points x which satisfy (12) constitute a half-plane bounded by the bisecting perpendicular $\mathcal{B}(\mathfrak{g})$ of the segment $x_0, \mathfrak{g}x_0$, containing x_0, and closed on $\mathcal{D}$. The normal domain $\mathcal{N}(\mathfrak{G}; x_0)$ is the intersection of all these half-planes for $\mathfrak{g}$ ranging over $\mathfrak{G}$, hence a convex region closed on $\mathcal{D}$. Any circle $\mathcal{C}(x_0; \rho)$ is only cut by a finite number of the $\mathcal{B}(\mathfrak{g})$, since $[x_0, \mathfrak{g}x_0] < 2\rho$ only holds for finitely many elements $\mathfrak{g} \in \mathfrak{G}$. Hence the $\mathcal{B}(\mathfrak{g})$ do not accumulate in $\mathcal{D}$, and the same holds for the set of their intersection points. This implies that any boundary point of $\mathcal{N}(x_0)$ in $\mathcal{D}$ is situated on at least one of the $\mathcal{B}(\mathfrak{g})$ and on not more than a finite number of them, while an inner point of $\mathcal{N}(x_0)$ is situated on none of them. This implies that $\mathcal{N}(x_0)$ satisfies condition 1) in the definition of fundamental polygons in Section 5.

Since any element $\mathfrak{f} \in \mathfrak{G}$ reproduces $\mathfrak{G}x_0$ and preserves distances, it follows from (12) that the normal domain $\mathcal{N}(\mathfrak{G}; \mathfrak{f}x_0)$ with central point $\mathfrak{f}x_0$ is the image of $\mathcal{N}(\mathfrak{G}; x_0)$ by $\mathfrak{f}$:

$$\mathcal{N}(\mathfrak{f}x_0) = \mathfrak{f}\mathcal{N}(x_0).$$

Let y denote an arbitrary point in $\mathcal{D}$ and $\mu = [y, \mathfrak{G}x_0]$ its distance from the point set $\mathfrak{G}x_0$. There is then at least one, and not more than a finite number of points of $\mathfrak{G}x_0$ distant μ from y, say the points $\mathfrak{f}_\nu x_0$, $1 \leqq \nu \leqq \sigma$. Then y belongs to the normal domains $\mathcal{N}(\mathfrak{f}_\nu x_0) = \mathfrak{f}_\nu \mathcal{N}(x_0)$ and to no other domain of the collection $\mathfrak{G}\mathcal{N}(x_0)$. If $\sigma = 1$, then y is an inner point of $\mathfrak{f}_1\mathcal{N}(x_0)$. If $\sigma > 1$, then y is a common boundary point of the domains $\mathfrak{f}_\nu\mathcal{N}(x_0)$. Hence, if $\mathfrak{g}$ ranges over $\mathfrak{G}$, the totality of domains $\mathfrak{g}\mathcal{N}(x_0)$, any two of which are congruent or symmetric, cover $\mathcal{D}$ completely, any point of $\mathcal{D}$ being either an inner point of exactly one domain, or a common boundary of a finite number. This shows that $\mathcal{N}(x_0)$ satisfies condition 2) of Section 5.

Every point y of $\mathcal{D}$ is embedded in a neighbourhood which has points in common with a finite number only of the domains of the collection $\mathfrak{G}\mathcal{N}(x_0)$. In the above notation, this is evident if $\sigma = 1$, and also if $\sigma = 2$. If $\sigma \geqq 3$, the σ domains on whose boundary y lies are separated by σ sides issuing from y. These sides form a certain *star*, and in virtue of the convexity of the domains the convex hull of this star is a neighbourhood of y as required. This implies that any bounded subset of $\mathcal{D}$ has points in common only with a finite number of domains of the collection $\mathfrak{G}\mathcal{N}(x_0)$. For, otherwise this collection of point sets would have at least one accumulation point in $\mathcal{D}$ (Section 1), and accumulation point would fail to have the above neighbourhood property. Hence $\mathcal{N}(x_0)$ satisfies condition 3) of Section 5.

It has thus been seen that a normal domain is a special case of a fundamental polygon. All results concerning the latter are thus applicable to normal domains. On the other hand, normal domains have special properties not applicable to all fundamental polygons. One of them is convexity. We point out another one:

Let y be a boundary point of $\mathcal{N}(x_0)$, and assume that it is a boundary point of $\sigma \geqq 3$ domains of the collection $\mathfrak{G}\mathcal{N}(x_0)$. A circle with centre y and passing through x_0 then contains σ points of the set $\mathfrak{G}x_0$. Let $\mathfrak{g}_1 x_0$ and $\mathfrak{g}_2 x_0$ be the two next to x_0 on that circle. Then $\mathcal{B}(\mathfrak{g}_1)$ and $\mathcal{B}(\mathfrak{g}_2)$ meet in y. Together with the lines x_0, $\mathfrak{g}_1 x_0$ and x_0, $\mathfrak{g}_2 x_0$ they bound a quadrangle with two right angles (Fig. 10.1). Hence the angle of the quadrangle at y is $< \pi$. The angular sector subtended at

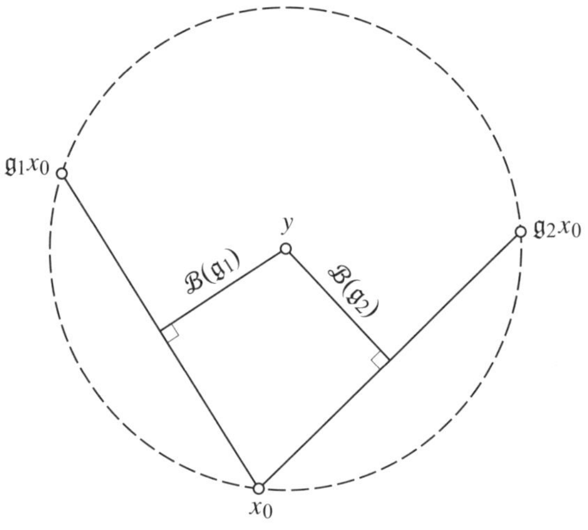

Figure 10.1

y by $\mathcal{B}(\mathfrak{g}_1)$ and $\mathcal{B}(\mathfrak{g}_2)$ contains $\mathcal{N}(x_0)$. This shows that y is a vertex of $\mathcal{N}(x_0)$. Hence the extension introduced in Section 5, namely to include among the vertices of a fundamental polygon such boundary points which are common to three or more polygons, lapses in the case of normal domains, because such points are vertices from the outset. The other convention, namely to count centres of half-turns on the boundary of $\mathcal{N}(x_0)$ as vertices even if they correspond to an interior angle π, is maintained. Such a centre constitutes by itself a complete vertex cycle.

§11 Groups with invariant points or lines

11.1 Quasi-abelian groups. It is the aim of the following paragraphs to characterize and to investigate all those groups of motions and reversions which are discontinuous in $\mathcal{D}$. Among these groups a particular category which is exceptional by its rather simple structure is treated beforehand in this paragraph, viz. those groups whose elements all leave invariant a point, or a line. These groups include all discontinuous abelian groups of motions and reversions. For an element $\mathfrak{f} \neq 1$ leaves invariant one point (proper or infinite), if it is a rotation or limit-rotation, or one line (its end-points individually) if it is a translation or reversed translation or (all its points) if it is a reflection in that line); according to §4.5, every element permutable with $\mathfrak{f}$ must leave invariant in the first case that same point and in the second case that line as a whole. Moreover, it will be seen in the sequel that a group of the category considered, if it is not abelian, results from an abelian group by the adjunction of an involutory element, thus has an abelian subgroup with index 2.

All groups of the category considered will be called *quasi-abelian*, this term including in particular, abelian groups. The letter $\mathfrak{A}$ is set apart for denotation of groups with invariant points or lines, thus for quasi-abelian groups.

11.2 Groups with a proper invariant point. Let a group $\mathfrak{A}_c$ discontinuous in $\mathcal{D}$ leave invariant the point c of $\mathcal{D}$ and no other point of $\overline{\mathcal{D}}$ being left invariant by *all* elements of the group. All its elements are then rotations with centre c or reflections in lines passing through c.

First, let all elements of $\mathfrak{A}_c$ be rotations. Then $\mathfrak{A}_c$ is abelian. Let x_0 be an arbitrary point of $\mathcal{D}$ different from c. All points of the equivalence class $\mathfrak{A}_c x_0$ are then situated on the circle $\mathfrak{C}$ with centre c and radius $[c, x_0]$. Since $\mathfrak{A}_c$ is assumed to be discontinuous in $\mathcal{D}$, their number is finite, say ν. Let x_1 be the one next to x_0 in the positive sense on $\mathfrak{C}$, and put $\mathfrak{g}_0 x_0 = x_1$, $\mathfrak{g}_0 \in \mathfrak{A}_c$. Since the equivalence class $\mathfrak{A}_c x_0$ is reproduced by $\mathfrak{g}_0$, and no interior point of the arc $x_0 x_1$ of $\mathfrak{C}$ belongs to $\mathfrak{A}_c x_0$, it follows that the ν points of $\mathfrak{A}_c x_0$ are equidistant on $\mathfrak{C}$, that the angle of rotation $\varphi(\mathfrak{g}_0)$ of $\mathfrak{g}_0$ is $\frac{2\pi}{\nu}$, and that $\mathfrak{g}_0$ is of the order ν and generates $\mathfrak{A}_c$. Hence $\mathfrak{A}_c$ is a cyclical group of order ν. The elements $\mathfrak{g}_0$ and $\mathfrak{g}_0^{-1}$ are called *primary elements* of $\mathfrak{A}_c$. We denote this group by the symbol $\mathfrak{A}_c(\circlearrowleft_\nu)$. In the case $\nu = 2$ we also write $\mathfrak{A}_c(\cdot)$ instead of $\mathfrak{A}_c(\circlearrowleft_2)$. It should be noticed that $\mathfrak{A}_c(\cdot)$ leaves every line through c invariant as a whole, but it leaves only one point invariant, namely c, and it is therefore appropriate to include it among the groups with invariant points.

The centre of rotation c is the only singular point of $\mathfrak{A}_c(\circlearrowleft_\nu)$. If x_0 is any point of $\mathcal{D}$ other than c, thus regular, the normal domain $\mathcal{N}(\mathfrak{A}_c(\circlearrowleft_\nu); x_0)$ is the angular sector of magnitude $\frac{2\pi}{\nu}$ which has c as its vertex and is symmetric with respect to the line cx_0 (compare Fig. 13.4). From this normal domain a fundamental domain is obtained by omitting one of the bounding half-lines except for its end-point c.

Secondly, let $\mathfrak{A}_c$ contain reflections. Let $\mathfrak{s}$ be such a reflection, thus $\mathcal{L}(\mathfrak{s})$ passing through c. The product of two reflections in lines through c is a rotation about c.

Thus, if $\mathfrak{A}_c$ contains no rotation, then it consists only of $\mathfrak{s}$ and the identity. In this case the group leaves all points of $\mathcal{L}(\mathfrak{s})$ invariant individually and therefore is not comprised under the present case of groups with only one invariant point in $\overline{\mathcal{D}}$. This group belongs to the case of an invariant line and will be mentioned in Section 5. We therefore have to investigate those $\mathfrak{A}_c$ which also contain rotations.

Let $\mathfrak{g}_0$ be a primary element of the subgroup of rotations contained in $\mathfrak{A}_c$, say the one with positive angle of rotation $\frac{2\pi}{\nu}$, ν being the order of the rotation subgroup. Then $\mathfrak{s}\mathfrak{g}_0 = \mathfrak{s}'$ is a reflection contained in $\mathfrak{A}_c$, $\mathfrak{g}_0 = \mathfrak{s}\mathfrak{s}'$, and the angle from $\mathcal{L}(\mathfrak{s}')$ to $\mathcal{L}(\mathfrak{s})$ is $\frac{\pi}{\nu}$; compare §7.3. The two reflections $\mathfrak{s}$ and $\mathfrak{s}'$ generate $\mathfrak{A}_c$, for they generate $\mathfrak{A}_c(\circledcirc_\nu)$ and its coset by $\mathfrak{s}$. All ν elements of this coset are reflections. They can be obtained in the following way: Since

$$\mathfrak{s}\mathfrak{g}_0\mathfrak{s} = \mathfrak{s}'\mathfrak{s} = \mathfrak{g}_0^{-1},$$

one gets

$$\mathfrak{g}_0^{-\sigma}\mathfrak{s}\mathfrak{g}_0^{\sigma} = \mathfrak{s}\mathfrak{g}_0^{2\sigma}.$$

If ν is odd, this yields the ν reflections for σ ranging from 0 to $\nu - 1$. If ν is even, the ν reflections are

$$\mathfrak{g}_0^{-\sigma}\mathfrak{s}\mathfrak{g}_0^{\sigma}, \quad \mathfrak{g}_0^{-\sigma}\mathfrak{s}'\mathfrak{g}_0^{\sigma}$$

for σ ranging from 0 to $\frac{1}{2}\nu - 1$.

The group $\mathfrak{A}_c$ is of order 2ν and will be denoted by the symbol $\mathfrak{A}_c(\times_\nu)$. In the particular case $\nu = 2$ the group $\mathfrak{A}_c(\times_2)$ consists besides the identity of two reflections in mutually orthogonal reflection lines and one half-turn; it is an abelian group. For $\nu > 2$ the group $\mathfrak{A}_c(\times_\nu)$ is a dihedral group; it is non-abelian, but with an abelian subgroup of index 2, thus quasi-abelian.

Let x_0 be a point in $\mathcal{D}$ which is not situated on any of the ν reflection lines of the group $\mathfrak{A}_c(\times_\nu)$, thus a regular point. These ν lines divide the plane into 2ν angular sectors of magnitude $\frac{\pi}{\nu}$. The angular sector closed on $\mathcal{D}$ which contains x_0 is the normal domain $\mathcal{N}(\mathfrak{A}_c(\times_\nu); x_0)$. It is at the same time a fundamental domain for $\mathfrak{A}_c(\times_\nu)$.

11.3 Groups with an infinite invariant point. Let $\mathfrak{A}_u$ denote a group discontinuous in $\mathcal{D}$ which leaves invariant the infinite point u, no other point of $\overline{\mathcal{D}}$ being left invariant by *all* elements of the group. The elements of $\mathfrak{A}_u$ are thus either limit-rotations about u, or they are reflections in lines terminating at u, or translations or reversed translations with axes terminating at u. If $\mathfrak{A}_u$ were of order 2, it would consist of the identity and one reflection and would leave all points of the reflection line invariant. This case therefore is excluded as in the case of an $\mathfrak{A}_c$. Hence the order of $\mathfrak{A}_u$ is > 2, and $\mathfrak{A}_u$ thus contains a non-identical subgroup of motions.

It can then be shown that $\mathfrak{A}_u$ contains limit-rotations. For, let $\mathfrak{f}$ be a (non-identical) motion in $\mathfrak{A}_u$. If $\mathfrak{f}$ is not a limit-rotation, it is a translation with u as one of its invariant points and another v. Since v is not an invariant point of the group considered, let $\mathfrak{g}$ be an element of $\mathfrak{A}_u$ which does not leave v invariant. Then the translations $\mathfrak{f}$ and $\mathfrak{g}\mathfrak{f}\mathfrak{g}^{-1}$ have different axes (with common end-point at u) and equal displacements. Then $\mathfrak{f}^{-1}\mathfrak{g}\mathfrak{f}\mathfrak{g}^{-1}$ has only u as invariant point and is by §9.1 a limit-rotation about u. Since limit-rotations are of infinite order, any $\mathfrak{A}_u$ is of infinite order.

If all elements of $\mathfrak{A}_u$ are limit-rotations, the group is an abelian group and it is denoted by $\mathfrak{A}_u(\hookrightarrow)$. All points of the equivalence class $\mathfrak{A}_u(\hookrightarrow)x$ of a point x of $\mathcal{D}$ are situated on a horocycle $\mathcal{O}$ with centre u and are different, since the elements of $\mathfrak{A}_u(\hookrightarrow)$ leave no point of $\mathcal{D}$ invariant. In virtue of the discontinuity of $\mathfrak{A}_u(\hookrightarrow)$ there must be two points of the class which are the next to x on that horocycle $\mathcal{O}$. Let one of these be $\mathfrak{g}_0 x$, the other then is $\mathfrak{g}_0^{-1}x$. Since $\mathfrak{g}_0$ displaces all points of $\mathcal{O}$ in the same sense, the sequence $\mathfrak{g}_0^{\nu}x$ is monotonous on the horocycle according to the sequence of exponents ν. Thus this sequence of points must be convergent both for $\nu \to \infty$ and for $\nu \to -\infty$, and the limit point can only be u in both cases, since it obviously must be an invariant point for $\mathfrak{g}_0$. The same is true if x is a point on $\mathcal{E}$. From this follows that $\mathfrak{A}_u(\hookrightarrow)$ is generated by $\mathfrak{g}_0$, for $\mathfrak{g}_0$ reproduces $\mathfrak{A}_u(\hookrightarrow)x$, and since the arc $x, \mathfrak{g}_0 x$ of $\mathcal{O}$ contains no point of this equivalence class in its interior, the same holds for its image under $\mathfrak{g}_0^{\nu}$. Hence $\mathfrak{A}_u(\hookrightarrow)$ is a free group generated by $\mathfrak{g}_0$ or by $\mathfrak{g}_0^{-1}$. In a free group with one generator this generator is uniquely determined except for replacement by its inverse. Thus $\mathfrak{g}_0$ and $\mathfrak{g}_0^{-1}$ do not depend on the choice of x. They are called the *primary elements* of the group.

An arbitrary point x_0 of $\mathcal{D}$ is regular, and the normal domain $\mathcal{N}(\mathfrak{A}_u(\hookrightarrow); x_0)$ is a strip bounded by two lines issuing from u which are symmetric with respect to the line ux_0 and which correspond by $\mathfrak{g}_0$ and its inverse (compare Fig. 13.5). A fundamental domain arises if the points on one of the bounding lines are omitted.

It is now assumed that $\mathfrak{A}_u$ contains, besides limit-rotations, reflections with respect to lines issuing from u, but no translations or reversed translations. The case then is similar to the case of an $\mathfrak{A}_c(\times_\nu)$ considered in Section 2. The limit-rotations contained in $\mathfrak{A}_u$ form an abelian subgroup $\mathfrak{A}_u(\hookrightarrow)$ with index 2 the structure of which has just been investigated. Let $\mathfrak{g}_0$ be a primary element of $\mathfrak{A}_u(\hookrightarrow)$ and $\mathfrak{s}$ a reflection contained in $\mathfrak{A}_u$. Then $\mathfrak{s}' = \mathfrak{s}\mathfrak{g}_0$ belongs to $\mathfrak{A}_u$ and is a reflection, since it carries a horocycle with centre u into itself. The reflections $\mathfrak{s}$ and $\mathfrak{s}'$ generate $\mathfrak{A}_u$; for their product $\mathfrak{s}\mathfrak{s}' = \mathfrak{g}_0$ generates $\mathfrak{A}_u(\hookrightarrow)$, and every reflection contained in $\mathfrak{A}_u$ is contained in the coset $\mathfrak{s}\mathfrak{A}_u(\hookrightarrow)$ of $\mathfrak{A}_u(\hookrightarrow)$ in $\mathfrak{A}_u$. We denote the group by $\mathfrak{A}_u(\wedge)$. The lines of reflection $\mathcal{L}(\mathfrak{s})$ and $\mathfrak{s}'\mathcal{L}(\mathfrak{s}) = \mathfrak{s}'\mathfrak{s}\mathcal{L}(\mathfrak{s}) = \mathfrak{g}_0^{-1}\mathcal{L}(\mathfrak{s}) = \mathcal{L}(\mathfrak{g}_0^{-1}\mathfrak{s}\mathfrak{g}_0)$ are symmetric with respect to $\mathcal{L}(\mathfrak{s}')$. All lines of reflection in $\mathfrak{A}_u(\wedge)$ are the images of $\mathcal{L}(\mathfrak{s})$ and $\mathcal{L}(\mathfrak{s}')$ by the elements of $\mathfrak{A}_u(\hookrightarrow)$. They are equidistant in the sense that any two neighbouring cut out congruent arcs on a fixed horocycle about u.

Any point x_0 of $\mathcal{D}$ which is not situated on any of these lines of reflections is regular. The normal domain $\mathcal{N}(\mathfrak{A}_u(\wedge); x_0)$ is the strip bounded by two neighbouring lines of reflections, containing x_0 and closed on $\mathcal{D}$. This strip is at the same time a fundamental domain.

11.4 Non discontinuous groups with an infinite invariant point. Let $\mathfrak{A}_u$ be a group leaving invariant the infinite point u but no other point of $\overline{\mathcal{D}}$. It will now be shown that *if* $\mathfrak{A}_u$ *contains a translation, then it cannot be discontinuous in* $\mathcal{D}$. The same then is true if $\mathfrak{A}_u$ contains a reversed translation, since the square of a reversed translation is a translation.

For an indirect proof, let it be assumed that $\mathfrak{A}_u$ is discontinuous in $\mathcal{D}$ and that $\mathfrak{f}$ is a translation in $\mathfrak{A}_u$ with u as its negative fundamental point (§7.2). It is known from Section 3 that $\mathfrak{A}_u$ contains limit-rotations. If $\mathfrak{A}_u(\hookrightarrow)$ denotes the subgroup consisting of all limit-rotations of $\mathfrak{A}_u$, even $\mathfrak{A}_u(\hookrightarrow)$ is discontinuous; let $\mathfrak{g}_0$ be a primary element of $\mathfrak{A}_u(\hookrightarrow)$. The motion $\mathfrak{f}\mathfrak{g}_0\mathfrak{f}^{-1}$ is a limit-rotation, thus contained in $\mathfrak{A}_u(\hookrightarrow)$. The image of the axis $\mathcal{A}(\mathfrak{f})$ of $\mathfrak{f}$ by $\mathfrak{f}\mathfrak{g}_0\mathfrak{f}^{-1}$ is the same as its image by $\mathfrak{f}\mathfrak{g}_0$, since $\mathfrak{f}^{-1}$ carries $\mathcal{A}(\mathfrak{f})$ into itself. Now the line $\mathfrak{f}\mathfrak{g}_0\mathcal{A}(\mathfrak{f})$ is situated between $\mathcal{A}(\mathfrak{f})$ and $\mathfrak{g}_0\mathcal{A}(\mathfrak{f})$ (Fig. 11.1) This does not comply with the fact that $\mathfrak{g}_0$ is a primary element of $\mathfrak{A}_u(\hookrightarrow)$.

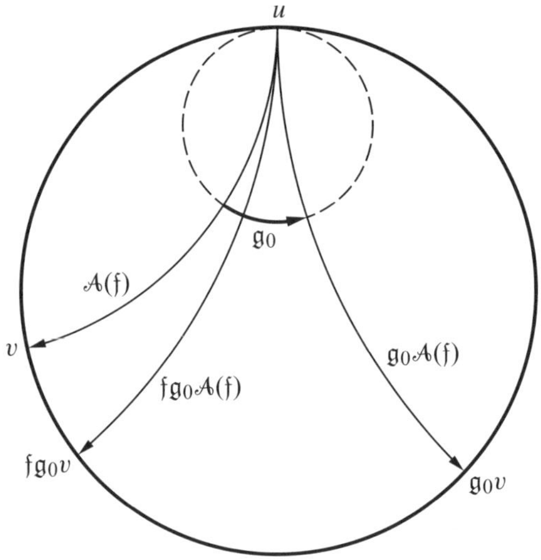

Figure 11.1

Hence the groups considered in Section 3 are the only discontinuous groups with one infinite invariant point. Moreover the following important consequence results.

In a group $\mathfrak{G}$ *discontinuous in* $\mathcal{D}$ *a fundamental point of a translation or a reversed translation cannot coincide with the centre* u *of a limit-rotation.* This follows from the theorem just proved, since the subgroup $\mathfrak{A}_u$ of $\mathfrak{G}$ leaving u invariant must also be discontinuous. Likewise, *two translations (direct or reversed) in* $\mathfrak{G}$ *cannot have one fundamental point in common without having also the other in common and thus being permutable.* For, if $\mathfrak{f}$ and $\mathfrak{g}$ are translations with u as common fundamental point and not coinciding axes, then $\mathfrak{f}$ and $\mathfrak{g}\mathfrak{f}^{-1}\mathfrak{g}^{-1}$ are translations with equal displacements and oppositely directed, not coinciding axes, and their product, the commutator $\mathfrak{fgf}^{-1}\mathfrak{g}^{-1}$, is then a limit-rotation with centre u (§9.1). Also, *the axis of a translation in $\mathfrak{G}$ cannot have one end-point in common with the line of a reflection in $\mathfrak{G}$ without coinciding with that line.* For, if the fundamental point u of the translation $\mathfrak{f}$ were an end-point of the line $\mathcal{L}$ of the reflection $\mathfrak{s}$, this line $\mathcal{L}$ not coinciding with the axis $\mathcal{A}$ of $\mathfrak{f}$, then $\mathfrak{s}\mathcal{A}$, which is the axis of the translation $\mathfrak{sfs}$ of $\mathfrak{G}$, and $\mathcal{A}$ would have u in common without coinciding.

11.5 Groups with an invariant line. Let u and v be two infinite points and $\mathcal{A}$ the line joining them. Let $\mathfrak{A}_{\mathcal{A}}$ be a group discontinuous in $\mathcal{D}$ which leaves $\mathcal{A}$ invariant. The elements which such a group can contain are determined as follows: They can either leave u and v invariant individually and are then translations or reversed translations with $\mathcal{A}$ as axis or the reflection in $\mathcal{A}$, or they can interchange u and v and are then reflections in normals of $\mathcal{A}$ or half-turns about centres on $\mathcal{A}$. The cases in which the group apart from the identity only contains one half-turn, or two reflections and one half-turn, have been treated in Section 2 and are now left aside. They are the groups $\mathfrak{A}_c(\cdot)$ or $\mathfrak{A}_c(\times 2)$ and are comprised among the groups with a proper invariant point.

We now consider the case in which the group $\mathfrak{A}_{\mathcal{A}}$ apart from the identity only contains one reflection $\mathfrak{s}$. If $\mathcal{L}(\mathfrak{s})$ is $\mathcal{A}$ itself, then this group is denoted by $\mathfrak{A}_{\mathcal{A}}(-)$. However, $\mathcal{L}(\mathfrak{s})$ may also be a normal of $\mathcal{A}$, and we then denote the group by $\mathfrak{A}(|)$. Both groups are of the same kind in as far as they are generated by a reflection. But *in relation to a given line* $\mathcal{A}$ it is advantageous to have the two symbols in order to distinguish between the two ways in which they carry $\mathcal{A}$ into itself. All points not on the reflection line are regular. The normal domain $\mathcal{N}(\mathfrak{A}_{\mathcal{A}}(-); x_0)$ is that half-plane bounded by $\mathcal{A}$ and closed on $\mathcal{D}$ which contains x_0. It is also a fundamental domain. Compare Fig. 13.7.

In all remaining cases translations must occur among the elements of the group $\mathfrak{A}_{\mathcal{A}}$. We now consider these general cases.

At first the groups leaving u and v invariant individually are considered. These are all abelian, since any two of the elements in question are permutable. If reversions do not occur, all elements of $\mathfrak{A}_{\mathcal{A}}$ are translations (apart from the identity). Let $\mathcal{A}$ be directed from u towards v and a sign accordingly attached to all displacements. The points of the equivalence class $\mathfrak{A}_{\mathcal{A}}x$ of a point x on $\mathcal{A}$ are all situated on $\mathcal{A}$ and are different, since the elements of $\mathfrak{A}_{\mathcal{A}}$ have no invariant point in $\mathcal{D}$. Since $\mathfrak{A}_{\mathcal{A}}$ is assumed discontinuous, there are in this equivalence class two points next to x. Let $\mathfrak{g}_0 x$ be the one between x and v; the other then is $\mathfrak{g}_0^{-1}x$. The sequence of points $\mathfrak{g}_0^{\nu}x$, $-\infty < \nu < \infty$, is monotonous on $\mathcal{A}$ and it converges towards v for $\nu \to \infty$ and towards u for $\nu \to -\infty$. The statement about convergence also holds, if x is any point of $\overline{\mathcal{D}}$ other than u and v. In the same way as in the case of groups of limit-rotations it is seen that $\mathfrak{A}_{\mathcal{A}}$ is a free group generated by $\mathfrak{g}_0$ or $\mathfrak{g}_0^{-1}$. The smallest positive displacement occurring in $\mathfrak{A}_{\mathcal{A}}$ is the displacement $\lambda(\mathfrak{g}_0)$. The elements $\mathfrak{g}_0$ and $\mathfrak{g}_0^{-1}$ are called the *primary elements of the group* and $\lambda(\mathfrak{g}_0)$ *its primary displacement*. This group is denoted by $\mathfrak{A}_{\mathcal{A}}(\rightarrow)$. No point of $\mathcal{D}$ is singular with respect to $\mathfrak{A}_{\mathcal{A}}(\rightarrow)$. If x_0

is an arbitrary point of $\mathcal{D}$, the normal domain $\mathcal{N}(\mathfrak{A}_{\mathcal{A}}(\rightarrow); x_0)$ is the strip symmetric with respect to the perpendicular from x_0 on $\mathcal{A}$ and bounded by two normals of $\mathcal{A}$ one of which is carried into the other by $\mathfrak{g}_0$ (compare Fig. 13.6). A fundamental domain arises if the points on one of these bounding lines are omitted.

Now, let reversions occur in $\mathfrak{A}_{\mathcal{A}}$. Let $\mathfrak{h}_0$ be the one with the smallest non-negative displacement. If $\mathfrak{h}_0$ is a reversed translation, it is seen in the same way as above that $\mathfrak{A}_{\mathcal{A}}$ is generated by $\mathfrak{h}_0$ (or $\mathfrak{h}_0^{-1}$). The elements $\mathfrak{h}_0$ and $\mathfrak{h}_0^{-1}$ are again called *primary* and $\lambda(\mathfrak{h}_0)$ the *primary displacement*. We denote this group by $\mathfrak{A}_{\mathcal{A}}(\rightleftharpoons)$. The subgroup of motions in $\mathfrak{A}_{\mathcal{A}}$ is the group $\mathfrak{A}_{\mathcal{A}}(\rightarrow)$, generated by $\mathfrak{h}_0{}^2$.

If $\mathfrak{h}_0$ is the reflection in $\mathcal{A}$, then the group of motions contained in $\mathfrak{A}_{\mathcal{A}}$ is a group of translations of the above kind $\mathfrak{A}(\rightarrow)$. If its primary elements are denoted by $\mathfrak{g}_0$ and $\mathfrak{g}_0^{-1}$, $\mathfrak{A}_{\mathcal{A}}$ consists apart from the identity, of the reflection $\mathfrak{h}_0$ and all translations and reversed translations with displacements $\nu\lambda(\mathfrak{g}_0)$, $\nu = \pm 1, \pm 2, \ldots$. The group is generated by $\mathfrak{h}_0$ and $\mathfrak{g}_0$ but cannot be generated by one single element. We denote it by $\mathfrak{A}_{\mathcal{A}}(-\rightarrow)$; one has $\mathfrak{A}_{\mathcal{A}}(-\rightarrow) = \mathfrak{A}_{\mathcal{A}}(\rightarrow) \cup \mathfrak{h}_0\mathfrak{A}_{\mathcal{A}}(\rightarrow)$.

To construct normal domains, consider first $\mathfrak{A}_{\mathcal{A}}(\rightleftharpoons)$ with the primary element $\mathfrak{h}_0$. No point in $\mathcal{D}$ is singular. If x_0 is a point on $\mathcal{A}$, the normal domain $\mathcal{N}(\mathfrak{A}_{\mathcal{A}}(\rightleftharpoons); x_0)$ is bounded by the bisecting perpendiculars of the segments $x_0, \mathfrak{h}_0 x_0$ and $x_0, \mathfrak{h}_0^{-1} x_0$. The same is true if x_0 is not on $\mathcal{A}$, provided it is sufficiently near to $\mathcal{A}$; exactly speaking, as long as the bisecting perpendiculars of the segments $x_0, \mathfrak{h}_0 x_0$ and $x_0, \mathfrak{h}_0^{-1} x_0$ are divergent or parallel. If they intersect, as they will do if x_0 is further removed from $\mathcal{A}$, then the normal domain is a half-strip bounded by two half-lines on the bisecting perpendiculars of $x_0, \mathfrak{h}_0^2 x_0$ and $x_0, \mathfrak{h}_0^{-2} x_0$ and two segments on those of $x_0, \mathfrak{h}_0 x_0$ and $x_0, \mathfrak{h}_0^{-1} x_0$; the first part of the boundary corresponds by $\mathfrak{h}_0^2$, the second by $\mathfrak{h}_0$ (see Fig. 11.2). In both cases a fundamental domain is obtained by omitting the points on one half of the boundary.

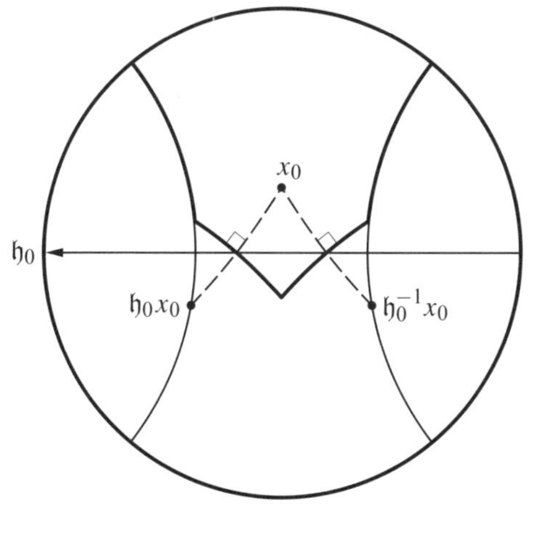

Figure 11.2

With respect to $\mathfrak{A}_{\mathcal{A}}(-\ \rightarrow)$ all points of $\mathcal{A}$ are singular, since $\mathfrak{A}(-\ \rightarrow)$ contains the reflection $\mathfrak{h}_0$ in $\mathcal{A}$; if $\mathfrak{g}_0$ denotes a primary element of the subgroup $\mathfrak{A}_{\mathcal{A}}(\rightarrow)$ consisting of translations, the normal domain $\mathcal{N}(\mathfrak{A}_{\mathcal{A}}(-\ \rightarrow); x_0)$ with central point x_0 not on $\mathcal{A}$ is obtained in the following way: The normal domain $\mathcal{N}(\mathfrak{A}_{\mathcal{A}}(\rightarrow); x_0)$ is divided by $\mathcal{A}$ into two half-strips; $\mathcal{N}(\mathfrak{A}_{\mathcal{A}}(-\ \rightarrow); x_0)$ is the closed half-strip containing x_0. A fundamental domain is obtained by omitting the points on one of the bounding half-lines.

It remains to discuss those groups $\mathfrak{A}_{\mathcal{A}}$ which interchange u and v. Such group must contain a group $\mathfrak{A}^0_{\mathcal{A}}$ of the kind hitherto considered in this section as a subgroup with index 2; for, if $\mathfrak{s}$ is an element of $\mathfrak{A}_{\mathcal{A}}$ which interchanges u and v, thus either the reflection in a normal of $\mathcal{A}$ or the half-turn about a point of $\mathcal{A}$, then every element of $\mathfrak{A}_{\mathcal{A}}$ can be written $\mathfrak{g}$ or $\mathfrak{s}\mathfrak{g}$, where $\mathfrak{g}$ belongs to $\mathfrak{A}_{\mathcal{A}}$ and leaves u and v invariant individually. These elements $\mathfrak{g}$ form a subgroup $\mathfrak{A}^0_{\mathcal{A}}$ which can belong to each of the three types $\mathfrak{A}_{\mathcal{A}}(\rightarrow)$, $\mathfrak{A}_{\mathcal{A}}(\rightleftharpoons)$, $\mathfrak{A}_{\mathcal{A}}(-\ \rightarrow)$ considered above.

First, let $\mathfrak{A}^0_{\mathcal{A}} = \mathfrak{A}_{\mathcal{A}}(\rightarrow)$ be a group of translations and $\mathfrak{s}$ a reflection in a normal of $\mathcal{A}$. If $\mathfrak{g}_0$ is a primary element of $\mathfrak{A}_{\mathcal{A}}(\rightarrow)$ then $\mathfrak{s}\mathfrak{g}_0 = \mathfrak{s}'$ is the reflection in another normal of $\mathcal{A}$. The group $\mathfrak{A}_{\mathcal{A}}$ is generated by $\mathfrak{s}$ and $\mathfrak{s}'$ and we denote it by $\mathfrak{A}_{\mathcal{A}}(\|)$. There are an infinity of reflections in $\mathfrak{A}_{\mathcal{A}}(\|)$ with equidistant reflection lines at right angles to $\mathcal{A}$. Two neighbouring ones cut out a segment of length $\frac{1}{2}\lambda(\mathfrak{g}_0)$ on $\mathcal{A}$, and the strip bounded by two such lines and closed on $\mathcal{D}$ is the normal domain for each point interior to the strip. The strip is also a fundamental domain.

Then let $\mathfrak{A}^0_{\mathcal{A}} = \mathfrak{A}_{\mathcal{A}}(\rightarrow)$ be a group of translations with primary element $\mathfrak{g}_0$, and $\mathfrak{s}$ a half-turn with its centre on $\mathcal{A}$. Then $\mathfrak{s}' = \mathfrak{s} \cdot \mathfrak{g}_0$ is another half-turn with centre on $\mathcal{A}$. The group $\mathfrak{A}_{\mathcal{A}}$ then is generated by $\mathfrak{s}$ and $\mathfrak{s}'$; it is denoted by $\mathfrak{A}_{\mathcal{A}}(\cdot\cdot)$. There are an infinity of half-turns in $\mathfrak{A}_{\mathcal{A}}(\cdot\cdot)$ with centres in $\mathcal{A}$ at equal distances $\frac{1}{2}\lambda(\mathfrak{g}_0)$. If x_0 is a regular point on $\mathcal{A}$, thus not a centre of rotation, then $\mathcal{N}(\mathfrak{A}_{\mathcal{A}}(\cdot\cdot); x_0)$ is bounded by two lines at right angles to $\mathcal{A}$ through the centres of rotation next to x_0 on $\mathcal{A}$. If x_0 is not on $\mathcal{A}$, different forms arise which need not be discussed in detail. A fundamental domain is obtained by omitting appropriate points of the boundary of the normal domain.

Then let $\mathfrak{A}^0_{\mathcal{A}}$ be a group $\mathfrak{A}_{\mathcal{A}}(\rightleftharpoons)$ generated by a reversed translation $\mathfrak{h}_0$. The element $\mathfrak{h}_0$ can be represented as the product of a half-turn with centre on $\mathcal{A}$ and a reflection in a normal of $\mathcal{A}$, and in this representation either the centre of rotation or the line of reflection can be chosen at will. From this it is seen that in the present case the adjunction of a half-turn with centre in $\mathcal{A}$ yields the same result as the adjunction of a reflection in a normal of $\mathcal{A}$. In both cases the resulting group can be generated by a half-turn together with a reflection in a normal of $\mathcal{A}$ distant $\frac{1}{2}\lambda(\mathfrak{h}_0)$ from the centre of rotation. It will be denoted by $\mathfrak{A}_{\mathcal{A}}(\cdot|)$. The group contains an infinity of half-turns with centres at equal distances on $\mathcal{A}$ and an infinity of reflections in normals of $\mathcal{A}$ bisecting the segments joining two neighbouring centres of rotation. If x_0 is a regular point in $\mathcal{D}$ thus neither a centre of rotation nor situated on a reflection line, the normal domain $\mathcal{N}(\mathfrak{A}_{\mathcal{A}}(\cdot|); x_0)$ is obtained in the following way: The strip bounded by two neighbouring lines of reflection and containing x_0 contains exactly one centre of rotation c. The perpendicular on x_0c in c divides the strip into two congruent parts, whether it cuts the boundaries of the strip or not. The part containing x_0 is the normal domain. A fundamental domain arises if one half of this perpendicular, terminating at the centre of rotation, is omitted.

Finally, let $\mathfrak{A}^0_{\mathcal{A}} = \mathfrak{A}_{\mathcal{A}}(- \rightarrow)$ be a group generated by a reflection $\mathfrak{h}_0$ in $\mathcal{A}$ and a translation $\mathfrak{g}_0$. Since $\mathfrak{h}_0$ can be represented as the product of a half-turn with centre on $\mathcal{A}$ and the reflection in a normal of $\mathcal{A}$ in the centre of rotation, the adjunction of a half-turn or of a reflection produce equal effects even in this case. The group $\mathfrak{A}_{\mathcal{A}}$ is generated by $\mathfrak{h}_0$ and two reflections in normals of $\mathcal{A}$, and also by $\mathfrak{h}_0$ and two half-turns with centres on $\mathcal{A}$, the distance of the two normals, or of the two centres, being $\frac{1}{2}\lambda(\mathfrak{g}_0)$. This group can be denoted by $\mathfrak{A}_{\mathcal{A}}(- \,\|)$ and also by $\mathfrak{A}_{\mathcal{A}}(- \cdot \cdot)$. Besides $\mathfrak{h}_0$ and an infinity of translations and reversed translations it contains an infinity of reflections in equidistant normals of $\mathcal{A}$ and of half-turns with centres at the intersection points of these normals with $\mathcal{A}$. If x_0 is a regular point in $\mathcal{D}$, thus not situated on any of the reflection lines, $\mathcal{N}(\mathfrak{A}_{\mathcal{A}}(- \,\|); x_0)$ is a half-strip bounded by $\mathcal{A}$ and two neighbouring reflection lines, containing x_0, and closed on $\mathcal{D}$. It is also a fundamental domain.

11.6 List of quasi-abelian groups. It has been seen in the preceding sections that there are twelve types of quasi-abelian groups discontinuous in $\mathcal{D}$. For convenience we list them in the following table:

Table 1

	abelian		non-abelian	
	finite	infinite	finite	infinite
inv. point	$\mathfrak{A}(\circlearrowleft_\nu)$	$\mathfrak{A}(\hookrightarrow)$	$\mathfrak{A}(\times_\nu)$	$\mathfrak{A}(\wedge)$
inv. line	$\mathfrak{A}(-)$ $\mathfrak{A}(\vert)$	$\mathfrak{A}(\rightarrow)$ $\mathfrak{A}(\rightleftharpoons)$ $\mathfrak{A}(- \rightarrow)$		$\mathfrak{A}(\Vert)$ $\mathfrak{A}(\cdot\cdot)$ $\mathfrak{A}(\cdot\vert)$ $\mathfrak{A}(- \,\Vert) = \mathfrak{A}(- \cdot \cdot)$

The invariant point or line of the different groups has not been indicated in this table, since it results from the different symbols, except in the case $\mathfrak{A}(-)$, where both the reflection line and all its normals are invariant. Thus $\mathfrak{A}(-)$ and $\mathfrak{A}(|)$ represent the same type. It should be noticed that a quasi-abelian group possesses at most one axis. Of course $\mathfrak{A}(\times_\nu)$ is abelian for $\nu = 2$.

11.7 Conclusion. This enumeration of all groups with an invariant point or an invariant line and discontinuous in $\mathcal{D}$ makes it evident, as stated in Section 1, that all these groups are abelian or derived from an abelian subgroup $\mathfrak{A}^0_{\mathcal{A}}$ by the adjunction of an element of order 2.

It is worth noticing that Table 1 contains all those groups which can be generated by one of their elements, thus all cyclic groups and all free groups with one generator;

furthermore all groups which can be generated by two involutory elements. Only two of the groups do not belong to any of these types, viz. the group $\mathfrak{A}(-\rightarrow)$ generated by a translation and the reflection in its axis, and the group $\mathfrak{A}(-\,\|)$ generated by three reflections, one reflection line being the common perpendicular of the two others.

Four of the twelve groups consist of motions only, namely

$$\mathfrak{A}(\circlearrowleft_\nu), \quad \mathfrak{A}(\hookrightarrow), \quad \mathfrak{A}(\rightarrow), \quad \mathfrak{A}(\cdot\cdot).$$

For those five groups which can be generated by reflections only, i.e.

$$\mathfrak{A}(-), \quad \mathfrak{A}(\times_\nu), \quad \mathfrak{A}(\wedge), \quad \mathfrak{A}(\|), \quad \mathfrak{A}(-\,\|),$$

normal domains can be characterized by the following general rule:

The normal domain $\mathcal{N}(x_0)$ for a group generated by reflections and an arbitrary regular point x_0 is the closure on $\mathcal{D}$ of that component of the set of regular points to which x_0 belongs. $\mathcal{N}(x_0)$ is also a fundamental domain. □

§12 A discontinuity theorem

12.1 An auxiliary theorem. After the special groups of §11 have been treated, the following investigations are concerned with groups of motions and reversions without invariant points or lines. It is the aim of this paragraph to establish a necessary and sufficient condition for the discontinuity of such groups. In order to facilitate the proofs given later, the following is established beforehand:

If in a groups of motions without invariant point two different axes of translations are parallel, then the group contains rotations. The same is true if an axis of translation terminates at the centre of a limit-rotation.

Proof. Let u be a common end-point of two different axes of translations in a group $\mathfrak{G}$ of motions. Let $\mathfrak{A}_u$ be the subgroup of $\mathfrak{G}$ consisting of those elements of $\mathfrak{G}$ which leave u invariant. This subgroup contains translations along the two axes and therefore evidently has no other invariant point than u. It is known from §11.3 that $\mathfrak{A}_u$, and thus $\mathfrak{G}$, contains limit-rotations with centre u. On the other hand, if the axis of a translation $\mathfrak{f}$ terminates at the centre of a limit-rotation $\mathfrak{g}$, then the axis of the translation $\mathfrak{f}' = \mathfrak{g}\mathfrak{f}\mathfrak{g}^{-1}$ is parallel to the axis of $\mathfrak{f}$. Thus the conditions in the two parts of the theorem are equivalent. – Now it is known from §11.4 that the subgroup $\mathfrak{A}_u(\hookrightarrow)$ consisting of all limit-rotations with centre u in $\mathfrak{G}$ is not discontinuous in $\mathcal{D}$. It has been pointed out in §9.1 that limit-rotations have no essential invariant, but that arcs of an arbitrary horocycle $\mathcal{O}$ touching $\mathcal{E}$ in the limit-centre represent displacements on $\mathcal{O}$, the displacement on $\mathcal{O}$ of a product being the sum of the displacements on $\mathcal{O}$ of the factors. Consider now a point x on an arbitrary horocycle $\mathcal{O}$ with u as limit-centre. Since $\mathfrak{A}_u(\hookrightarrow)$ is not discontinuous in $\mathcal{D}$, the point x is an accumulation point of the equivalence class $\mathfrak{A}_u(\hookrightarrow)x$, hence $\mathfrak{A}_u(\hookrightarrow)$ contains limit-rotations with arbitrarily small displacements on $\mathcal{O}$. Since $\mathfrak{G}$ has not u as an invariant point, let $\mathfrak{g}$ be an element of $\mathfrak{G}$ for which u is not invariant. Let $\mathfrak{f}_1$ and $\mathfrak{f}_2$ be translations belonging to two given axes respectively. Then at least one of the two translations $\mathfrak{g}\mathfrak{f}_1\mathfrak{g}^{-1}$ and $\mathfrak{g}\mathfrak{f}_2\mathfrak{g}^{-1}$ does not leave u invariant; for $\mathfrak{g}$ displaces u and can at most carry one of the other end-points of the axes into u. Thus there exists in $\mathfrak{G}$ a translation $\mathfrak{f}$ which does not leave u invariant. Let the above horocycle $\mathcal{O}$ with centre at u be so chosen as to intersect $\mathcal{A}(\mathfrak{f})$, and let $\mathfrak{h}$ denote an element of $\mathfrak{A}_u(\hookrightarrow)$. For sufficiently small displacements on $\mathcal{O}$ of $\mathfrak{h}$ the axes of the translations $\mathfrak{f}$ and $\mathfrak{f}' = \mathfrak{h}\mathfrak{f}\mathfrak{h}^{-1}$ intersect, and they even intersect at an arbitrarily small angle φ (see Fig. 12.1).

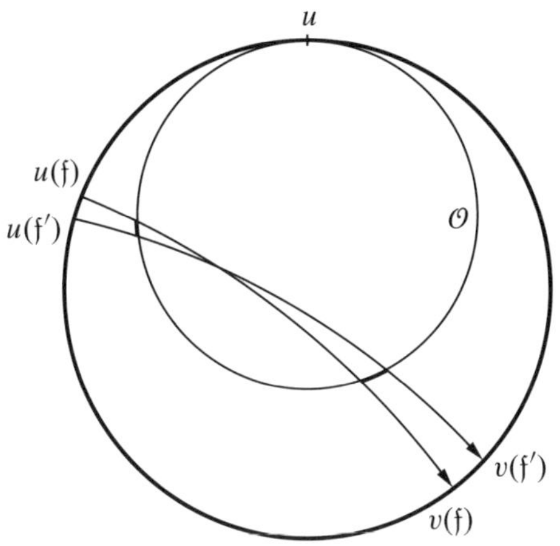

Figure 12.1

In particular, let $\mathfrak{h}$ be chosen with a so small displacement on $\mathcal{O}$ that

$$\sin\varphi < \frac{1}{\sinh^2 \frac{1}{2}\lambda(\mathfrak{f})}.$$

If $\mathfrak{f}$ and $\mathfrak{f}'$ are represented as products of half-turns (which need not belong to $\mathfrak{G}$), using in both cases the intersection point of their axes as one of the centres, $\sin\varphi \sinh^2 \frac{1}{2}\lambda(\mathfrak{f})$ is the sine amplitude (§8.4) of the triangle of centres. Hence by §9.8 the commutator $\mathfrak{k}_{\mathfrak{f}\mathfrak{f}'}$, which is an element of $\mathfrak{G}$, is a rotation. This proves the theorem. □

12.2 The discontinuity theorem. In this and the next two sections we will prove the following theorem:

Consider a group $\mathfrak{G}$ of motions and reversions which leaves invariant no point and no line. Then $\mathfrak{G}$ is discontinuous in $\mathfrak{D}$ if and only if the centres of rotations in $\mathfrak{G}$, if any, do not accumulate in $\mathfrak{D}$.

As a first step of the proof we make the following reduction:

It is sufficient to prove this theorem for groups of motions.

Proof. For let $\mathfrak{G}$ be a group which contains reversions and $\mathfrak{H}$ the subgroup of motions in $\mathfrak{G}$; this is a normal subgroup. At the end of §10.2 it has been shown that $\mathfrak{G}$ and $\mathfrak{H}$ are simultaneously discontinuous in $\mathfrak{D}$ or not. So it only remains to be proved that, if $\mathfrak{G}$ has no invariant point or line, the same is true for $\mathfrak{H}$. Let it be assumed that $\mathfrak{H}$ leaves invariant the proper or finite point c but no other point of $\overline{\mathfrak{D}}$. Since $\mathfrak{G}$ has no invariant point, let $\mathfrak{g}$ be an element of $\mathfrak{G}$ which displaces c. Then the subgroup $\mathfrak{g}\mathfrak{H}\mathfrak{g}^{-1}$ of $\mathfrak{G}$ leaves invariant the point $\mathfrak{g}c$ but no other point of $\mathfrak{D}$, thus not c, in contradiction to the fact that this subgroup is $\mathfrak{H}$ itself. It is proved in exactly the same way that $\mathfrak{H}$ leaves no line invariant. □

Remark. The condition of the non-existence of an invariant point or line is obviously necessary for the validity of the theorem. For if $\mathfrak{G}$ is the group consisting of *all* limit-rotations about a given infinite point, or the group consisting of *all* translations along a given line, the group contains no rotations at all and nevertheless is not discontinuous in $\mathfrak{D}$.

12.3 Proof for groups containing rotations. In virtue of the preceding section it can be assumed that $\mathfrak{G}$ is a group of motions without invariant point or line. It has already been proved in §10.4 that centres of rotations of a group discontinuous in $\mathfrak{D}$ do not accumulate in $\mathfrak{D}$. The condition of the theorem, therefore, is necessary for the discontinuity. It remains to be seen that it is sufficient: Let c be a centre of rotation in $\mathfrak{G}$. Since $\mathfrak{G}$ has no invariant point, let $\mathfrak{g}$ be an element of $\mathfrak{G}$ which displaces c. Then $c' = \mathfrak{g}c$ is also a centre of rotation in $\mathfrak{G}$. From this it is inferred that the abelian subgroup of $\mathfrak{G}$ consisting of all rotations with centre c must be a group of finite order. For these rotations carry c' into other centres of rotation of $\mathfrak{G}$, and these centres would accumulate on the circle $\mathfrak{C}(c; [c, c'])$ if the order were infinite. Hence the equivalence class $\mathfrak{G}c$ of c consists of centres of rotation for which the corresponding subgroups of rotations have all one and the same finite order. Since, by assumption, the centres do not accumulate, and each is counted only a finite number of times in $\mathfrak{G}c$, this equivalence class does not accumulate in $\mathfrak{D}$. This proves the discontinuity of $\mathfrak{G}$.

12.4 Proof for groups not containing rotations. It remains to prove the discontinuity theorem under the assumption that no rotations occur in $\mathfrak{G}$. It is then sufficient to put the first condition in the form that $\mathfrak{G}$ leaves no point of $\mathfrak{E}$ invariant; for it can leave neither a point in $\mathfrak{D}$ nor a line invariant, since the only motions which can interchange the end-points of a line are half-turns, and these are excluded by the assumption. So the theorem which has yet to be proved can be thus formulated:

A group of motions which leaves invariant no point of $\mathfrak{E}$ and which contains no rotations is discontinuous in $\mathfrak{D}$.

Proof. It is seen from §11 that a group $\mathfrak{G}$ which satisfies these conditions cannot be abelian. Furthermore, translations must occur in $\mathfrak{G}$. For otherwise all elements of $\mathfrak{G}$ would be limit-rotations and, by assumption, would not all have the same centre; but it is known from §9.6 that the commutator of two limit-rotations with different centres is a translation (case $\mathrm{I}(\rightarrow\hookrightarrow\hookrightarrow)$ or $\mathrm{I}(\hookrightarrow\hookrightarrow\hookrightarrow)$).

Let $\mathfrak{f}$ be an arbitrary translation of $\mathfrak{G}$, let $\mathcal{A}(\mathfrak{f})$ be its axis and $\lambda(\mathfrak{f})$ its displacement. Let $\mathfrak{g}$ be an element of $\mathfrak{G}$ not permutable with $\mathfrak{f}$. The element $\mathfrak{f}' = \mathfrak{g}\mathfrak{f}^{\pm 1}\mathfrak{g}^{-1}$ then is a translation with the same displacement $\lambda(\mathfrak{f})$ and with an axis $\mathcal{A}(\mathfrak{f}') = \mathfrak{g}\mathcal{A}(\mathfrak{f})$ different from $\mathcal{A}(\mathfrak{f})$. According to Section 1 $\mathcal{A}(\mathfrak{f})$ and $\mathcal{A}(\mathfrak{f}')$ cannot be parallel. If they are divergent, let δ denote their distance, and let the exponent of $\mathfrak{f}$ in $\mathfrak{f}'$ be so chosen that $\mathcal{A}(\mathfrak{f})$ and $\mathcal{A}(\mathfrak{f}')$ have opposite directions. The invariant of the product of two such translations has been calculated in §9.4. In virtue of the condition imposed on $\mathfrak{G}$ the product $\mathfrak{f}\mathfrak{f}'$ is not a rotation; therefore, from §9.4 one gets the inequality

$$\sinh \tfrac{1}{2}\delta \sinh \tfrac{1}{2}\lambda(\mathfrak{f}) \geqq 1. \tag{1}$$

If $\mathcal{A}(\mathfrak{f})$ and $\mathcal{A}(\mathfrak{f}')$ are concurrent, let φ be the angle at which they intersect. The invariant of the commutator of two such translations has been calculated in §9.8. The sine amplitude of the triangle made up of the centres of half-turns by which the two translations can be generated takes the value $\sin\varphi \ \sinh^2 \tfrac{1}{2}\lambda(\mathfrak{f})$, compare §8.4. Since the commutator $\mathfrak{k}_{\mathfrak{f}\mathfrak{f}'}$ is not a rotation, one gets from §9.8 the inequality

$$\sin\varphi \sinh^2 \tfrac{1}{2}\lambda(\mathfrak{f}) \geqq 1. \tag{2}$$

These inequalities are first applied for a fixed $\mathfrak{g}$ and variable $\mathfrak{f}$: Let $\mathcal{A}$ be an axis of $\mathfrak{G}$. The translations of $\mathfrak{G}$ pertaining to $\mathcal{A}$ form an abelian subgroup $\mathfrak{A}_{\mathcal{A}}(\rightarrow)$ of $\mathfrak{G}$. Let $\mathfrak{g}$ be a fixed element of $\mathfrak{G}$ not leaving $\mathcal{A}$ invariant, thus not permutable with the elements of $\mathfrak{A}(\rightarrow)$. Then $\mathfrak{g}\mathcal{A}$ is different from $\mathcal{A}$ and has a definite distance δ from $\mathcal{A}$ or cuts $\mathcal{A}$ at a definite angle φ according as $\mathcal{A}$ and $\mathfrak{g}\mathcal{A}$ are divergent or concurrent. According to these two cases inequalities (1) or (2) apply. If $\mathfrak{f}$ ranges over all elements of $\mathfrak{A}_{\mathcal{A}}(\rightarrow)$, (1) or (2) yield a positive lower bound for the displacements of the elements of $\mathfrak{A}_{\mathcal{A}}(\rightarrow)$. This implies that $\mathfrak{A}_{\mathcal{A}}(\rightarrow)$ is discontinuous. Let λ_0 denote the primary displacement of $\mathfrak{A}_{\mathcal{A}}(\rightarrow)$.

Then the same inequalities are applied for a fixed element $\mathfrak{f}$ of $\mathfrak{A}_{\mathcal{A}}(\rightarrow)$ and variable $\mathfrak{g}$ ranging over a set of representatives $\mathfrak{g}_1, \mathfrak{g}_2, \ldots$ of the cosets of $\mathfrak{A}_{\mathcal{A}}(\rightarrow)$ in $\mathfrak{G}$. Then the axes $\mathfrak{g}_\nu\mathcal{A}$ of the translations $\mathfrak{f}_\nu = \mathfrak{g}_\nu\mathfrak{f}\mathfrak{g}_\nu^{-1}$ range exactly over all different axes of the equivalence class $\mathfrak{G}\mathcal{A}$ of $\mathcal{A}$. Since the displacements of all translations $\mathfrak{f}_\nu$ equal the displacement $\lambda(\mathfrak{f})$ of $\mathfrak{f}$, inequality (1) or (2) is applicable to any two of these axes. There is thus a positive lower bound for the mutual distance δ of any two divergent and for the acute angle of intersection φ of any two concurrent among the axes $\mathfrak{g}_\nu\mathcal{A}$. This implies that these axes do not accumulate in $\mathcal{D}$.

Now let x be an arbitrary point on $\mathcal{A}$. All points of the equivalence class $\mathfrak{G}x$ are situated on the collection of axes $\mathfrak{g}_\nu\mathcal{A}$. On each of these axes the points of $\mathfrak{G}x$ which are situated on that axis form a sequence at equal distances λ_0. A circle with x as its centre and $\frac{1}{2}\lambda_0$ as its radius cannot contain more than one point of such a sequence in its interior. Thus the number of points of $\mathfrak{G}x$ inside this circle is at most equal to the number of axes $\mathfrak{g}_\nu \mathcal{A}$ which cut the circle. This number is finite, since the axes $\mathfrak{g}_\nu \mathcal{A}$ do no accumulate in $\mathcal{D}$. Hence x is not an accumulation point of its equivalence class. This completes the proof of the discontinuity theorem. □

§13 $\mathfrak{F}$-groups. Fundamental set and limit set

13.1 Quasi-abelian subgroups of $\mathfrak{F}$. The investigations to follow are concerned with a group of motions and reversions satisfying the conditions of the discontinuity theorem, thus a group $\mathfrak{F}$ discontinuous in $\mathcal{D}$ and without invariant point or line. Such a group will shortly be spoken of as an *$\mathfrak{F}$-group*.

A point c in $\mathcal{D}$ is called a *centre of* $\mathfrak{F}$, if it is a centre of some rotation contained in $\mathfrak{F}$. The subgroup $\mathfrak{A}_c$ consisting of those elements of $\mathfrak{F}$ which leave c invariant is either a finite group of rotations or is obtained from such a group by the adjunction of a reflection (cf. §11.2); thus $\mathfrak{A}_c$ has a finite order. It is an $\mathfrak{A}_c(\circlearrowright_\nu)$ or an $\mathfrak{A}_c(\times_\nu)$.

A point u on $\mathcal{E}$ is called a *limit-centre of* $\mathfrak{F}$, if it is the centre of some limit-rotation contained in $\mathfrak{F}$. The subgroup $\mathfrak{A}_u$ consisting of those elements of $\mathfrak{F}$ which leave u invariant is either a group of limit-rotations or is obtained from such a group by the adjunction of a reflection (cf. §11.3 and §11.4); thus $\mathfrak{A}_u$ has an infinite order. It is an $\mathfrak{A}_u(\hookrightarrow)$ or an $\mathfrak{A}_u(\wedge)$.

A line $\mathcal{A}$ is called an *axis of* $\mathfrak{F}$, if it is the axis of some translation contained in $\mathfrak{F}$. The subgroup $\mathfrak{A}_\mathcal{A}$ consisting of those elements of $\mathfrak{F}$ which leave $\mathcal{A}$ invariant is a group of one of the types described in §11.5, thus of infinite order. Its corresponding symbol is one of the following:

$$\mathfrak{A}_\mathcal{A}(\rightarrow),\quad \mathfrak{A}_\mathcal{A}(\rightleftharpoons),\quad \mathfrak{A}_\mathcal{A}(-\rightarrow),\quad \mathfrak{A}_\mathcal{A}(\|),\quad \mathfrak{A}_\mathcal{A}(\cdot\cdot),\quad \mathfrak{A}_\mathcal{A}(\cdot|),\quad \mathfrak{A}_\mathcal{A}(-\|) = \mathfrak{A}_\mathcal{A}(-\cdot\cdot).$$

All end-points of axes of $\mathfrak{F}$ are called *fundamental points of* $\mathfrak{F}$. As pointed out in §11.4 two different axes cannot have one end-point in common; thus the fundamental points of $\mathfrak{F}$ belong together in pairs in a unique way; such pair will be called a *joined pair*. – Moreover, by §11.4, a limit-centre cannot at the same time be a fundamental point.

A straight line $\mathcal{L}$ is called a *reflection line of* $\mathfrak{F}$, if $\mathfrak{F}$ contains the reflection in $\mathcal{L}$. A reflection line can at the same time be an axis of $\mathfrak{F}$. In that case its end-points are a joined pair of fundamental points, and the subgroup $\mathfrak{A}_\mathcal{L}$ of $\mathfrak{F}$ is either $\mathfrak{A}_\mathcal{L}(-\rightarrow)$ or $\mathfrak{A}_\mathcal{L}(-\|)$. If $\mathcal{L}$ is not an axis, its corresponding subgroup is $\mathfrak{A}_\mathcal{L}(-)$ consisting of the reflection in $\mathcal{L}$ and the identity. In that case neither of the end-points of $\mathcal{L}$ is a fundamental point (§11.4).

13.2 Equivalence classes with respect to $\mathfrak{F}$. An $\mathfrak{F}$-group is not quasi-abelian, thus not abelian. If it contains reversions, its subgroup of motions, which has index 2 in $\mathfrak{F}$, is not abelian.

The equivalence class $\mathfrak{F}\mathcal{M}$ of a bounded subset $\mathcal{M}$ of $\mathcal{D}$ has no accumulation point in $\mathcal{D}$.

Proof. Let δ denote the diameter of $\mathcal{M}$, i.e. the least upper bound for the distance of any two points of $\mathcal{M}$. If the point x of $\mathcal{D}$ were an accumulation point for the collection of point sets $\mathfrak{F}\mathcal{M}$ (§10.1), then for infinitely many $\mathfrak{f} \in \mathfrak{F}$ and an arbitrary $\rho > 0$ the set $\mathfrak{f}\mathcal{M}$ would have points in common with the circular disc bounded by $\mathfrak{C}(x; \rho)$. If q is an arbitrary point of $\mathcal{M}$, then $[x, \mathfrak{f}q] < \rho + \delta$ for those $\mathfrak{f}$ in contradiction to the discontinuity of $\mathfrak{F}$. □

Let $\mathfrak{A}$ denote any such subgroup of $\mathfrak{F}$ as described in the preceding section, pertaining to a centre c, a limit-centre u, an axis $\mathcal{A}$, or a reflection line $\mathcal{L}$, and let $\mathfrak{g}$ be an element of $\mathfrak{F}$ not in $\mathfrak{A}$. Then the conjugate subgroup $\mathfrak{g}\mathfrak{A}\mathfrak{g}^{-1}$ belongs to the same type as $\mathfrak{A}$, and corresponding elements of the two subgroups have the same displacement, if they are translations or reversed translations, and the same angle of rotation, possibly except for sign, if they are rotations; in particular, primary elements correspond to primary elements. The group $\mathfrak{g}\mathfrak{A}\mathfrak{g}^{-1}$ leaves invariant $\mathfrak{g}c$, $\mathfrak{g}u$, $\mathfrak{g}\mathcal{A}$ or $\mathfrak{g}\mathcal{L}$. If $\mathfrak{g}$ ranges over a set of representatives of the cosets of $\mathfrak{A}$ in $\mathfrak{F}$, then $\mathfrak{g}c$, $\mathfrak{g}u$, $\mathfrak{g}\mathcal{A}$ or $\mathfrak{g}\mathcal{L}$ range over all different elements of the equivalence classes $\mathfrak{F}c$, $\mathfrak{F}u$, $\mathfrak{F}\mathcal{A}$ or $\mathfrak{F}\mathcal{L}$ respectively. According to the definition of the equivalence class of a point set given in §10.1, it may happen that an element of the class has to be counted a finite or infinite number of times in its class. As far as *classes of centres, classes of limit-centres, classes of axes* or *classes of reflection lines* are concerned, it is convenient to stipulate from now on that each such point or line be counted only once in its class. The same agreement applies to *equivalence classes of fundamental points.*

In every $\mathfrak{F}$-group there are axes, and every equivalence class of axes contains an infinite number of axes.

Proof. To prove this assertion, let $\mathfrak{f}_1$ and $\mathfrak{f}_2$ be any two non-permutable motions of $\mathfrak{F}$. If one of them is a translation, there exists an axis; if not, they are rotations or limit-rotations, and according to §9.1–9.3 one of the cases I occurs. Then by §9.6 the commutator of $\mathfrak{f}_1$ and $\mathfrak{f}_2$ is known to be a translation. Thus axes occur in $\mathfrak{F}$. Let $\mathcal{A}$ be one of them. If $\mathfrak{g}$ denotes an element of $\mathfrak{F}$ not in $\mathfrak{A}_{\mathcal{A}}$, it carries $\mathcal{A}$ into another axis $\mathcal{A}' = \mathfrak{g}\mathcal{A}$. The infinitely many elements of $\mathfrak{A}_{\mathcal{A}}$ carry $\mathcal{A}'$ into infinitely many different axes all contained in $\mathfrak{F}\mathcal{A}$. □

The axes of a class $\mathfrak{F}\mathcal{A}$ do not accumulate in $\mathcal{D}$.

Proof. If they did, there would exist a point x in $\mathcal{D}$ the distance of which from infinitely many axes of the class would be smaller than an assigned positive number ρ. If λ is the displacement of a fixed translation pertaining to $\mathcal{A}$, the conjugate translations on these infinitely many axes have the same displacement λ. These infinitely many translations then would displace x by amounts smaller than $2\rho + \lambda$ in contradiction to the discontinuity of $\mathfrak{F}$. □

According to §10.4 neither the reflection lines of $\mathfrak{F}$ nor the centres of $\mathfrak{F}$ accumulate in $\mathcal{D}$. All the more this is true of an equivalence class of reflection lines, and of an equivalence class of centres.

13.3 The fundamental set $\mathcal{G}(\mathfrak{F})$. The set of all fundamental points of $\mathfrak{F}$ is enumerable, since the elements of $\mathfrak{F}$ are enumerable (§10.2). It will be denoted by $\mathcal{G}(\mathfrak{F})$ and called the *fundamental set of* $\mathfrak{F}$. As has already been pointed out, it falls into joined pairs. Every fundamental point is the positive fundamental point of certain translations and the negative of their inverses.

Every fundamental point is an accumulation point of any prescribed class of fundamental points.

Proof. Given a single fundamental point p and an equivalence class $\mathfrak{F}q$ of fundamental points, let $\mathfrak{f}$ be a translation in $\mathfrak{F}$ for which $p = v(\mathfrak{f})$. The class $\mathfrak{F}q$ may or may not contain one or both of the fundamental points of $\mathfrak{f}$. In any case a representative q of $\mathfrak{F}q$ can be chosen which does not coincide with either of the fundamental points of $\mathfrak{f}$. Let $\mathfrak{g}$ be a translation in $\mathfrak{F}$ such that $q = v(\mathfrak{g})$. Then $\mathfrak{g}$ is not permutable with $\mathfrak{f}$. The fundamental points $\mathfrak{f}^{\nu} v(\mathfrak{g})$ of the translation $\mathfrak{f}^{\nu}\mathfrak{g}\mathfrak{f}^{-\nu}$ are all different and converge to $v(\mathfrak{f}) = p$ for $\nu \to \infty$. This proves the assertion. □

If $\mathcal{J}$ and $\mathcal{J}'$ are any two open intervals on $\mathcal{E}$ not necessarily disjoint and if there are fundamental points in each of them, then there exists an axis of $\mathfrak{F}$ with one of its fundamental points in $\mathcal{J}$ the other in $\mathcal{J}'$.

Proof. Since there are fundamental points in both intervals, let $\mathfrak{f}$ and $\mathfrak{f}'$ be translations in $\mathfrak{F}$ with their positive fundamental points in $\mathcal{J}$ and $\mathcal{J}'$ respectively. If $\mathfrak{f}$ and $\mathfrak{f}'$ belong to the same axis, the assertion comes true. If $\mathcal{A}(\mathfrak{f})$ and $\mathcal{A}(\mathfrak{f}')$ are different, then for some suitable positive integer μ the element $\mathfrak{f}^{\mu}\mathfrak{f}'^{-\mu}$ will prove to be a translation the axis of which has the required property: At first, let $\mathcal{A}(\mathfrak{f})$ and $\mathcal{A}(\mathfrak{f}')$ be concurrent. According to §9.2 the element $\mathfrak{f}^{\mu}\mathfrak{f}'^{-\mu}$ is a translation and can be obtained as the product of two half-turns with centres on $\mathcal{A}(\mathfrak{f})$ and $\mathcal{A}(\mathfrak{f}')$ at distances $\frac{\mu}{2}\lambda(\mathfrak{f})$ and $\frac{\mu}{2}\lambda(\mathfrak{f}')$ respectively from the intersection point. (These half-turns need not belong to $\mathfrak{F}$, of course.) These centres converge to $v(\mathfrak{f})$ and $v(\mathfrak{f}')$ respectively, if $\mu \to \infty$. Hence for μ sufficiently large the first centre is in the half-plane bounded by $\mathcal{J}$ and the other in the half-plane bounded by $\mathcal{J}'$. The line joining them, which is the axis of the translation $\mathfrak{f}^{\mu}\mathfrak{f}'^{-\mu}$, must then have its end-points situated in $\mathcal{J}$ and $\mathcal{J}'$ respectively (Fig. 13.1).

If $\mathcal{A}(\mathfrak{f})$ and $\mathcal{A}(\mathfrak{f}')$ are divergent, the element $\mathfrak{f}^{\mu}\mathfrak{f}'^{-\mu}$ can according to §9.3 be obtained as the product of two reflections at lines orthogonal to $\mathcal{A}(\mathfrak{f})$ and $\mathcal{A}(\mathfrak{f}')$ and distant $\frac{\mu}{2}\lambda(\mathfrak{f})$ and $\frac{\mu}{2}\lambda(\mathfrak{f}')$ respectively from the common perpendicular of $\mathcal{A}(\mathfrak{f})$ and $\mathcal{A}(\mathfrak{f}')$. (These reflections need not belong to $\mathfrak{F}$, of course.) Their lines of reflection converge to $v(\mathfrak{f})$ and $v(\mathfrak{f}')$ respectively, if $\mu \to \infty$. Hence, since $v(\mathfrak{f})$ and $v(\mathfrak{f}')$ are different, for sufficiently large values of μ they are divergent and both end-points of the first are in $\mathcal{J}$, and both of the second in $\mathcal{J}'$. Thus $\mathfrak{f}^{\mu}\mathfrak{f}'^{-\mu}$ is a translation, and its

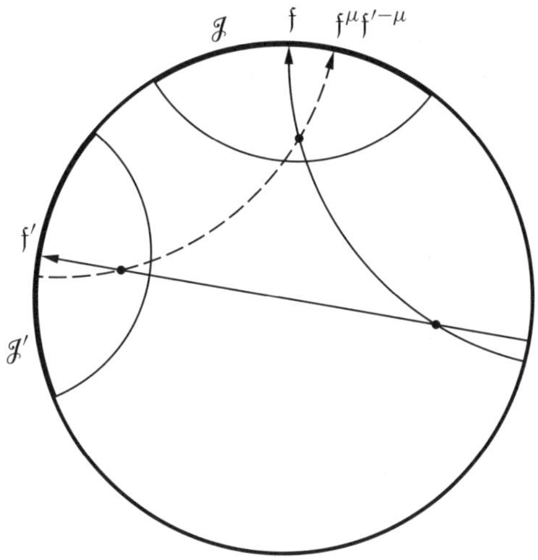

Figure 13.1

fundamental points are evidently situated in $\mathcal{J}$ and $\mathcal{J}'$ respectively (Fig. 13.2). Finally, the case where $\mathcal{A}(\mathfrak{f})$ and $\mathcal{A}'(\mathfrak{f})$ are parallel (and different) cannot occur by §11.4. □

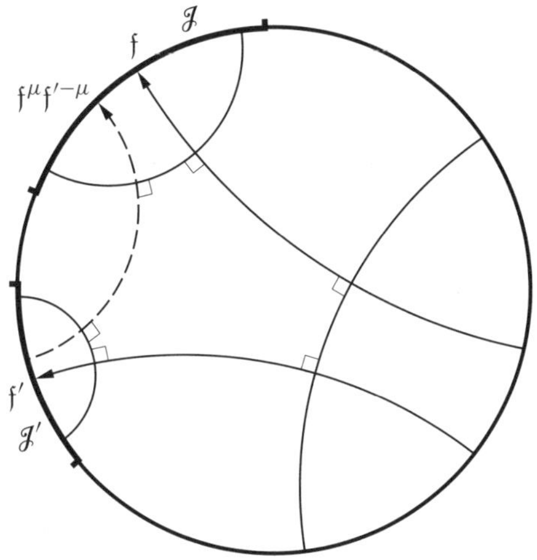

Figure 13.2

13.4 The limit set $\overline{\mathcal{G}}(\mathfrak{F})$. The closure $\overline{\mathcal{G}}(\mathfrak{F})$ of the fundamental set $\mathcal{G}(\mathfrak{F})$ is called the *limit set* of $\mathfrak{F}$; its points are called *limit points* of $\mathfrak{F}$. (Thus the limit set contains the fundamental set.) According to the preceding section every fundamental point is an accumulation point of a class of fundamental points; hence:

The limit set can be obtained as the closure of an arbitrary equivalence class of fundamental points. □

The limit set is a perfect subset of $\mathcal{E}$ (i.e. closed and without isolated points); for particular groups $\mathfrak{F}$ it can be identical with the whole of $\mathcal{E}$. If $\overline{\mathcal{G}}(\mathfrak{F})$ is a proper subset of $\mathcal{E}$, its complemental set with respect to $\mathcal{E}$ consists of open intervals no two of which can have a common end-point, since no point of $\overline{\mathcal{G}}(\mathfrak{F})$ is isolated. These intervals are called *intervals of discontinuity* of $\mathfrak{F}$, and their points *discontinuity points* on $\mathcal{E}$. It will come out later on that $\mathfrak{F}$ is discontinuous in every discontinuity point on $\mathcal{E}$.

Every element $\mathfrak{f}$ of $\mathfrak{F}$ effects a one-to-one mapping of $\mathcal{G}(\mathfrak{F})$ onto itself, since it effects a one-to-one mapping of $\mathcal{E}$ onto itself and both $\mathfrak{f}$ and $\mathfrak{f}^{-1}$ carry fundamental points into fundamental points. Thus, for reasons of continuity, it also effects a one-to-one mapping of $\overline{\mathcal{G}}(\mathfrak{F})$ onto itself. This implies that the same is true of the complementary set on $\mathcal{E}$. Hence $\mathfrak{f}$ carries every complete interval of discontinuity into a complete interval of discontinuity.

If intervals of discontinuity exist at all, they are everywhere dense on $\mathcal{E}$. For if p denotes a discontinuity point on $\mathcal{E}$ and $\mathfrak{f}$ an arbitrary translation in $\mathfrak{F}$, then the points $\mathfrak{f}^{\nu}p$, $0 < \nu < \infty$, are discontinuity points and converge to the positive fundamental point of $\mathfrak{f}$ for $\nu \to \infty$. Thus every fundamental point, and consequently every limit point, is an accumulation point of discontinuity points on $\mathcal{E}$, which proves the statement. Since a limit point is not the common end-point of two intervals of discontinuity, these intervals must accumulate to the point from at least one side. Hence:

If $\overline{\mathcal{G}}(\mathfrak{F})$ is not identical with the whole of $\mathcal{E}$, the intervals of discontinuity are (enumerably) infinite in number and everywhere dense in $\mathcal{E}$. □

13.5 Accumulation in limit points. If centres exist in an $\mathfrak{F}$-group, they are infinite in number. For a given centre is carried into an infinity of others by the translations contained in the group. Since the centres of $\mathfrak{F}$ do not accumulate in $\mathcal{D}$, their accumulation points are on $\mathcal{E}$.

The accumulation points of the set of centres are points of the limit set.

Proof. Let p be an accumulation point of centres of $\mathfrak{F}$ and $\mathcal{B}$ an arbitrary arc of $\mathcal{E}$ containing p as an inner point. The assertion is that there are fundamental points on $\mathcal{B}$. Since p is an accumulation point of centres, there is a centre c of $\mathfrak{F}$ inside the half-plane $\mathcal{H}$ bounded by $\mathcal{B}$. Let ν be the order of the subgroup $\mathfrak{A}_c$ of $\mathfrak{F}$ consisting of all rotations of $\mathfrak{F}$ with centre c. Let an angular sector of magnitude $\frac{2\pi}{\nu}$ and vertex c be drawn inside $\mathcal{H}$; this is possible even in the extreme case $\nu = 2$, since c is inside $\mathcal{H}$. This angular sector cuts out on $\mathcal{E}$ an arc $\mathcal{B}_1$ contained in $\mathcal{B}$, and the elements of $\mathfrak{A}_c$ carry $\mathcal{B}_1$ into other arcs $\mathcal{B}_2, \ldots, \mathcal{B}_\nu$, making up the whole of $\mathcal{E}$. These ν

arcs are interchanged by the elements of $\mathfrak{A}_c$, and since these elements reproduce the fundamental set, there must be fundamental points on each of the ν arcs, thus on $\mathcal{B}_1$ and hence on $\mathcal{B}$. □

It is inferred in the same way as for centres that there is an infinity of limit-centres in $\mathfrak{F}$, if any. It has been pointed out in Section 1 that a limit-centre cannot be a fundamental point; but the following theorem holds:

Every limit-centre is a limit point.

Proof. For if $v(\mathfrak{f})$ is a fundamental point of the translation $\mathfrak{f}$ of $\mathfrak{F}$ and $\mathfrak{g}$ a limit-rotation of $\mathfrak{F}$ pertaining to the limit-centre u, then the fundamental points $\mathfrak{g}^{\nu}v(\mathfrak{f})$ of the translations $\mathfrak{g}^{\nu}\mathfrak{f}\mathfrak{g}^{-\nu}$ converge to u for $\nu \to \infty$. □

There are an infinity of reflection lines of $\mathfrak{F}$, if any. As already mentioned in Section 2, they do not accumulate in $\mathcal{D}$. Let $\mathcal{L}_\nu$, $\nu = 1, 2, \ldots$, be an infinite sequence of reflection lines of $\mathfrak{F}$, and let an arbitrary point p_ν be chosen on $\mathcal{L}_\nu$. Then the accumulation points of the sequence of points p_ν are on $\mathcal{E}$. Let it be assumed that the sequence p_ν converges to a point p on $\mathcal{E}$ for $\nu \to \infty$. Since the $\mathcal{L}_\nu$ do not accumulate in $\mathcal{D}$, the euclidean lengths of the circular arcs $\mathcal{L}_\nu$ must converge to zero. Thus both end-points of the $\mathcal{L}_\nu$ converge to p and the same is true of the lines $\mathcal{L}_\nu$ in their whole extent. Now the following theorem holds:

Every accumulation point of reflection lines is a limit point.

Proof. Let p be such an accumulation point, $\mathcal{B}$ an arbitrary arc of $\mathcal{E}$ containing p as an inner point, and $\mathcal{H}$ the half-plane bounded by $\mathcal{B}$. It is sufficient to show that $\mathcal{B}$ contains limit points. In consequence of the above statement there exists a reflection line $\mathcal{L}$ of $\mathfrak{F}$ which in its full extent belongs to $\mathcal{H}$. The reflection in $\mathcal{L}$ belongs to $\mathfrak{F}$ and thus reproduces the limit set. Hence both arcs of $\mathcal{E}$ determined by $\mathcal{L}$ contain limit points. One of them is part of $\mathcal{B}$. This proves the assertion. □

13.6 A theorem on sequences of elements of $\mathfrak{F}$. Let $\mathfrak{f}_\nu$, $1 \leqq \nu < \infty$, be an arbitrary sequence of elements of $\mathfrak{F}$ with the only restriction that no element of $\mathfrak{F}$ occurs an infinite number of times in the sequence. With this sequence $\mathfrak{f}_\nu$ two sequences u_ν and v_ν of points of $\overline{\mathcal{D}}$ are associated in the following way: If $\mathfrak{f}_\nu$ is a rotation or limit-rotation, both u_ν and v_ν are taken to mean the centre or limit-centre respectively. If $\mathfrak{f}_\nu$ is a reflection, u_ν means one of the end-points of the reflection line, chosen at will, and v_ν likewise; u_ν and v_ν may thus be different end-points or the same. If $\mathfrak{f}_\nu$ is a translation or reversed translation, u_ν is the negative and v_ν the positive fundamental point of $\mathfrak{f}_\nu$. If $\mathfrak{f}_\nu = 1$ then u_ν and v_ν are not defined. The set of accumulation points of the sequences u_ν and v_ν are denoted by $\mathcal{U}$ and $\mathcal{V}$ respectively. It results from the preceding section that both $\mathcal{U}$ and $\mathcal{V}$ are closed subsets of $\overline{\mathcal{G}}(\mathfrak{F})$; they may well have common points. With these definitions the following theorem holds:

If $\mathcal{M} \neq \emptyset$ denotes an arbitrary subset of $\overline{\mathfrak{D}}$ the closure $\overline{\mathcal{M}}$ of which has no point in common with $\mathcal{U}$, then the set of accumulation points of the sequence of sets $\mathfrak{f}_\nu \mathcal{M}$, $\nu = 1, 2, \ldots$, coincides with $\mathcal{V}$.

In preparation for the proof a lemma is established in advance:

Let $\mathcal{H}$ and $\mathcal{H}_0 \subset \mathcal{H}$ be two open half-planes and let the end-points h' and h'' respectively h_0' and h_0'' of their bounding lines be four different points on $\mathcal{E}$. Let x_0 be a fixed point in $\mathfrak{D}$ chosen arbitrarily. Then there exists a positive number ρ such that every line distant more than ρ from x_0 and having some of its points in $\mathcal{H}_0$ is in its whole extent situated in $\mathcal{H}$.

Proof. This may be seen as follows: Let h' and h'' be the end-points of the line bounding $\mathcal{H}$, and h_0' and h_0'' the end-points of the line bounding $\mathcal{H}_0$, and let h', h_0', h_0'', h'' indicate a cyclical order on $\mathcal{E}$ (Fig. 13.3). Let a circle $\mathcal{K}$ be drawn with centre x_0 and intersecting

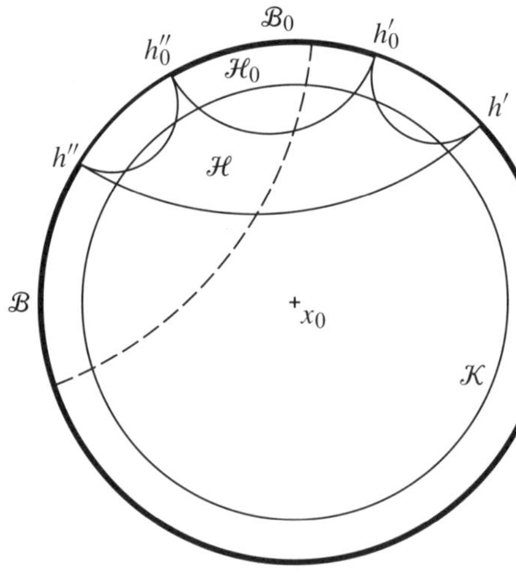

Figure 13.3

both the line $h'h_0'$ and the line $h''h_0''$. The radius of $\mathcal{K}$ is a number ρ such as required; for a line which has some of its points in $\mathcal{H}_0$ has one at least of its end-points on the arc $\mathcal{B}_0$ bounding $\mathcal{H}_0$ on $\mathcal{E}$; if it is not wholly inside $\mathcal{H}$, the other end-point must be on the arc $\mathcal{B}$ bounding the half-plane complementary to $\mathcal{H}$; it thus separates the lines $h'h_0'$ and $h''h_0''$ and hence intersects $\mathcal{K}$. This proves the lemma. □

Proof of the theorem. Let x_0 denote a fixed regular point in $\mathfrak{D}$ chosen arbitrarily. With every element $\mathfrak{f}_\nu$ of the prescribed sequence of elements of $\mathfrak{F}$ a certain domain $\mathcal{N}_\nu$ is associated in the following way: $\mathfrak{f}_\nu$ generates an abelian subgroup $\mathfrak{A}_\nu$ of $\mathfrak{F}$ which belongs to one of the types considered in §11. If $\mathfrak{f}_\nu$ is a rotation or a limit-rotation, or a translation, or a reflection (Fig. 13.4–13.7), the closed normal domain $\mathcal{N}(\mathfrak{A}_\nu; x_0)$ is taken as $\mathcal{N}_\nu$. If $\mathfrak{f}_\nu$ is a reversed translation, let x_0' denote the projection of x_0 on the axis of $\mathfrak{f}_\nu$ (Fig. 13.8); then the closed normal domain $\overline{\mathcal{N}}(\mathfrak{A}_\nu; x_0')$ is taken to mean

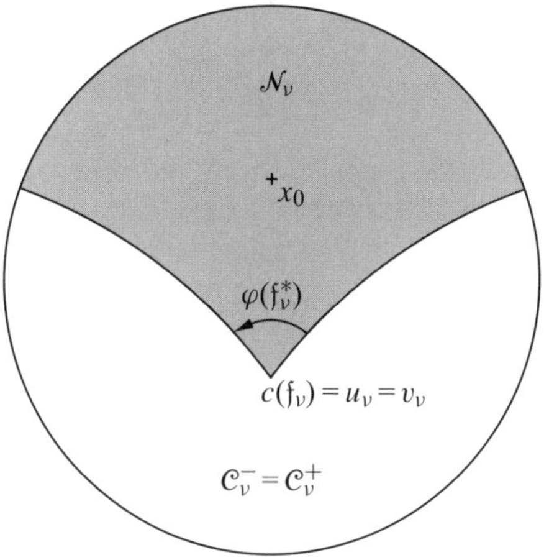

Figure 13.4

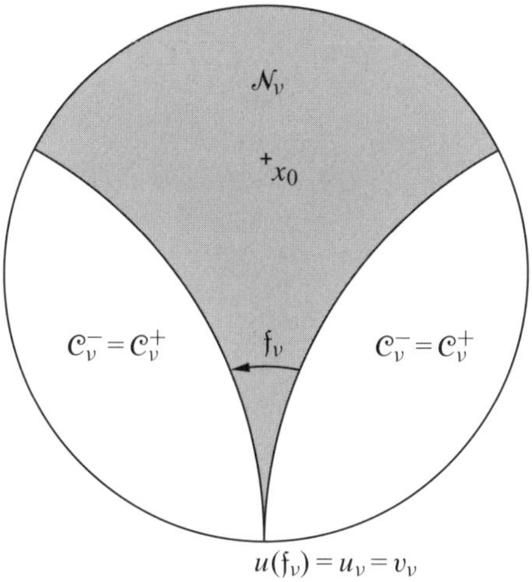

Figure 13.5

$\mathcal{N}_\nu$ (cf. §11.5). Now, let ρ_ν denote the radius of the greatest circle with centre x_0 contained in $\mathcal{N}_\nu$. It will first be shown that $\rho_\nu \to \infty$ for $\nu \to \infty$. It results from §10.3 and §10.8 that, if $\varepsilon(\mathfrak{A}_\nu)$ denotes the distance function of the group $\mathfrak{A}_\nu$, the greatest

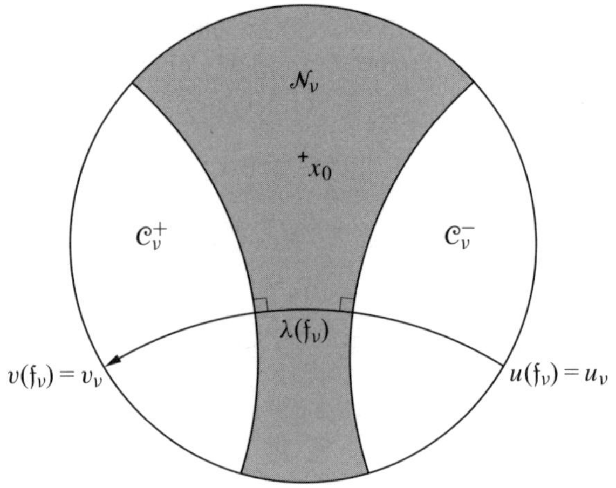

Figure 13.6

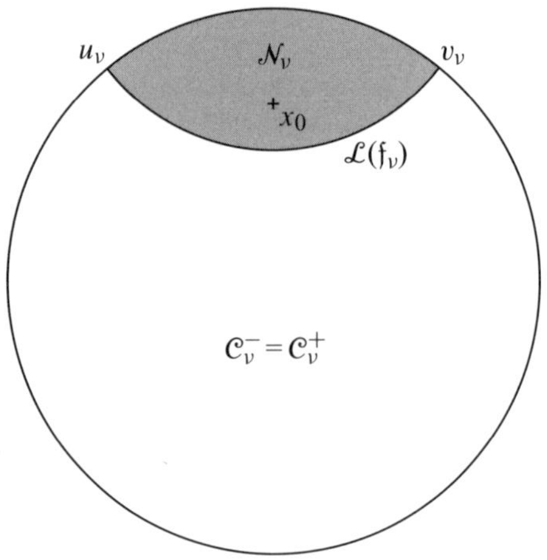

Figure 13.7

circle with centre x_0 contained in the normal domain $\mathcal{N}(\mathfrak{A}_\nu; x_0)$ has its radius equal to $\frac{1}{2}\varepsilon(\mathfrak{A}_\nu; x_0)$. In the case of a group $\mathfrak{A}_\nu$ generated by an element $\mathfrak{f}_\nu$ which is either a translation, or a limit-rotation, or a reflection one evidently gets $\varepsilon(\mathfrak{A}_\nu; x_0) = [x_0, \mathfrak{f}_\nu x_0]$ and hence

$$\rho_\nu = \tfrac{1}{2}[x_0, \mathfrak{f}_\nu x_0]. \tag{1}$$

In the case of a rotation $\mathfrak{f}_\nu$ let $\mathfrak{f}_\nu^*$ denote the element with the smallest positive angle

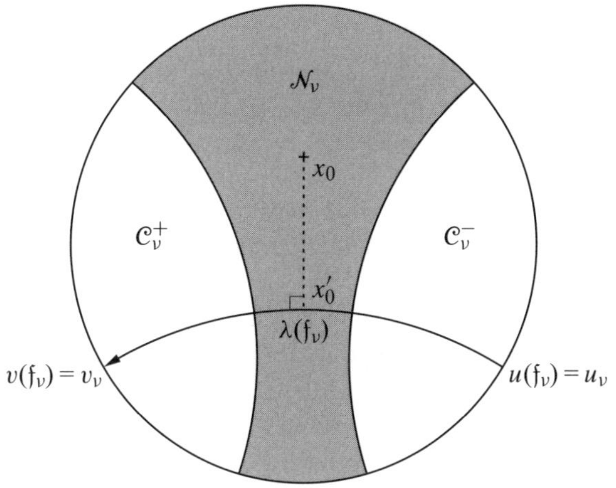

Figure 13.8

of rotation in the group $\mathfrak{A}_\nu$ generated by $\mathfrak{f}_\nu$; then $\varepsilon(\mathfrak{A}_\nu; x_0) = [x_0, \mathfrak{f}_\nu^* x_0]$ and hence

$$\rho_\nu = \tfrac{1}{2}[x_0, \mathfrak{f}_\nu^* x_0]. \tag{2}$$

In the case of a reversed translation $\mathfrak{f}_\nu$, let $\mathfrak{f}_\nu'$ denote the translation with the same axis and the same displacement as $\mathfrak{f}_\nu$. Then $\mathcal{N}_\nu$ coincides with what would be the normal domain with central point x_0 for a group generated by $\mathfrak{f}_\nu'$. Hence $\rho = \frac{1}{2}[x_0; \mathfrak{f}_\nu' x_0]$. Now $\mathfrak{f}_\nu'^2 = \mathfrak{f}_\nu^2$ and

$$[x_0, \mathfrak{f}_\nu'^2 x_0] \leqq [x_0, \mathfrak{f}_\nu' x_0] + [f_\nu' x_0, \mathfrak{f}_\nu'^2 x_0] = 2[x_0, \mathfrak{f}_\nu' x_0] = 4\rho_\nu.$$

Hence in this case

$$\rho_\nu \geqq \tfrac{1}{4}[x_0, \mathfrak{f}_\nu^2 x_0]. \tag{3}$$

Since the orders of all rotation groups in $\mathfrak{F}$ are finite, in the replacement of a rotation $\mathfrak{f}_\nu$ by the corresponding $\mathfrak{f}_\nu^*$ in (2) a definite element $\mathfrak{f}_\nu^*$ (which need not be itself an element of the sequence $\mathfrak{f}_\nu$) is only used a finite number of times, since elements of its rotation group only occur a finite number of times in the sequence $\mathfrak{f}_\nu$. Now it is known from §10.2 that there are only a finite number of elements $\mathfrak{f}$ in $\mathfrak{F}$ for which the distance $[x_0, \mathfrak{f}x_0]$ is smaller than an assigned positive number. From this and from (1), (2), (3) the assertion $\rho_\nu \to \infty$ is inferred.

Now the domain complementary to $\mathcal{N}_\nu$ with respect to $\overline{\mathcal{D}}$ is considered. If $\mathfrak{f}_\nu$ is a translation or reversed translation, this complementary domain is made up of two half-planes. One of these half-planes contains $u(\mathfrak{f}_\nu)$ as a boundary point on $\mathcal{E}$; its closure is denoted by $\mathfrak{C}_\nu^-$; the closure of the other half-plane, which contains $v(\mathfrak{f}_\nu)$ as a boundary point on $\mathcal{E}$, is denoted by $\mathfrak{C}_\nu^+$. If $\mathfrak{f}_\nu$ is a rotation, or a limit-rotation, or a reflection, the closure of the complementary domain of $\mathcal{N}_\nu$ is denoted by $\mathfrak{C}_\nu^- = \mathfrak{C}_\nu^+$ without regard to the fact that in the case of a limit-rotation even this complementary domain consists of two half-planes. See Fig. 13.4-8. Then in all five cases the following statement holds: $\mathfrak{C}_\nu^-$ contains u_ν and $\mathfrak{C}_\nu^+$ contains v_ν; any point set in $\overline{\mathcal{D}}$ which has no point in common with $\mathfrak{C}_\nu^-$ is, by the application of $\mathfrak{f}_\nu$, carried into a subset of $\mathfrak{C}_\nu^+$. In the case of a rotation $\mathfrak{f}_\nu$ this is due to the fact that $\mathfrak{f}_\nu$ is a power of $\mathfrak{f}_\nu^*$.

It will now be proved that *the set of accumulation points of the sequence of sets* $\mathfrak{C}_\nu^+$ *coincides with* $\mathcal{V}$ and that every neighbourhood of an arbitrary point of $\mathcal{V}$ contains at least one and thus infinitely many of the sets $\mathfrak{C}_\nu^+$ completely. The same is valid for $\mathfrak{C}_\nu^-$ and $\mathcal{U}$ instead of $\mathfrak{C}_\nu^+$ and $\mathcal{V}$. Since $\mathfrak{C}_\nu^+$ contains the point v_ν, it is evident that every point of $\mathcal{V}$ is an accumulation point of the sequence of sets $\mathfrak{C}_\nu^+$. Conversely, let p be an arbitrary accumulation point of the sequence of sets $\mathfrak{C}_\nu^+$. First, p must be a point on $\mathcal{E}$; for since $\mathcal{N}_\nu$ contains a circle with centre x_0 and radius ρ_ν, and $\rho_\nu \to \infty$, any point of $\mathcal{D}$ is inside $\mathcal{N}_\nu$ for all sufficiently large values of ν and thus cannot be an accumulation point of the $\mathfrak{C}_\nu^+$. Now let $\mathcal{H}$ denote an arbitrary open half-plane containing p as an interior point of its bounding arc of $\mathcal{E}$. It has to be shown that this half-plane $\mathcal{H}$, however near to p, contains one of the sets $\mathfrak{C}_\nu^+$ completely. Without restriction $\mathcal{H}$ may be so chosen as to exclude x_0, which from the outset was a fixed point of $\mathcal{D}$. Let $\mathcal{H}_0$ and $\mathcal{H}_0'$ denote two open half-planes, both containing p as a boundary point, and chosen such that $\mathcal{H}_0'$ is contained in $\mathcal{H}_0$ and $\mathcal{H}_0$ is contained in $\mathcal{H}$ (Fig. 13.9).

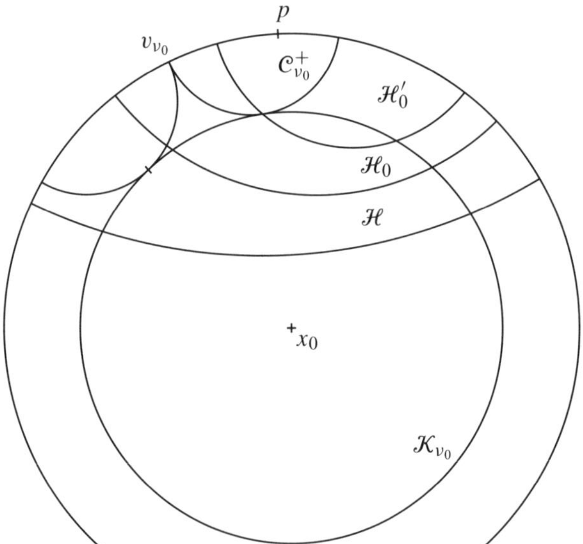

Figure 13.9

The values of the index ν considered will all be taken above a certain lower bound chosen such that the corresponding ρ_ν possesses the property of the number ρ in the above lemma, both in respect to the pair of half-planes $\mathcal{H}_0'$, $\mathcal{H}_0$ and to the pair $\mathcal{H}_0$, $\mathcal{H}$, and that moreover a circle $\mathcal{K}_\nu$ with centre x_0 and radius ρ_ν has points in common with $\mathcal{H}_0'$; in virtue of $\rho_\nu \to \infty$ this is possible. Since p is an accumulation point of the $\mathfrak{C}_\nu^+$, there is among these large values of ν one, ν_0 say, for which $\mathfrak{C}_{\nu_0}^+$ has points in common with $\mathcal{H}_0'$. On the other hand even $\mathcal{N}_{\nu_0}$ has points in common with $\mathcal{H}_0'$, namely points of the circle $\mathcal{K}_\nu$, and hence the boundary of $\mathfrak{C}_{\nu_0}^+$ in $\mathcal{D}$ has points in common with $\mathcal{H}_0'$. This boundary of $\mathfrak{C}_{\nu_0}^+$ consists either of one line (translation, reversed translation, reflection) or of two lines with common end-point on $\mathcal{E}$ (limit-rotation; this case has been chosen for illustration in Fig. 13.9), or of two half-lines with common end-point in $\mathcal{D}$ (rotation). In consequence of the lemma this line or, in the latter two cases, one of these lines or half-lines, is completely contained in $\mathcal{H}_0$ since it does not cut $\mathcal{K}_{\nu_0}$. In the latter two cases one more application of the lemma yields that the other line or half-line is completely contained in $\mathcal{H}$ since it does not cut $\mathcal{K}_{\nu_0}$. Thus in all five cases the complete boundary of $\mathfrak{C}_{\nu_0}^+$ in $\mathcal{D}$ is contained in $\mathcal{H}$. Then $\mathfrak{C}_{\nu_0}^+$ itself must be contained in $\mathcal{H}$, for the point x_0, contained in $\mathcal{N}_{\nu_0}$, is outside $\mathcal{H}$. Since the point v_{ν_0} is contained in $\mathfrak{C}_{\nu_0}^+$, thus in $\mathcal{H}$, it is inferred that p is a point of $\mathcal{V}$. – The same facts pertaining to $\mathcal{U}$ and the sequence of sets $\mathfrak{C}_\nu^-$ are established in a completely analogous way.

Now, let $\mathcal{M} \neq \emptyset$ be any subset of $\overline{\mathcal{D}}$ the closure $\overline{\mathcal{M}}$ of which has no point in common with $\mathcal{U}$. Then $\mathcal{M}$ cannot have points in common with the sets $\mathfrak{C}_\nu^-$ but for a finite number of values of ν. Thus for all sufficiently large values of ν the set $\mathfrak{f}_\nu\mathcal{M}$ is contained in $\mathfrak{C}_\nu^+$. Hence the set X of accumulation points of the sequence $\mathfrak{f}_\nu\mathcal{M}$ is contained in the set of accumulation points of the sequence $\mathfrak{C}_\nu^+$, thus in $\mathcal{V}$. On the other hand, every neighbourhood of an arbitrary point p of $\mathcal{V}$ contains a $\mathfrak{C}_{\nu_0}^+$ completely, and thus also $\mathfrak{f}_{\nu_0}\mathcal{M}$. Hence $p \in X$, and $X = \mathcal{V}$. This completes the proof of the main theorem. □

In the next two sections two important particular cases of this theorem are considered.

13.7 Accumulation points for equivalence classes. Let the sequence $\mathfrak{f}_\nu$ be an enumeration of all elements of $\mathfrak{F}$. Every fundamental point of $\mathfrak{F}$ is the positive fundamental point for some translation and the negative for its inverse. Hence $\mathcal{U} = \mathcal{V} = \overline{\mathcal{G}}(\mathfrak{F})$. The theorem of the preceding section then yields:

If $\mathcal{M} \neq \emptyset$ is a subset of $\mathcal{D}$ the closure of which has no point in common with $\overline{\mathcal{G}}(\mathfrak{F})$, then $\mathfrak{F}\mathcal{M}$ has exactly $\overline{\mathcal{G}}(\mathfrak{F})$ as set of accumulation points. □

In particular, let $\mathcal{M}$ consist of one point x not belonging to $\overline{\mathcal{G}}(\mathfrak{F})$, thus situated either in $\mathcal{D}$ or in an interval of discontinuity. The theorem then yields:

For every point x of $\overline{\mathfrak{D}}$ not belonging to $\overline{\mathfrak{G}}(\mathfrak{F})$ the set of accumulation points of its equivalence class $\mathfrak{F}x$ coincides with the limit set $\overline{\mathfrak{G}}(\mathfrak{F})$. □

Hence such a point is not itself an accumulation point of its equivalence class. This implies that $\mathfrak{F}$ is discontinuous in every point of an interval of discontinuity, a fact which explains the notation chosen for these intervals.

It has been pointed out in §13.4 that the limit set is the set of accumulation points of an arbitrary class of fundamental points. It is also the set of accumulation points of an arbitrary class of limit points. For if p is an arbitrary limit point and $\mathcal{J}$ an open interval of $\mathcal{E}$ containing p, let $\mathfrak{f}$ be a translation in $\mathfrak{F}$ which has its positive fundamental point in $\mathcal{J}$; if then q is an arbitrary limit point which is not a fundamental point, the image of q by a sufficiently high power of $\mathfrak{f}$ will belong to $\mathcal{J}$. Moreover, the domain $\mathcal{D}^*$ outside $\mathcal{E}$ evidently plays a rôle equal to that of $\mathcal{D}$, as may be seen by an inversion with respect to $\mathcal{E}$. Hence the above statement of this section may be thus formulated:

The limit set of an $\mathfrak{F}$-group coincides with the set of accumulation points of the equivalence class of an arbitrary point of the complex sphere. □

13.8 Fundamental sequences. If the set $\mathcal{V}$ of Section 6 pertaining to a sequence $\mathfrak{f}_\nu$ in which no element of $\mathfrak{F}$ occurs an infinite number of times consists of one single point p, then the sequence $\mathfrak{f}_\nu$ is called a *fundamental sequence* and p its *point of convergence*. For any given limit point p a fundamental sequence with p as its point of convergence can easily be formed, since the fundamental set is dense in the limit set. From the theorem of Section 6 is inferred:

The necessary and sufficient condition for a sequence $\mathfrak{f}_\nu$ in which no element occurs an infinite number of times being a fundamental sequence is that for one point x of $\overline{\mathcal{D}} \setminus \mathcal{U}$ the sequence of points $\mathfrak{f}_\nu x$ is convergent. □

If convergence occurs for one point of $\overline{\mathcal{D}} \setminus \mathcal{U}$, it also occurs for every other point of $\overline{\mathcal{D}} \setminus \mathcal{U}$. Since $\mathcal{U}$ is contained in $\overline{\mathfrak{G}}(\mathfrak{F})$, one can always choose an arbitrary point x of $\overline{\mathcal{D}} \setminus \overline{\mathfrak{G}}(\mathfrak{F})$. – The criterion can be put into the following slightly more general form:

Let a sequence of elements $\mathfrak{f}_\nu$ with its corresponding set $\mathcal{U}$ be given, and let x_ν be a sequence of points from $\overline{\mathcal{D}} \setminus \mathcal{U}$ all accumulation points of which are outside $\mathcal{U}$. If, and only if, $\mathfrak{f}_\nu$ is a fundamental sequence, the sequence of points $\mathfrak{f}_\nu x_\nu$ is convergent, namely to the point of convergence of the fundamental sequence.

Proof. Let X denote the set of all points x_ν. If $\mathfrak{f}_\nu$ is a fundamental sequence, even the sequence of sets $\mathfrak{f}_\nu X$ converges to the point of convergence of the sequence $\mathfrak{f}_\nu$; all the more this is true of the sequence of points $\mathfrak{f}_\nu x_\nu$. On the other hand, let it be assumed that $\mathfrak{f}_\nu x_\nu$ is convergent and that $\mathfrak{f}_\nu$ is not a fundamental sequence. Then from the sequence $\mathfrak{f}_\nu$ two fundamental sequences $\mathfrak{f}_{\nu'}$, and $\mathfrak{f}_{\nu''}$ with different points of convergence p' and p'' can be selected. Then the sequence of sets $\mathfrak{f}_{\nu'} X$ and thus the sequence of points $\mathfrak{f}_{\nu'}x_{\nu'}$ are convergent to p' and the sequence of sets $\mathfrak{f}_{\nu''}X$ and thus the sequence of points $\mathfrak{f}_{\nu''}x_{\nu''}$ are convergent to p'' in contradiction to the assumed convergence of the complete sequence $\mathfrak{f}_\nu x_\nu$. □

§14 The convex domain of an $\mathfrak{F}$-group. Characteristic and isometric neighbourhood

14.1 The convex domain $\mathcal{K}(\mathfrak{F})$. Given any $\mathfrak{F}$-group $\mathfrak{F}$, the intersection of the convex hull (§6.1) of the fundamental set $\mathcal{G}(\mathfrak{F})$ with $\mathcal{D}$ will be denoted by $\mathcal{K}(\mathfrak{F})$ and called the *convex domain of* $\mathfrak{F}$. Its closure on $\mathcal{D}$ is denoted by $\tilde{\mathcal{K}}(\mathfrak{F})$, and its complete closure by $\overline{\mathcal{K}}(\mathfrak{F})$. The latter contains the limit set $\overline{\mathcal{G}}(\mathfrak{F})$ and is the convex hull of $\overline{\mathcal{G}}(\mathfrak{F})$. Also $\overline{\mathcal{K}}(\mathfrak{F}) \cap \mathcal{D} = \tilde{\mathcal{K}}(\mathfrak{F})$.

If $\overline{\mathcal{G}}(\mathfrak{F}) = \mathcal{E}$, thus $\mathcal{G}(\mathfrak{F})$ dense on $\mathcal{E}$, then $\mathcal{K}(\mathfrak{F}) = \tilde{\mathcal{K}}(\mathfrak{F}) = \mathcal{D}$ and $\overline{\mathcal{K}}(\mathfrak{F}) = \overline{\mathcal{D}}$. Hence the investigation of the convex domain presents an interest only in the case in which $\overline{\mathcal{G}}(\mathfrak{F})$ is a proper subset of $\mathcal{E}$, thus in which intervals of discontinuity occur.

Let $\mathcal{I}$ denote any interval of discontinuity. The line joining its end-points cuts off from $\mathcal{D}$ an open half-plane $\mathcal{H}(\mathcal{I})$ the boundary of which on $\mathcal{E}$ are the points of $\overline{\mathcal{I}}$. This open half-plane $\mathcal{H}(\mathcal{I})$ is the intersection with $\mathcal{D}$ of the convex hull of the (open) interval $\mathcal{I}$, thus $\mathcal{H}(\mathcal{I})$ and $\tilde{\mathcal{K}}(\mathfrak{F})$ are disjoint. The closures of two such half-planes $\overline{\mathcal{H}}(\mathcal{I}_1)$ and $\overline{\mathcal{H}}(\mathcal{I}_2)$ have no point in common, since two intervals of discontinuity $\mathcal{I}_1$ and $\mathcal{I}_2$ have no end-point in common.

Now, $\tilde{\mathcal{K}}(\mathfrak{F})$ is obtained by omitting from $\mathcal{D}$ those open half-planes $\mathcal{H}(\mathcal{I})$ pertaining to all intervals of discontinuity $\mathcal{I}$. For, the set obtained in this way has the following properties. Its closure contains $\overline{\mathcal{G}}(\mathfrak{F})$. It is convex, since it is the intersection of the convex sets formed by the half-planes closed on $\mathcal{D}$ complementary to the half-planes $\mathcal{H}(\mathcal{I})$. Thus it contains $\tilde{\mathcal{K}}(\mathfrak{F})$. On the other hand, the end-points of each interval of discontinuity are limit points, thus the whole boundary of the set inside $\mathcal{D}$ belongs to $\tilde{\mathcal{K}}(\mathfrak{F})$. Hence the set coincides with $\tilde{\mathcal{K}}(\mathfrak{F})$. If $\mathcal{K}(\mathfrak{F})$ is not identical with the whole of $\mathcal{D}$, $\tilde{\mathcal{K}}(\mathfrak{F})$ is a region closed on $\mathcal{D}$ the boundary of which on $\mathcal{E}$ is the limit set and the boundary of which in $\mathcal{D}$ consists of an enumerable infinity of lines, which will be called the *sides of* $\mathcal{K}(\mathfrak{F})$.

Any circle $\mathcal{C}$ in $\mathcal{D}$ only meets a finite number of sides of $\mathcal{K}(\mathfrak{F})$, if any.

Proof. Any interval of discontinuity subtends at the centre p of $\mathcal{C}$ a certain angle ψ, and any two of these angles ψ do not overlap, since the corresponding intervals of discontinuity have no point in common. (In case p is outside $\tilde{\mathcal{K}}(\mathfrak{F})$, thus in one of the half-planes $\mathcal{H}(\mathcal{I})$, the angle ψ pertaining to the corresponding interval $\mathcal{I}$ exceeds π which makes no difference.) Thus the sum of all these angles ψ does not exceed 2π, and only a finite number of them can exceed an assigned positive amount. For every ψ let σ denote the distance from p to the side of $\mathcal{K}(\mathfrak{F})$ pertaining to the corresponding interval of discontinuity. σ is the side in a right-angled triangle with angles 0 and $\frac{\psi}{2}$, or $\pi - \frac{\psi}{2}$, and opposite to the angle 0. Thus from formula (III.4) in §8.1 one gets $\cosh\sigma \sin\frac{\psi}{2} = 1$. Hence $\psi \to 0$ implies $\sigma \to \infty$. Thus the distances σ of the sides of $\mathcal{K}(\mathfrak{F})$ from p all exceed the radius of $\mathfrak{C}$ but for a finite number. □

Since every element of $\mathfrak{F}$ preserves convexity and carries $\mathcal{G}(\mathfrak{F})$ into itself, it also carries $\mathcal{K}(\mathfrak{F})$ into itself, and the same holds for $\tilde{\mathcal{K}}(\mathfrak{F})$ and $\overline{\mathcal{K}}(\mathfrak{F})$. It will now be shown that $\overline{\mathcal{K}}(\mathfrak{F})$ is the *smallest* closed convex set in $\overline{\mathcal{D}}$ reproduced by every element of $\mathfrak{F}$:

If $\mathcal{M} \neq \emptyset$ is a closed convex subset of $\overline{\mathcal{D}}$, and if $\mathfrak{f}\mathcal{M} \subset \mathcal{M}$ for every $\mathfrak{f}$ in $\mathfrak{F}$, then $\overline{\mathcal{K}}(\mathfrak{F})$ is contained in $\mathcal{M}$.

Proof. $\mathcal{M}$ must contain more than one point, since $\mathfrak{F}$ leaves no single point invariant. Therefore, $\mathcal{M} \cap \mathcal{D}$ is not empty for, in virtue of the convexity of $\mathcal{M}$, a segment, a half-line, or a line of $\mathcal{D}$ is contained in $\mathcal{M}$. If x is a point of $\mathcal{D}$ contained in $\mathcal{M}$, the equivalence class $\mathfrak{F}x$ belongs to $\mathcal{M}$, since $\mathcal{M}$ is reproduced by $\mathfrak{F}$, and so does the set of accumulation points of $\mathfrak{F}x$, thus $\overline{\mathcal{G}}(\mathfrak{F})$, since $\mathcal{M}$ is closed. Then the convexity of $\mathcal{M}$ implies that $\overline{\mathcal{K}}(\mathfrak{F})$ is contained in $\mathcal{M}$. □

If $\mathfrak{F}$ and $\mathfrak{F}'$ are two $\mathfrak{F}$-groups, if $\mathfrak{F}'$ possesses intervals of discontinuity and if every side of $\mathcal{K}(\mathfrak{F}')$ is at the same time a side of $\mathcal{K}(\mathfrak{F})$, then $\tilde{\mathcal{K}}(\mathfrak{F}') = \tilde{\mathcal{K}}(\mathfrak{F})$.

Proof. An arbitrary interval of discontinuity $\mathcal{I}$ of $\mathfrak{F}$ contains a discontinuity point p' of $\mathfrak{F}'$, since the intervals of discontinuity of $\mathfrak{F}'$ are everywhere dense on $\mathcal{E}$. By assumption, the interval of discontinuity $\mathcal{I}'$ of $\mathfrak{F}'$ containing p' is also an interval of discontinuity of $\mathfrak{F}$, thus identical with $\mathcal{I}$. Hence the collections of intervals of discontinuity of $\mathfrak{F}$ and $\mathfrak{F}'$ coincide and, consequently, $\tilde{\mathcal{K}}(\mathfrak{F}) = \tilde{\mathcal{K}}(\mathfrak{F}')$. □

If the $\mathfrak{F}$-group $\mathfrak{F}'$ is a subgroup of $\mathfrak{F}$, then $\mathcal{K}(\mathfrak{F}')$ is contained in $\mathcal{K}(\mathfrak{F})$.

Proof. This follows immediately from the fact that the fundamental set $\mathcal{G}(\mathfrak{F}')$ of $\mathfrak{F}'$ is contained in the fundamental set $\mathcal{G}(\mathfrak{F})$ of $\mathfrak{F}$. It implies that $\tilde{\mathcal{K}}(\mathfrak{F}') \subset \tilde{\mathcal{K}}(\mathfrak{F})$ and that $\overline{\mathcal{K}}(\mathfrak{F}') \subset \overline{\mathcal{K}}(\mathfrak{F})$. □

If the $\mathfrak{F}$-group $\mathfrak{F}'$ is a normal subgroup of $\mathfrak{F}$, then $\overline{\mathcal{K}}(\mathfrak{F}') = \overline{\mathcal{K}}(\mathfrak{F})$.

Proof. In virtue of the preceding theorem it is sufficient to prove that $\overline{\mathcal{K}}(\mathfrak{F})$ is contained in $\overline{\mathcal{K}}(\mathfrak{F}')$ or, which amounts to the same, that $\overline{\mathcal{G}}(\mathfrak{F})$ is contained in $\overline{\mathcal{G}}(\mathfrak{F}')$. Let $\mathfrak{f}'$ be a translation in $\mathfrak{F}'$ with $v(\mathfrak{f}')$ as its positive fundamental point and $\mathfrak{f}$ an arbitrary element of $\mathfrak{F}$. Then $\mathfrak{f}v(\mathfrak{f}')$ is the positive fundamental point of $\mathfrak{f}\mathfrak{f}'\mathfrak{f}^{-1}$, and this element belongs to $\mathfrak{F}'$, since $\mathfrak{F}'$ is self conjugate. Hence $\mathfrak{F}v(\mathfrak{f}')$ is contained in $\mathcal{G}(\mathfrak{F}')$, which implies that the closure of $\mathfrak{F}v(\mathfrak{f}')$ is contained in $\overline{\mathcal{G}}(\mathfrak{F}')$. Now it is known from §13.4 that the closure of the class of fundamental points $\mathfrak{F}v(\mathfrak{f}')$ coincides with $\overline{\mathcal{G}}(\mathfrak{F})$. This proves the assertion. □

The result $\overline{\mathcal{K}}(\mathfrak{F}') = \overline{\mathcal{K}}(\mathfrak{F})$ implies that $\overline{\mathcal{K}}(\mathfrak{F}') \cap \mathcal{D} = \overline{\mathcal{K}}(\mathfrak{F}) \cap \mathcal{D}$, thus $\tilde{\mathcal{K}}(\mathfrak{F}') = \tilde{\mathcal{K}}(\mathfrak{F})$. The two convex domains $\mathcal{K}(\mathfrak{F}')$ and $\mathcal{K}(\mathfrak{F})$ have thus the same closure on

$\mathcal{D}$, thus the same sides. It should be noticed that this does not necessarily imply the equation $\mathcal{K}(\mathfrak{F}') = \mathcal{K}(\mathfrak{F})$, because a side of $\mathcal{K}(\mathfrak{F})$ which belongs to $\mathcal{K}(\mathfrak{F})$ need not belong to $\mathcal{K}(\mathfrak{F}')$ although it is a side of $\mathcal{K}(\mathfrak{F}')$.

14.2 Boundary axes and limit sides. Any element $\mathfrak{f} \in \mathfrak{F}$ transforms translations of $\mathfrak{F}$ into translations of $\mathfrak{F}$. It therefore maps $\mathcal{G}(\mathfrak{F})$ onto itself. It then also maps the convex hull of $\mathcal{G}(\mathfrak{F})$ onto itself and, since it also maps $\mathcal{D}$ onto itself, it maps the intersection $\mathcal{K}(\mathfrak{F})$ of the two point sets onto itself. We consider the case in which $\mathcal{K}(\mathfrak{F})$ is not the whole of $\mathcal{D}$. Then $\mathfrak{f}$ maps the closure $\tilde{\mathcal{K}}(\mathfrak{F})$ of $\mathcal{K}(\mathfrak{F})$ on $\mathcal{D}$ onto itself, and thus also its complement in $\mathcal{D}$, i.e. the collection of infinitely many open half-planes $\mathcal{H}(\mathcal{I})$ considered in Section 1. As stated there, the closure $\overline{\mathcal{H}}(\mathcal{I}_1)$ and $\overline{\mathcal{H}}(\mathcal{I}_2)$ of any two such half-planes are disjoint.

Assume now that the element $\mathfrak{f} \neq 1$ of $\mathfrak{F}$ carries a particular point x_0 of $\overline{\mathcal{H}}(\mathcal{I})$ into a point $\mathfrak{f}x_0$ which also belongs to $\overline{\mathcal{H}}(\mathcal{I})$. Then $\mathfrak{f}$ must map $\overline{\mathcal{H}}(\mathcal{I})$ onto itself and thus, in particular, the side $\mathcal{S}$ of $\mathcal{K}(\mathfrak{F})$ bounding $\mathcal{H}(\mathcal{I})$ onto itself. Let $\mathfrak{A}_{\mathcal{S}}$ denote the subgroup of $\mathfrak{F}$ consisting of all those elements of $\mathfrak{F}$ which leave $\mathcal{S}$ invariant. We then have $\mathfrak{f} \in \mathfrak{A}_{\mathcal{S}}$. On the other hand, let $\mathfrak{g} \neq 1$ be any element of $\mathfrak{A}_{\mathcal{S}}$. Since $\mathcal{S}$ is a side of $\mathcal{K}(\mathfrak{F})$, and $\mathfrak{g}$ maps $\mathcal{K}(\mathfrak{F})$ onto itself, $\mathfrak{g}$ can neither be the reflection in $\mathcal{S}$, nor a half-turn around a point of $\mathcal{S}$, nor a reversed translation along $\mathcal{S}$, thus only a translation along $\mathcal{S}$ or the reflection in a normal of $\mathcal{S}$. – Instead of $\mathcal{H}(\mathcal{I})$ we also write $\mathcal{H}(\mathcal{S})$ for the open half-plane adjacent to $\mathcal{K}(\mathfrak{F})$ along its side $\mathcal{S}$. With the notation introduced in §11.6 and §10.1 one thus gets the result:

The subgroup $\mathfrak{A}_{\mathcal{S}}$ of $\mathfrak{F}$ leaving the side $\mathcal{S}$ of $\mathcal{K}(\mathfrak{F})$ invariant takes one of the four forms

$$\mathfrak{A}_{\mathcal{S}} = 1, \quad \mathfrak{A}_{\mathcal{S}} = \mathfrak{A}(|), \quad \mathfrak{A}_{\mathcal{S}} = \mathfrak{A}(\rightarrow), \quad \mathfrak{A}_{\mathcal{S}} = \mathfrak{A}(\|). \tag{1}$$

In $\overline{\mathcal{H}}(\mathcal{S})$ the groups $\mathfrak{F}$ and $\mathfrak{A}_{\mathcal{S}}$ are of equal effect. □

If $\mathfrak{A}_{\mathcal{S}} = \mathfrak{A}(\rightarrow)$ or $\mathfrak{A}_{\mathcal{S}} = \mathfrak{A}(\|)$, then $\mathcal{S}$ is an axis of $\mathfrak{F}$, its end-points belong to $\mathcal{G}(\mathfrak{F})$ as a joined pair, and $\mathcal{S}$ belongs to $\mathcal{K}(\mathfrak{F})$. In this case we call $\mathcal{S}$ a *boundary axis of* $\mathfrak{F}$ and the corresponding interval $\mathcal{I}$ a *periodic interval of discontinuity*. The whole equivalence class $\mathfrak{F}\mathcal{S}$ then consists of boundary axes of $\mathfrak{F}$, and all intervals of the equivalence class $\mathfrak{F}\mathcal{I}$ are periodic. – All axes of $\mathfrak{F}$ which are not boundary axes are called *inner axes of* $\mathfrak{F}$.

Consider the cases $\mathfrak{A}_{\mathcal{S}} = 1$ or $\mathfrak{A}_{\mathcal{S}} = \mathfrak{A}(|)$. Since $\mathfrak{F}$ and $\mathfrak{A}_{\mathcal{S}}$ have equal effect in $\overline{\mathcal{H}}(\mathcal{S})$ and thus on $\mathcal{S}$, an element of $\mathfrak{F}$ leaving an end-point of $\mathcal{S}$ fixed can only be the identity. Thus these end-points are not fundamental points of $\mathfrak{F}$. Hence $\mathcal{G}(\mathfrak{F})$ is contained in the open arc $\mathcal{I}$ of $\mathcal{E}$ bounding the open half-plane $\mathcal{H}'(\mathcal{S})$ determined by $\mathcal{S}$ and different from $\mathcal{H}(\mathcal{S})$. Thus $\mathcal{K}(\mathfrak{F})$ is contained in $\mathcal{H}'(\mathcal{S})$, and $\mathcal{S}$ does not belong to $\mathcal{K}(\mathfrak{F})$. The end-points of $\mathcal{S}$ are limit points of $\mathfrak{F}$, since they bound $\mathcal{I}$, and are thus accumulation points of $\mathcal{G}(\mathfrak{F})$. In this case we call $\mathcal{S}$ a *limit side* of $\mathcal{K}(\mathfrak{F})$ and $\mathcal{I}$ an *aperiodic interval of discontinuity*. The whole equivalence class $\mathfrak{F}\mathcal{S}$ consists of limit sides, and all intervals in $\mathfrak{F}\mathcal{I}$ are aperiodic.

Let $\mathfrak{I}$ be an arbitrary interval of discontinuity of $\mathfrak{F}$ and x an arbitrary point in $\mathfrak{I}$. In the two first cases of (1) the set $\mathfrak{A}_{\mathfrak{I}}x$ consists of one or two points, respectively; in the last two cases it consists of a sequence of infinitely many points converging to the end-points of $\mathfrak{I}$. Any element of $\mathfrak{F}$ not in $\mathfrak{A}_{\mathfrak{I}}$ maps x into a point of another interval of discontinuity, since $\mathfrak{F}$ and $\mathfrak{A}_{\mathfrak{I}}$ are of equal effect in $\mathfrak{I}$. This shows once more the fact already stated in §13.7 that $\mathfrak{F}$ is discontinuous in every point of an interval of discontinuity.

14.3 Further properties of the convex domain. It is an immediate consequence of the definition of the convex domain that every axis of $\mathfrak{F}$ is contained in $\mathcal{K}(\mathfrak{F})$; for its end-points belong to $\mathcal{G}(\mathfrak{F})$. Concerning the disposition of axes in the convex domain the following statement holds:

Every half-line issuing from a point of $\mathcal{K}(\mathfrak{F})$ meets an axis of $\mathfrak{F}$.

Proof. This is evident if the point is a boundary point of $\mathcal{K}(\mathfrak{F})$ on $\mathcal{D}$, because it is then itself situated on an axis, namely a boundary axis. So let p be an inner point of $\mathcal{K}(\mathfrak{F})$, let $\mathcal{T}$ denote a half-line issuing from p, and let q be the end-point of $\mathcal{T}$ on $\mathcal{E}$. Let $\mathcal{L}$ denote a line through p, and let m and n be its end-points on $\mathcal{E}$. If q belongs to a closed interval of discontinuity $\overline{\mathfrak{I}}$ as an inner point or end-point of $\overline{\mathfrak{I}}$, then $\mathcal{L}$ is so chosen as not to meet $\overline{\mathfrak{I}}$; this is possible, since $\overline{\mathfrak{I}}$ subtends at the inner point p of $\mathcal{K}(\mathfrak{F})$ an angle $< \pi$ (Fig. 14.1). If q does not belong to such an $\overline{\mathfrak{I}}$, then the line $\mathcal{L}$ is

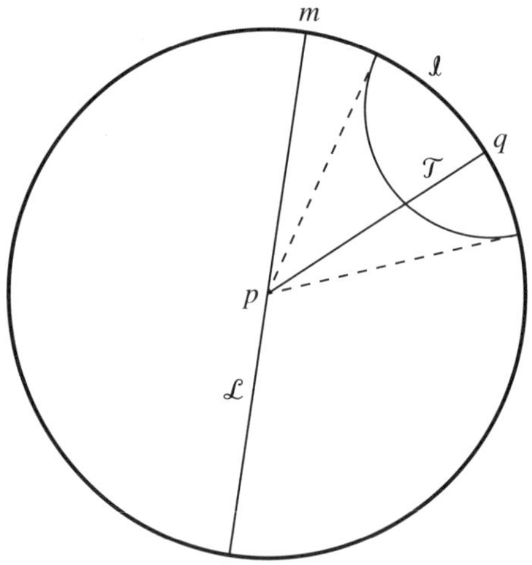

Figure 14.1

arbitrary, except that it shall not contain $\mathcal{T}$. Under these conditions the open intervals mq and nq of the arc mqn of $\mathcal{E}$ both contain fundamental points. Then it has been proved in §13.3 that there is an axis in $\mathfrak{F}$ with one of its fundamental points in mq, the other in nq. This axis meets $\mathcal{T}$. □

If $\mathcal{L}$ is an arbitrary line contained in $\tilde{\mathcal{K}}(\mathfrak{F})$, $\mathcal{L}_0$ a finite subsegment of $\mathcal{L}$ and ε a positive number, then there is in $\mathfrak{F}$ an axis $\mathcal{A}$ such that the distances of all points of $\mathcal{L}_0$ from $\mathcal{A}$ are smaller than ε.

Proof. This is easily seen by the same theorem in §13.3 concerning the existence of axes with end-points in prescribed intervals: Since $\mathcal{L}$ is contained in $\tilde{\mathcal{K}}(\mathfrak{F})$, its end-points belong to $\overline{\mathfrak{G}}(\mathfrak{F})$. If $\mathcal{J}$ and $\mathcal{J}'$ are arbitrary intervals on $\mathcal{E}$ each containing one of the end-points of $\mathcal{L}$, then there are fundamental points both in $\mathcal{J}$ and $\mathcal{J}'$, and there is, therefore, an axis $\mathcal{A}$ in $\mathfrak{F}$ with one end-point in $\mathcal{J}$, the other in $\mathcal{J}'$. It is evident that $\mathcal{A}$ has the said property, if $\mathcal{J}$ and $\mathcal{J}'$ are chosen sufficiently small. □

All centres of $\mathfrak{F}$ are inner points of $\mathcal{K}(\mathfrak{F})$.

Proof. Let $\mathfrak{A}_c(\circlearrowleft_\nu)$ be the subgroup of $\mathfrak{F}$ consisting of all rotations in $\mathfrak{F}$ with centre c and ν its order. If $\mathcal{E}$ is divided into ν arcs which are cyclically interchanged by $\mathfrak{A}_c(\circlearrowleft_\nu)$, then each of these arcs contains in its interior points of $\mathfrak{G}(\mathfrak{F})$, because $\mathfrak{G}(\mathfrak{F})$ is reproduced by $\mathfrak{A}_c(\circlearrowleft_\nu)$. Since each of these arcs subtends an angle $\frac{2\pi}{\nu}$ in c, any interval of discontinuity must subtend a smaller angle in c. This implies that c can neither be a boundary point of $\mathcal{K}(\mathfrak{F})$ in $\mathcal{D}$, nor situated in an open half-plane $\mathcal{H}(\mathcal{I})$. □

More precisely, the calculation indicated in Section 1 shows that the distance of a centre of a rotation subgroup of order ν from a side of $\mathcal{K}(\mathfrak{F})$ exceeds the value σ_ν calculated from

$$\cosh \sigma_\nu \sin \frac{\pi}{\nu} = 1.$$

This yields $\sigma_\nu > 0$ for $\nu > 2$ and $\sigma_\nu \to \infty$ for $\nu \to \infty$. Hence, if c is a centre of $\mathfrak{F}$ and $\nu > 2$ is the order of the rotation subgroup of $\mathfrak{F}$ belonging to c, then σ_ν is a positive lower bound for the distance of c from the boundary of $\mathcal{K}(\mathfrak{F})$.

Centres of rotation subgroups of order 2 may be situated arbitrarily close to the boundary of $\mathcal{K}(\mathfrak{F})$, as shown by suitably constructed examples for $\mathfrak{F}$.

If c is a centre of $\mathfrak{F}$ with corresponding subgroup $\mathfrak{A}_c(\times_\nu)$ and if the ν reflection lines of $\mathfrak{F}$ passing through c are wholly contained in $\mathcal{K}(\mathfrak{F})$ (thus, if their end-points are limit points), then the positive root τ_ν of the equation

$$\cosh \tau_\nu \sin \frac{\pi}{2\nu} = 1$$

is a lower bound for the distance of c from the boundary of $\mathcal{K}(\mathfrak{F})$. This is seen from the fact that an arbitrary interval of discontinuity of $\mathfrak{F}$ subtends at c an angle smaller than $\frac{\pi}{\nu}$. Here τ_2, determined by $\cosh \tau_2 = \sqrt{2}$, yields a positive lower bound even for $\nu = 2$.

Remark. Any half-line issuing from c together with its images by $\mathfrak{A}_c(\circlearrowleft_\nu)$ divides $\mathcal{K}(\mathfrak{F})$ into ν congruent parts.

All limit-centres of $\mathfrak{F}$ are known to be limit points of $\mathfrak{F}$ (§13.5). They are thus boundary points of $\mathcal{K}(\mathfrak{F})$ on $\mathfrak{E}$.

Every reflection line of $\mathfrak{F}$ divides $\mathcal{K}(\mathfrak{F})$ into two symmetric parts. If it is not wholly contained in $\mathcal{K}(\mathfrak{F})$, it is orthogonal to one or two sides of $\mathcal{K}(\mathfrak{F})$. If the line of the reflection $\mathfrak{s}$ cuts the side $\mathcal{S}$ of $\mathcal{K}(\mathfrak{F})$, then $\mathfrak{s}$ is contained in the group $\mathfrak{A}_{\mathcal{S}}$ considered in Section 2.

14.4 Characteristic neighbourhood of a point in $\mathcal{D}$. Let x be an arbitrary point of $\mathcal{D}$, and let $\mathfrak{A}_x$ denote the subgroup of $\mathfrak{F}$ consisting of all those elements of $\mathfrak{F}$ which leave x invariant. If x is a regular point, then $\mathfrak{A}_x = 1$. If x is a centre of $\mathfrak{F}$, then $\mathfrak{A}_x = \mathfrak{A}(\circlearrowleft_\nu)$ or $\mathfrak{A}_x = \mathfrak{A}(\times_\nu)$. If x is situated on a reflection line of $\mathfrak{F}$ without being a centre, then $\mathfrak{A}_x = \mathfrak{A}(-)$. In all cases let $\mathfrak{F} \setminus \mathfrak{A}_x$ denote the collection of all elements of $\mathfrak{F}$ not in $\mathfrak{A}_x$. Then $(\mathfrak{F} \setminus \mathfrak{A}_x)x$ is the set of those points of the equivalence class $\mathfrak{F}x$ which are different from x. There is one such point for each of the left cosets of $\mathfrak{A}_x$ in $\mathfrak{F}$ which make up $\mathfrak{F} \setminus \mathfrak{A}_x$. We then put

$$\eta(\mathfrak{F}; x) = [x, (\mathfrak{F} \setminus \mathfrak{A}_x)x] = \min_{\mathfrak{g} \in \mathfrak{F} \setminus \mathfrak{A}_x} [x, \mathfrak{g}x]. \tag{2}$$

Thus $\eta(\mathfrak{F}; x)$, which we simply write $\eta(x)$ if there can be no doubt as to the group in question, denotes the distance from x to the nearest point of its equivalence class not coinciding with x. In the set of regular points $\eta(x)$ is the same as the distance function $\varepsilon(x)$ introduced in §10.3, hence $\eta(x)$ is continuous in this set. In the singular set $\varepsilon(x) = 0$, while $\eta(x) > 0$ in the whole of $\mathcal{D}$. Thus $\eta(x)$ is not continuous in the singular set.

The interior of the circle $\mathfrak{C}(x; \frac{1}{2}\eta(x))$ will be called the *characteristic neighbourhood of* x and denoted by $\mathcal{D}(\mathfrak{F}; x)$, or in short $\mathcal{D}(x)$. It is reproduced by $\mathfrak{A}_x$ and has no point in common with any of its images by $\mathfrak{F} \setminus \mathfrak{A}_x$, since any element of $\mathfrak{F} \setminus \mathfrak{A}_x$ displaces x at least at the distance $\eta(x)$. This implies that $\mathfrak{F}$ and $\mathfrak{A}_x$ have equal effect in $\mathcal{D}(x)$. This fact, and the following considerations, are generalizations of those presented in §10.4. As a special consequence, all points of $\mathcal{D}(x)$ are regular except possibly x itself or points on reflection lines passing through x. Evidently, $\eta(x)$ takes the same value in all points of the equivalence class $\mathfrak{F}x$, thus all these points have congruent characteristic neighbourhoods.

Let $\mathfrak{g}$ be an element of $\mathfrak{F} \setminus \mathfrak{A}_x$ chosen such that $[x, \mathfrak{g}x] = \eta(x)$. Then $\mathfrak{C}(x; \frac{1}{2}\eta(x))$ and $\mathfrak{g}C$ touch each other, say in the point z. Thus $\mathfrak{g}^{-1}z$ is a point of $\mathfrak{C}$. Whether z coincides with $\mathfrak{g}^{-1}z$ or not, this fact shows that $\mathfrak{F}$ and $\mathfrak{A}_x$ are no more of equal effect on $\mathfrak{C}$, for $\mathfrak{g}^{-1}$ is not an element of $\mathfrak{A}_x$. Thus no open circular disc with centre x and a radius exceeding $\frac{1}{2}\eta(x)$ has the property that $\mathfrak{F}$ and $\mathfrak{A}_x$ have equal effect in it. $\mathcal{D}(x)$ is thus characterized by this maximum property.

In the special case in which z and $\mathfrak{g}^{-1}z$ coincide the element $\mathfrak{g}^{-1}$ carries $\mathfrak{g}x$ into x and leaves the bisecting point z of the segment x, $\mathfrak{g}x$ invariant. It is thus either a half-turn about z or the reflection in the line touching $\mathfrak{C}$ in z. In both cases $\mathfrak{g}^2 = 1$. Conversely, if $\mathfrak{g}^2 = 1$, for the above element $\mathfrak{g} \in \mathfrak{F} \setminus \mathfrak{A}_x$, then $\mathfrak{g}z = z$, since x and $\mathfrak{g}x$ are interchanged by $\mathfrak{g}$.

14.5 Distance modulo $\mathfrak{F}$. For any two points x and y of $\mathcal{D}$ let $\xi(\mathfrak{F}; x, y)$ denote the distance of the point sets $\mathfrak{F}x$ and $\mathfrak{F}y$. We call ξ *the distance modulo* $\mathfrak{F}$ *of* x *and* y. It depends only on the equivalence classes of the points, and $\xi(\mathfrak{F}; x, y) = \xi(\mathfrak{F}; y, x)$. Since $\mathfrak{F}$ preserves distances, one may write

$$\xi(\mathfrak{F}; x, y) = \min_{\mathfrak{g}, \mathfrak{h} \in \mathfrak{F}} [\mathfrak{g}x, \mathfrak{h}y] = \min_{\mathfrak{f} \in \mathfrak{F}} [x, \mathfrak{f}y] = \min_{\mathfrak{f} \in \mathfrak{F}} [\mathfrak{f}x, y]. \tag{3}$$

$\xi = 0$ if and only if x and y are equivalent with respect to $\mathfrak{F}$. Since for any pair $\mathfrak{g}, \mathfrak{h}$ of elements of $\mathfrak{F}$

$$\begin{aligned} \xi(\mathfrak{F}; x', y') &\leqq [\mathfrak{g}x', \mathfrak{h}y'] \leqq [\mathfrak{g}x', \mathfrak{g}x] + [\mathfrak{g}x, \mathfrak{h}y] + [\mathfrak{h}y, \mathfrak{h}y'] \\ &= [\mathfrak{g}x, \mathfrak{h}y] + [x, x'] + [y, y'] \end{aligned}$$

and $\mathfrak{g}$ and $\mathfrak{h}$ can be chosen such that $[\mathfrak{g}x, \mathfrak{h}y] = \xi(\mathfrak{F}; x, y)$, it is inferred that ξ is a continuous function of x and y.

14.6 Isometric neighbourhood of a point in $\mathcal{D}$. If, as in Section 4, x is any point of $\mathcal{D}$ and $\mathfrak{A}_x$ denotes the subgroup of $\mathfrak{F}$ leaving x invariant, we define as *isometric neighbourhood of* x and denote by $\mathcal{I}(\mathfrak{F}; x)$, or in short $\mathcal{I}(x)$, the largest open circular disc with centre x for which the equation

$$\xi(\mathfrak{F}; x_1, x_2) = \xi(\mathfrak{A}_x; x_1, x_2) \tag{4}$$

holds for any two points x_1 and x_2 of the disc. Let $\mathcal{K}(x)$ denote the bounding circle of $\mathcal{I}(x)$ and $\iota(x)$ its radius. Evidently $\iota(\mathfrak{f}x) = \iota(x)$ for any $\mathfrak{f} \in \mathfrak{F}$. – We now show:

For, every $x \in \mathcal{D}$ *the interior of the circle* $\mathfrak{C}(x; \frac{1}{4}\eta(x))$ *belongs to* $\mathcal{I}(x)$.

Proof. Let x_1 and x_2 be any two points inside that circle. Then

$$\xi(\mathfrak{A}_x; x_1, x_2) \leqq [x_1, x_2] < \tfrac{1}{2}\eta(x),$$

while for any $\mathfrak{f} \in \mathfrak{F} \setminus \mathfrak{A}_x$ in virtue of $[x, \mathfrak{f}x] \geqq \eta(x)$ one gets

$$[x_1, \mathfrak{f}x_2] \geqq -[x_1, x] + [x, \mathfrak{f}x] - [\mathfrak{f}x_1, \mathfrak{f}x_2] > -\tfrac{1}{4}\eta(x) + \eta(x) - \tfrac{1}{4}\eta(x) = \tfrac{1}{2}\eta(x).$$

This shows the theorem. □

Hence

$$\iota(x) \geqq \tfrac{1}{4}\eta(x) \tag{5}$$

for all x in $\mathcal{D}$. This implies that an isometric neighbourhood exists for all x in $\mathcal{D}$, since $\eta(x)$ is positive throughout $\mathcal{D}$.

The lower bound $\frac{1}{4}\eta(x)$ for $\iota(x)$ established in (5) cannot be replaced by a larger one valid for all $\mathfrak{F}$-groups. To see this, consider a group $\mathfrak{F}$ consisting besides the identity of translations only. (As will be seen in §21.9, a simple example of such a group is a group generated by three translations $\mathfrak{f}$, $\mathfrak{g}$, $\mathfrak{h}$ with $\mathfrak{fgh} = 1$ and mutually divergent axes which do not separate each other; compare Fig. 9.3). Moreover, it will result from §15 that among the displacements for the translations in an $\mathfrak{F}$-group there is a smallest one, say λ, if the group can be generated by a finite number of its elements (as is the case in the example given). All points in $\mathcal{D}$ are regular, since neither rotations nor reflections occur in $\mathfrak{F}$. Let $\mathfrak{f}$ be a translation with $\lambda(\mathfrak{f}) = \lambda$. Then $\eta(x) = \lambda$ for all points x on $\mathcal{A}(\mathfrak{f})$. Let $\mathcal{A}(\mathfrak{f})$ be directed (Fig. 14.2), and choose on

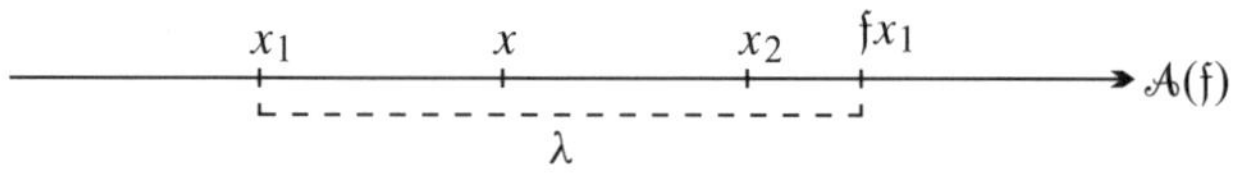

Figure 14.2

$\mathcal{A}(\mathfrak{f})$ three points x_1, x, x_2 in this order such that

$$\tfrac{1}{4}\lambda < [x_1, x] = [x, x_2] < \tfrac{1}{2}\lambda.$$

Then

$$\tfrac{1}{2}\lambda < [x_1, x_2] < \lambda.$$

Since $[x_1, \mathfrak{f}x_1] = \lambda$, one gets, counting distances with sign included,

$$[x_2, \mathfrak{f}x_1] = [x_2, x_1] + [x_1, \mathfrak{f}x_1] = -[x_1, x_2] + \lambda$$

thus

$$0 < [x_2, \mathfrak{f}x_1] < \tfrac{1}{2}\lambda$$

and, since $\mathfrak{A}_x = 1$,

$$\xi(\mathfrak{A}_x; x_1, x_2) = [x_1, x_2] > [x_2, \mathfrak{f}x_1].$$

This shows that the radius of $\mathcal{I}(x)$ cannot exceed $\frac{1}{4}\eta(x)$ for this particular $\mathfrak{F}$-group.

For, every $\mathfrak{F}$-group $\mathfrak{F}$, and every point x in $\mathcal{D}$, $\mathcal{I}(\mathfrak{F}; x)$ is contained in $\mathcal{D}(\mathfrak{F}; x)$.

Proof. Since every $\mathfrak{h} \in \mathfrak{A}_x$, $\mathfrak{h} \neq 1$, is either a rotation with centre x, or the reflection in a line through x, one gets $[\mathfrak{h}x_1, x] = [x_1, x]$ for all $\mathfrak{h} \in \mathfrak{A}_x$, and any point $x_1 \in \mathcal{D}$. Hence

$$\xi(\mathfrak{A}_x; x_1, x) = \min_{\mathfrak{h} \in \mathfrak{A}_x} [\mathfrak{h}x_1, x] = [x_1, x]$$

and thus $\xi(\mathfrak{A}_x; x_1, x) > \frac{1}{2}\eta(x)$, if x_1 is outside $\mathfrak{C}(x; \frac{1}{2}\eta(x))$. Let $\mathfrak{g}$ be an element of $\mathfrak{F} \setminus \mathfrak{A}_x$ such that $[x, \mathfrak{g}x] = \eta(x)$, and let x_1 be any point on the segment x, gx outside $\mathfrak{C}(x; \frac{1}{2}\eta(x))$. Then

$$\xi(\mathfrak{F}; x_1, x) \leqq [x_1, \mathfrak{g}x] < \tfrac{1}{2}\eta(x).$$

This shows that (4) does not hold for any open circular disc containing $\mathfrak{C}(x; \frac{1}{2}\eta(x))$, and the assertion is proved. □

It yields the inequality

$$\iota(x) \leqq \tfrac{1}{2}\eta(x). \tag{6}$$

The upper bound for $\iota(x)$ established in (6) cannot be replaced by a smaller one valid for all $\mathfrak{F}$-groups. To see this, consider the following example (Fig. 14.3). Let

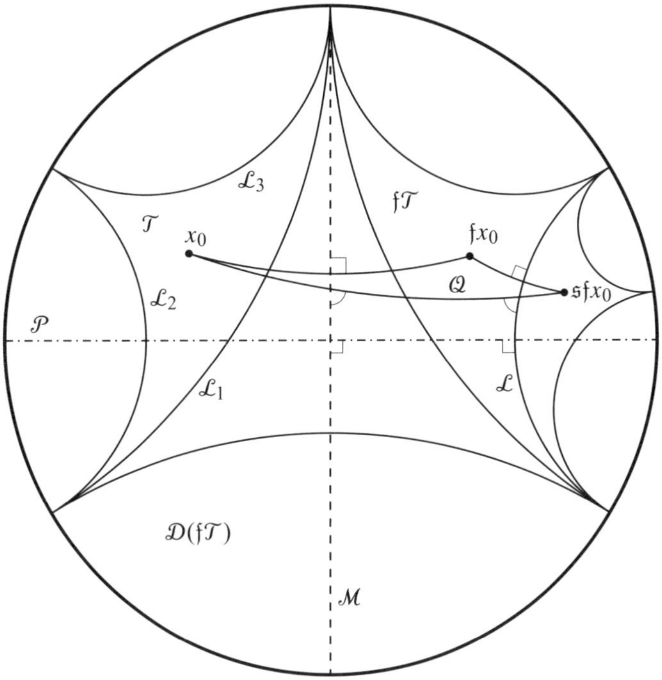

Figure 14.3

three lines $\mathcal{L}_1, \mathcal{L}_2, \mathcal{L}_3$ form an asymptotic triangle; its interior is called $\mathcal{T}$. The reflections $\mathfrak{s}_1, \mathfrak{s}_2, \mathfrak{s}_3$ in $\mathcal{L}_1, \mathcal{L}_2, \mathcal{L}_3$ respectively generate a group $\mathfrak{F}$ which produces a tesselation of $\mathcal{D}$ by asymptotic triangles. It is discontinuous in $\mathcal{D}$, since $\tilde{\mathcal{T}}$ is a fundamental polygon, and it is not quasi-abelian, thus an $\mathfrak{F}$-group, since it contains limit-rotations with different limit-centres, e.g. the vertices of $\mathcal{T}$. Since no two points in $\mathcal{T}$ are equivalent, one gets $\mathfrak{A}_x = 1$ for every point x in $\mathcal{T}$. The groups $\mathfrak{F}$ and $\mathfrak{A}_x$

have equal effect in $\mathcal{T}$. Hence: *for any x in $\mathcal{T}$ $\mathcal{D}(\mathfrak{F}; x)$ is the largest open circular disc with centre x contained in $\mathcal{T}$.*

Now let x_0 be any fixed point in $\mathcal{T}$. We first want to show that the normal domain $\mathcal{N}(\mathfrak{F}; x_0)$ is $\tilde{\mathcal{T}}$. Since $\mathcal{L}_\nu$, $\nu = 1, 2, 3$, is the bisecting perpendicular of $x_0, \mathfrak{s}_\nu x_0$, it follows that $\tilde{\mathcal{T}}$ contains $\mathcal{N}(\mathfrak{F}; x_0)$. The assertion then is that for any element $\mathfrak{f} \neq 1$ of $\mathfrak{F}$ the bisecting perpendicular of $x_0, \mathfrak{f}x_0$ does not intersect $\mathcal{T}$. The point $\mathfrak{f}x_0$ is contained in the triangle $\mathfrak{f}\mathcal{T}$ of the tesselation. For any $\mathfrak{f} \in \mathfrak{F}$ let $\rho(\mathfrak{f}\mathcal{T})$ denote the number of lines out of the collection $\mathcal{U} = \bigcup_{1 \leqq \nu \leqq 3} \mathfrak{F}\mathcal{L}_\nu$ which separate $\mathcal{T}$ and $\mathfrak{f}\mathcal{T}$. It is the number of points in which a segment joining an arbitrary point of $\mathcal{T}$ with an arbitrary point of $\mathfrak{f}\mathcal{T}$ cuts $\mathcal{U}$. We may call ρ the *rank* of $\mathfrak{f}\mathcal{T}$. The only triangle of rank 0 is $\mathcal{T}$, the only triangles of rank 1 are $\mathfrak{s}_\nu\mathcal{T}$, $\nu = 1, 2, 3$. For any $\mathfrak{f}\mathcal{T}$ of positive rank let $\mathcal{D}(\mathfrak{f}\mathcal{T})$ denote the subset of $\mathcal{D}$ separating $\mathcal{T}$ from $\mathfrak{f}\mathcal{T}$. If $\rho(\mathfrak{f}\mathcal{T}) = 1$, then $\mathcal{D}(\mathfrak{f}\mathcal{T})$ is one of the lines $\mathcal{L}_\nu$. If $\rho(\mathfrak{f}\mathcal{T}) > 1$, then $\mathcal{D}(\mathfrak{f}\mathcal{T})$ is a region closed on $\mathcal{D}$. – Let it now be assumed that for any value of the rank ρ of $\mathfrak{f}\mathcal{T}$ satisfying $1 \leqq \rho \leqq \rho^*$ the bisecting perpendicular of $x_0, \mathfrak{f}x_0$ belongs to $\mathcal{D}(\mathfrak{f}\mathcal{T})$. This holds for $\rho^* = 1$. We want to show that it holds for $\rho^* + 1$. Let $\mathfrak{f}\mathcal{T}$ be of rank ρ^*. The bisecting perpendicular $\mathcal{M}$ of $x_0, \mathfrak{f}_{x_0}$ is contained in $\mathcal{D}(\mathfrak{f}\mathcal{T})$. Let $\mathcal{L}$ denote one of the two sides of $\mathfrak{f}\mathcal{T}$ not belonging to $\mathcal{D}(\mathfrak{f}\mathcal{T})$ (Fig. 14.3). The reflection $\mathfrak{s}$ in $\mathcal{L}$ is an element of $\mathfrak{F}$. Then $\mathfrak{s}\mathfrak{f}\mathcal{T}$ is of rank $\rho^* + 1$. The bisecting perpendicular of $\mathfrak{f}x_0$, $\mathfrak{s}\mathfrak{f}x_0$ is $\mathcal{L}$, which is outside $\mathcal{D}(\mathfrak{f}\mathcal{T})$. Thus $\mathcal{L}$ and $\mathcal{M}$ are either parallel or divergent. Let $\mathcal{P}$ denote their common end-point on $\mathcal{E}$ or their common normal, as the case may be. Consider now the triangle or quadrangle bounded by the straight line $\mathcal{Q} = x_0, \mathfrak{s}\mathfrak{f}x_0$ together with $\mathcal{L}$, $\mathcal{M}$ and the point or line $\mathcal{P}$. The two angles at $\mathcal{Q}$ of this triangle or quadrangle are acute, since for each of them its vertical angle is contained in a right-angled triangle. In virtue of the last theorem in §9.9 the bisecting perpendicular of $x_0, \mathfrak{s}\mathfrak{f}x_0$ belongs to the pencil determined by $\mathcal{L}$ and $\mathcal{M}$. It is thus the perpendicular from $\mathcal{P}$ on $\mathcal{Q}$, or the common normal of $\mathcal{P}$ and $\mathcal{Q}$, in the two cases respectively. In both cases its foot on $\mathcal{Q}$ is situated within the triangle or quadrangle, thus the perpendicular is situated in $\mathcal{D}(\mathfrak{s}\mathfrak{f}\mathcal{T})$, and hence does not intersect $\mathcal{T}$. Induction now shows that $\mathcal{N}(\mathfrak{F}; x_0) = \tilde{\mathcal{T}}$ for any point x_0 of $\mathcal{T}$.

Let x, x_1, x_2 be any three points of $\mathcal{T}$. Since $\mathfrak{A}_x = 1$ and $\mathcal{N}(\mathfrak{F}; x_1) = \tilde{\mathcal{T}}$, one gets by (3)

$$\xi(\mathfrak{F}; x_1, x_2) = \min_{\mathfrak{f} \in \mathfrak{F}}[x_1, \mathfrak{f}x_2] = [x_1, x_2] = \xi(\mathfrak{A}_x; x_1, x_2).$$

This shows that (4) holds for any three points x, x_1, x_2 of $\mathcal{T}$. Hence $\mathcal{I}(\mathfrak{F}; x)$ *is the largest open circular disc with centre x contained in $\mathcal{T}$* and thus coincides with $\mathcal{D}(\mathfrak{F}; x)$. This shows that the sign of equality in (6) holds for the present group $\mathfrak{F}$.

We restate the main results of Sections 4-6:

For any $\mathfrak{F}$-group $\mathfrak{F}$ and any point x of $\mathcal{D}$, if $\mathfrak{A}_x$ denotes the subgroup of $\mathfrak{F}$ leaving x fixed, the characteristic neighbourhood $\mathcal{D}(\mathfrak{F}; x)$ of x is the largest open circular disc with centre x in which $\mathfrak{F}$ and $\mathfrak{A}_x$ have equal effect, and the isometric neighbourhood $\mathcal{I}(\mathfrak{F}; x)$ of x is the largest open circular disc with centre x such that distance modulo $\mathfrak{F}$ coincides with distance modulo $\mathfrak{A}_x$ for any two of its points. If $\chi(x)$ and $\iota(x)$ denote *the radii of these two discs respectively, then in virtue of* (5) *and* (6)

$$\tfrac{1}{2}\chi(x) \leqq \iota(x) \leqq \chi(x). \tag{7}$$

This inequality (7) *cannot be sharpened to more narrow limits valid for all* $\mathfrak{F}$*-groups.*

14.7 Characteristic neighbourhood of a limit-centre. The investigation contained in the preceding sections concerning the neighbourhoods of point in $\mathcal{D}$, will now be followed by a similar investigation concerning limit-centres in case such are contained in $\mathfrak{F}$. The results prove to be similar.

Let u be a limit-centre of $\mathfrak{F}$ and $\mathfrak{A}_u$ the subgroup of $\mathfrak{F}$ leaving u invariant. Then $\mathfrak{A}_u$ belongs to the type $\mathfrak{A}_u(\hookrightarrow)$ or $\mathfrak{A}_u(\wedge)$. Let $\mathcal{O}$ be a horocycle with centre u. For horocycles with the same centre we define their mutual distance as the length of the segment cut out by them on any line through the centre. If $\mathcal{O}_1$ is a horocycle with centre $u_1 \neq u$ (u_1 need not be a limit-centre of $\mathfrak{F}$), we define the distance $\delta(\mathcal{O}, \mathcal{O}_1)$ of $\mathcal{O}$ and $\mathcal{O}_1$ as the length of the segment cut out on the line uu_1 by $\mathcal{O}$ and $\mathcal{O}_1$, this distance being negative, zero or positive, according as $\mathcal{O}$ and $\mathcal{O}_1$ intersect or touch each other or have no point in common. Likewise, if $\mathcal{L}$ is a line with end-points different from u, the distance $\delta(\mathcal{O}, \mathcal{L})$ means the length of the segment cut out by $\mathcal{O}$ and $\mathcal{L}$ on the normal of $\mathcal{L}$ terminating at u with the sign of δ fixed accordingly. Obviously, $\delta(\mathcal{O}, \mathcal{L}) = \frac{1}{2}\delta(\mathcal{O}, \mathfrak{s}\mathcal{O})$, if $\mathfrak{s}$ denotes the reflection in $\mathcal{L}$ ($\mathfrak{s}$ need not be an element of $\mathfrak{F}$), compare Fig. 14.4.

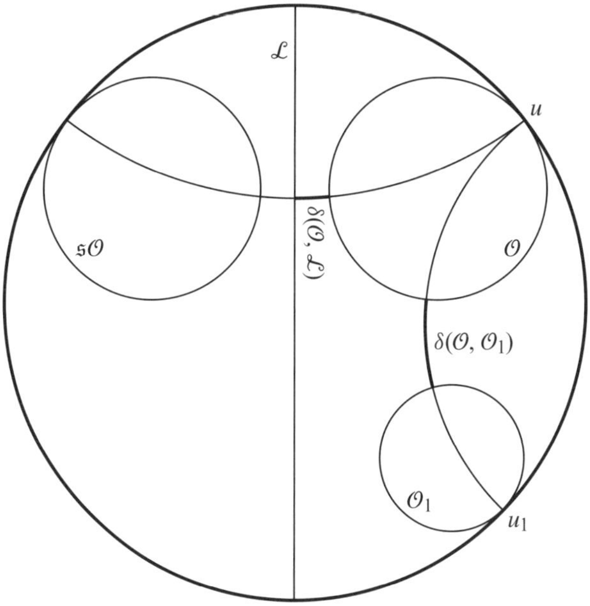

Figure 14.4

If a horocycle or line is replaced by its image by some element of $\mathfrak{A}_u$, its distance from $\mathcal{O}$ is not altered.

In the case $\mathfrak{A}_u(\wedge)$ let p and q denote the intersections of $\mathcal{O}$ with the lines of two reflections generating $\mathfrak{A}_u$. In the case $\mathfrak{A}_u(\hookrightarrow)$ let p be an arbitrary point of $\mathcal{O}$ and q its image by a primary limit-rotation in $\mathfrak{A}_u$. Then the arc $\mathcal{J} = pq$ of $\mathcal{O}$ is a fundamental arc for $\mathfrak{A}_u$ on $\mathcal{O}$; (exactly: if the point q is omitted in the case $\mathfrak{A}_u(\hookrightarrow)$).

We now prove two auxiliary theorems.

The distances from $\mathcal{O}$ of all horocycles equivalent with $\mathcal{O}$ with respect to $\mathfrak{F}$ have no accumulation value.

Proof. Let $\mathcal{O}_1$ with centre u_1 be the image of $\mathcal{O}$ by some element of $\mathfrak{F} \setminus \mathfrak{A}_u$; thus $u_1 \neq u$. Without affecting the distance $\delta(\mathcal{O}, \mathcal{O}_1)$ one can replace $\mathcal{O}_1$ by its image by some element of $\mathfrak{A}_u$. It can therefore be assumed that uu_1 intersects $\mathcal{J}$. There is on $\mathcal{O}_1$ an arc equivalent with $\mathcal{J}$ and containing the intersection point of $\mathcal{O}_1$ and uu_1. Let this arc be $\mathfrak{g}\mathcal{J}$ (Fig. 14.5). Then $\mathcal{O}_1 = \mathfrak{g}\mathcal{O}$. For different horocycles $\mathcal{O}_1$ of the

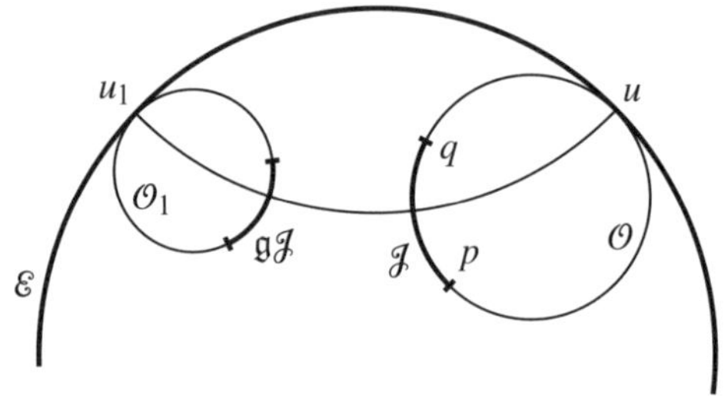

Figure 14.5

equivalence class $\mathfrak{F}\mathcal{O}$ the corresponding elements $\mathfrak{g}$ belong to different left cosets of $\mathfrak{A}_u$ in the collection $\mathfrak{F} \setminus \mathfrak{A}_u$. Now, since chords of a horocycle increase with increasing arcs,

$$[p, \mathfrak{g}p] \leqq [p, q] + |\delta(\mathcal{O}, \mathcal{O}_1)| + [p, q].$$

For values of $|\delta(\mathcal{O}, \mathcal{O}_1)|$ smaller than a prescribed positive bound, this inequality can – on account of the discontinuity of $\mathfrak{F}$ – only hold for a finite number of elements $\mathfrak{g}$. This proves the assertion. □

There can only be a finite number of negative values of $\delta(\mathcal{O}, \mathfrak{g}\mathcal{O})$, if any.

Proof. Let the equivalent horocycle $\mathcal{O}_1$ intersect $\mathcal{O}$ (Fig. 14.6). As in the preceding proof we arrange that one of the intersection points, say r, belongs to $\mathcal{J}$. Let $\mathfrak{g}\mathcal{J}$ be the arc of $\mathcal{O}_1$ to which r belongs. Then $\mathcal{O}_1 = \mathfrak{g}\mathcal{O}$. We now get

$$[p, \mathfrak{g}p] \leqq [p, r] + [r, \mathfrak{g}p] \leqq 2[p, q].$$

This only holds for a finite number of elements $\mathfrak{g}$, which proves the assertion. □

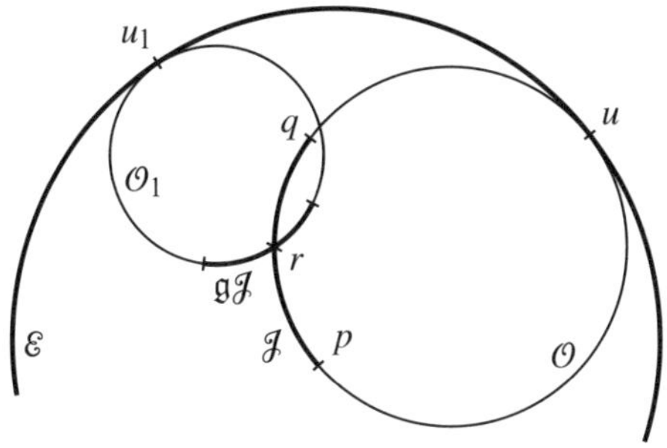

Figure 14.6

From these two auxiliary theorems is inferred that the values $\delta(\mathcal{O}, \mathfrak{g}\mathcal{O})$ have a definite minimum as $\mathfrak{g}$ ranges over a collection of representatives of the left cosets of $\mathfrak{A}_u$ in $\mathfrak{F}$ which make up $\mathfrak{F} \setminus \mathfrak{A}_u$. This minimum is called δ_0. Let now $\mathcal{O}_u$ denote the horocycle with centre u and at the distance $\frac{1}{2}|\delta_0|$ from $\mathcal{O}$ (measured on the lines through u) and situated inside $\mathcal{O}$ or outside $\mathcal{O}$ according as δ_0 is negative or positive. If $\delta = 0$, then $\mathcal{O}_u = \mathcal{O}$. Then $\mathcal{O}_u$ *touches some of its equivalents with respect to* $\mathfrak{F}$*, but is not intersected by any of them.* This implies that $\mathcal{O}_u$ depends only on u and $\mathfrak{F}$, not on the initial choice of a horocycle $\mathcal{O}$. The interior of $\mathcal{O}_u$ is denoted by $\mathcal{D}(\mathfrak{F}; u)$ and called the *characteristic neighbourhood of* u. As in the case of points in $\mathcal{D}$, the groups $\mathfrak{F}$ and $\mathfrak{A}_u$ have equal effect in $\mathcal{D}(\mathfrak{F}; u)$, and $\mathcal{D}(\mathfrak{F}; u)$ is the largest open horocyclical disc belonging to u with that property. As a special consequence $\mathcal{D}(\mathfrak{F}; u)$ cannot contain any centre of rotation of $\mathfrak{F}$, and it is not cut by any reflection line of $\mathfrak{F}$ not terminating at u. Hence $\mathcal{D}(\mathfrak{F}; u)$ consists of regular points except possibly for points on reflection lines terminating at u. It is evident that for any element $\mathfrak{f}$ of $\mathfrak{F}$ one gets $\mathcal{D}(\mathfrak{F}; \mathfrak{f}u) = \mathfrak{f}\mathcal{D}(\mathfrak{F}; u)$.

14.8 Isometric neighbourhood of a limit-centre. Consider any fundamental arc $p_u q_u$ for $\mathfrak{A}_u$ on the horocycle $\mathcal{O}_u$ determined in the preceding section. Let $\mathcal{P}$ and $\mathcal{Q}$ be the straight lines joining u with p_u and q_u respectively, and $p_{\mathcal{E}}$ and $q_{\mathcal{E}}$ their other infinite points. Let $\mathcal{O}$ be any horocycle with centre u situated inside $\mathcal{O}_u$, and p and q the intersection points of $\mathcal{O}$ with $\mathcal{P}$ and $\mathcal{Q}$ respectively. Then qp is a fundamental arc for $\mathfrak{A}_u$ on $\mathcal{O}$. As the distance $\delta(\mathcal{O}, \mathcal{O}_u)$ increases continuously from zero to infinity, $[p, q]$ decreases continuously from $[p_u, q_u]$ to zero. Let $\mathcal{O}_0$ denote the uniquely determined horocycle with centre u for which $[p_0, q_0] = \delta(\mathcal{O}, \mathcal{O}_u)$ (Fig. 14.7), and let $\mathcal{I}_0$ denote the open horocyclical disc bounded by $\mathcal{O}_0$.

Let $\mathfrak{f}$ be any element of $\mathfrak{F} \setminus \mathfrak{A}_u$, thus $\mathfrak{f}u \neq u$. In the above construction, the fundamental arc $p_u q_u$ can be replaced by any of its images by $\mathfrak{A}_u$. We may therefore assume it to be so chosen that $p_{\mathcal{E}} q_{\mathcal{E}}$, which is a fundamental arc for $\mathfrak{A}_u$ on $\mathcal{E}$, contains $\mathfrak{f}u$. From the construction of $\mathcal{O}_u$ in the preceding section follows that $\delta(\mathcal{O}_u, \mathfrak{f}\mathcal{O}_u) \geqq 0$.

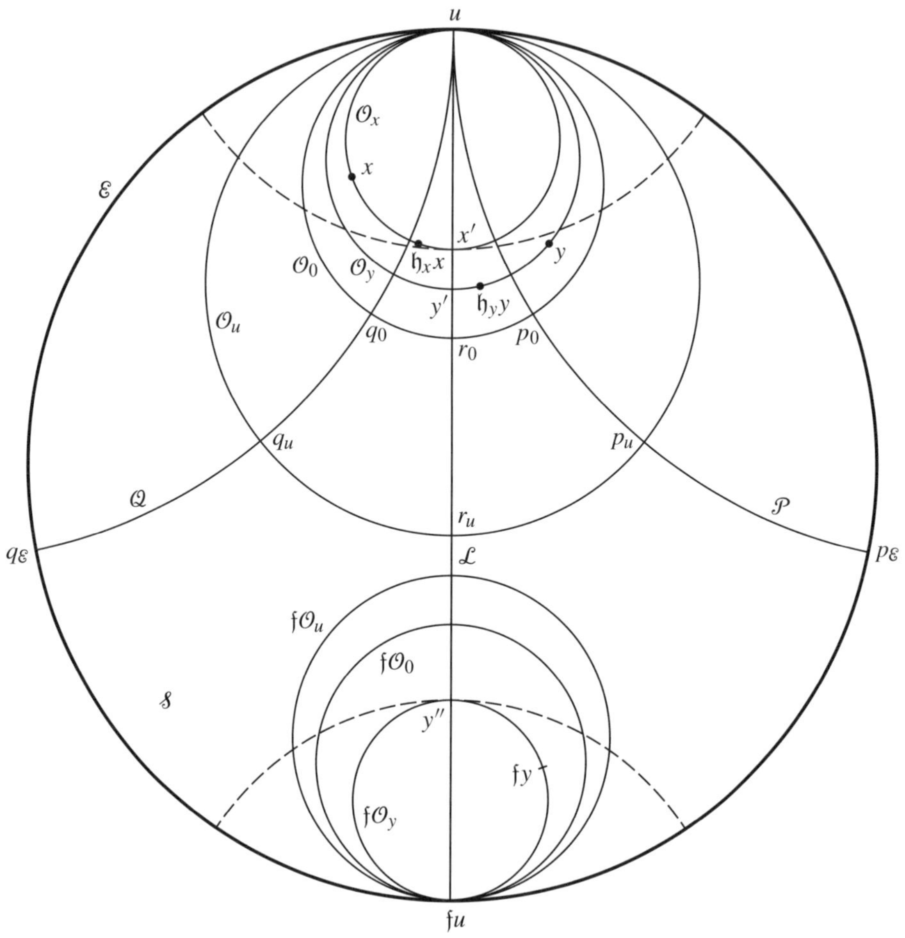

Figure 14.7

Also $\delta(\mathfrak{f}\mathcal{O}_0, \mathfrak{f}\mathcal{O}_u) = \delta(\mathcal{O}_0, \mathcal{O}_u) = [p_0, q_0]$.

Let x and y be any two points in $\mathcal{I}_0$, $\mathcal{O}_x$ and $\mathcal{O}_y$ the horocycles with centre u passing through x and y respectively, and x', y', r_0, r_u the intersection points of the line $\mathcal{L} = u, \mathfrak{f}u$ with $\mathcal{O}_x, \mathcal{O}_y, \mathcal{O}_0, \mathcal{O}_u$ respectively, thus $[r_0, r_u] = \delta(\mathcal{O}_0, \mathcal{O}_u) = [p_0, q_0]$. We choose the notation x and y for the two given points such that $x'r_0 \geqq y'r_0$. Let $\mathfrak{h}_x$ and $\mathfrak{h}_y$ be two elements of $\mathfrak{A}_u$ for which $\mathfrak{h}_x x$ and $\mathfrak{h}_y y$ belong to the sector $\mathcal{S}$ closed on $\mathcal{D}$ bounded by $\mathcal{P}$ and $\mathcal{Q}$. Then

$$\xi(\mathfrak{A}_u; x, y) \leqq [\mathfrak{h}_x x, \mathfrak{h}_y y] \leqq [\mathfrak{h}_x x, x'] + [x', y'] + [y', \mathfrak{h}_y y] < [x', y'] + 2[p_0, q_0]. \tag{8}$$

The point $\mathfrak{f}y$ is situated in $\mathfrak{f}\mathcal{I}_0$, thus inside $\mathfrak{f}\mathcal{O}_0$. Let y'' be the intersection of $\mathcal{L}$ with the horocycle $\mathfrak{f}\mathcal{O}_y$; this horocycle has its centre in $\mathfrak{f}u$ and passes through $\mathfrak{f}y$. Consider the open strip $\mathcal{T}$ of $\mathcal{D}$ bounded by the normals of $\mathcal{L}$ in x' and y''. The distance of its bounding lines is $[x', y'']$. Neither x nor $\mathfrak{f}y$ belongs to $\mathcal{T}$. Hence

$$[x, \mathfrak{f}y] \geqq [x', y''] > [x', y'] + 2[p_0, q_0] \tag{9}$$

since $x'y''$ contains the segments $x'y'$ and r_0r_u and the one between $\mathfrak{f}\mathcal{O}_u$ and $\mathfrak{f}\mathcal{O}_0$, thus of length $[r_0, r_u]$, and these three segments are disjoint.

The last term of the inequality (9) does not depend on $\mathfrak{f}$, since the lengths x', y' and $[p_0, q_0]$ depend on $\mathfrak{A}_u$ and x and y only. Comparison of (8) and (9) now shows that the equation

$$\xi(\mathfrak{F}; x, y) = \xi(\mathfrak{A}_u; x, y) \tag{10}$$

holds for any pair of points x, y in $\mathcal{I}_0$. In analogy to (4) we establish the following definition:

If u is any limit-centre of $\mathfrak{F}$, and $\mathfrak{A}_u$ denotes the subgroup of $\mathfrak{F}$ leaving u invariant, we define as *isometric neighbourhood of* u and denote by $\mathcal{I}(\mathfrak{F}; u)$ the largest open horocyclical disc with centre u for which the equation

$$\xi(\mathfrak{F}; x_1, x_2) = \xi(\mathfrak{A}_u; x_1, x_2) \tag{11}$$

holds for any two points x_1 and x_2 of the disc. Let $\mathcal{K}_u$ denote its bounding horocycle. Equation (10) shows that such a neighbourhood exists: The above $\mathcal{I}_0$ is part of $\mathcal{I}(\mathfrak{F}; u)$.

For every $\mathfrak{F}$-group $\mathfrak{F}$, and every limit-centre u of it, $\mathcal{I}(\mathfrak{F}; u)$ is contained in $\mathcal{D}(\mathfrak{F}; u)$.

Proof. Since $\mathcal{O}_u$ touches some horocycles of its equivalence class, let $\mathfrak{g}'$ be such an element of $\mathfrak{F} \setminus \mathfrak{A}_u$ that $\mathfrak{g}'\mathcal{O}_u$ touches $\mathcal{O}_u$, say in z. Then $\mathfrak{g}'^{-1}z$ is a point of $\mathcal{O}_u$. For every element $\mathfrak{h}$ of $\mathfrak{A}_u$ one gets $\mathfrak{g}'\mathfrak{h}\mathcal{O}_u = \mathfrak{g}'\mathcal{O}_u$, and $\mathfrak{h}^{-1}\mathfrak{g}'^{-1}z$ is thus a point of $\mathcal{O}_u$. For suitably chosen $\mathfrak{h}$ then $\mathfrak{h}^{-1}\mathfrak{g}'^{-1}z$ and z represent the distance of their equivalence classes with respect to $\mathfrak{A}_u$. In other words: There is in the left coset of $\mathfrak{A}_u$ in $\mathfrak{F}$ to which $\mathfrak{g}'$ belongs, an element $\mathfrak{g}$ such that $\mathfrak{g}\mathcal{O}_u$ touches $\mathcal{O}_u$ in z, and such that

$$\xi(\mathfrak{A}_u; z, g^{-1}z) = [z, \mathfrak{g}^{-1}z]. \tag{12}$$

Assume first that $\mathfrak{g}^{-1}z$ and z do not coincide; they are then not equivalent with respect to $\mathfrak{A}_u$. Since they are equivalent with respect to $\mathfrak{F}$, one gets

$$\xi(\mathfrak{F}; z, \mathfrak{g}^{-1}z) = 0, \tag{13}$$

while $\xi(\mathfrak{A}_u; z, \mathfrak{g}^{-1}z) > 0$. Hence (11), (12), (13) show that $\mathcal{O}_u$ is not contained in $\mathcal{I}(\mathfrak{F}; u)$. Moreover, since the functions ξ in (12) and (13) are continuous, if $x_{1\nu}$ and $x_{2\nu}$ are sequences of points in $\mathcal{D}(\mathfrak{F}; u)$ convergent to z and $\mathfrak{g}^{-1}z$, respectively, then (11) cannot hold throughout the sequences, since its left side tends to zero and the right one to a positive value. From this follows that the bounding horocycle $\mathcal{K}(u)$ of $\mathcal{I}(\mathfrak{F}; u)$ belongs to $\mathcal{D}(\mathfrak{F}; u)$, thus that $\mathcal{I}(\mathfrak{F}; u)$ is a proper subset of $\mathcal{D}(\mathfrak{F}; u)$.

We now assume that z and $\mathfrak{g}^{-1}z$ coincide, thus $\mathfrak{g} \in \mathfrak{A}_z$. If z is not situated on a reflection line of $\mathfrak{F}$, then $\mathfrak{A}_z = \mathfrak{A}(\odot 2)$, since $\mathcal{O}_u$ is not cut by any of its equivalents. Thus $\mathfrak{g}$ is the half-turn contained in $\mathfrak{A}_z$. Since z is not on a reflection line of $\mathfrak{F}$, one may choose two points x and y on $\mathcal{O}_u$ such that they are separated by z but by no reflection line of $\mathfrak{A}_u$, if any, that they represent the distance of their equivalence classes with respect to $\mathfrak{A}_u$, and that the angle $\measuredangle(zx,\ zy)$ is obtuse. In fact, these conditions are satisfied, if both x and y are sufficiently near to z and separated by z. Then $\mathfrak{g}y$ is on $\mathfrak{g}\mathcal{O}_u$, and $[z, \mathfrak{g}y] = [z, y]$. Moreover, the angle $\measuredangle(z(\mathfrak{g}y),\ zx) = \pi - \measuredangle(zx,\ zy)$ is acute (Fig. 14.8).

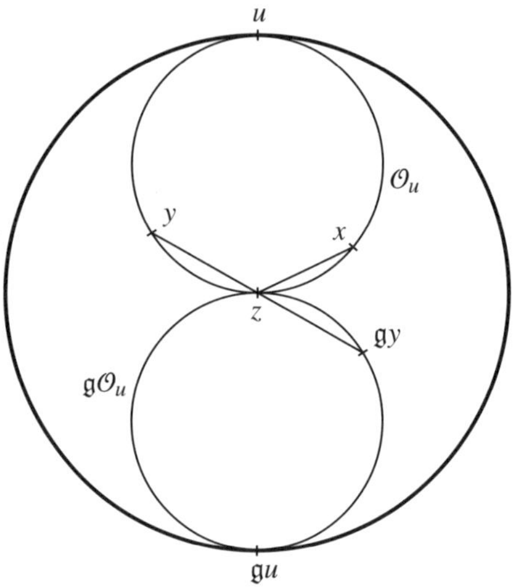

Figure 14.8

Hence

$$\xi(\mathfrak{A}_u;\, x,\, y) = [x,\, y],$$

while

$$\xi(\mathfrak{F};\, x,\, y) \leqq [x,\, \mathfrak{g}y] < [x,\, y]$$

in virtue of formula (IIIa) of §8.7. Again continuity shows that (11) cannot continue to hold for sequences $x_{1\nu}$ and $x_{2\nu}$ of points in $\mathcal{D}(\mathfrak{F};\, u)$ convergent to x and y, respectively, since $\xi(\mathfrak{A}_u;\, x_{1\nu},\, x_{2\nu})$ and $\xi(\mathfrak{F};\, x_{1\nu},\, x_{2\nu})$ tend to different limits for $\nu \to \infty$. Hence even in this case $\mathcal{I}(\mathfrak{F};\, u)$ is a proper subset of $\mathcal{D}(\mathfrak{F};\, u)$.

Finally, let $z = \mathfrak{g}^{-1}z$ be situated on a reflection line of $\mathfrak{F}$. If $\mathcal{M} = zu$ is not a reflection line of $\mathfrak{F}$, then the reflection line of $\mathfrak{F}$ containing z must be the tangent $\mathcal{L}$ of $\mathcal{O}_u$ in z, because $\mathcal{O}_u$ is not cut by any reflection line not terminating at u. If $\mathcal{M}$ is a reflection line of $\mathfrak{F}$, then the element $\mathfrak{g} \in \mathfrak{A}_z$ which carries $\mathcal{O}_u$ into $\mathfrak{g}\mathcal{O}_u$ is not the reflection in $\mathcal{M}$, because $\mathfrak{g}\mathcal{O}_u$ is different from $\mathcal{O}_u$; hence $\mathfrak{g}$ is either the reflection in

$\mathcal{L}$ or the half-turn about z; but even in the latter case, $\mathfrak{A}_z$ contains the reflection in $\mathcal{L}$. One can thus in all cases take $\mathfrak{g}$ to be the reflection in $\mathcal{L}$, and $\mathfrak{A}_z$ is either $\mathfrak{A}(-)$ or $\mathfrak{A}(\times 2)$. Let x (Fig. 14.9) be a point on the half-line uz, and y a point on the half-line

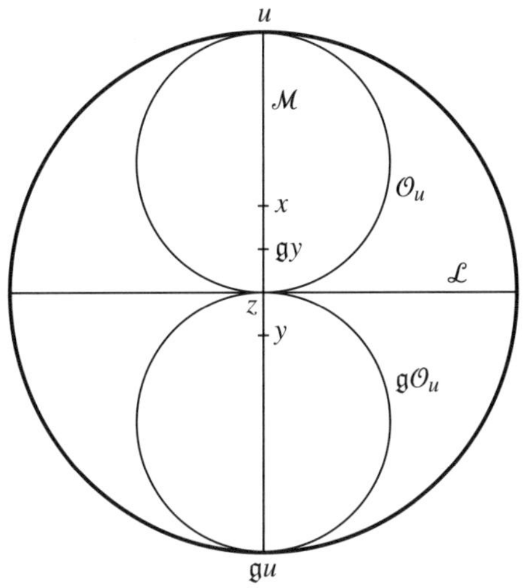

Figure 14.9

$z, \mathfrak{g}u$. Then $\mathfrak{g}y$ is on uz. Hence

$$\xi(\mathfrak{F}; x, y) \leqq [x, \mathfrak{g}y] < [x, y] = \xi(\mathfrak{A}_u; x, y),$$

showing that $\mathcal{O}_u$ cannot be contained in the open disc $\mathcal{I}(\mathfrak{F}; u)$. Hence $\mathcal{I}(\mathfrak{F}; u) \subset \mathcal{D}(\mathfrak{F}; u)$. This completes the proof of the theorem. □

$\mathcal{I}(\mathfrak{F}; u)$ *and* $\mathcal{D}(\mathfrak{F}; u)$ *may coincide.*

Proof. Consider the group $\mathfrak{F}$ illustrated in Fig. 14.3. Let u be the common end-point of $\mathcal{L}_1$ and $\mathcal{L}_2$, thus a limit-centre of $\mathfrak{F}$. The corresponding subgroup of $\mathfrak{F}$ is $\mathfrak{A}_u(\wedge)$ generated by $\mathfrak{s}_1$ and $\mathfrak{s}_2$. Consider the horocycle with limit-centre u touching $\mathcal{L}_3$ (Fig. 14.10) and the open horocyclical disc bounded by it. The horocycle touches its image by $\mathfrak{s}_3$, and thus all those derived from it by $\mathfrak{A}_u$, but the horocycle is not cut by any of its equivalents with respect to $\mathfrak{F}$, since $\mathcal{T}$ is a fundamental polygon for $\mathfrak{F}$. Hence $\mathfrak{F}$ and $\mathfrak{A}_u$ have equal effect in the disc, the disc is $\mathcal{D}(\mathfrak{F}; u)$, and its bounding horocycle is $\mathcal{O}_u$. – Let $\mathcal{S}$ denote the sector cut out from $\mathcal{D}(\mathfrak{F}; u)$ by $\mathcal{L}_1$ and $\mathcal{L}_2$. Let x_1 and x_2 be any two points of $\mathcal{D}(\mathfrak{F}; u)$. Let x_1' and x_2', respectively, be the points in $\mathcal{S}$ derived from them by $\mathfrak{A}_u(\wedge)$; they are uniquely determined, because $\mathcal{S}$ is a fundamental domain for $\mathfrak{A}_u$ in $\mathcal{D}(\mathfrak{F}; u)$. Then

$$\xi(\mathfrak{F}; x_1, x_2) = \xi(\mathfrak{F}; x_1', x_2') = [x_1', x_2'],$$

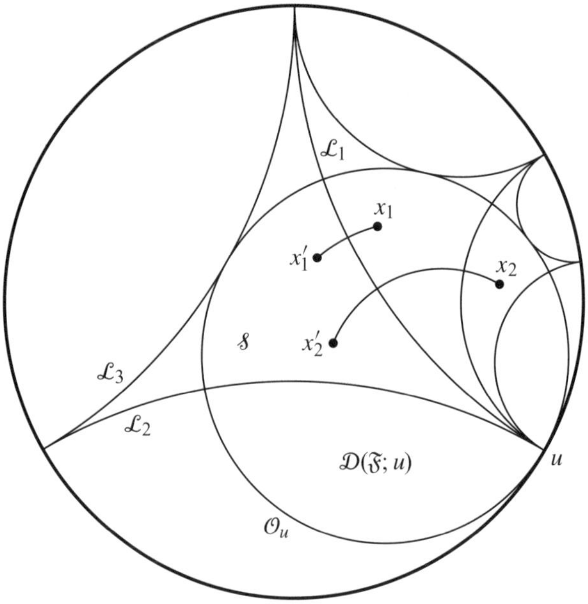

Figure 14.10

because x_2' is contained in $\mathcal{N}(\mathfrak{F}; x_1') = \tilde{\mathcal{T}}$ (Section 6). This shows that $\xi(\mathfrak{F}; x_1, x_2)$ is not smaller than $\xi(\mathfrak{A}_u; x_1, x_2)$ for any two points of $\mathcal{D}(\mathfrak{F}; u)$. Hence $\mathcal{I}(\mathfrak{F}; u) = \mathcal{D}(\mathfrak{F}; u)$, and $\mathcal{O}_u = \mathcal{K}_u$. □

We restate the main results of the last two sections:

For any $\mathfrak{F}$-group $\mathfrak{F}$ containing limit-rotations, and any limit-centre u of $\mathfrak{F}$, if $\mathfrak{A}_u$ is the subgroup of $\mathfrak{F}$ leaving u fixed, then the characteristic neighbourhood $\mathcal{D}(\mathfrak{F}; u)$ of u is the largest open horocyclical disc with centre u in which $\mathfrak{F}$ and $\mathfrak{A}_u$ have equal effect, and the isometric neighbourhood $\mathcal{I}(\mathfrak{F}; u)$ of u is the largest open horocyclical disc with centre u such that distance modulo $\mathfrak{F}$ coincides with distance modulo $\mathfrak{A}_u$ for any two of its points. $\mathcal{I}(\mathfrak{F}; u)$ is contained in $\mathcal{D}(\mathfrak{F}; u)$ and may, in particular cases, coincide with $\mathcal{D}(\mathfrak{F}; u)$. □

14.9 Centre free part of $\mathcal{K}(\mathfrak{F})$. Let c be any centre of $\mathfrak{F}$. As shown in Section 1, a circle with centre c only meets a finite number of sides of $\mathcal{K}(\mathfrak{F})$. Thus, if at all $\mathcal{K}(\mathfrak{F})$ has sides in $\mathcal{D}$, there is a side of $\mathcal{K}(\mathfrak{F})$ nearest to c, thus determining the distance $\delta(c)$ of c from the boundary of $\mathcal{K}(\mathfrak{F})$ in $\mathcal{D}$. Let $\rho(c)$ be chosen as a positive number smaller than $\frac{1}{4}\eta(c)$ and also smaller than $\delta(c)$, and this for every centre c of $\mathfrak{F}$. Since both $\eta(c)$ and $\delta(c)$ only depend on the equivalence class of c, we choose the same ρ

for all centres in the equivalence class of c. Let $\mathcal{Q}(c)$ denote the open circular disc bounded by the circle $\mathcal{C}(c; \rho(c))$. In virtue of (5) $\mathcal{Q}(c)$ is contained in $\mathcal{I}(\mathfrak{F}; c)$ as a proper subset, thus also in $\mathcal{D}(\mathfrak{F}; c)$.

No two of the discs $\tilde{\mathcal{Q}}$ have common points.

Proof. Assume that the closed discs $\tilde{\mathcal{Q}}(c_1)$ and $\tilde{\mathcal{Q}}(c_2)$ had points in common. Let the notation be such that $\eta(c_1) \geqq \eta(c_2)$. Then

$$[c_1, c_2] \leqq \rho(c_1) + \rho(c_2) < \tfrac{1}{4}\eta(c_1) + \tfrac{1}{4}\eta(c_2) \leqq \tfrac{1}{2}\eta(c_1),$$

thus $c_2 \in \mathcal{D}(\mathfrak{F}; c_1)$ in contradiction to the fact that the characteristic neighbourhood $\mathcal{D}(\mathfrak{F}; c_1)$ contains no centre except c_1. □

Hence the collection of discs $\tilde{\mathcal{Q}}(c)$ for all centres c of $\mathfrak{F}$ are disjoint. If the $\mathcal{Q}(c)$ are omitted from $\tilde{\mathcal{K}}(\mathfrak{F})$ for all centres c of $\mathfrak{F}$, the rest of $\tilde{\mathcal{K}}(\mathfrak{F})$, is a region closed on $\mathcal{D}$ with all boundaries of $\tilde{\mathcal{K}}(\mathfrak{F})$ intact, and it is reproduced by $\mathfrak{F}$. Also those reflection lines of $\mathfrak{F}$ which contain no centre are left intact, since a reflection line which has points in common with a $\tilde{\mathcal{Q}}(c)$, and thus with the corresponding $\mathcal{D}(\mathfrak{F}; c)$, is known to pass through c.

14.10 Truncated domain of an $\mathfrak{F}$-group. For $\mathfrak{F}$-groups containing limit-rotations, a construction similar to the one carried out for centres in Section 9 can be performed for limit-centres. Let u be a limit-centre of $\mathfrak{F}$, $\mathcal{O}_u$ the bounding horocycle for $\mathcal{D}(\mathfrak{F}; u)$, $p_u q_u$ a fundamental arc for $\mathfrak{A}_u$ on $\mathcal{O}_u$ as in Fig. 14.7. Besides $\mathcal{O}_u$ there exists exactly one more horocycle passing through p_u and q_u, namely the horocycle $\mathcal{O} = \mathfrak{s}\mathcal{O}_u$ obtained from $\mathcal{O}_u$ by the reflection $\mathfrak{s}$ in the straight line $p_u q_u$ (Fig. 14.11). It touches $\mathcal{E}$ in $\mathfrak{s}u$. The reflection $\mathfrak{s}$ does not belong to $\mathfrak{F}$, because its line cuts $\mathcal{D}(\mathfrak{F}; u)$ without terminating at u. The line $u, \mathfrak{s}u$ is the bisecting normal of the segment $p_u q_u$, since the reflection in it reproduces $\mathcal{O}_u$ and $\mathcal{O}'$. The distance of $\mathcal{O}_u$ and $\mathcal{O}'$ is negative according to the definition in Section 7; it is measured on the line $u, \mathfrak{s}u$. Denote its absolute value by $\delta(u) = -\delta(\mathcal{O}_u, \mathcal{O}')$.

Any horocycle which cuts out on $\mathcal{O}_u$ an arc smaller than $p_u q_u$ has a numerical distance smaller than $\delta(u)$ from $\mathcal{O}_u$. A line which cuts out on $\mathcal{O}_u$ an arc smaller than or equal to $p_u q_u$ has a numerical distance $\leqq \frac{1}{2}\delta(u)$ from $\mathcal{O}_u$.

For every limit-centre u of $\mathfrak{F}$ let $\mathcal{O}^*(u)$ denote a horocycle with u as its centre, situated inside $\mathcal{O}_u$ at a distance numerically greater than $\frac{1}{2}\delta(u)$, the same distance being chosen for equivalent limit-centres. Then *$\mathcal{O}^*(u)$ does not meet any side of $\mathcal{K}(\mathfrak{F})$*. For, let $\mathcal{S}$ be such a side, and suppose that $\mathcal{S}$ intersects $\mathcal{O}_u$. Since $\mathcal{S}$ does not terminate at u, it has a certain negative distance from $\mathcal{O}_u$, and the absolute value of that distance is $\leqq \frac{1}{2}\delta(u)$. This follows immediately from the fact (§14.2) that no two points of the closed half-plane cut off by $\mathcal{S}$ are equivalent by an element of $\mathfrak{A}_u$, and therefore the arc of $\mathcal{O}_u$ cut off by $\mathcal{S}$ is smaller than the arc $p_u q_u$.

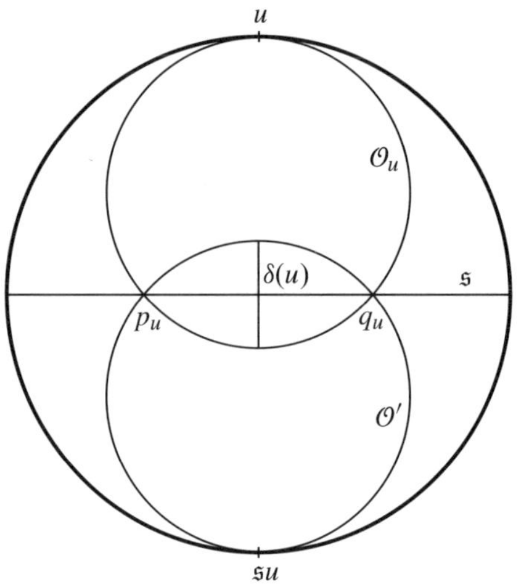

Figure 14.11

Moreover,

Any two of these horocycles $\mathcal{O}^*$ have no point in common.

Proof. For an indirect proof, suppose that $\mathcal{O}^*(u_1)$ and $\mathcal{O}^*(u_2)$ have a common point and let $\delta(u_1) \geqq \delta(u_2)$; the limit-centres u_1 and u_2 may or may not be equivalent. Then (Fig. 14.12)

$$|\delta(\mathcal{O}_{u_1}, \mathcal{O}_{u_2})| > \tfrac{1}{2}\delta(u_1) + \tfrac{1}{2}\delta(u_2) \geqq \delta(u_2).$$

Therefore the arc of $\mathcal{O}_{u_2}$ cut out by $\mathcal{O}_{u_1}$ contains a fundamental arc of $\mathcal{O}_{u_2}$ as a proper subset, and there are therefore inside $\mathcal{O}_{u_1}$ two points equivalent with respect to $\mathfrak{A}_{u_2}$ in contradiction to the fact that $\mathcal{O}_{u_1}$ bounds the characteristic neighbourhood $\mathcal{D}(\mathfrak{F}; u_1)$ of u_1. □

We summarize these result:

The system of horocycles $\mathcal{O}^*(u)$ is situated in the interior of $\mathcal{K}(\mathfrak{F})$ and any two horocycles of the system have no point in common. No centre of $\mathfrak{F}$ is situated on or inside any $\mathcal{O}^*$, and no reflection line of $\mathfrak{F}$ meets any $\mathcal{O}^*$, except when it ends in its limit-centre. □

If the interior $\mathcal{Q}(u)$ of all the horocycles $\mathcal{O}^*(u)$ is omitted from $\mathcal{K}(\mathfrak{F})$, one gets a domain bounded in $\mathcal{D}$ by lines and horocycles. It will be called a *truncated domain of* $\mathfrak{F}$ and denoted by $\mathcal{K}^*(\mathfrak{F})$. In case there are no limit-rotations in $\mathfrak{F}$, $\mathcal{K}^*(\mathfrak{F})$ means the same as $\mathcal{K}(\mathfrak{F})$. All elements of $\mathfrak{F}$ carry $\mathcal{K}^*(\mathfrak{F})$ into itself. $\mathcal{K}^*(\mathfrak{F})$ contains all centres

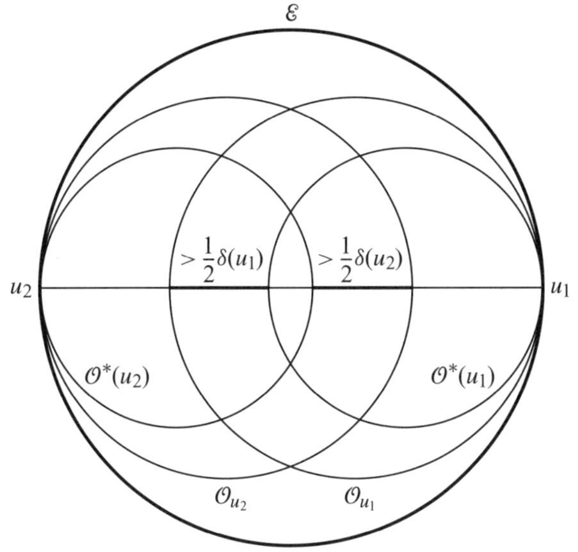

Figure 14.12

of $\mathfrak{F}$. A reflection line of $\mathfrak{F}$ is either wholly situated in $\mathcal{K}^*(\mathfrak{F})$ or cuts the boundary of $\mathcal{K}^*(\mathfrak{F})$ at right angles; a reflection of $\mathfrak{F}$ carries a bounding horocycle into itself or into another bounding horocycle; in the first case the reflection line cuts the horocycle at right angles and thus terminates at its centre. Every axis of $\mathfrak{F}$ has points in common with $\mathcal{K}^*(\mathfrak{F})$; if it cuts a bounding horocycle, the intersection cannot be at right angles, since the axis cannot terminate at a limit-centre (§11.4).

A circle $\mathcal{C}$ in $\mathcal{D}$ can only meet a finite number of bounding horocycles of $\mathcal{K}^*(\mathfrak{F})$.

Proof. This is most easily shown in the language of euclidean geometry: The bounding horocycles then are circles which touch the circle $\mathcal{E}$ inside and are situated outside each other. Those which meet a fixed circle $\mathcal{C}$ situated in $\mathcal{D}$ have a positive lower bound for their radius. Thus their number must be finite. □

§15 Quasi-compactness modulo $\mathfrak{F}$ and finite generation of $\mathfrak{F}$

15.1 Quasi-compactness and compactness modulo $\mathfrak{F}$. The convex domain $\mathcal{K}(\mathfrak{F})$ of $\mathfrak{F}$ is called *quasi-compact modulo* $\mathfrak{F}$, if there exists a polygon $\mathcal{P}$ with a finite number of vertices situated in $\mathcal{D}$ or in limit-centres of $\mathfrak{F}$ such that for any point $x \in \mathcal{K}(\mathfrak{F})$ there is in the interior of $\mathcal{P}$ at least one point equivalent with x with respect to $\mathfrak{F}$. If there are no limit-rotations in $\mathfrak{F}$, then all vertices of $\mathcal{P}$ are in $\mathcal{D}$; the convex domain

$\mathcal{K}(\mathfrak{F})$ is then called *compact modulo* $\mathfrak{F}$. In this latter case the condition may be thus stated: There exists in $\mathcal{D}$ an open circular disc $\mathcal{Q}$ containing for every point of $\mathcal{K}(\mathfrak{F})$ at least one equivalent point.

A truncated domain $\mathcal{K}^*(\mathfrak{F})$ of $\mathfrak{F}$ is called *compact modulo* $\mathfrak{F}$, if there exists in $\mathcal{D}$ an open circular disc $\mathcal{Q}^*$ such that for any point $x \in \tilde{\mathcal{K}}^*(\mathfrak{F})$ there is in $\mathcal{Q}^*$ at least one point equivalent with x with respect to $\mathfrak{F}$.

If there are no limit-rotations in $\mathfrak{F}$, this definition only means that $\tilde{\mathcal{K}}(\mathfrak{F})$ is compact modulo $\mathfrak{F}$. It is the aim of this section it show that *the compactness modulo $\mathfrak{F}$ of $\tilde{\mathcal{K}}^*(\mathfrak{F})$ and the quasi-compactness modulo $\mathfrak{F}$ of $\tilde{\mathcal{K}}(\mathfrak{F})$ are equivalent concepts*; the existence of limit-rotations in $\mathfrak{F}$ is, therefore, assumed. For brevity, the addition "modulo $\mathfrak{F}$" is omitted if that can cause no mistake. The attention is drawn to the fact that, whereas the convex domain of $\mathfrak{F}$ is uniquely determined, the concept of truncated domain carries with it a certain arbitrariness, since the bounding horocycles $\mathcal{O}^*(u)$ of $\mathcal{K}^*(\mathfrak{F})$ are not uniquely determined.

If the convex domain $\tilde{\mathcal{K}}(\mathfrak{F})$ is quasi-compact, then a truncated domain $\tilde{\mathcal{K}}^*(\mathfrak{F})$ is compact.

Proof. The intersection of $\tilde{\mathcal{K}}^*(\mathfrak{F})$ with a polygon $\mathcal{P}$ derived from the definition of quasi-compactness contains an equivalent point to every point of $\tilde{\mathcal{K}}^*(\mathfrak{F})$; for there is such point inside $\mathcal{P}$ and, since $\tilde{\mathcal{K}}^*(\mathfrak{F})$ is reproduced by $\mathfrak{F}$, it must belong to the said intersection. Since the infinite cusps of $\mathcal{P}$ at limit-centres are finite in number and are cut off by some of the bounding horocycles of $\mathcal{K}^*(\mathfrak{F})$, the intersection of $\mathcal{K}^*(\mathfrak{F})$ and $\mathcal{P}$ is bounded, thus contained in an open circular disc $\mathcal{Q}^*$, which proves the assertion. □

If a truncated domain is compact, then the convex domain is quasi-compact.

Proof. $\mathcal{Q}^*$ being an open circular disc corresponding to the definition of compactness of $\mathcal{K}^*(\mathfrak{F})$, a polygon $\mathcal{P}$ corresponding to the definition of quasi-compactness of $\mathcal{K}(\mathfrak{F})$ has to be constructed. Since the collection of bounding horocycles of $\mathcal{K}^*(\mathfrak{F})$ is reproduced by $\mathfrak{F}$, some of these horocycles must have an arc inside $\mathcal{Q}^*$; on the other hand, this can only be the case for a finite number of bounding horocycles, since the circle bounding $\mathcal{Q}^*$ can only meet a finite number of these horocycles (§14.10). Let $\mathcal{O}^*(u)$ with limit-centre u be one of the bounding horocycles of $\mathcal{K}^*(\mathfrak{F})$ which meets the boundary of $\mathcal{Q}^*$ in two points p and q. If half-lines are drawn from p and q to u, a cuspidal sector inside $\mathcal{O}^*(u)$ arises. The part of it which is outside $\mathcal{Q}^*$ is added to $\mathcal{Q}^*$. This construction is performed for the finite number of bounding horocycles intersecting $\mathcal{Q}^*$. Let $\mathcal{M}$ denote the point set consisting of $\mathcal{Q}^*$ and the added cuspidal sectors. *Then $\mathcal{M}$ contains an equivalent point to every point of $\mathcal{K}(\mathfrak{F})$.* From the outset this is evident for every point of $\mathcal{K}^*(\mathfrak{F})$, since $\mathcal{M}$ contains $\mathcal{Q}^*$. If x is a point of $\mathcal{K}(\mathfrak{F})$ outside $\mathcal{K}^*(\mathfrak{F})$, let $\mathcal{O}^*(u)$ be the bounding horocycle in the interior of which it is situated, u being the centre of $\mathcal{O}^*(u)$ (Fig. 15.1). Let x' be the intersection point of the line ux with $\mathcal{O}^*(u)$. Since x' belongs to $\mathcal{K}^*(\mathfrak{F})$, there is an element $\mathfrak{f}$ of $\mathfrak{F}$ for which $\mathfrak{f}x'$ belongs to $\mathcal{Q}^*$. Now $\mathfrak{f}$ carries $\mathcal{K}^*(\mathfrak{F})$ into itself, and since x' is a boundary point of

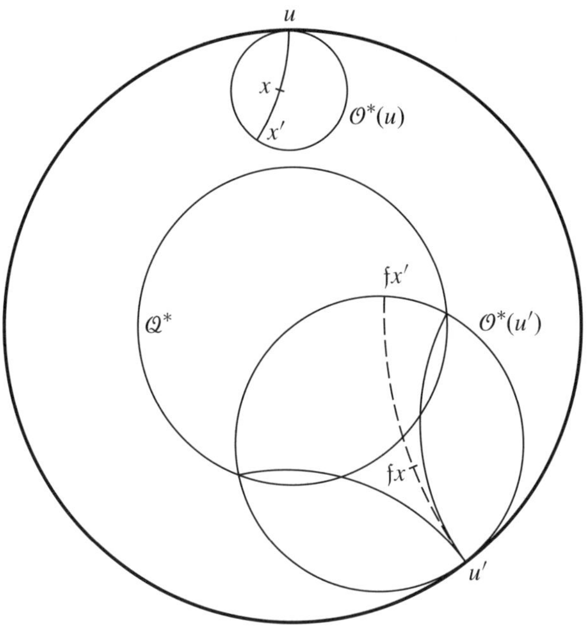

Figure 15.1

$\mathcal{K}^*(\mathfrak{F})$ on one of its bounding horocycles the same is true of $\mathfrak{f}x'$. Hence $\mathfrak{f}x'$ is situated on a bounding horocycle $\mathcal{O}^*(u')$ of $\mathcal{K}^*(\mathfrak{F})$ which may be the same as $\mathcal{O}^*(u)$ or not. Then $\mathfrak{f}\mathcal{O}^*(u) = \mathcal{O}^*(u')$, for $\mathfrak{f}$ carries $\mathcal{O}^*(u)$ into a bounding horocycle, and $\mathfrak{f}\mathcal{O}^*(u)$ and $\mathcal{O}^*(u')$ have the point $\mathfrak{f}x'$ in common. $\mathfrak{f}$ carries the half-line ux' into the half-line joining the centre $\mathfrak{f}u = u'$ of $\mathcal{O}^*(u')$ with the point $\mathfrak{f}x'$ of $\mathcal{Q}^*$, and thus situated in the cuspidal sector determined in $\mathcal{O}^*(u')$. The point $\mathfrak{f}x$ is situated on that half-line, thus in the sector, and thus in $\mathcal{M}$. Evidently, the set $\mathcal{M}$ can be included in a polygon $\mathcal{P}$ with a finite number of vertices among which are the limit-centres used as vertices of the sectors. Hence $\mathcal{K}(\mathfrak{F})$ is quasi-compact. □

The equivalence of the two definitions at the head of this section has thus been proved. Moreover, it is seen that the property of a truncated domain of being compact does not depend on the choice of the bounding horocycles $\mathcal{O}^*(u)$.

15.2 Some consequences of quasi-compactness. In this section some consequences of the property of compactness of $\mathcal{K}^*(\mathfrak{F})$ are pointed out. Let a group $\mathfrak{F}$ be given, and let $\mathcal{K}^*(\mathfrak{F})$ be a truncated domain compact modulo $\mathfrak{F}$, every point to $\mathcal{K}^*(\mathfrak{F})$ having at least one equivalent in the open circular disc $\mathcal{Q}^*$. It is remembered that $\mathcal{K}^*(\mathfrak{F})$ means $\mathcal{K}(\mathfrak{F})$ itself in case there are no limit-rotations in $\mathfrak{F}$.

Since $\mathcal{K}^*(\mathfrak{F})$ contains all centres of $\mathfrak{F}$, every class of centres of $\mathfrak{F}$ is represented in $\mathcal{Q}^*$ by at least one centre. Since the centres do not accumulate in $\mathcal{D}$, this implies:

The number of equivalence classes of centres of $\mathfrak{F}$ *is finite.* □

There is a bounding horocycle of $\mathcal{K}^*(\mathfrak{F})$ corresponding to every limit-centre of $\mathfrak{F}$, and equivalent limit-centres correspond to equivalent bounding horocycles. In every equivalence class of bounding horocycles there must be at least one intersecting $\mathcal{Q}^*$. On the other hand only a finite number of bounding horocycles intersect $\mathcal{Q}^*$ (§14.10). Hence:

The number of equivalence classes of limit-centres of $\mathfrak{F}$ *is finite.*

Every reflection line of $\mathfrak{F}$ has points in common with $\mathcal{K}^*(\mathfrak{F})$; for it is either entirely contained in $\mathcal{K}^*(\mathfrak{F})$, or it intersects the boundary of $\mathcal{K}^*(\mathfrak{F})$ at right angles (§14.10). Thus in every class of reflection lines there is at least one which intersects $\mathcal{Q}^*$. Since the reflection lines do not accumulate in $\mathcal{D}$, we conclude that:

The number of equivalence classes of reflection lines of $\mathfrak{F}$ *is finite.* □

Let $\mathcal{T}_1, \mathcal{T}_2, \ldots, \mathcal{T}_n$ denote the segments situated in $\mathcal{Q}^*$ of the finite number of reflection lines which intersect $\mathcal{Q}^*$. The part of a reflection line $\mathcal{L}$ situated in $\mathcal{K}^*(\mathfrak{F})$ can be completely covered with segments each of which is equivalent to one of the $\mathcal{T}_\nu$, $1 \leqq \nu \leqq n$. (This is even true if the $\mathcal{T}_\nu$ are replaced by the part which they have in common with the intersection of $\mathcal{K}^*(\mathfrak{F})$ and $\mathcal{Q}^*$.) Now two possibilities occur: Either at least one of the $\mathcal{T}_\nu$ is used more than once in this covering, its equivalent being placed in two different parts of $\mathcal{L}$ and in the same direction on $\mathcal{L}$, or this is not the case. In the first case there is an element of $\mathfrak{F}$ which makes $\mathcal{L}$ slide in itself, and $\mathcal{L}$ is then an axis of $\mathfrak{F}$, and thus wholly inside $\mathcal{K}(\mathfrak{F})$. Moreover, $\mathcal{L}$ is wholly inside $\mathcal{K}^*(\mathfrak{F})$, for otherwise it would cut a bounding horocycle of $\mathcal{K}^*(\mathfrak{F})$ at right angles, since it is a reflection line, and an axis cannot terminate in a limit-centre. Conversely, if $\mathcal{L}$ is an axis of $\mathfrak{F}$, the first case must occur, for $\mathcal{L}$ is wholly inside $\mathcal{K}^*(\mathfrak{F})$, and the total length of $\mathcal{T}_1, \ldots, \mathcal{T}_n$ is finite. – In the second case, therefore, $\mathcal{L}$ is not an axis of $\mathfrak{F}$, and the intersection of $\mathcal{L}$ with $\mathcal{K}^*(\mathfrak{F})$ must be finite. The line $\mathcal{L}$ is then the common perpendicular of two sides, or two bounding horocycles, or of a side and a bounding horocycle of $\mathcal{K}^*(\mathfrak{F})$.

Every side of $\mathcal{K}(\mathfrak{F})$ is at the same time a side of $\mathcal{K}^*(\mathfrak{F})$. For each side there must, therefore, be an equivalent intersecting $\mathcal{Q}^*$. Since this can only occur for a finite number of sides (§14.1), we conclude that:

The number of equivalence classes of sides of $\mathcal{K}(\mathfrak{F})$ *is finite.* □

Let $\mathcal{T}_1, \mathcal{T}_2, \ldots, \mathcal{T}_n$ denote the segments situated in $\mathcal{Q}^*$ of the finite number of sides of $\mathcal{K}(\mathfrak{F})$ which intersect $\mathcal{Q}^*$. Every side $\mathcal{S}$ of $\mathcal{K}(\mathfrak{F})$ can be completely covered with segments each of which is equivalent to one of the $\mathcal{T}_\nu$. Since $\mathcal{S}$ is infinite, at least one of the $\mathcal{T}_\nu$ must be used repeatedly. There is, therefore, an element of $\mathfrak{F}$ which makes $\mathcal{S}$ slide in itself. Hence:

Every side of $\mathcal{K}(\mathfrak{F})$ is a boundary axis. Thus $\tilde{\mathcal{K}}(\mathfrak{F}) = \mathcal{K}(\mathfrak{F})$. □

Let λ be a positive number and $\mathfrak{f}$ a translation in $\mathfrak{F}$ with a displacement $\lambda(\mathfrak{f}) < \lambda$. Since $\mathcal{A}(\mathfrak{f})$ has points in common with $\mathcal{K}^*(\mathfrak{F})$, there is in the class of elements conjugate to $\mathfrak{f}$ at least one element, say $\mathfrak{f}'$, the axis $\mathcal{A}(\mathfrak{f}')$ of which intersects $\mathcal{Q}^*$. By $\mathfrak{f}'$ the centre x^* of $\mathcal{Q}^*$ is displaced by the amount

$$[x^*, \mathfrak{f}'x^*] < 2\rho^* + \lambda(\mathfrak{f}') = 2\rho^* + \lambda(\mathfrak{f}) < 2\rho^* + \lambda,$$

ρ^* being the radius of the circle bounding $\mathcal{Q}^*$. Since only a finite number of elements of $\mathfrak{F}$ displace x^* less than $2\rho^*+\lambda$, only a finite number of classes of conjugate translations with displacements smaller than λ can exist, and there can be no finite accumulation value for the displacements belonging to $\mathfrak{F}$. This remains true if reversed translations are taken into account, since the square of a reversed translation is a translation. In particular among the displacements of the translations occurring in $\mathfrak{F}$ there is a smallest one.

In summing up the main results of this section, the following theorem can be stated:

If a group $\mathfrak{F}$ satisfies the condition of quasi-compactness, i.e. if a truncated domain $\mathcal{K}^*(\mathfrak{F})$ is compact modulo $\mathfrak{F}$, then the number of equivalence classes of centres, of limit-centres, of reflection lines and of sides of $\mathcal{K}(\mathfrak{F})$ are finite. Every side of $\mathcal{K}(\mathfrak{F})$ is a boundary axis. The displacements of all translations and reversed translations contained in $\mathfrak{F}$ have no accumulation value; in particular, they have a positive minimum.

□

15.3 Quasi-compactness and finite generation. It is the aim of this and the next two sections to prove the following theorem:

The condition of quasi-compactness is a necessary and sufficient condition for a group $\mathfrak{F}$ to be finitely generated.

It will first be shown in this section that the condition is sufficient, thus:

If a truncated domain $\mathcal{K}^*(\mathfrak{F})$ of a group $\mathfrak{F}$ is compact modulo $\mathfrak{F}$, then $\mathfrak{F}$ can be generated by a finite number of its elements.

Proof. Let $\mathcal{Q}^*$ denote an open circular disc containing at least one equivalent to every point of $\mathcal{K}^*(\mathfrak{F})$, and let x^* be the centre and ρ^* the radius of $\mathcal{Q}^*$. Let x be an arbitrary point of $\mathcal{K}^*(\mathfrak{F})$ and $\mathfrak{f}$ such an element of $\mathfrak{F}$ that $\mathfrak{f}^{-1}x$ is situated in $\mathcal{Q}^*$. This can be expressed by the fact that $\mathfrak{f}\mathcal{Q}^*$ covers the point x. The point set $\mathfrak{F}\mathcal{Q}^*$ made up of the equivalence class of $\mathcal{Q}^*$ thus covers the whole of $\mathcal{K}^*(\mathfrak{F})$; it is an open set in $\mathcal{D}$, since it is a union of open sets. For an arbitrary element $\mathfrak{f}$ of $\mathfrak{F}$ the sets $\mathcal{Q}^*$ and $\mathfrak{f}\mathcal{Q}^*$ have points in common if

$$[x^*, \mathfrak{f}x^*] < 2\rho^*$$

and in that case only. In virtue of the discontinuity of $\mathfrak{F}$ this inequality only holds for a finite number of elements $\mathfrak{f}$, say for $\mathfrak{f}_1, \mathfrak{f}_2, \dots, \mathfrak{f}_\alpha$ besides the identity. (It is evident that such elements $\mathfrak{f} \neq 1$ exist; for if x is a boundary point of $\mathcal{Q}^*$ belonging to $\mathcal{K}^*(\mathfrak{F})$, then there is in $\mathfrak{F}$ an element $\mathfrak{f}$ for which $\mathfrak{f}^{-1}x$ belongs to $\mathcal{Q}^*$, and $\mathcal{Q}^*$ and $\mathfrak{f}\mathcal{Q}^*$ then have points in common; thus such an $\mathfrak{f}$ belongs to the above collection of elements, and so does $\mathfrak{f}^{-1}$.) An arbitrary point of $\mathcal{Q}^*$ belongs to a certain number among the $\alpha+1$ circular discs $\mathcal{Q}^*, \mathfrak{f}_1\mathcal{Q}^*, \dots, \mathfrak{f}_\alpha\mathcal{Q}^*$, and any of its equivalent points then belongs to the same number of discs in the collection $\mathfrak{F}\mathcal{Q}^*$.

It will now be shown that $\mathfrak{F}$ can be generated by the collection of elements $\mathfrak{f}_\nu$, $\nu = 1, 2, \dots, \alpha$. The $\mathfrak{f}_\nu$ generate a subgroup $\mathfrak{H}$ of $\mathfrak{F}$. Let $\mathcal{M}$ denote the point set $\mathfrak{H}\mathcal{Q}^*$, i.e. the set of points equivalent to points of $\mathcal{Q}^*$ with respect to $\mathfrak{H}$. $\mathcal{M}$ is an open set, and it will first be proved that it is closed on $\mathcal{K}^*(\mathfrak{F})$, i.e. that if a point x of $\mathcal{K}^*(\mathfrak{F})$ is an accumulation point of $\mathcal{M}$, then x belongs itself to $\mathcal{M}$. Let $\mathcal{C}$ be a circular disc with centre x and radius ε. Since x is an accumulation point of $\mathcal{M}$, the disc $\mathcal{C}$ has points in common with some discs of the collection $\mathfrak{H}\mathcal{Q}^*$. As this can only be the case with a finite number of discs from that collection there must be at least one of these discs with which $\mathcal{C}$ keeps points in common as ε tends to 0. Let this, for instance, be the case with $\mathfrak{h}\mathcal{Q}^*$, $\mathfrak{h}$ being an element of $\mathfrak{H}$. If x belongs itself to $\mathfrak{h}\mathcal{Q}^*$, it is evident that x belongs to $\mathcal{M}$. If not, x is a boundary point of the (open) disc $\mathfrak{h}\mathcal{Q}^*$. Then $\mathfrak{h}^{-1}x$ is a boundary point of $\mathcal{Q}^*$ and belongs to $\mathcal{K}^*(\mathfrak{F})$, since x does. Therefore, $\mathfrak{h}^{-1}x$ belongs to at least one of the α discs $\mathfrak{f}_\nu\mathcal{Q}^*$. Then x belongs to $\mathfrak{h}\mathfrak{f}_\nu\mathcal{Q}^*$, and since $\mathfrak{h}\mathfrak{f}_\nu$ is an element of $\mathfrak{H}$, the accumulation point x belongs to $\mathcal{M}$.

Since $\mathcal{M}$ contains points of $\mathcal{K}^*(\mathfrak{F})$ and is both an open set and a set closed on $\mathcal{K}^*(\mathfrak{F})$, it must contain the whole of $\mathcal{K}^*(\mathfrak{F})$. Thus for every point of $\mathcal{K}^*(\mathfrak{F})$ there is in $\mathcal{Q}^*$ at least one point equivalent to it with respect to $\mathfrak{H}$. Now, if $\mathfrak{f}$ is an arbitrary element of $\mathfrak{F}$ and x a point belonging to the intersection of $\mathcal{Q}^*$ and $\mathcal{K}^*(\mathfrak{F})$, then $\mathfrak{f}x$ is a point of $\mathcal{K}^*(\mathfrak{F})$, and there is an element $\mathfrak{h}$ of $\mathfrak{H}$ for which $\mathfrak{h}\mathfrak{f}x$ belongs to $\mathcal{Q}^*$. Thus $\mathcal{Q}^*$ and $\mathfrak{h}\mathfrak{f}\mathcal{Q}^*$ have points in common. Then $\mathfrak{h}\mathfrak{f}$ must be one of the elements $1, \mathfrak{f}_1, \dots, \mathfrak{f}_\alpha$, thus $\mathfrak{f}$ one of the elements $\mathfrak{h}^{-1}, \mathfrak{h}^{-1}\mathfrak{f}_1, \dots, \mathfrak{h}^{-1}\mathfrak{f}_\alpha$, which all are elements of $\mathfrak{H}$. Hence $\mathfrak{H} = \mathfrak{F}$, which proves the assertion. □

15.4 Generation by translations and reversed translations. It remains to be proved that the condition of the main theorem at the head of the preceding section is necessary. In preparation for the proof the following auxiliary theorem is established in advance:

Every $\mathfrak{F}$-group can be generated by a finite or enumerable collection of translations and reversed translations the axes of which are situated in the interior of the convex domain, this collection being finite if the group admits a finite system of generators.

Proof. First, every reflection $\mathfrak{s}$ in $\mathfrak{F}$ is obtainable as the product of a translation and a reversed translation in $\mathfrak{F}$. For there is in $\mathfrak{F}$ a translation $\mathfrak{f}$ the axes $\mathcal{A}(\mathfrak{f})$ of which is not orthogonal to the reflection line $\mathcal{L}(\mathfrak{s})$; this may be seen from the fact that $\mathcal{L}(\mathfrak{s})$ is not a side of $\mathcal{K}(\mathfrak{F})$ (§14.2) together with the fact (§13.3) that there is in $\mathfrak{F}$ a translation the fundamental points of which are situated in any two assigned intervals of $\mathcal{E}$ containing fundamental points. The product $\mathfrak{s}\mathfrak{f} = \mathfrak{g}$ cannot be a reflection, for otherwise $\mathfrak{f}$ would be the product of two reflections $\mathfrak{s}$ and $\mathfrak{g}$, which is impossible, since $\mathcal{A}(\mathfrak{f})$ is not orthogonal to $\mathcal{L}(\mathfrak{s})$. The reversion $\mathfrak{g}$ must then be a reversed translation. Hence $\mathfrak{s} = \mathfrak{g}\mathfrak{f}^{-1}$ proves the assertion.

Secondly, every rotation or limit-rotation $\mathfrak{f}$ is obtainable as the product of two translations from $\mathfrak{F}$. Let $c(\mathfrak{f})$ denote the centre of $\mathfrak{f}$ (infinite if $\mathfrak{f}$ is a limit-rotation). Evidently there is a translation axis $\mathcal{A}$ in $\mathfrak{F}$ which does not pass through $c(\mathfrak{f})$. Let $\mathfrak{s}'$ denote the reflection at the perpendicular from $c(\mathfrak{f})$ on $\mathcal{A}$ (Fig. 15.2). Then $\mathfrak{f} = \mathfrak{s}\mathfrak{s}'$,

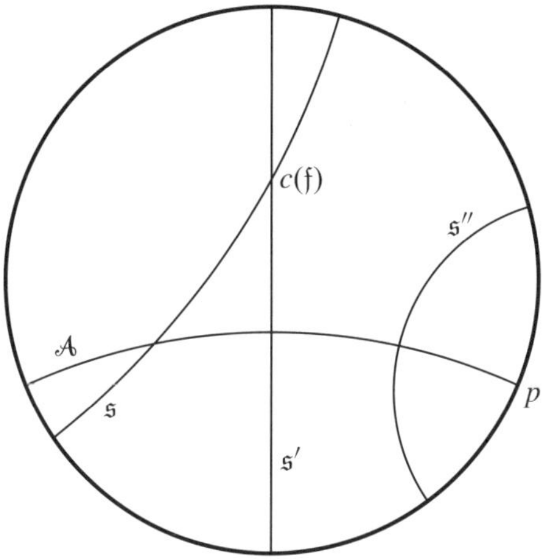

Figure 15.2

if $\mathfrak{s}$ denotes the reflection in a certain line through $c(\mathfrak{f})$. An end-point p of $\mathcal{A}$ which is not situated on the reflection line $\mathcal{L}(\mathfrak{s})$ (at most one of them can be on $\mathcal{L}(\mathfrak{s})$) is taken as the negative fundamental point of a translation $\mathfrak{g}$ with axis $\mathcal{A}$. If $\mathfrak{g} = \mathfrak{s}'\mathfrak{s}''$, the reflection line $\mathcal{L}(\mathfrak{s}'')$ is arbitrarily near to p, if the displacement of $\mathfrak{g}$ is chosen sufficiently large, and $\mathcal{L}(\mathfrak{s}')$ and $\mathcal{L}(\mathfrak{s}'')$ are then divergent. Finally $\mathfrak{f}\mathfrak{g} = \mathfrak{s}\mathfrak{s}'' = \mathfrak{h}^{-1}$ is an element of $\mathfrak{F}$, since $\mathfrak{f}$ and $\mathfrak{g}$ are, and $\mathfrak{h}^{-1}$ is a translation, since $\mathcal{L}(\mathfrak{s})$ and $\mathcal{L}(\mathfrak{s}'')$ are divergent. Now $\mathfrak{f} = \mathfrak{h}^{-1}\mathfrak{g}^{-1}$ proves the assertion.

Finally, a translation $\mathfrak{f}$ in $\mathfrak{F}$ belonging to a boundary axis $\mathcal{A}(\mathfrak{f})$ is obtainable as the product of two translations with inner axes, i.e. axes in the interior of $\mathcal{K}(\mathfrak{F})$. For let $\mathcal{A}$ be an inner axis; the existence of such axes is evident, e.g. from §13.3. Then $\mathcal{A}$ and $\mathcal{A}(\mathfrak{f})$ are divergent. Let $\mathfrak{g}$ be a translation with axis $\mathcal{A}$ such that the direction of $\mathfrak{f}$ and $\mathfrak{g}$ are in accordance. If $\mathfrak{s}'$ denotes the reflection at the common perpendicular of $\mathcal{A}(\mathfrak{f})$

and $\mathcal{A}(\mathfrak{g})$, and if $\mathfrak{f} = \mathfrak{s}\mathfrak{s}'$ and $\mathfrak{g} = \mathfrak{s}'\mathfrak{s}''$, then $\mathcal{L}(\mathfrak{s})$ and $\mathcal{L}(\mathfrak{s}'')$ are divergent (Fig. 15.3) and $\mathfrak{f}\mathfrak{g} = \mathfrak{s}\mathfrak{s}'' = \mathfrak{h}^{-1}$ is a translation in $\mathfrak{F}$ the axis of which is situated between $\mathcal{A}(\mathfrak{f})$

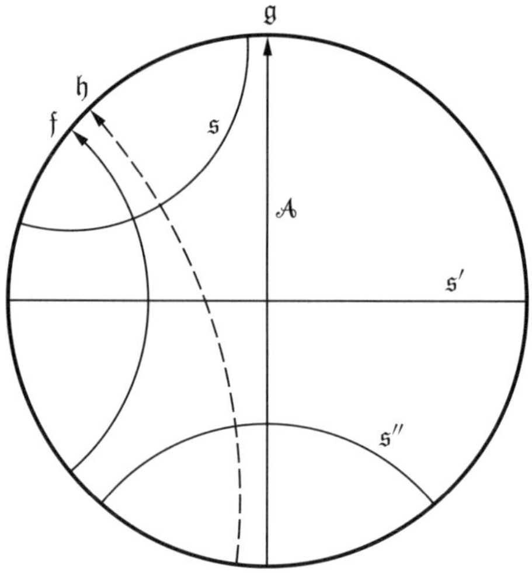

Figure 15.3

and $\mathcal{A}(\mathfrak{g})$, thus in the interior of $\mathcal{K}(\mathfrak{F})$. Now $\mathfrak{f} = \mathfrak{h}^{-1}\mathfrak{g}^{-1}$ proves the assertion. – It has been remarked in §14.2 that the axis of a reversed translation cannot be a boundary axis.

From these three statements the auxiliary theorem is readily obtained: Let $\mathfrak{f}_1, \mathfrak{f}_2, \dots$ be an arbitrary system of generators of $\mathfrak{F}$. If the generator $\mathfrak{f}_\nu$ is a reflection, it is replaced by two new generators, namely a translation and a reversed translation having $\mathfrak{f}_\nu$ as their product. In the same way every rotation or limit-rotation $\mathfrak{f}_\nu$ is replaced by two translations having $\mathfrak{f}_\nu$ as their product. The new system of generators obtained in this way consists only of translations and reversed translations. If some of the translations belong to boundary axes, each of them is replaced by two translations with inner axes having it as their product. If the system of generators was finite from the outset, it remains finite. Thus the proof of the auxiliary theorem is complete. □

15.5 Necessity of the condition in the main theorem of Section 3. This will result from a more general theorem which we are going to establish in this section. We first introduce two definitions relating to an $\mathfrak{F}$-group $\mathfrak{F}$:

A point set $\mathcal{M}$ of $\mathcal{D}$ is called *simple modulo* $\mathfrak{F}$, if for any element $\mathfrak{f} \in \mathfrak{F}$ the point sets $\mathcal{M}$ and $\mathfrak{f}\mathcal{M}$ are disjoint, except if they coincide.

A collection of point sets $\mathcal{M}_\nu$ in $\mathcal{D}$, ν ranging over a finite or infinite set of values, is called *disjoint modulo* $\mathfrak{F}$, if the point sets $\mathfrak{F}\mathcal{M}_\mu$ and $\mathfrak{F}\mathcal{M}_\rho$ are disjoint for any two different values μ and ρ of ν.

Remark. To the first of these two definitions we attach the following remark: Let $\mathcal{M}$ be simple modulo $\mathfrak{F}$. Those elements $\mathfrak{f}$ of $\mathfrak{F}$ for which $\mathcal{M}$ and $\mathfrak{f}\mathcal{M}$ coincide, i.e. which map $\mathcal{M}$ onto itself, constitute a subgroup $\mathfrak{M}$ of $\mathfrak{F}$. If an element $\mathfrak{g}$ of $\mathfrak{F}$ carries a point of $\mathcal{M}$ into a point of $\mathcal{M}$, then $\mathfrak{g}$ must belong to $\mathfrak{M}$. Hence $\mathfrak{M}$ and $\mathfrak{F}$ have equal effect in $\mathcal{M}$. The collection $\mathfrak{F}\mathcal{M}$ consists of the disjoint sets $\mathfrak{f}_\nu\mathcal{M}$, $\nu = 1, 2, \dots$, where $\mathfrak{f}_\nu$ ranges over a set of representatives of the left cosets of $\mathfrak{M}$ in $\mathfrak{F}$.

The theorem in question then reads as follows:

Let $\mathfrak{G}$ be an $\mathfrak{F}$-subgroup of $\mathfrak{F}$, and let $\mathfrak{G}$ be generated by a finite number of its elements. Then there is a subgroup $\mathfrak{U}$ of $\mathfrak{F}$ which contains $\mathfrak{G}$ and satisfies the condition of quasi-compactness, and for which the interior of $\mathcal{K}(\mathfrak{U})$ is simple modulo $\mathfrak{F}$.

Remark. Assume this theorem to be valid. If $\mathfrak{F}$ can be generated by a finite number of its elements, one can take $\mathfrak{F}$ itself as $\mathfrak{G}$. Then also $\mathfrak{U} = \mathfrak{F}$. Thus $\mathfrak{F}$ satisfies the condition of quasi-compactness. This proves the necessity of the condition in the theorem of Section 3.

Proof. Since $\mathfrak{G}$ can be generated by a finite number of its elements, it can, in virtue of Section 4, be generated by a finite collection $\mathfrak{g}_\mu$, $\mu = 1, 2, \dots, m$, of translations and (possibly) reversed translations, such that the axes of the $\mathfrak{g}_\mu$ are situated in the interior of $\mathcal{K}(\mathfrak{G})$. If the axes of $\mathfrak{g}_1$ and $\mathfrak{g}_\rho$, $\rho > 1$, diverge, their end-points determine four intervals on $\mathcal{E}$, and each of these intervals contains limit points of $\mathfrak{G}$, since $\mathfrak{g}_1$ and $\mathfrak{g}_\rho$ do not belong to boundary axes of $\mathcal{K}(\mathfrak{G})$. Thus, in virtue of §13.3, there is in $\mathfrak{G}$ a translation $\mathfrak{g}'_\rho$ whose axis cuts the axes of $\mathfrak{g}_1$ and $\mathfrak{g}_\rho$; it is therefore situated in the interior of $\mathcal{K}(\mathfrak{G})$. Let the $\mathfrak{g}'_\rho$ be added to the $\mathfrak{g}_\mu$ for all those values of ρ for which the axes of $\mathfrak{g}_1$ and $\mathfrak{g}_\rho$ diverge, and let this extended system of generators of $\mathfrak{G}$ be denoted by $\mathfrak{f}_\nu$, $\nu = 1, 2, \dots, n$, $n \geqq m$. The effect of this extension is that the collection of axes $\mathcal{A}_\nu$ of the $\mathfrak{f}_\nu$ constitutes a connected point set in the interior of $\mathcal{K}(\mathfrak{G})$. – It is recalled that the term *axis* refers to a straight line in $\mathcal{D}$, *not including its end-points on* $\mathcal{E}$.

Let $\mathcal{Z}$ denote the collection of axes obtained by adding to the $\mathcal{A}_\nu$ all their equivalents with respect to $\mathfrak{F}$, thus the collection $\mathcal{Z} = \bigcup_\nu \mathfrak{F}\mathcal{A}_\nu$. This is a collection of inner axes of $\mathcal{K}(\mathfrak{F})$, because $\mathcal{K}(\mathfrak{G}) \subset \mathcal{K}(\mathfrak{F})$ (§14.1) and the $\mathcal{A}_\nu$ are in the interior of $\mathcal{K}(\mathfrak{G})$ and thus in the interior of $\mathcal{K}(\mathfrak{F})$. They do not accumulate in $\mathcal{D}$, since $\mathcal{Z}$ consists of a finite number of equivalence classes of axes and the axes of a class do not accumulate in $\mathcal{D}$ (§13.2). We consider the components into which the point set $\mathcal{Z}$ falls. The subset $\bigcup_\nu \mathcal{A}_\nu$ of $\mathcal{Z}$ is connected according to the above construction. Let $\mathcal{Z}_0$ denote the component of $\mathcal{Z}$ to which it belongs, and let $\mathcal{Y}$ denote the convex hull of $\mathcal{Z}_0$.

The open convex set $\mathcal{D}$ contains $\mathcal{Z}$ and thus also $\mathcal{Y}$. Assume that $\mathcal{Y}$ is not the whole of $\mathcal{D}$, and let p be a boundary point of $\mathcal{Y}$ in $\mathcal{D}$. Let $\mathcal{L}$ be a supporting line of $\mathcal{Y}$ through p, i.e. such that one of the open half-planes determined by $\mathcal{L}$ is disjoint with $\mathcal{Y}$; it is then, in particular, disjoint with $\mathcal{Z}_0 \subset \mathcal{Y}$. Also $\mathcal{L}$ itself does not belong to $\mathcal{Z}_0$, since every line of $\mathcal{Z}_0$ is cut by other lines of $\mathcal{Z}_0$. Hence $\mathcal{Z}_0$, and thus also $\mathcal{Y}$, is contained in the other open half-plane determined by $\mathcal{L}$. Hence p does not belong to $\mathcal{Y}$. It follows that $\mathcal{Y}$ is an open set.

Any element of $\mathfrak{F}$ carries $\mathcal{Z}$ into itself and thus a component of $\mathcal{Z}$ into a component of $\mathcal{Z}$. Let $\mathfrak{U}$ denote the subgroup of $\mathfrak{F}$ whose elements carry $\mathcal{Z}_0$, and thus also $\mathcal{Y}$, into itself. Since $\mathfrak{f}_\nu$ carries $\mathcal{A}_\nu$, and thus $\mathcal{Z}_0$, into itself, it belongs to $\mathfrak{U}$. This being the case for all the generators $\mathfrak{f}_\nu$ of $\mathfrak{G}$, it follows that $\mathfrak{G}$ is contained in $\mathfrak{U}$. Thus $\mathfrak{U}$ is an $\mathfrak{F}$-group.

The closure $\overline{\mathcal{Y}}$ of $\mathcal{Y}$ is a convex set in $\overline{\mathcal{D}}$ which is carried into itself by all elements of $\mathfrak{U}$. It thus contains $\mathcal{K}(\mathfrak{U})$ (§14.1). On the other hand, the translations and reversed translations along any axis of $\mathcal{Z}_0$ carry $\mathcal{Z}_0$ into itself and thus belong to $\mathfrak{U}$. Hence $\mathcal{Z}_0$ is contained in $\mathcal{K}(\mathfrak{U})$, and $\overline{\mathcal{Y}}$ is contained in $\overline{\mathcal{K}}(\mathfrak{U})$. Thus $\overline{\mathcal{Y}} = \overline{\mathcal{K}}(\mathfrak{U})$, and $\mathcal{Y}$ is the interior of $\mathcal{K}(\mathfrak{U})$. The axes of $\mathcal{Z}_0$ are all situated in $\mathcal{Y}$.

Assume that $\mathfrak{U}$ does not exhaust the whole of $\mathfrak{F}$. Let $\mathfrak{f}$ be an element of $\mathfrak{F}$ outside $\mathfrak{U}$. Then $\mathfrak{f}$ carries $\mathcal{Z}_0$ into another component of $\mathcal{Z}$, say $\mathfrak{f}\mathcal{Z}_0 = \mathcal{Z}_1$, and $\mathfrak{f}\mathcal{Y}$ is the convex hull of $\mathcal{Z}_1$. A straight line in $\mathcal{D}$ which does not intersect $\mathcal{Z}_0$ has no point in common with $\mathcal{Y}$, because otherwise it would divide $\mathcal{Y}$ into two convex parts, and, since it does not meet $\mathcal{Z}_0$, the whole of $\mathcal{Z}_0$ would be contained in one of these parts, which contradicts the fact that $\mathcal{Y}$ is the convex hull of $\mathcal{Z}_0$. This applies in particular to the axes of $\mathcal{Z}_1$. Also an axis of $\mathcal{Z}_1$ cannot be a boundary axis of $\mathcal{K}(\mathfrak{U})$, since it is cut by other axes of $\mathcal{Z}_1$. The whole of $\mathcal{Z}_1$ is thus contained in one of the open half-planes outside $\mathcal{K}(\mathfrak{U})$, and $\mathcal{Y}$ and $\mathfrak{f}\mathcal{Y}$ are disjoint. It follows that $\mathcal{Y}$ is simple modulo $\mathfrak{F}$. Hence $\mathfrak{U}$ and $\mathfrak{F}$ have equal effect in $\mathcal{Y}$. We may thus write $\mathcal{Z}_0 = \bigcup_\nu \mathfrak{U}\mathcal{A}_\nu$.

It has so far been shown that $\mathfrak{U}$ is an $\mathfrak{F}$-subgroup of $\mathfrak{F}$ which contains $\mathfrak{G}$, and that the interior of $\mathcal{K}(\mathfrak{U})$ is simple modulo $\mathfrak{F}$, this being trivial in case $\mathfrak{U}$ coincides with $\mathfrak{F}$. It remains to prove that $\mathfrak{U}$ satisfies the condition of quasi-compactness.

Consider the axis $\mathcal{A}_\nu$ of $\mathfrak{f}_\nu$. It is cut by other axes of $\mathcal{Z}_0$, and these axes divide it into segments. These segments do not accumulate on $\mathcal{A}_\nu$, because $\mathcal{Z}$ does not accumulate in $\mathcal{D}$. Let $\mathfrak{h}_\nu$ be a primary translation or reversed translation of $\mathfrak{F}$, and thus of $\mathfrak{U}$, belonging to the axis $\mathcal{A}_\nu$; thus $\mathfrak{f}_\nu$ is a power of $\mathfrak{h}_\nu$. Since $\mathfrak{h}_\nu$ reproduces $\mathcal{Z}_0$, it reproduces the division of $\mathcal{A}_\nu$ into segments. Let α_ν be the number of segments of $\mathcal{A}_\nu$ contained in a part of $\mathcal{A}_\nu$ corresponding to the displacement of $\mathfrak{h}_\nu$. The totality of segments of $\mathcal{A}_\nu$ is then derived from these α_ν segments by the powers of $\mathfrak{h}$. It should be noticed that these α_ν segments need not be inequivalent with respect to $\mathfrak{F}$, and thus to $\mathfrak{U}$: Two of the α_ν segments may be equivalent with their direction on $\mathcal{A}_\nu$ reversed, if $\mathfrak{F}$ contains an element which reverses $\mathcal{A}_\nu$, thus a half-turn with centre on $\mathcal{A}_\nu$ or a reflection in a normal of $\mathcal{A}_\nu$. If two segments are equivalent with their direction on $\mathcal{A}_\nu$ preserved, it must be a power of $\mathfrak{h}_\nu$.

Every axis of $\mathcal{Z}_0$ is equivalent to one of the axes $\mathcal{A}_\nu$ with respect to $\mathfrak{F}$ and thus also with respect to $\mathfrak{U}$. Even some of the $\mathcal{A}_\nu$ may be equivalent. Hence the total number α of equivalence classes with respect to $\mathfrak{F}$ and thus also with respect to $\mathfrak{U}$ into which the totality of segments of $\mathcal{Z}_0$ falls is finite and at most equal to $\sum_{\nu=1}^{n} \alpha_\nu$. Let $\mathcal{S}_\rho$, $\rho = 1, 2, \dots, \alpha$, denote α inequivalent segments selected arbitrarily on $\mathcal{Z}_0$.

Consider the decomposition of $\mathcal{Y}$ by Z_0. The regions of decomposition are convex, because $\mathcal{Y}$ is convex and Z_0 consists of complete straight lines. Z_0 contributes to the boundary of any of these regions, the part of the boundary of the region inside $\mathcal{Y}$ consisting of chains of segments of Z_0. Besides these even boundaries of $\mathcal{Y}$, thus sides of $\mathcal{K}(\mathfrak{U})$, may contribute to the boundary of the region. Let $\mathcal{B}$ be a region of decomposition and $\mathcal{S}$ a segment of Z_0 on its boundary. There is an element $\mathfrak{f}$ of $\mathfrak{U}$ which carries $\mathcal{S}$ into one of the $\mathcal{S}_\rho$, and $\mathfrak{f}\mathcal{B}$ is then one of the two regions adjacent to that $\mathcal{S}_\rho$. Hence the number of equivalence classes into which the regions of decomposition fall (with respect to $\mathfrak{F}$ and thus to $\mathfrak{U}$) is a finite number $\beta \leqq 2\alpha$. Let $\mathcal{B}_\mu$, $\mu = 1, 2, \dots, \beta$, be a set of representatives of these β equivalence classes.

Consider a region $\mathcal{B}_\mu$ to whose boundary only a finite number of segments contribute. Its boundary must be a simple, closed polygon consisting of segments of Z_0, thus situated in $\mathcal{Y}$. Let δ_μ denote the non-euclidean radius of a circle containing $\mathcal{B}_\mu$.

Consider a region $\mathcal{B}_\mu$ whose boundary is not a closed polygon, thus to whose boundary an infinite number of segments contribute. Every boundary component of $\mathcal{B}_\mu$ consisting of segments must then contain infinitely many segments. Let $\mathcal{C}$ be such a boundary component. At least one of the α equivalence classes of segments must be represented infinitely often on $\mathcal{C}$. Let $\mathcal{S}$ be a segment of such a class on $\mathcal{C}$, and let $\mathfrak{h}_\sigma$, $0 < \sigma < \infty$, be such elements of $\mathfrak{U}$ that the $\mathfrak{h}_\sigma\mathcal{S}$ are all the other segments of that class on $\mathcal{C}$. For some values of σ the element $\mathfrak{h}_\sigma$ may map $\mathcal{S}$ on $\mathfrak{h}_\sigma\mathcal{S}$ with its direction on $\mathcal{C}$ preserved, for some other values of σ with its direction on $\mathcal{C}$ reversed. At least one of the two must happen infinitely often. If $\mathfrak{h}_\omega$ and $\mathfrak{h}_\tau$ map $\mathcal{S}$ in the same way, then $\mathfrak{h}_\tau\mathfrak{h}_\omega^{-1}$ maps $\mathfrak{h}_\omega\mathcal{S}$ on $\mathfrak{h}_\tau\mathcal{S}$ with its direction on $\mathcal{C}$ preserved. One can thus choose $\mathcal{S}$ in its class in such a way that more than one element among the $\mathfrak{h}_\sigma$ maps it with its direction on $\mathcal{C}$ preserved. Let $\mathfrak{h}_1$ and $\mathfrak{h}_2$ map $\mathcal{S}$ on $\mathfrak{h}_1\mathcal{S}$ and $\mathfrak{h}_2\mathcal{S}$ with its direction on $\mathcal{C}$ preserved. If both $\mathfrak{h}_1$ and $\mathfrak{h}_2$ happen to be reversions, then $\mathfrak{h}_1\mathfrak{h}_2^{-1}$ is a motion which maps $\mathfrak{h}_2\mathcal{S}$ on $\mathfrak{h}_1\mathcal{S}$ with its direction on $\mathcal{C}$ preserved. In all, one can therefore find a segment $\mathcal{S}$ on $\mathcal{C}$ and a motion $\mathfrak{h}$ in $\mathfrak{U}$ such that $\mathfrak{h}$ maps $\mathcal{S}$ on the segment $\mathfrak{h}\mathcal{S}$ of $\mathcal{C}$ with its direction on $\mathcal{C}$ preserved. This motion $\mathfrak{h}$ then carries $\mathcal{B}_\mu$ into itself, advancing the boundary component $\mathcal{C}$ of $\mathcal{B}_\mu$ in itself a certain number of steps. The finite sequence of segments of $\mathcal{C}$ from $\mathcal{S}$ included to $\mathfrak{h}\mathcal{S}$ excluded together with its images by all powers of $\mathfrak{h}$ then make up the whole of $\mathcal{C}$. The motion $\mathfrak{h}$ is of infinite order, thus either a translation or a limit-rotation.

If $\mathfrak{h}$ is a translation, the ends of $\mathcal{C}$ converge to the end-points of the axis $\mathcal{A}$ of $\mathfrak{h}$. Owing to the convexity of $\mathcal{B}_\mu$ the axis $\mathcal{A}$ then belongs to the closure of $\mathcal{B}_\mu$ on $\mathcal{D}$. Hence the region bounded by $\mathcal{C}$ and $\mathcal{A}$ belongs to $\mathcal{Y}$. Since Z_0 is connected, the half-plane determined by $\mathcal{A}$ and containing $\mathcal{C}$ contains the whole of Z_0. Thus $\mathcal{A}$ is a boundary of $\mathcal{B}_\mu$ and of $\mathcal{Y}$, and the complete boundary of $\mathcal{B}_\mu$ then consists of $\mathcal{C}$ and $\mathcal{A}$ and the end-points of $\mathcal{A}$. Since $\mathcal{C}$ is reproduced by $\mathfrak{h}$, the region $\mathcal{B}_\mu$ is situated in the distance strip between $\mathcal{A}$ and a suitable hypercycle belonging to $\mathcal{A}$. Let δ_μ denote the width of that strip.

If $\mathfrak{h}$ is a limit-rotation, both ends of $\mathcal{C}$ converge to the limit-centre u of $\mathfrak{h}$. Hence the region bounded by $\mathcal{C}$ and u belongs to $\mathcal{Y}$ and is $\mathcal{B}_\mu$ itself, since Z_0 is connected.

Since $\mathfrak{C}$ is reproduced by $\mathfrak{h}$, it is situated between two suitably chosen horocycles belonging to u. Let $\mathcal{O}_1$ and $\mathcal{O}_2$ be two horocycles belonging to u chosen in such a way that they include between them both $\mathfrak{C}$ and the bounding horocycle $\mathcal{O}^*(u)$ of $\mathcal{K}^*(\mathfrak{U})$ belonging to u and thus also the intersection of $\mathcal{B}_\mu$ and $\mathcal{K}^*(\mathfrak{U})$. Let δ_μ denote the distance between $\mathcal{O}_1$ and $\mathcal{O}_2$.

Let δ denote a number greater than all the β values δ_μ. Taking into account that every region of decomposition is congruent to one of the $\mathcal{B}_\mu$, one may state the result as follows: The distance of any point of $\mathcal{K}^*(\mathfrak{U})$ from $\mathcal{Z}_0$, thus from a suitable segment of $\mathcal{Z}_0$ is smaller than δ.

Finally, let $\mathcal{Q}$ denote a circular disc in $\mathcal{D}$ containing the α segments $\mathcal{S}_\rho$, and $\mathcal{Q}^*$ the circular disc obtained by increasing the non-euclidean radius of $\mathcal{Q}$ by the amount δ. Since every segment of $\mathcal{Z}_0$ is equivalent with respect to $\mathfrak{U}$ with one of the $\mathcal{S}_\rho$, any point of $\mathcal{K}^*(\mathfrak{U})$ has at least one equivalent in $\mathcal{Q}^*$. Thus $\mathcal{K}^*(\mathfrak{U})$ is compact modulo $\mathfrak{U}$, and $\mathfrak{U}$ satisfies the condition of quasi-compactness. □

15.6 The hull of a finitely generated subgroup. Let $\mathfrak{V}$ denote any subgroup of $\mathfrak{F}$ which contains $\mathfrak{G}$, and for which the interior of $\mathcal{K}(\mathfrak{V})$ is simple modulo $\mathfrak{F}$. Then $\mathcal{K}(\mathfrak{V})$ contains $\mathcal{K}(\mathfrak{G})$ (§14.1). Therefore the above axes $\mathcal{A}_\nu$ of the $\mathfrak{f}_\nu$, $1 \leqq \nu \leqq n$, which are inner axes of $\mathfrak{G}$, are also inner axes of $\mathfrak{V}$. Consider again the above collection $\mathcal{Z} = \bigcup_\nu \mathfrak{F}\mathcal{A}_\nu$. Since $\mathfrak{V}$ and $\mathfrak{F}$ have equal effect in the interior of $\mathcal{K}(\mathfrak{V})$, those axes of $\mathcal{Z}$ which have points in common with the interior of $\mathcal{K}(\mathfrak{V})$ belong to the subcollection $\mathcal{Z}' = \bigcup_\nu \mathfrak{V}\mathcal{A}_\nu$. The whole of $\mathcal{Z}'$ is situated in the interior of $\mathcal{K}(\mathfrak{V})$, since the $\mathcal{A}_\nu$ are. Also no boundary of $\mathcal{K}(\mathfrak{V})$ belongs to $\mathcal{Z}$, because every axis of $\mathcal{Z}$ is cut by other axes of $\mathcal{Z}$. It therefore follows that $\mathcal{K}(\mathfrak{V})$ contains one or more complete components of $\mathcal{Z}$, and in particular the component $\mathcal{Z}_0$, whose convex hull is the interior of $\mathcal{K}(\mathfrak{U})$. Owing to the convexity of $\mathcal{K}(\mathfrak{V})$ then $\mathcal{K}(\mathfrak{U})$ is contained in $\mathcal{K}(\mathfrak{V})$, and since $\mathfrak{V}$ and $\mathfrak{F}$ have equal effect in the interior of $\mathcal{K}(\mathfrak{V})$, it follows that $\mathfrak{U}$ is contained in $\mathfrak{V}$. We call $\mathfrak{U}$ *the hull of* $\mathfrak{G}$ *in* $\mathfrak{F}$ and denote it by $\mathfrak{H}'(\mathfrak{G})$. The definition can thus be rephrased in the following way: *The hull* $\mathfrak{H}'(\mathfrak{G})$ *in* $\mathfrak{F}$ *of a finitely generated* $\mathfrak{F}$*-subgroup* $\mathfrak{G}$ *of* $\mathfrak{F}$ *is the intersection of all subgroups of* $\mathfrak{F}$ *which contain* $\mathfrak{G}$ *and for which the interior of their convex domain is simple modulo* $\mathfrak{F}$.

$\mathfrak{H}'(\mathfrak{G})$ is the *smallest* subgroup of $\mathfrak{F}$ with this property. This shows that $\mathfrak{H}'(\mathfrak{G})$ does not depend on the choice of the $\mathfrak{f}_\nu$ used as generators for $\mathfrak{G}$. The hull $\mathfrak{H}'(\mathfrak{G})$ is uniquely determined by $\mathfrak{G}$.

Since it was shown that $\mathcal{K}^*(\mathfrak{H}')$ is compact modulo $\mathfrak{H}'$, one gets in virtue of Section 3:

The hull $\mathfrak{H}'(\mathfrak{G})$ *of a finitely generated* $\mathfrak{F}$*-subgroup* $\mathfrak{G}$ *of* $\mathfrak{F}$ *is itself finitely generated.* □

If, in particular, the interior of $\mathcal{K}(\mathfrak{G})$ is simple modulo $\mathfrak{F}$, thus if $\mathfrak{F}$ and $\mathfrak{G}$ have equal effect in the interior of $\mathcal{K}(\mathfrak{G})$, then $\mathfrak{H}'(\mathfrak{G}) = \mathfrak{G}$.

Chapter III
Surfaces associated with discontinuous groups

§16 The surfaces $\mathcal{D}$ modulo $\mathfrak{G}$ and $\mathcal{K}(\mathfrak{F})$ modulo $\mathfrak{F}$

16.1 The surface $\mathcal{D}$ mod $\mathfrak{G}$. Let $\mathfrak{G}$ be an arbitrary group discontinuous in $\mathcal{D}$. It is then possible to define a *surface $\mathcal{D}$ mod $\mathfrak{G}$* by taking as its points the equivalence classes $X = \mathfrak{G}x$ of points x of $\mathcal{D}$. The distance $[X, Y]$ of two points X and Y of $\mathcal{D}$ mod $\mathfrak{G}$ means the distance of the point sets $\mathfrak{G}x$ and $\mathfrak{G}y$ in $\mathcal{D}$, thus according to §14.5

$$[X, Y] = [Y, X] = \min_{\substack{\mathfrak{g}\in\mathfrak{G}\\ \mathfrak{f}\in\mathfrak{G}}} [\mathfrak{g}x, \mathfrak{f}y] = \min_{\mathfrak{g}\in\mathfrak{G}}[\mathfrak{G}x, y] = \xi(\mathfrak{G}; x, y).$$

This concept of distance satisfies the usual requirements: $[X, Y] \geqq 0$, the sign of equality occurring only if $X = Y$; for since y is not an accumulation point of $\mathfrak{G}x$, the distance $[X, Y]$ can only vanish, if y belongs to $\mathfrak{G}x$, thus if $\mathfrak{G}x = \mathfrak{G}y$. Moreover any three points $X = \mathfrak{G}x$, $Y = \mathfrak{G}y$, $Z = \mathfrak{G}z$, satisfy the relation

$$[X, Y] + [Y, Z] = \min_{\mathfrak{g}\in\mathfrak{G}}[\mathfrak{g}x, y] + \min_{\mathfrak{f}\in\mathfrak{G}}[y, \mathfrak{f}z] \tag{1}$$

$$= \min_{\substack{\mathfrak{g}\in\mathfrak{G}\\ \mathfrak{f}\in\mathfrak{G}}} ([\mathfrak{g}x, y] + [y, \mathfrak{f}z]) \geqq \min_{\substack{\mathfrak{g}\in\mathfrak{G}\\ \mathfrak{f}\in\mathfrak{G}}} [\mathfrak{g}x, \mathfrak{f}z] = [X, Z]. \tag{2}$$

Hence $\mathcal{D}$ mod $\mathfrak{G}$ is a metric space.

We now take the results of §14.4 into account. If x is an arbitrary point of $\mathcal{D}$ and $\mathfrak{A}_x$ the subgroup of $\mathfrak{G}$ leaving x invariant, then $\mathfrak{G}$ and $\mathfrak{A}_x$ are of equal effect in the characteristic neighbourhood $\mathcal{D}(\mathfrak{G}; x)$ of x. Therefore, there is a one-to-one correspondence between the parts of the surfaces $\mathcal{D}$ mod $\mathfrak{G}$ and $\mathcal{D}$ mod $\mathfrak{A}_x$ derived from points of $\mathcal{D}(\mathfrak{G}; x)$. This correspondence is isometric in the parts of the two surfaces derived from the disc $\mathcal{I}(\mathfrak{G}; x)$ determining the isometric neighbourhood of x; compare (4), §14.6.

We denote as *regular and singular points of $\mathcal{D}$ mod $\mathfrak{G}$* points corresponding to regular and singular points, respectively, for $\mathfrak{G}$ in $\mathcal{D}$ according to the definition given in §10.4. Thus singular points of the surface are points derived from centres of $\mathfrak{G}$ or from points situated on reflection lines of $\mathfrak{G}$.

The part of the surface $\mathcal{D}$ mod $\mathfrak{G}$ consisting of regular points is an open surface endowed with non-euclidean metric, in other words a two-dimensional Riemann manifold with constant negative curvature to which all local metric concepts of non-euclidean geometry are applicable.

The singular points of the surface fall into three possible types according as the corresponding subgroup takes the form $\mathfrak{A}(-)$, $\mathfrak{A}(\odot)$, or $\mathfrak{A}(\times)$. A more detailed description will result from the discussion in the next section.

16.2 Surfaces derived from quasi-abelian groups. In this section we deal with the surfaces $\mathfrak{D}$ mod $\mathfrak{A}$, where $\mathfrak{A}$ denotes one of the 12 quasi-abelian groups treated in §11. In this respect it is noted that the concepts of characteristic and isometric neighbourhood defined in §14 equally well apply to quasi-abelian groups. It is in general not possible to embed these surfaces in three-dimensional space in such a way that the euclidean metric of the space induces the non-euclidean metric of the surfaces. Therefore the following descriptions only aim at characterizing the structure of the different surfaces in its main features.

As a general remark, one may visualize a surface $\mathfrak{D}$ mod $\mathfrak{G}$ by operating on a fundamental polygon of $\mathfrak{G}$ in such a way as to make equivalent sides coincide and join them together. By taking the fundamental polygons described in §11 as starting point, the surfaces corresponding to the 12 types of quasi-abelian groups present the following aspect:

$\mathfrak{A}_{\mathcal{L}}(-)$: A half-plane together with its bounding line; compare Fig. 13.7. This line will be called a *reflection edge* of the surface. In a way to be specified later the same denotation will be used for images on other surfaces derived from a reflection line.

$\mathfrak{A}_c(\circlearrowleft_\nu)$: A cone shaped surface with enlarging aperture, the image of c on the surface being its vertex. The full angle round the vertex has the magnitude $\frac{2\pi}{\nu}$. The vertex is called a *conical point of order* ν. Compare Fig. 13.4.

$\mathfrak{A}_c(\times_\nu)$: An angular sector in $\mathfrak{D}$ of magnitude $\frac{\pi}{\nu}$. The bounding half-lines are reflection edges. The vertex is called an *angular point of order* ν. The surface can also be thought of as one half of the preceding surface.

$\mathfrak{A}_u(\hookrightarrow)$: A tube shaped surface which has an enlarging aperture at one side and reduces at the other like a pseudosphere. A horocycle with centre u yields a closed curve as its image on the surface. The reducing part cut off by such a curve will be called a *mast* of the surface. Compare Fig. 13.5.

$\mathfrak{A}_u(\wedge)$: The part of $\mathfrak{D}$ bounded by two parallel lines of reflection pertaining to two generators of the group; they are reflection edges of the surface, which is one half of the preceding surface. The reducing part cut off by the image of a horocycle with centre u will be called a *half-mast*.

$\mathfrak{A}_{\mathcal{A}}(\rightarrow)$: A tube shaped surface with enlarging aperture at both sides. The image of the axis $\mathcal{A}$ divides the surface into two symmetric parts, each of which will be called a *funnel*. Compare Fig. 13.6.

$\mathfrak{A}_{\mathcal{A}}(\|)$: A strip in $\mathfrak{D}$ bounded by two divergent lines, reflection edges of the surface. The image of the axis $\mathcal{A}$ decomposes the surface into two symmetric parts, each of which can be thought of as one half of a funnel and therefore will be called a *half-funnel*.

$\mathfrak{A}_{\mathcal{A}}(\cdot\cdot)$: A surface which is closed at one side and has an enlarging aperture at the other. It possesses two conical points of order 2.

$\mathfrak{A}_{\mathcal{A}}(\rightleftharpoons)$: This surface is obtained from a strip in $\mathfrak{D}$ bounded by two divergent lines by making these two lines coincide with opposite directions. Compare

Fig. 13.6. The surface part derived from a strip of $\mathcal{D}$ bounded by two hypercycles at equal distances from the axis $\mathcal{A}$ has the structure of a Möbius band and will be called a *cross cap*. It is not decomposed by the image of the axis.

$\mathfrak{A}_{\mathcal{A}}(\cdot\,|)$: One half of the surface $\mathcal{D}$ mod $\mathfrak{A}_{\mathcal{A}}(\cdot\cdot)$. It has one conical point of order 2 and one reflection edge.

$\mathfrak{A}_{\mathcal{A}}(-\rightarrow)$: A funnel on which the bounding curve, the image of $\mathcal{A}$, is a reflection edge.

$\mathfrak{A}_{\mathcal{A}}(-\,||)$: A half-funnel on which all three bounding lines are reflection edges and which possesses two angular points of order 2.

$\mathfrak{A}_{\mathcal{A}}(-\cdot\cdot)$: As $\mathfrak{A}_{\mathcal{A}}(-\,||)$.

16.3 Geodesics. Now let $\mathfrak{G}$ again denote an arbitrary group discontinuous in $\mathcal{D}$. The mapping of $\mathcal{D}$ onto $\mathcal{D}$ mod $\mathfrak{G}$ by the correspondence $x \mapsto \mathfrak{G}x$ is obviously one-valued, and it is uniformly continuous in $\mathcal{D}$ in consequence of the inequality

$$[X, Y] = \min_{\mathfrak{g}\in\mathfrak{G}}[\mathfrak{g}x, y] \leqq [x, y].$$

Let $X_1, X_2, X_3, \ldots$ denote a bounded sequence of points of $\mathcal{D}$ mod $\mathfrak{G}$. Thus for some point X_0 of $\mathcal{D}$ mod $\mathfrak{G}$ and some positive number ρ:

$$[X_0, X_\nu] < \rho, \quad \nu = 1, 2, \ldots .$$

If $X_0 = \mathfrak{G}x_0$, $X_\nu = \mathfrak{G}x_\nu$, then such elements $\mathfrak{g}_\nu$ of $\mathfrak{G}$ exists that

$$[x_0, \mathfrak{g}_\nu x_\nu] < \rho, \quad \nu = 1, 2, \ldots .$$

Therefore the sequence $\mathfrak{g}_\nu x_\nu$ of points in $\mathcal{D}$ is bounded and has at least one accumulation point, say x. Then according to the uniform continuity just mentioned the point $X = \mathfrak{G}x$ of $\mathcal{D}$ mod $\mathfrak{G}$ is an accumulation point of the sequence X_ν. Hence the theorem:

Every closed bounded subset of $\mathcal{D}$ mod $\mathfrak{G}$ is compact in $\mathcal{D}$ mod $\mathfrak{G}$. □

A *geodetic arc* of $\mathcal{D}$ mod $\mathfrak{G}$ means the image of a straight segment of $\mathcal{D}$ and a *geodesic* or *half-geodesic* (*geodetic ray*) the image of a line or half-line of $\mathcal{D}$ respectively by the correspondence $x \mapsto \mathfrak{G}x$. From these definitions the following theorem is evident:

Any two points of the surface can be joined by at least one geodetic arc. The shortest one of these is the shortest path between the two points of the surface. □

Every geodetic arc is contained in exactly one geodesic. In other words, it can be extended at both ends to arbitrary lengths, it being observed that the prolongation may well, in special cases, lead to a closed geodesic. What the prolongation of an arc looks like on the surface is evident as far as the regular part of the surface is concerned. As to the singular points, the following has to be observed:

At the image of a reflection line a geodesic is reflected according to the usual law of reflection: *The two arcs issuing from a point on the image of the reflection line subtend equal angles with that image.*

The image on $\mathfrak{D}$ mod $\mathfrak{G}$ of a centre of $\mathfrak{G}$ with corresponding subgroup $\mathfrak{A}(\odot_\nu)$ is called a *conical point of order* ν of $\mathfrak{D}$ mod $\mathfrak{G}$.

To clarify the course of a geodesic passing through a conical point of order ν, the following has to be observed: Let c be a centre of order ν in $\mathfrak{D}$. Let the half-lines issuing from c be identified by the angle φ which they subtend with an individual half-line issuing from c. A fundamental domain of the subgroup $\mathfrak{A}_c(\odot_\nu)$ is marked by $0 \leqq \varphi < \frac{2\pi}{\nu}$. Let a geodesic run into c at an angle $\varphi_0, 0 \leqq \varphi_0 < \frac{2\pi}{\nu}$. Then its prolongation leaves c at the angle $\varphi_0 + \pi$. If this angle is reduced modulo $\frac{2\pi}{\nu}$, one gets φ_0, if ν is even, and $\varphi_0 \pm \frac{\pi}{\nu}$, if ν is odd. Hence:

A geodesic which runs into a conical point of order ν is reflected in itself, if ν is even. If ν is odd, it goes straight through in the sense that together with its prolongation it divides the full angle $\frac{2\pi}{\nu}$ around the conical point into two equal parts. □

The image on $\mathfrak{D}$ mod $\mathfrak{G}$ of a centre of $\mathfrak{G}$ with corresponding subgroup $\mathfrak{A}(\times_\nu)$ is called an *angular point* of order ν of $\mathfrak{D}$ mod $\mathfrak{G}$.

By an analogous consideration of the fundamental domain $0 \leqq \varphi < \frac{\pi}{\nu}$ of the group $\mathfrak{A}_c(\times_\nu)$ pertaining to an angular point of order ν, or by the adjunction of a reflection to $\mathfrak{A}_c(\times_\nu)$, one gets:

A geodesic which runs into an angular point of order ν is reflected in itself, if ν is even. If ν is odd, it is reflected in such a way that the geodesic and its prolongation subtend equal angles with the images of the reflection lines bounding the fundamental domain. □

An axis $\mathcal{A}$ of the group $\mathfrak{G}$ yields a closed geodesic. For if λ denotes the primary displacement of $\mathcal{A}$ (pertaining to a translation of reversed translation), then any two points of $\mathcal{A}$ at distance λ correspond to one and the same point of the surface. λ is called the *length of the closed geodesic*. Conversely, every closed geodesic is the image of some axis of $\mathfrak{G}$. For if a certain part of the geodesic of length λ is traversed, one gets back to the same points in the same order; this means that any two points at distance λ on a line corresponding to the geodesic are equivalent. There must then be a translation or reversed translation with displacement λ which carries the line into itself.

On the surfaces corresponding to quasi-abelian groups of the types

$$\mathfrak{A}(-), \quad \mathfrak{A}(\odot), \quad \mathfrak{A}(\times), \quad \mathfrak{A}(\hookrightarrow), \quad \mathfrak{A}(\wedge),$$

there is no closed geodesic. On the surfaces corresponding to the 7 other quasi-abelian groups there is exactly one. On four of the latter, viz. those corresponding to the non-abelian groups of type

$$\mathfrak{A}(||), \quad \mathfrak{A}(\cdot\cdot), \quad \mathfrak{A}(\cdot\,|), \quad \mathfrak{A}(-\,||),$$

this closed geodesic consists of an arc traversed in both directions. Its length is then twice the length of that arc.

On every surface $\mathcal{D}$ mod $\mathfrak{F}$, where $\mathfrak{F}$ is an $\mathfrak{F}$-group, thus not quasi-abelian, there are infinitely many closed geodesics. For according to §14.3 there exists for every line $\mathcal{L}$ in $\mathcal{K}(\mathfrak{F})$ a sequence of axes of $\mathfrak{F}$ which converge to $\mathcal{L}$. Since the axes of an equivalence class do not accumulate in $\mathcal{D}$, they must belong to infinitely many equivalence classes of axes and thus represent infinitely many closed geodesics of $\mathcal{D}$ mod $\mathfrak{F}$.

16.4 Description of $\mathcal{D}$ mod $\mathfrak{F}$. The concepts introduced in the preceding sections enable us to give in its essential features, if not completely a description of the surface $\mathcal{D}$ mod $\mathfrak{F}$, where $\mathfrak{F}$ is an $\mathfrak{F}$-group.

To every equivalence class $\mathfrak{F}c$ of centres corresponds a conical point of order ν, if the subgroup pertaining to a centre c of the class belongs to the type $\mathfrak{A}_c(\circlearrowleft_\nu)$, and an angular point of order ν, if the subgroup belongs to the type $\mathfrak{A}_c(\times_\nu)$. According to Section 1, $\mathcal{D}(\mathfrak{F}; c)$ mod $\mathfrak{A}_c$ is part of $\mathcal{D}$ mod $\mathfrak{F}$.

If $\mathfrak{F}u$ is an equivalence class of limit-centres, the result obtained in §14.7 implies that there is a horocycle with centre u whose image on the surface cuts off a mast or half-mast, according as the subgroup pertaining to u belongs to the type $\mathfrak{A}_u(\hookrightarrow)$ or $\mathfrak{A}_u(\wedge)$. In fact, every horocycle with centre u inside $\mathcal{O}_u$ has this property. In particular, this is the case for the horocycles $\mathcal{O}^*(u)$ bounding a truncated domain $\mathcal{K}^*(\mathfrak{F})$ (§14.10).

Let $\mathcal{S}$ denote a side of $\mathcal{K}(\mathfrak{F})$ and $\mathfrak{A}_\mathcal{S}$ the subgroup of $\mathfrak{F}$ leaving $\mathcal{S}$ invariant. According to §14.2 the groups $\mathfrak{F}$ and $\mathfrak{A}_\mathcal{S}$ are of equal effect in the half-plane cut off by $\mathcal{S}$ and not belonging to $\mathcal{K}(\mathfrak{F})$. This means that on one side of the geodesic corresponding to $\mathcal{S}$ the surface $\mathcal{D}$ mod $\mathfrak{F}$ coincides with the surface $\mathcal{D}$ mod $\mathfrak{A}_\mathcal{S}$. Now, according to (1), §14.2, there are four possibilities for the group $\mathfrak{A}_\mathcal{S}$. If $\mathfrak{A}_\mathcal{S} = 1$, the corresponding part of the surface $\mathcal{D}$ mod $\mathfrak{F}$ is a half-plane. If $\mathfrak{A}_\mathcal{S} = \mathfrak{A}(|)$, the part is a quarter of a plane bounded by two orthogonal lines; the part of its boundary not corresponding to $\mathcal{S}$ corresponds to a reflection line. In these two cases $\mathcal{S}$ is a limit side of $\mathcal{K}(\mathfrak{F})$. If $\mathfrak{A}_\mathcal{S} = \mathfrak{A}(\rightarrow)$ or $\mathfrak{A}_\mathcal{S} = \mathfrak{A}(||)$, then $\mathcal{S}$ is a boundary axis; its image is a closed geodesic which cuts off from the surface $\mathcal{D}$ mod $\mathfrak{F}$ a funnel or a half-funnel respectively.

We now consider the image on $\mathcal{D}$ mod $\mathfrak{F}$ of the totality of reflection lines of $\mathfrak{F}$. If a reflection line $\mathcal{L}$ of $\mathfrak{F}$ is not cut by any other reflection line, then it contains no centre of $\mathfrak{F}$. The line may or may not be an axis of $\mathfrak{F}$.

Let the reflection line $\mathcal{L}$ of $\mathfrak{F}$ be cut by other reflection lines of $\mathfrak{F}$. It thus contains centres of $\mathfrak{F}$. The image on $\mathcal{D}$ mod $\mathfrak{F}$ of a centre with corresponding subgroup $\mathfrak{A}(\times_\nu)$ of $\mathfrak{F}$ was called an angular point of order ν of $\mathcal{D}$ mod $\mathfrak{F}$. If there is only one centre on $\mathcal{L}$, it divides $\mathcal{L}$ into two half-lines. If there is more than one centre on $\mathcal{L}$, then the centres on $\mathcal{L}$ divide $\mathcal{L}$ into segments and possibly half-lines. If the reflection line $\mathcal{L}$ is an axis of $\mathfrak{F}$, then the centres on $\mathcal{L}$ divide the whole of $\mathcal{L}$ into segments, this sequence of segments on $\mathcal{L}$ being periodic: The segments contained in a primary displacement on $\mathcal{L}$ are carried into all segments on $\mathcal{L}$ be the powers of a primary translation or reversed translation along $\mathcal{L}$.

In this way the totality of reflection lines of $\mathfrak{F}$ consists of a collection of segments, half-lines, and full lines, and this collection is reproduced by $\mathfrak{F}$. The image on $\mathfrak{D}$ mod $\mathfrak{F}$ of an equivalence class of segments, half-lines, or full lines, of this collection is called a *reflection edge of $\mathfrak{D}$ mod $\mathfrak{F}$.*

If $\mathcal{L}$ contains no centre of $\mathfrak{F}$, then its image on $\mathfrak{D}$ mod $\mathfrak{F}$ is a geodesic which runs into no angular point. The geodesic is closed or open according as $\mathcal{L}$ is an axis of $\mathfrak{F}$ or not.

The above statement concerning centres of odd or even order now leads to the following addition to the considerations of Section 3:

A geodesic of $\mathfrak{D}$ mod $\mathfrak{F}$ running into an angular point of order ν along a reflection edge leaves the angular point along the other edge, if ν is odd, whereas it is reflected in itself, if ν is even (*Fig.* 16.1–16.2). □

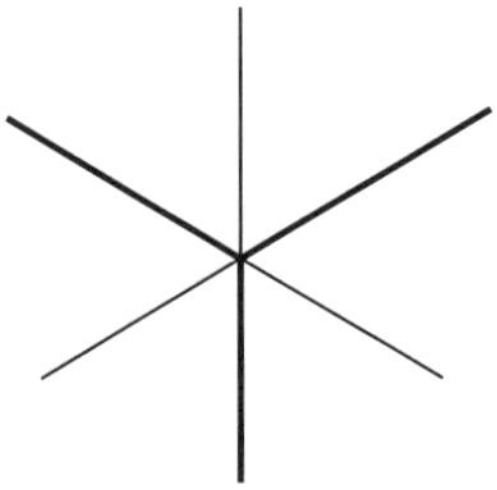

Figure 16.1

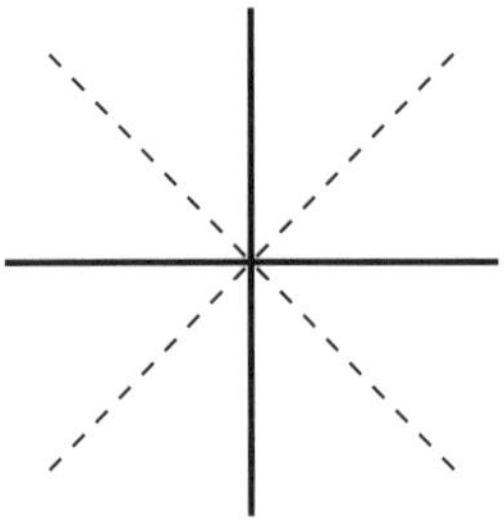

Figure 16.2

16.5 Reflection chains and reflection rings. We now especially consider the case in which $\mathfrak{F}$ contains reflections. It is recalled that the reflection lines of $\mathfrak{F}$ do not accumulate in $\mathfrak{D}$ (§13.2), that every accumulation point of reflection lines is a limit point (§13.5) and that any two reflection lines determine a centre, or a limit-centre, or an axis of $\mathfrak{F}$, according as they are concurrent, parallel, or divergent.

An intersection point of reflection lines is the centre c of a group $\mathfrak{A}_c(\times_\nu)$ with some $\nu \geqq 2$. The smallest angle subtended at c by two reflection lines is $\frac{\pi}{\nu}$. Every reflection line passing through c is equivalent to one of the two lines subtending this angle. If ν is odd, these two lines are equivalent among themselves (Fig. 16.1). If ν is even, they may or may not be equivalent (Fig. 16.2). Thus:

All reflection lines passing through a centre of odd order belong to one equivalence class. The reflection lines passing through a centre can at most belong to two classes, and if they do belong to two distinct classes, then the centre has an even order. □

Let $\mathcal{R}$ denote a reflection edge on the surface $\mathcal{D}$ mod $\mathfrak{F}$. One can determine a sequence of reflection edges on the surface in the following way: Traversing $\mathcal{R}$ in a definite sense, one may run into an angular point, or a half-mast, or a half-funnel. Then the other reflection edge pertaining to that angular point, or half-mast, or half-funnel, is taken as the next edge in the sequence, and one can then in the same way continue from that edge. This process can come to an end in three different ways:

1) After a certain number of steps one is lead back to $\mathcal{R}$ and thus gets a cyclical sequence. In this case this finite cyclical sequence of reflection edges is called a *reflection ring* of $\mathcal{D}$ mod $\mathfrak{F}$. This includes the case in which $\mathcal{R}$ itself is a closed reflection edge.

2) After a certain number of steps one arrives at an open reflection edge (it may be $\mathcal{R}$ itself) which continues on the surface without running into an angular point or half-mast or half-funnel.

3) The process of determining new edges in the sequence goes on indefinitely without leading back to $\mathcal{R}$.

In the two last cases one may start a new by traversing $\mathcal{R}$ in the opposite sense and determine the sequence corresponding to that start. The two sequences of reflection edges taken together are called a *reflection chain* on $\mathcal{D}$ mod $\mathfrak{F}$. Parts of the chain corresponding to traversing the chain from an arbitrary of its points in one direction or the other are spoken of as *ends* of the chain. A chain may, in particular, consist of an open reflection edge, if neither of its ends runs into an angular point, or a half-mast, or a half-funnel.

The reflection edges considered in the sequences are called *links* of the reflection chain or reflection ring. An end of a reflection chain may contain a finite or infinite number of links. The surface $\mathcal{D}$ mod $\mathfrak{F}$ may contain any number, finite or infinite, of reflection chains and reflection rings. The determination of a chain or ring does obviously not depend on the choice of the link $\mathcal{R}$ of the chain or ring, which is used as the starting element.

16.6 The surface $\mathcal{K}(\mathfrak{F})$ mod $\mathfrak{F}$. The reduction modulo $\mathfrak{F}$ is now considered for the convex domain only: The surfaces $\mathcal{K}(\mathfrak{F})$ mod $\mathfrak{F}$ is the aggregate of all equivalence classes $\mathfrak{F}x$, where x belongs to $\mathcal{K}(\mathfrak{F})$. This definition is consistent, since an arbitrary equivalence class of points of $\mathcal{D}$ either belongs to $\mathcal{K}(\mathfrak{F})$ as a whole or has no point in common with $\mathcal{K}(\mathfrak{F})$. The same holds, if one considers $\tilde{\mathcal{K}}(\mathfrak{F})$ mod $\mathfrak{F}$.

Since $\mathcal{D}$ mod $\mathfrak{F}$ has already been considered, we are only concerned with the case where $\mathcal{K}(\mathfrak{F})$ is not the whole of $\mathcal{D}$. From the definition of $\mathcal{K}(\mathfrak{F})$ in §14.1, and from what has been said in Section 4 concerning the images of sides of $\mathcal{K}(\mathfrak{F})$, it is evident that $\mathcal{K}(\mathfrak{F})$ mod $\mathfrak{F}$ is obtained from $\mathcal{D}$ mod $\mathfrak{F}$ by cutting off such half-planes, quarter-planes, funnels, or half-funnels, as are bounded by the geodesics corresponding to the sides of $\mathcal{K}(\mathfrak{F})$. In the first two cases the half-planes or quarter-planes removed are closed on $\mathcal{D}$, because the corresponding sides of $\mathcal{K}(\mathfrak{F})$ are limit sides and thus do not belong to $\mathcal{K}(\mathfrak{F})$; they belong to $\tilde{\mathcal{K}}(\mathfrak{F})$.

The surfaces $\tilde{\mathcal{K}}(\mathfrak{F})$ mod $\mathfrak{F}$ is thus in general bordered by two essentially different types of curves, both of them geodetic, namely on the one hand by reflection edges (if any), on the other hand by the images of sides $\mathcal{K}(\mathfrak{F})$. In the sequel only the latter will be spoken of as *boundaries of the surface* $\tilde{\mathcal{K}}(\mathfrak{F})$ mod $\mathfrak{F}$. Thus if $\mathcal{K}(\mathfrak{F}) = \mathcal{D}$, the surface has no boundaries at all. If $\mathcal{K}(\mathfrak{F})$ does not coincide with $\mathcal{D}$, then the boundaries of $\tilde{\mathcal{K}}(\mathfrak{F})$ mod $\mathfrak{F}$ are those geodesics which bound that surface on the surface $\mathcal{D}$ mod $\mathfrak{F}$. When speaking of $\tilde{\mathcal{K}}(\mathfrak{F})$ mod $\mathfrak{F}$, there is no question of prolongating such geodesics which run into a point on the boundary, whereas reflection edges do not, as has been seen, interrupt the course of a geodesic. A boundary cannot contain conical or angular points. It is either an open geodesic, or a geodetic ray starting at right angles from a reflection edge, or a simple closed geodesic, or a geodetic arc which is the common perpendicular of two reflection edges.

All closed geodesics of $\mathcal{D}$ mod $\mathfrak{F}$ are completely contained in $\mathcal{K}(\mathfrak{F})$ mod $\mathfrak{F}$, since all axes of $\mathfrak{F}$ are situated in $\mathcal{K}(\mathfrak{F})$. In virtue of a theorem in §14.3 the closed geodesics are dense in the set of all geodesics belonging to $\tilde{\mathcal{K}}(\mathfrak{F})$ mod $\mathfrak{F}$: If a geodesic is contained in $\mathcal{K}(\mathfrak{F})$ mod $\mathfrak{F}$ then to every finite arc of it, and every positive number ε there is on $\mathcal{K}(\mathfrak{F})$ mod $\mathfrak{F}$ a closed geodesic which in some part of its course accompanies the arc within a distance smaller than ε.

16.7 The surface $\mathcal{K}^*(\mathfrak{F})$ mod $\mathfrak{F}$. In case $\mathfrak{F}$ contains limit-rotations the reduction modulo $\mathfrak{F}$ can even be applied to a truncated domain $\mathcal{K}^*(\mathfrak{F})$ (§14.10) since $\mathcal{K}^*(\mathfrak{F})$ is reproduced by $\mathfrak{F}$. The surface $\mathcal{K}^*(\mathfrak{F})$ mod $\mathfrak{F}$ is part of $\mathcal{K}(\mathfrak{F})$ mod $\mathfrak{F}$ and is obtained from $\mathcal{K}(\mathfrak{F})$ mod $\mathfrak{F}$ by cutting off its masts and half-masts along the images of the horocycles $\mathcal{O}^*(u)$ constructed in §14.10.

If $\mathcal{K}^*(\mathfrak{F})$ is compact modulo $\mathfrak{F}$ (§15.1), the distance of any two points of $\mathcal{K}^*(\mathfrak{F})$ mod $\mathfrak{F}$ is smaller than the diameter of the circular disc $\mathcal{Q}^*$ used in the definition of compactness. Moreover, $\tilde{\mathcal{K}}^*(\mathfrak{F}) = \mathcal{K}^*(\mathfrak{F})$ (§15.2). In other words, $\mathcal{K}^*(\mathfrak{F})$ mod $\mathfrak{F}$ is a bounded subset of $\mathcal{D}$ mod $\mathfrak{F}$, thus according to Section 3 compact on $\mathcal{D}$ mod $\mathfrak{F}$, hence compact in itself, since it is a closed set. Conversely, if $\mathcal{K}^*(\mathfrak{F})$ mod $\mathfrak{F}$ is compact in itself, then it is bounded, thus the distance of any two of its points smaller than some quantity ρ. Then a circle with radius ρ and an arbitrary point of $\mathcal{K}^*(\mathfrak{F})$ as its centre contains for every point of $\mathcal{K}^*(\mathfrak{F})$ at least one equivalent with respect to $\mathfrak{F}$, which means that $\mathcal{K}^*(\mathfrak{F})$ is compact modulo $\mathfrak{F}$. Hence:

The two statements: $\mathcal{K}^*(\mathfrak{F})$ mod $\mathfrak{F}$ is compact in itself, and: $\mathcal{K}^*(\mathfrak{F})$ is compact modulo $\mathfrak{F}$ are equivalent.

From the definition of quasi-compactness modulo $\mathfrak{F}$ of $\mathcal{K}(\mathfrak{F})$ (§15.1) and from the equivalence of this concept with the compactness modulo $\mathfrak{F}$ of $\mathcal{K}^*(\mathfrak{F})$ the following conclusion is reached:

It is a necessary and sufficient condition for $\mathcal{K}(\mathfrak{F})$ to be quasi-compact modulo $\mathfrak{F}$ that every infinite subset of $\mathcal{K}(\mathfrak{F})$ mod $\mathfrak{F}$ contains a subsequence which is either convergent or tends to infinity on a mast or half-mast of the surface. □

We now consider a surface $\mathcal{D}$ mod $\mathfrak{F}$ for which $\mathcal{K}(\mathfrak{F})$ is quasi-compact modulo $\mathfrak{F}$, thus for which $\mathcal{K}^*(\mathfrak{F})$ mod $\mathfrak{F}$ is bounded. There can then only be a finite number of reflection edges on $\mathcal{K}^*(\mathfrak{F})$ mod $\mathfrak{F}$, because they do not accumulate on the surface. Thereby case 3) of Section 5 is excluded. Every reflection edge on $\mathcal{D}$ mod $\mathfrak{F}$ has points in common with $\mathcal{K}^*(\mathfrak{F})$ mod $\mathfrak{F}$. Suppose a particular reflection edge of $\mathcal{D}$ mod $\mathfrak{F}$, when traversed in a given sense, never runs into an angular point, or half-mast, or half-funnel. It then stays on the surface $\mathcal{K}^*(\mathfrak{F})$ mod $\mathfrak{F}$. Since this surface is bounded, the edge must be closed. Thereby case 2) of Section 5 is excluded. We thus get the following result:

If $\mathfrak{F}$ satisfies the condition of quasi-compactness, then the surface $\mathcal{D}$ mod $\mathfrak{F}$ contains no open reflection chain but only reflection rings. Moreover, the number of reflection rings is finite. This number is positive, if $\mathfrak{F}$ at all contains reflections. □

§17 Area and type numbers

17.1 Properties of normal domains. We consider a normal domain $\mathcal{N}(\mathfrak{F}; x_0) = \mathcal{N}(x_0) = \mathcal{N}$ (§10.8) of an $\mathfrak{F}$-group with some regular point $x_0 \in \mathcal{D}$ as central point and denote its closure by $\overline{\mathcal{N}}(x_0)$. Then $\overline{\mathcal{N}}(x_0)$ contains no fundamental point, for if $\mathcal{A}$ is an arbitrary axis of $\mathfrak{F}$, the normal domain $\mathcal{N}(\mathfrak{A}_{\mathcal{A}}; x_0)$ contains $\mathcal{N}(\mathfrak{F}; x_0)$ and its boundary does not contain the end-points of $\mathcal{A}$.

Since $\mathcal{N}(x_0)$ is a fundamental polygon, every class of axes is represented in $\mathcal{N}(x_0)$ by at least one segment. We prove:

Only a finite number of members of an equivalence class of axes has points in common with $\mathcal{N}(x_0)$.

Proof. Let $\mathcal{A}$ be an axis and $\mathfrak{A}^*_{\mathcal{A}}(\rightarrow)$ the subgroup of all translations of $\mathfrak{F}$ belonging to $\mathcal{A}$. The points $\mathfrak{A}^*_{\mathcal{A}}x_0$ form an equidistant sequence at constant distance from $\mathcal{A}$. Thus all points of $\mathcal{A}$ are at a bounded distance from this point set, say at a distance smaller than ρ. Then for an arbitrary element $\mathfrak{f}$ of $\mathfrak{F}$ the points of $\mathfrak{f}\mathcal{A}$ are at a distance smaller than ρ from the point set $\mathfrak{f}\mathfrak{A}^*_{\mathcal{A}}x_0$. It now follows from the definition of normal domain that an axis of the class of $\mathcal{A}$ cannot have points in common with $\mathcal{N}(x_0)$, if its distance from x_0 exceeds ρ, which is the case for all axes of the class but for a finite number. □

Similarly, every class of horocycles belonging to limit-centres of $\mathfrak{F}$ is represented in $\mathcal{N}(x_0)$ by at least one arc. We prove:

Only a finite number of members of an equivalence class of horocycles belonging to limit-centres of $\mathfrak{F}$ have points in common with $\mathcal{N}(x_0)$.

Proof. Let $\mathcal{O}$ be a horocycle belonging to the limit-centre u of $\mathfrak{F}$ and $\mathfrak{A}_u^*(\hookrightarrow)$ the subgroup of all limit-rotations of $\mathfrak{F}$ belonging to u. The points $\mathfrak{A}_u^* x_0$ form an equidistant sequence at constant distance from $\mathcal{O}$, thus all points of $\mathcal{O}$ are at a bounded distance from this point set, and the proof is completed in exactly the same way as for axes. □

Let $n(x_0)$ denote the complete boundary of $\mathcal{N}(x_0)$ and put $\mathcal{Z} = \mathfrak{F}n(x_0) \cap \mathcal{D}$. Then $\mathcal{Z}$ consists of the collection of all segments, half-lines, or lines derived from the boundary of $\mathcal{N}(x_0)$ in $\mathcal{D}$ by all elements of $\mathfrak{F}$. Thus $\mathcal{Z}$ determines the tesselation $\mathfrak{F}\mathcal{N}(x_0)$ of $\mathcal{D}$. Let $\mathcal{O}$ be a horocycle belonging to a limit-centre u of $\mathfrak{F}$, and let $\mathfrak{g}_\nu$, $0 \leqq \nu \leqq \mu$, denote elements of $\mathfrak{F}$ which carry $\mathcal{O}$ into that finite collection of horocycles which meet $\mathcal{N}(x_0)$ and belong to the class of $\mathcal{O}$. These elements are only determined up to an element of the group $\mathfrak{A}_u$ pertaining u, because that group carries $\mathcal{O}$ into itself. So they stand as representatives of certain left cosets of $\mathfrak{A}_u$ in $\mathfrak{F}$. Let p be such a point on $\mathcal{O}$ that the line determined by p and u does not contain any segment, half-line, or line of $\mathcal{Z}$, and consider that part pu of that line (Fig. 17.1). If pu meets $\mathcal{Z}$,

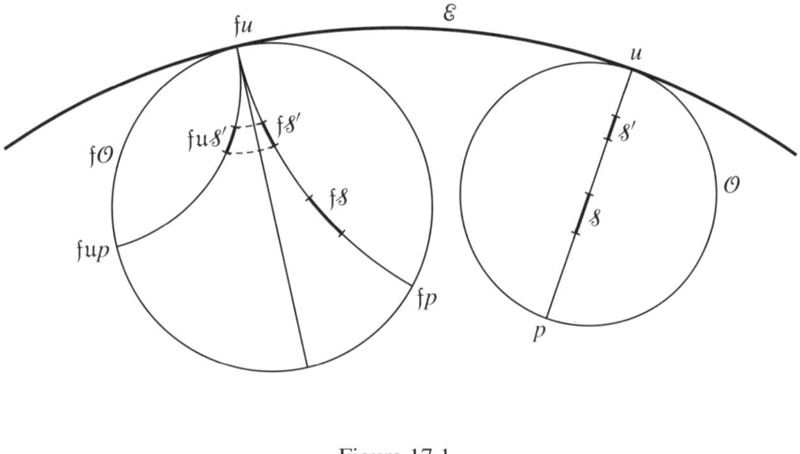

Figure 17.1

let $\mathcal{S}$ be an open segment of the decomposition of pu by $\mathcal{Z}$. It belongs to the interior of a certain domain $\mathfrak{f}^{-1}\mathcal{N}$, say, of the tesselation $\mathfrak{F}\mathcal{N}(x_0)$. Then $\mathfrak{f}\mathcal{S}$ belongs to $\mathcal{N}$ and is inside $\mathfrak{f}\mathcal{O}$, and thus $\mathfrak{f}$ belongs to one of the left cosets determined by the $\mathfrak{g}_\nu$. Let $\mathcal{S}'$ be another open segment of the decomposition of pu by $\mathcal{Z}$. Then $\mathcal{S}'$ cannot belong to the same domain $\mathfrak{f}^{-1}\mathcal{N}$ owing to the convexity of that domain. Let it be assumed that $\mathcal{S}'$ belongs to the domain $\mathfrak{u}^{-1}\mathfrak{f}^{-1}\mathcal{N}$, where $\mathfrak{u}$ is an element of $\mathfrak{A}_u$. Then both $\mathfrak{f}\mathcal{S}$ and $\mathfrak{f}\mathfrak{u}\mathcal{S}' = \mathfrak{f}\mathfrak{u}\mathfrak{f}^{-1}(\mathfrak{f}\mathcal{S}')$ belong to the interior of $\mathcal{N}$ and are inside $\mathfrak{f}\mathcal{O}$. They are situated on two lines with common end-point $\mathfrak{f}u$ which correspond by the element $\mathfrak{f}\mathfrak{u}\mathfrak{f}^{-1}$ of the group $\mathfrak{A}_{\mathfrak{f}u} = \mathfrak{f}\mathfrak{A}_u\mathfrak{f}^{-1}$. This contradicts the fact that $\mathcal{N}(\mathfrak{F}; x_0)$ is contained in the normal domain $\mathcal{N}(\mathfrak{A}_{\mathfrak{f}u}; x_0)$. Thus the domains to which the different segments cut out on pu by $\mathcal{Z}$ belong are carried into $\mathcal{N}$ by elements belonging to different of the μ cosets of $\mathfrak{A}_u$ in $\mathfrak{F}$ determined by the $\mathfrak{g}_\nu$. Hence the number of these segments is finite, and the last one has u as its end-points and is thus a half-line. If there is more than one segment, all but the last domain do not have u as a boundary point; this is immediately seen from the convexity of the domains.

Let $\mathfrak{u}$ denote a primary limit-rotation of $\mathfrak{A}_u$. The arc of $\mathcal{O}$ from p to $\mathfrak{u}p$ meets a finite number of domains of the collection $\mathfrak{F}\mathcal{N}(x_0)$. Let α of these have u as a boundary point and β not. Alle other domains of the collection $\mathfrak{F}\mathcal{N}$ cut by $\mathcal{O}$ are derived from these $\alpha + \beta$ domains by powers of $\mathfrak{u}$. Let $\mathfrak{C}$ be a horocycle with limit-centre u and situated inside $\mathcal{O}$ which does not meet the β domains for which u is not a boundary point. Then $\mathfrak{C}$ cuts the α domains for which u is a boundary point and all their images by powers of $\mathfrak{u}$, but no other domain of the collection $\mathfrak{F}\mathcal{N}$. If we join the intersection points of $\mathfrak{C}$ and $\mathcal{Z}$ with u by half-lines, these half-lines belong to $\mathcal{Z}$. No intersection of lines of $\mathcal{Z}$ takes place inside $\mathfrak{C}$. Hence:

After x_0 has been chosen, the horocycles bounding a truncated domain can be so chosen that the collection $\mathfrak{F}n(x_0)$ decomposes their interior into simple sectors as illustrated by Fig. 17.2. □

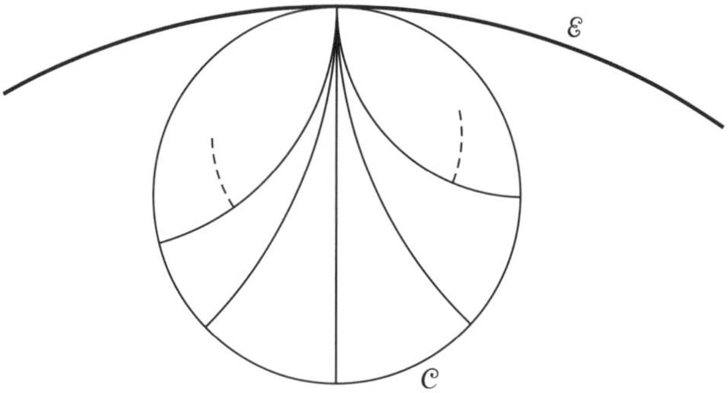

Figure 17.2

It is inferred that every equivalence class of limit-centres is represented in $n(x_0)$ at least once, but only a finite number of times, and that each limit-centre belonging to $\overline{\mathcal{N}}(x_0)$ is a vertex of $\overline{\mathcal{N}}(x_0)$.

Similarly, let $\mathcal{A}$ be a boundary axis of $\mathfrak{F}$, and let $\mathfrak{g}_\nu$, $1 \leqq \nu \leqq \mu$, denote elements of $\mathfrak{F}$ which carry $\mathcal{A}$ into that finite collection of boundary axes of $\mathfrak{F}$ which meet $\mathcal{N}(x_0)$ and belong to the class of $\mathcal{A}$. These elements stand as representatives of certain left cosets of $\mathfrak{A}_{\mathcal{A}}$ in $\mathfrak{F}$. Let $\mathcal{H}$ denote the half-plane outside $\mathcal{K}(\mathfrak{F})$ determined by $\mathcal{A}$, and $\mathcal{J}$ the periodic interval of discontinuity of $\mathcal{E}$ belonging to $\mathcal{A}$. Let p be such a point on $\mathcal{A}$ that the normal of $\mathcal{A}$ in p does not contain any segment, half-line or line of $\mathcal{Z}$, and let q be the end-point of that normal in $\mathcal{J}$ (Fig. 17.3). The half-line pq belongs

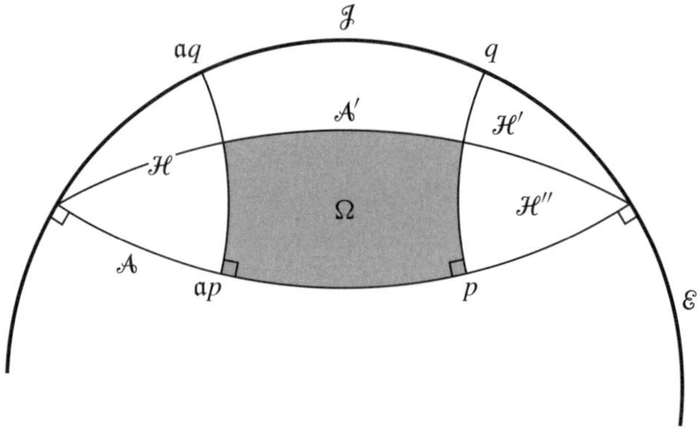

Figure 17.3

to $\mathcal{H}$. If pq meets $\mathcal{Z}$, let $\mathcal{S}$ be an open segment determined on pq by $\mathcal{Z}$. Then for a certain $\mathfrak{f} \in \mathfrak{F}$ the segment $\mathfrak{f}\mathcal{S}$ belongs to the interior of $\mathcal{N}$ and is situated in $\mathfrak{f}\mathcal{H}$. Thus $\mathfrak{f}$ belongs to one of the left cosets of $\mathfrak{A}_{\mathcal{A}}$ in $\mathfrak{F}$ determined by the $\mathfrak{g}_{\nu}$. It is seen in an analogous way as above that another open segment $\mathcal{S}'$ determined on pq by $\mathcal{Z}$ leads to a coset different from that belonging to $\mathcal{S}$, and hence that the number of intersections of pq with $\mathcal{Z}$ is finite. Thus pq ends with a half-line which is wholly contained in the interior of one of the domains $\mathcal{N}(x_0)$, and q is a boundary point of that domain.

Let $\mathfrak{a}$ be a primary translation in $\mathfrak{A}_{\mathcal{A}}$ and consider the arc of $\mathcal{A}$ from p to $\mathfrak{a}p$. It intersects a finite number of domains of the collection $\mathfrak{F}\mathcal{N}$. Let α of these have boundary points on $\mathcal{J}$ and β not. All domains of the collection $\mathfrak{F}\mathcal{N}$ which have points in common with $\mathcal{H}$ are derived from these $\alpha + \beta$ domains by powers of $\mathfrak{a}$. Let $\mathcal{A}'$ denote a hypercycle in $\mathcal{H}$ belonging to $\mathcal{A}$ which does not meet the β domains which have no boundary point on $\mathcal{J}$, and let $\mathcal{H}'$ denote the part of $\mathcal{H}$ bounded by $\mathcal{A}'$ and $\mathcal{J}$. Then $\mathcal{H}'$ contains no vertices of $\mathfrak{F}\mathcal{N}(x_0)$, but only half-lines belonging to $\mathcal{Z}$. No intersection of lines of $\mathcal{Z}$ takes place in $\mathcal{H}'$.

Assume that two half-lines of $\mathcal{Z}$ in $\mathcal{H}'$ are parallel. Let r be their common end-point in $\mathcal{J}$. We consider the collection of lines or half-lines of $\mathcal{Z}$ which have r as an end-point. Let m of these lines, consecutive in their order around r, be selected. They contribute to the boundary of $m-1$ domains, each of which has r as a boundary point, its only boundary point on $\mathcal{J}$. If $\mathcal{N}_1$ is one of these domains, then $\mathfrak{a}^{\rho}\mathcal{N}_1$ has $\mathfrak{a}^{\rho}r$ as its only boundary point on $\mathcal{J}$. Therefore no two of the selected domains bounded by r can be equivalent by a power of $\mathfrak{a}$. Since the number of domains of the collection $\mathfrak{F}\mathcal{N}$ which enter into $\mathcal{H}'$ does not exceed α modulo the powers of $\mathfrak{a}$, it follows that

$m-1 \leqq \alpha$. Hence any point of $\mathcal{J}$ can be boundary point for a finite number of domains only. The first and the last domain in the order around r are thus bounded by an arc of $\mathcal{J}$. Thus every domain of the collection $\mathfrak{F}\mathcal{N}$ which has points in common with $\mathcal{H}'$ has on its boundary either a single point of $\mathcal{J}$ or an arc of $\mathcal{J}$, and every point of $\mathcal{J}$ is a boundary point of at least one domain.

It is inferred that every equivalence class of periodic intervals of discontinuity is represented in $n(x_0)$ at least once, but only a finite number of times, by closed arcs of that class of intervals besides possibly single points, even such single points being finite in number.

With a common denotation we call such arcs and single points *boundary components of $\mathcal{N}(x_0)$ in periodic intervals of discontinuity.* Thus the intersection of $\overline{\mathcal{N}}(x_0)$ with a given equivalence class of periodic intervals of discontinuity consists of a finite number of boundary components of $\mathcal{N}(x_0)$; they are called a *cycle of boundary components of $\mathcal{N}(x_0)$*. Occasionally, treating periodic intervals of discontinuity and limit-centres together, we may also speak of a *cycle of boundary components* of $\mathcal{N}(x_0)$ in limit-centres instead of saying *cycle of vertices*.

Since $\mathcal{N}(x_0)$ is a fundamental polygon, it is evident that each equivalence class of centres of $\mathfrak{F}$ is represented in the boundary $n(x_0)$ of $\mathcal{N}(x_0)$ by at least one vertex, but only by a finite number owing to the finiteness of the cycle of vertices (§10.5).

17.2 Normal domains in the case of quasi-compactness. In this section we specify the results of the preceding section for a group $\mathfrak{F}$ satisfying the condition of quasi-compactness.

As seen in §15.2 all intervals of discontinuity are periodic, and both the intervals of discontinuity and the limit-centres of $\mathfrak{F}$ fall into a finite number of equivalence classes. Also each cycle of boundary components of $\mathcal{N}(x_0)$ belonging to these equivalence classes (Section 1) consists of a finite number of components. Hence we get from Section 1 the statement:

The intersection of $\overline{\mathcal{N}}(\mathfrak{F}; x_0)$ with the totality $\mathcal{E} \setminus \overline{\mathfrak{G}}(\mathfrak{F})$ of intervals of discontinuity consists of a finite number of components each of which in either a single point or a closed arc of $\mathcal{E}$. The same holds for the intersection of $\overline{\mathcal{N}}(\mathfrak{F}; x_0)$ with the set of all limit-centres of $\mathfrak{F}$, the components being single points. □

Consider the construction illustrated by Fig. 17.3 for an arbitrary periodic interval of discontinuity. Let $\mathcal{H}'$ denote the part of $\mathcal{H}$ between $\mathcal{J}$ and $\mathcal{A}'$, and $\mathcal{H}''$ the part of $\mathcal{H}$ between $\mathcal{A}'$ and $\mathcal{A}$. Let Ω denote the part of $\mathcal{H}''$ between the normals of $\mathcal{A}$ in p and $\mathfrak{a}p$ (shaded in Fig. 17.3). Its images by the powers of $\mathfrak{a}$ make up the whole of $\mathcal{H}''$. If the number of equivalence classes of boundary axes of $\mathfrak{F}$ is m, let the axes $\mathcal{A}_i$, $1 \leqq i \leqq m$, represent these classes, and apply notations $\mathcal{A}_i'$, $\mathcal{J}_i$, $\mathcal{H}_i'$, $\mathcal{H}_i''$, Ω_i accordingly. The truncated domain $\mathcal{K}^*(\mathfrak{F})$ is fixed by the horocycles called $\mathfrak{C}$ in Section 1. If they fall into ℓ equivalence classes, let $\mathfrak{C}_k$, $1 \leqq k \leqq \ell$, represent these classes.

Let $\mathcal{Q}^*$ be a circular disc containing for every point of $\mathcal{K}^*(\mathfrak{F})$ at least one equivalent point. Let $\mathcal{Q}$ denote a circular disc containing $\mathcal{Q}^*$ and the m bounded regions Ω_i, $1 \leqq i \leqq m$. Let $\mathcal{M}$ denote the part of $\mathcal{D}$ obtained from $\mathcal{D}$ by omitting the interior of the horocycles $\mathfrak{F}\mathfrak{C}_k$, $1 \leqq k \leqq \ell$, and the regions $\mathfrak{F}\mathcal{H}_i'$, $1 \leqq i \leqq m$; in other words, one gets $\mathcal{M}$ by adding to $\mathcal{K}^*(\mathfrak{F})$ the strips $\mathfrak{F}\mathcal{H}_i''$, $1 \leqq i \leqq m$. Then $\mathcal{Q}$ contains for every point of $\mathcal{M}$ at least one equivalent point. According to its construction $\mathcal{M}$ contains all those vertices of the tesselation $\mathfrak{F}\mathcal{N}(x_0)$ which belong to $\mathcal{D}$, since there is none inside the $\mathfrak{C}_k$ nor in the $\mathcal{H}_i'$. Since these vertices do not accumulate in $\mathcal{D}$, and since $\mathcal{Q}$ is bounded, it follows that they fall into a finite number of equivalence classes, thus that $\mathcal{N}(x_0)$ has a finite number of cycles of vertices in $\mathcal{D}$. Since every vertex cycle contains a finite number of vertices, *the number of vertices of $\mathcal{N}(x_0)$ in $\mathcal{D}$ is finite.*

Assume that $\mathcal{N}(x_0)$ has boundary points on $\mathcal{E}$, thus that it is not simply a closed polygon in $\mathcal{D}$. Let q be a boundary point of $\mathcal{N}(x_0)$ on $\mathcal{E}$. Then the half-line x_0q is contained in $\mathcal{N}(x_0)$ owing to the convexity of $\mathcal{N}(x_0)$. Since x_0 is an inner point of $\mathcal{N}(x_0)$, the half-line x_0q consists of inner points of $\mathcal{N}(x_0)$. Hence two different points of x_0q cannot be equivalent. Every mesh of the tesselation $\mathfrak{F}\mathcal{N}(x_0)$ contains exactly one half-line equivalent with x_0q. The disc $\mathcal{Q}$ has points in common only with a finite number of meshes of the tesselation $\mathfrak{F}\mathcal{N}(x_0)$, since it is bounded. Thus if the intersection $\mathcal{Q} \cap \mathfrak{F}(x_0q)$ is not empty, it consists of a finite number of segments $\mathcal{S}_j$, $1 \leqq j \leqq r$. If x_0q intersects $\mathcal{M}$, the intersection $(x_0q) \cap \mathcal{M}$ is covered by equivalents of the $\mathcal{S}_j$, since every point of $\mathcal{M}$ has an equivalent in $\mathcal{Q}$. No segment $\mathcal{S}_j$ can be used more than once in this covering, since x_0q contains no two equivalent points. Hence from a certain point p of x_0q onwards the half-line pq is outside $\mathcal{M}$. It follows that q is either a limit-centre or belongs to an interval of discontinuity.

In particular, the last sentence of the above statement can be sharpened to read:

The intersection of $\overline{\mathcal{N}}(\mathfrak{F}; x_0)$ with $\overline{\mathfrak{G}}(\mathfrak{F})$ consists of a finite number of limit-centres, if $\mathfrak{F}$ at all contains limit-rotations.

In all then, $\mathcal{N}(\mathfrak{F}; x_0)$ has a finite number of sides in $\mathcal{D}$. On the other hand, it has been shown in §10.6 that, if a fundamental polygon for an $\mathfrak{F}$-group has a finite number of sides in $\mathcal{D}$, then the group can be generated by a finite number of its elements; in virtue of §15 it then satisfies the condition of quasi-compactness. Thus one gets the theorem:

A necessary and sufficient condition for an $\mathfrak{F}$-group $\mathfrak{F}$ satisfying the condition of quasi-compactness is that the normal domain $\mathcal{N}(\mathfrak{F}; x_0)$ with an arbitrary regular point x_0 of $\mathcal{D}$ as central point has a finite number of sides in $\mathcal{D}$. $\mathcal{N}(\mathfrak{F}; x_0)$ then has a finite number of vertices in limit-centres of $\mathfrak{F}$, if $\mathfrak{F}$ contains limit-rotations, and a finite number of boundary arcs on $\mathcal{E}$, if $\mathfrak{F}$ contains boundary translations. Every boundary point of $\mathcal{N}(\mathfrak{F}; x_0)$ on $\mathcal{E}$ is either a limit-centre or belongs to an interval of discontinuity of $\mathfrak{F}$, and to every limit-centre or point of an interval of discontinuity there is at least one equivalent point on the boundary of $\mathcal{N}(\mathfrak{F}; x_0)$. □

17.3 Area of $\mathcal{K}(\mathfrak{F})$ mod $\mathfrak{F}$. For an arbitrary $\mathfrak{F}$-group $\mathfrak{F}$ we want to introduce a quantity $\Phi(\mathfrak{F})$, which will be called the *area of the surface* $\mathcal{K}(\mathfrak{F})$ mod $\mathfrak{F}$.

Let $\mathcal{N} = \mathcal{N}(\mathfrak{F}; x_0)$ be the normal domain constructed with an arbitrary regular point x_0 of $\mathcal{D}$ as central point. We put $\mathcal{N}' = \mathcal{N} \cap \tilde{\mathcal{K}}(\mathfrak{F})$. Being the intersection of two convex domains, $\mathcal{N}'$ is convex. The boundary of $\mathcal{N}'$ on $\mathcal{E}$ is the intersection of the boundary of $\mathcal{N}$ on $\mathcal{E}$, which contains no fundamental point, and the boundary of $\tilde{\mathcal{K}}(\mathfrak{F})$ on $\mathcal{E}$, which is $\overline{G}(\mathfrak{F})$. Hence no interval of $\mathcal{E}$ belongs to the boundary of $\mathcal{N}'$, because every interval of $\mathcal{E}$ containing points of $\overline{G}(\mathfrak{F})$ contains fundamental points. Thus boundary points of $\mathcal{N}'$ on $\mathcal{E}$, if any, are limit points of $\mathfrak{F}$ which are not fundamental.

We first consider the case of non-quasi-compactness, thus the case in which, according to Section 2, $\mathcal{N}$ has an infinite number of vertices.

First, let the number of vertices of $\mathcal{N}'$ on $\mathcal{E}$ be infinite. If we select an arbitrary number n of these vertices and join them in their cyclical order on $\mathcal{E}$ by straight lines, we get a polygon $\mathcal{P}$ inside $\mathcal{N}'$, because $\mathcal{N}'$ is convex. This polygon $\mathcal{P}$ has n sides and all its angles equal to zero. It is itself convex and thus simply connected. Hence we get for the area of $\mathcal{P}$ from §8.3 the general formula

$$\Phi(\mathcal{P}) = n - 2 - \frac{1}{\pi}\sum_{\mu=1}^{n} \varphi_\mu, \tag{1}$$

n denoting the number of vertices and φ_μ the angles. Since in our present case the angles vanish, we get $\Phi(\mathcal{P}) = n - 2$, thus arbitrary large values, since n is arbitrary.

Secondly, let the number of boundary points of $\mathcal{N}'$ on $\mathcal{E}$ be finite. We want to show:

Even $\mathcal{N}'$ has infinitely many vertices (and thus infinitely many of them in $\mathcal{D}$).

Proof. Since any point in $\mathcal{D}$, and thus also an inner point of $\mathcal{K}(\mathfrak{F})$, belongs to at least one of the domains $\mathfrak{F}\mathcal{N}$, and since $\mathcal{K}(\mathfrak{F})$ is reproduced by $\mathfrak{F}$, the normal domain $\mathcal{N}$ has points in common with the interior of $\mathcal{K}(\mathfrak{F})$, and so has every of the domains of the tesselation $\mathfrak{F}\mathcal{N}$. If $\mathcal{K}(\mathfrak{F}) = \mathcal{D}$, then $\mathcal{N}' = \mathcal{N}$, and $\mathcal{N}'$ thus has infinitely many vertices. So let $\mathcal{K}(\mathfrak{F})$ have sides in $\mathcal{D}$. If $\mathcal{N}$ has points in common with the open half-plane $\mathcal{H}$ outside $\tilde{\mathcal{K}}(\mathfrak{F})$ and bounded by the side $\mathcal{S}$ of $\mathcal{K}(\mathfrak{F})$, then $\mathcal{S}$ cuts off from $\mathcal{N}$ the part $\mathcal{N} \cap \mathcal{H}$, and $\mathcal{N}'$ then has two vertices on $\overline{\mathcal{S}}$, each of them being an inner point or end-point of $\mathcal{S}$. Thus if $\mathcal{N}$ has points in common with infinitely many of the half-planes outside $\tilde{\mathcal{K}}(\mathfrak{F})$, then the assertion that $\mathcal{N}'$ has infinitely many vertices holds. So let $\mathcal{N}$ have points in common with only a finite number of the half-planes outside $\tilde{\mathcal{K}}(\mathfrak{F})$. It then follows that the number of equivalence classes of these half-planes, and thus the number of equivalence classes of sides of $\mathcal{K}(\mathfrak{F})$, is finite, since to any point of any of these half-planes there is at least one equivalent in $\mathcal{N}$. If the number of vertices of $\mathcal{N}$ in each of the finitely many half-planes with which it has points in common is finite, then one looses only a finite number of vertices of $\mathcal{N}$ when passing to $\mathcal{N}'$, hence $\mathcal{N}'$ has infinitely many vertices. We therefore assume that there is a half-plane $\mathcal{H}$ in which $\mathcal{N}$ has infinitely many vertices, and thus infinitely many different elements $\mathfrak{f}_i$ of $\mathfrak{F}$ such that $\mathcal{N}$ and $\mathfrak{f}_i^{-1}\mathcal{N}$ have a common side in $\mathcal{H}$. Since $\mathfrak{f}_i^{-1}\mathcal{N}$ has points in common with $\mathcal{H}$, $\mathcal{N}$ has points in common with $\mathfrak{f}_i\mathcal{H}$. Now $\mathfrak{f}_i\mathcal{H} = \mathcal{H}$ if, and only if, $\mathfrak{f}_i$ belongs to the subgroup $\mathfrak{A}_{\mathcal{S}}$ of $\mathfrak{F}$, where $\mathcal{S}$ bounds $\mathcal{H}$. Since $\mathcal{N}$ has points in common only with a finite number of half-planes, the $\mathfrak{f}_i$ must belong to a finite number of left cosets of $\mathfrak{A}_{\mathcal{S}}$ in $\mathfrak{F}$. This excludes the possibility of $\mathcal{S}$ being a limit side of $\mathcal{K}(\mathfrak{F})$, because $\mathfrak{A}_{\mathcal{S}}$ then is of finite order. Hence $\mathcal{S}$ is boundary axis, however, it results from Section 1 that $\mathcal{N}$ can only have a common side in $\mathcal{H}$ with a finite number of its equivalents by $\mathfrak{A}_{\mathcal{S}}$. – This finishes the proof of the fact that $\mathcal{N}'$ has infinitely many vertices. □

Since it was assumed that $\mathcal{N}'$ had only a finite number of vertices on $\mathcal{E}$, and since it is bounded by no interval of $\mathcal{E}$, these vertices of $\mathcal{N}'$ on $\mathcal{E}$ are linked in turn by chains of boundary segments of $\mathcal{N}'$ in $\mathcal{D}$. Since $\mathcal{N}'$ has infinitely many vertices in $\mathcal{D}$, at least one of these chains must contain infinitely many segments, so that there is at one end, at least, of the chain no half-line as the last link of the chain, but an infinity of segments converging to the point of $\mathcal{E}$. We now cut off the points at infinity by joining any two consecutive chain-ends by a straight segment; it is interior to $\mathcal{N}'$ owing to the convexity of $\mathcal{N}'$. In that way we get a convex polygon $\mathcal{P}$ inside $\mathcal{N}'$ with a finite number n of sides, and we apply (1) to find its area. This formula may be written

$$\Phi(\mathcal{P}) = -2 + \sum_{\mu=1}^{n}\left(1 - \frac{\varphi_\mu}{\pi}\right) \tag{2}$$

Assume that there are, among the vertices of $\mathcal{P}$, a complete cycle of vertices of $\mathcal{N}'$; we compute the contribution Γ to the sum in (2) which this cycle accounts for. It depends on the character of the point $\mathfrak{C}$ of the surface $\tilde{\mathcal{K}}(\mathfrak{F})$ mod $\mathfrak{F}$ corresponding to the cycle c, and we have to examine the different possibilities. Let $\gamma = \gamma(c)$ denote the number of vertices in the cycle. Then

$$\Gamma = \sum_{\mu=1}^{\gamma}\left(1 - \frac{\varphi_\mu}{\pi}\right) = \gamma - \frac{1}{\pi}\sum_{\mu=1}^{\gamma}\varphi_\mu .$$

If $\mathfrak{C}$ is an inner regular point of the surface, the sum of the angles of the cycle is 2π, thus $\Gamma = \gamma - 2$. In this case we have $\gamma \geqq 3$, thus $\Gamma \geqq 1$.

If $\mathfrak{C}$ is a regular point on the boundary of the surface $\tilde{\mathcal{K}}(\mathfrak{F})$ mod $\mathfrak{F}$, the sum of the angles of the cycle is π, thus $\Gamma = \gamma - 1$. In this case we have $\gamma \geqq 2$, thus $\Gamma \geqq 1$.

If $\mathfrak{C}$ is a conical point of order ν of the surface, the sum of the angles of the cycle is $\frac{2\pi}{\nu}$, thus $\Gamma = \gamma - \frac{2}{\nu}$. In this case we have $\gamma \geqq 1$, and there is a possibility for a vanishing Γ: If $\nu = 2$ together with $\gamma = 1$, then $\Gamma = 0$. In all other cases we get $\Gamma \geqq \frac{1}{3}$.

If $\mathfrak{C}$ is an angular point of order ν of the surface, the sum of the angles of the cycle is $\frac{\pi}{\nu}$, thus $\Gamma = \gamma - \frac{1}{\nu} \geqq \frac{1}{2}$.

If $\mathfrak{C}$ is an inner point of a reflection edge of the surface, the sum of the angles of the cycle is π, thus $\Gamma = \gamma - 1$. In this case $\gamma \geqq 2$, thus $\Gamma \geqq 1$.

Finally, if $\mathfrak{C}$ is situated on a reflection edge and at the same time on the boundary curve of the surface, the sum of the angles of the cycle is $\frac{\pi}{2}$, since a reflection line can only cut a side of $\mathcal{K}(\mathfrak{F})$ at right angles. Thus $\Gamma = \gamma - \frac{1}{2}$. In this case $\gamma \geqq 1$, thus $\Gamma \geqq \frac{1}{2}$.

From this survey it is seen that $\Gamma = 0$ only occurs in the case of a cycle consisting of one centre of a half-turn with the interior angle of $\mathcal{N}'$ equal to π. It has been proved in §10.5 that there is, on the boundary of $\mathcal{N}'$, between any two vertices of this type at least one vertex of another type. Thus the number of vertices of $\mathcal{N}'$ not belonging to this special type is infinite, and they give rise to an infinite number of cycles of vertices, since every cycle comprises a finite number of vertices. For all these cycles, then, we have as a common lower bound $\Gamma \geqq \frac{1}{3}$.

We now select an arbitrary number m of cycles of this latter type on the boundary of $\mathcal{N}'$ and then cut off the points at infinity by segments inside $\mathcal{N}'$, as previously described, in such a way, that all the vertices of the m cycles are among the vertices of the resulting closed polygon $\mathcal{P}$. They contribute at least $\frac{m}{3}$ to the sum in (2). Therefore, we get from (2) for the polygon $\mathcal{P}$

$$\Phi(\mathcal{P}) \geqq \tfrac{1}{3}m - 2,$$

which is arbitrarily large, since m is at our disposal.

Hence, in all, in the case of non-quasi-compactness to every assigned quantity κ we can construct, as a part of $\mathcal{N}'$, a polygon with an area greater than κ. In this case we write $\Phi(\mathcal{N}') = \infty$ and likewise $\Phi(\mathfrak{F}) = \infty$.

If $\mathcal{K}(\mathfrak{F})$ is not quasi-compact modulo $\mathfrak{F}$, the area of the surface $\mathcal{K}(\mathfrak{F})$ mod $\mathfrak{F}$ *is infinite.* □

It is evident that the result does not depend on the choice of the point x_0 used as central point of the normal domain $\mathcal{N}(x_0)$: For every choice of x_0 we get an $\mathcal{N}'(x_0)$ with an infinite number of sides in virtue of the non-quasi-compactness of $\mathfrak{F}$, and that enables us to carry out the above construction.

If $\mathcal{K}(\mathfrak{F})$ is quasi-compact modulo $\mathfrak{F}$, the boundary of $\mathcal{N}$ is a closed polygon, as results from the preceding section. Its area is found by (2).

If $\mathcal{K}(\mathfrak{F})$ is quasi-compact modulo $\mathfrak{F}$, the area $\Phi(\mathfrak{F})$ of the surface $\mathcal{K}(\mathfrak{F})$ mod $\mathfrak{F}$ is defined as the area of $\mathcal{N}' = \mathcal{N}(x_0) \cap \mathcal{K}(\mathfrak{F})$ as computed from formula (2).

The calculation will be carried out in detail in Section 7 of this paragraph. In this section we prove the consistency of the definition, i.e. the independence of $\Phi(\mathfrak{F})$ from x_0. This proof will be given in a slightly more general form.

Let $\mathcal{P}$ denote an arbitrary fundamental polygon of $\mathfrak{F}$ with the only restriction that $\mathcal{P}$ has a finite number of sides. We put $\mathcal{P}' = \mathcal{P} \cap \mathcal{K}(\mathfrak{F})$. The sides and vertices of the network $\mathfrak{F}\mathcal{N}' = \mathfrak{F}(\mathcal{N}(x_0) \cap \mathcal{K}(\mathfrak{F}))$ do not accumulate in $\mathcal{D}$ nor do the sides and vertices of the network $\mathfrak{F}\mathcal{P}' = \mathfrak{F}(\mathcal{P} \cap \mathcal{K}(\mathfrak{F}))$, and therefore the same holds for the network produced by their superposition, because all boundaries are straight. In case there are limit-rotations in $\mathfrak{F}$, a limit-centre bounding $\mathcal{P}'$ must be the common end-point of two half-lines bounding $\mathcal{P}'$ in virtue of the finiteness of the number of sides of $\mathcal{P}'$, just as was the case with $\mathcal{N}'$. Therefore $\mathcal{N}'$ is decomposed into a finite number of "small" regions by the network $\mathfrak{F}\mathcal{P}'$ and vice versa. The small regions into which $\mathcal{K}(\mathfrak{F})$ is decomposed by the two superposed networks fall into a finite number λ of equivalence classes modulo $\mathfrak{F}$. The small regions produced in $\mathcal{N}'$ by $\mathfrak{F}\mathcal{P}'$ comprise exactly one representative of each of these λ classes, and so do the small regions produced in $\mathcal{P}'$ by $\mathfrak{F}\mathcal{N}'$. Therefore, in pairs, they are isometric and thus have the same area. Finally one gets $\Phi(\mathcal{P}') = \Phi(\mathcal{N}')$ from the additivity of areas, and this proves the assertion. Hence:

In the case of quasi-compactness the area $\Phi(\mathfrak{F})$ of the surface $\mathcal{K}(\mathfrak{F})$ mod $\mathfrak{F}$ can be obtained by measuring the area of the intersection of $\mathcal{K}(\mathfrak{F})$ with an arbitrary fundamental polygon with a finite number of sides. □

17.4 Type numbers. The considerations to follow aim at the establishment of a system of characteristic numbers suitable for classifying the groups treated under certain types. In so doing quasi-abelian groups can be left aside, because the classification of these groups carried out in §11 will prove sufficient. We are thus only concerned with $\mathfrak{F}$-groups, and the geometric meaning of the numbers in question is most easily deducted from the surface $\mathcal{D}$ mod $\mathfrak{F}$. In regarding the surface the attention is immediately drawn to the number of funnels, masts and singular points of different kind. For these numbers the notation $\vartheta()$ with the corresponding symbol as argument will be used, and they will be called *type numbers* of $\mathfrak{F}$.

The type numbers are chiefly important for groups satisfying the condition of quasi-compactness. Assuming $\mathcal{K}(\mathfrak{F})$ quasi-compact modulo $\mathfrak{F}$, it is known from §15.2 that the numbers indicated below are finite and that only a finite number of them do not vanish.

If $\mathfrak{F}$ contains no reversions, the following numbers are introduced:

$$\vartheta(\circlearrowleft_\nu) = \text{number of (equivalence) classes (with respect to } \mathfrak{F}\text{) of centres of order } \nu \text{ in } \mathfrak{F}$$
$$= \text{number of conical points of order } \nu \text{ of } \mathcal{D} \bmod \mathfrak{F} \text{ or } \mathcal{K}(\mathfrak{F}) \bmod \mathfrak{F}.$$

$$\vartheta(\hookrightarrow) = \text{number of classes of limit-centres of } \mathfrak{F}$$
$$= \text{number of masts of } \mathcal{D} \bmod \mathfrak{F} \text{ or } \mathcal{K}(\mathfrak{F}) \bmod \mathfrak{F}.$$

$$\vartheta(\rightarrow) = \text{number of classes of boundary axes of } \mathfrak{F}$$
$$= \text{number of funnels of } \mathcal{D} \bmod \mathfrak{F}$$
$$= \text{number of boundaries of } \mathcal{K}(\mathfrak{F}) \bmod \mathfrak{F}.$$

When taken together with the area $\Phi(\mathfrak{F})$, these numbers will prove sufficient for the classification of groups without reversions.

If $\mathfrak{F}$ contains reversions, the following numbers are introduced in addition to those preceding:

$$\begin{aligned}\vartheta(\times_\nu) &= \text{number of classes of centres of } \mathfrak{F} \text{ pertaining to subgroups of the} \\ &\quad\ \text{type } \mathfrak{A}(\times_\nu) \\ &= \text{number of angular points of order } \nu \text{ of } \mathcal{D} \text{ mod } \mathfrak{F} \text{ or } \mathcal{K}(\mathfrak{F}) \text{ mod } \mathfrak{F}.\end{aligned}$$

$$\begin{aligned}\vartheta(\wedge) &= \text{number of classes of limit-centres of } \mathfrak{F} \text{ pertaining to subgroups of} \\ &\quad\ \text{the type } \mathfrak{A}(\wedge) \\ &= \text{number of half-masts of } \mathcal{D} \text{ mod } \mathfrak{F} \text{ or } \mathcal{K}(\mathfrak{F}) \text{ mod } \mathfrak{F}.\end{aligned}$$

$$\begin{aligned}\vartheta(\|) &= \text{number of classes of boundary axes of } \mathfrak{F} \text{ pertaining to subgroups of} \\ &\quad\ \text{the type } \mathfrak{A}(\|) \\ &= \text{number of half-funnels of } \mathcal{D} \text{ mod } \mathfrak{F} \\ &= \text{number of such geodetic boundaries of } \mathcal{K}(\mathfrak{F}) \text{ mod } \mathfrak{F} \\ &\quad\ \text{as join two reflection edges .}\end{aligned}$$

$$\begin{aligned}\vartheta(-\to) &= \text{number of classes of reflection lines of } \mathfrak{F} \text{ pertaining to subgroups} \\ &\quad\ \text{of the type } \mathfrak{A}(-\to) \text{ and containing no centre} \\ &= \text{number of closed geodetic reflection edges of } \mathcal{D} \text{ mod } \mathfrak{F} \text{ or } \mathcal{K}(\mathfrak{F}) \\ &\quad\ \text{mod } \mathfrak{F} \text{ passing through no angular point .}\end{aligned}$$

It is remarked that any reflection line belonging to these $\vartheta(-\to)$ classes is simple modulo $\mathfrak{F}$ and that any two of them are disjoint modulo $\mathfrak{F}$ (§15.5), since an intersection point of any two such lines would be a centre of $\mathfrak{F}$.

A more complete classification of $\mathfrak{F}$-groups containing reflections would furthermore require some indication about the cyclical order of angular points, half-masts and half-funnels on the reflection rings of the surface (§16.5). In the present context, however, we may restrict ourselves to the introduction of the type number:

$$\begin{aligned}\vartheta(-) &= \text{number of reflection rings of } \mathcal{D} \text{ mod } \mathfrak{F}, \text{ this number thus including} \\ &\quad\ \text{the } \vartheta(\to -) \text{ closed reflection edges .}\end{aligned}$$

With a view to later use it is observed that the total number of bounding curves of the surface $\mathcal{K}^*(\mathfrak{F})$ mod $\mathfrak{F}$, including boundaries properly speaking as well as reflection rings and simple reflection edges is equal to

$$\vartheta(\to) + \vartheta(\hookrightarrow) + \vartheta(-).$$

If $\mathcal{K}(\mathfrak{F})$ is not quasi-compact modulo $\mathfrak{F}$, some or all of these numbers may be ∞, and moreover the numbers $\vartheta(\circledcirc_\nu)$ and $\vartheta(\times_\nu)$ may be different from zero for infinitely many ν. Moreover, $\mathcal{K}(\mathfrak{F})$ may, in this case, have limit sides, and these also contribute to the boundary of $\tilde{\mathcal{K}}(\mathfrak{F})$ mod $\mathfrak{F}$.

17.5 Orientability. Finally we introduce a type number $\vartheta(\rightleftharpoons)$ in order to indicate the property of the surface $\mathcal{D}$ mod $\mathfrak{F}$ of being orientable or not. We put $\vartheta(\rightleftharpoons) = 1$, if there is in $\mathfrak{F}$ an axis $\mathcal{A}$ pertaining to a subgroup $\mathfrak{A}_{\mathcal{A}}(\rightleftharpoons)$ and not intersecting any axis of reflection, and $\vartheta(\rightleftharpoons) = 0$ otherwise.

Let $\mathcal{N} = \mathcal{N}(\mathfrak{F}; x_0)$ be a normal domain for $\mathfrak{F}$. Then $\vartheta(\rightleftharpoons) = 1$, if there is at least one pair of (different) sides of $\mathcal{N}$ which correspond by a reversed translation of $\mathfrak{F}$, and in that case only.

Proof. In order to prove this, we first assume $\vartheta(\rightleftharpoons) = 1$. Then $\mathfrak{F}$ contains axes of reversed translations which are not reflection lines and which meet no reflection lines. At least one of them, $\mathcal{A}$ say, has points in common with $\mathcal{N}$. Let $\mathfrak{f}$ denote a reversed translation belonging to $\mathcal{A}$, and let it be assumed that no reversed translation occurs among those elements of $\mathfrak{F}$ which make the sides of $\mathcal{N}$ correspond in pairs. Let $\mathcal{A}$ be traversed from an inner point p of $\mathcal{N}$ to the equivalent point $\mathfrak{f}p$ of $\mathfrak{f}\mathcal{N}$. If this segment passes through some vertex of the network $\mathfrak{F}\mathcal{N}$, this vertex may be avoided by passing around it on a small circular arc in such a way as not to meet any reflection line, because in virtue of the assumption no reflection line can pass through this vertex. This path from p to $\mathfrak{f}p$ leads through a finite number of domains of the collection $\mathfrak{F}\mathcal{N}$, passing from one domain into the next by crossing a side. Every crossing of the path corresponds to an element of the group, and this element is, in virtue of the assumption, neither a reversed translation nor a reflection. That does not comply with the fact that the product of these elements is the reversed translation $\mathfrak{f}$.

Conversely, let it be assumed that the sides $\mathcal{S}$ and $\mathfrak{f}\mathcal{S}$ of $\mathcal{N}$ correspond by the reversed translation $\mathfrak{f}$ of $\mathfrak{F}$. Then $\mathcal{S}$ and $\mathfrak{f}\mathcal{S}$ are different sides of $\mathcal{N}$, because $\mathfrak{f}$ leaves no point of $\mathcal{D}$ fixed. Hence $\mathcal{S}$ is not situated on a reflection line of $\mathfrak{F}$, since it does not coincide with its equivalent side, and the same holds for $\mathfrak{f}\mathcal{S}$. Let p be an inner point of $\mathcal{S}$, and consider the straight segment $\mathfrak{T} = p, \mathfrak{f}p$. It belongs to $\mathcal{N}$ owing to the convexity of $\mathcal{N}$. It is either situated on $\mathcal{A}(\mathfrak{f})$ or intersect $\mathcal{A}(\mathfrak{f})$. Hence $\mathcal{A}(\mathfrak{f})$ intersects $\mathcal{N}$; thus it cannot be a reflection line of $\mathfrak{F}$. Also, since no reflection line of $\mathfrak{F}$ has points in common with the interior of $\mathcal{N}$, the segment $\mathfrak{T}$ meets no reflection line of $\mathfrak{F}$. The same then holds for the whole equivalence class $\mathfrak{F}\mathfrak{T}$. The sequence of segments $\mathfrak{f}^r\mathfrak{T}$, $-\infty < r < \infty$, is a straight or broken line joining the fundamental points of $\mathfrak{f}$. Since it meets no reflection line, the same holds for $\mathcal{A}(\mathfrak{f})$. □

It follows from what has been proved that *the surface $\mathcal{D}$* mod $\mathfrak{F}$ *is orientable, if* $\vartheta(\rightleftharpoons) = 0$. For, if $\mathcal{N}$ is decomposed into triangles in an arbitrary way and all triangles are oriented according to the positive orientation of the plane, then the orientation of the triangles is concordant even on the surface $\mathcal{D}$ mod $\mathfrak{F}$, since any two distinct equivalent sides of $\mathcal{N}$ correspond with opposite directions on the boundary of $\mathcal{N}$.

On the other hand, *if* $\vartheta(\rightleftharpoons) = 1$, *then* $\mathcal{D}$ mod $\mathfrak{F}$ *is not orientable.* For, if a chain of triangles in $\mathcal{N}$ is formed connecting two sides of $\mathcal{N}$ which correspond by a reversed translation, and if this chain is concordantly oriented in $\mathcal{N}$, the orientation of the triangles of the chain is not concordant on $\mathcal{D}$ mod $\mathfrak{F}$, since the two sides correspond with the same direction on the boundary of $\mathcal{N}$.

17.6 Characteristic and genus. The term *characteristic* $\chi(\mathfrak{F})$, which in short we write χ if only one group is involved, means the Euler characteristic of the surface $\mathcal{K}^*(\mathfrak{F})$ mod $\mathfrak{F}$, derived from a truncated domain $\mathcal{K}^*(\mathfrak{F})$ of $\mathfrak{F}$. Here we restrict ourselves to the case of quasi-compactness for $\mathfrak{F}$, in which the surface $\mathcal{K}^*(\mathfrak{F})$ mod $\mathfrak{F}$ can be decomposed into a finite number of simply connected polygonal parts. Let α_0 denote the number of vertices, α_1 the number of sides and α_2 the number of simply connected parts. Then we have

$$\chi = \alpha_0 - \alpha_1 + \alpha_2.$$

It will result from the next section that χ does not depend on the choice of the truncated domain. The same holds for the independence of χ of the decomposition to the extent in which it will be used.

Let $\mathcal{N}$ be a normal domain for $\mathfrak{F}$. If there are limit-rotations in $\mathfrak{F}$, and thus a finite number of infinite vertices of $\mathcal{N}$, let the horocycle used for the determination of a truncated domain $\mathcal{K}^*(\mathfrak{F})$ be so chosen that the closure of their interior contains no other vertex of $\mathcal{N}$ than the corresponding limit-centre; see Fig. 17.2. The said decomposition may now be achieved by choosing a point p in the interior of $\mathcal{N} \cap \mathcal{K}^*(\mathfrak{F})$ and joining it with all vertices of that polygonal domain. Since that domain is not convex if there are bounding horocycles, it may be necessary to use curved lines of decomposition emanating from p. In this way the surface $\mathcal{K}^*(\mathfrak{F})$ mod $\mathfrak{F}$ is decomposed into m triangles with common vertex p, where m denotes the total number of sides of $\mathcal{N} \cap \mathcal{K}^*(\mathfrak{F})$. Hence

$$\alpha_2 = m.$$

Let ξ denote the number of cycles of vertices of $\mathcal{N} \cap \mathcal{K}^*(\mathfrak{F})$ and η the number of pairs of equivalent sides of $\mathcal{N} \cap \mathcal{K}^*(\mathfrak{F})$ which are different. Then we have

$$\alpha_0 = 1 + \xi, \quad \alpha_1 = 2m - \eta,$$

and thus

$$\chi = 1 + \xi - m + \eta. \tag{3}$$

Consider the reflection rings of $\mathcal{D}$ mod $\mathfrak{F}$. All reflection edges of $\mathcal{D}$ mod $\mathfrak{F}$ belong in their whole extent to $\mathcal{K}^*(\mathfrak{F})$ mod $\mathfrak{F}$ except those which run into a half-mast or a half-funnel. In the first case a half-mast is cut off from $\mathcal{D}$ mod $\mathfrak{F}$ by a boundary of $\mathcal{K}^*(\mathfrak{F})$ mod $\mathfrak{F}$, this boundary being the image on $\mathcal{K}^*(\mathfrak{F})$ mod $\mathfrak{F}$ of a bounding horocycle of $\mathcal{K}^*(\mathfrak{F})$ with corresponding subgroup $\mathfrak{A}(\wedge)$. In the second case a half-funnel is cut off from $\mathcal{D}$ mod $\mathfrak{F}$ by a boundary of $\mathcal{K}(\mathfrak{F})$ mod $\mathfrak{F}$, and thus also of $\mathcal{K}^*(\mathfrak{F})$ mod $\mathfrak{F}$, this boundary being the image on $\mathcal{K}(\mathfrak{F})$ mod $\mathfrak{F}$ of a boundary axis of $\mathcal{K}(\mathfrak{F})$ with corresponding subgroup $\mathfrak{A}(\|)$. If a reflection ring of $\mathcal{D}$ mod $\mathfrak{F}$ is contained in $\mathcal{K}^*(\mathfrak{F})$ mod $\mathfrak{F}$, then it is a closed curve on $\mathcal{K}^*(\mathfrak{F})$ mod $\mathfrak{F}$. If a reflection ring of $\mathcal{D}$ mod $\mathfrak{F}$ is not wholly contained in $\mathcal{K}^*(\mathfrak{F})$ mod $\mathfrak{F}$, then a closed curve on $\mathcal{K}^*(\mathfrak{F})$ mod $\mathfrak{F}$ is derived from it by adding certain boundaries of $\mathcal{K}^*(\mathfrak{F})$ mod $\mathfrak{F}$ as described above. One can thus in a uniquely determined way speak of the closed curve on $\mathcal{K}^*(\mathfrak{F})$ mod $\mathfrak{F}$ corresponding to a given reflection ring. These closed curves together with such boundaries as are derived from boundary axes with corresponding subgroup $\mathfrak{A}(\rightarrow)$ and bounding horocycles with corresponding subgroup $\mathfrak{A}(\hookrightarrow)$ constitute the complete border of $\mathcal{K}^*(\mathfrak{F})$ mod $\mathfrak{F}$. The number of cycles of vertices of $\mathcal{N} \cap \mathcal{K}^*(\mathfrak{F})$ which correspond to boundary points or to points on reflection edges of $\mathcal{K}^*(\mathfrak{F})$ mod $\mathfrak{F}$ is, therefore, equal to the number of sides of $\mathcal{N} \cap \mathcal{K}^*(\mathfrak{F})$ which belong to boundary axes, bounding horocycles, or reflection lines; they are those $m-2\eta$ sides of $\mathcal{N} \cap \mathcal{K}^*(\mathfrak{F})$ which either have no equivalent side, or which are self-equivalent, i.e. situated on a reflection line. All other cycles of vertices of $\mathcal{N} \cap \mathcal{K}^*(\mathfrak{F})$ yield inner points of $\mathcal{K}^*(\mathfrak{F})$ mod $\mathfrak{F}$. Let ξ_r denote the number of those cycles which correspond to regular inner points, and ξ_c the number of those cycles which correspond to conical points of $\mathcal{K}^*(\mathfrak{F})$ mod $\mathfrak{F}$. One thus gets for the total number of cycles of vertices of $\mathcal{N} \cap \mathcal{K}^*(\mathfrak{F})$

$$\xi = m - 2\eta + \xi_r + \xi_c,$$

and hence from (3)

$$\chi = 1 + \xi_r + \xi_c - \eta. \tag{4}$$

Observe that the right-hand side in (4) depends only on $\mathcal{N} \cap \mathcal{K}(\mathfrak{F})$ thus not on the choice of horocycles bounding $\mathcal{K}^*(\mathfrak{F})$ inside $\mathcal{K}(\mathfrak{F})$.

We now define a number p, which is called the *genus* of the surface $\mathcal{D}$ mod $\mathfrak{F}$ (or of $\mathcal{K}(\mathfrak{F})$ mod $\mathfrak{F}$, or of $\mathcal{K}^*(\mathfrak{F})$ mod $\mathfrak{F}$) by the equation

$$(2 - \vartheta(\rightleftharpoons))p = 2 - \chi - \vartheta(\rightarrow) - \vartheta(\hookrightarrow) - \vartheta(-). \tag{5}$$

It follows from (4) that

$$(2 - \vartheta(\rightleftharpoons))p = 1 - \xi_r - \xi_c - \vartheta(\rightarrow) - \vartheta(\hookrightarrow) - \vartheta(-) + \eta.$$

We set out to prove that this number p is not negative. The boundary of $\mathcal{N} \cap \mathcal{K}^*(\mathfrak{F})$ gives rise to a coherent one-dimensional complex on the surface. If every closed boundary curve, and every closed curve derived above from a reflection ring, is contracted to a point and represented by that point, then the complex has

$$\beta_0 = \xi_r + \xi_c + \vartheta(\rightarrow) + \vartheta(\hookrightarrow) + \vartheta(-)$$

points and $\beta_1 = \eta$ segments. It is thus sufficient to prove that

$$1 - \beta_0 + \beta_1 \geqq 0$$

for an arbitrary connected one-dimensional complex, and this can be done by induction: The statement is true for a complex consisting of one segment with its end-points. Let a complex be given for which it holds, and let a segment be added which has at least one end-point in common with the complex. Then β_1 is increased by unity, and β_0 is either increased by unity or left unchanged. This proves the assertion.

From $p \geqq 0$ and (5) one gets the inequality

$$\chi \leqq 2 - \vartheta(\rightarrow) - \vartheta(\hookrightarrow) - \vartheta(-) \leqq 2. \tag{6}$$

17.7 Relation between area and type numbers. Let $\mathfrak{F}$ be an $\mathfrak{F}$-group whose convex domain $\mathcal{K}(\mathfrak{F})$ is quasi-compact modulo $\mathfrak{F}$, and let $\mathcal{N} = \mathcal{N}(\mathfrak{F}; x_0)$ be a normal domain. According to Section 2 the intersection $\mathcal{N} \cap \mathcal{K}(\mathfrak{F})$ is a polygonal convex region with a finite number n of sides, and according to Section 3 the area of $\mathcal{K}(\mathfrak{F})$ mod $\mathfrak{F}$ is given by a formula analogous to (2):

$$\Phi(\mathfrak{F}) = -2 + \sum_{\mu=1}^{n} \left(1 - \frac{\varphi_\mu}{\pi}\right), \tag{7}$$

the sum extending to the n vertices of $\mathcal{N} \cap \mathcal{K}(\mathfrak{F})$ and φ_μ denoting the polygonal angle at a vertex. In particular, this angle is zero at an infinite vertex, in case $\mathfrak{F}$ contains limit-rotations.

In order to compute $\Phi(\mathfrak{F})$ we first enumerate the contribution Γ of a single cycle of vertices to the sum in (7) which has already been established in Section 3, the symbol γ denoting the number of vertices in the cycle. For our present task we only have to add the cycles of infinite vertices, if any. Moreover, Table 2 below contains a symbol indicating the number of cycles of the type in question, which depends on the nature of the corresponding point on the surface $\mathcal{K}(\mathfrak{F})$ mod $\mathfrak{F}$.

Table 2

Type of cycle	Γ of the cycle	number of cycles
regular, inner	$\gamma - 2$	ξ_r
regular, boundary	$\gamma - 1$	ξ_r'
conical	$\gamma - \frac{2}{\nu}$	ξ_c
angular	$\gamma - \frac{1}{\nu}$	ξ_a
reflection edge, inner	$\gamma - 1$	ξ_i
reflection edge, boundary	$\gamma - \frac{1}{2}$	ξ_b
limit-centre without reflection	γ	$\vartheta(\hookrightarrow)$
limit-centre with reflection	γ	$\vartheta(\wedge)$

It has to be remembered that both γ and ν depend on the individual cycle concerned, and that for the number of cycles

$$\xi_c = \sum_\nu \vartheta(\odot_\nu), \quad \xi_a = \sum_\nu \vartheta(\times_\nu), \quad \xi_b = 2\vartheta(\|). \tag{8}$$

We now compute $\Phi(\mathfrak{F})$ by adding all these contributions and inserting in (7), bearing (8) in mind and remembering that the sum of the γ of all cycles is n:

$$\begin{aligned}
\Phi(\mathfrak{F}) &= -2 + n - 2\xi_r - \xi_r' - 2\sum_\nu \frac{1}{\nu}\vartheta(\odot_\nu) - \sum_\nu \frac{1}{\nu}\vartheta(\times_\nu) - \xi_i - \vartheta(\|) \\
&= -2 + n - 2\xi_r - \xi_r' - 2\xi_c + 2\sum_\nu \left(1 - \frac{1}{\nu}\right)\vartheta(\odot_\nu) - \xi_a \\
&\quad + \sum_\nu \left(1 - \frac{1}{\nu}\right)\vartheta(\times_\nu) - \xi_i - 2\vartheta(\|) + \vartheta(\|).
\end{aligned}$$

Now the number of those sides of $\mathcal{N} \cap \mathcal{K}(\mathfrak{F})$ which are situated on boundary axes or on lines of reflection is equal to the number of vertices on boundary axes and lines of reflection, since all boundaries and curves derived from reflection rings of the surface $\mathcal{K}(\mathfrak{F})$ mod $\mathfrak{F}$ are closed; there are

$$\xi_r' + \xi_a + \xi_i + 2\vartheta(\|) + \vartheta(\wedge)$$

vertices on boundary axes and reflection lines. On the other hand, the number of these sides is $n - 2\eta$, if η denotes the number of pairs of distinct equivalent sides of $\mathcal{N} \cap \mathcal{K}(\mathfrak{F})$. Hence

$$n - 2\eta = \xi_r' + \xi_a + \xi_i + 2\vartheta(\|) + \vartheta(\wedge).$$

In case $\mathfrak{F}$ contains limit-rotations, let the truncated domain $\mathcal{K}^*(\mathfrak{F})$ be chosen as in the preceding section, i.e. such that the closure of the interior of a bounding horocycle contains no other vertex than the corresponding limit-centre. Then the numbers ξ_r, ξ_c and η are the same for $\mathcal{N} \cap \mathcal{K}^*(\mathfrak{F})$ and $\mathcal{N} \cap \mathcal{K}(\mathfrak{F})$ and these numbers as used in this section can therefore be inserted in (4) in order to determine the characteristic $\chi(\mathfrak{F})$. We then get from two of the formulae above and (4)

$$\Phi(\mathfrak{F}) = -2\chi + 2\sum_\nu \left(1 - \frac{1}{\nu}\right)\vartheta(\odot_\nu) + \sum_\nu \left(1 - \frac{1}{\nu}\right)\vartheta(\times_\nu) + \vartheta(\wedge) + \vartheta(\|). \tag{9}$$

If the genus as defined by (5) is introduced in (9), one gets

$$\begin{aligned}
\Phi(\mathfrak{F}) &= 2p(2 - \vartheta(\overset{\rightharpoonup}{=})) - 4 + 2\vartheta(\rightarrow) + 2\vartheta(\hookrightarrow) + 2\vartheta(-) \\
&\quad + 2\sum_\nu \left(1 - \frac{1}{\nu}\right)\vartheta(\odot_\nu) + \sum_\nu \left(1 - \frac{1}{\nu}\right)\vartheta(\times_\nu) + \vartheta(\wedge) + \vartheta(\|).
\end{aligned} \tag{10}$$

In particular, one gets for such groups which contain reversed translations but no reflections

$$\Phi = 2p - 4 + 2\vartheta(\rightarrow) + 2\vartheta(\hookrightarrow) + 2\sum_\nu \left(1 - \frac{1}{\nu}\right) \vartheta(\circlearrowright_\nu) \tag{11}$$

and for groups without reversions

$$\Phi = 4p - 4 + 2\vartheta(\rightarrow) + 2\vartheta(\hookrightarrow) + 2\sum_\nu \left(1 - \frac{1}{\nu}\right) \vartheta(\circlearrowright_\nu). \tag{12}$$

It is evident from (9) and (10) that the characteristic and the genus only depend on the group and not on the choice of the normal domain used for their definitions in the preceding section. For, all the other quantities entering into these formulae only depend on the group.

It will now be inferred from (10) that there is a positive minimum for the area Φ of an $\mathfrak{F}$-group.

If there are no reflections in $\mathfrak{F}$, one gets from (10)

$$\tfrac{1}{2}\Phi = p(2 - \vartheta(\rightleftharpoons)) - 2 + \vartheta(\rightarrow) + \vartheta(\hookrightarrow) + \sum_\nu \left(1 - \frac{1}{\nu}\right) \vartheta(\circlearrowright_\nu),$$

which takes the form

$$\tfrac{1}{2}\Phi = i + \sum_\nu \vartheta(\circlearrowright_\nu) \left(1 - \frac{1}{\nu}\right), \tag{13}$$

when using i to denote the integer

$$i = p(2 - \vartheta(\rightleftharpoons)) - 2 + \vartheta(\rightarrow) + \vartheta(\hookrightarrow) \geqq -2.$$

If there are reflections in $\mathfrak{F}$, one gets from (10)

$$\Phi = i + \sum_\nu (2\vartheta(\circlearrowright_\nu) + \vartheta(\times_\nu)) \left(1 - \frac{1}{\nu}\right), \tag{14}$$

where i denotes the integer

$$i = 2p(2 - \vartheta(\rightleftharpoons)) - 4 + 2\vartheta(\rightarrow) + 2\vartheta(\hookrightarrow) + 2\vartheta(-) + \vartheta(\wedge) + \vartheta(\|) \geqq -2,$$

since in this case $\vartheta(-) \geqq 1$. Thus both in (13) and (14) the problem is to determine the smallest positive value of the quantity

$$\Psi = i + \sum_\nu a_\nu \left(1 - \frac{1}{\nu}\right),$$

where i denotes an integer not inferior to -2 and where ν ranges over the values $\nu = 2, 3, \ldots$ and the a_ν are non-negative integers.

In looking for the smallest positive value of Ψ one can restrict oneself to assuming $a_\nu < \nu$, for, otherwise it would be possible to transfer a positive integer from the sum to i. Consequently, there is in the sum at most one contribution $\frac{1}{2}$, at most two contributions $\frac{2}{3}$, at most three contributions $\frac{3}{4}$, and so on.

If $i \geqq 0$, the smallest positive value of Ψ is obviously $\frac{1}{2}$.

If $i = -1$, there must be at least two terms in the sum. The two smallest contributions are $\frac{1}{2}$ and $\frac{2}{3}$, which yields $\Psi \geqq \frac{1}{6}$.

If $i = -2$, there must be at least three terms in the sum. If there are four terms or more in the sum, the four smallest possible contributions would be one $\frac{1}{2}$, two $\frac{2}{3}$ and one $\frac{3}{4}$; thus

$$\Psi \geqq -2 + \frac{1}{2} + \frac{2}{3} + \frac{2}{3} + \frac{3}{4} = \frac{7}{12}.$$

We now consider the case of three terms in the sum. If $a_2 = 0$, then

$$\Psi \geqq -2 + \frac{2}{3} + \frac{2}{3} + \frac{3}{4} = \frac{1}{12}.$$

If $a_2 = 1,\ a_3 = 0$, one gets

$$\Psi \geqq -2 + \frac{1}{2} + \frac{3}{4} + \frac{4}{5} = \frac{1}{20}.$$

If $a_2 = 1,\ a_3 \geqq 1$, the inequality

$$\frac{1}{2} + \frac{2}{3} + \left(1 - \frac{1}{\nu}\right) > 2$$

requires $\nu \geqq 7$. In this case

$$\Psi \geqq -2 + \frac{1}{2} + \frac{2}{3} + \frac{6}{7} = \frac{1}{42},$$

which thus is the smallest positive value for Ψ. By inserting this in (13) and (14) one gets the following theorem:

There is a positive lower bound for the area $\Phi(\mathfrak{F})$ of the surface $\mathcal{K}(\mathfrak{F})$ mod $\mathfrak{F}$ *given by $\Phi \geqq \frac{1}{21}$ in case there are no reflections in $\mathfrak{F}$, and by $\Phi \geqq \frac{1}{42}$ in case there are reflections in $\mathfrak{F}$.* □

Later on it will be verified by examples that both these lower bounds actually occur.

Chapter IV
Decompositions of groups

§18 Composition of groups

18.1 Generalized free products. The considerations of the present chapter are essentially based on the following concept treated in abstract group theory.

Let $\mathfrak{G}_i$, where the index i is drawn from a certain index range I, denote a collection of subgroups of a group $\mathfrak{G}$. By the *multiplication table* M_i of $\mathfrak{G}_i$ we mean the totality of the equations $\mathfrak{x}_1\mathfrak{x}_2 = \mathfrak{x}_3$ for all pairs of (different or equal) elements $\mathfrak{x}_1, \mathfrak{x}_2$ in $\mathfrak{G}_i$. Let Γ denote the collection of all those elements of $\mathfrak{G}$ which appear in at least one of the subgroups $\mathfrak{G}_i$, $i \in I$; an element in Γ may belong to more than one $\mathfrak{G}_i$. If the collection of subgroups $\mathfrak{G}_i$ generates $\mathfrak{G}$, and if, taking Γ as a system of generators for $\mathfrak{G}$, the union of all the M_i, $i \in I$, is a system of defining relations for $\mathfrak{G}$ in these generators, then $\mathfrak{G}$ is called the *generalized free product of the* $\mathfrak{G}_i$, $i \in I$.

Of course, the unit element of $\mathfrak{G}$ can be omitted as a generator, and also those equations in a multiplication table where at least one element of the pair $\mathfrak{x}_1, \mathfrak{x}_2$ is the unit element. Moreover, if $\mathfrak{G}_j \subset \mathfrak{G}_i$, $i, j \in I$, then $\mathfrak{G}_j$ can be omitted from the collection of subgroups, since the omission of $\mathfrak{G}_j$ does not affect Γ, and $M_j \subset M_i$.

Let the intersection of any two of the subgroups be denoted by

$$\mathfrak{G}_i \cap \mathfrak{G}_j = \mathfrak{H}_{ij} = \mathfrak{H}_{ji}, \quad i, j \in I.$$

The union of the subgroups $\mathfrak{H}_{ij}$ of $\mathfrak{G}$ is called the *amalgam* of the product.

Let Φ_i denote any system of generators of $\mathfrak{G}_i$, and Ψ_i a system of defining relations for $\mathfrak{G}_i$ in these generators. These relations are, of course, a consequence of the multiplication table M_i of $\mathfrak{G}_i$. Conversely, if in any equation $\mathfrak{x}_1\mathfrak{x}_2 = \mathfrak{x}_3$ of M_i the three elements of $\mathfrak{G}_i$ are expressed as products of elements from $\mathfrak{G}_i$, then the relation obtained is a consequence of the system Ψ_i; it may, of course, become an identical relation between elements from Φ_i. If $\mathfrak{h} \neq 1$ is an element of the amalgam, say $\mathfrak{h} \in \mathfrak{H}_{ij}$, and if

$$\mathfrak{h} = \varphi_i(\Phi_i), \quad \mathfrak{h} = \varphi_j(\Phi_j),$$

are expressions for $\mathfrak{h}$ by generators drawn from Φ_i and drawn from Φ_j, respectively, then

$$\varphi_i(\Phi_i) = \varphi_j(\Phi_j)$$

is a relation in the system of generators $\Phi = \bigcup_{i \in I} \Phi_i$ of $\mathfrak{G}$; it does not, in general, follow from the system of relations $\Psi = \bigcup_{i \in I} \Psi_i$ valid in $\mathfrak{G}$. If χ denotes the system of relations obtained in this way be letting $\mathfrak{h}$ range over the whole amalgam, then the union of Ψ and χ is a system of defining relations for $\mathfrak{G}$ in the generators Φ. For, one can extend Φ_i to a system of generators of $\mathfrak{G}_i$ including all elements of $\mathfrak{G}_i$, each element of $\mathfrak{G}_i$ being denoted by a definite symbol, and Ψ_i then yields M_i as a system of defining relations for $\mathfrak{G}_i$ in this extended system of generators of $\mathfrak{G}_i$. In consequence of the relations χ common elements of any two subgroups $\mathfrak{G}_i$ and $\mathfrak{G}_j$ to be denoted by the same symbol in this process. One thus gets the system Γ of generators of $\mathfrak{G}$ and $\bigcup_{i \in I} M_i$ is a system of defining relations for $\mathfrak{G}$ in them.

In establishing the system χ of relations valid in $\mathfrak{G}$ it is, of course, sufficient to let $\mathfrak{h}$ range over any system of generators of $\mathfrak{H}_{ij}$ for every pair $i, j \in I, i \neq j$. If, in particular, $\mathfrak{h} \in \mathfrak{H}_{ij}$ is used as a generator both in the system Φ_i for $\mathfrak{G}_i$ and in the system Φ_j for $\mathfrak{G}_j$ and denoted by the same symbol in both groups, then the corresponding relations in the system χ lapses.

$\mathfrak{G}$ is determined by the $\mathfrak{G}_i$ and the $\mathfrak{H}_{ij}$. We express this fact by the notation

$$\mathfrak{G} = \prod\nolimits^{*} \mathfrak{G}_i \operatorname{am} \mathfrak{H}_{ij}, \quad i, j \in I.$$

More precisely then $\mathfrak{G}$ is called the (generalized) *free product of the $\mathfrak{G}_i$ with amalgamated subgroups $\mathfrak{H}_{ij}$*. The group $\mathfrak{G}$ is said to be *decomposed into a generalized free product*. If I is a finite range, one may spell the product out and write

$$\mathfrak{G} = \mathfrak{G}_1 * \cdots * \mathfrak{G}_s \operatorname{am} \mathfrak{H}_{ij}; \quad i, j = 1, 2, \ldots, s; \quad i \neq j.$$

In the special case in which the amalgam consists only the unit element of $\mathfrak{G}$, thus where any two of the subgroups $\mathfrak{G}_i$, $\mathfrak{G}_j$, have only the unit element in common, $\mathfrak{G}$ is called the (ordinary) *free product of its subgroups $\mathfrak{G}_i$*, and we simply write

$$\mathfrak{G} = \prod\nolimits^{*} \mathfrak{G}_i, \quad i \in I.$$

If I' is a subrange of I and the subgroup $\mathfrak{G}'$ of $\mathfrak{G}$ is the generalized free product of the $\mathfrak{G}_i$, $i \in I'$, then $\mathfrak{G}$ is the generalized free product of $\mathfrak{G}'$ and the $\mathfrak{G}_i$, $i \in I$, $i \notin I'$. For, $\mathfrak{G}'$ is generated by the $\mathfrak{G}_i$, $i \in I'$, and the multiplication of $\mathfrak{G}'$ can be derived from the multiplication tables M_i, $i \in I'$.

Conversely, if one of the $\mathfrak{G}_i$ is decomposable into a generalized free product, then in the decomposition of $\mathfrak{G}$ this subgroup $\mathfrak{G}_i$ may be replaced by the collection of factors into which it is decomposed, thus yielding a *refinement* of the decomposition of $\mathfrak{G}$.

In the application to follow the amalgam of the generalized free product will satisfy a special condition, namely that any two subgroups $\mathfrak{H}_{ij}$ and $\mathfrak{H}_{i\ell}$ of $\mathfrak{G}_i$ have only the unit element in common, and this for every $i \in I$. The groups considered will be groups of motions and (possibly) reversions in $\mathcal{D}$.

18.2 Generalized free product of two groups operating on two mutually adjacent regions. Let $\mathcal{S}$ be a straight line in $\mathcal{D}$, and let $\mathcal{D}_1$ and $\mathcal{D}_2$ denote the two open half-planes determined by $\mathcal{S}$. Consider a group $\mathfrak{G}_1$ of motions and (possibly) reversions satisfying the following conditions:

1) $\mathfrak{G}_1$ is discontinuous in $\mathcal{D}$;

2) $\mathcal{S}$ is not an invariant line of $\mathfrak{G}_1$;

3) For every element $\mathfrak{f}$ of $\mathfrak{G}_1$ for which $\mathfrak{f}\mathcal{S} \neq \mathcal{S}$, the line $\mathfrak{f}\mathcal{S}$ is situated in $\mathcal{D}_1$.

Note that these conditions are in particular satisfied if $\mathfrak{G}_1$ is an $\mathfrak{F}$-group and $\mathcal{S}$ is either a side of $\mathcal{K}(\mathfrak{G}_1)$ or contained in any of the half-planes outside $\mathcal{K}(\mathfrak{G}_1)$ and simple modulo that subgroup of $\mathfrak{G}_1$ which carries that half-plane into itself. The following conclusions immediately result from the conditions 1), 2), 3). Let $\mathfrak{h}$ be an element of $\mathfrak{G}_1$ such that $\mathfrak{h}\mathcal{S} \neq \mathcal{S}$ (condition 2)). Then $\mathfrak{h}\mathcal{S} \subset \mathcal{D}_1$ (condition 3)). There is in $\mathfrak{G}_1$ no subgroup $\mathfrak{A}(\odot_\nu)$ or $\mathfrak{A}(\times_\nu)$ with $\nu > 2$ and centre on $\mathcal{S}$ (condition 3)). $\mathcal{S}$ contains no centre of order $\nu = 2$ of $\mathfrak{G}_1$, nor is $\mathcal{S}$ a reflection line of $\mathfrak{G}_1$, nor is there in $\mathfrak{G}_1$ a reversed translation with axis $\mathcal{S}$; for, if $\mathfrak{k}$ were an element of $\mathfrak{G}_1$ corresponding to one of these cases, then $\mathfrak{k}\mathfrak{h}\mathcal{S}$ would be situated in $\mathcal{D}_2$ contrary to condition 3). Let $\mathfrak{A}_\mathcal{S}$ denote the subgroup of $\mathfrak{G}_1$ which carries $\mathcal{S}$ into itself. $\mathfrak{A}_\mathcal{S}$ is discontinuous in virtue of condition 1). It then follows that $\mathfrak{A}_\mathcal{S} = 1$, $\mathfrak{A}_\mathcal{S} = \mathfrak{A}(|)$, $\mathfrak{A}_\mathcal{S} = \mathfrak{A}(\rightarrow)$, and $\mathfrak{A}_\mathcal{S} = \mathfrak{A}(\|)$ are the only possibilities for this subgroup.

If $\mathfrak{A}_\mathcal{S}$ contains reflections, the reflection lines of $\mathfrak{A}_\mathcal{S}$ cut $\mathcal{S}$ at right angles.

If $\mathfrak{A}_\mathcal{S}$ contains translations, then both $\mathcal{S}$ and $\mathfrak{h}\mathcal{S}$ are axes of $\mathfrak{G}_1$ and $\mathfrak{G}_1$ is thus an $\mathfrak{F}$-group. In this case $\mathcal{S}$ is a boundary axis of $\mathcal{K}(\mathfrak{G}_1)$, for, if $\mathcal{S}$ had points in common with the interior of $\mathcal{K}(\mathfrak{G}_1)$, there would be axes of $\mathfrak{G}_1$ intersecting $\mathcal{S}$, and that would lead to a violation of condition 3). $\mathcal{S}$ is simple modulo $\mathfrak{G}_1$ (§15.5), since it is not cut by any of its equivalents by $\mathfrak{G}_1$ (condition 3)). The collection $\mathfrak{G}_1\mathcal{S}$ consists of lines which are mutually disjoint and which, except for $\mathcal{S}$ itself, are situated in $\mathcal{D}_1$. This is reproduced by every element of $\mathfrak{G}_1$. Hence, if $\mathfrak{f}$ is any element of $\mathfrak{G}_1$ not in $\mathfrak{A}_\mathcal{S}$, the collection $\mathfrak{G}_1\mathcal{S}$ is, except for $\mathfrak{f}\mathcal{S}$, situated in $\mathfrak{f}\mathcal{D}_1$, which is one of the half-planes determined by $\mathfrak{f}\mathcal{S}$, and thus the one which contains $\mathcal{S}$. It is inferred that the intersection of the collection of half-planes $\mathfrak{G}_1\mathcal{D}_1$ is a convex subregion $\mathfrak{C}_1$ of $\mathcal{D}_1$ which is bounded on $\mathcal{D}$ by the complete collection $\mathfrak{G}_1\mathcal{S}$. The collection $\mathfrak{G}_1\mathcal{D}_2$ (strictly speaking the collection $\mathfrak{f}_\nu\mathcal{D}_2$, where $\mathfrak{f}_\nu$ ranges over a set of representatives of the left cosets of $\mathfrak{A}_\mathcal{S}$ in $\mathfrak{G}_1$) consists of disjoint half-planes, and their union is the complement relative to $\mathcal{D}$ of the closure $\tilde{\mathfrak{C}}_1$ of $\mathfrak{C}_1$ on $\mathcal{D}$. The region $\mathfrak{C}_1$ is carried into itself by all elements of $\mathfrak{G}_1$. The number of its boundary lines in $\mathcal{D}$ may be finite or infinite.

If $\mathfrak{G}_1$ is an $\mathfrak{F}$-group, the closure $\overline{\mathfrak{C}}_1$ (on $\overline{\mathcal{D}}$) of $\mathfrak{C}_1$ comprises $\overline{\mathcal{K}}(\mathfrak{G}_1)$, since $\overline{\mathfrak{C}}_1$ is convex and closed and is reproduced by all elements of $\mathfrak{G}_1$ (§14.1). Since $\mathfrak{G}_1$ is discontinuous in $\mathcal{D}$ (condition 1)), the surface $\mathcal{D}$ mod $\mathfrak{G}_1$ exists. The contribution of $\mathcal{D}_2$ to $\mathcal{D}$ mod $\mathfrak{G}_1$ is a half-plane, or a quarter-plane, or a funnel, or a half-funnel, according to the four possibilities for $\mathfrak{A}_\mathcal{S}$.

Let $\mathfrak{G}_2$ be a group satisfying the same conditions as $\mathfrak{G}_1$ with $\mathcal{D}_2$ substituted for $\mathcal{D}_1$ in condition 3). Moreover, the subgroup of $\mathfrak{G}_2$ carrying $\mathcal{S}$ into itself shall consist of the same motions and reversions as the subgroup $\mathfrak{A}_\mathcal{S}$ of $\mathfrak{G}_1$. The collection $\mathfrak{G}_2\mathcal{S}$ then bounds in $\mathcal{D}_2$ a convex region $\mathfrak{C}_2$ adjacent to $\mathfrak{C}_1$ along $\mathcal{S}$. The half-planes adjacent to $\mathfrak{C}_2$ along the lines of $\mathfrak{G}_2\mathcal{S}$ are the half-planes $\mathfrak{G}_2\mathcal{D}_1$. The region $\mathfrak{C}_2$ is carried into itself by all elements of $\mathfrak{G}_2$.

The elements of $\mathfrak{A}_{\mathcal{S}}$ carry both $\mathfrak{C}_1$ and $\mathfrak{C}_2$ into themselves; no other element of $\mathfrak{G}_1$ or $\mathfrak{G}_2$ has this property.

We are going to study the group $\mathfrak{G}$ generated by $\mathfrak{G}_1$ and $\mathfrak{G}_2$. It will result from the following considerations that the lines of the collection $\mathfrak{G}\mathcal{S}$ are mutually disjoint and determine a collection of convex regions $\mathfrak{G}\mathfrak{C}_1$ and $\mathfrak{G}\mathfrak{C}_2$ in $\mathcal{D}$. If $\mathcal{S}'$ and $\mathcal{S}''$ designate two lines of the collection $\mathfrak{G}\mathcal{S}$, the straight segment joining a point of $\mathcal{S}'$ with a point of $\mathcal{S}''$ crosses a certain number ν of these regions, and this number is independent of the points chosen on $\mathcal{S}'$ and $\mathcal{S}''$ owing to the convexity of the regions and to the fact that they in $\mathcal{D}$ bounded by complete lines. It will then be said that $\mathcal{S}'$ and $\mathcal{S}''$ are *separated by* ν *regions*.

Every element of $\mathfrak{G}$ is either an element of $\mathfrak{A}_{\mathcal{S}}$, or it can be written in the form

$$\mathfrak{g} = \mathfrak{g}_\ell \mathfrak{g}_{\ell-1} \cdots \mathfrak{g}_2 \mathfrak{g}_1, \tag{1}$$

where the factors of this product are alternately drawn from the two generating groups and none from $\mathfrak{A}_{\mathcal{S}}$, because a factor from $\mathfrak{A}_{\mathcal{S}}$ can always be combined with a neighbouring factor. Let i denote one of the indices 1 and 2 and j the other. If $\mathfrak{g}_1 \in \mathfrak{G}_i$, then $\mathfrak{g}_1\mathfrak{C}_i = \mathfrak{C}_i$, the line $\mathfrak{g}_1\mathcal{S}$ is a side of $\mathfrak{C}_i$ different from $\mathcal{S}$, thus in $\mathcal{D}_i$, and separated from $\mathcal{S}$ by $\mathfrak{C}_i$; the region $\mathfrak{g}_1\mathfrak{C}_j$ is adjacent to $\mathfrak{C}_i$ in $\mathcal{D}_i$ along $\mathfrak{g}_1\mathcal{S}$ (Fig. 18.1). Now let it be assumed that $\mathfrak{g}$ in (1) satisfies the following condition:

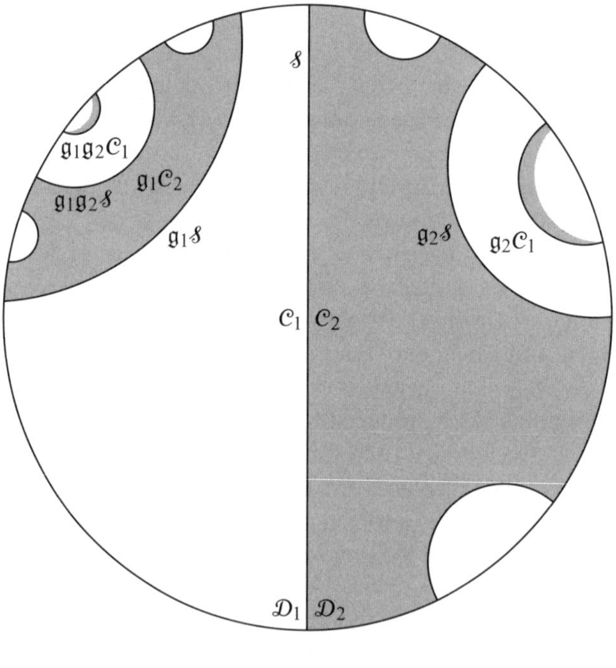

Figure 18.1

If $\mathfrak{g}_\ell$ belongs to $\mathfrak{G}_i$, then $\mathfrak{g}\mathcal{S}$ belongs to $\mathcal{D}_i$ and is separated from $\mathcal{S}$ by ℓ regions which in order from $\mathcal{S}$ to $\mathfrak{g}\mathcal{S}$ belong alternately to $\mathfrak{G}\mathfrak{C}_i$ and $\mathfrak{G}\mathfrak{C}_j$.

This condition holds for $\ell = 1$.

We then consider an element $\mathfrak{g}_{\ell+1}\mathfrak{g}$, where $\mathfrak{g}_{\ell+1}$ belongs to $\mathfrak{G}_j$, but not to $\mathfrak{A}_\mathcal{S}$. We then get $\mathfrak{g}_{\ell+1}\mathfrak{C}_j = \mathfrak{C}_j$, the line $\mathfrak{g}_{\ell+1}\mathcal{S}$ is a side of $\mathfrak{C}_j$ different from $\mathcal{S}$, thus in $\mathcal{D}_j$, and $\mathfrak{g}_{\ell+1}\mathcal{D}_i \subset \mathcal{D}_j$. Hence $\mathfrak{g}_{\ell+1}\mathfrak{g}\mathcal{S}$ belongs to $\mathcal{D}_j$ and is separated from $\mathcal{S}$ by $\ell+1$ regions which in the order from $\mathcal{S}$ to $\mathfrak{g}_{\ell+1}\mathfrak{g}\mathcal{S}$ belongs alternately to $\mathfrak{G}\mathfrak{C}_j$ and $\mathfrak{G}\mathfrak{C}_i$. Induction thus shows that the condition holds for any number of factors in (1).

The collection of lines $\mathfrak{G}\mathcal{S}$ is reproduced by all elements of $\mathfrak{G}$, and the same holds for the collection of regions $\mathfrak{G}\mathfrak{C}_1$ and likewise for $\mathfrak{G}\mathfrak{C}_2$.

Assume that

$$\mathfrak{g} = \mathfrak{g}'_m\mathfrak{g}'_{m-1}\cdots\mathfrak{g}'_2\mathfrak{g}'_1 \tag{1'}$$

is another representation of the same element $\mathfrak{g} \in \mathfrak{G}$ as in (1) by a product of factors alternately drawn from the two generating groups and none from $\mathfrak{A}_\mathcal{S}$. The same property as just described for the product (1) then holds for (1′). Since $\mathfrak{g}\mathcal{S}$ is separated from $\mathcal{S}$ by ℓ regions, it is inferred that $m = \ell$. Thus the number of factors in a product of the form (1) depends only on the element $\mathfrak{g} \in \mathfrak{G}$ represented. We shall call it the *length of* $\mathfrak{g}$, and we shall speak of elements of $\mathfrak{A}_\mathcal{S}$ as elements of $\mathfrak{G}$ of *length zero*. We may now write ℓ instead of m. If $\ell = 1$, we immediately get $\mathfrak{g}_1 = \mathfrak{g}'_1$. Assume $\ell > 1$. If $\mathfrak{g}\mathcal{S} \subset \mathcal{D}_i$, then both $\mathfrak{g}_\ell$ and $\mathfrak{g}'_\ell$ belong to $\mathfrak{G}_i$. We then get from (1) and (1′)

$$\mathfrak{g}_{\ell-1}\cdots\mathfrak{g}_1 = \left(\mathfrak{g}_\ell^{-1}\mathfrak{g}'_\ell\right)\mathfrak{g}'_{\ell-1}\cdots\mathfrak{g}'_1,$$

where $\mathfrak{g}_\ell^{-1}\mathfrak{g}'_\ell \in \mathfrak{G}_i$. Here both $\mathfrak{g}_{\ell-1}$ and $\mathfrak{g}'_{\ell-1}$ belongs to $\mathfrak{G}_j$. It then follows that $\mathfrak{g}_\ell^{-1}\mathfrak{g}'_\ell$ must belong to $\mathfrak{A}_\mathcal{S}$, since otherwise the left- and right-hand terms in the last equation would have different length. Hence $\mathfrak{g}'_\ell = \mathfrak{g}_\ell\mathfrak{a}_\ell$, where $\mathfrak{a}_\ell \in \mathfrak{A}_\mathcal{S}$. Putting $\mathfrak{a}_\ell\mathfrak{g}'_{\ell-1} = \mathfrak{g}''_{\ell-1}$, which is an element of $\mathfrak{G}_j$ but not of $\mathfrak{A}_\mathcal{S}$, one may continue the same process on the equation

$$\mathfrak{g}_{\ell-1}\mathfrak{g}_{\ell-2}\cdots\mathfrak{g}_1 = \mathfrak{g}''_{\ell-1}\mathfrak{g}'_{\ell-2}\cdots\mathfrak{g}'_1$$

thus again reducing the number of factors.

On the other hand, if $\ldots\mathfrak{g}_{t+1}\mathfrak{g}_t\ldots$ are two consecutive factors in (1), they may obviously be replaced by the two factors $\ldots(\mathfrak{g}_{t+1}\mathfrak{a})(\mathfrak{a}^{-1}\mathfrak{g}_t)\ldots$, $\mathfrak{a} \in \mathfrak{A}_\mathcal{S}$, without violating the conditions imposed on the product. Hence this is the only arbitrariness in the representation of elements of $\mathfrak{G}$, but outside $\mathfrak{A}_\mathcal{S}$, in the form (1).

This arbitrariness may be removed by choosing fixed representatives for the cosets of $\mathfrak{A}_\mathcal{S}$ in $\mathfrak{G}_1$ and equally in $\mathfrak{G}_2$. Let $\bar{\mathfrak{g}}$ denote the chosen representative of the left coset of $\mathfrak{A}_\mathcal{S}$ in $\mathfrak{G}_1$ or $\mathfrak{G}_2$ to which $\mathfrak{g}_n$ belongs, thus $\mathfrak{g}_n = \bar{\mathfrak{g}}_n\mathfrak{a}_n$, where $\mathfrak{a}_n$ is an element of $\mathfrak{A}_\mathcal{S}$. We then replace $\mathfrak{g}_\ell \in \mathfrak{G}_i$ by $\bar{\mathfrak{g}}_\ell\mathfrak{a}_\ell$ and put $\mathfrak{a}_\ell\mathfrak{g}_{\ell-1} = \mathfrak{g}^*_{\ell-1}$. This is an element of $\mathfrak{G}_j$ outside $\mathfrak{A}_\mathcal{S}$. We replace it by $\bar{\mathfrak{g}}^*_{\ell-1}\mathfrak{a}^*_{\ell-1}$, combine $\mathfrak{a}^*_{\ell-1}$ with $\mathfrak{g}_{\ell-2}$ and continue this process. In this way $\mathfrak{g}$ is transformed into a *normal form*

$$\mathfrak{g} = \mathfrak{k}_\ell\mathfrak{k}_{\ell-1}\ldots\mathfrak{k}_2\mathfrak{k}_1\mathfrak{a}, \tag{1''}$$

where the $\mathfrak{k}_n$ are taken from the set of fixed representatives of the left cosets of $\mathfrak{A}_{\mathcal{S}}$ in $\mathfrak{G}_1$ or $\mathfrak{G}_2$, and $\mathfrak{a}$ is an element of $\mathfrak{A}_{\mathcal{S}}$. The number of $\mathfrak{k}$-factors is the length of $\mathfrak{g}$. The elements of $\mathfrak{A}_{\mathcal{S}}$ are included in the normal form (1″) as elements of length zero. By repeating the consideration applied above to two representations (1) and (1′) of the same element of $\mathfrak{G}$, and now assuming moreover that both representations are in normal form, it is immediately seen that they must be identical. Hence every element of $\mathfrak{G}$ is represented by one, and only one, product (1″) in normal form in relation to a fixed choice of a system of representatives of the left cosets of $\mathfrak{A}_{\mathcal{S}}$ in $\mathfrak{G}_1$ and $\mathfrak{G}_2$, the representative of $\mathfrak{A}_{\mathcal{S}}$ itself being 1.

Let Γ denote the collection of those elements of $\mathfrak{G}$ which belong to $\mathfrak{G}_1$ or to $\mathfrak{G}_2$; the elements of $\mathfrak{A}_{\mathcal{S}} = \mathfrak{G}_1 \cap \mathfrak{G}_2$ belong to both. Since $\mathfrak{G}$ is generated by $\mathfrak{G}_1$ and $\mathfrak{G}_2$, one may take Γ as a set of generators of $\mathfrak{G}$. Consider any relation $\varphi(\Gamma) = 1$ in these generators. Since the representation of elements of $\mathfrak{G}$ in normal form was seen to be unique, the relation becomes an identity, if the product $\varphi(\Gamma)$ is brought into normal form. It results from the above considerations that in every single step of this process of bringing $\varphi(\Gamma)$ into normal form only $\mathfrak{G}_1$ or $\mathfrak{G}_2$ is involved, hence that the step is carried out either by means of the multiplication table M_1 of $\mathfrak{G}_1$ or by means of the M_2 of $\mathfrak{G}_2$. Thus the union of M_1 and M_2 is a system of defining relations for $\mathfrak{G}$ in the system Γ of generators. Hence one gets by Section 1:

$$\mathfrak{G} = \mathfrak{G}_1 * \mathfrak{G}_2 \text{ am } \mathfrak{A}_{\mathcal{S}}. \tag{2}$$

If $\mathfrak{A}_{\mathcal{S}} = 1$, one simply gets

$$\mathfrak{G} = \mathfrak{G}_1 * \mathfrak{G}_2. \tag{3}$$

If the group $\mathfrak{A}_{\mathcal{S}}$ does not contain reflections, i.e. if $\mathfrak{A}_{\mathcal{S}} = 1$ or $\mathfrak{A}_{\mathcal{S}} = \mathfrak{A}(\rightarrow)$, and if $\mathfrak{t}$ is an arbitrary translation with axis $\mathcal{S}$, then we can use the group $\mathfrak{t}\mathfrak{G}_2\mathfrak{t}^{-1}$ instead of the group $\mathfrak{G}_2$ to build the amalgamated product $\mathfrak{G}(\mathfrak{t}) = \mathfrak{G}_1 * \mathfrak{t}\mathfrak{G}_2\mathfrak{t}^{-1} \text{ am } \mathfrak{A}_{\mathcal{S}}$.

Therefore in this case one has a whole family of amalgamated products, parametrized by the group of all translations with axis $\mathcal{S}$. In particular one has $\mathfrak{G}(1) = \mathfrak{G}$. If the line $\mathcal{S}$ is directed, then every translation with axis $\mathcal{S}$ is uniquely determined by its displacement and a parametrization of the family of amalgamated products by real numbers results. This displacement of $\mathfrak{t}$ will then be called the *amalgamation parameter* of the group $\mathfrak{G}(\mathfrak{t})$.

If $\mathfrak{A}_{\mathcal{S}}$ contains reflections, then the conjugate of a reflection $\mathfrak{s}$ in $\mathfrak{A}_{\mathcal{S}}$ with reflection line $\ell(\mathfrak{s})$ by a translation $\mathfrak{t}$ with axis $\mathcal{S}$ is the reflection $\mathfrak{t}\mathfrak{s}\mathfrak{t}^{-1}$ with reflection line $\mathfrak{t}\ell(\mathfrak{s})$. Therefore one can build only those amalgamated products $\mathfrak{G}(\mathfrak{t})$, for which $\mathfrak{t} \in \mathfrak{A}_{\mathcal{S}}$, and nothing new arises.

In order to arrive at other abstract characterizations of $\mathfrak{G}$ we carry out in detail a construction which corresponds to a general remark in Section 1. Let Φ_1 be any system of generators and Ψ_1 a system of defining relations in them for $\mathfrak{G}_1$, and likewise Φ_2 and Ψ_2 for $\mathfrak{G}_2$. Let further fixed representatives for the left cosets of $\mathfrak{A}_{\mathcal{S}}$ in $\mathfrak{G}_1$ and in $\mathfrak{G}_2$ be chosen, $\mathfrak{A}_{\mathcal{S}}$ itself being represented by the unit element. Consider an arbitrary product $\Pi = \Pi(\Phi_1, \Phi_2)$ of generators drawn from Φ_1 and Φ_2. If it contains generators from both systems, it may be subdivided into partial products formed alternately by factors from Φ_1 and from Φ_2

$$\Pi_1\Pi_2\Pi_3\dots\Pi_s,$$

say $\Pi_1 = \Pi_1(\Phi_1)$, $\Pi_2 = \Pi_2(\Phi_2)$, $\Pi_3 = \Pi_3(\Phi_1)$, and so on. Let

$$\Pi_1(\Phi_1) = \mathfrak{k}_1(\Phi_1)\mathfrak{a}_1(\Phi_1),$$

where $\mathfrak{k}_1$ is the representative of the left coset of $\mathfrak{A}_\mathfrak{g}$ in $\mathfrak{G}_1$ to which $\Pi_1(\Phi_1)$ belongs, and $\mathfrak{a}_1(\Phi_1)$ is an element of $\mathfrak{A}_\mathfrak{g}$. This is a relation between generators from Φ, thus a consequence of the system Ψ_1. Since $\mathfrak{a}_1$ is also an element in $\mathfrak{G}_2$, there exists an expression of the element $\mathfrak{a}_1(\Phi_1)$ of $\mathfrak{A}_\mathfrak{g}$ as a product $\mathfrak{a}'_1(\Phi_2)$, thus an equation

$$\mathfrak{a}_1(\Phi_1) = \mathfrak{a}'_1(\Phi_2).$$

We then put

$$\mathfrak{a}'_1(\Phi_2)\Pi_2(\Phi_2) = \Pi'(\Phi_2).$$

If $\mathfrak{k}_1 = 1$, a reduction in the number of partial products results. We now put accordingly

$$\Pi'_2(\Phi_2) = \mathfrak{k}_2(\Phi_2)\mathfrak{a}_2(\Phi_2),$$
$$\mathfrak{a}_2(\Phi_2) = \mathfrak{a}'_2(\Phi_1),$$

and continue the process in the same way provided $\mathfrak{k}_2 \neq 1$. If $\mathfrak{k}_2 = 1$, then

$$\begin{aligned}\Pi_1(\Phi_1)\Pi_2(\Phi_2)\Pi_3(\Phi_1) &= \mathfrak{k}_1(\Phi_1)\Pi'_2(\Phi_2)\Pi_3(\Phi_1)\\ &= \mathfrak{k}_1(\Phi_1)\mathfrak{a}'_2(\Phi_1)\Pi_3(\Phi_1)\\ &= \Pi'_3(\Phi_1)\end{aligned}$$

is an element of $\mathfrak{G}_1$, and we have then reached a reduction of the number of factors. By continuing this process, $\Pi(\Phi_1, \Phi_2)$ is brought into a normal form

$$\mathfrak{k}'_1\mathfrak{k}'_2\dots\mathfrak{k}'_r\mathfrak{a}, \quad r \leqq s, \quad \mathfrak{a} \in \mathfrak{A}_\mathfrak{g},$$

all the $\mathfrak{k}'$ belonging to the chosen set of representatives. The relations between the generators from Φ_1 or Φ_2 applied in this process are on the one hand the Ψ_1 from $\mathfrak{G}_1$ and Ψ_2 from $\mathfrak{G}_2$, and on the other hand relations expressing that elements of $\mathfrak{A}_\mathfrak{g}$ can be represented both by products of generators from Φ_1 and by products of generators from Φ_2. It is obviously sufficient to do this for generators of $\mathfrak{A}_\mathfrak{g}$. Thus to each generator of $\mathfrak{A}_\mathfrak{g}$ corresponds a relation

$$\Pi'(\Phi_1) = \Pi''(\Phi_2). \tag{4}$$

If $\Pi(\Phi_1, \Phi_2) = 1$, then $\Pi(\Phi_1, \Phi_2)$ reduces to the empty product when brought into normal form, and the relation $\Pi(\Phi_1, \Phi_2) = 1$ is thus a consequence of the relations brought into play by that process. In all then, we get the result:

If Φ_1 and Φ_2 are a system of generators and $\Psi_1(\Phi_1)$, $\Psi_2(\Phi_2)$ a system of defining relations of $\mathfrak{G}_1$ and $\mathfrak{G}_2$, respectively, then the Φ_1 together with Φ_2 generate $\mathfrak{G}$, and as a system of defining relations of $\mathfrak{G}$, in these generators one may take $\Psi_1(\Phi_1)$ and $\Psi_2(\Phi_2)$ together with two relations (4), *if $\mathfrak{A}_{\mathfrak{g}} = \mathfrak{A}(\|)$, one relation* (4), *if $\mathfrak{A}_{\mathfrak{g}} = \mathfrak{A}(|)$ or $\mathfrak{A}_{\mathfrak{g}} = \mathfrak{A}(\rightarrow)$, and none, if $\mathfrak{A}_{\mathfrak{g}} = 1$.* □

If a generator of $\mathfrak{A}_{\mathfrak{g}}$ is included in Φ_1 and also in Φ_2 and denoted by the same symbol, in both systems then the corresponding relation (4) lapses.

18.3 Properties of generalized free products. *The groups $\mathfrak{G}_1$* and *$\mathfrak{G}$ have equal effect in $\tilde{\mathfrak{C}}_1$.*

Proof. Let x and $\mathfrak{g}x$, $\mathfrak{g} \in \mathfrak{G}$, both belong to the closure $\tilde{\mathfrak{C}}_1$ of $\mathfrak{C}_1$ on $\mathfrak{D}$. If $\mathfrak{g}$ belongs to $\mathfrak{A}_{\mathfrak{g}}$, it belongs to $\mathfrak{G}_1$. If not, let it have the length $\ell > 0$. If $\ell > 1$, or if $\ell = 1$ and $\mathfrak{g} \in \mathfrak{G}_2$, then $\mathfrak{g}\tilde{\mathfrak{C}}_1$ would be separated from $\tilde{\mathfrak{C}}_1$ by at least one region in contradiction to the fact that $\tilde{\mathfrak{C}}_1$ and $\mathfrak{g}\tilde{\mathfrak{C}}_1$ have the point $\mathfrak{g}x$ in common. Thus $\mathfrak{g} \in \mathfrak{G}_1$, which proves the assertion. – Likewise $\mathfrak{G}_2$ and $\mathfrak{G}$ have equal effect in $\tilde{\mathfrak{C}}_2$. □

The group $\mathfrak{G}$ is discontinuous in $\mathfrak{D}$.

Proof. This immediately follows from the preceding theorem: If $x \in \mathfrak{C}_1$, the equivalence class $\mathfrak{G}x$ does not accumulate in x, because $\mathfrak{G}_1 x$ does not (condition 1)). □

Every element of $\mathfrak{G}$ of positive and even length is of infinite order.

Proof. If the length ℓ of the element $\mathfrak{g}$ of $\mathfrak{G}$ is positive and even, the first and the last factors in its representation in the form (1) belong to different of the groups $\mathfrak{G}_1$ and $\mathfrak{G}_2$, and $\mathfrak{g}^n$ then has the length $n\ell$ for all positive n and this is not the identity. □

Every element of finite order of $\mathfrak{G}$ is conjugate to some element of finite order in $\mathfrak{G}_1$ or $\mathfrak{G}_2$.

Proof. Let the element $\mathfrak{g} \neq 1$ of $\mathfrak{G}$ be of finite order. If $\ell = 0$, then $\mathfrak{g}$ is a reflection contained in $\mathfrak{A}_{\mathfrak{g}}$, thus in both $\mathfrak{G}_1$ and $\mathfrak{G}_2$. If the length ℓ of $\mathfrak{g}$ is positive, then it must be an odd number, as has just been shown. If $\ell = 1$, $\mathfrak{g}$ belongs either to $\mathfrak{G}_1$ or $\mathfrak{G}_2$.

Assume $\ell > 1$, thus $\ell \geqq 3$. Let $\mathfrak{g}$, written in the form (1), be $\mathfrak{g} = \mathfrak{g}_1\mathfrak{g}_2 \ldots \mathfrak{g}_\ell$. Since ℓ is odd, $\mathfrak{g}_1$ and $\mathfrak{g}_\ell$ belong to the same of the two generating groups, say to $\mathfrak{G}_1$. We then consider the element

$$\mathfrak{g}' = \mathfrak{g}_1^{-1}\mathfrak{g}\mathfrak{g}_1 = \mathfrak{g}_2 \ldots \mathfrak{g}_{\ell-1}(\mathfrak{g}_\ell\mathfrak{g}_1).$$

Here $\mathfrak{g}_\ell\mathfrak{g}_1$ is an element of $\mathfrak{G}_1$. If it were not in $\mathfrak{A}_{\mathfrak{g}}$, $\mathfrak{g}'$ would have a positive and even length in contradiction to the fact that also $\mathfrak{g}'$ is of finite order. Hence $\mathfrak{g}_\ell\mathfrak{g}_1$ is an element of $\mathfrak{A}_{\mathfrak{g}}$, and $\mathfrak{g}_{\ell-1}\mathfrak{g}_\ell\mathfrak{g}_1 = \mathfrak{g}'_{\ell-1}$ is thus an element of $\mathfrak{G}_2$ not in $\mathfrak{A}_{\mathfrak{g}}$. Therefore $\mathfrak{g}' = \mathfrak{g}_2 \ldots \mathfrak{g}_{\ell-2}\mathfrak{g}'_{\ell-1}$ is again an element of finite order of the form (1) with length

$\ell-2$. If $\ell-2>1$, then $\mathfrak{g}''=\mathfrak{g}_2^{-1}\mathfrak{g}'\mathfrak{g}_2$ becomes in the same way an element of length $\ell-4$, and so on until we reach an element of length 1, which then is an element of finite order either in $\mathfrak{G}_1$ or $\mathfrak{G}_2$. This proves the assertion. □

If $\mathfrak{f}$ is any element of $\mathfrak{G}$, the centres of $\mathfrak{G}$ in $\mathfrak{f}\mathcal{C}_1$ are the images by $\mathfrak{f}$ of the centres in $\mathcal{C}_1$ and no other points. Similarly, the reflection lines passing through $\mathfrak{f}\mathcal{C}_1$ are images by $\mathfrak{f}$ of the reflection lines passing through $\mathcal{C}_1$; likewise for $\mathcal{C}_2$. There are no centres of $\mathfrak{G}$ on the set $\mathfrak{G}\mathcal{S}$, because there are no centres on $\mathcal{S}$. All lines of $\mathfrak{G}\mathcal{S}$ which are cut by reflection lines of $\mathfrak{G}$ are cut at right angles. If $\mathfrak{G}_1$ and $\mathfrak{G}_2$ contain no rotation and no reflection, then $\mathfrak{G}$ has the same property.

18.4 Quasi-abelian groups as generalized free products. Suppose that $\mathfrak{G}$ is not an $\mathfrak{F}$-group. Since $\mathfrak{G}$ is discontinuous in $\mathcal{D}$ and contains more than the unit element, it must be quasi-abelian. We consider the different possibilities.

a) $\mathfrak{G}$ has an *invariant point* c, thus a centre or limit-centre. Wherever c is situated in $\overline{\mathcal{D}}$, it belongs to the closure of one of the half-planes $\mathcal{D}_1$ and $\mathcal{D}_2$, say to $\overline{\mathcal{D}}_1$. An element $\mathfrak{f}$ of $\mathfrak{G}_2$ outside $\mathfrak{A}_\mathcal{S}$ carries $\mathcal{S}$ into a line $\mathfrak{f}\mathcal{S}$ situated in $\mathcal{D}_2$ (condition 3)) and $\mathcal{D}_1$ into the open half-plane $\mathfrak{f}\mathcal{D}_1$, which is part of $\mathcal{D}_2$. This excludes the possibility of c being a centre. Moreover, $\mathfrak{f}\overline{\mathcal{D}}_1$ has no point in common with $\overline{\mathcal{D}}_1$ except possibly one of the end-points of $\mathcal{S}$. Hence we have only to consider the case that c is an end-point of $\mathcal{S}$. Now let $\mathfrak{g}_1$ be an element of $\mathfrak{G}_1$ such that $\mathfrak{g}_1\mathcal{S}\neq\mathcal{S}$ (condition 2)). Then $\mathfrak{g}_1 c=c$, because c is an invariant point of $\mathfrak{G}$. Hence $\mathfrak{g}_1$ is not a rotation. Also $\mathfrak{g}_1$ cannot be a limit-rotation, because it would have to have c as its limit-centre and a suitable power of $\mathfrak{g}_1$ would carry $\mathcal{S}$ into $\mathcal{D}_2$ contrary to condition 3). Similarly, $\mathfrak{g}_1$ cannot be a translation, because c would have to be one of the end-points of its axis and, again, a suitable power of $\mathfrak{g}_1$ would carry $\mathcal{S}$ into $\mathcal{D}_2$. Finally, $\mathfrak{g}_1$ cannot be a reversed translation because $\mathfrak{g}_1^2$ would be a translation. Hence $\mathfrak{g}_1$ can only be a reflection with its line of reflection ending in c. Moreover, there can only be one such reflection line belonging to $\mathfrak{G}_1$, because otherwise $\mathfrak{G}_1$ would contain limit-rotations with c as limit-centre, which we had to exclude. Therefore, even $\mathfrak{A}_\mathcal{S}$ contains no translation, because that would carry the reflection line of $\mathfrak{G}_1$ into another one. Since a reflection belonging to $\mathfrak{A}_\mathcal{S}$ would interchange the end-points of $\mathcal{S}$, we conclude that $\mathfrak{A}_\mathcal{S}=1$. Thus, in all, $\mathfrak{G}_1$ can only consist of the identity and one reflection $\mathfrak{g}_1$, and likewise $\mathfrak{G}_2$ consists of the identity and one reflection $\mathfrak{g}_2$. It follows that $\mathfrak{G}$ is the group $\mathfrak{A}_c(\wedge)$ generated by $\mathfrak{g}_1$ and $\mathfrak{g}_2$ (Fig. 18.2).

b) Suppose that $\mathcal{A}$ is an *invariant line for* $\mathfrak{G}$ and thus also for $\mathfrak{G}_1$ and $\mathfrak{G}_2$. Hence $\mathcal{A}$ is different from $\mathcal{S}$ (condition 2)). If $\mathcal{A}$ did not intersect $\mathcal{S}$, it would belong to $\overline{\mathcal{D}}_1$, say, and we would get the same contradiction as for a centre in case a) above. We therefore have to assume that $\mathcal{A}$ intersects $\mathcal{S}$ (Fig. 18.3). No translation along $\mathcal{A}$ is contained in $\mathfrak{G}_1$ or $\mathfrak{G}_2$, for this would evidently lead to a violation of condition 3). Therefore $\mathfrak{G}_1$, besides the identity and possibly the reflection in $\mathcal{A}$, contains only either a half-turn around a point of $\mathcal{A}$ or a reflection in a normal of $\mathcal{A}$, and this point or normal is then inside $\mathcal{C}_1$. The same holds for $\mathfrak{G}_2$ and $\mathcal{C}_2$ independently of $\mathfrak{G}_1$. According to the

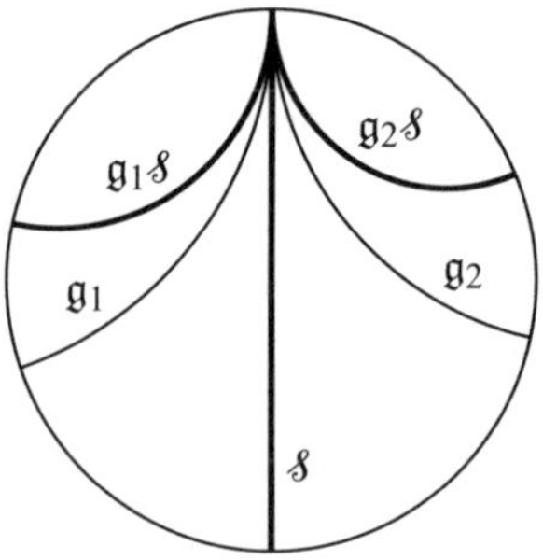

Figure 18.2

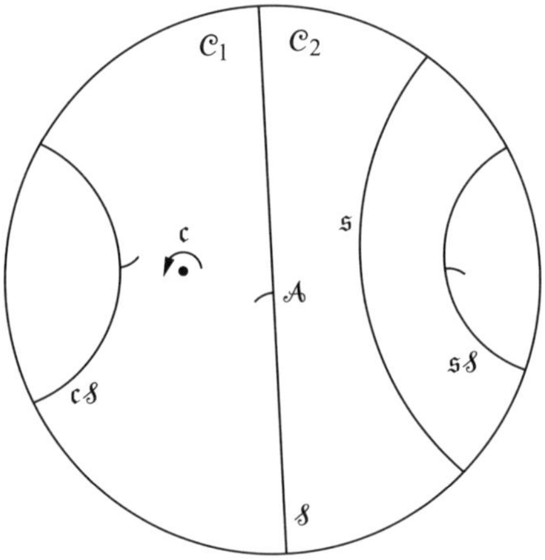

Figure 18.3

possible combinations it therefore results that $\mathfrak{G}$ belongs to one of the types $\mathfrak{A}(\cdot\cdot)$, $\mathfrak{A}(\|)$ or $\mathfrak{A}(\cdot|)$, if $\mathfrak{A}_{\mathcal{S}}$ does not contain the reflection in $\mathcal{A}$, and $\mathfrak{A}(-\ \|) = \mathfrak{A}(-\cdot\cdot)$, if $\mathfrak{A}_{\mathcal{S}}$ contains the reflection in $\mathcal{A}$. In the latter case, $\mathcal{A}$ is a normal of $\mathcal{S}$ and $\mathfrak{A}_{\mathcal{S}} = \mathfrak{A}(|)$. In the first three cases $\mathfrak{A}_{\mathcal{S}} = 1$, and $\mathcal{A}$ and $\mathcal{S}$ need not be orthogonal.

In all cases a) and b) the region $\mathfrak{C}_1$ is bounded in $\mathcal{D}$ by only two lines of the set $\mathfrak{G}\mathcal{S}$, and likewise $\mathfrak{C}_2$.

In all cases except these very special ones a) and b) the group $\mathfrak{G}$ is an $\mathfrak{F}$-group.

18.5 Tesselation of $\mathcal{D}$ by the collection of domains. $\tilde{\mathfrak{C}}_1$ and $\tilde{\mathfrak{C}}_2$ denoting the closure of $\mathfrak{C}_1$ and $\mathfrak{C}_2$ on $\mathcal{D}$, consider the union

$$\mathfrak{C} = \mathfrak{G}\tilde{\mathfrak{C}}_1 \cup \mathfrak{G}\tilde{\mathfrak{C}}_2.$$

If p is any point of $\mathfrak{C}$, it is contained in some $\mathfrak{g}\tilde{\mathfrak{C}}_1$ or $\mathfrak{g}\tilde{\mathfrak{C}}_2$, where $\mathfrak{g} \in \mathfrak{G}$. If p_0 is a point of $\mathcal{S}$, the segment pp_0 is contained in the union of a finite number of the convex regions closed on $\mathcal{D}$ which make up $\mathfrak{C}$, namely in at most $n+1$ regions, if $n \geqq 0$ is the length of $\mathfrak{g}$. Thus, if q is another point of $\mathfrak{C}$, both p and q are contained in a finite connected subset of the regions of $\mathfrak{C}$. Now, every finite connected subset of regions of $\mathfrak{C}$ is convex, because it is obtained by omitting from $\mathcal{D}$ a certain collection of disjoint half-planes, namely all those half-planes complementary to the single region in which it is not adjacent to another region of the connected collection. Therefore any two points of $\mathfrak{C}$ are contained in a convex subset of $\mathfrak{C}$ and hence $\mathfrak{C}$ is convex. Moreover, $\mathfrak{C}$ and hence even its closure $\overline{\mathfrak{C}}$ is reproduced by $\mathfrak{G}$. Thus, if $\mathfrak{G}$ is an $\mathfrak{F}$-group, it is inferred from §14.1 that its closed convex domain $\overline{\mathcal{K}}(\mathfrak{G})$ is contained in $\overline{\mathfrak{C}}$. Also $\mathfrak{C}$ is an open set: for, any point of $\mathfrak{C}$ is either an inner point of one of the regions in $\mathfrak{G}\mathfrak{C}_1$ or in $\mathfrak{G}\mathfrak{C}_2$, or it is a common boundary point of two such regions.

That $\overline{\mathcal{K}}(\mathfrak{G})$ is contained in $\overline{\mathfrak{C}}$ is obvious, if $\mathfrak{C} = \mathcal{D}$. We want to establish a condition which is sufficient for this to be the case. Assume that there is a positive lower bound ρ for the distance between $\mathcal{S}$ and any other side of at least one of the two adjacent regions, say $\mathfrak{C}_1$. (This is for instance the case, if the sides of $\mathfrak{C}_1$ are finite in number and no two of them are parallel; and also if $\mathfrak{A}_{\mathcal{S}} = \mathfrak{A}(\rightarrow)$ or $\mathfrak{A}(\|)$, because the projection of any other side of $\mathfrak{C}_1$ on $\mathcal{S}$ cannot exceed the primary displacement on $\mathcal{S}$. But the assumption holds even under other conditions.) Then $\mathfrak{C}_1$ cuts off from every line intersecting it a segment of length $\geqq \rho$. Now consider any half-line issuing from a point of $\mathcal{S}$. It intersects a finite ($\geqq 0$) or infinite number of lines of the collection $\mathfrak{G}\mathcal{S}$. In the first case it is from a certain of its points onwards wholly contained in one of the domains of $\mathfrak{C}$. In the second case, $\mathfrak{G}\mathcal{S}$ cuts out an infinite number of segments on the half-line, and these belong alternately to $\mathfrak{G}\tilde{\mathfrak{C}}_1$ and $\mathfrak{G}\tilde{\mathfrak{C}}_2$. Those of $\mathfrak{G}\tilde{\mathfrak{C}}_1$, at least, are of length $\geqq \rho$, and the whole half-line therefore belongs to $\mathfrak{C}$. Hence $\mathfrak{C} = \mathcal{D}$.

Consider the case that $\mathfrak{C}$ is not the whole of $\mathcal{D}$. Then there exists a half-line h, issuing from a point of $\mathcal{S}$, which does not wholly belong to $\mathfrak{C}$. It must then intersect an infinite number of lines of the collection $\mathfrak{G}\mathcal{S}$, and the points of intersection converge to a point of h. Thus $\mathfrak{G}\mathcal{S}$ accumulates in $\mathcal{D}$ in this case.

18.6 Surface corresponding to a generalized free product. If $\mathfrak{A}_{\mathcal{S}} = \mathfrak{A}(\rightarrow)$ or $\mathfrak{A}_{\mathcal{S}} = \mathfrak{A}(\|)$, in which case we have $\mathfrak{C} = \mathcal{D}$ and both $\mathfrak{G}_1$ and $\mathfrak{G}_2$ are $\mathfrak{F}$-groups, the surfaces $\mathcal{D}$ mod $\mathfrak{G}_1$ and $\mathcal{D}$ mod $\mathfrak{G}_2$ contain a simple closed geodesic corresponding to $\mathcal{S}$, and this geodesic cuts off a funnel or half-funnel from each of the two surfaces. Moreover, this closed geodesic has the same length on both surfaces. The surface $\mathcal{D}$ mod $\mathfrak{G}$ arises, if these two funnels or half-funnels are removed and the rest of the two surfaces are joined together along the boundaries thereby arising. Boundaries of $\mathcal{K}(\mathfrak{G})$ mod $\mathfrak{G}$ are those boundaries of $\mathcal{K}(\mathfrak{G})$ mod $\mathfrak{G}_1$ and $\mathcal{K}(\mathfrak{G})$ mod $\mathfrak{G}_2$ which are not derived from $\mathcal{S}$, if any.

If $\mathfrak{A}_{\mathcal{S}} = 1$ or $\mathfrak{A}_{\mathcal{S}} = (|)$, the surfaces $\mathcal{D}$ mod $\mathfrak{G}_1$ and $\mathcal{D}$ mod $\mathfrak{G}_2$ contain a simple open geodesic or half-geodesic corresponding to $\mathcal{S}$, and this cuts off a half-plane or quarter-plane from each of the two surfaces. The surface $\mathcal{D}$ mod $\mathfrak{G}$ arises, if these two half-planes or quarter-planes are removed and the rest of the two surfaces are joined together along the boundaries thereby arising. In this case none of the groups $\mathfrak{G}_1$ and $\mathfrak{G}_2$ need be an $\mathfrak{F}$-group, and even if $\mathfrak{G}_1$, say, is an $\mathfrak{F}$-group, $\mathcal{S}$ need not be a side of $\mathcal{K}(\mathfrak{G}_1)$. However, $\mathcal{S}$ cannot intersect $\mathcal{K}(\mathfrak{G}_1)$ (condition 3)). It can be any line in one of the half-planes outside $\mathcal{K}(\mathfrak{G}_1)$ provided it is not intersected by any of its equivalents.

If $\mathfrak{A}_\mathcal{S}$ does not contain reflections and $\mathfrak{t}$ is a translation with axis $\mathcal{S}$ then the surface of $\mathcal{D}$ mod $\mathfrak{G}(\mathfrak{t})$ results from the surface $\mathcal{D}$ mod $\mathfrak{G}$ by a *shift* (in the case $\mathfrak{A}_\mathcal{S} = 1$) or a *twist* (in the case $\mathfrak{A}_\mathcal{S} = \mathfrak{A}(\rightarrow)$) of length $\lambda(\mathfrak{t})$ of $\mathcal{D}_2$ mod $\mathfrak{G}_2$ along the geodesic corresponding to $\mathcal{S}$ with $\mathcal{D}_1$ mod $\mathfrak{G}_1$ held fixed. Therefore we will speak instead of the amalgamation parameter of the *shift parameter* resp. the *twist parameter* of the surface $\mathcal{D}$ mod $\mathfrak{G}(\mathfrak{t})$.

18.7 Generalized free product of a group operating on a region with a quasi-abelian group operating on a boundary of that region. Let $\mathcal{S}$, $\mathcal{D}_1$, $\mathcal{D}_2$, $\mathfrak{G}_1$, $\mathfrak{C}_1$ and $\mathfrak{A}_\mathcal{S}$ have the same meaning as in the preceding sections. Let $\mathfrak{H}$ denote a quasi-abelian group carrying $\mathcal{S}$ into itself. We want to study the group $\mathfrak{G}$ generated by $\mathfrak{G}_1$ and $\mathfrak{H}$. In particular, we want $\mathfrak{G}$ to become discontinuous in $\mathcal{D}$ and such that $\mathfrak{C}_1$ mod $\mathfrak{G}_1$ becomes part of $\mathcal{D}$ mod $\mathfrak{G}$. Hence $\mathfrak{G}$ and $\mathfrak{G}_1$ have to have equal effect in $\mathfrak{C}_1$.

If $\mathfrak{H}$ coincides with $\mathfrak{A}_\mathcal{S}$ or is a subgroup of $\mathfrak{A}_\mathcal{S}$, then $\mathfrak{G}$ coincides with $\mathfrak{G}_1$. We therefore assume that $\mathfrak{H}$ contains certain elements not in $\mathfrak{A}_\mathcal{S}$. Let $\mathfrak{a}$ be an element of $\mathfrak{H}$ carrying $\mathcal{D}_1$ into itself. Thus $\mathfrak{a}\mathfrak{C}_1$ has points in common with $\mathfrak{C}_1$; their intersection is an open set. Let $\mathcal{B}$ be an open subset of $\mathfrak{C}_1$ such that $\mathfrak{a}\mathcal{B}$ belongs to $\mathfrak{C}_1$. Since $\mathfrak{G}$ and $\mathfrak{G}_1$ shall have equal effect in $\mathfrak{C}_1$, it follows that $\mathfrak{a}$ must belong to $\mathfrak{G}_1$ and hence to $\mathfrak{A}_\mathcal{S}$. Thus the subgroup $\mathfrak{H}_1$ of $\mathfrak{H}$ carrying $\mathcal{D}_1$ into itself is a subgroup of $\mathfrak{A}_\mathcal{S}$. Any element of $\mathfrak{H}$ not in $\mathfrak{G}_1$ must then interchange $\mathcal{D}_1$ and $\mathcal{D}_2$ and is thus

α) the reflection in $\mathcal{S}$, or

β) a half-turn around a centre on $\mathcal{S}$, or

γ) a reversed translation along $\mathcal{S}$.

It follows that $\mathfrak{H}_1$ has index 2 in $\mathfrak{H}$. On the other hand, if we extend $\mathfrak{H}$ by the adjunction of elements of $\mathfrak{A}_\mathcal{S}$ not in $\mathfrak{H}_1$, the resulting group $\mathfrak{G}$ remains unchanged.

In all, therefore, we can assume that $\mathfrak{H}$ is a quasi-abelian group containing $\mathfrak{A}_\mathcal{S}$ as a subgroup with index 2 and resulting from $\mathfrak{A}_\mathcal{S}$ by the adjunction of an element α) or β) or γ).

Let $\mathfrak{h}$ be an element of $\mathfrak{H}$ outside $\mathfrak{A}_\mathcal{S}$. Then $\mathfrak{h}\mathfrak{A}_\mathcal{S}\mathfrak{h}^{-1} = \mathfrak{A}_\mathcal{S}$. Hence $\mathfrak{h}$ must carry the totality of those reflection lines of $\mathfrak{G}_1$ which intersect $\mathcal{S}$ into itself. This condition is fulfilled, if $\mathfrak{h}$ is the reflection in $\mathcal{S}$ (case α)), since those reflection lines, if any, cut $\mathcal{S}$ at right angles. – If $\mathfrak{h}$ is a half-turn (case β)) and $\mathfrak{A}_\mathcal{S} = \mathfrak{A}(|)$, the centre of $\mathfrak{h}$ must be the intersection point of the reflection line with $\mathcal{S}$. If $\mathfrak{h}$ is a half-turn and $\mathfrak{A}_\mathcal{S} = \mathfrak{A}(\|)$, the centre of $\mathfrak{h}$ is either the intersection point of one of the reflection lines with $\mathcal{S}$, or the centre bisects the segment between two consecutive intersection points. – If $\mathfrak{h}$ is a reversed translation (case γ)), the case $\mathfrak{A}_{\mathcal{S}} = \mathfrak{A}(|)$ cannot occur; if $\mathfrak{A}_{\mathcal{S}} = \mathfrak{A}(\|)$, the displacement $\lambda(\mathfrak{h})$ of $\mathfrak{h}$ must be a multiple of the distance between two consecutive lines of reflection.

Moreover, $\mathfrak{h}^2$ belongs to $\mathfrak{A}_{\mathcal{S}}$. In the cases α) and β) we have $\mathfrak{h}^2 = 1$. In the case γ) the element $\mathfrak{h}^2$ is a translation along $\mathcal{S}$, thus a power of the primary translation along $\mathcal{S}$ in $\mathfrak{A}_{\mathcal{S}}$.

We resume these *conditions of compatibility*: Case α) is compatible with all four forms for $\mathfrak{A}_{\mathcal{S}}$. The same holds in the case β) provided the location of the centre is as indicated above for $\mathfrak{A}(|)$ and $\mathfrak{A}(\|)$. Case γ) is only compatible with $\mathfrak{A}(\rightarrow)$ and $\mathfrak{A}(\|)$ and requires that $2\lambda(\mathfrak{h})$ is among the displacements of $\mathfrak{A}_{\mathcal{S}}$.

Now, let $\mathfrak{h}$ denote a fixed element of $\mathfrak{H}$ outside $\mathfrak{A}_{\mathcal{S}}$. We put $\mathfrak{G}_2 = \mathfrak{h}\mathfrak{G}_1\mathfrak{h}^{-1}$. This is a subgroup of $\mathfrak{G}$ carrying $\mathcal{D}_2$, and in particular the region $\mathcal{C}_2 = \mathfrak{h}\mathcal{C}_1$ of $\mathcal{D}_2$, into itself. The subgroup of $\mathfrak{G}_2$ carrying $\mathcal{S}$ into itself is $\mathfrak{h}\mathfrak{A}_{\mathcal{S}}\mathfrak{h}^{-1}$, which is $\mathfrak{A}_{\mathcal{S}}$ itself. Thus $\mathfrak{G}_1$ and $\mathfrak{G}_2$ satisfy the conditions imposed on the groups $\mathfrak{G}_1$ and $\mathfrak{G}_2$ of Section 2. We can thus form the subgroup

$$\mathfrak{G}^* = \mathfrak{G}_1 * \mathfrak{G}_2 \operatorname{am} \mathfrak{A}_{\mathcal{S}} \tag{5}$$

of $\mathfrak{G}$, and $\mathfrak{G}^*$ is then discontinuous in $\mathcal{D}$.

On the other hand, the group $\mathfrak{G}$ generated by $\mathfrak{G}_1$ and $\mathfrak{H}$ is also generated by $\mathfrak{G}_1$ and $\mathfrak{h}$, since every element of $\mathfrak{H}$ not in $\mathfrak{G}_1$ belongs to the coset $\mathfrak{h}\mathfrak{A}_{\mathcal{S}} = \mathfrak{A}_{\mathcal{S}}\mathfrak{h}$ of $\mathfrak{A}_{\mathcal{S}}$ in $\mathfrak{H}$. Also, since $\mathfrak{h}^2$ belongs to $\mathfrak{G}_1$ and likewise to $\mathfrak{G}_2$, every element of $\mathfrak{G}$ which contains an even number of factors $\mathfrak{h}$ or $\mathfrak{h}^{-1}$ can be written as a product of factors alternately drawn from $\mathfrak{G}_1$ and $\mathfrak{G}_2$, and if it contains an odd number of factors $\mathfrak{h}$ or $\mathfrak{h}^{-1}$, it can be written as such a product preceded or followed by $\mathfrak{h}$. Hence $\mathfrak{G}^*$ has index 2 in $\mathfrak{G}$. It follows that $\mathfrak{G}$ is discontinuous in $\mathcal{D}$ (§10.1). Also $\mathfrak{G}^*$ is a normal subgroup of $\mathfrak{G}$.

$\mathfrak{G}$ is the generalized free product of $\mathfrak{G}_1$ and $\mathfrak{H}$ with amalgamated subgroup $\mathfrak{A}_{\mathcal{S}}$, and we write

$$\mathfrak{G} = \mathfrak{G}_1 * \mathfrak{H} \operatorname{am} \mathfrak{A}_{\mathcal{S}}. \tag{6}$$

It results from Section 5 that, if $\mathfrak{A}_{\mathcal{S}}$ contains translations, $\mathfrak{G}\tilde{\mathcal{C}}_1$ is the whole of $\mathcal{D}$. The same is the case, if $\mathfrak{H}$ contains the reflection in $\mathcal{S}$, because $\mathfrak{G}\mathcal{S}$ then consists of lines of reflection of a discontinuous group and thus cannot accumulate in $\mathcal{D}$.

Remark. If $\mathfrak{H}$ does not contain reflections, i.e. if we are in the cases β) or γ), and if $\mathfrak{t}$ is an arbitrary translation with axis $\mathcal{S}$, then we can use the group $\mathfrak{t}\mathfrak{H}\mathfrak{t}^{-1}$ instead of the group $\mathfrak{H}$ to form the amalgamated product $\mathfrak{G}(\mathfrak{t}) = \mathfrak{G}_1 * \mathfrak{t}\mathfrak{H}\mathfrak{t}^{-1} \operatorname{am} \mathfrak{A}_{\mathcal{S}}$. In case γ) this produces nothing new since then $\mathfrak{H} = \mathfrak{t}\mathfrak{H}\mathfrak{t}^{-1}$. But in case β) we get as in Section 2 a whole family of groups, parametrized by the displacement of the translation $\mathfrak{t}$.

18.8 Abstract characterization of the generalized free product. Let a set of generators $\mathfrak{g}_\nu$, $\nu = 1, 2, \dots$, and defining relations $R_\mu = 1$, $\mu = 1, 2, \dots$, for $\mathfrak{G}_1$ be known. In order first to characterize $\mathfrak{G}^*$ on the model of Section 2 we introduce as generators of $\mathfrak{G}_2$ the elements $\mathfrak{g}'_\nu = \mathfrak{h}\mathfrak{g}_\nu\mathfrak{h}^{-1}$ and as defining relations of $\mathfrak{G}_2$ the relations $R'_\mu = 1$, where R'_μ is the same product of the $\mathfrak{g}'_\nu$ as R_μ of the $\mathfrak{g}_\nu$. Then $\mathfrak{G}^*$ has a presentation with the $\mathfrak{g}_\nu$ and the $\mathfrak{g}'_\nu$ as generators and defining relations $R_\mu = 1$, $R'_\mu = 1$ and, moreover, none or one or two relations corresponding to (4), which in short may be written

$$\Pi'(\mathfrak{g}_\nu) = \Pi''(\mathfrak{g}'_\nu). \tag{7}$$

In order to extend $\mathfrak{G}^*$ to $\mathfrak{G}$ we introduce $\mathfrak{h}$ as a new generator and add a relation

$$\mathfrak{h}^2 = \Psi(\mathfrak{g}_\nu) \tag{8}$$

based on the fact that $\mathfrak{h}^2$ belongs to $\mathfrak{G}_1$. We then remark that the generators $\mathfrak{g}'_\nu$ can be eliminated, and also the relations $R'_\mu = 1$, because they reduce to $\mathfrak{h}R_\mu\mathfrak{h}^{-1} = 1$, hence $R_\mu = 1$. As to the relations (7) they take the form

$$\Pi''(\mathfrak{g}_\nu) = \mathfrak{h}\,\Pi'(\mathfrak{g}_\nu)\mathfrak{h}^{-1}. \tag{7'}$$

Thus $\mathfrak{G}$ has a presentation with generators $\mathfrak{g}_\nu$ and $\mathfrak{h}$ and defining relations $R_\mu = 1$ together with (8) and none, or one, or two relations (7′), namely none if $\mathfrak{A}_\mathfrak{s} = 1$, one if $\mathfrak{A}_\mathfrak{s} = \mathfrak{A}(|)$ or $\mathfrak{A}_\mathfrak{s} = \mathfrak{A}(\rightarrow)$, and two if $\mathfrak{A}_\mathfrak{s} = \mathfrak{A}(\|)$. The relations (8) and (7′) may be brought into a simple form by including generators of $\mathfrak{A}_\mathfrak{s}$ among the $\mathfrak{g}_\nu$.

We now survey in detail the different possibilities by examining in Table 1, §11.6, of quasi-abelian groups those groups which leave a line invariant and contain elements interchanging the two half-planes determined by that line. The groups $\mathfrak{A}(\hookrightarrow)$ and $\mathfrak{A}(\wedge)$ leave no line invariant. The groups $\mathfrak{A}(\rightarrow)$ and $\mathfrak{A}(\|)$, while leaving a line invariant, contain no element interchanging the half-planes determined by it. The remaining eight groups can occur as groups $\mathfrak{H}$. They are listed in the following table:

Table 3

Case	$\mathfrak{H}$	$\mathfrak{A}_\mathfrak{s}$	$\mathfrak{h}$	(8)	(7′)
1)	$\mathfrak{A}(-)$	1	α	$\mathfrak{h}^2 = 1$	none
2)	$\mathfrak{A}(\times_2)$	$\mathfrak{A}_\mathfrak{s}(\vert)$	α	$\mathfrak{h}^2 = 1$	$(\mathfrak{h}\mathfrak{s})^2 = 1$
3)	$\mathfrak{A}(- \rightarrow)$	$\mathfrak{A}_\mathfrak{s}(\rightarrow)$	α	$\mathfrak{h}^2 = 1$	$\mathfrak{h}\mathfrak{t}\mathfrak{h}\mathfrak{t}^{-1} = 1$
4)	$\mathfrak{A}(- \Vert)$	$\mathfrak{A}_\mathfrak{s}(\Vert)$	α	$\mathfrak{h}^2 = 1$	$(\mathfrak{h}\mathfrak{s}_1)^2 = 1$, $(\mathfrak{h}\mathfrak{s}_2)^2 = 1$
5)	$\mathfrak{A}(\circlearrowleft\!\!\!\!\odot_2)$	1	β	$\mathfrak{h}^2 = 1$	none
6)	$\mathfrak{A}(\cdot\cdot)$	$\mathfrak{A}_\mathfrak{s}(\rightarrow)$	β	$\mathfrak{h}^2 = 1$	$(\mathfrak{h}\mathfrak{t})^2 = 1$
7)	$\mathfrak{A}(\cdot\vert)$	$\mathfrak{A}_\mathfrak{s}(\Vert)$	β	$\mathfrak{h}^2 = 1$	$\mathfrak{h}\mathfrak{s}_1\mathfrak{h}\mathfrak{s}_2 = 1$, $\mathfrak{h}\mathfrak{s}_2\mathfrak{h}\mathfrak{s}_1 = 1$
8)	$\mathfrak{A}(\rightleftharpoons)$	$\mathfrak{A}_\mathfrak{s}(\rightarrow)$	γ	$\mathfrak{h}^2 = \mathfrak{t}$	$\mathfrak{h}\mathfrak{t}\mathfrak{h}^{-1}\mathfrak{t}^{-1} = 1$

In each of the eight cases the subgroup $\mathfrak{A}_\mathfrak{s}$ of $\mathfrak{H}$ is indicated in the table. In the cases marked α we choose for $\mathfrak{h}$ the reflection in $\mathfrak{s}$, in the cases marked β the half-turn around a point of $\mathfrak{s}$, in the case marked γ the primary reversed translation along $\mathfrak{s}$. In the case $\mathfrak{A}_{\mathcal{S}}(|)$ let $\mathfrak{s}$ denote the reflection contained in $\mathfrak{A}_{\mathcal{S}}$ (it is a reflection in a normal of $\mathcal{S}$). In the case $\mathfrak{A}_{\mathcal{S}}(\|)$ let $\mathfrak{s}_1$ and $\mathfrak{s}_2$ denote reflections in two consecutive normals of $\mathcal{S}$ and $\mathfrak{t}$ their product. In the case $\mathfrak{A}_{\mathcal{S}}(\rightarrow)$ let $\mathfrak{t}$ denote a primary translation along $\mathcal{S}$. The relation (8) takes the form $\mathfrak{h}^2 = 1$ in all cases marked α and β, and $\mathfrak{h}^2 = \mathfrak{t}$ in the case γ. The relations (7′) reduce to the forms listed in the table when expressed by generators of $\mathfrak{A}_{\mathcal{S}}$. In the case 7) one of the relations listed is redundant in view of $\mathfrak{h}^2 = 1$. In the case 8) the relation (7′) is redundant, because it follows from $\mathfrak{h}^2 = \mathfrak{t}$.

18.9 Surface corresponding to the generalized free product. In order to describe how $\mathcal{D}$ mod $\mathfrak{G}$ is derived from $\mathcal{D}$ mod $\mathfrak{G}_1$, we remove from $\mathcal{D}$ mod $\mathfrak{G}_1$

I) a half-plane, if $\mathfrak{A}_{\mathcal{S}} = 1$,

II) a quarter-plane, if $\mathfrak{A}_{\mathcal{S}} = \mathfrak{A}(|)$,

III) a funnel, if $\mathfrak{A}_{\mathcal{S}} = \mathfrak{A}(\rightarrow)$,

IV) a half-funnel, if $\mathfrak{A}_{\mathcal{S}} = \mathfrak{A}(\|)$,

these parts of the surface $\mathcal{D}$ mod $\mathfrak{G}_1$ resulting in all cases from $\mathcal{D}_2$ mod $\mathfrak{G}_1$. It is recalled that $\mathfrak{G}_1$ is in the cases III and IV an $\mathfrak{F}$-group, and $\mathcal{S}$ is a boundary axis of $\mathcal{K}(\mathfrak{G}_1)$. In the cases I and II the group $\mathfrak{G}_1$ may be quasi-abelian and, even if it is an $\mathfrak{F}$-group, $\mathcal{S}$ may be contained in one of the half-planes outside $\mathcal{K}(\mathfrak{G}_1)$ provided the three conditions at the beginning of Section 2 are satisfied.

It now depends on the structure of $\mathfrak{H}$ what happens to the boundary arising on $\mathcal{D}$ mod $\mathfrak{G}_1$ from the cut.

In the cases 1)–4) of Table 3, marked α, and in these cases only, the reflection in $\mathcal{S}$ is contained in $\mathfrak{H}$. Hence, passing from $\mathcal{D}$ mod $\mathfrak{G}_1$ to $\mathcal{D}$ mod $\mathfrak{G}$ means that the boundary arising from the cut is turned into a reflection edge. In the case 1) the surface $\mathcal{D}$ mod $\mathfrak{G}$ has an open reflection edge, in the case 2) an angular point of order 2, in the case 3) a closed reflection edge, in the case 4) two angular points of order 2.

In the cases 5)–7) the element $\mathfrak{h}$ is a half-turn around a point of $\mathcal{S}$. In the case 5) the boundary arising from the cut along $\mathcal{S}$ is bent together to produce a conical point of order 2. In the cases 6) and 7) the line $\mathcal{S}$ is a boundary axis of $\mathcal{K}(\mathfrak{G}_1)$. In the case 6) the cut is made along the simple, closed, bounding geodesic of $\mathcal{K}(\mathfrak{G}_1)$ mod $\mathfrak{G}_1$, and the effect of passing to $\mathcal{D}$ mod $\mathfrak{G}$ is to squeeze that bounding geodesic together to produce two conical points of order 2. In the case 7) the cut is made along the geodetic arc bounding $\mathcal{K}(\mathfrak{G}_1)$ mod $\mathfrak{G}_1$; this arc is bent together to produce a conical point of order 2, and the reflection edges which on $\mathcal{K}(\mathfrak{G}_1)$ mod $\mathfrak{G}_1$ terminated at that arc become prolongations of each other on $\mathcal{D}$ mod $\mathfrak{G}$.

Finally, in the case 8) the surface $\mathcal{D}$ mod $\mathfrak{G}$ arises by making the two halves of the closed bounding geodesic of $\mathcal{K}(\mathfrak{G}_1)$ mod $\mathfrak{G}_1$ which correspond to the cut coincide with concordant directions. A strip of the surface $\mathcal{K}(\mathfrak{G}_1)$ mod $\mathfrak{G}_1$ along that geodesic is then turned into a Möbius band (a *cross cap*).

18.10 Quasi-abelian generalized free products. Finally we ask under what conditions $\mathfrak{G}$ can be quasi-abelian. Assume it is. Then $\mathfrak{G}^*$, being a subgroup of $\mathfrak{G}$, is also quasi-abelian, thus one of the possible cases enumerated in Section 4.

If $\mathfrak{G}^*$ is the group $\mathfrak{A}(\wedge)$ with limit-centre c, case a) of Section 4, generated by reflections $\mathfrak{g}_1$ and $\mathfrak{g}_2$ of $\mathfrak{G}_1$ and $\mathfrak{G}_2$, respectively, then $\mathfrak{G}^*$ leaves no straight line fixed and leaves no other point than c fixed. Thus $\mathfrak{G}$ cannot have an invariant line and can only have c as its invariant point. Hence $\mathfrak{h}$ cannot be a half-turn. Also $\mathfrak{h}$ cannot be a reversed translation because $\mathfrak{h}^2$ would be a translation belonging to $\mathfrak{A}_{\mathcal{S}}$, and it is known that $\mathfrak{A}_{\mathcal{S}}$ consists only of the identity in the case a). Thus $\mathfrak{h}$ is the reflection in $\mathcal{S}$, and $\mathfrak{g}_2 = \mathfrak{h}\mathfrak{g}_1\mathfrak{h}$. It follows that $\mathfrak{G}$ is the group $\mathfrak{A}(\wedge)$ generated by $\mathfrak{g}_1$ and $\mathfrak{h}$.

Then let $\mathfrak{G}^*$ correspond to the case b) of Section 4. Here $\mathfrak{G}^*$ leaves $\mathcal{A}$ fixed, but no point or end-point of $\mathcal{A}$. Hence $\mathfrak{G}$ must also leave $\mathcal{A}$ fixed. Therefore $\mathfrak{h}$ cannot be a reversed translation. So $\mathfrak{h}$ is either a half-turn around the intersection point of $\mathcal{A}$ and $\mathcal{S}$, or the reflection in $\mathcal{S}$. In the latter case, and also if $\mathfrak{A}_{\mathcal{S}}$ contains the reflection in $\mathcal{A}$, the lines $\mathcal{A}$ and $\mathcal{S}$ are orthogonal. Moreover, $\mathfrak{G}^*$ cannot be of the type $\mathfrak{A}(\cdot|)$, since $\mathfrak{G}_1$ and $\mathfrak{G}_2$ are conjugate. $\mathfrak{G}$ itself can belong to any of the four types indicated in case b) of Section 1.

18.11 Generalized free product of infinitely many groups operating on congruent regions. Let $\mathcal{S}$ and $\mathcal{T}$ denote two non-concurrent straight lines and $\mathcal{D}_0$ the open part of $\mathcal{D}$ situated between them (Fig. 18.4 and 18.5).

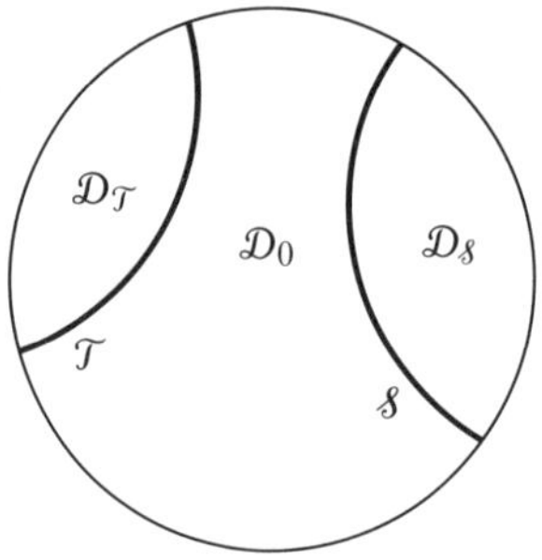

Figure 18.4

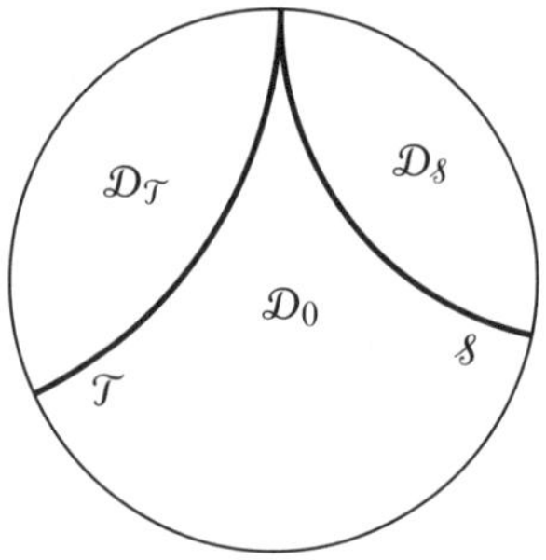

Figure 18.5

Consider a group $\mathfrak{G}_0$ of motions and reversions satisfying the following conditions:

1) $\mathfrak{G}_0$ is discontinuous in $\mathcal{D}$.

2) Neither $\mathcal{S}$ nor $\mathcal{T}$ is an invariant line of $\mathfrak{G}_0$.

3) If $\mathfrak{f}\mathcal{S} \neq \mathcal{S}$, $\mathfrak{f} \in \mathfrak{G}_0$, then $\mathfrak{f}\mathcal{S} \subset \mathcal{D}_0$.

4) If $\mathfrak{f}\mathcal{T} \neq \mathcal{T}$, $\mathfrak{f} \in \mathfrak{G}_0$, then $\mathfrak{f}\mathcal{T} \subset \mathcal{D}_0$.

5) The subgroups $\mathfrak{A}_{\mathcal{S}}$ and $\mathfrak{A}_{\mathcal{T}}$ of $\mathfrak{G}_0$ which leave invariant $\mathcal{S}$ and $\mathcal{T}$, respectively, are isomorphic. If they contain translations, then the primary displacements are equal.

6) $\mathfrak{A}_{\mathcal{S}}$ and $\mathfrak{A}_{\mathcal{T}}$ have only the identity in common.

The line $\mathcal{S}$ fulfills the conditions 1), 2), 3) of Section 2, if the group there called $\mathfrak{G}_1$ is our present $\mathfrak{G}_0$ and the half-plane there called $\mathcal{D}_1$ means the half-plane determined by $\mathcal{S}$ and containing $\mathcal{D}_0$; for, these then follow from our present conditions 1), 2), 3). Hence all consequences derived from these conditions in Section 2 are valid. In particular, if $\mathcal{D}_{\mathcal{S}}$ is the half-plane determined by $\mathcal{S}$ and not containing $\mathcal{D}_0$, and thus corresponds to the $\mathcal{D}_2$ of Section 2, and if we write $\mathfrak{C}_{\mathcal{S}}$ for the region there called $\mathfrak{C}_1$, then the collection of disjoint half-planes $\mathfrak{G}_0\mathcal{D}_{\mathcal{S}}$ is the complement relative to $\mathcal{D}$ of the closure $\tilde{\mathfrak{C}}_{\mathcal{S}}$ of $\mathfrak{C}_{\mathcal{S}}$ on $\mathcal{D}$.

Since the analogous conditions 1), 2), 4) are imposed on $\mathcal{T}$ instead of $\mathcal{S}$, the same conclusions hold in relation to $\mathcal{T}$. In particular, if $\mathcal{D}_{\mathcal{T}}$ is the half-plane determined by $\mathcal{T}$ and not containing $\mathcal{D}_0$, then $\mathfrak{G}_0\mathcal{D}_{\mathcal{T}}$ is the complement relative to $\mathcal{D}$ of the closure $\tilde{\mathfrak{C}}_{\mathcal{T}}$ on $\mathcal{D}$ of the region $\mathfrak{C}_{\mathcal{T}}$ bounded on $\mathcal{D}$ by the collection $\mathfrak{G}_0\mathcal{T}$. No line of the collection $\mathfrak{G}_0\mathcal{S} \cup \mathfrak{G}_0\mathcal{T}$ separates two lines of that collection; for, if $\mathfrak{f} \in \mathfrak{G}_0$, and $\mathfrak{f}\mathcal{S}$, say, separated two lines $\mathcal{L}$ and $\mathcal{M}$ of that collection, then one of the lines $\mathfrak{f}^{-1}\mathcal{L}, \mathfrak{f}^{-1}\mathcal{M}$ would lie in $\mathcal{D}_{\mathcal{S}}$, thus outside $\tilde{\mathcal{D}}_0$.

Since $\mathcal{S}$ and $\mathcal{T}$ are non-concurrent, and the whole collection $\mathfrak{G}_0\mathcal{T}$ is, except for $\mathcal{T}$ itself, contained in $\mathcal{D}_0$, it follows that $\mathcal{S}$ and $\mathcal{T}$ are disjoint modulo $\mathfrak{G}_0$ (§15.5). It is recalled that each of the lines $\mathcal{S}$ and $\mathcal{T}$ is simple modulo $\mathfrak{G}_0$ (Section 2).

The collection $\mathfrak{G}_0\mathcal{T}$ is contained in $\tilde{\mathcal{D}}_0$, hence disjoint with $\mathcal{D}_{\mathcal{S}}$, hence disjoint with $\mathfrak{G}_0\mathcal{D}_{\mathcal{S}}$, since it is reproduced by $\mathfrak{G}_0$. It is thus contained in $\mathfrak{C}_{\mathcal{S}}$. Equally $\mathfrak{G}_0\mathcal{S}$ is contained in $\mathfrak{C}_{\mathcal{T}}$. Thus the intersection $\mathfrak{C}_0 = \mathfrak{C}_{\mathcal{S}} \cap \mathfrak{C}_{\mathcal{T}}$ is a convex region bounded on $\mathcal{D}$ by the union $\mathfrak{G}_0\mathcal{S} \cup \mathfrak{G}_0\mathcal{T}$. It is reproduced by $\mathfrak{G}_0$, since both $\mathfrak{C}_{\mathcal{S}}$ and $\mathfrak{C}_{\mathcal{T}}$ are reproduced by $\mathfrak{G}_0$. The complement of $\tilde{\mathfrak{C}}_0$ on $\mathcal{D}$ is the union $\mathfrak{G}_0\mathcal{D}_{\mathcal{S}} \cup \mathfrak{G}_0\mathcal{D}_{\mathcal{T}}$. If $\mathfrak{G}_0$ is an $\mathfrak{F}$-group, then $\overline{\mathfrak{C}}_0$ contains $\mathcal{K}(\mathfrak{G}_0)$ in virtue of §14.4, for $\overline{\mathfrak{C}}_0$ is closed and convex and is reproduced by $\mathfrak{G}_0$.

We now take conditions 5) and 6) into account. It was seen in Section 2 that $\mathfrak{A}_{\mathcal{S}}$, and thus also $\mathfrak{A}_{\mathcal{T}}$, belongs to one of the types

$$\text{I) } 1, \quad \text{II) } \mathfrak{A}(|), \quad \text{III) } \mathfrak{A}(\rightarrow), \quad \text{IV) } \mathfrak{A}(\|).$$

No two of these four groups are isomorphic. Hence condition 5) implies that $\mathfrak{A}_{\mathcal{T}}$ belongs to the same of these four types as $\mathfrak{A}_{\mathcal{S}}$. There exists a motion, and also a reversion, carrying $\mathcal{S}$ into $\mathcal{T}$ and the half-plane $\mathcal{D}_{\mathcal{S}}$ into the half-plane complementary to $\mathcal{D}_{\mathcal{T}}$; such a motion, and such a reversion, is determined up to an arbitrary translation along $\mathcal{T}$. – In the cases I) and III) let $\mathfrak{h}$ denote any such motion or reversion. – In the case II) we impose on $\mathfrak{h}$ the additional condition that it carries the reflection line of $\mathfrak{G}_0$ cutting $\mathcal{S}$ into the reflection line of $\mathfrak{G}_0$ cutting $\mathcal{T}$. This is possible since these reflection lines are normals of $\mathcal{S}$ and $\mathcal{T}$, and the motion or reversion is then uniquely determined. The two reflection lines do not coincide, since otherwise the reflection in them would be a common element of $\mathfrak{A}_{\mathcal{S}}$ and $\mathfrak{A}_{\mathcal{T}}$ contrary to the condition 6). – In the case IV) we impose on $\mathfrak{h}$ the additional condition that it carries one of the reflection lines of $\mathfrak{G}_0$ cutting $\mathcal{S}$ into one of the reflection lines of $\mathfrak{G}_0$ cutting $\mathcal{T}$; owing to the fact that the primary displacements in $\mathfrak{A}_{\mathcal{S}}$ and $\mathfrak{A}_{\mathcal{T}}$ are equal (condition 5)), $\mathfrak{h}$ then carries the whole collection of reflection lines of $\mathfrak{G}_0$ cutting $\mathcal{S}$ into the whole collection of reflection lines of $\mathfrak{G}_0$ cutting $\mathcal{T}$. These two collections have no reflection line in common in virtue of condition 6).

In consequence of these conditions any two points of $\mathcal{S}$ which are equivalent with respect to $\mathfrak{A}_{\mathcal{S}}$ are carried into two points of $\mathcal{T}$ which are equivalent with respect to $\mathfrak{A}_{\mathcal{T}}$; in the case I) no two such points exist. The motion or reversion $\mathfrak{h}^{-1}$ has the same properties as $\mathfrak{h}$ with the interchange of the rôle of $\mathcal{S}$ and $\mathcal{T}$.

If $\mathcal{S}$ and $\mathcal{T}$ are parallel (Fig. 18.5), they cannot be axes of $\mathfrak{G}_0$. Thus only the cases I) and II) occur. In these cases we take for $\mathfrak{h}$ the limit-rotation which carries $\mathcal{S}$ into $\mathcal{T}$. In the case II) this requires the additional assumption that it is this limit-rotation which carries the reflection line of $\mathfrak{G}_0$ cutting $\mathcal{S}$ into the reflection line of $\mathfrak{G}_0$ cutting $\mathcal{T}$; this additional assumption will hold in the applications to be made later (cf. §19.2). Condition 6) holds, since the two reflection lines cannot coincide, because $\mathcal{S}$ and $\mathcal{T}$ have no common normal.

If $\mathcal{S}$ and $\mathcal{T}$ are divergent (Fig. 18.4), the half-plane $\mathcal{D}_{\mathcal{S}}$ is by $\mathfrak{h}$ carried into the complement of $\tilde{\mathcal{D}}_{\mathcal{T}}$. It thus contains an invariant point of $\mathfrak{h}$ on its boundary (Fig. 18.4). Since $\mathcal{D}_{\mathcal{T}}$ is the image by $\mathfrak{h}$ of the complement of $\tilde{\mathcal{D}}_{\mathcal{S}}$, it also contains an invariant point of $\mathfrak{h}$ on its boundary. These two points are different, hence $\mathfrak{h}$ is a translation or reversed translation, and the two points are, in the order indicated, the negative and positive fundamental points of $\mathfrak{h}$.

In the sequel now $\mathfrak{h}$ denotes a *fixed* translation, or limit-rotation, or reversed translation satisfying the additional conditions imposed above regarding reflection lines of $\mathfrak{G}_0$, if any.

In all cases $\mathfrak{h}$ is of infinite order, and it does not belong to $\mathfrak{G}_0$, since it carries $\mathfrak{C}_0$ into a region situated in $\mathcal{D}_{\mathcal{T}}$.

We now write $\mathcal{S}_0$ for $\mathcal{S}$, $\mathfrak{A}_0$ for $\mathfrak{A}_{\mathcal{S}}$, and for any n, $-\infty < n < \infty$, we put

$$\mathcal{S}_n = \mathfrak{h}^n \mathcal{S}_0, \quad \mathfrak{C}_n = \mathfrak{h}^n \mathfrak{C}_0, \quad \mathfrak{G}_n = \mathfrak{h}^n \mathfrak{G}_0 \mathfrak{h}^{-n}, \quad \mathfrak{A}_n = \mathfrak{h}^n \mathfrak{A}_0 \mathfrak{h}^{-n}.$$

Thus, in particular, $\mathcal{S}_1$ means $\mathcal{T}$, and $\mathfrak{A}_1$ means $\mathfrak{A}_{\mathcal{T}}$. Then any two of the regions $\mathfrak{C}_n$ are disjoint. $\mathfrak{C}_{n-1}$ and $\mathfrak{C}_n$ are adjacent along $\mathcal{S}_n$. The group $\mathfrak{G}_n$ carries $\mathfrak{C}_n$ into itself, because $\mathfrak{G}_0$ carries $\mathfrak{C}_0$ into itself. Let $\mathcal{H}_n$ denote the half-plane determined by $\mathcal{S}_n$ and containing $\mathcal{S}_{n+1}$, and $\mathcal{H}'_n$ the complement of $\tilde{\mathcal{H}}_n$ (Fig. 18.6). Then $\mathfrak{C}_n \subset \tilde{\mathcal{H}}_n \cap \tilde{\mathcal{H}}'_{n+1}$. Let $\mathfrak{g}$ be an arbitrary element of $\mathfrak{G}_n$. If $\mathfrak{g} \notin \mathfrak{A}_{n+1}$, then $\mathfrak{g}\mathcal{H}_{n+1} \subset \mathcal{H}'_{n+1}$, hence for all $m \geqq 1$ we get $\mathfrak{g}\mathfrak{C}_{n+m} \subset \mathcal{H}'_{n+1}$, hence $\mathfrak{C}_{n+m}$ and $\mathfrak{g}\mathfrak{C}_{n+m}$ are disjoint, hence $\mathfrak{g} \notin \mathfrak{G}_{n+m}$, $m \geqq 1$.

If $\mathfrak{g} \in \mathfrak{A}_{n+1}$, then it follows from condition 6), that $\mathfrak{g} \notin \mathfrak{A}_{n+2}$, hence $\mathfrak{g}\mathcal{H}_{n+2} \subset \mathcal{H}'_{n+2}$, hence for all $m \geqq 2$, $\mathfrak{g}\mathcal{H}_{n+m} \subset \mathcal{H}'_{n+2}$, hence $\mathfrak{g}\mathfrak{C}_{n+m}$ and $\mathfrak{C}_{n+m}$ are disjoint, hence $\mathfrak{g} \notin \mathfrak{G}_{n+m}$, $m \geqq 2$. This shows that any two groups $\mathfrak{G}_n$ and $\mathfrak{G}_{n+m}$, $m \geqq 2$, have no element in common, and that $\mathfrak{G}_n \cap \mathfrak{G}_{n+1} = \mathfrak{A}_{n+1}$.

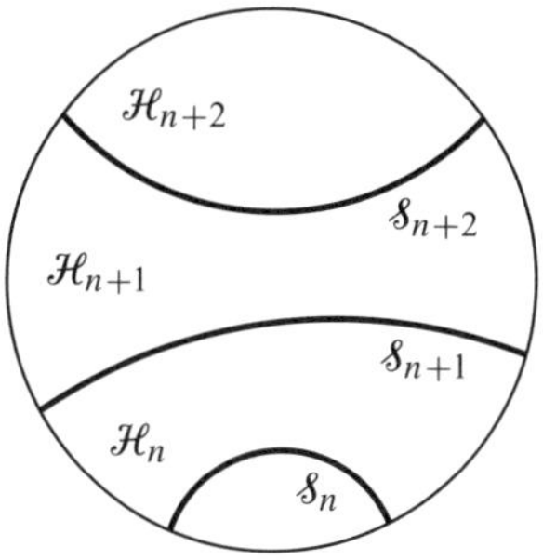

Figure 18.6

The line $\mathcal{S}_0$ (= $\mathcal{S}$) and the two groups $\mathfrak{G}_0$ and $\mathfrak{G}_{-1}$ satisfy the conditions imposed in Section 2 on the line there called $\mathcal{S}$ and the groups there called $\mathfrak{G}_1$ and $\mathfrak{G}_2$. One can, therefore, derive the generalized free product

$$\mathfrak{G}_{0,-1} = \mathfrak{G}_0 * \mathfrak{G}_{-1} \text{ am } \mathfrak{A}_0.$$

Let $\mathfrak{C}'_0$ and $\mathfrak{C}'_{-1}$ denote the regions corresponding to those called $\mathfrak{C}_1$ and $\mathfrak{C}_2$ in Section 2; thus $\mathfrak{C}'_0$ is bounded on $\mathcal{D}$ by $\mathfrak{G}_0\mathcal{S}_0$ and $\mathfrak{C}'_{-1}$ by $\mathfrak{G}_{-1}\mathcal{S}_0$. The region here called $\mathfrak{C}_0$ is bounded on $\mathcal{D}$ by $\mathfrak{G}_0\mathcal{S}_0 \cup \mathfrak{G}_0\mathcal{S}_1$, since $\mathcal{S}_1 = \mathcal{T}$. Thus $\mathfrak{C}_0$ is the part of $\mathfrak{C}'_0$ bounded on $\mathfrak{C}'_0$ by $\mathfrak{G}_0\mathcal{S}_1$. Likewise $\mathfrak{C}_{-1}$ is the part of $\mathfrak{C}'_{-1}$ bounded on $\mathfrak{C}'_{-1}$ by $\mathfrak{G}_{-1}\mathcal{S}_{-1}$. It follows from Section 3 that $\mathfrak{G}_{0,-1}$ is discontinuous in $\mathcal{D}$, and further that $\mathfrak{C}'_0$ and $\mathfrak{C}'_{-1}$ are disjoint modulo $\mathfrak{G}_{0,-1}$. Hence, in particular, $\mathcal{S}_1$ and $\mathcal{S}_{-1}$ are disjoint modulo $\mathfrak{G}_{0,-1}$, since they are contained in $\mathfrak{C}'_0$ and $\mathfrak{C}'_{-1}$, respectively. Let $\mathfrak{C}_{0,-1}$ denote the region bounded on $\mathcal{D}$ by $\mathfrak{G}_{0,-1}\mathcal{S}_1 \cup \mathfrak{G}_{0,-1}\mathcal{S}_{-1}$.

It also follows from Section 3 that $\mathfrak{G}_{0,-1}$ has the same effect as $\mathfrak{G}_0$ in $\mathfrak{C}'_0$, thus, in particular, in $\mathfrak{C}_0$.

We now proceed to the next step. The line $\mathcal{S}_1$ (= $\mathcal{T}$) and the two groups $\mathfrak{G}_1$ and $\mathfrak{G}_{0,-1}$ again satisfy the starting conditions of Section 2. One can therefore derive the generalized free product

$$\mathfrak{G}_{1,0,-1} = \mathfrak{G}_1 * \mathfrak{G}_{0,-1} \text{ am } \mathfrak{A}_1.$$

Let, correspondingly, $\mathfrak{C}'_1$ denote the region bounded on $\mathcal{D}$ by $\mathfrak{G}_1\mathcal{S}_1$, and $\mathfrak{C}'_{0,-1}$ the region bounded on $\mathcal{D}$ by $\mathfrak{G}_{0,-1}\mathcal{S}_1$. Then $\mathfrak{C}_1$ is part of $\mathfrak{C}'_1$ and bounded on $\mathfrak{C}'_1$ by $\mathfrak{G}_1\mathcal{S}_2$; and $\mathfrak{C}_{0,-1}$ is the part of $\mathfrak{C}'_{0,-1}$ bounded on $\mathfrak{C}'_{0,-1}$ by $\mathfrak{G}_{0,-1}\mathcal{S}_{-1}$. Again $\mathfrak{G}_{1,0,-1}$ is discontinuous in $\mathcal{D}$ and has in $\mathfrak{C}_{0,-1}$ the same effect as $\mathfrak{G}_{0,-1}$.

One may now continue this process by constructing in turn

$$\mathfrak{G}_{1,0,-1,-2} = \mathfrak{G}_{1,0,-1} * \mathfrak{G}_{-2} \text{ am } \mathfrak{A}_{-1},$$
$$\mathfrak{G}_{2,1,0,-1,-2} = \mathfrak{G}_2 * \mathfrak{G}_{1,0,-1,-2} \text{ am } \mathfrak{A}_2,$$

and so on. Each group so constructed contains the preceding one as a subgroup. The union of this ascending sequence of groups is the group

$$\mathfrak{G}^* = \prod_n{}_* \mathfrak{G}_n \operatorname{am} \mathfrak{A}_n, \qquad -\infty < n < \infty.$$

Let p be a point of $\mathfrak{C}_0$, and assume that $\mathfrak{g}^* p$, where $\mathfrak{g}^*$ is an element of $\mathfrak{G}^*$, belongs to $\mathfrak{C}_0$. The element $\mathfrak{g}^* \in \mathfrak{G}^*$ being the product of a finite number of factors drawn from the groups $\mathfrak{G}_n$, there exists a number m such that $\mathfrak{g}^* \in \mathfrak{G}_{m.m-1,\dots,-m}$. Now $\mathfrak{G}_{m,\dots,-m}$ has the same effect in $\mathfrak{C}_{m-1,\dots,-m}$ as $\mathfrak{G}_{m-1,\dots,-m}$, thus in $\mathfrak{C}_{m-1,\dots,-(m-1)}$ the same effect as $\mathfrak{G}_{m-1,\dots,-(m-1)}$, and so on until: in $\mathfrak{C}_0$ the same effect as $\mathfrak{G}_0$. Hence $\mathfrak{g}^* \in \mathfrak{G}_0$. Hence $\mathfrak{G}^*$ and $\mathfrak{G}_0$ have equal effect in $\mathfrak{C}_0$. Since $\mathfrak{G}_0 p$ does not accumulate in p (condition 1)), it is inferred that $\mathfrak{G}^*$ is discontinuous in $\mathcal{D}$.

It was seen in Section 2 – bearing the change of notations in mind – that the collection $\mathfrak{G}_{0,-1}\tilde{\mathfrak{C}}'_0 \cup \mathfrak{G}_{0,-1}\tilde{\mathfrak{C}}'_{-1}$ is a convex and open set reproduced by $\mathfrak{G}_{0,-1}$. In Section 2 it was denoted by $\mathfrak{C}$; let it here be called $\mathcal{U}_{0,-1}$. Any two adjacent of the domains making up $\mathcal{U}_{0,-1}$ have a line from $\mathfrak{G}_{0,-1}\mathcal{S}_0$ in common, and one of the domains belongs to $\mathfrak{G}_{0,-1}\tilde{\mathfrak{C}}'_0$, the other to $\mathfrak{G}_{0,-1}\tilde{\mathfrak{C}}'_{-1}$. We now consider the collection $\mathcal{V}_{0,-1} = \mathfrak{G}_{0,-1}\tilde{\mathfrak{C}}_0 \cup \mathfrak{G}_{0,-1}\tilde{\mathfrak{C}}_{-1}$. It is obtained from $\mathcal{U}_{0,-1}$ by omitting the half-planes bounded by $\mathfrak{G}_{0,-1}\mathcal{S}_1$, all contained in $\mathfrak{G}_{0,-1}\mathfrak{C}'_0$, and also the half-planes bounded by $\mathfrak{G}_{0,-1}\mathcal{S}_{-1}$, all contained in $\mathfrak{G}_{0,-1}\mathfrak{C}'_{-1}$. Hence also $\mathcal{V}_{0,-1}$ is convex. We may call the collection $\mathfrak{G}_{0,-1}\mathcal{S}_1 \cup \mathfrak{G}_{0,-1}\mathcal{S}_{-1}$ the *free sides* of $\mathcal{V}_{0,-1}$. Proceeding now to $\mathfrak{G}_{1,0,-1}$ one gets in the same way the collection $\mathcal{V}_{1,0,-1}$ consisting of $\mathfrak{G}_{1,0,-1}\tilde{\mathfrak{C}}_1 \cup \mathfrak{G}_{1,0,-1}\mathcal{V}_{0,-1}$. It is composed of the union $\mathcal{V}_{1,0,-1} = \mathfrak{G}_{1,0,-1}\tilde{\mathfrak{C}}_1 \cup \mathfrak{G}_{1,0,-1}\tilde{\mathfrak{C}}_0 \cup \mathfrak{G}_{1,0,-1}\tilde{\mathfrak{C}}_{-1}$, and it contains $\mathcal{V}_{0,-1}$. Any two adjacent domains of this collection $\mathcal{V}_{1,0,-1}$ are either adjacent along a line of the collection $\mathfrak{G}_{1,0,-1}\mathcal{S}_1$ and are then equivalent with respect to $\mathfrak{G}_{1,0,-1}$ with $\tilde{\mathfrak{C}}_1$ and $\tilde{\mathfrak{C}}_0$, respectively, or they are adjacent along a line of the collection $\mathfrak{G}_{1,0,-1}$ and are then equivalent with respect to $\mathfrak{G}_{1,0,-1}$ with $\tilde{\mathfrak{C}}_0$ and $\tilde{\mathfrak{C}}_{-1}$, respectively. $\mathcal{V}_{1,0,-1}$ has $\mathfrak{G}_{1,0,-1}\mathcal{S}_2 \cup \mathfrak{G}_{1,0,-1}\mathcal{S}_{-1}$ as free sides. This may now be extended in turn to the group $\mathfrak{G}_{1,0,-1,-2}$ (and $\mathcal{V}_{1,0,-1,-2}$ then contains $\mathcal{V}_{1,0,-1}$), from there to the group $\mathfrak{G}_{2,1,0,-1,-2}$ and so on. The union of all the domains obtained by this infinite process is

$$\mathcal{V} = \bigcup_n \mathfrak{G}^*\tilde{\mathfrak{C}}_n, \qquad -\infty < n < \infty.$$

It contains no free sides, and it is a convex, open set. Any two adjacent of the domains making up $\mathcal{V}$ have in common a line of the collection $\mathfrak{G}^*\mathcal{S}_n$ for some value n. They are then equivalent with respect to $\mathfrak{G}^*$ with $\tilde{\mathfrak{C}}_n$ and $\tilde{\mathfrak{C}}_{n-1}$, respectively. If $\mathfrak{G}^*$ is an $\mathfrak{F}$-group, then $\mathcal{K}(\mathfrak{G}^*) \subset \overline{\mathcal{V}}$, since $\overline{\mathcal{V}}$ is closed and convex and is reproduced by $\mathfrak{G}^*$.

If for any line intersecting $\mathfrak{C}_0$ there is a positive lower bound ρ for the length of the segment cut out on the line by $\mathfrak{C}_0$, then $\mathcal{V}$ is the whole of $\mathcal{D}$. For, the same then holds with the same ρ for all the domains making up $\mathcal{V}$, since the $\mathfrak{C}_n$, and hence all domains of $\mathcal{V}$, are congruent, and it then follows by the same consideration as in Section 5 that

$\mathcal{V}$ is a tesselation of the whole of $\mathcal{D}$. The condition is, in particular, satisfied, if $\mathcal{S}$ is an axis of $\mathfrak{G}_0$; for, then the projection of any other side of $\mathfrak{C}_0$ on $\mathcal{S}$ cannot exceed the primary displacement on $\mathcal{S}$, and this implies the existence of a positive lower bound for the distance of any two sides of $\mathfrak{C}_0$. – If $\mathcal{V}$ is not the whole of $\mathcal{D}$, then the collection $\bigcup_n \mathfrak{G}^* \mathcal{S}_n$ accumulates in $\mathcal{D}$.

As in Section 2 an arbitrary translation $\mathfrak{t}$ with axis $\mathcal{S}$ provides in the cases I and III a group $\mathfrak{G}^*(\mathfrak{t})$; it is obtained if one substitutes in the preceding construction the element $\mathfrak{h}$ by $\mathfrak{h}\mathfrak{t}$. Therefore one has, again, a whole family $\mathfrak{G}^*(\mathfrak{t})$ of groups parametrized by the translations with axis $\mathcal{S}$. If moreover $\mathfrak{h}$ is the translation along the common normal of $\mathcal{S}$ and $\mathcal{T}$ which maps $\mathcal{S}$ onto $\mathcal{T}$, then a canonical normalization (relative to $\mathcal{S}$ and $\mathcal{T}$) of the amalgamation parameter arises.

An abstract characterization of $\mathfrak{G}^*$ is obtained by using the results of Section 2. Let Φ_0 denote any system of generators of $\mathfrak{G}_0$, and Ψ_0 a system of defining relations between them. If λ_0 is an element in Φ_0, then $\lambda_n = \mathfrak{h}^n \lambda_0 \mathfrak{h}^{-n}$ belongs to $\mathfrak{G}_n$. If λ_0 ranges over Φ_0, then λ_n ranges over a system Φ_n of generators of $\mathfrak{G}_n$, and a system Ψ_n of defining relations for $\mathfrak{G}_n$ in these generators is obtained from Ψ_0 by replacing every generator λ_0 in every relation of Ψ_0 by the corresponding λ_n. Then $\bigcup_n \Phi_n$ generates $\mathfrak{G}^*$, and all relations $\bigcup_n \Psi_n$ are valid in $\mathfrak{G}^*$. For any generator of $\mathfrak{A}_0$ one gets an additional relation

$$\varphi'(\Phi_0) = \varphi''(\Phi_{-1}),$$

and this induces a copy in $\mathfrak{A}_n$:

$$\varphi'(\Phi_n) = \varphi''(\Phi_{n-1}).$$

If these relations, for a system of generators of $\mathfrak{A}_0$ and for all n, are added to $\bigcup_n \Psi_n$, then a system of defining relations for $\mathfrak{G}^*$ in the generators $\bigcup_n \Phi_n$ is obtained. For, it was shown above that $\mathfrak{G}_m \cap \mathfrak{G}_n = 1$, if $|m-n| \geqq 2$, and that $\mathfrak{G}_{n-1} \cap \mathfrak{G}_n = \mathfrak{A}_n$.

18.12 Extension of a group by an adjunction. $\mathfrak{G}_0$ and $\mathfrak{h}$ being determined as in the preceding section, consider the group $\mathfrak{G} = \langle \mathfrak{G}_0, \mathfrak{h} \rangle$ generated by $\mathfrak{h}$ and $\mathfrak{G}_0$. It contains all the $\mathfrak{G}_n$, thus also $\mathfrak{G}^*$ as a subgroup. Φ_0 together with $\mathfrak{h}$ is a system Φ of generators for $\mathfrak{G}$. As a system of defining relations for $\mathfrak{G}$, one gets the collection of relations Ψ_0 together with the relations

$$\varphi'(\Phi_0) = \varphi''(\mathfrak{h}^{-1}\Phi_0\mathfrak{h}),$$

where $\mathfrak{h}^{-1}\Phi_0\mathfrak{h}$ denotes the system of elements of $\mathfrak{G}$ derived from the elements in Φ_0 by transformation with $\mathfrak{h}^{-1}$. One can restrict oneself to two, or one, or no relation of this latter kind according as $\mathfrak{A}_{\mathcal{S}}$ belongs to the type IV, to the types III or II, or to the type I.

The group $\mathfrak{G}$ is said to be derived from $\mathfrak{G}_0$ by the *adjunction of the element* $\mathfrak{h}$. It contains $\mathfrak{G}^*$ as a normal subgroup, since conjugation with $\mathfrak{h}$ replaces $\mathfrak{G}_n$ by $\mathfrak{G}_{n+1}$ and thus leaves $\Pi_n^* \mathfrak{G}_n$ invariant. The index of $\mathfrak{G}^*$ in $\mathfrak{G}$ is infinite, the cosets of $\mathfrak{G}^*$ in $\mathfrak{G}$ being $\mathfrak{G}^*\mathfrak{h}^n, \quad -\infty < n < \infty$.

It was seen above that $\mathfrak{G}^*$ and $\mathfrak{G}_0$ have equal effect in $\mathfrak{C}_0$. An element of $\mathfrak{G}$ outside $\mathfrak{G}^*$, being of the form $\mathfrak{g}^*\mathfrak{h}^n$, $\mathfrak{g}^* \in \mathfrak{G}^*$, $n \neq 0$, carries $\mathfrak{C}_0$ into the region $\mathfrak{g}^*\mathcal{C}_n$, which is disjoint with $\mathfrak{C}_0$ modulo $\mathfrak{G}^*$. Hence $\mathfrak{C}_0$ and $\mathfrak{g}^*\mathfrak{h}^n\mathfrak{C}_0$, $n \neq 0$, have no point in common. Thus $\mathfrak{G}$ and $\mathfrak{G}_0$ have equal effect in $\mathfrak{C}_0$. It follows that $\mathfrak{G}$ is discontinuous in $\mathcal{D}$. It should be noticed that $\mathfrak{G}$ and $\mathfrak{G}_0$ do not have equal effect in the closure $\tilde{\mathfrak{C}}_0$ of $\mathfrak{C}_0$ on $\mathcal{D}$, since $\mathcal{S}_0$ and $\mathcal{S}_1$ are equivalent with respect to $\mathfrak{G}$, but not with respect to $\mathfrak{G}_0$.

All the domains making up the collection $\mathcal{V} = \bigcup_n \mathfrak{G}^*\tilde{\mathfrak{C}}_n$ are equivalent with respect to $\mathfrak{G}$, since all the $\tilde{\mathfrak{C}}_n$ are equivalent by powers of $\mathfrak{h}$, thus with respect to $\mathfrak{G}$, and $\mathfrak{G}^* \subset \mathfrak{G}$. Since $\mathfrak{G}$ and $\mathfrak{G}_0$ have equal effect in $\mathfrak{C}_0$, it is easily inferred that for any $\mathfrak{g} \in \mathfrak{G}$ the groups $\mathfrak{G}$ and $\mathfrak{g}\mathfrak{G}_0\mathfrak{g}^{-1}$ have equal effect in $\mathfrak{g}\mathfrak{C}_0$.

The boundaries of $\mathfrak{C}_0$ in $\mathcal{D}$ fall into two equivalence classes with respect to $\mathfrak{G}_0$, one being $\mathfrak{G}_0\mathcal{S}_0$, the other $\mathfrak{G}_0\mathcal{S}_1$ (in the notation of Section 11 we called them $\mathfrak{G}_0\mathcal{S}$ and $\mathfrak{G}_0\mathcal{T}$). They belong to one equivalence class with respect to $\mathfrak{G}$, since $\mathcal{S}_1 = \mathfrak{h}\mathcal{S}_0$. Hence the surface $\mathcal{D}$ mod $\mathfrak{G}$ is obtained by removing from $\mathcal{D}$ mod $\mathfrak{G}_0$ the two half-planes, quarter-planes, funnels, or half-funnels outside $\tilde{\mathfrak{C}}_0$ mod $\mathfrak{G}_0$, and joining the two cuts thereby arising in such a way that points corresponding by $\mathfrak{h}$ coincide.

§19 Decomposition of groups

19.1 Decomposition of an $\mathfrak{F}$-group. Let $\mathcal{S}$ be a straight line in $\mathcal{D}$ which is simple modulo an $\mathfrak{F}$-group $\mathfrak{F}$. If $\mathcal{S}$ is an axis of $\mathfrak{F}$, or a reflection line of $\mathfrak{F}$, then it is known from §13.2 that the collection $\mathfrak{F}\mathcal{S}$ does not accumulate in $\mathcal{D}$. In the general case we *assume* $\mathcal{S}$ to satisfy this condition. Since $\mathcal{S}$ is simple modulo $\mathfrak{F}$, it cannot pass through any centre of $\mathfrak{F}$ of order higher than 2, and it can only cut reflection lines of $\mathfrak{F}$ at right angles.

Since $\mathfrak{F}\mathcal{S}$ does not accumulate in $\mathcal{D}$, the collection $\mathfrak{F}\mathcal{S}$ decomposes the whole of $\mathcal{D}$ into convex regions bounded on $\mathcal{D}$ by lines of $\mathfrak{F}\mathcal{S}$ only. The number of these regions is infinite. This is evident, if $\mathcal{S}$ is a side of $\mathcal{K}(\mathfrak{F})$, or is wholly contained in one of the half-planes outside $\mathcal{K}(\mathfrak{F})$; and if $\mathcal{S}$ has points in common with the interior of $\mathcal{K}(\mathfrak{F})$, then there are axes of $\mathfrak{F}$ cutting $\mathcal{S}$ (§13.3), and translations along such an axis carry $\mathcal{S}$ into lines of the collection $\mathfrak{F}\mathcal{S}$ which belong to the boundary of an infinity of regions. Since $\mathfrak{F}\mathcal{S}$ is reproduced by $\mathfrak{F}$, the same holds for the totality of the regions of decomposition.

Let $\mathcal{G}_1$ and $\mathcal{G}_2$ denote the two regions of decomposition adjacent to $\mathcal{S}$, and $\tilde{\mathcal{G}}_1$ and $\tilde{\mathcal{G}}_2$ their closures on $\mathcal{D}$, and $\mathcal{D}_1$ and $\mathcal{D}_2$ the half-planes bounded by $\mathcal{S}$ and containing $\mathcal{G}_1$ and $\mathcal{G}_2$ respectively. If $\mathcal{G}'$ is any of the regions of decomposition and $\mathcal{S}'$ any of its sides, then there is an element of $\mathfrak{F}$ carrying $\mathcal{S}'$ into $\mathcal{S}$ and thus $\mathcal{G}'$ into either $\mathcal{G}_1$ or $\mathcal{G}_2$. Thus the collection of regions falls into one or two equivalence classes with respect to $\mathfrak{F}$, according as $\mathcal{G}_1$ and $\mathcal{G}_2$ are or are not equivalent with respect to $\mathfrak{F}$. In the first case $\mathcal{S}$ will be called *non-dividing modulo* $\mathfrak{F}$, in the second case *dividing modulo* $\mathfrak{F}$. If $\mathcal{S}$ is a side of $\mathcal{K}(\mathfrak{F})$, or is contained in one of the half-planes outside $\mathcal{K}(\mathfrak{F})$, then it is obviously dividing modulo $\mathfrak{F}$.

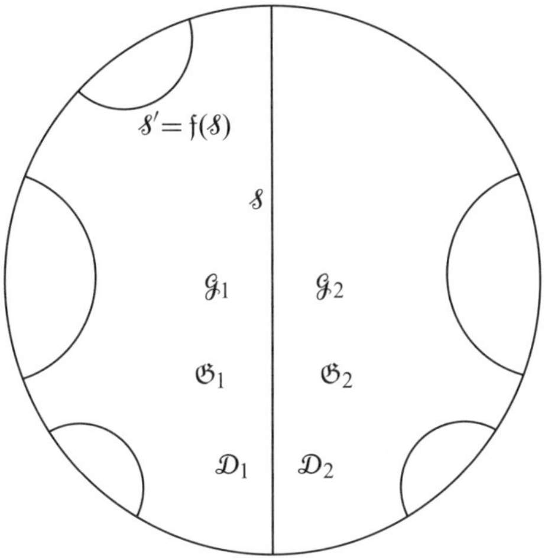

Figure 19.1

Let $\mathfrak{A}_{\mathcal{S}}$ be the subgroup of $\mathfrak{F}$ which leaves $\mathcal{S}$ invariant. If α) the reflection in $\mathcal{S}$, or β) a half-turn about a point of $\mathcal{S}$, or γ) a reversed translation along $\mathcal{S}$ occurs in $\mathfrak{A}_{\mathcal{S}}$, then $\mathcal{S}$ is obviously non-dividing modulo $\mathfrak{F}$. But $\mathcal{S}$ may be non-dividing modulo $\mathfrak{F}$ even if none of these cases occur, as we shall see in Section 2.

Let $\mathfrak{G}_1$ denote the subgroup of $\mathfrak{F}$ which carries $\mathcal{G}_1$ into itself, and correspondingly $\mathfrak{G}_2$ for $\mathcal{G}_2$. Both $\mathfrak{G}_1$ and $\mathfrak{G}_2$ are discontinuous in $\mathcal{D}$, because they are subgroups of $\mathfrak{F}$. If $\mathcal{S}$ is non-dividing modulo $\mathfrak{F}$, then they are conjugate subgroups of $\mathfrak{F}$. It is evident that $\mathfrak{G}_1$ and $\mathfrak{F}$ have equal effect in $\mathcal{G}_1$ and likewise $\mathfrak{G}_2$ and $\mathfrak{F}$ in $\mathcal{G}_2$.

The group $\mathfrak{A}_{\mathcal{S}}$ contains a subgroup $\mathfrak{A}^*_{\mathcal{S}}$ consisting of those elements of $\mathfrak{A}_{\mathcal{S}}$ which, while carrying $\mathcal{S}$ into itself, do not interchange $\mathcal{G}_1$ and $\mathcal{G}_2$. This group $\mathfrak{A}^*_{\mathcal{S}}$ is a subgroup both of $\mathfrak{G}_1$ and $\mathfrak{G}_2$, since it carries each of the regions $\mathcal{G}_1$ and $\mathcal{G}_2$ into itself. It can take one of the four forms

(1) $\mathfrak{A}^*_{\mathcal{S}} = 1$,

(2) $\mathfrak{A}^*_{\mathcal{S}} = \mathfrak{A}(|)$,

(3) $\mathfrak{A}^*_{\mathcal{S}} = \mathfrak{A}(\rightarrow)$,

(4) $\mathfrak{A}^*_{\mathcal{S}} = \mathfrak{A}(\|)$.

Only those reflection lines of $\mathfrak{F}$ pertaining to $\mathfrak{A}_{\mathcal{S}}$ intersect $\mathcal{S}$; they are normals of $\mathcal{S}$ (cases (2) and (4)). Since there is more than one line in $\mathfrak{F}\mathcal{S}$, at least one of the regions $\mathcal{G}_1$ and $\mathcal{G}_2$ has more than one bounding line in $\mathcal{D}$. Suppose that one of them, $\mathcal{G}_2$ say, has only $\mathcal{S}$ as boundary in $\mathcal{D}$. It is then a half-plane and cannot be equivalent with $\mathcal{G}_1$, hence $\mathfrak{G}_2 = \mathfrak{A}^*_{\mathcal{S}} = \mathfrak{A}_{\mathcal{S}}$. No axis of $\mathfrak{F}$ can then intersect $\mathcal{S}$, and $\mathcal{S}$ can thus not have points in common with the interior of $\mathcal{K}(\mathfrak{F})$. Hence in the cases (3) and (4) the line $\mathcal{S}$ is a boundary axis of $\mathfrak{F}$. In the cases (1) and (2) $\mathcal{S}$ is a limit side of $\mathcal{K}(\mathfrak{F})$ or it is contained in the half-plane cut off by a side of $\mathcal{K}(\mathfrak{F})$.

Conversely, if $\mathcal{S}$ is a side of $\mathcal{K}(\mathfrak{F})$ bounding a half-plane $\mathcal{H}$ outside $\mathcal{K}(\mathfrak{F})$, or if $\mathcal{S}$ is wholly contained in such a half-plane $\mathcal{H}$, then one of the half-planes determined by $\mathcal{S}$ belongs to $\mathcal{H}$ and contains no line of the collection $\mathfrak{F}\mathcal{S}$, because $\mathfrak{F}$ has in $\mathcal{H}$ the same effect as the subgroup of $\mathfrak{F}$ pertaining to the side of $\mathcal{K}(\mathfrak{F})$ which bounds $\mathcal{H}$.

In this case, in which one of the regions of decomposition of $\mathcal{D}$, say $\mathcal{G}_2$, is a half-plane, $\tilde{\mathcal{G}}_1$ contains $\mathcal{K}(\mathfrak{F})$, and $\mathfrak{G}_1$ is then $\mathfrak{F}$ itself. Therefore no decomposition of $\mathfrak{F}$ arises. We are thus only concerned with the case in which $\mathcal{S}$ has points in common with the interior of $\mathcal{K}(\mathfrak{F})$. Both $\tilde{\mathcal{G}}_1$ and $\tilde{\mathcal{G}}_2$ then possess more than one line of the collection $\mathfrak{F}\mathcal{S}$ as bounding lines in $\mathcal{D}$.

If $\mathcal{S}$ is an axis of $\mathfrak{F}$, then it is an axis of $\mathfrak{G}_1$. Another side of $\mathcal{G}_1$ is then also an axis of $\mathfrak{G}_1$. Thus $\mathfrak{G}_1$ admits more than one axis and is therefore an $\mathfrak{F}$-group. The same holds for $\mathfrak{G}_2$. Since the closure $\overline{\mathcal{G}}_1$ of $\mathcal{G}_1$ on $\overline{\mathcal{D}}$ is convex and is reproduced by $\mathfrak{G}_1$, it contains $\overline{\mathcal{K}}(\mathfrak{G}_1)$ (§14.1). Being an *axis* of $\mathfrak{G}_1$, the line $\mathcal{S}$ belongs to $\mathcal{K}(\mathfrak{G}_1)$ and is thus a boundary axis of $\mathfrak{G}_1$, because it is a boundary of $\mathcal{G}_1$. Likewise, all boundaries of $\mathcal{G}_1$ are boundary axes of $\mathfrak{G}_1$. The same holds for $\mathfrak{G}_2$ and $\mathcal{G}_2$.

Neither $\mathfrak{G}_1$ nor $\mathfrak{G}_2$ coincides with $\mathfrak{A}^*_{\mathcal{S}}$.

Proof. This is evident in the cases (3) and (4), because $\mathfrak{G}_1$ and $\mathfrak{G}_2$ are then $\mathfrak{F}$-groups, as we have just seen. We consider the cases (1) and (2). Assume that $\mathfrak{G}_1 = \mathfrak{A}^*_{\mathcal{S}}$, thus either the identity or $\mathfrak{A}(|)$. Since $\mathcal{G}_1$ has at least one more boundary in $\mathcal{D}$, let $\mathfrak{f}\mathcal{S}$, $\mathfrak{f} \in \mathfrak{F}$, be another side of $\mathcal{G}_1$. Since $\mathfrak{f}$ cannot belong to $\mathfrak{G}_1$, it must carry $\mathcal{G}_2$ into $\mathcal{G}_1$, $\mathfrak{f}\mathcal{G}_2 = \mathcal{G}_1$, thus all regions are equivalent with respect to $\mathfrak{F}$. If $\mathfrak{G}_1 = 1$, $\mathfrak{h} \in \mathfrak{F}$, and $\mathfrak{h}\mathcal{S} \neq \mathcal{S}$ is also a side of $\mathcal{G}_1$, then also $\mathfrak{h}\mathcal{G}_2 = \mathcal{G}_1$ and $\mathfrak{f}\mathfrak{h}^{-1}\mathcal{G}_1 = \mathcal{G}_1$. Since $\mathfrak{G}_1 = 1$, one gets $\mathfrak{h} = \mathfrak{f}$. Thus $\mathcal{G}_1$ has exactly two sides. If $\mathfrak{G}_1 = \mathfrak{A}(|)$, there is exactly one reflection line in $\mathcal{G}_1$. Since every side of $\mathcal{G}_1$ is cut by a reflection line, the number of sides must also in this case be 2, the reflection line being their common normal. In both cases it is inferred that the $\mathfrak{f}^{\nu}\mathcal{S}$, $-\infty < \nu < \infty$, constitute the whole of $\mathfrak{F}\mathcal{S}$ and that they have at most two accumulation points on $\mathcal{E}$ in contradiction to the fact that the limit set of $\mathfrak{F}$ contains more than two points. This proves the assertion. □

It is thus seen that $\mathfrak{G}_1$ satisfies the three conditions listed §18.1: 1) $\mathfrak{G}_1$, being a subgroup of $\mathfrak{F}$, is discontinuous in $\mathcal{D}$. 2) $\mathfrak{G}_1$ has not $\mathcal{S}$ as an invariant line, because $\mathfrak{G}_1$ contains elements outside $\mathfrak{A}^*_{\mathcal{S}}$, and these are also outside $\mathfrak{A}_{\mathcal{S}}$ because elements of $\mathfrak{A}_{\mathcal{S}}$ outside $\mathfrak{A}^*_{\mathcal{S}}$ carry $\mathcal{G}_1$ into $\mathcal{G}_2$ and thus do not belong to $\mathfrak{G}_1$. 3) Such elements of $\mathfrak{G}_1$ carry $\mathcal{S}$ into another side of $\mathcal{G}_1$, thus situated in $\mathcal{D}_1$. – The same holds for $\mathfrak{G}_2$ with $\mathcal{D}_2$ instead of $\mathcal{D}_1$. One can thus form the group corresponding to (2) of §18.2:

$$\mathfrak{G}_{1,2} = \mathfrak{G}_1 * \mathfrak{G}_2 \text{ am } \mathfrak{A}^*_{\mathcal{S}}.$$

At this point an important distinction has to be made. In §18.2 the region $\mathfrak{C}_1$ was defined as the region adjacent to $\mathcal{S}$ which is bounded by the collection $\mathfrak{G}_1\mathcal{S}$. In our present consideration the region $\mathcal{G}_1$ is determined by $\mathfrak{F}$, being one of the two regions adjacent to $\mathcal{S}$ in the decomposition of $\mathcal{D}$ by the collection $\mathfrak{F}\mathcal{S}$. Since $\mathfrak{G}_1$ is a subgroup of $\mathfrak{F}$ and carries $\mathcal{G}_1$ into itself, the collection $\mathfrak{G}_1\mathcal{S}$ is contained in the collection of boundary lines of $\mathcal{G}_1$. If $\mathfrak{G}_1\mathcal{S}$ is not the whole boundary of $\mathcal{G}_1$ in $\mathcal{D}$, then $\mathcal{G}_1$ is only part of the region bounded by $\mathfrak{G}_1\mathcal{S}$. Only if $\mathfrak{G}_1\mathcal{S}$ is the whole boundary of $\mathcal{G}_1$ in $\mathcal{D}$ does $\mathcal{G}_1$ correspond to the $\mathcal{C}_1$ of §18.2 and only if that is the case and the same holds for $\mathfrak{G}_2$ will the images of $\tilde{\mathcal{G}}_1$ and $\tilde{\mathcal{G}}_2$ by $\mathfrak{G}_{1,2}$ cover $\mathcal{D}$ completely.

19.2 Decomposition of $\mathfrak{F}$ in the case I. The investigation to follow will be split into three cases denoted by I, II, III, which together will cover all possibilities completely.

We first investigate the **case** I, in which $\mathfrak{G}_1\mathcal{S}$ *is not the whole boundary of* $\mathcal{G}_1$ *in* $\mathcal{D}$. Let $\mathcal{S}'$ be a side of $\mathcal{G}_1$ not belonging to $\mathfrak{G}_1\mathcal{S}$. Let $\mathfrak{f}$ be an element of $\mathfrak{F}$ such that $\mathcal{S}' = \mathfrak{f}\mathcal{S}$. Since $\mathfrak{f}$ does not belong to $\mathcal{G}_1$, it does not carry $\mathcal{G}_1$ into itself. Therefore $\mathfrak{f}\mathcal{G}_1$ is adjacent to $\mathcal{G}_1$ along $\mathcal{S}$ and we must have $\mathfrak{f}\mathcal{G}_2 = \mathcal{G}_1$. *Thus* $\mathcal{S}$ *must be non-dividing modulo* $\mathfrak{F}$. Moreover, if $\mathfrak{h}$ were an element of $\mathfrak{A}_\mathcal{S}$ outside $\mathfrak{A}^*_\mathcal{S}$, then $\mathfrak{h}\mathcal{G}_1 = \mathcal{G}_2$ and $\mathfrak{f}\mathfrak{h}\mathcal{G}_1 = \mathfrak{f}\mathcal{G}_2 = \mathcal{G}_1$, while $\mathfrak{f}\mathfrak{h}\mathcal{S} = \mathfrak{f}\mathcal{S} = \mathcal{S}'$. Thus $\mathfrak{f}\mathfrak{h}$ would be an element of $\mathfrak{G}_1$ and carry $\mathcal{S}$ into $\mathcal{S}'$ contrary to the assumption that $\mathcal{S}'$ does not belong to $\mathfrak{G}_1\mathcal{S}$. *Hence* $\mathfrak{A}_\mathcal{S}$ *must coincide with* $\mathfrak{A}^*_\mathcal{S}$.

Conversely, let it be assumed that $\mathcal{S}$ is non-dividing modulo $\mathfrak{F}$ and that $\mathfrak{A}_\mathcal{S} = \mathfrak{A}^*_\mathcal{S}$. Since $\mathcal{S}$ is non-dividing modulo $\mathfrak{F}$, let $\mathfrak{f}$ be an element of $\mathfrak{F}$ such that $\mathfrak{f}\mathcal{G}_2 = \mathcal{G}_1$. Since $\mathfrak{A}_\mathcal{S} = \mathfrak{A}^*_\mathcal{S}$, the element $\mathfrak{f}$ cannot interchange $\mathcal{G}_2$ and $\mathcal{G}_1$, thus $\mathcal{S}' = \mathfrak{f}\mathcal{S}$ is a side of $\mathcal{G}_1$ different from $\mathcal{S}$. Then $\mathcal{S}'$ does not belong to $\mathfrak{G}_1\mathcal{S}$; for if there were some $\mathfrak{g} \in \mathfrak{G}_1$ such that $\mathcal{S}' = \mathfrak{g}\mathcal{S}$, then $\mathfrak{g}^{-1}\mathfrak{f}\mathcal{S} = \mathfrak{g}^{-1}\mathcal{S}' = \mathcal{S}$ while $\mathfrak{g}^{-1}\mathfrak{f}\mathcal{G}_2 = \mathfrak{g}^{-1}\mathcal{G}_1 = \mathcal{G}_1$, which contradicts the fact that $\mathfrak{A}_\mathcal{S} = \mathfrak{A}^*_\mathcal{S}$.

If also $\mathcal{S}'' = \mathfrak{h}\mathcal{S}$ is a side of $\mathcal{G}_1$ outside $\mathfrak{G}_1\mathcal{S}$, then $\mathfrak{h}\mathfrak{f}^{-1}\mathcal{S}' = \mathcal{S}''$ while $\mathfrak{h}\mathfrak{f}^{-1}\mathcal{G}_1 = \mathfrak{h}\mathcal{G}_2 = \mathcal{G}_1$, and thus $\mathfrak{h}\mathfrak{f}^{-1} \in \mathfrak{G}_1$. The whole collection of boundaries of $\mathcal{G}_1$ outside $\mathfrak{G}_1\mathcal{S}$ is thus the collection $\mathfrak{G}_1\mathfrak{f}\mathcal{S}$ derived from $\mathcal{S}$ by a certain right coset of $\mathfrak{G}_1$ in $\mathfrak{F}$.

Hence, in all, we have the statement:

In order that $\mathfrak{G}_1\mathcal{S}$ *be not the whole boundary of* $\mathcal{G}_1$ *in* $\mathcal{D}$ *it is necessary and sufficient that* $\mathcal{S}$ *is non-dividing modulo* $\mathfrak{F}$ *and that* $\mathfrak{A}_\mathcal{S} = \mathfrak{A}^*_\mathcal{S}$. *In that case the lines bounding* $\mathcal{G}_1$ *fall into two equivalence classes with respect to* $\mathfrak{G}_1$. □

We now prove the following

Lemma. *There exists a side of* $\mathcal{G}_1$ *outside* $\mathfrak{G}_1\mathcal{S}$ *not left invariant by any element* $\neq 1$ *of* $\mathfrak{A}_\mathcal{S}$.

Proof. Let $\mathfrak{f}\mathcal{S}$, $\mathfrak{f} \in \mathfrak{F}$, be a side of $\mathcal{G}_1$ outside $\mathfrak{G}_1\mathcal{S}$. A translation in $\mathfrak{A}_\mathcal{S}$, if any, leaves no side of $\mathcal{G}_1$ invariant except $\mathcal{S}$. Hence if $\mathfrak{f}\mathcal{S}$ is left invariant by an element of $\mathfrak{A}_\mathcal{S}$, it must be by a certain reflection $\mathfrak{s}$ contained in $\mathfrak{A}_\mathcal{S}$, and the reflection line $\mathcal{L}(\mathfrak{s})$ is then the common normal of $\mathcal{S}$ and $\mathfrak{f}\mathcal{S}$. If $\mathfrak{A}_\mathcal{S} = \mathfrak{A}(|)$, let $\mathfrak{g}$ be an element of $\mathfrak{G}_1$ outside $\mathfrak{A}_\mathcal{S}$; such an element exists, since $\mathfrak{G}_1$ does not coincide with $\mathfrak{A}_\mathcal{S}$. Then $\mathfrak{g}\mathfrak{f}\mathcal{S}$ is a side of $\mathcal{G}_1$ outside $\mathfrak{G}_1\mathcal{S}$, and it is not left invariant by $\mathfrak{s}$. – If $\mathfrak{A}_\mathcal{S} = \mathfrak{A}(\|)$, and $\mathcal{S}$ thus is an axis of $\mathfrak{F}$, let $\mathfrak{s}_1$ and $\mathfrak{s}_2$ be two reflections in $\mathfrak{A}_{\mathcal{S}}$ generating $\mathfrak{A}_{\mathcal{S}}$. Since $\mathcal{S}$ is an axis, $\mathfrak{G}_1$ is an $\mathfrak{F}$-group, $\mathcal{S}$ being one of the sides of $\mathcal{K}(\mathfrak{G}_1)$; no other side of $\mathcal{K}(\mathfrak{G}_1)$ is cut both by $\mathcal{L}(\mathfrak{s}_1)$ and $\mathcal{L}(\mathfrak{s}_2)$. Select in the strip between $\mathcal{L}(\mathfrak{s}_1)$ and $\mathcal{L}(\mathfrak{s}_2)$ a side of $\mathcal{K}(\mathfrak{G}_1)$ equivalent with $\mathfrak{f}\mathcal{S}$ with respect to $\mathfrak{G}_1$. Its common normal with $\mathcal{S}$ belongs to the strip and is thus not a reflection line of $\mathfrak{A}_{\mathcal{S}}$. Hence it is not left invariant by any element of $\mathfrak{A}_{\mathcal{S}}$. This completes the proof of the lemma. □

In virtue of this lemma we now select a side $\mathcal{T}$ of $\mathcal{G}_1$ outside $\mathfrak{G}_1\mathcal{S}$ which is not left invariant by any element of $\mathfrak{A}_{\mathcal{S}}$. Since $\mathcal{S}$ is simple modulo $\mathfrak{F}$, and $\mathcal{S}$ and $\mathcal{T}$ are equivalent with respect to $\mathfrak{F}$, the lines $\mathcal{S}$ and $\mathcal{T}$ are non-concurrent. Let $\mathcal{D}_0$ denote the open part of $\mathcal{D}$ situated between them. The group here called $\mathfrak{G}_1$ then satisfies the six conditions imposed on the group $\mathfrak{G}_0$ of §18.11:

1) $\mathfrak{G}_1$ is discontinuous in $\mathcal{D}$, because it is a subgroup of $\mathfrak{F}$.

2) $\mathcal{S}$ is not an invariant line of $\mathfrak{G}_1$, because it was shown above that $\mathfrak{A}^*_{\mathcal{S}}$ is not the whole of $\mathfrak{G}_1$; and $\mathfrak{A}^*_{\mathcal{S}}$ is here the same as $\mathfrak{A}_{\mathcal{S}}$. The same holds for $\mathcal{T}$, because $\mathfrak{A}_{\mathcal{S}}$ and $\mathfrak{A}_{\mathcal{T}}$ are conjugate subgroups of $\mathfrak{F}$.

3) and 4) $\mathfrak{G}_1\mathcal{S} \cup \mathfrak{G}_1\mathcal{T}$ is the collection of sides of the convex region $\mathcal{G}_1$, which is part of $\mathcal{D}_0$. Thus $\mathfrak{G}_1\mathcal{S} \cup \mathfrak{G}_1\mathcal{T}$ belongs to $\mathcal{D}_0$ except that $\mathcal{S}$ and $\mathcal{T}$ themselves bound $\mathcal{D}_0$.

5) $\mathfrak{A}_{\mathcal{S}}$ and $\mathfrak{A}_{\mathcal{T}}$ are conjugate subgroups of $\mathfrak{F}$, hence isomorphic. The primary displacements along $\mathcal{S}$ and $\mathcal{T}$ are equal, if $\mathcal{S}$ and $\mathcal{T}$ are axes, because they belong to conjugate elements of $\mathfrak{F}$.

6) $\mathcal{T}$ was so selected that this condition holds.

Let $\mathfrak{f}$ be an element of $\mathfrak{F}$ such that $\mathcal{T} = \mathfrak{f}\mathcal{S}$. The totality of elements of $\mathfrak{F}$ which carry $\mathcal{S}$ into $\mathcal{T}$ is then the right coset $\mathfrak{A}_{\mathcal{T}}\mathfrak{f}$ of $\mathfrak{A}_{\mathcal{T}}$ in $\mathfrak{F}$. For the element $\mathfrak{h}$ of §18.11 one may take any element of that coset. The additional conditions we had to impose on $\mathfrak{h}$ in §18.11 regarding reflection lines of $\mathfrak{A}_{\mathcal{S}}$ and $\mathfrak{A}_{\mathcal{T}}$, if any, are here satisfied from the outset, since any $\mathfrak{h} \in \mathfrak{A}_{\mathcal{T}}\mathfrak{f}$ is an element of $\mathfrak{F}$ and thus carries reflection lines into reflection lines.

If $\mathcal{S}$ and $\mathcal{T}$ are parallel, they are not axes of $\mathfrak{F}$, thus $\mathfrak{A}_{\mathcal{T}} = 1$ or $\mathfrak{A}_{\mathcal{T}} = \mathfrak{A}(|)$. Let u be the common end-point of $\mathcal{S}$ and $\mathcal{T}$, the other end-points being $u_{\mathcal{S}}$ and $u_{\mathcal{T}}$ respectively. If $\mathfrak{f}u = u$, $\mathfrak{f}u_{\mathcal{S}} = u_{\mathcal{T}}$, then $\mathfrak{f}$ must be a limit-rotation, for if $\mathfrak{f}$ were a translation or reversed translation with some point v as its positive and u as its negative fundamental point, then $\mathfrak{f}^n\mathcal{S}$ would converge to the line uv for $n \to \infty$ contrary to the assumption that $\mathfrak{F}\mathcal{S}$ does not accumulate in $\mathcal{D}$. If $\mathfrak{f}u_{\mathcal{S}} = u$, $\mathfrak{f}u = u_{\mathcal{T}}$ together with $\mathfrak{A}_{\mathcal{T}} = \mathfrak{A}(|)$, let $\mathfrak{s}$ be the reflection in $\mathfrak{A}_{\mathcal{T}}$; then $\mathfrak{s}\mathfrak{f}$ is a limit-rotation with centre u as before. Finally, if $\mathfrak{f}u_{\mathcal{S}} = u$, $\mathfrak{f}u = u_{\mathcal{T}}$ together with $\mathfrak{A}_{\mathcal{T}} = 1$, then $\mathfrak{f}$ is seen to be a reversed translation, if one bears in mind that $\mathfrak{f}\mathcal{G}_1$ is adjacent to $\mathcal{G}_1$ along $\mathcal{T}$. The situation then is the same as for $\mathcal{S}$ and $\mathcal{T}$ divergent and is, in the sequel, included in that case without mentioning.

If $\mathcal{S}$ and $\mathcal{T}$ are divergent, then $\mathfrak{f}$ is a translation or reversed translation, as seen in §18.11.

We now choose in the coset $\mathfrak{A}_{\mathcal{T}}\mathfrak{f}$ a *fixed* element which is either a limit-rotation or a translation or reversed translation according to the different cases just examined. In order to adapt the notation to that of §18.11–12, we call this fixed element $\mathfrak{h}$ and write $\mathfrak{G}_0$ instead of $\mathfrak{G}_1$ and $\mathfrak{C}_0$ instead of $\mathcal{G}_1$ as well as $\mathcal{S}_0$ instead of $\mathcal{S}$. As in §18.11, we put for $-\infty < n < \infty$

$$\mathcal{S}_n = \mathfrak{h}^n \mathcal{S}_0, \quad \mathfrak{C}_n = \mathfrak{h}^n \mathfrak{C}_0, \quad \mathfrak{G}_n = \mathfrak{h}^n \mathfrak{G}_0 \mathfrak{h}^{-n}, \quad \mathfrak{A}_n = \mathfrak{h}^n \mathfrak{A}_0 \mathfrak{h}^{-n},$$

where $\mathfrak{A}_0$ means $\mathfrak{A}_{\mathcal{S}}$. We can then construct the group

$$\mathfrak{G}^* = \prod_n^* \mathfrak{G}_n \operatorname{am} \mathfrak{A}_n, \quad -\infty < n < \infty,$$

and also the group

$$\mathfrak{G} = \langle \mathfrak{G}_0, \mathfrak{h} \rangle \tag{5}$$

obtained by the adjunction of $\mathfrak{h}$ to $\mathfrak{G}_0$ (§18.12); it contains $\mathfrak{G}^*$ as a normal subgroup with infinite index.

We set out to show that $\mathfrak{G}$ coincides with $\mathfrak{F}$. Since $\mathfrak{G}_0$ is a subgroup of $\mathfrak{F}$ and $\mathfrak{h}$ an element of $\mathfrak{F}$, the group $\mathfrak{G}$ derived by the adjunction of $\mathfrak{h}$ to $\mathfrak{G}_0$ is a subgroup of $\mathfrak{F}$. We thus have to show that an arbitrary element $\mathfrak{f} \in \mathfrak{F}$ belongs to $\mathfrak{G}$. Let p be an arbitrary regular point of the region $\mathfrak{C}_0$. Then $\mathfrak{f}p$ belongs to $\mathfrak{f}\mathfrak{C}_0$. In virtue of §18.11–12 the collection $\mathfrak{G}\mathfrak{C}_0$ is a tesselation of $\mathcal{D}$, the one produced by the collection of lines $\mathfrak{F}\mathcal{S}$. Hence there is an element $\mathfrak{g} \in \mathfrak{G}$ such that $\mathfrak{g}\mathfrak{f}\mathfrak{C}_0 = \mathfrak{C}_0$, and thus $\mathfrak{g}\mathfrak{f}p \in \mathfrak{C}_0$.

Now $\mathfrak{F}$ has in $\mathfrak{C}_0$ the same effect as its subgroup $\mathfrak{G}_0$, and $\mathfrak{G}_0$ is also a subgroup of $\mathfrak{G}$. Hence $\mathfrak{g}\mathfrak{f}$ belongs to $\mathfrak{G}$, and thus $\mathfrak{f}$ belongs to $\mathfrak{G}$. Hence $\mathfrak{G} = \mathfrak{F}$.

The effect of the decomposition of $\mathcal{D}$ by the collection $\mathfrak{F}\mathcal{S}$ in this case I is thus to reduce the investigation of $\mathfrak{F}$ to the investigation of its subgroup $\mathfrak{G}_0$. From this subgroup one regains $\mathfrak{F}$ by the adjunction of the element $\mathfrak{h}$ on the model of §18.12.

19.3 Decomposition of $\mathfrak{F}$ in the case II. For the further consideration of decompositions of $\mathcal{D}$ by $\mathfrak{F}\mathcal{S}$ we again apply the notations introduced in Section 1. The necessary and sufficient conditions for the case I dealt with in Section 2, i.e. for $\mathfrak{G}_1\mathcal{S}$ not being the whole boundary of $\mathcal{G}_1$ in $\mathcal{D}$, were that $\mathcal{S}$ was non-dividing modulo $\mathfrak{F}$ and that $\mathfrak{A}_{\mathcal{S}} = \mathfrak{A}^*_{\mathcal{S}}$. We shall arrive at the two remaining cases II and III by cancelling one or the other of these two conditions, and $\mathfrak{G}_1\mathcal{S}$ will then be the whole boundary of $\mathcal{G}_1$ in $\mathcal{D}$. It should be noted that the two conditions cannot be cancelled simultaneously: If $\mathcal{S}$ is dividing modulo $\mathfrak{F}$, then clearly $\mathfrak{A}_{\mathcal{S}} = \mathfrak{A}^*_{\mathcal{S}}$ must hold.

The present **case II** deals with the assumption that $\mathfrak{A}^*_{\mathcal{S}}$ *is not the whole of* $\mathfrak{A}_{\mathcal{S}}$. Thus if one bears in mind that our present $\mathfrak{A}_{\mathcal{S}}$ corresponds to the group called $\mathfrak{H}$, and our present $\mathfrak{A}^*_{\mathcal{S}}$ to the group called $\mathfrak{A}_{\mathcal{S}}$ in §18.7–18.9, then our present $\mathfrak{A}_{\mathcal{S}}$ falls under one of the eight types denoted by $\mathfrak{H}$ in Table 3 of §18.8. Also our present $\mathcal{G}_1$ corresponds to the $\mathfrak{C}_1$ of §18.

When in §18.7 the group $\mathfrak{H}$ was introduced as a quasi-abelian group leaving a side $\mathcal{S}$ of $\mathfrak{C}_1$ invariant, certain conditions of compatibility had to be observed. It is easily verified that they are now satisfied from the outset, since our $\mathfrak{A}_{\mathcal{S}}$ is a quasi-abelian subgroup of $\mathfrak{F}$ operating on the side $\mathcal{S}$ of the region $\mathcal{G}_1$ left invariant by the subgroup $\mathfrak{G}_1$ of $\mathfrak{F}$.

We now form the group

$$\mathfrak{G} = \mathfrak{G}_1 * \mathfrak{A}_{\mathcal{S}} \operatorname{am} \mathfrak{A}^*_{\mathcal{S}} \tag{6}$$

corresponding to (6), §18.

In order to show that $\mathfrak{G}$ coincides with $\mathfrak{F}$ we follow the same procedure as at the end of the preceding section: $\mathfrak{G}$ is a subgroup of $\mathfrak{F}$, since both factors of the generalized free product (6) are. If p is a regular point in $\mathcal{G}_1$ and $\mathfrak{f} \in \mathfrak{F}$, then $\mathfrak{f}p$ is situated in $\mathfrak{f}\mathcal{G}_1$, which is one of the regions of decomposition of $\mathcal{D}$ by $\mathfrak{F}\mathcal{S}$. There is thus a $\mathfrak{g} \in \mathfrak{G}$ such that $\mathfrak{g}\mathfrak{f}\mathcal{G}_1 = \mathcal{G}_1$ and thus $\mathfrak{g}\mathfrak{f}p \in \mathcal{G}_1$. Since $\mathfrak{F}$ and its subgroup $\mathfrak{G}_1$ have equal effect in $\mathcal{G}_1$, and $\mathfrak{G}_1$ is a subgroup of $\mathfrak{G}$, the proof is completed as in Section 2.

The effect of the decomposition in this case II is to reduce the investigation of $\mathfrak{F}$ to the investigation of its subgroup $\mathfrak{G}_1$. From this subgroup one regains $\mathfrak{F}$ by the generalized free product of $\mathfrak{G}_1$ and $\mathfrak{A}_{\mathcal{S}}$ according to (6).

19.4 Decomposition of $\mathfrak{F}$ in the case III. This case deals with the assumption that *$\mathcal{S}$ is dividing modulo $\mathfrak{F}$*. As already stated it implies that $\mathfrak{A}_{\mathcal{S}} = \mathfrak{A}^*_{\mathcal{S}}$, because $\mathcal{G}_1$ and $\mathcal{G}_2$ are not equivalent with respect to $\mathfrak{F}$. The case corresponds to §18.2–7. In this case $\mathfrak{F}$ coincides with the group

$$\mathfrak{G}_{1,2} = \mathfrak{G}_1 * \mathfrak{G}_2 \operatorname{am} \mathfrak{A}_{\mathcal{S}}. \tag{7}$$

This is shown in an analogous way as in the cases I and II: $\mathfrak{G}_{1,2}$ is a subgroup of $\mathfrak{F}$, since both $\mathfrak{G}_1$ and $\mathfrak{G}_2$ are. Let p be a regular point in $\mathcal{G}_1$, and $\mathfrak{f} \in \mathfrak{F}$, thus $\mathfrak{f}p \in \mathfrak{f}\mathcal{G}_1$. This is one of the regions of decomposition of $\mathcal{D}$ by $\mathfrak{F}\mathcal{S}$ and not equivalent with $\mathcal{G}_2$ with respect to $\mathfrak{F}$, thus belonging to $\mathfrak{G}_{1,2}\mathcal{G}_1$. There is thus a $\mathfrak{g} \in \mathfrak{G}_{1,2}$ such that $\mathfrak{g}\mathfrak{f}p \in \mathcal{G}_1$. In $\mathcal{G}_1$ the group $\mathfrak{F}$ has the same effect as its subgroup $\mathfrak{G}_1$ which is also a subgroup of $\mathfrak{G}_{1,2}$. This again completes the proof.

The effect of the decomposition in this case III is to reduce the investigation of $\mathfrak{F}$ to the investigation of its two subgroups $\mathfrak{G}_1$ and $\mathfrak{G}_2$. From these two subgroups one regains $\mathfrak{F}$ by the generalized free product (7).

19.5 Effect on the surface. The line $\mathcal{S}$ in $\mathcal{D}$, simple modulo $\mathfrak{F}$, corresponds on the surface $\mathcal{D}$ mod $\mathfrak{F}$ to a geodesic s which does not intersect itself, and whose form depends on the subgroup $\mathfrak{A}_{\mathcal{S}}$ of $\mathfrak{F}$. The decomposition of $\mathcal{D}$ by $\mathfrak{F}\mathcal{S}$ corresponds to introducing s as a new boundary component on the surface $\mathcal{D}$ mod $\mathfrak{F}$. We are only concerned with the case in which a decomposition of $\mathfrak{F}$ arises, thus where s does not only cut off from $\mathcal{D}$ mod $\mathfrak{F}$ a half-plane, quarter-plane, funnel, or half-funnel or part of these surface parts outside $\mathcal{K}(\mathfrak{F})$ mod $\mathfrak{F}$. Thus s has points in common with the interior of $\mathcal{K}(\mathfrak{F})$ mod $\mathfrak{F}$ (Section 1).

In the case III the surface $\mathfrak{D}$ mod $\mathfrak{F}$ is divided into two parts, one corresponding to $\mathfrak{g}_1$ mod $\mathfrak{G}_1$, the other to $\mathfrak{g}_2$ mod $\mathfrak{G}_2$. According to the structure (1)–(4) of $\mathfrak{A}^*_{\mathfrak{s}}$, which in this case is $\mathfrak{A}_{\mathfrak{s}}$ itself, the geodetic cut s is an open geodesic, a geodetic ray issuing at right angles from a reflection edge, a closed geodesic, or a geodetic segment joining two reflection edges at right angles. The cutting along s results in one boundary component on each of the two new surfaces.

In the case II the surface $\mathfrak{D}$ mod $\mathfrak{F}$ remains connected, when s is introduced as a new boundary component. The character of this new component depends on the group $\mathfrak{A}_{\mathfrak{s}}$. We get the following possibilities in accordance with Table 3 of §18.8. For four forms of $\mathfrak{A}_{\mathfrak{s}}$ the boundary component of the new surface arises out of:

$\mathfrak{A}_{\mathfrak{s}} = \mathfrak{A}(-)$: an open reflection edge,

$\mathfrak{A}_{\mathfrak{s}} = \mathfrak{A}(\times_2)$: a reflection edge issuing from an angular point of order 2,

$\mathfrak{A}_{\mathfrak{s}} = \mathfrak{A}(- \rightarrow)$: a closed reflection edge,

$\mathfrak{A}_{\mathfrak{s}} = \mathfrak{A}(- \parallel)$: a reflection edge connecting two angular points of order 2.

For the following three forms of $\mathfrak{A}_{\mathfrak{s}}$ the surface is cut up along:

$\mathfrak{A}_{\mathfrak{s}} = \mathfrak{A}(\odot_2)$: a geodetic ray issuing from a conical point of order 2,

$\mathfrak{A}_{\mathfrak{s}} = \mathfrak{A}(\cdot\cdot)$: a geodetic segment connecting two conical points of order 2,

$\mathfrak{A}_{\mathfrak{s}} = \mathfrak{A}(\cdot|)$: a geodetic segment issuing from a conical point of order 2 and terminating at right angles at a reflection edge.

In all these three cases a conical point becomes an ordinary point of a geodetic boundary component of the new surface. This boundary component is in the three cases respectively an open geodesic, a closed geodesic, a geodetic segment joining two reflection edges. – One last form for $\mathfrak{A}_{\mathfrak{s}}$ remains:

$\mathfrak{A}_{\mathfrak{s}} = \mathfrak{A}(\rightleftharpoons)$: The surface $\mathfrak{D}$ mod $\mathfrak{F}$ is cut along the middle line of a Möbius band forming part of the surface, the boundary component arising on the new surface being a simple closed geodesic.

This last case may be described as "cutting up a cross cap".

In the case I the nature of the cut s is the same as in the case III, because the possible groups $\mathfrak{A}_{\mathfrak{s}}$ are the same. The difference is that the surface $\mathfrak{D}$ mod $\mathfrak{F}$ remains connected after the cut, being now the surface $\mathfrak{g}_1$ mod $\mathfrak{G}_1$, and that the cutting along s results in *two* new boundary components of the surface. They correspond to the two equivalence classes with respect to $\mathfrak{G}_1$ into which the sides of $\mathfrak{g}_1$ fall.

In §18 the effect on the surface of a group composition was described. It is seen that the decomposition here considered has the inverse effect.

19.6 Orientation of the decomposing line. We are going to consider the cases in which $\mathfrak{F}\mathfrak{s}$ can be oriented unambiguously. This excludes the cases in which $\mathfrak{A}_\mathfrak{s}$ contains transformations which interchange the end-points of $\mathfrak{s}$, thus half-turns around centres on $\mathfrak{s}$ and reflections in normals of $\mathfrak{s}$. The only possible forms of $\mathfrak{A}_\mathfrak{s}$ are then

$$\mathfrak{A}_\mathfrak{s} = 1, \ \mathfrak{A}(\rightarrow), \ \mathfrak{A}(-), \ \mathfrak{A}(- \rightarrow) \quad \text{and} \quad \mathfrak{A}(\rightleftharpoons).$$

Let $\mathfrak{s}$ be so oriented that it bounds $\mathcal{G}_1$ positively and $\mathcal{G}_2$ negatively. If the orientation of $\mathfrak{s}$ is carried by the application of any element of $\mathfrak{F}$, then every line of $\mathfrak{F}\mathfrak{s}$ is provided with unambiguous orientation. We then call the collection $\mathfrak{F}\mathfrak{s}$ *coherently oriented*.– We now discuss the three cases considered in Sections 2–4.

Case III. Here only $\mathfrak{A}_\mathfrak{s} = 1$ and $\mathfrak{A}_\mathfrak{s} = \mathfrak{A}(\rightarrow)$ occur under our present assumptions. If $\mathfrak{f}$ is an element of $\mathfrak{G}_1$, $\mathfrak{f}\mathfrak{s}$ bounds $\mathcal{G}_1$ positively or negatively according as $\mathfrak{f}$ is a motion or a reversion. Thus if $\mathfrak{G}_1$ contains motions only, then all sides of $\mathcal{G}_1$ bound $\mathcal{G}_1$ positively. If $\mathfrak{G}_1$ contains reversions, the sides of $\mathcal{G}_1$, while all equivalent with respect to $\mathfrak{G}_1$, fall into two equivalence classes with respect to the subgroup of motions in $\mathfrak{G}_1$, one bounding $\mathcal{G}_1$ positively, the other negatively. The two classes are interchanged by a reversion in $\mathfrak{G}_1$. The same considerations apply to $\mathcal{G}_2$ and $\mathfrak{G}_2$ independently of $\mathcal{G}_1$ and $\mathfrak{G}_1$.

Case II. Here there are for $\mathfrak{A}_\mathfrak{s}$ the three possibilities $\mathfrak{A}(-)$ or $\mathfrak{A}(- \rightarrow)$ or $\mathfrak{A}(\rightleftharpoons)$. Even in this case all sides of $\mathcal{G}_1$ are equivalent with respect to $\mathfrak{G}_1$. Thus the considerations of case III with respect to $\mathfrak{G}_1$ and $\mathcal{G}_1$ hold true also in case II. Here $\mathfrak{G}_2$ is in $\mathfrak{F}$ conjugate to $\mathfrak{G}_1$. Two different elements of $\mathfrak{F}$ carrying $\mathcal{G}_1$ into $\mathcal{G}_2$ belong to the same left coset of $\mathfrak{G}_1$ in $\mathfrak{F}$. Any transformation $\mathfrak{h}$ in $\mathfrak{F}$ for which $\mathfrak{h}\mathcal{G}_1 = \mathcal{G}_2$ is thus a transformation in $\mathfrak{G}_1$ followed by a reflection in $\mathfrak{s}$ or, if $\mathfrak{A}_\mathfrak{s}$ does not contain this reflection, by a reversed translation along $\mathfrak{s}$. One thus gets: If $\mathfrak{G}_1$ consists only of motions, the same holds true for $\mathfrak{G}_2$, and every $\mathfrak{h}$ is a reversion; all sides of $\mathfrak{G}_1$ then bound $\mathcal{G}_1$ positively, and all sides of $\mathcal{G}_2$ bound $\mathcal{G}_2$ negatively. If, however, $\mathfrak{G}_1$ contains reversions, there are both motions and reversions carrying $\mathcal{G}_1$ into $\mathcal{G}_2$, and $\mathcal{G}_2$ is, like $\mathcal{G}_1$, bounded by its sides in both senses.

Case I. Since no element interchanges $\mathcal{G}_1$ and $\mathcal{G}_2$, we have only the same two possibilities $\mathfrak{A}_\mathfrak{s} = 1$ and $\mathfrak{A}_\mathfrak{s} = \mathfrak{A}(\rightarrow)$ as in case III. Here the sides of $\mathcal{G}_1$, while all equivalent with respect to $\mathfrak{F}$, are known to fall into two equivalence classes with respect to $\mathfrak{G}_1$. The same holds for $\mathcal{G}_2$ and $\mathfrak{G}_2$. Let $\mathfrak{f}$ be an element of $\mathfrak{F}$ such that $\mathfrak{f}\mathcal{G}_2 = \mathcal{G}_1$. Then $\mathfrak{f}\mathfrak{s}$ is known to be a side of $\mathcal{G}_1$ not equivalent with $\mathfrak{s}$ with respect to $\mathfrak{G}_1$, and $\mathfrak{f}$ is either a translation or a limit-rotation about an end-point of $\mathfrak{s}$, or it is a reversed translation. Accordingly $\mathfrak{f}\mathfrak{s}$ bounds $\mathcal{G}_1$ negatively or positively, because $\mathfrak{s}$ bounds $\mathcal{G}_2$ negatively.

In this case I we may distinguish between three possible subcases. In these $\mathfrak{f}$ continues to denote any element of $\mathfrak{F}$ for which $\mathfrak{f}\mathcal{G}_2 = \mathcal{G}_1$.

Ia) $\mathfrak{F}$ (and hence also $\mathfrak{G}_1$ and $\mathfrak{G}_2$) consists of motions only. All $\mathfrak{f}$ are translations or limit-rotations. All sides $\mathfrak{G}_1\mathfrak{f}\mathcal{S}$ bound $\mathcal{G}_1$ negatively, and all sides $\mathfrak{G}_1\mathcal{S}$ bound $\mathcal{G}_1$ positively. This distinguishes the two classes.

Ib) $\mathfrak{G}_1$ (and hence also $\mathfrak{G}_2$ and $\mathfrak{F}$) contains reversions. Each of the two classes with respect to $\mathfrak{G}_1$ of sides of $\mathcal{G}_1$ then contains sides bounding positively as well as sides bounding negatively. Among the $\mathfrak{f}$ there are both motions and reversions.

Ic) $\mathfrak{G}_1$ (and hence also $\mathfrak{G}_2$) consists of motions only, but $\mathfrak{F}$ contains reversions. Assume that a motion $\mathfrak{f}$ exists in $\mathfrak{F}$ with $\mathfrak{f}\mathcal{G}_2 = \mathcal{G}_1$. Then every region of the collection $\mathfrak{F}\mathcal{G}_1$ can be carried into any of its adjacent regions by a motion. Thus all regions $\mathfrak{F}\mathcal{G}_1$ are equivalent with respect to the subgroup of motions in $\mathfrak{F}$. Let $\mathfrak{g}$ be a reversion in $\mathfrak{F}$ and $\mathfrak{h}$ a motion carrying $\mathcal{G}_1$ into $\mathfrak{g}\mathcal{G}_1$. Then $\mathfrak{h}^{-1}\mathfrak{g}$ carries $\mathcal{G}_1$ into itself and thus belongs to $\mathfrak{G}_1$. Since $\mathfrak{h}^{-1}\mathfrak{g}$ is a reversion, this contradicts the assumption. Hence all elements $\mathfrak{f}$ are reversed translations. They then carry $\mathcal{S}$ into sides $\mathfrak{f}\mathcal{S}$ bounding $\mathcal{G}_1$ positively, because $\mathcal{S}$ bounds $\mathcal{G}_2$ negatively. Hence in this case all sides of $\mathcal{G}_1$ bound positively.

From the considerations of this section the following statement results. Let $\mathcal{S}$ be a straight line, simple modulo $\mathfrak{F}$, and not inverted by any element of $\mathfrak{F}$; let the collection $\mathfrak{F}\mathcal{S}$, which is supposed not to accumulate in $\mathcal{D}$, be coherently oriented; let $\mathcal{G}$ denote any of the regions of decomposition of $\mathcal{D}$ by $\mathfrak{F}\mathcal{S}$, and $\mathfrak{G}$ the subgroup of $\mathfrak{F}$ carrying $\mathcal{G}$ into itself.

If all sides of $\mathcal{G}$ are equivalent with respect to $\mathfrak{G}$, then $\mathcal{S}$ is non-dividing or dividing modulo $\mathfrak{F}$, according as $\mathfrak{F}$ contains reversions leaving $\mathcal{S}$ fixed or does not contain such elements. All sides of $\mathcal{G}$ bound $\mathcal{G}$ in the same sense or not, according as $\mathfrak{G}$ does not or does contain reversions.

If the sides of $\mathcal{G}$ fall into two equivalence classes with respect to $\mathfrak{G}$, then $\mathcal{S}$ is non-dividing modulo $\mathfrak{F}$. If the two classes bound $\mathcal{G}$ in opposite senses, then $\mathfrak{F}$ contains no reversions. If the two classes bound $\mathcal{G}$ in the same sense, then $\mathfrak{G}$ contains no reversions, but $\mathfrak{F}$ does. If in one of the classes some sides bound $\mathcal{G}$ in one sense and others in the opposite sense, then the same holds for the other class and $\mathfrak{G}$ contains reversions. An element of $\mathfrak{F}$ carrying $\mathcal{G}$ into an adjacent region is in the three cases respectively: a translation or limit-rotation; a reversed translation; either of the two.

Independently of the question of orientation the following statement holds: If $\mathcal{S}$ is dividing modulo $\mathfrak{F}$, the surface $\mathcal{D}$ mod $\mathfrak{F}$ is divided into two parts by the geodesic s corresponding to $\mathcal{S}$; each of these parts has one boundary derived from s. If $\mathcal{S}$ is non-dividing modulo $\mathfrak{F}$, the surface $\mathcal{D}$ modulo $\mathfrak{F}$ is not divided by s. According as $\mathfrak{F}$ contains reversions leaving $\mathcal{S}$ fixed, or not, the surface $\mathcal{G}$ mod $\mathfrak{G}$ has one or two boundary components derived from s.

19.7 Simultaneous decomposition. Let $\mathcal{U}$ denote a collection of straight lines in $\mathcal{D}$ satisfying the following conditions:

1) If $\mathcal{S}$ is a line of $\mathcal{U}$, then the whole equivalence class $\mathfrak{F}\mathcal{S}$ belongs to $\mathcal{U}$.

2) Every line of $\mathcal{U}$ is simple modulo $\mathfrak{F}$.

3) Any two non-equivalent lines of $\mathcal{U}$ are disjoint modulo $\mathfrak{F}$.

4) Every accumulation point of $\mathcal{U}$ in $\mathcal{D}$, if any, is situated on a limit side of $\mathcal{K}(\mathfrak{F})$.

5) No line of $\mathcal{U}$ is cut by a reflection line of $\mathfrak{F}$.

6) Any line of $\mathcal{U}$ is either dividing modulo $\mathfrak{F}$, or $\mathfrak{F}$ contains an element interchanging the two half-planes determined by it.

Let p be an accumulation point of $\mathcal{U}$ in $\mathcal{D}$ (§10.1), thus situated on a limit side $\mathcal{L}$ of $\mathcal{K}(\mathfrak{F})$. Consider a finite segment $\mathcal{L}_0$ of $\mathcal{L}$ and a bounded quadrangle neighbourhood $abcd$ of p in $\mathcal{D}$ which contains no point of any other limit side of $\mathcal{K}(\mathfrak{F})$. Those lines of $\mathcal{U}$, if any, which intersect ab or cd, are finite in number, because the quadrangle contains no accumulation point of $\mathcal{U}$ outside $\mathcal{L}$ (condition 4)). Moreover, all lines of $\mathcal{U}$ are mutually disjoint (condition 2) and 3)). From this is inferred that there is a sequence of lines from $\mathcal{U}$ converging to $\mathcal{L}$, and that $\mathcal{L}$ is not cut by any line from $\mathcal{U}$. Condition 5) then implies that $\mathcal{L}$ is not cut by any reflection line of $\mathfrak{F}$.

We now introduce the following definition: Let $\tilde{\mathcal{H}}^*$ be the designation for any half-plane, closed on $\mathcal{D}$ and outside $\mathcal{K}(\mathfrak{F})$, for which the bounding line consists of accumulation points of $\mathcal{U}$. *We denote by $\mathcal{D}_{\mathcal{U}}$ the part of $\mathcal{D}$ obtained by the omission of all the $\tilde{\mathcal{H}}^*$.* This is an open, convex set, bounded on $\mathcal{D}$ by complete straight lines, namely the above lines $\mathcal{L}$. The region $\mathcal{D}_{\mathcal{U}}$ is reproduced by $\mathfrak{F}$ (condition 1)), since we are only concerned with its decomposition by $\mathcal{U}$, it can be assumed that the whole of $\mathcal{U}$ is situated in $\mathcal{D}_{\mathcal{U}}$. The collection $\mathcal{U}$ does not accumulate in $\mathcal{D}_{\mathcal{U}}$. Hence in virtue of conditions 2) and 3) the collection $\mathcal{U}$ decomposes $\mathcal{D}_{\mathcal{U}}$ into a collection of regions; they will be spoken of as *$\mathcal{B}$-regions*. Each $\mathcal{B}$-region is bounded on $\mathcal{D}$ by complete lines from $\mathcal{U}$, called its *sides*; it is thus convex. In virtue of condition 1) the collection $\mathcal{U}$, and thus also the collection of $\mathcal{B}$-regions, is reproduced by $\mathfrak{F}$. – If $\mathfrak{F}$ satisfies the condition of quasi-compactness, then $\mathcal{D}_{\mathcal{U}} = \mathcal{D}$, because there are no limit sides of $\mathcal{K}(\mathfrak{F})$.

In preparation for subsequent theorems we first prove the following

Lemma. *For any $\mathfrak{F}$-group $\mathfrak{F}$ containing limit-rotations the horocycles $\mathcal{O}_u^*$ which bound a truncated domain $\mathcal{K}^*(\mathfrak{F})$ (§14.10) on the convex domain $\mathcal{K}(\mathfrak{F})$ can be so chosen that every straight line which is simple modulo $\mathfrak{F}$ and situated in $\mathcal{K}(\mathfrak{F})$ without terminating in a limit-centre is situated in $\mathcal{K}^*(\mathfrak{F})$.*

Proof. Let u be a limit-centre of $\mathfrak{F}$, $\mathfrak{g}$ a primary limit-rotation in $\mathfrak{F}$ with centre u, and $p \neq u$ a point of $\mathcal{E}$. Let $\mathcal{O}_u'$ be that horocycle with centre u which touches the straight line $p, \mathfrak{g}p$ (Fig. 19.2). $\mathcal{O}_u'$ does not depend on the choice of p, because $\mathfrak{g}$ is permutable with any limit-rotation with centre u whether belonging to $\mathfrak{F}$ or not. Let $\mathcal{S}$ be a straight line not terminating in u, and p one of its end-points, the other, q, being situated on the arc $p, \mathfrak{g}p, u$ of $\mathcal{E}$. If q is on the arc $\mathfrak{g}p, u$, then the pair p, q is separated on $\mathcal{E}$

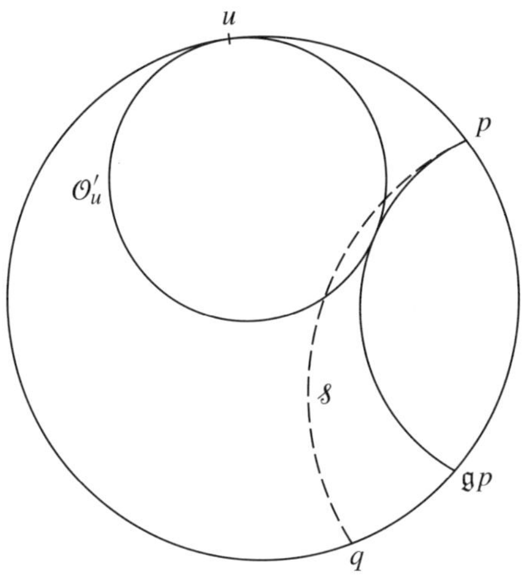

Figure 19.2

by the pair $\mathfrak{g}p$, $\mathfrak{g}q$, and $\mathcal{S}$ is then cut by $\mathfrak{g}\mathcal{S}$. This shows that any horocycle $\mathcal{O}^*_u$ inside $\mathcal{O}'_u$ has the required property; one can obviously choose $\mathcal{O}^*_{\mathfrak{f}u} = \mathfrak{f}\mathcal{O}^*_u$ as the horocycle belonging to $\mathfrak{f}u$ for every $\mathfrak{f} \in \mathfrak{F}$. The proof is independent of whether $\mathfrak{A}_u$ is $\mathfrak{A}(\hookrightarrow)$ or $\mathfrak{A}(\wedge)$. □

We now prove:

If $\mathfrak{F}$ satisfies the condition of quasi-compactness, then the number of equivalence classes with respect to $\mathfrak{F}$ into which the lines of $\mathcal{U}$ fall is finite; also the number of equivalence classes of $\mathcal{B}$-regions is finite.

Proof. Since $\mathcal{K}(\mathfrak{F})$ has no limit sides, $\mathcal{D}_{\mathcal{U}}$ coincides with $\mathcal{D}$. If $\mathfrak{F}$ contains limit-rotations, let u be a limit-centre of $\mathfrak{F}$ with corresponding subgroup $\mathfrak{A}_u$. If some lines of $\mathcal{U}$ terminate in u, conditions 1) and 4) imply that they fall into a finite number of equivalence classes with respect to $\mathfrak{A}_u$ and thus with respect to $\mathfrak{F}$. Also the number of equivalence classes with respect to $\mathfrak{F}$ of limit-centres of $\mathfrak{F}$ is finite, since $\mathcal{K}(\mathfrak{F})$ is quasi-compact modulo $\mathfrak{F}$. Hence those lines of $\mathcal{U}$ which terminate in limit-centres of $\mathfrak{F}$ fall into a finite number of equivalence classes with respect to $\mathfrak{F}$. Those lines of $\mathcal{U}$ which do not terminate in limit-centres of $\mathfrak{F}$ are situated in a truncated domain $\mathcal{K}^*(\mathfrak{F})$ chosen in accordance with the above lemma. Let Q^* be a circular disc in $\mathcal{D}$ containing at least one equivalent to every point of $\mathcal{K}^*(\mathfrak{F})$. Condition 4) and the fact that $\mathcal{D}_{\mathcal{U}} = \mathcal{D}$ imply that only a finite number of lines of $\mathcal{U}$ can have points in common with Q^*. All the more a finite number of equivalence classes occur. – If $\mathfrak{F}$ contains no limit-rotations, then $\mathcal{K}(\mathfrak{F})$ itself takes the place of $\mathcal{K}^*(\mathfrak{F})$.

Let n be the number of equivalence classes of lines from $\mathcal{U}$, and let $\mathcal{S}_1, \ldots, \mathcal{S}_n$ be the lines from $\mathcal{U}$ representing them. If $\mathcal{B}$ is any $\mathcal{B}$-region and $\mathcal{S}$ any of its sides, there is an element of $\mathfrak{F}$ carrying $\mathcal{S}$ into one of the lines $\mathcal{S}_1, \ldots, \mathcal{S}_n$, and thus $\mathcal{B}$ into one of the two $\mathcal{B}$-regions adjacent to that line. This finishes the proof of the theorem. □

The lines of $\mathcal{U}$ constitute an enumerable collection.

Proof. If $\mathfrak{F}$ is not finitely generated, and thus does not satisfy the conditions of quasi-compactness, it is the union of an ascending sequence of finitely generated subgroups $\mathfrak{F}_i$, $1 \leqq i < \infty$. It follows from the preceding theorem that the lines of $\mathcal{U}$ which have points in common with a fundamental domain for $\mathfrak{F}_i$ in $\mathcal{K}(\mathfrak{F}_i)$ fall into a finite number of equivalence classes with respect to $\mathfrak{F}_i$, and thus all the more with respect to $\mathfrak{F}$. As $\mathfrak{F}$ is exhausted by the sequence of subgroups $\mathfrak{F}_i$ the number of equivalence classes is, therefore, at most enumerable. Since the lines in every equivalence class are enumerable, $\mathcal{U}$ is in all an enumerable collection.

This implies that also the $\mathcal{B}$-regions constitute an enumerable collection. □

The decomposition of $\mathcal{D}_{\mathcal{U}}$ by $\mathcal{U}$ will now be studied in the two steps treated in the next three sections.

19.8 Decomposition by non-dividing lines. Let $\mathcal{U}'$ denote the subcollection consisting of those lines of $\mathcal{U}$ which are non-dividing modulo $\mathfrak{F}$. *We denote by $\mathcal{D}_{\mathcal{U}'}$, the part of $\mathcal{D}$ corresponding to the $\mathcal{D}_{\mathcal{U}}$ of the preceding section, if $\mathcal{U}'$ takes the place of $\mathcal{U}$.* We speak of the regions of decomposition of $\mathcal{D}_{\mathcal{U}'}$ by the lines of $\mathcal{U}'$ as *$\mathcal{C}$-regions.* For every line of $\mathcal{U}'$ there is in $\mathfrak{F}$ an element which interchanges the two half-planes determined by that line (condition 6) of Section 7). This implies that every $\mathcal{C}$-region is equivalent with all its adjacent regions, and hence that all $\mathcal{C}$-regions constitute one equivalence class with respect to $\mathfrak{F}$.

We select one particular $\mathcal{C}$-region, denote it by $\mathcal{C}$, and the subgroup of $\mathfrak{F}$ carrying $\mathcal{C}$ into itself by $\mathfrak{C}$. Then $\mathfrak{F}$ and $\mathfrak{C}$ have equal effect in $\mathcal{C}$, though in general not in the closure $\tilde{\mathcal{C}}$ of $\mathcal{C}$ on $\mathcal{D}$. Since all $\mathcal{C}$-regions are equivalent, every equivalence class with respect to $\mathfrak{F}$ of lines of $\mathcal{U}'$ is represented on the boundary of $\mathcal{C}$. Let the sides $\mathcal{W}_r$ of $\mathcal{C}$ represent these equivalence classes; here $r = 1, 2, \ldots$ is a finite or enumerable index range. If two different sides of $\mathcal{C}$ are equivalent by an element of $\mathfrak{F}$, this element can be so chosen that it carries $\mathcal{C}$ into itself, and thus belongs to $\mathfrak{C}$. Hence the boundary of $\mathcal{C}$ on $\mathcal{D}$ is $\bigcup_r \mathfrak{C}\mathcal{W}_r$. This boundary belongs to $\mathcal{D}_{\mathcal{U}'}$.

Let $\mathfrak{W}_r$ denote the subgroup of $\mathfrak{F}$ carrying $\mathcal{W}_r$ into itself, and $\mathfrak{W}_r^*$ the subgroup of $\mathfrak{W}_r$ not interchanging the half-planes determined by $\mathcal{W}_r$. Then $\mathfrak{W}_r^*$ has index 2 in $\mathfrak{W}_r$. Any two of the $\mathcal{W}_r$ have only the identity in common (condition 5) of Section 7). This condition also implies that out of the eight types in Table 3 of §18.8 for the group there called $\mathfrak{H}$ only five can occur for the groups $\mathfrak{W}_r$, namely

$$\mathfrak{A}(-), \quad \mathfrak{A}(\circlearrowleft_2), \quad \mathfrak{A}(- \to), \quad \mathfrak{A}(\cdot\cdot), \quad \mathfrak{A}(\rightleftharpoons).$$

The two first types yield $\mathfrak{W}_r^* = 1$, the three last types yield $\mathfrak{W}_r^* = \mathfrak{A}(\to)$.

One now gets by simultaneous application of the generalized free product (6) of Section 3 to all the factors $\mathfrak{W}_r$

$$\mathfrak{F} = \prod_r^* (\mathfrak{C}, \mathfrak{W}_r) \text{ am } \mathfrak{W}_r^*. \tag{8}$$

The proof that the right hand product in (8) coincides with $\mathfrak{F}$ is analogous to the corresponding proofs in Sections 2, 3, 4: The right hand product in (8) reproduces the decomposition of $\mathcal{D}_{\mathcal{U}'}$, into $\mathcal{C}$-regions just as $\mathfrak{F}$ does, and in $\mathcal{C}$ the groups $\mathfrak{F}$ and $\mathfrak{C}$ have equal effect.

The formula (8) reduces the investigation of $\mathfrak{F}$ to the investigation of its subgroup $\mathfrak{C}$. From this one regains $\mathfrak{F}$ be the generalized free product (8). This reduction supposes, of course, that there are in $\mathcal{U}$ lines which are non-dividing modulo $\mathfrak{F}$. If all lines of $\mathcal{U}$ are non-dividing modulo $\mathfrak{F}$, then (8) is the final result of the decomposition of $\mathcal{D}_{\mathcal{U}}$ by the collection $\mathcal{U}$.

19.9 Decomposition by dividing lines. We now assume that $\mathcal{U}$ contains lines which are dividing modulo $\mathfrak{F}$. Let $\mathcal{U}''$ denote the subcollection of $\mathcal{U}$ consisting of those lines of $\mathcal{U}$ which are dividing modulo $\mathfrak{F}$ *and situated in the $\mathcal{C}$-region $\mathcal{C}$ considered in the preceding section.* Any line of $\mathcal{U}$ which is dividing modulo $\mathfrak{F}$ has an equivalent line in $\mathcal{C}$, because all $\mathcal{C}$-regions are equivalent with respect to $\mathfrak{F}$. If $\mathcal{S}$ is a line of $\mathcal{U}''$, then $\mathfrak{C}\mathcal{S}$ is situated in $\mathcal{C}$, and no other line of $\mathfrak{F}\mathcal{S}$ is situated in $\mathcal{C}$. Any line of $\mathcal{U}''$ is dividing modulo $\mathfrak{C}$, since it is dividing modulo $\mathfrak{F}$.

If $\mathcal{U}''$ accumulates in $\mathcal{C}$, then the accumulation points make up limit sides of $\mathcal{K}(\mathfrak{F})$ (Section 7). They are then also limit sides of $\mathcal{K}(\mathfrak{C})$. Let $\tilde{\mathcal{H}}^{**}$ be the designation for any half-plane, closed on $\mathcal{C}$ and outside $\mathcal{K}(\mathfrak{C})$ for which the bounding line consists of accumulation points of $\mathcal{U}''$ in $\mathcal{C}$. *We denote by $\mathcal{C}_{\mathcal{U}''}$ the region obtained from $\mathcal{C}$ by the omission of all the $\tilde{\mathcal{H}}^{**}$.* Then $\mathcal{U}''$ brings about a decomposition of $\mathcal{C}_{\mathcal{U}''}$ into regions which are $\mathcal{B}$-regions as defined in Section 7, because $\mathfrak{F}$ and $\mathfrak{C}$ have equal effect in $\mathcal{C}$. For any region of the decomposition of $\mathcal{D}_{\mathcal{U}}$ by $\mathcal{U}$, i.e. for any $\mathcal{B}$-region, there is in $\mathcal{C}_{\mathcal{U}''}$ a region $\mathcal{B}$ equivalent to it with respect to $\mathfrak{F}$, and the whole collection $\mathfrak{C}\mathcal{B}$ belongs to $\mathcal{C}_{\mathcal{U}''}$.

It should be borne in mind that the investigation of the present section also comprises the case in which the subcollection $\mathcal{U}'$ in Section 8 is empty. In that case $\mathfrak{C}$ means $\mathfrak{F}$, $\mathcal{C}$ means $\mathcal{D}$, $\mathcal{D}_{\mathcal{U}''}$ means $\mathcal{D}_{\mathcal{U}}$.

Only if, in the general case, $\mathcal{U}$ contains both dividing and non-dividing lines, is the decomposition into $\mathcal{B}$-regions a refinement of the decomposition into $\mathcal{C}$-regions with due regard to the open sets to which the decomposition refers, namely $\mathcal{D}_{\mathcal{U}'}$ for the $\mathcal{C}$-regions, and $\mathcal{C}_{\mathcal{U}''}$ for the $\mathcal{B}$-regions contained in a particular $\mathcal{C}$-region.

It is emphasized that, as long as we operate in $\mathcal{C}$, equivalence with respect to $\mathfrak{C}$ and with respect to $\mathfrak{F}$ are synonymous, and likewise disjointness modulo these two groups. This is to be borne in mind in the sequel.

Since $\mathfrak{C}$ is the subgroup of $\mathfrak{F}$ reproducing $\mathcal{C}$, the region $\mathcal{C}_{\mathcal{U}''}$ is reproduced by $\mathfrak{C}$, and so is its decomposition into $\mathcal{B}$-regions. The $\mathcal{B}$-regions in $\mathcal{C}$ – all situated in $\mathcal{C}_{\mathcal{U}''}$ – fall into equivalence classes with respect to $\mathfrak{C}$. We are going to select a representative for each one of these equivalence classes in such a way that the collection of these representatives, together with those lines from $\mathcal{U}''$ which are the common boundary of any two of these chosen representatives, forms a connected point set. For a line from $\mathcal{U}''$ which is the common boundary of two adjacent $\mathcal{B}$-regions in this selected collection we reserve the letter $\mathcal{V}$ with subscripts from an index range to be described in detail. If $\mathcal{V}_N$ stands for such a line, the two adjacent $\mathcal{B}$-regions will be called $\mathcal{B}_N$ and $\mathcal{B}'_N$, and the subgroups of $\mathfrak{C}$ carrying these $\mathcal{B}$-regions into themselves will be called $\mathfrak{B}_N$ and $\mathfrak{B}'_N$ respectively. If one considers the decomposition of $\mathcal{C}_{\mathcal{U}''}$ by $\mathfrak{C}\mathcal{V}_N$ for this particular line $\mathcal{V}_N$ only, the regions of decomposition adjacent to $\mathcal{V}_N$ will be called $\mathcal{G}_N$ and $\mathcal{G}'_N$; they correspond to the $\mathcal{G}_1$ and $\mathcal{G}_2$ of Section 1, and they are disjoint modulo $\mathfrak{C}$, because $\mathcal{V}_N$ is dividing modulo $\mathfrak{F}$ and thus also modulo $\mathfrak{C}$. Therefore also their subregions $\mathcal{B}_N$ and $\mathcal{B}'_N$ are disjoint modulo $\mathfrak{C}$. The notation will be such that $\mathcal{B}_N \subset \mathcal{G}_N$ and $\mathcal{B}'_N \subset \mathcal{G}'_N$.

We now select an arbitrary one of the $\mathcal{B}$-regions in $\mathcal{C}$ and denote it by $\mathcal{B}$. It is bounded on $\mathcal{D}$ by some lines from $\mathcal{U}''$, and possibly also by some lines from $\mathcal{U}'$ contributing to the boundary of $\mathcal{C}$ on $\mathcal{D}$. The subgroup $\mathfrak{B}$ of $\mathfrak{C}$ which carries $\mathcal{B}$ into itself has in $\mathcal{B}$ the same effect as $\mathfrak{C}$ and $\mathfrak{F}$. Even on the closure $\tilde{\mathcal{B}}$ of $\mathcal{B}$ on $\mathcal{D}$ the groups $\mathfrak{B}$ and $\mathfrak{C}$ have equal effect. (It is recalled that on a line of $\mathcal{U}'$ the groups $\mathfrak{C}$ and $\mathfrak{F}$ do not, in general, have the same effect; see Section 8.) Thus if $\mathcal{S}$ is any side of $\mathcal{B}$, then all the lines $\mathfrak{B}\mathcal{S}$, and no other line of $\mathfrak{C}\mathcal{S}$ (or of $\mathfrak{F}\mathcal{S}$), are sides of $\mathcal{B}$. (The cases I and II treated in Sections 2 and 3 do not occur, because $\mathcal{S}$ is dividing modulo $\mathfrak{F}$.)

Those sides of $\mathcal{B}$ which belong to $\mathcal{U}''$ fall into a finite or infinite number of equivalence classes with respect to $\mathfrak{B}$. We denote this number by ν, $1 \leqq \nu \leqq \infty$, and denote by $\mathcal{V}_i$, $i = 1, 2, \dots$, a complete collection of representatives of these ν equivalence classes; thus every side of $\mathcal{B}$ belonging to $\mathcal{U}''$ is equivalent with respect to $\mathfrak{B}$ (and thus also with respect to $\mathfrak{C}$ and with respect to $\mathfrak{F}$) with exactly one of the $\mathcal{V}_i$. According to the above stipulation the two $\mathcal{B}$-regions adjacent to $\mathcal{V}_i$ are to be called $\mathcal{B}_i$ and $\mathcal{B}'_i$. This notation is now so fixed that $\mathcal{B}'_i$ is our region $\mathcal{B}$ for all values of i.

The regions $\mathcal{G}_i$ and $\mathcal{G}'_i$ were defined above as the two regions adjacent to $\mathcal{V}_i$ in the decomposition of $\mathcal{C}_{\mathcal{U}''}$ by $\mathfrak{C}\mathcal{V}_i$. Now $\mathfrak{B}\mathcal{V}_i$ is contained in $\mathfrak{C}\mathcal{V}_i$, and all lines of $\mathfrak{B}\mathcal{V}_i$ are sides of $\mathcal{B}$. Since it was agreed that $\mathcal{B} = \mathcal{B}'_i$, the region $\mathcal{B}$ is contained in $\mathcal{G}'_i$. This holds for all ν values of i. Thus $\mathcal{B}$ is contained in the intersection of all the $\mathcal{G}'_i$. On the other hand, any side of $\mathcal{B}$ belonging to $\mathcal{U}''$ belongs to $\mathfrak{B}\mathcal{V}_i$ for a certain value of i. Hence if $\mathcal{U}'$ does not contribute to the boundary of $\mathcal{B}$, one gets $\mathcal{B}$ as the intersection of all the $\mathcal{G}'_i$, and this intersection is contained in $\mathcal{C}$. Since in the general case sides of $\mathcal{C}$, i.e. lines from $\mathcal{U}'$, may contribute to the boundary of $\mathcal{B}$, the region $\mathcal{B}$ is only the part of that intersection contained in $\mathcal{C}$. Thus in all cases the general formula

$$\mathcal{B} = \bigcap_i \mathcal{G}'_i \cap \mathcal{C} \tag{9}$$

holds, where i ranges over ν values, this being a finite or infinite range as the case may be.

In our agreed notation, since $\mathcal{B} = \mathcal{B}'_i$, the $\mathcal{B}$-region adjacent to $\mathcal{B}$ along $\mathcal{V}_i$ is called $\mathcal{B}_i$, and the subgroup of $\mathfrak{C}$ reproducing it is called $\mathfrak{B}_i$. The region $\mathcal{B}_i$ is contained in $\mathcal{G}_i$. For any two different values i_1 and i_2 of i the regions $\mathcal{G}_{i_1}$ and $\mathcal{G}_{i_2}$ are disjoint modulo $\mathfrak{C}$. For $\mathcal{G}_{i_2}$, being adjacent to $\mathcal{B}$ along $\mathcal{V}_{i_2}$, is not separated from $\mathcal{B}$ by any line of $\mathfrak{C}\mathcal{V}_{i_1}$ and thus belongs to $\mathcal{G}'_{i_1}$, and $\mathcal{G}_{i_1}$ and $\mathcal{G}'_{i_1}$, were seen to be disjoint modulo $\mathfrak{C}$. Hence $\mathcal{B}_{i_1}$ and $\mathcal{B}_{i_2}$ are not equivalent with respect to $\mathfrak{C}$, since they are contained in $\mathcal{G}_{i_1}$ and $\mathcal{G}_{i_2}$ respectively. Moreover, none of the $\mathcal{B}_i$ is equivalent with $\mathcal{B}$, since all lines of $\mathcal{U}''$ are dividing modulo $\mathfrak{C}$.

For any value of i we now select a complete collection $\mathcal{V}_{ij}$ of representatives of the equivalence classes with respect to $\mathfrak{B}_i$ of the sides of $\mathcal{B}_i$ belonging to $\mathcal{U}''$ *and different from the equivalence class of* $\mathcal{V}_i$. The number of these equivalence classes is called ν_i, $0 \leqq \nu_i \leqq \infty$. Identifying $\mathcal{B}_i$ with the notation $\mathcal{B}'_{ij}$ for all values of j, and remembering that $\mathcal{B}_i \subset \mathcal{G}_i$, one gets, in an analogous way as in (9),

$$\mathcal{B}_i = \bigcap_j (\mathcal{G}_i \cap \mathcal{G}_{ij} \cap \mathcal{C}). \tag{10}$$

Here the number ν_i may, in particular, be zero, which signifies that all sides of $\mathcal{B}_i$ drawn from $\mathcal{U}''$ are equivalent with $\mathcal{V}_i$. In this particular case formula (10) simply means

$$\mathcal{B}_i = \mathcal{G}_i \cap \mathcal{C}.$$

(This again simply reads $\mathcal{B}_i = \mathcal{G}_i$, if $\nu_i = 0$ and no side of $\mathcal{C}$, i.e. no line from $\mathcal{U}'$, contributes to the boundary of $\mathcal{B}_i$.)

$\mathcal{B}_{ij}$ being (for $\nu_i > 0$) the $\mathcal{B}$-region adjacent to $\mathcal{B}_i$ along $\mathcal{V}_{ij}$, and $\mathfrak{B}_{ij}$ the subgroup of $\mathfrak{C}$ reproducing it, we now select in the same way a complete collection of representatives $\mathcal{V}_{ijk}$ for the equivalence classes with respect to $\mathfrak{B}_{ij}$ of sides of $\mathcal{B}_{ij}$ drawn from $\mathcal{U}''$ and different from the class of $\mathcal{V}_{ij}$. Their number is denoted by ν_{ij}, $0 \leqq \nu_{ij} \leqq \infty$. Identifying $\mathcal{B}_{ij}$ with $\mathcal{B}'_{ijk}$ one gets $\mathcal{B}_{ijk}$ as the $\mathcal{B}$-region adjacent to $\mathcal{B}_{ij}$ along $\mathcal{V}_{ijk}$, and $\mathfrak{B}_{ijk}$ as the subgroup of $\mathfrak{C}$ reproducing it.

Continuing in this way one gets a collection of $\mathcal{B}$-regions denoted by

$$\mathcal{B}, \mathcal{B}_i, \mathcal{B}_{ij}, \mathcal{B}_{ijk}, \dots, \mathcal{B}_{ijk\dots r}, \dots \tag{11}$$

The regions $\mathcal{B}_{ijk\dots r}$ arrived at after κ steps have κ subscripts, each of which has a certain range of values. The subgroup of $\mathfrak{C}$ reproducing one of the $\mathcal{B}$-regions (11) is denoted by the letter $\mathfrak{B}$ with the same sequence of subscripts. If $\mathcal{B}_{ij\dots r}$ has κ subscripts, then $\mathcal{V}_{ij\dots rs}$, $s = 1, 2, \dots$, is a complete collection of representatives of the equivalence classes with respect to $\mathfrak{B}_{ij\dots r}$ of sides of $\mathcal{B}_{ij\dots r}$ drawn from $\mathcal{U}''$ and different from the class of $\mathcal{V}_{ij\dots r}$. The number of these classes, i.e. the number of values for the subscript s, is denoted by $\nu_{ij\dots r}$, $0 \leqq \nu_{ij\dots r} \leqq \infty$. If for specified values of $i, j, \dots, r$ the number $\nu_{ij\dots r} = 0$, then there is in (11) no region with more than κ subscripts for which the first κ subscripts take these specified values.

Any region of the collection (11), say $\mathcal{B}_{ij\dots r}$, is adjacent to a preceding one along $\mathcal{V}_{ij\dots r}$. Since the $\mathcal{B}$-regions are convex regions bounded by complete straight lines, this implies that any segment joining a point of $\mathcal{B}$ with a point of $\mathcal{B}_{ij\dots r}$ cuts the lines $\mathcal{V}_i, \mathcal{V}_{ij}, \dots, \mathcal{V}_{ij\dots r}$ in this order and no other line of $\mathcal{U}''$. The region $\mathcal{B}_{ij\dots r}$ is thus separated from $\mathcal{B}$ by exactly κ lines of the collection

$$\mathcal{V}_i, \mathcal{V}_{ij}, \dots, \mathcal{V}_{ij\dots r}, \dots \tag{12}$$

(12) is for all values which the subscripts take the totality of all $\mathcal{V}$-lines brought into play by the above construction.

If for any particular $\mathcal{V}$-line $\mathcal{V}_{ij\dots r}$ one decomposes $\mathcal{C}_{\mathcal{U}''}$ by the collection $\mathfrak{C}\mathcal{V}_{ij\dots r}$, the regions of decomposition adjacent to $\mathcal{V}_{ij\dots r}$ were called $\mathcal{G}_{ij\dots r}$ and $\mathcal{G}'_{ij\dots r}$, and it was stipulated that $\mathcal{B}_{ij\dots r} \subset \mathcal{G}_{ij\dots r}$. This implies that $\mathcal{B}_{ij\dots r} \subset \mathcal{G}'_{ij\dots rs}$ for all values which s takes. The boundary of $\mathcal{B}_{ij\dots r}$ on $\mathcal{C}$ consists of the equivalence classes with respect to $\mathfrak{B}_{ij\dots r}$ of $\mathcal{V}_{ij\dots r}$ and $\mathcal{V}_{ij\dots rs}$ for all values which s takes. Hence

$$\mathcal{B}_{ij\dots r} = \bigcap_s (\mathcal{G}_{ij\dots r} \cap \mathcal{G}'_{ij\dots rs} \cap \mathcal{C}). \tag{13}$$

If $\nu_{ij\dots r} = 0$, this reduces to

$$\mathcal{B}_{ij\dots r} = \mathcal{G}_{ij\dots r} \cap \mathcal{C}.$$

Let $\mathfrak{G}_{ij\dots r}$ denote the subgroup of $\mathfrak{C}$ which carries the region $\mathcal{G}_{ij\dots r}$, obtained by the decomposition of $\mathcal{C}_{\mathcal{U}''}$ by $\mathfrak{C}\mathcal{V}_{ij\dots r}$ and containing $\mathcal{B}_{ij\dots r}$, into itself. Then $\mathfrak{G}_{ij\dots r}\mathcal{V}_{ij\dots r}$ is the complete boundary of $\mathcal{G}_{ij\dots r}$ on $\mathcal{C}$ (Section 4). Any $\mathcal{V}$-line $\mathcal{V}_{ij\dots rs}$ with the same values for the first κ subscripts $i, j, \dots, r$ is a side of $\mathcal{B}_{ij\dots r}$, not equivalent with $\mathcal{V}_{ij\dots r}$, hence situated in $\mathcal{G}_{ij\dots r}$; it is dividing modulo $\mathfrak{C}$ and thus also modulo $\mathfrak{G}_{ij\dots r}$ which in $\mathcal{G}_{ij\dots r}$ has the same effect as $\mathfrak{C}$. Consider the decomposition of $\mathcal{C}_{\mathcal{U}''}$ by the collection $\mathfrak{C}\mathcal{V}_{ij\dots rs}$. The part of this collection contained in $\mathcal{G}_{ij\dots r}$ is $\mathfrak{G}_{ij\dots r}\mathcal{V}_{ij\dots rs}$. In the decomposition of $\mathcal{G}_{ij\dots r}$ by this subcollection, the $\mathcal{B}$-region $\mathcal{B}_{ij\dots rs}$, which is adjacent to $\mathcal{B}_{ij\dots r}$ along $\mathcal{V}_{ij\dots rs}$, belongs to $\mathcal{G}_{ij\dots rs}$, while $\mathcal{B}_{ij\dots r}$ belongs to $\mathcal{G}'_{ij\dots rs}$; compare (13).

Thus $\mathcal{G}_{ij\dots rs}$ is a subregion of $\mathcal{G}_{ij\dots r}$, and $\mathfrak{G}_{ij\dots rs}$ is a subgroup of $\mathfrak{G}_{ij\dots r}$. This implies that $\mathcal{G}'_{ij\dots r}$ is a subregion of $\mathcal{G}'_{ij\dots rs}$. Since $\mathcal{B}_{ij\dots r}$ is situated in $\mathcal{G}'_{ij\dots rs}$, the same holds for any of its sides $\mathcal{V}_{ij\dots rs'}$ with $s' \neq s$, and thus also for $\mathcal{G}_{ij\dots rs'}$.

This shows that any two of the $\mathcal{B}$-regions (11), *say $\mathcal{B}^*$ and $\mathcal{B}^{**}$ are disjoint modulo* $\mathfrak{C}$: Let κ be the maximum number such that the first κ subscripts $i, j, \dots r$ take the same values for $\mathcal{B}^*$ and $\mathcal{B}^{**}$. In at least one of them, say in $\mathcal{B}^*$, the number of subscripts must exceed κ. If $\mathcal{B}^* = \mathcal{B}_{ij\dots rs\dots}$, then $\mathcal{B}^* \subset \mathcal{G}_{ij\dots rs\dots} \subset \mathcal{G}_{ij\dots rs\dots}$. The region $\mathcal{B}^{**}$ is either $\mathcal{B}_{ij\dots r}$ or some $\mathcal{B}_{ij\dots rs'\dots}$ with $s' \neq s$, and thus belongs to $\mathcal{G}'_{ij\dots rs}$; and $\mathcal{G}_{ij\dots rs}$ and $\mathcal{G}'_{ij\dots rs}$ are known to be disjoint modulo $\mathfrak{C}$.

From this follows that *no two of the $\mathcal{V}$-lines* (12) *can be equivalent with respect to* $\mathfrak{C}$; for the pair of $\mathcal{B}$-regions adjacent to the first one cannot be equivalent with respect to $\mathfrak{C}$ with the pair of $\mathcal{B}$-regions adjacent to the second one.

19.10 $\mathfrak{C}$ and $\mathfrak{F}$ as generalized free products. At this point it is convenient to introduce a change of notation. The construction in Section 9 of the $\mathcal{B}$-regions (11) started with the arbitrary choice of a particular region $\mathcal{B}$ in the decomposition of $\mathcal{C}_{\mathcal{U}''}$ by the collection $\mathcal{U}''$, and this is reflected in the systems of subscripts used. We are now concerned with the (finite or enumerable) totality of $\mathcal{B}$-regions (11), and we denote them *in an arbitrary order* by

$$\mathcal{B}_m, \quad m = 1, 2, \dots. \tag{14}$$

The subgroup of $\mathfrak{C}$ reproducing $\mathcal{B}_m$ is denoted by

$$\mathfrak{B}_m, \quad m = 1, 2, \dots. \tag{15}$$

Any $\mathcal{V}$-line, i.e. any line of the collection (12), is the common boundary of two of the regions (14). If these are $\mathcal{B}_m$ and $\mathcal{B}_n$, we denote the $\mathcal{V}$-line by

$$\mathcal{V}_{mn} = \mathcal{V}_{nm} = \tilde{\mathcal{B}}_m \cap \tilde{\mathcal{B}}_n. \tag{16}$$

The union of the $\mathcal{B}$-regions (14) and the $\mathcal{V}$-lines (16) is a connected, open subset of $\mathcal{C}_{\mathcal{U}''}$. We denote it by $\mathcal{Y}$. The boundaries of $\mathcal{Y}$ on $\mathcal{C}$ are, on the one hand, certain limit sides of $\mathcal{K}(\mathfrak{C})$ which consist of accumulation points of $\mathcal{U}''$, if any; on the other hand all those sides of the $\mathcal{B}$-regions (14) which do not belong to the collection (16). The latter will be called *free sides* of the $\mathcal{B}_m$ and also *free sides* of $\mathcal{Y}$, and the half-planes, closed on $\mathcal{C}$, cut off by them and not containing $\mathcal{Y}$, *free half-planes* for the $\mathcal{B}_m$, and also *free half-planes* for $\mathcal{Y}$.

Thus $\mathcal{Y}$ is obtained from $\mathcal{C}$ by the omission of all free half-planes for $\mathcal{Y}$ together with the half-planes called $\tilde{\mathcal{H}}^{**}$ in the beginning of the preceding section, if any. Thus $\mathcal{Y}$ is convex. (All the omitted half-planes are closed on $\mathcal{C}$.)

For any two $\mathcal{B}$-regions $\mathcal{B}'$ and $\mathcal{B}''$ in $\mathcal{C}_{\mathcal{U}''}$ let $\Delta(\mathcal{B}', \mathcal{B}'')$ denote the number of lines from $\mathcal{U}''$ separating them; it is called the *distance* of $\mathcal{B}'$ and $\mathcal{B}''$. The distance $\Delta(\mathcal{B}', \mathcal{B}'')$ may be obtained as the number of intersection points of the collection $\mathcal{U}''$ with a segment joining an arbitrary point of $\mathcal{B}'$ with an arbitrary point of $\mathcal{B}''$, because the $\mathcal{B}$-regions are bounded by complete straight lines; it is a finite number, since the lines of $\mathcal{U}''$ do not accumulate in $\mathcal{C}_{\mathcal{U}''}$. The distance is zero, if the two regions coincide, and 1, if they are adjacent. For any $\mathfrak{c} \in \mathfrak{C}$ one gets

$$\Delta(\mathfrak{c}\mathcal{B}', \mathfrak{c}\mathcal{B}'') = \Delta(\mathcal{B}', \mathcal{B}''), \tag{17}$$

because $\mathcal{U}''$ is reproduced by $\mathfrak{C}$.

The number of $\mathcal{B}$-regions separating $\mathcal{B}'$ and $\mathcal{B}''$ is $\Delta(\mathcal{B}', \mathcal{B}'') - 1$.

To every point of $\mathcal{C}_{\mathcal{U}''}$ there is in $\mathcal{Y}$ an equivalent with respect to $\mathfrak{C}$.

Proof. Let $\mathcal{B}'$ be an arbitrary $\mathcal{B}$-region in $\mathcal{C}$ outside $\mathcal{Y}$, thus situated in one of the free half-planes for $\mathcal{Y}$. Let $\mathcal{S}$ denote the free side of $\mathcal{Y}$ cutting off that free half-plane, and $\mathcal{B}_m$ the $\mathcal{B}$-region of $\mathcal{Y}$ adjacent to $\mathcal{S}$. The distance $\Delta(\mathcal{B}', \mathcal{B}_m)$ may then also be denoted by $\Delta(\mathcal{B}', \mathcal{Y})$ and called *the distance of $\mathcal{B}'$ and $\mathcal{Y}$*. Now $\mathcal{S}$ is a free side of $\mathcal{B}_m$; it is equivalent with respect to $\mathfrak{B}_m$ with a $\mathcal{V}$-line (16) bounding $\mathcal{B}_m$, say $\mathfrak{b}\mathcal{S} = \mathcal{V}_{mn}$, $\mathfrak{b} \in \mathfrak{B}_m$. Then (17) yields

$$\Delta(\mathfrak{b}\mathcal{B}', \mathcal{B}_m) = \Delta(\mathfrak{b}\mathcal{B}', \mathfrak{b}\mathcal{B}_m) = \Delta(\mathcal{B}', \mathcal{B}_m).$$

One of the lines from $\mathcal{U}''$ separating $\mathfrak{b}\mathcal{B}'$ and $\mathcal{B}_m$ is $\mathcal{V}_{mn}$, and $\mathcal{B}_m \subset \mathcal{Y}$. Hence $\Delta(\mathfrak{b}\mathcal{B}', \mathcal{Y}) \leqq \Delta(\mathcal{B}', \mathcal{Y}) - 1$. If $\Delta(\mathfrak{b}\mathcal{B}', \mathcal{Y}) > 0$, the process is repeated until a region $\mathcal{B}_r$ of $\mathcal{Y}$ equivalent with $\mathcal{B}'$ is reached. – Every point of $\mathcal{C}$ on a side of $\mathcal{B}'$ has then an equivalent with respect to $\mathfrak{C}$ on a side of $\mathcal{B}_r$ on $\mathcal{C}$, and this again has an equivalent with respect to $\mathfrak{B}_r$ on a $\mathcal{V}$-line bounding $\mathcal{B}_r$, thus in $\mathcal{Y}$. This finishes the proof. □

$\mathfrak{C}$ *is generated by the collection of groups* (15).

Proof. All the $\mathfrak{B}_m$ are subgroups of $\mathfrak{C}$. Let $\mathfrak{K}$ denote the subgroup of $\mathcal{C}$ generated by them. In the above construction of a region $\mathcal{B}_r$ in $\mathcal{Y}$ equivalent with respect to $\mathfrak{C}$ with a given $\mathcal{B}$-region of $\mathcal{C}$ only elements of the subgroups (15) of $\mathfrak{C}$ were brought into play. Let p be a regular point in a particular $\mathcal{B}_m$ and $\mathfrak{c}$ an arbitrary element of $\mathfrak{C}$. According to the preceding theorem there exists an element $\mathfrak{k} \in \mathfrak{K}$ carrying the region $\mathfrak{c}\mathcal{B}_m$ into a $\mathcal{B}$-region of $\mathcal{Y}$. Now $\mathfrak{k}\mathfrak{c}\mathcal{B}_m$ must coincide with $\mathcal{B}_m$, because the $\mathcal{B}$-regions of $\mathcal{Y}$ are disjoint modulo $\mathfrak{C}$. Hence both p and $\mathfrak{k}\mathfrak{c}p$ belong to $\mathcal{B}_m$ and are thus equivalent by an element $\mathfrak{k}'$ of $\mathfrak{B}_m$, because $\mathfrak{C}$ and $\mathfrak{B}_m$ have equal effect in $\mathcal{B}_m$. The equation $p = \mathfrak{k}'\mathfrak{k}\mathfrak{c}p$ yields $\mathfrak{c} = \mathfrak{k}^{-1}\mathfrak{k}'^{-1}$, since p is regular. Hence $\mathfrak{K} = \mathfrak{C}$.

In virtue of the condition 5) in Section 7 and the fact that the lines of $\mathcal{U}''$ are dividing modulo $\mathfrak{C}$ the subgroup of $\mathfrak{C}$ leaving invariant a line of $\mathcal{U}''$ is either a translation group $\mathfrak{A}(\rightarrow)$ or the identity, according as the line is an axis of $\mathfrak{C}$ or not. Let

$$\mathfrak{V}_{mn} = \mathfrak{B}_m \cap \mathfrak{B}_n = \mathfrak{V}_{nm} \tag{18}$$

denote the intersection of the groups $\mathfrak{B}_m$ and $\mathfrak{B}_n$ belonging to any two $\mathcal{B}$-regions $\mathcal{B}_m$ and $\mathcal{B}_n$ of $\mathcal{Y}$, whether adjacent or not. Then $\mathfrak{V}_{mn} = 1$ unless $\mathcal{B}_m$ and $\mathcal{B}_n$ are adjacent and their common side $\mathcal{V}_{mn}$ is an axis of $\mathfrak{C}$.

Let $\mathfrak{b} \neq 1$ be any element of $\mathfrak{B}_m$, thus $\mathfrak{b}\mathfrak{B}_m = \mathfrak{B}_m$. Let $\mathcal{V}_{mn}$ denote the (finite or infinite) subcollection of (16) situated on the boundary of this particular $\mathcal{B}_m$; thus m is a fixed number, and n ranges over those values for which $\mathcal{B}_m$ and $\mathcal{B}_n$ are adjacent.

Assume first that $\mathfrak{b}$ does not belong to any of the groups (18), thus is not a translation along a $\mathcal{V}$-line. It then leaves none of the $\mathcal{V}_{mn}$ invariant. Since the $\mathcal{V}_{mn}$ represent the different equivalence classes with respect to $\mathfrak{B}_m$ of sides of $\mathcal{B}_m$ drawn from $\mathcal{U}''$, and no two of them thus are equivalent by $\mathfrak{b}$, the sides $\mathfrak{b}\mathcal{V}_{mn}$ of $\mathcal{B}_m$ are free sides of $\mathcal{B}_m$ for all pertinent values of n. The part of $\mathcal{Y}$ belonging to the half-plane determined by $\mathcal{V}_{mn}$ and not containing $\mathcal{B}_m$ has its image by $\mathfrak{b}$ belonging to the free half-plane for $\mathcal{Y}$ determined by the free side $\mathfrak{b}\mathcal{V}_{mn}$ of $\mathcal{B}_m$. This part of $\mathcal{Y}$ and its image by $\mathfrak{b}$ are thus disjoint. Since this holds for all pertinent values of n, one gets

$$\mathcal{Y} \cap \mathfrak{b}\mathcal{Y} = \mathcal{B}_m. \tag{19}$$

Any $\mathcal{B}$-region $\neq \mathcal{B}_m$ of $\mathcal{Y}$ and any $\mathcal{B}$-region $\neq \mathfrak{b}\mathcal{B}_m$ of $\mathfrak{b}\mathcal{Y}$ are separated by $\mathcal{B}_m$, since they belong to half-planes outside $\mathcal{B}_m$ cut off by different sides of $\mathcal{B}_m$.

Secondly, let $\mathfrak{b}$ be a translation along the side $\mathcal{V}_{mn}$ of $\mathcal{B}_m$ thus $\mathfrak{b}\mathcal{B}_m = \mathcal{B}_m$ and $\mathfrak{b}\mathcal{B}_n = \mathcal{B}_n$. No $\mathcal{V}$-line on the boundary of $\mathcal{B}_m$ or of $\mathcal{B}_n$ except $\mathcal{V}_{mn}$ is left invariant by $\mathfrak{b}$. Therefore, the part of $\mathcal{Y}$ belonging to the half-plane determined by any $\mathcal{V}_{mn'}$, $n' \neq n$, and not containing $\mathcal{B}_m$ has its image by $\mathfrak{b}$ belonging to the free half-plane for $\mathcal{Y}$ determined by the free side $\mathfrak{b}\mathcal{V}_{mn'}$ of $\mathcal{B}_m$. Likewise, the part of $\mathcal{Y}$ belonging to the half-plane determined by an $\mathcal{V}_{nm'}$, $m' \neq m$, and not containing $\mathcal{B}_n$ has its image by $\mathfrak{b}$ belonging to the free half-plane for $\mathcal{Y}$ determined by the free side $\mathfrak{b}\mathcal{V}_{nm'}$ of $\mathcal{B}_n$. Hence in this case one gets

$$\mathcal{Y} \cap \mathfrak{b}\mathcal{Y} = \mathcal{B}_m \cup \mathcal{V}_{mn} \cup \mathcal{B}_n \tag{20}$$

instead of (19).

Let $\mathcal{B}_r$ and $\mathcal{B}_s$ be any two $\mathcal{B}$-regions of $\mathcal{Y}$ both different from $\mathcal{B}_m$ and from $\mathcal{B}_n$; the case of coinciding regions, $r = s$, is not excluded. $\mathcal{B}_r$ belongs to a half-plane determined by some $\mathcal{V}_{mn'}$, $n' \neq n$, or to a half-plane determined by some $\mathcal{V}_{nm'}$, $m' \neq m$, while $\mathfrak{b}\mathcal{B}_s$ belongs either to a free half-plane for $\mathcal{B}_m$ or to a free half-plane for $\mathcal{B}_n$. A straight segment joining a point of $\mathcal{B}_r$ with a point of $\mathfrak{b}\mathcal{B}_s$ passes either through $\mathcal{B}_m$ or $\mathcal{B}_n$, namely if $\mathcal{B}_r$ and $\mathfrak{b}\mathcal{B}_s$ are not separated by $\mathcal{V}_{mn}$; or through both $\mathcal{B}_m$ and $\mathcal{B}_n$, if $\mathcal{B}_r$ and $\mathfrak{b}\mathcal{B}_s$ are separated by $\mathcal{V}_{mn}$. □

Taking the two cases corresponding to (19) and (20) together, we formulate the

Auxiliary theorem. *Any $\mathcal{B}$-region of $\mathcal{Y}$ and any $\mathcal{B}$-region of $\mathfrak{b}\mathcal{Y}$ both not belonging to $\mathcal{Y} \cap \mathfrak{b}\mathcal{Y}$, are separated by $\mathcal{Y} \cap \mathfrak{b}\mathcal{Y}$.* □

Since $\mathfrak{C}$ is generated by the groups (15), any non-identical element $\mathfrak{c} \in \mathfrak{C}$ can be written in the form

$$\mathfrak{c} = \mathfrak{b}_\ell \mathfrak{b}_{\ell-1} \dots \mathfrak{b}_2 \mathfrak{b}_1 \tag{21}$$

where every right hand factor is $\neq 1$ and drawn from one of the groups (15), and where no two neighbouring factors are drawn from the same group $\mathfrak{B}_m$. If, in particular, a factor $\mathfrak{b}_i$ in (21) belongs to one of the groups (18), then, since $\mathfrak{b}_i \neq 1$, this $\mathfrak{V}_{mn}$ is the translation group belonging to the common side $\mathcal{V}_{mn}$ of two adjacent regions $\mathcal{B}_m$ and $\mathcal{B}_n$, and none of the factors $\mathfrak{b}_{i-1}$ and $\mathfrak{b}_{i+1}$ neighbouring $\mathfrak{b}_i$ in (21) then belongs to $\mathfrak{B}_m$ or to $\mathfrak{B}_n$.

Assume $\ell = 2$ in (21), thus $\mathfrak{c} = \mathfrak{b}_2\mathfrak{b}_1$. If neither $\mathfrak{b}_1$ nor $\mathfrak{b}_2$ is a translation along a $\mathcal{V}$-line, then they belong to uniquely determined groups (15), say $\mathfrak{b}_1 \in \mathfrak{B}_m$, $\mathfrak{b}_2 \in \mathfrak{B}_n$, $\Delta(\mathcal{B}_m, \mathcal{B}_n) \geqq 1$. If $\mathfrak{b}_1 \in \mathfrak{V}_{mn}$, then $\mathfrak{b}_2$ belongs neither to $\mathfrak{B}_m$ nor to $\mathfrak{B}_n$; let $\mathfrak{B}_r$ be the group (15), or one of the two groups (15), to which $\mathfrak{b}_2$ belongs. Since $\mathcal{B}_m$ and $\mathcal{B}_n$ are adjacent, one of them separates the other from $\mathcal{B}_r$. We assign $\mathfrak{b}_1$ to $\mathcal{B}_m$, if $\mathcal{B}_m$ separates $\mathcal{B}_r$ and $\mathcal{B}_n$ and to $\mathfrak{B}_n$, if $\mathcal{B}_n$ separates $\mathcal{B}_r$ and $\mathcal{B}_m$. Likewise, if $\mathfrak{b}_2 \in \mathfrak{V}_{rs}$, we assign $\mathfrak{b}_2$ to $\mathfrak{B}_r$, if $\mathcal{B}_r$ separates $\mathcal{B}_m$ and $\mathcal{B}_s$, and to $\mathfrak{B}_s$, if $\mathcal{B}_s$ separates

$\mathcal{B}_m$ and $\mathcal{B}_r$. In other words: Whenever the inclusions $\mathfrak{b}_1 \in \mathfrak{B}_m$, $\mathfrak{b}_2 \in \mathfrak{B}_r$ are not uniquely determined, they are so chosen that $\Delta(\mathcal{B}_m, \mathcal{B}_r)$ gets its smallest value; this value is $\geqq 1$.

$\mathcal{B}_m$ and $\mathcal{B}_r$ both belong to $\mathcal{Y}$ and are thus separated by $\mathcal{V}$-lines only, say by $\nu = \Delta(\mathcal{B}_m, \mathcal{B}_r) \geqq 1$ lines. The first and the last of these $\mathcal{V}$-lines is a side $\mathcal{V}_{rt}$ of $\mathcal{B}_r$ and a side $\mathcal{V}_{mu}$ of $\mathcal{B}_m$; they coincide if $\nu = 1$, thus if $t = u$. If $\mathfrak{b}_1 \in \mathfrak{V}_{mn}$, then $\mathcal{B}_m$ separates $\mathcal{B}_r$ and $\mathcal{B}_n$ according to our choice of notation, hence $n \neq u$, and $\Delta(\mathcal{B}_r, \mathcal{B}_n) = \Delta(\mathcal{B}_r, \mathcal{B}_m) + 1$. Let $\mathcal{H}_m$ denote the half-plane determined by $\mathcal{V}_{mu}$ and containing $\mathcal{B}_m$, and $\mathcal{H}_r$ the half-plane determined by $\mathcal{V}_{rt}$ and containing $\mathcal{B}_n$. Thus $\mathcal{H}_m \subset \mathcal{H}_r$; they coincide, if $\nu = 1$. We now apply the above auxiliary theorem: $\mathcal{H}_m$ contains $\mathcal{Y} \cap \mathfrak{b}_1\mathcal{Y}$, since it contains $\mathcal{B}_m$ or both $\mathcal{B}_m$ and $\mathcal{B}_n$; see (19) and (20). The rest of $\mathfrak{b}_1\mathcal{Y}$ is separated by this intersection from any $\mathcal{B}$-region of $\mathcal{Y}$ not belonging to it, thus in particular from $\mathcal{B}_r$. This shows that $\mathfrak{b}_1\mathcal{Y} \subset \mathcal{H}_m$. If $\mathfrak{b}_2 \in \mathfrak{V}_{rs}$, then $s \neq t$ according to our choice of notations. Hence $\mathfrak{b}_2\mathcal{V}_{rt}$ is a free side of $\mathcal{B}_r$, and thus $\mathfrak{b}_2\mathcal{H}_r$ a free half-plane for $\mathcal{Y}$. Since $\mathcal{H}_m \subset \mathcal{H}_r$, it follows that *$\mathcal{Y}$ and $\mathfrak{b}_2\mathfrak{b}_1\mathcal{Y}$ are disjoint.*

The number of lines from $\mathcal{U}''$ separating $\mathcal{B}_m$ and $\mathcal{B}_r$ is not changed by the application of $\mathfrak{b}_2$; see (17). Since $\mathcal{V}_{mn}$ is a side of $\mathfrak{b}_1\mathcal{Y}$, one gets

$$\Delta(\mathcal{Y}, \mathfrak{b}_2\mathfrak{b}_1\mathcal{Y}) = \nu. \tag{22}$$

We now proceed by induction with respect to the index ℓ in (21). We put

$$\mathfrak{c}' = \mathfrak{b}_{\ell+1}\mathfrak{c} = \mathfrak{b}_{\ell+1}\mathfrak{b}_\ell \dots \mathfrak{b}_2\mathfrak{b}_1 \tag{23}$$

maintaining the conditions imposed on (21). We put $\mathfrak{b}_\ell \in \mathfrak{B}_m$, $\mathfrak{b}_{\ell+1} \in \mathfrak{B}_r$. If the assignment to these groups is not uniquely determined, because one or both of the elements $\mathfrak{b}_\ell$ and $\mathfrak{b}_{\ell+1}$ belong to a group (18), then their assignment to groups $\mathfrak{B}_m$ and $\mathfrak{B}_r$ is such that $\Delta(\mathcal{B}_m, \mathcal{B}_r)$ gets its minimum value. This value is $\geqq 1$. We put $\Delta(\mathcal{B}_m, \mathcal{B}_r) = \nu$.

The assumptions of the induction are for $\ell \geqq 2$:

1) $\mathfrak{c}\mathcal{Y}$ is contained in one of the free half-planes for $\mathcal{B}_m$,

2) $\sigma = \Delta(\mathcal{Y}, \mathfrak{c}\mathcal{Y}) \geqq \ell - 1$.

These assumptions have been seen to hold for $\ell = 2$. (In the above proof of this fact for $\ell = 2$ the $\mathcal{B}$-region there called $\mathcal{B}_r$ corresponds to our present $\mathcal{B}_m$.)

We again use the following denotations: $\mathcal{V}_{mu}$ is the $\mathcal{V}$-side of $\mathcal{B}_m$ separating it from $\mathcal{B}_r$, and $\mathcal{V}_{rt}$ the $\mathcal{V}$-side of $\mathcal{B}_r$ separating it from $\mathcal{B}_m$; they coincide if $\nu = 1$. The half-plane determined by $\mathcal{V}_{mu}$ and containing $\mathcal{B}_m$ is called $\mathcal{H}_m$. The half-plane determined by $\mathcal{V}_{rt}$ and containing $\mathcal{B}_m$ is called $\mathcal{H}_r$. Thus $\mathcal{H}_m \subset \mathcal{H}_r$; they coincide if $\nu = 1$.

Since the $\mathcal{V}$-side $\mathcal{V}_{mu}$ of $\mathcal{B}_m$ bounds $\mathcal{H}_m$, every free half-plane for $\mathcal{B}_m$ is contained in $\mathcal{H}_m$. Thus, in particular, $\mathfrak{c}\mathcal{Y}$ is contained in $\mathcal{H}_m$ in virtue of assumption 1. The element $\mathfrak{b}_{\ell+1}$ is not a translation along $\mathcal{V}_{rt}$, since otherwise $\Delta(\mathcal{B}_m, \mathcal{B}_r)$ would not have its minimum value. Hence $\mathfrak{b}_{\ell+1}\mathcal{V}_{rt}$ is a free side of $\mathcal{B}_r$, and $\mathfrak{b}_{\ell+1}\mathcal{H}_r$ is a free half-plane for $\mathcal{B}_r$. Since $\mathcal{H}_m \subset \mathcal{H}_r$, and $\mathfrak{c}\mathcal{Y} \subset \mathcal{H}_m$, the image $\mathfrak{b}_{\ell+1}\mathfrak{c}\mathcal{Y} = \mathfrak{c}'\mathcal{Y}$ is contained in a free half-plane for $\mathcal{B}_r$. Thus the assumption 1 continues to hold for $\ell + 1$ factors.

Since $\mathfrak{c}\mathcal{Y}$ is contained in a free half-plane for $\mathcal{B}_m$, the distance $\sigma = \Delta(\mathcal{Y}, \mathfrak{c}\mathcal{Y})$ is equal to the number of lines from $\mathcal{U}''$ which separate $\mathfrak{c}\mathcal{Y}$ from $\mathcal{B}_m$, and since $\mathfrak{c}\mathcal{Y} \subset \mathcal{H}_m$, one gets

$$\Delta(\mathfrak{c}\mathcal{Y}, \mathcal{B}_r) = \Delta(\mathfrak{c}\mathcal{Y}, \mathcal{B}_m) + \Delta(\mathcal{B}_m, \mathcal{B}_r).$$

This number is preserved under the mapping $\mathfrak{b}_{\ell+1}$, which carries $\mathcal{V}_{rt}$ into a free side of $\mathcal{B}_r$. Hence by (17)

$$\Delta(\mathfrak{c}'\mathcal{Y}, \mathcal{Y}) = \Delta(\mathfrak{b}_{\ell+1}\mathfrak{c}\mathcal{Y}, \mathcal{B}_r) = \Delta(\mathfrak{c}\mathcal{Y}, \mathcal{B}_r) = \Delta(\mathfrak{c}\mathcal{Y}, \mathcal{B}_m) + \Delta(\mathcal{B}_m, \mathcal{B}_r) = \sigma + \nu \geqq \ell,$$

since $\sigma \geqq \ell - 1$ and $\nu \geqq 1$. Thus assumption 2 continues to hold for $\ell + 1$ factors.

We have thus arrived at the following result: Let an arbitrary element $\mathfrak{c}$ of $\mathfrak{C}$ be written in the form (21). If all factors are $\neq 1$, no two neighbouring factors belong to the same group (15), and if $\ell \geqq 2$, then $\mathcal{Y}$ and $\mathfrak{c}\mathcal{Y}$ are disjoint. If $\ell = 1$ and $\mathfrak{b}_1 \neq 1$, then the intersection of $\mathcal{Y}$ and $\mathfrak{c}\mathcal{Y}$ is given by (19) or (20). Hence $\mathfrak{c} = 1$ requires that every factor in (21) is the identity. Since the single factors are drawn from the generating groups $\mathfrak{B}_m$ of $\mathfrak{C}$, the relation $\mathfrak{b}_m = 1$ is a consequence of the multiplication table for $\mathfrak{B}_m$. Hence if all elements of all the $\mathfrak{B}_m$ are taken as a system of generators for $\mathfrak{C}$, then any relation for $\mathfrak{C}$ in these generators is a consequence of the multiplication tables for all the $\mathfrak{B}_m$. One thus gets by Section 1

$$\mathfrak{C} = \prod_m{}^{*} \mathfrak{B}_m \text{ am } \mathfrak{W}_{mn}. \tag{24}$$

By inserting (24) in (8) one finally gets

$$\mathfrak{F} = \prod_{r,m}{}^{*} (\mathfrak{W}_r, \mathfrak{B}_m) \text{ am } \mathfrak{W}_r^*, \ \mathfrak{W}_{mn}. \tag{25}$$

The sides $\mathcal{W}_r$ of $\mathcal{C}$ were chosen as representatives of their equivalence classes with respect to $\mathfrak{C}$. Every equivalence class is represented by a side of exactly one of the $\mathcal{B}$-regions $\mathcal{B}_m$, thus by a side of $\mathcal{Y}$. We may therefore assume that all the $\mathcal{W}_r$ are chosen among the sides of $\mathcal{Y}$. In §18.2–6 it was assumed that the two factors of the generalized free product $\mathfrak{G} = \mathfrak{G}_1 * \mathfrak{G}_2$ operated on two mutually adjacent regions; and in §18.7–10 we considered the generalized free product of a group $\mathfrak{G}_1$ operating on a certain region with a quasi-abelian group $\mathfrak{H}$ operating on a boundary of that region. In order to comply with these assumptions we arrange that any factor in (25) is applied only when the partial product so far established operates on a region to which the new factor is adjacent as a group operating on a region (factors $\mathfrak{B}_m$) or as a quasi-abelian group (factors $\mathfrak{W}_r$) operating on a boundary of the region.

If $\mathcal{K}(\mathfrak{F})$ is quasi-compact modulo $\mathfrak{F}$, the collection $\mathcal{U}$ comprises only a finite number of equivalence classes with respect to $\mathfrak{F}$. Then (25) yields $\mathfrak{F}$ as a generalized free product of a finite number of factors with given amalgamations.

§20 Decompositions of $\mathfrak{F}$-groups containing reflections

20.1 The reflection subgroup $\mathfrak{R}$. In this paragraph we consider an $\mathfrak{F}$-group $\mathfrak{F}$ which contains reflections. Let $\mathcal{Z}$ denote the point set made up of all reflection lines of $\mathfrak{F}$ not including their end-points on $\mathcal{E}$. The lines of $\mathcal{Z}$ are infinite in number, for, if $\mathcal{S}$ is a reflection line and $\mathfrak{f}$ is a translation in $\mathfrak{F}$ whose axis is different from $\mathcal{S}$, then the powers of $\mathfrak{f}$ carry $\mathcal{S}$ into infinitely many other reflection lines. Centres of $\mathfrak{F}$ on $\mathcal{Z}$ are those, and only those points of $\mathcal{Z}$ in which reflection lines intersect. The complement of $\mathcal{Z}$ in $\mathcal{D}$ is open and is made up of convex regions, since the decomposition of $\mathcal{D}$ is effected by complete straight lines. This collection of regions is reproduced by every element of $\mathfrak{F}$, since both $\mathcal{D}$ and $\mathcal{Z}$ are.

We denote by *reflection segment*,

1) a segment of a line of $\mathcal{Z}$ joining two consecutive centres on that line,

2) a half-line of $\mathcal{Z}$ issuing from a centre and containing no other centre,

3) a complete line of $\mathcal{Z}$ containing no centre.

Owing to the convexity of the regions two adjacent regions are only adjacent along one reflection segment.

We denote by $\mathfrak{R}$ the subgroup of $\mathfrak{F}$ generated by all reflections contained in $\mathfrak{F}$. Since every element of $\mathfrak{F}$ transforms a reflection into a reflection, $\mathfrak{R}$ is a normal subgroup of $\mathfrak{F}$. Hence $\mathcal{K}(\mathfrak{R}) = \mathcal{K}(\mathfrak{F})$ (§14.1). The group $\mathfrak{R}$ is called the *reflection subgroup of $\mathfrak{F}$*.

Let $\mathfrak{s}$ be a reflection in $\mathfrak{F}$ and $\mathcal{B}$ any of the regions of decomposition of $\mathcal{D}$ by $\mathcal{Z}$. Let p be a point of $\mathcal{B}$ and $\mathcal{P}$ the perpendicular from p on $\mathcal{L}(\mathfrak{s})$. Let p be so chosen that $\mathcal{P}$ passes through no centre of $\mathcal{Z}$; these centres are enumerable, if any. Then $\mathcal{P}$ intersects on its way from p to $\mathcal{L}(\mathfrak{s})$ a finite number $r \geqq 0$ of reflection lines, because these do not accumulate in $\mathcal{D}$ (§10.4). Let them in the order in which $\mathcal{P}$ meets them be

$$\mathcal{L}(\mathfrak{s}_1),\ \mathcal{L}(\mathfrak{s}_2),\ \ldots,\ \mathcal{L}(\mathfrak{s}_r),\ \mathcal{L}(\mathfrak{s}).$$

The prolongation of $\mathcal{P}$ from $\mathcal{L}(\mathfrak{s})$ to $\mathfrak{s}p$ then intersects

$$\mathfrak{s}\mathcal{L}(\mathfrak{s}_r),\ \mathfrak{s}\mathcal{L}(\mathfrak{s}_{r-1}),\ \ldots,\ \mathfrak{s}\mathcal{L}(\mathfrak{s}_1)$$

in this order and no other reflection line. The product of the reflections in these $2r+1$ lines in the given order then is

$$\mathfrak{s}\mathfrak{s}_1\mathfrak{s}\cdot\mathfrak{s}\mathfrak{s}_2\mathfrak{s}\ldots\mathfrak{s}\mathfrak{s}_{r-1}\mathfrak{s}\cdot\mathfrak{s}\mathfrak{s}_r\mathfrak{s}\cdot\mathfrak{s}\cdot\mathfrak{s}_r\cdot\mathfrak{s}_{r-1}\ldots\mathfrak{s}_2\cdot\mathfrak{s}_1=\mathfrak{s}. \tag{1}$$

($r = 0$ corresponds to replacing $\mathfrak{s}_1, \mathfrak{s}_2, \ldots, \mathfrak{s}_r$ by 1.) Thus the reflection $\mathfrak{s}$ can be replaced by this product (1). The factors of this product have the following effect: $\mathfrak{s}_1$ carries $\mathcal{B}$ into $\mathfrak{s}_1\mathcal{B}$, which is adjacent to $\mathcal{B}$ along a reflection segment on $\mathcal{L}(\mathfrak{s}_1)$, then

$\mathfrak{s}_2$ carries $\mathfrak{s}_1\mathcal{B}$ into $\mathfrak{s}_2\mathfrak{s}_1\mathcal{B}$, which is adjacent to $\mathfrak{s}_1\mathcal{B}$ along a reflection segment on $\mathcal{L}(\mathfrak{s}_2)$, and so on.

We now define a *chain of regions* in the decomposition of $\mathcal{D}$ by $\mathcal{Z}$ as an ordered sequence $\mathcal{B}_1, \mathcal{B}_2, \ldots, \mathcal{B}_s$ of regions such that $\mathcal{B}_n$ and $\mathcal{B}_{n+1}$, $n = 1, 2, \ldots, s-1$, are adjacent along a reflection segment. If $\mathcal{B}_1 = \mathcal{B}_s$, then the chain is said to be *closed*. If the sequence contains no subsequence of the form $\mathcal{B}'\mathcal{B}''\mathcal{B}'$, then the chain is said to be *reduced*.

Equation (1) shows that any reflection, and thus also any product of reflections, hence any element $\mathfrak{r}$ of $\mathfrak{R}$, can be represented by a product of reflections pertaining to a chain of regions connecting any region $\mathcal{B}$ with $\mathfrak{r}\mathcal{B}$. On the other hand, given any two regions $\mathcal{B}$ and $\mathcal{B}'$, if one joins a point of $\mathcal{B}$ with a point of $\mathcal{B}'$ by a straight segment which passes through no centre on $\mathcal{Z}$, the regions containing this segment form a chain of regions connecting $\mathcal{B}$ with $\mathcal{B}'$. Hence any two regions $\mathcal{B}$ and $\mathcal{B}'$ are equivalent with respect to $\mathfrak{R}$, and thus also with respect to $\mathfrak{F}$.

20.2 A fundamental domain for $\mathfrak{R}$. We now want to show that any two chains of regions connecting two given regions $\mathcal{B}$ and $\mathcal{B}'$ lead to the same element of $\mathfrak{R}$. This is equivalent to stating that a closed chain leads to the identical element.

First, consider the case that there are no centres on $\mathcal{Z}$, thus all reflection segments are complete reflection lines. If one joins a point of $\mathcal{B}$ with a point of $\mathcal{B}'$ by a straight segment, this segment determines a reduced chain of regions, and this chain is independent of the points chosen. This is the only reduced chain connecting $\mathcal{B}$ and $\mathcal{B}'$. The process of reducing a chain by replacing a subsequence $\mathcal{B}'\mathcal{B}''\mathcal{B}'$ by $\mathcal{B}'$ obviously does not alter the element concerned, because it consists in omitting $\mathfrak{s}^2$, where $\mathfrak{s}$ is the reflection in the reflection segment between $\mathcal{B}'$ and $\mathcal{B}''$, and $\mathfrak{s}^2 = 1$. Hence there is one, and only one element of $\mathfrak{R}$ carrying $\mathcal{B}$ into $\mathcal{B}'$. Every closed chain is reducible and reduces to one single region; the corresponding element is the identity.

Secondly, assume that $\mathcal{Z}$ contains centres, and consider a closed chain. Let $\mathcal{M}$ denote the open point set of $\mathcal{D}$ consisting of all regions of the chain and of all those open reflection segments which connect any two consecutive regions of the chain; and let $\mathfrak{C}_{\mathcal{M}}$ denote the complement of $\mathcal{M}$ in $\mathcal{D}$. We now distribute the centres on $\mathcal{Z}$ into two classes: A centre c on $\mathcal{Z}$ belongs to class I, if there is at least one half-line on $\mathcal{Z}$ issuing from c and belonging to $\mathfrak{C}_{\mathcal{M}}$, otherwise to class II. The centres of class II, if any, form a bounded point set: Since the chain is finite, $\mathcal{M}$ contains only a finite number of reflection segments, and these end in a finite number of centres, namely two, one or none each. Let $\mathcal{K}$ be a circle in $\mathcal{D}$ containing all these centres in its interior. Let c be a centre outside $\mathcal{K}$. Then there exists a half-line on $\mathcal{Z}$ issuing from c and not intersecting $\mathcal{K}$. This half-line contains no point of $\mathcal{M}$, because for every finite reflection segment or half-line segment of $\mathcal{M}$ the centre or centres in which it ends are inside $\mathcal{K}$; thus c belongs to class I.

Being bounded, class II contains only a finite number of centres. Consider a centre of class II and a half-line on $\mathcal{Z}$ issuing from that point. It contains a last centre of class II, say c. Beyond c it begins with a reflection segment $\mathcal{S}$ contained in $\mathcal{M}$, namely either a finite segment connecting c with a centre of class I or a half-line segment. Let $\mathfrak{s}$ denote the reflection in $\mathcal{S}$ and $\mathcal{B}_1$ and $\mathcal{B}_2$ the regions adjacent to $\mathcal{S}$ (Fig. 20.1). We then insert between $\mathcal{B}_1$ and $\mathcal{B}_2$ the sequence of the other regions adjacent to c

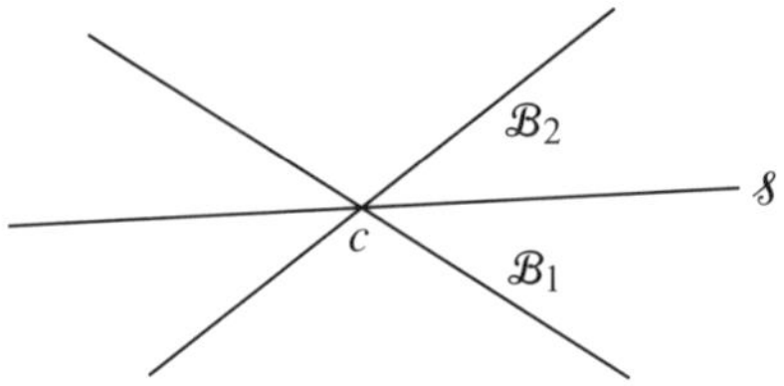

Figure 20.1

in the order in which they are met with from $\mathcal{B}_1$ to $\mathcal{B}_2$. In the corresponding group element this means replacing $\mathfrak{s}$ by the product of the other reflections at lines issuing from c in their cyclical order, and this product is equal to $\mathfrak{s}$, because the product of all reflections at lines issuing from a centre in their cyclical order is the full power of a rotation and thus equal to 1. If this is done wherever the subsequence $\mathcal{B}_1\mathcal{B}_2$ occurs in the chain, and likewise for all subsequences $\mathcal{B}_2\mathcal{B}_1$, then c changes from class II to class I. No centre of class I is by this process converted into a centre of class II, for, even if a segment $\mathcal{S}' \neq \mathcal{S}$ issuing from c did not belong to $\mathcal{M}$ before the change, and now belongs to $\mathcal{M}$ after the change, the half-line issuing from c on which it is situated, contained points of $\mathcal{M}$ before the change, since c belonged to class II.

By repeating this process all points of class II can be changed into class I. Then the situation is exactly as if there were no centres at all on $\mathcal{Z}$. (Imagine for a chain of this kind the plane $\mathcal{D}$ being slit along a half-line of $\mathcal{Z}$ in $\mathfrak{C}_{\mathcal{M}}$ from every centre on $\mathcal{Z}$.) Thus the closed chain is completely reducible, and the corresponding element of $\mathfrak{N}$ is 1.

We denote by b the boundary of $\mathcal{B}$ on $\overline{\mathcal{D}}$, by $b_{\mathcal{D}} = b \cap \mathcal{D}$ the boundary of $\mathcal{B}$ on $\mathcal{D}$, by $\overline{\mathcal{B}} = \mathcal{B} \cup b$ the closure of $\mathcal{B}$ on $\overline{\mathcal{D}}$ by $\tilde{\mathcal{B}} = \mathcal{B} \cup b_{\mathcal{D}}$ the closure of $\mathcal{B}$ on $\mathcal{D}$. In then follows from what has just been shown that $\tilde{\mathcal{B}}$ does not contain two different points equivalent with respect to $\mathfrak{N}$. On the other hand, all regions of the decomposition of $\mathcal{D}$ by $\mathcal{Z}$ are equivalent with respect to $\mathfrak{N}$. Thus $\tilde{\mathcal{B}}$ is a fundamental polygon and even a *fundamental domain for* $\mathfrak{N}$ (§10.3). Also $\tilde{\mathcal{B}}$ is the normal domain (§10.5) with respect to $\mathfrak{N}$ for any point x_0 of $\mathcal{B}$, since, on the one hand, $\tilde{\mathcal{B}}$ is a fundamental polygon for $\mathfrak{N}$ and, on the other hand, every side of $\mathcal{B}$ is the bisecting normal for x_0 and $\mathfrak{s}x_0$, if $\mathfrak{s}$ is the reflection in that side.

We denote by $\mathfrak{B}$ the subgroup of $\mathfrak{F}$ carrying $\mathcal{B}$ into itself. It then follows that:

$\mathcal{D}$ mod $\mathfrak{B}$ *is the same surface as* $\tilde{\mathcal{B}}$ mod $\mathfrak{B}$ *with the understanding that the boundaries of* $\tilde{\mathcal{B}}$ mod $\mathfrak{B}$ *derived from* $b_{\mathcal{D}}$ mod $\mathfrak{B}$ *are considered as reflection edges of* $\mathcal{D}$ mod $\mathfrak{F}$. □

The group $\mathfrak{B}$ contains no reflections, since there are no reflection lines in $\mathcal{B}$.

20.3 Abstract presentation of the reflection group. Since $\tilde{\mathcal{B}}$ is a fundamental polygon for $\mathfrak{R}$, it follows immediately from §10.6–7 that the reflections $\mathfrak{s}_r$, $r = 1, 2, \ldots$, in the sides of $b_{\mathcal{D}}$ generate $\mathfrak{R}$ and that a complete system of defining relations for $\mathfrak{R}$ in these generators can be taken to consist of the relations

$$\mathfrak{s}_r^2 = 1, \quad r = 1, 2, \ldots, \tag{2}$$

together with the relations belonging to centres on $b_{\mathcal{D}}$, if any. If c is such a centre of order ν, and if $\mathfrak{s}_r$ and $\mathfrak{s}_s$ are reflections in the sides of $\mathcal{B}$ meeting in c, then the relation is

$$(\mathfrak{s}_r \mathfrak{s}_s)^\nu = 1. \tag{3}$$

The angle of $\mathcal{B}$ at c is $\frac{\pi}{\nu}$.

20.4 Reflection chains and reflection rings. Let $\mathcal{Y}$ denote the point set of $\overline{\mathcal{D}}$ obtained by adding to $\mathcal{Z}$ those points of $\mathcal{E}$ and intervals of discontinuity on $\mathcal{E}$ in which more than one reflection line terminates, i.e. all limit-centres u of $\mathfrak{F}$ with corresponding group $\mathfrak{A}_u = \mathfrak{A}(\wedge)$ and all intervals of discontinuity cut off by boundary axes $\mathcal{A}$ of $\mathfrak{F}$ with corresponding group $\mathfrak{A}_{\mathcal{A}} = \mathfrak{A}(\|)$. This point set $\mathcal{Y}$ contains no fundamental point, since none of the points of $\mathcal{E}$ added to $\mathcal{Z}$ is fundamental. $\mathcal{Y}$ may be composed of a finite or infinite number of components. In general more than one, even an infinite number of these components may contribute to the boundary of one of the regions into which $\mathcal{D}$ is decomposed by $\mathcal{Z}$.

We now consider one of the regions, $\mathcal{B}$, its boundary b, and the corresponding subgroup $\mathfrak{B}$ of $\mathfrak{F}$. Let b_1 be the part contributed to b by a component $\mathcal{Y}_1$ of $\mathcal{Y}$, thus $b_1 = b \cap \mathcal{Y}_1$. This will be called a *boundary chain* of $\mathcal{B}$. We put $b_1 \cap \mathcal{D} = b_{1\mathcal{D}}$. Any element of $\mathfrak{B}$ carries a boundary chain of $\mathcal{B}$ into a boundary chain of $\mathcal{B}$, because it carries both b and $\mathcal{Y}$ into themselves. Let $\mathfrak{B}_1$ denote the subgroup of $\mathfrak{B}$ whose elements carry the boundary chain b_1 into itself. This *polygonal* chain b_1 may contain points on $\mathcal{E}$ (in certain limit-centres) among its vertices, and arcs of $\mathcal{E}$ (in certain intervals of discontinuity) among its sides. It may be closed and then contains only a finite number of vertices and sides. If it is open, we may traverse it in a definite sense and then speak of its *end*. This end must converge to a point e of $\mathcal{E}$, either by having a finite number of sides, the last one being on a reflection line ending in e, which is then not a limit-centre, or by having an infinite number of sides with e as their only accumulation point. Traversing the chain b_1 in the opposite sense, its end converges in the same way to a point e' of $\mathcal{E}$. The two points e and e' belonging to the two ends of b_1 may even coincide. In that case the number of sides of b_1 must be infinite, for, otherwise that common point of convergence would be the common end-point of two reflection lines and would belong to $\mathcal{Y}$, and b_1 would be closed.

Assume that there is an element $\mathfrak{f} \neq 1$ of $\mathfrak{F}$ carrying a side $\mathcal{S}$ of $b_{1\mathcal{D}}$ into a side $\mathcal{S}'$ of $b_{1\mathcal{D}}$. It then carries $\mathcal{B}$ into a region $\mathfrak{f}\mathcal{B}$ adjacent to $\mathcal{S}'$. By applying the reflection in $\mathcal{S}'$, if necessary, one can obtain that $\mathfrak{f}\mathcal{B} = \mathcal{B}$, hence $\mathfrak{f}b_1 = b_1$ and $\mathfrak{f} \in \mathfrak{B}_1$. We direct $\mathcal{S}$ and carry its direction by $\mathfrak{f}$. If $\mathfrak{f}\mathcal{S}$ and $\mathcal{S}$ were oppositely directed on b_1, then a point of b_1 would be left fixed by $\mathfrak{f}$, while $\mathfrak{f} \neq 1$. Now, $\mathfrak{f}$ cannot be a reflection, because there are no reflections in $\mathfrak{B}$. Also $\mathfrak{f}$ cannot be a rotation or limit-rotation, because a rotation or limit-rotation around a centre or limit-centre on b_1 carries $\mathcal{B}$ into another region. Nor can $\mathfrak{f}$ be a translation or reversed translation, because there is no fundamental point on b_1. Hence $\mathcal{S}$ and $\mathfrak{f}\mathcal{S}$ cannot be oppositely directed on b_1, and it follows that $\mathfrak{f}$ is a motion. If $\mathfrak{f}\mathcal{S} = \mathcal{S}$, then $\mathfrak{f}$ is a translation along $\mathcal{S}$, and $\mathcal{S}$ is then a complete line of reflection, which is also an axis of $\mathfrak{F}$ and contains no centres; thus in this case $\mathcal{S} = b_1$. – We now assume that $\mathfrak{f}\mathcal{S} \neq \mathcal{S}$. Then $\mathfrak{f}$ advances b_1 in itself a certain number of steps. Therefore $\mathfrak{f}$ is a rotation, if b_1 is closed, and a translation or limit-rotation, if b_1 is open. In both cases we call b_1 *periodic*, and there exists a primary rotation or translation or limit-rotation, respectively, determining the shortest period on b_1. This primary motion then generates $\mathfrak{B}_1$. – This also applies to the case $\mathcal{S} = b_1$, $\mathcal{S}$ being an axis of $\mathfrak{F}$. – If no element of $\mathfrak{F}$ satisfies the given condition, we call b_1 *aperiodic*, and we then get $\mathfrak{B}_1 = 1$.

If b_1 is closed, whether periodic or aperiodic, and also if b_1 is open and periodic, b_1 mod $\mathfrak{B}_1$ determines on the surface $\mathcal{D}$ mod $\mathfrak{F}$ a *reflection ring* (§16.4). If b_1 is open and aperiodic, b_1 determines on $\mathcal{D}$ mod $\mathfrak{F}$ a *reflection chain*. The links of a reflection ring or reflection chain are reflection edges on $\mathcal{D}$ mod $\mathfrak{F}$ which are connected either by angular points or by half-masts or half-funnels. In the case of a closed reflection edge or an open infinite reflection edge there is only one link.

Let $\mathfrak{Y}_1$ denote the subgroup of $\mathfrak{F}$ generated by all those reflections whose reflection lines belong to the component $\mathcal{Y}_1$ of $\mathcal{Y}$. This group $\mathfrak{Y}_1$ is a subgroup of the group $\mathfrak{R}$ considered in Section 1, and it coincides with $\mathfrak{R}$, if $\mathcal{Y}$ consists of only one component. $\mathcal{Y}_1$ is reproduced by all elements of $\mathfrak{Y}_1$, and so is the totality of all (convex, open) regions into which $\mathcal{D}$ is decomposed by the reflection lines contained in $\mathcal{Y}_1$. It is then seen as in Section 1 that any two regions $\mathcal{C}$ and $\mathcal{C}'$ of that decomposition are equivalent with respect to $\mathfrak{Y}_1$, as in Section 2 that the closure $\tilde{\mathcal{C}}$ of $\mathcal{C}$ on $\mathcal{D}$ is a fundamental domain for $\mathfrak{Y}_1$, and as in Section 3 that (2) and (3) form a set of defining relations, if the reflections $\mathfrak{s}_r$ in the reflection lines of $\mathcal{Y}_1$ on the boundary of $\mathcal{C}$ are taken as generators of $\mathfrak{Y}_1$.

We now proceed to a survey of the possible special cases and consider the types of surfaces or surface parts arising therefrom. We first consider the cases in which b_1 is closed and then the cases in which b_1 is open. In all cases $\mathfrak{F}$ is obtained by the adjunction of a set of generators of $\mathfrak{B}$ to $\mathfrak{R}$.

20.5 The finite polygonal disc. We assume b_1 to be closed. It is then the complete boundary of $\mathcal{B}$ on $\overline{\mathcal{D}}$, thus $b_1 = b$ and $\mathfrak{B}_1 = \mathfrak{B}$. Moreover we assume b to be aperiodic, thus $\mathfrak{B} = 1$. We thus get $\mathfrak{F} = \mathfrak{R}$. Here $\mathcal{B}$ mod $\mathfrak{B}$ is $\mathcal{B}$ itself. Hence the surface $\mathcal{D}$ mod $\mathfrak{F}$ is a plane region $\mathcal{B}$ determined by a finite number of reflection edges, which are connected by angular points or half-masts or half-funnels, thus forming a reflection ring. We call this surface a *finite polygonal disc*.

Let the reflections in the lines of $b_{\mathfrak{D}}$ be denoted by

$$\mathfrak{s}_1, \mathfrak{s}_2, \ldots, \mathfrak{s}_s \tag{4}$$

in their cyclical order on b. The reflection line of $\mathfrak{s}_r$ is called $\mathcal{L}_r$. Then $\mathfrak{F}$ is generated by the elements (4), and a set of defining relations in these generators is, in virtue of Section 3,

$$\mathfrak{s}_r^2 = 1, \quad r = 1, 2, \ldots, s, \tag{5}$$

together with the relations corresponding to (3)

$$(\mathfrak{s}_{r+1}\mathfrak{s}_r)^{\nu_r} = 1, \quad r = 1, 2, \ldots, s, \tag{6}$$

where $\mathfrak{s}_{s+1}$ stands for $\mathfrak{s}_1$, and where the relation (6) lapses for all r for which $\mathfrak{s}_{r+1}\mathfrak{s}_r$ is a limit-rotation or a translation. It is convenient to denote this case by putting $\nu_r = \infty$. This notation holds throughout this paragraph. Thus the number of effective relations (6) is equal to the number of angular points on the disc, if any.

Here the *condition of quasi-compactness* (($\mathcal{K}^*(\mathfrak{F})$ is compact modulo $\mathfrak{F}$) is satisfied, since (4) is a finite set of generators of $\mathfrak{F}$ (§15). The Euler-characteristic χ of $\mathcal{K}^*(\mathfrak{F})$ mod $\mathfrak{F}$ (§17.5) is 1, since this surface is a simple polygonal area. Hence we get from (9), §17, for the area of $\mathcal{K}(\mathfrak{F})$ mod $\mathfrak{F}$

$$\Phi(\mathfrak{F}) = -2 + \sum_{\nu}\left(1 - \frac{1}{\nu}\right)\vartheta(\times_\nu) + \vartheta(\wedge) + \vartheta(\|). \tag{7}$$

In this simple case this can, of course, easily be computed directly from the formula (1), §17, for the area of a polygon, possibly with vertices on $\mathcal{E}$. If we use the notation $\nu_r = \infty$ for the cases in which $\mathfrak{s}_{r+1}\mathfrak{s}_r$ is a limit-rotation or a translation, and if we interpret $\frac{1}{\nu_r}$ as meaning zero for these values, then (7) may be written in the simpler form

$$\Phi(\mathfrak{F}) = \sum_{r=1}^{s}\left(1 - \frac{1}{\nu_r}\right) - 2. \tag{7'}$$

For convenience we denote by ε the quantity

$$\varepsilon = \sum_{r=1}^{s}\left(1 - \frac{1}{\nu_r}\right) \tag{8}$$

with the above interpretation of the ν_r. This quantity plays a rôle also in the following sections. Then (7′) reads $\Phi(\mathfrak{F}) = \varepsilon - 2$.

By inserting $\chi = 1$, $\vartheta(-) = 1$, $\vartheta(\rightarrow) = \vartheta(\hookrightarrow) = 0$ in (5), §17, one gets $p = 0$. Also the surface is orientable (§17.4), since $\vartheta(\rightleftharpoons) = 0$. For, if $\mathfrak{f}$ is a reversed translation in $\mathfrak{F}$, the end-points of its axis do not belong to $\mathcal{Y}$, because $\mathcal{Y}$ contains no fundamental point; this axis must therefore intersect $\mathcal{Z}$ on leaving a region $\mathcal{B}$ with which is has points in common.

If $s = 2$, then $\mathcal{L}_1$ and $\mathcal{L}_2$ must be divergent in order to produce a closed reflection ring, and the resulting group is then the quasi-abelian group $\mathfrak{A}(\|)$. Also, if $s = 3$ and $\nu_1 = \nu_2 = 2$, the quasi-abelian group $\mathfrak{A}(-\ \|)$ arises. Thus these two cases have to be excluded, if we only consider $\mathfrak{F}$-groups.

The surface $\mathcal{D}$ mod $\mathfrak{F}$ arises by adding to $\mathcal{K}(\mathfrak{F})$ mod $\mathfrak{F}$ half-funnels corresponding to those values of r, if any, for which $\mathcal{L}_r$ and $\mathcal{L}_{r+1}$ are divergent. Their common normal $\mathcal{A}_r^*$ is a boundary axis of $\mathfrak{F}$ belonging to a subgroup $\mathfrak{A}_r(\|)$. For, one of the arcs determined by $\mathcal{A}_r^*$ belongs to $\mathcal{Y}$, being a periodic interval of discontinuity of $\mathfrak{F}$ of the required type. – This remark also applies in the Sections 6–8, 13–17, 19–20.

20.6 The polygonal cone. Next, b_1 is assumed to be closed and periodic, thus again $b_1 = b$ and $\mathfrak{B}_1 = \mathfrak{B}$. Let the shortest period of b correspond to the rotation $\mathfrak{c}$ of order ν and centre c. The number of lines of $b_{\mathcal{D}}$ then is divisible by ν, say $s\nu$. We then have $\mathfrak{B} = \mathfrak{A}_c(\circledcirc_\nu)$. The surface $\mathcal{D}$ mod $\mathfrak{F}$ has a conical point of order ν in its interior. We call it a *polygonal cone*. It contains s reflection edges connected by angular points or half-masts or half-funnels, thus forming one reflection ring b mod $\mathfrak{A}_{\mathfrak{c}}$, thus $\vartheta(-) = 1$. Here we are only concerned with the case $s\nu > 2$, since for $s = 1$, $\nu = 2$, we would get the quasi-abelian group $\mathfrak{A}(\cdot|)$.

As generators of $\mathfrak{F}$ one can take $\mathfrak{c}$ together with the reflections (4) in a period of b in the order corresponding to the sense of the rotation $\mathfrak{c}$, and as a set of defining relations between them

$$\mathfrak{c}^\nu = 1 \tag{9}$$

together with (5) and (6), where $\mathfrak{s}_{s+1}$ now means $\mathfrak{c}\mathfrak{s}_1\mathfrak{c}^{-1}$ and the ν_r are interpreted as in the preceding section.

Again the condition of quasi-compactness is satisfied, since $\mathfrak{F}$ can be generated by a finite number of elements. Also $\chi = 1$. From (9), §17, the area of $\mathcal{K}(\mathfrak{F})$ mod $\mathfrak{F}$ is found to be

$$\begin{aligned}\Phi(\mathfrak{F}) &= -2 + \sum_\tau \Big(1 - \frac{1}{\tau}\Big)\vartheta(\times_\tau) + \vartheta(\wedge) + \vartheta(\|) + 2\Big(1 - \frac{1}{\nu}\Big)\\ &= \sum_{r=1}^{s}\Big(1 - \frac{1}{\nu_r}\Big) - \frac{2}{\nu} = \varepsilon - \frac{2}{\nu},\end{aligned} \tag{10}$$

if again we use the form corresponding to (7′) and (8). For the same reasons as in the preceding section one gets $\vartheta(\rightleftharpoons) = 0$, and from (5), §17, one gets $p = 0$.

20.7 The infinite polygonal disc. Next, b_1 is assumed to be open and aperiodic, both ends of b_1 converging to the same point e of $\mathcal{E}$. This point e does not belong to $\mathcal{Y}$, and $b_{1\mathcal{D}}$ consists of infinitely many reflection segments. Here $b = b_1 \cup e$ and $\mathfrak{B}_1 = \mathfrak{B} = 1$. Hence $\mathcal{D}$ mod $\mathfrak{F}$ is a plane region $\mathcal{B}$ determined by an infinite number of reflection edges connected by angular points, half-masts, or half-funnels, thus forming a reflection chain. The links of the chain converge to a point at infinity. We call this surface an *infinite polygonal disc*. Since no two points of $b_{\mathcal{D}}$ are equivalent with respect to $\mathfrak{F}$, the infinitely many lines of $b_{\mathcal{D}}$ are represented on $\mathcal{K}^*(\mathfrak{F})$ mod $\mathfrak{F}$ by infinitely many arcs of reflection edges. Since these do not accumulate on $\mathcal{K}^*(\mathfrak{F})$ mod $\mathfrak{F}$, this surface must be unbounded. Here then $\mathcal{K}(\mathfrak{F})$ is not quasi-compact modulo $\mathfrak{F}$, hence $\Phi(\mathfrak{F}) = \infty$. Since b contains only one point outside $\mathcal{Y}$, namely e, we get $\vartheta(\rightleftharpoons) = 0$ by the same consideration as in the preceding section.

As generators of $\mathfrak{R} = \mathfrak{F}$ we take the infinitely many reflections $\mathfrak{s}_r$ in the lines of $b_{\mathcal{D}}$ in their order on b, where now $1 \leqq r < \infty$ or $-\infty < r < \infty$ according as one or none of the ends of b_1 is finite, i.e. ends in e with a line of $\mathcal{Z}$. As a set of defining relations between them we get (5) and (6) for infinitely many values of r (and with the usual interpretation of the ν_r).

20.8 The polygonal mast. Next, b_1 is assumed to be open and periodic, both ends of b_1 converging to the same point of $\mathcal{E}$, which here is called u. Thus $b = b_1 \cup u$. Here none of the ends of b_1 can have a finite number of sides owing to the periodicity of b_1, since u does not belong to $\mathcal{Y}$. Again $\mathfrak{B}_1 = \mathfrak{B}$. The motion of $\mathfrak{B}$ which corresponds to the shortest period of b_1 must leave u fixed and is thus a limit-rotation $\mathfrak{u}$ with u as its limit-centre. It generates $\mathfrak{B} = \mathfrak{A}_u(\hookrightarrow)$. The surface $\mathcal{D}$ mod $\mathfrak{F}$ is called a *polygonal mast*. It has one reflection ring b_1 mod $\mathfrak{A}_u$, thus $\vartheta(-) = 1$. Also $\vartheta(\rightleftharpoons) = 0$.

As generators of $\mathfrak{F}$ we take $\mathfrak{u}$ together with the reflections (4) in a period of b_1 in the order of the limit-rotation $\mathfrak{u}$, and as a set of defining relations we get (5) and (6), where $\mathfrak{s}_{s+1}$ now means $\mathfrak{u}\mathfrak{s}_1\mathfrak{u}^{-1}$.

$\mathcal{K}^*(\mathfrak{F})$ mod $\mathfrak{F}$ has the characteristic $\chi = 0$. We thus get $p = 0$ from (5), §17, and from (9), §17,

$$\Phi(\mathfrak{F}) = \sum_{\tau}\left(1 - \frac{1}{\tau}\right)\vartheta(\times_\tau) + \vartheta(\wedge) + \vartheta(\|) = \sum_{r=1}^{s}\left(1 - \frac{1}{\nu_r}\right) = \varepsilon.$$

This corresponds to letting ν increase towards infinity in (10), thus considering a mast as a limit case for the cone.

20.9 Open boundary chain with different end-points. Finally, we consider an open boundary chain b_1 of $\mathcal{B}$ with different points u and v on $\mathcal{E}$ as points of convergence for the ends of b_1. Then u and v do not belong to $\mathcal{Y}$. The straight line uv is called $\mathcal{A}$, and $\mathfrak{A}_{\mathcal{A}}$ denotes the subgroup of $\mathfrak{F}$ carrying $\mathcal{A}$ into itself. The subgroup $\mathfrak{B}_1$ of $\mathfrak{B}$ which carries b_1 into itself must also carry the pair u, v and thus $\mathcal{A}$ into itself and is thus a subgroup of $\mathfrak{A}_{\mathcal{A}}$.

We denote by $\mathfrak{G}$ the group generated by $\mathfrak{B}_1$ and the reflections in the sides of $b_{1\mathcal{D}}$, thus by $\mathfrak{B}_1$ and $\mathfrak{Y}_1$. Then $\mathfrak{G}$ is a subgroup of $\mathfrak{F}$ and may, in special cases, coincide with $\mathfrak{F}$. We therefore take the surface $\mathcal{D}$ mod $\mathfrak{G}$ into consideration in order to cover all types of surfaces $\mathcal{D}$ mod $\mathfrak{F}$. In the general case the group $\mathfrak{G}$ will play a rôle for the complete determination of $\mathfrak{F}$.

Two different cases present themselves according as b_1 coincides with $\mathcal{A}$ or not. The first case will be considered in the next section. In the second case $\mathcal{A}$ is either situated in the interior of the (convex) region $\mathcal{B}$, or is itself a side of $\mathcal{B}$. We denote by $\mathcal{B}_1$ the open part of $\mathcal{B}$ whose complete boundary consists of b_1, $\mathcal{A}$, u and v, thus the interior of the convex hull of b_1. It may, or may not, coincide with $\mathcal{B}$. Any element of $\mathfrak{F}$ carrying $\mathcal{B}_1$ into itself must carry b_1 into itself and thus belongs to $\mathfrak{B}_1$, and every element of $\mathfrak{B}_1$ carries b_1 and thus $\mathcal{B}_1$ into itself. Inside $\mathcal{B}_1$ the groups $\mathfrak{F}$ and $\mathfrak{B}_1$ have equal effect.

Let $\mathfrak{f}$ be an arbitrary element of $\mathfrak{F}$ not in $\mathfrak{B}_1$. The continuous curve $\mathfrak{f}b_1$ joins $\mathfrak{f}u$ and $\mathfrak{f}v$ on $\overline{\mathcal{D}}$. The parts of $\mathfrak{f}b_1$ which are in $\mathcal{D}$ consist of reflection segments and thus cannot enter into the region $\mathcal{B}_1$, which is part of $\mathcal{B}$. The parts of $\mathfrak{f}b_1$ which are on $\mathcal{E}$ belong to $\mathcal{Y}$ and thus cannot contain u or v. It follows that the pair $\mathfrak{f}u$, $\mathfrak{f}v$ cannot separate the pair u, v on $\mathcal{E}$. Thus $\mathcal{A}$ is simple modulo $\mathfrak{F}$. Since $\mathcal{A}$ belongs to $\tilde{\mathcal{B}}_1$, it is not cut by any reflection line. The possible forms for $\mathfrak{A}_{\mathcal{A}}$ are then

$$1, \quad \mathfrak{A}(\rightarrow), \quad \mathfrak{A}(-), \quad \mathfrak{A}(-\rightarrow), \quad \mathfrak{A}(\circlearrowleft_2), \quad \mathfrak{A}(\cdot\cdot), \quad \mathfrak{A}(\rightleftharpoons). \tag{11}$$

20.10 The case of the full reflection line. Let b_1 coincide with $\mathcal{A}$. Being a complete boundary chain of $\mathcal{B}$, the reflection line $\mathcal{A}$ cannot be cut by any other reflection line, thus in particular by any line equivalent with $\mathcal{A}$ with respect to $\mathfrak{F}$. Hence $\mathcal{A}$ is simple modulo $\mathfrak{F}$. Since $\mathfrak{B}_1$ carries $\mathcal{B}$ and $\mathcal{A}$ into themselves, the only two possibilities for $\mathfrak{B}_1$ are $\mathfrak{B}_1 = 1$ and $\mathfrak{B}_1 = \mathfrak{A}_{\mathcal{A}}(\rightarrow)$. Corresponding to these two cases, we get $\mathfrak{G} = \mathfrak{A}_{\mathcal{A}} = \mathfrak{A}(-) = \mathfrak{Y}_1$ and $\mathfrak{G} = \mathfrak{A}_{\mathcal{A}} = \mathfrak{A}(-\rightarrow)$. These groups being quasi-abelian, $\mathfrak{G}$ does not coincide with $\mathfrak{F}$. The surface $\mathcal{D}$ mod $\mathfrak{G}$ is in the two cases a half-plane bordered by an open reflection edge and a funnel bordered by a closed reflection edge. These edges are special cases of reflection chains and reflections rings, respectively, for $\mathcal{D}$ mod $\mathfrak{F}$. It should be noticed that, in the first case, the open reflection edge does not terminate in a half-mast or half-funnel of $\mathcal{D}$ mod $\mathfrak{F}$, because u and v do not belong to $\mathcal{Y}$. In this first case, $\mathfrak{F}$ does not satisfy the condition of quasi-compactness because it contains an open reflection chain.

20.11 The incomplete reflection strip. In the following Sections 11–17, the boundary chain b_1 of $\mathcal{B}$ does not coincide with $\mathcal{A}$. The type of the surfaces $\mathcal{D}$ mod $\mathfrak{G}$ depends on the possible forms of $\mathfrak{A}_{\mathcal{A}}$. We consider the seven cases enumerated in (11) in the same order.

$\mathfrak{A}_{\mathcal{A}} = 1 = \mathfrak{B}_1$. The region $\mathcal{B}_1$ does not contain two points equivalent with respect to $\mathfrak{F}$. It is thus part of the surface $\mathcal{D}$ mod $\mathfrak{F}$. The group $\mathfrak{G}$ coincides with $\mathfrak{Y}_1$ and is generated by the reflections $\mathfrak{s}_r$ in the sides of $b_{1\mathcal{D}}$, which again we number in their order on b_1. A set of defining relations for $\mathfrak{G}$ for this choice of generators is given by (5) and (6), if the following is borne in mind. If the number of sides is finite, say $s(\geqq 2)$, then (5) holds for $r = 1, 2, \dots, s$, and (6) for $r = 1, 2, \dots, s-1$. If the number is infinite, then we either have $1 \leqq r < \infty$ or $-\infty < r < \infty$, and (5) and (6) hold for all values of r in the same way as in Section 7. If $\mathfrak{G}$ is an $\mathfrak{F}$-group, then in all cases the union $\tilde{\mathcal{G}} = \mathfrak{G}\tilde{\mathcal{B}}_1$ contains $\mathcal{K}(\mathfrak{G})$. In $\tilde{\mathcal{G}}$ the group $\mathfrak{G}$ and $\mathfrak{F}$ have equal effect.

The part of the surface $\mathcal{D}$ mod $\mathfrak{F}$ which consists of $\tilde{\mathcal{B}}_1$ will be called an *incomplete reflection strip*. It has an open simple geodesic as its boundary and is bordered by a finite or infinite reflection chain whose ends converge to the same point at infinity as the ends of the bounding geodesic. Since $\mathcal{D}$ mod $\mathfrak{F}$ contains a reflection chain which is not a reflection ring, $\mathfrak{F}$ cannot satisfy the condition of quasi-compactness.

If b_1 *is the only boundary chain of* $\mathcal{B}$, then even $\mathfrak{B} = 1$, and $\mathfrak{G} = \mathfrak{F}$. If $\mathcal{D}_1$ denotes the half-plane cut off by $\mathcal{A}$ and not containing $\mathcal{B}_1$, then $\mathcal{D}_1$ is part of $\mathcal{D}$ mod $\mathfrak{F}$, and $\mathcal{D}$ mod $\mathfrak{F} = \mathcal{B}_1 \cup b_{1\mathcal{D}} \cup \mathcal{A} \cup \mathcal{D}_1 = \mathcal{B} \cup b_{1\mathcal{D}} = \tilde{\mathcal{B}}$. In this case $b_{1\mathcal{D}}$ must contain an infinite number of sides. For, if this number were finite, say s, then $\mathfrak{s}_1$ and $\mathfrak{s}_s$ would be reflections in lines ending in u and v, respectively. Moreover $s > 2$, since two reflections do not generate an $\mathfrak{F}$-group. Thus $\mathfrak{f} = \mathfrak{s}_s\mathfrak{s}_1$ would be a translation in $\mathfrak{F}$ and generate a subgroup of the type $\mathfrak{A}(\|)$. Now the arc uv of $\mathcal{E}$ bounding $\mathcal{D}_1$ does not contain limit points of $\mathfrak{F}$, since $\mathcal{D}_1$ does not contain two equivalent points. The same then holds for $\mathfrak{f}^{\nu}(uv)$ for all ν. Hence u and v would be inside an interval of discontinuity belonging to the type $\mathfrak{A}(\|)$ in contradiction to the fact that they are outside $\mathcal{Y}$. Hence, at least one of the points u and v is an accumulation point of reflection lines of $\mathfrak{F}$ and thus belongs to the limit set $\overline{\mathcal{G}}(\mathfrak{F})$. If both do, then $\mathcal{A}$ is a limit side of $\mathcal{K}(\mathfrak{F})$, intersected by no reflection line. If v, say, is the end-point of the reflection line belonging to $\mathfrak{s}_1$, then $u' = \mathfrak{s}_1 u$ belongs to $\overline{\mathcal{G}}(\mathfrak{F})$, and uu' is a limit side of $\mathcal{K}(\mathfrak{F})$ intersected by one reflection line. It is inferred in both cases that $\mathfrak{F}$ does not satisfy the condition of quasi-compactness. Thus $\Phi(\mathfrak{F})$ is infinite. Moreover, $\vartheta(\rightleftharpoons) = 0$ for similar reasons as in Section 6.

20.12 The crown. $\mathfrak{A}_{\mathcal{A}} = \mathfrak{A}(\rightarrow) = \mathfrak{B}_1$. Let $\mathfrak{f}$ be a primary translation along $\mathcal{A}$ and $\mathfrak{s}_r$, $r = 1, 2, \ldots, s$, $s \geqq 1$, the reflections in the sides of $b_{1\mathcal{D}}$ within a period corresponding to $\mathfrak{f}$. Then $\tilde{\mathcal{B}}_1$ mod $\mathfrak{A}_{\mathcal{A}}$ will be called a *crown*. It has a simple closed geodesic as its boundary and is bordered by one reflection ring, namely b_1 mod $\mathfrak{A}_{\mathcal{A}}$.

$\mathfrak{G}$ is generated by $\mathfrak{f}$ together with (4), and a set of defining relations for $\mathfrak{G}$ in these generators is (5) and (6), where $\mathfrak{s}_{s+1}$ now means $\mathfrak{f}\mathfrak{s}_1\mathfrak{f}^{-1}$. The surface $\mathcal{D}$ mod $\mathfrak{G}$ arises by adding to the crown a funnel along the simple closed geodesic. This surface corresponds to the polygonal cone (Section 6) or the polygonal mast (Section 8), when replacing the conical point or the mast by a funnel.

Since $\mathcal{A}$ is an axis of $\mathfrak{G}$, also $\mathfrak{s}_1\mathcal{A}$ is an axis of $\mathfrak{G}$, and it follows that $\mathfrak{G}$ is an $\mathfrak{F}$-group. Moreover, $\mathcal{A}$ is a boundary axis of $\mathfrak{G}$. The set $\tilde{\mathcal{G}} = \mathfrak{G}\tilde{\mathcal{B}}_1$ is $\mathcal{K}(\mathfrak{G})$, and $\mathcal{K}(\mathfrak{G})$ is quasi-compact modulo $\mathfrak{G}$, since $\mathfrak{G}$ is generated by a finite number of its elements. Since $\mathcal{K}^*(\mathfrak{G})$ mod $\mathfrak{G}$ is ring-shaped, one gets $\chi = 0$. Since $\vartheta(-) = 1$, $\vartheta(\rightarrow) = 1$, and $\vartheta(\hookrightarrow) = 0$, equation (5), §17, yields $p = 0$. Also (9), §17, yields for the area of $\mathcal{K}(\mathfrak{G})$ mod $\mathfrak{G}$ with the notation (8) the same value $\Phi(\mathfrak{G}) = \varepsilon$ as for the polygonal mast. This then we call the *area of the crown*. Also $\vartheta(\rightleftharpoons) = 0$ for $\mathfrak{G}$. For, on the one hand, $\tilde{\mathcal{B}}$ contains a fundamental domain for $\mathfrak{G}$; on the other hand, the only fundamental points on the boundary of $\mathcal{B}$ are u and v, but $\mathfrak{A}_{\mathcal{A}}$ contains no reversed translation.

Thus the axis of a reversed translation in $\mathfrak{G}$, if any, must have points in common with $\mathcal{Y}_1$, and hence the conditions of non-orientability are not satisfied (§17.5).

If b_1 *is the only boundary chain of* $\mathcal{B}$, then $\mathfrak{B} = \mathfrak{A}_{\mathcal{A}}$ and $\mathfrak{G} = \mathfrak{F}$. Thus, in this case, $\mathcal{D}$ mod $\mathfrak{F}$ is composed of a crown and a funnel, and an abstract presentation of $\mathfrak{F}$ is given above.

20.13 The complete reflection strip. $\mathfrak{A}_{\mathcal{A}} = \mathfrak{A}(-)$, thus $\mathfrak{B}_1 = 1$. Since $\mathcal{A}$ is a reflection line of $\mathfrak{F}$, we get $\mathcal{B} = \mathcal{B}_1$ and $\mathfrak{B} = \mathfrak{B}_1 = 1$. The region $\mathcal{B}$ has two boundary chains, one being $\mathcal{A}$ itself, and $\tilde{\mathcal{B}}$ is a fundamental domain for $\mathfrak{F}$. No reflection line other than $\mathcal{A}$ can terminate in u or v, because these points do not belong to $\mathcal{Y}$. Hence both ends of b_1 contain infinitely many sides, and u and v are limit points of $\mathfrak{F}$. The fact that $\mathcal{K}(\mathfrak{F})$ mod $\mathfrak{F}$ contains reflection chains which are not reflection rings shows that $\mathcal{K}^*(\mathfrak{F})$ mod $\mathfrak{F}$ cannot be compact, hence that $\mathfrak{F}$ does not satisfy the condition of quasi-compactness. Hence $\Phi(\mathfrak{F}) = \infty$. Also $\vartheta(\rightleftharpoons) = 0$, since $\mathfrak{A}_{\mathcal{A}}$ does not contain reversed translations.

The surface $\mathcal{D}$ mod $\mathfrak{F}$ is $\tilde{\mathcal{B}}$ itself. It is bordered by two reflection chains, one of which is an open reflection edge. We call this surface a *complete reflection strip.* It has two points at infinity besides the half-masts and half-funnels which it may contain.

If $\mathfrak{s}$ denotes the reflection in $\mathcal{A}$, one gets for an abstract presentation of $\mathfrak{F}$ as generators $\mathfrak{s}$ and the $\mathfrak{s}_r$, $-\infty < r < \infty$, with defining relations $\mathfrak{s}^2 = 1$ together with (5) and (6) for all values of r.

20.14 The reflection crown. $\mathfrak{A}_{\mathcal{A}} = \mathfrak{A}(- \rightarrow)$, thus $\mathfrak{B}_1 = \mathfrak{A}(\rightarrow)$. Owing to the fact that the reflection $\mathfrak{s}$ in $\mathcal{A}$ is contained in $\mathfrak{F}$, we get, as in the preceding section, $\mathcal{B} = \mathcal{B}_1$, $\mathfrak{B} = \mathfrak{B}_1 = \mathfrak{A}(\rightarrow)$, $\vartheta(\rightleftharpoons) = 0$, and u and v are limit points of $\mathfrak{F}$ but not limit-centres. Further $\chi = 0$, $\vartheta(-) = 2$, $p = 0$; and $\Phi(\mathfrak{F}) = \varepsilon$ is the same as for a crown. The surface $\mathcal{D}$ mod $\mathfrak{F}$ is called a *reflection crown*, since it is obtained from a crown by turning the simple closed geodesic bounding the crown into a reflection edge. An abstract presentation of $\mathfrak{F}$ is obtained by adding to the presentation of $\mathfrak{G}$ in Section 12 the generator $\mathfrak{s}$ and the relations $\mathfrak{s}^2 = 1$ and $\mathfrak{s}\mathfrak{f} = \mathfrak{f}\mathfrak{s}$.

20.15 The conical reflection strip. $\mathfrak{A}_{\mathcal{A}} = \mathfrak{A}(\circlearrowleft_2)$, thus $\mathfrak{B}_1 = 1$. Let $\mathfrak{c}$ denote the half-turn around the centre c on $\mathcal{A}$. Then $\mathcal{B} = \mathcal{B}_1 \cup \mathfrak{c}\mathcal{B}_1 \cup \mathcal{A}$ and $\mathfrak{B} = \mathfrak{A}_{\mathcal{A}}$. Since u and v are the only points outside $\mathcal{Y}$ on the boundary of $\mathcal{B}$, and since $\mathfrak{A}_{\mathcal{A}}$ contains no reversed translation, one gets $\vartheta(\rightleftharpoons) = 0$. Moreover $v = \mathfrak{c}u$. If both u and v were end-points of reflection lines of b_1, then there would be a reflection line of b_1 and also one of $\mathfrak{c}b_1$ ending in u in contradiction to the fact that u does not belong to $\mathcal{Y}$. Hence, at least one end of b_1 has infinitely many sides, and since b_1 is aperiodic, it is inferred that $\mathfrak{F}$ does not satisfy the condition of quasi-compactness.

$\mathcal{D}$ mod $\mathfrak{F}$ is derived from an incomplete reflection strip by bending the open, simple, bounding geodesic of the strip together to form a conical point of order 2. It is called a *conical reflection strip.* The surface has one point at infinity besides possible half-masts and half-funnels.

An abstract presentation of $\mathfrak{F}$ is obtained by adding to the presentation of $\mathfrak{G}$ in Section 11 with infinitely many values of r one more generator $\mathfrak{c}$ and the relation $\mathfrak{c}^2 = 1$.

20.16 The conical crown. $\mathfrak{A}_{\mathcal{A}} = \mathfrak{A}(\cdot\cdot)$. Let $\mathfrak{c}_1$ and $\mathfrak{c}_2$ denote half-turns around neighbouring centres c_1 and c_2 on $\mathcal{A}$. Then $\mathfrak{c}_1\mathfrak{c}_2$ is a primary translation along $\mathcal{A}$ and generates $\mathfrak{B}_1 = \mathfrak{A}(\rightarrow)$. Again, $\mathcal{B} = \mathcal{B}_1 \cup \mathfrak{c}_1\mathcal{B}_1 \cup \mathcal{A}$, $\mathfrak{B} = \mathfrak{A}_{\mathcal{A}}$, $\vartheta(\rightleftharpoons) = 0$.

$\mathcal{D}$ mod $\mathfrak{F}$ is derived from a crown (Section 12) by squeezing the simple, closed, geodetic boundary of the crown into a doubly covered geodetic arc connecting two conical points of order 2. It is called a *conical crown.*

An abstract presentation of $\mathfrak{F}$ is obtained by adding to the presentation of $\mathfrak{G}$ in Section 12 two generators $\mathfrak{c}_1$ and $\mathfrak{c}_2$ and relations ${\mathfrak{c}_1}^2 = 1$, ${\mathfrak{c}_2}^2 = 1$, $\mathfrak{c}_1\mathfrak{c}_2 = \mathfrak{f}$. The generator $\mathfrak{f}$ may then be eliminated.

Here one gets $\chi = 1$, $\vartheta(-) = 1$, $\vartheta(\rightarrow) = \vartheta(\hookrightarrow) = 0$, thus from (5), §17, $p = 0$ and from (9), §17, the same area $\Phi(\mathfrak{F}) = \varepsilon$ as for the crown.

20.17 The cross cap crown. $\mathfrak{A}_{\mathcal{A}} = \mathfrak{A}(\rightleftharpoons)$. Let $\mathfrak{h}$ be a reversed translation generating $\mathfrak{A}_{\mathcal{A}}$. Then $\mathfrak{h}^2 = \mathfrak{f}$ is a primary translation along $\mathcal{A}$ and thus generates $\mathfrak{B}_1 = \mathfrak{A}(\rightarrow)$. Again one gets $\mathcal{B} = \mathcal{B}_1 \cup \mathfrak{h}\mathcal{B}_1 \cup \mathcal{A}$ and $\mathfrak{B} = \mathfrak{A}_{\mathcal{A}}$. In this case $\vartheta(\rightleftharpoons) = 1$, for, the axis $\mathcal{A}$ of $\mathfrak{h}$ is not a reflection line and it intersects no reflection line.

The surface $\mathcal{D}$ mod $\mathfrak{F}$ arises from a crown, if a strip along the simple closed geodesic bounding the crown is twisted into a Möbius band by making the two halves of that geodesic coincide with direction preserved. It will be called a *cross cap crown.*

An abstract presentation of $\mathfrak{F}$ is obtained by adding to the presentation of $\mathfrak{G}$ in Section 12 a generator $\mathfrak{h}$ and the relation $\mathfrak{h}^2 = \mathfrak{f}$, whereupon $\mathfrak{f}$ can be eliminated.

In this case one calculates $\chi = 0$, $\vartheta(\rightarrow) = \vartheta(\hookrightarrow) = 0$, $\vartheta(-) = 1$, thus $p = 1$ from (5), §17. Then (9), §17, shows that the area $\Phi(\mathfrak{F}) = \varepsilon$ is the same as for the corresponding crown.

20.18 The general case. Reviewing the Sections 1–17 of the present paragraph, we get the following picture concerning $\mathfrak{F}$-groups which contain reflections. Sections 1–4 are preparatory; they introduce the point sets $\mathcal{Z}$ and $\mathcal{Y}$, define the reflection subgroup $\mathfrak{R}$ of $\mathfrak{F}$ and the boundary chains of a region $\mathcal{B}$ of the decomposition of $\mathcal{D}$ by $\mathcal{Z}$, these boundary chains being the intersections of the boundary b of $\mathcal{B}$ with the different components of $\mathcal{Y}$.

In the Sections 5–17, the starting point then is a particular boundary chain b_1 of $\mathcal{B}$, the intersection of b with the particular component $\mathcal{Y}_1$ of $\mathcal{Y}$. The subgroup of $\mathfrak{R}$ generated by the reflections in the lines of $\mathcal{Y}_1$ is called $\mathfrak{Y}_1$ and is also generated by the reflections in the sides of $b_{1\mathcal{D}} = b_1 \cap \mathcal{D}$ (compare the end of Section 4). The subgroup of $\mathfrak{F}$ carrying $\mathcal{B}$ into itself is called $\mathfrak{B}$, and the subgroup of $\mathfrak{B}$ carrying b_1 into itself is called $\mathfrak{B}_1$. Finally, $\mathfrak{G}$ is the subgroup of $\mathfrak{F}$ generated by $\mathfrak{Y}_1$ and $\mathfrak{B}_1$.

In Section 5 and 6, the boundary chain b_1 is closed, in Sections 7 and 8 it is open, both ends converging to the same point of $\mathcal{E}$. In all four cases, $\mathfrak{F}$ has been completely determined by an abstract presentation, and the surface $\mathcal{D}$ mod $\mathfrak{F}$ has been described.

In the remaining Section 9–17, the boundary chain b_1 is open, its ends converging to two different points u and v on $\mathcal{E}$ outside $\mathcal{Y}$. Section 9 contains preparatory considerations for this case. Then the Sections 13–17 again lead to a complete determination of $\mathfrak{F}$ and of the surface $\mathcal{D}$ mod $\mathfrak{F}$.

Hence it is only in the cases of the Sections 10, 11 and 12 that a full determination of $\mathfrak{F}$ needs further considerations, provided that $\mathfrak{F}$ does not coincide with the group $\mathfrak{G}$ pertaining to these cases. To put it precisely: In the remaining sections of this paragraph the following statements hold:

I. $\mathcal{B}$ admits more than one boundary chain.

II. The subgroup of $\mathfrak{F}$ leaving the line uv invariant, is neither $\mathfrak{A}(\circlearrowleft_2)$, nor $\mathfrak{A}(\cdot\cdot)$, nor $\mathfrak{A}(\rightleftharpoons)$. Moreover, if uv is not a reflection line, then this group is neither $\mathfrak{A}(-)$ nor $\mathfrak{A}(- \rightarrow)$.

The statement I makes sure that $\mathfrak{F}$ is not exhausted by the group $\mathfrak{G}$ generated by $\mathfrak{Y}_1$ and $\mathfrak{B}_1$. It implies that the cases of the finite or infinite polygonal disc, the polygonal cone, and the polygonal mast (Sections 5, 6, 7, 8) are eliminated, since u and v do not coincide. Then II states that $\mathcal{D}$ mod $\mathfrak{F}$ is not a conical reflection strip, nor a conical crown, nor a cross cap crown (Sections 15, 16, 17) and that it is not a complete reflection strip, nor a reflection crown (Sections 13, 14). By comparison with the enumeration (11) it thus follows that the group leaving uv invariant is either $\mathfrak{A}(-)$ or $\mathfrak{A}(- \rightarrow)$, if uv is a reflection line of $\mathfrak{F}$, and 1 or $\mathfrak{A}(\rightarrow)$, if it is not.

We now take the whole of $\mathcal{B}$ with its different boundary chains into consideration. In $\mathcal{B}$ the groups $\mathfrak{B}$ and $\mathfrak{F}$ have equal effect. The boundary chains of $\mathcal{B}$ fall into complete equivalence classes with respect to $\mathfrak{B}$. If the number of these classes is 1, then $\mathfrak{B}$ must contain elements outside its subgroup $\mathfrak{B}_1$, since otherwise b_1 would be the only boundary chain of $\mathcal{B}$.

Let $\mathcal{W}_r$, $r = 1, 2, \ldots$, denote a system of representatives for those equivalence classes with respect to $\mathfrak{B}$ of boundary chains of $\mathcal{B}$ which consist of one full reflection line of $\mathfrak{F}$, thus of only one reflection segment. The corresponding subgroup $\mathfrak{W}_r$ of $\mathfrak{F}$ is either $\mathfrak{A}(-)$ or $\mathfrak{A}(- \rightarrow)$ (statement II); the subgroup $\mathfrak{W}_r^*$ of $\mathfrak{W}_r$ not interchanging the half-planes determined by $\mathcal{W}_r$ is then correspondingly 1 or $\mathfrak{A}(\rightarrow)$. Each of the end-points of $\mathcal{W}_r$ is either situated in an aperiodic interval of discontinuity, or is a limit point of $\mathfrak{F}$ which is not a limit-centre, because it does not belong to $\mathcal{Y}$. The $\mathcal{W}_r$ are obviously non-dividing modulo $\mathfrak{F}$. They determine a simple, open or closed, full reflection edge on $\mathcal{D}$ mod $\mathfrak{F}$.

Let b_m, $m = 1, 2, \ldots$, denote a system of representatives for those equivalence classes with respect to $\mathfrak{B}$ of boundary chains of $\mathcal{B}$ which contain more than one reflection segment. The ends of b_m converge to different points on $\mathcal{E}$. We denote by u_m and v_m the end-points of b_m on $\mathcal{E}$, by $\mathcal{V}_m$ the straight line $u_m v_m$ and by $\mathcal{B}_m$ the interior of the convex hull of b_m; thus the boundary of this convex hull consists of b_m, $\mathcal{V}_m$, u_m and v_m. The subgroup $\mathfrak{B}_m$ of $\mathfrak{F}$ leaving $\mathcal{V}_m$ invariant is either 1 or $\mathfrak{A}(\rightarrow)$

(statement II); for, $\mathcal{V}_m$ is not a reflection line of $\mathfrak{F}$, because the cases of the complete reflection strip and the reflection crown (Sections 13, 14) have been eliminated. $\mathfrak{B}_m$ is also the subgroup of $\mathfrak{B}$ the elements of which carry b_m and $\mathcal{B}_m$ into themselves; in $\mathcal{B}_m$ the groups $\mathfrak{B}_m$, $\mathfrak{B}$, and $\mathfrak{F}$ have equal effect.

If b_m has a side terminating in u_m, then u_m is either a point of an aperiodic interval of discontinuity of $\mathfrak{F}$ or a limit point of $\mathfrak{F}$ (not a limit-centre). If an end of b_m with infinitely many sides converges to u_m, then u_m is a limit point of $\mathfrak{F}$.

In order to base the full determination of $\mathfrak{F}$ on the results of §19.7–19.10, we consider the collection

$$\mathcal{U} = \bigcup_{r,m} (\mathfrak{F}\,\mathcal{W}_r, \mathfrak{F}\,\mathcal{V}_m) \tag{12}$$

and prove that it satisfies the six conditions imposed on the collection $\mathcal{U}$ of §19.7–9.

1) Evidently, $\mathcal{U}$ consists of complete equivalence classes of straight lines.

2) Every line of $\mathcal{U}$ is simple modulo $\mathfrak{F}$: For lines of the type $\mathcal{V}$ this was shown in Section 9, for lines of the type $\mathcal{W}$ in Section 10. (In both cases the line was called $\mathcal{A}$.)

3) Any two non-equivalent lines of $\mathcal{U}$ are disjoint modulo $\mathfrak{F}$: If both are of the type $\mathcal{W}$, then this follows from the fact that a line $\mathcal{W}$ is not cut by any other reflection line. If one is of the type $\mathcal{W}$, the other of the type $\mathcal{V}$, then the former bounds a region $\mathcal{B}$, while the second is situated in its interior. Let $\mathcal{V}'$ and $\mathcal{V}''$ by any two lines of the type $\mathcal{V}$; let u', v' and u'', v'' denote their end-points on $\mathcal{E}$. It is then seen in exactly the same way as in Section 9, that the pair $\mathfrak{f}u''$, $\mathfrak{f}v''$ cannot separate the pair u', v' on $\mathcal{E}$ for any $\mathfrak{f} \in \mathfrak{F}$.

4) Condition 4 is here satisfied in the trivial way that the lines of $\mathcal{U}$ do not at all accumulate in $\mathcal{D}$. For lines of the type $\mathcal{W}$ this follows from the fact that reflection lines of $\mathfrak{F}$ do not accumulate in $\mathcal{D}$ (§10). A line of the type $\mathcal{V}$, say $\mathcal{V}_m$, cuts off from $\mathcal{B}$ a part $\mathcal{B}_m$, and no other line of $\mathcal{U}$ enters into $\mathcal{B}_m$. Hence in at least one of the half-planes determined by $\mathcal{V}_m$, this line is separated from the rest of $\mathcal{U}$ by reflection lines, namely lines of $b_m \cap \mathcal{D}$. Since the reflection lines of $\mathfrak{F}$ do not accumulate in $\mathcal{D}$, the same holds for the lines of the type $\mathcal{V}$, and hence for $\mathcal{U}$.

5) No line of $\mathcal{U}$ is cut by a reflection line of $\mathfrak{F}$, since the $\mathcal{W}_r$ are simple modulo $\mathfrak{F}$, and the $\mathcal{V}_m$ are situated in a region $\mathcal{B}$ into which no reflection line of $\mathfrak{F}$ enters.

6) Every line of the type $\mathcal{W}$ admits an element of $\mathfrak{F}$ interchanging the two half-planes determined by it, namely the reflection in it. Consider a line of the type $\mathcal{V}$, say $\mathcal{V}_m$, and the decomposition of $\mathcal{D}$ brought about by $\mathfrak{F}\mathcal{V}_m$. Let $\mathcal{G}_m$ denote that region of this decomposition which is adjacent to $\mathcal{V}_m$ in the half-plane containing $\mathcal{B}_m$. The collection $\mathfrak{Y}_m\tilde{\mathcal{B}}_m$ is on $\mathcal{D}$ bounded by $\mathfrak{Y}_m\mathcal{V}_m$ and by no other line. No line equivalent with $\mathcal{V}_m$ has points in common with the interior

of $\mathfrak{Y}_m \tilde{\mathcal{B}}_m$, because none has points in common with the interior of $\mathcal{B}_m$. Thus $\mathfrak{Y}_m \tilde{\mathcal{B}}_m$ coincides with $\tilde{\mathcal{G}}_m$. The subgroup of $\mathfrak{F}$ carrying the region $\mathcal{G}_m$ of the decomposition of $\mathcal{D}$ by $\mathfrak{F}\mathcal{V}_m$ into itself, is the group $\mathfrak{G}_m$ generated by $\mathfrak{Y}_m$ and $\mathfrak{B}_m$. Since all sides of $\mathcal{G}_m$ are equivalent with respect to $\mathfrak{Y}_m$, hence belong to one equivalence class with respect to $\mathfrak{G}_m$, the case I of §19, dealt with in §19.2, does not arise. Since the subgroup of $\mathfrak{F}$ pertaining to $\mathcal{V}_m$ is either 1 or $\mathfrak{A}(\rightarrow)$, it is inferred that $\mathcal{V}_m$ is dividing modulo $\mathfrak{F}$. Thus condition 6) of §19.7 holds.

The structure of $\mathfrak{F}$ is thus determined by a generalized free product. However, in view of the fact that the present case is particularly simple, compared with the more general assumptions of §19, we carry the analysis a little further.

With notation following §19.7–10 the collection $\mathcal{U}$ is the union $\bigcup_r \mathfrak{F}\mathcal{W}_r$. Let $\mathcal{C}$ be that region of the decomposition of $\mathcal{D}$ by $\mathcal{U}'$ which contains the region $\mathcal{B}$ hitherto considered, and $\mathfrak{C}$ the subgroup of $\mathfrak{F}$ carrying $\mathcal{C}$ into itself. All the $\mathcal{W}_r$ are then among the sides of $\mathcal{C}$, because they are among the sides of $\mathcal{B}$. If the collection $\mathcal{W}_r$ is empty, then $\mathcal{C}$ means $\mathcal{D}$ itself, and $\mathfrak{C}$ means $\mathfrak{F}$ itself.

Next, we have to decompose $\mathcal{C}$ by that part $\mathcal{U}''$ of $\mathcal{U}$ which is situated in $\mathcal{C}$. The lines of $\mathcal{U}''$ belong to the union $\bigcup_m \mathfrak{F}\mathcal{V}_m$.

If $\mathcal{U}''$ is empty, then no lines $\mathcal{V}_m$ occur at all. In this case $\mathfrak{F}$ is determined by $\mathfrak{C}$ and the $\mathfrak{W}_r$, namely by (8), §19. This case arises if all reflection edges of $\mathcal{D}$ mod $\mathfrak{F}$ are simple and disjoint and no two of them are connected by a half-mast or half-funnel. Thus (8), §19, implies that the determination of $\mathfrak{F}$ is reduced to the determination of the subgroup $\mathfrak{C}$ of $\mathfrak{F}$, which contains no reflections.

The case in which neither lines $\mathcal{W}_r$ nor lines $\mathcal{V}_m$ occur, means that there are no reflections in $\mathfrak{F}$, so this case falls outside the scope of the present paragraph.

We now assume that some lines $\mathcal{V}_m$ actually occur. Thus $\mathcal{C}$ is decomposed by the collection $\mathfrak{C}\mathcal{V}_m$. For every value of m one of the regions of this decomposition is the region $\mathcal{G}_m$ adjacent to $\mathcal{V}_m$, which was considered above, and which is bounded by all $\mathfrak{Y}_m \mathcal{V}_m$. This region $\mathcal{G}_m$ has the subregion $\mathcal{B}_m$ in common with $\mathcal{B}$, but $\mathcal{B}_m$ is not the whole of $\mathcal{B}$, because we exclude the possibility of $\mathcal{V}_m$ coinciding with a $\mathcal{W}_r$, which was the case already dealt with in Sections 13 and 14. Moreover, any two of these regions, say $\mathcal{G}_1$ and $\mathcal{G}_2$, are disjoint modulo $\mathfrak{F}$; for, otherwise they would be equivalent with respect to $\mathfrak{F}$, thus with respect to $\mathfrak{C}$, because both are in $\mathcal{C}$. Hence b_1 and b_2 would be equivalent with respect to $\mathfrak{C}$; but then b_1 and b_2 would be equivalent with respect to $\mathfrak{B}$, because both are boundary chains of $\mathcal{B}$. This contradicts the fact that they have been chosen as representatives of different equivalence classes with respect to $\mathfrak{B}$.

We first consider the case in which the union $\bigcup_m \mathfrak{B}\mathcal{B}_m$ exhausts the whole of $\mathcal{B}$. Then $\mathcal{V}_1$, which bounds $\mathcal{B}_1$ inside $\mathcal{B}$, is also the boundary of another region of the collection $\bigcup_m \mathfrak{B}\mathcal{B}_m$. This region cannot be equivalent with $\mathcal{B}_1$ with respect to $\mathfrak{F}$, because $\mathcal{V}_1$ is dividing modulo $\mathfrak{F}$. If it is equivalent with $\mathcal{B}_2$ with respect to $\mathfrak{F}$, and thus also with respect to $\mathfrak{B}$ (which in $\mathcal{B}$ has the same effect as $\mathfrak{F}$), we may take it to be $\mathcal{B}_2$. Thus $\mathcal{V}_1$ and $\mathcal{V}_2$ coincide. In this case the complete boundary b of $\mathcal{B}$

consists of b_1 and b_2 together with the two points $u_1 = u_2$ and $v_1 = v_2$, and m takes only two values. $\mathcal{B}$ is the union of $\mathcal{B}_1$, $\mathcal{V}_1$, and $\mathcal{B}_2$. According as the subgroup $\mathfrak{B}_1$ of $\mathfrak{F}$ carrying $\mathcal{V}_1$ into itself is 1 or $\mathfrak{A}(\rightarrow)$ we get two cases. In both cases evidently $\vartheta(\rightleftharpoons) = 0$.

20.19 The double reflection strip. $\mathfrak{B}_1 = 1$. The surface $\mathcal{D}$ mod $\mathfrak{F}$ consists of two incomplete reflection strips (Section 11) adjacent along an open, simple geodesic of the surface. It is bordered by two reflection chains. We call this surface a *double reflection strip*. $\mathfrak{F}$ is the free product

$$\mathfrak{G}_1 * \mathfrak{G}_2$$

of the two groups generated by the reflections in the sides of $b_{1\mathcal{D}}$ and $b_{2\mathcal{D}}$ respectively. These groups have been characterized by generators and defining relations in Section 11. It should however be noticed that b_1 and b_2 cannot both have a finite number of sides. For, b_1 and b_2 cannot both have a reflection line ending in the same end-point of $\mathcal{V}_1$, because the end-points of $\mathcal{V}_1$ are outside $\mathcal{Y}$. An infinite number of sides in at least one of the two boundary chains b_1 and b_2, together with the fact that these chains are aperiodic on account of $\mathfrak{B}_1 = 1$, implies that $\mathfrak{F}$ cannot satisfy the condition of quasi-compactness.

20.20 The double crown. $\mathfrak{B}_1 = \mathfrak{A}(\rightarrow)$. The surface $\mathcal{D}$ mod $\mathfrak{F}$ consists of two crowns with their simple, closed, bounding geodesic coinciding. It is called a *double crown*. It is bordered by two reflection rings. In this case one gets

$$\mathfrak{F} = \mathfrak{G}_1 * \mathfrak{G}_2 \text{ am } \mathfrak{B}_1.$$

$\mathfrak{G}_1$ is generated by the reflections in the sides of $b_{1\mathcal{D}}$ together with a primary translation $\mathfrak{f}$ along $\mathcal{V}_1$; correspondingly, for $\mathfrak{G}_2$ with the same $\mathfrak{f}$. Both $\mathfrak{G}_1$ and $\mathfrak{G}_2$ are $\mathfrak{F}$-groups. The Euler characteristic of $\mathcal{K}^*(\mathfrak{F})$ mod $\mathfrak{F}$ is $\chi = 0$, moreover $\vartheta(-) = 2$ and $\vartheta(\rightarrow) = \vartheta(\hookrightarrow) = 0$, thus $p = 0$ from (5), §17. The area of $\mathcal{K}(\mathfrak{F})$ mod $\mathfrak{F}$ is

$$\Phi(\mathfrak{F}) = \sum_r \left(1 - \frac{1}{\nu_r}\right)$$

this sum being extended over both reflection rings. The area is thus equal to the sum of the areas of the two crowns.

20.21 Determination of $\mathfrak{F}$ by a free product with amalgamation. Reverting to the general case, we assume that the collection $\bigcup_m \mathfrak{B}\mathcal{B}_m$ does not exhaust $\mathcal{B}$. The regions of this collection are bounded on $\mathcal{B}$ by the collection of lines $\bigcup_m \mathfrak{B}\mathcal{V}_m$, and any two of these regions are either equivalent with respect to $\mathfrak{B}$ or disjoint modulo $\mathfrak{B}$. If all regions of this collection are omitted from $\mathcal{B}$, the remaining part of $\mathcal{B}$ is convex and has all the lines $\bigcup_m \mathfrak{B}\mathcal{V}_m$ on its boundary. We call the open part of this convex set $\mathcal{B}_0$. The boundary of $\mathcal{B}_0$ in $\mathcal{D}$ consists of the lines $\bigcup_m \mathfrak{B}\mathcal{V}_m$ together with the lines $\bigcup_r \mathfrak{B}\mathcal{W}_r$, if any. Any element of $\mathfrak{F}$ carrying $\mathcal{B}_0$ into itself, belongs to $\mathfrak{B}$ because it must carry $\mathcal{B}$ into itself. On the other hand, any element of $\mathfrak{B}$ carries the boundary $\bigcup_{r,m} (\mathfrak{B}\mathcal{W}_r, \mathfrak{B}_m\mathcal{V})$ of $\mathcal{B}_0$ and hence also $\mathcal{B}_0$ into itself. Hence $\mathfrak{B}$ is exactly the subgroup of $\mathfrak{F}$ carrying $\mathcal{B}_0$ into itself, while its elements in general interchange the regions of the collection $\bigcup_m \mathfrak{B}\mathcal{B}_m$.

We now follow the pattern of §19.7–10. There, in the decomposition of $\mathfrak{C}$ by $\mathcal{U}''$, we chose arbitrarily one of the regions of decomposition for the start of the construction (it was denoted by $\mathcal{B}$ in §19.9). In our present case we choose the above region $\mathcal{B}_0$. A complete collection of representatives of equivalence classes with respect to $\mathfrak{B}$ of boundaries of $\mathcal{B}_0$ drawn from $\mathcal{U}''$ are our present lines $\mathcal{V}_m$, where the number ν of values of m may be finite or infinite. For, a region $\mathcal{G}_i$ adjacent to $\mathcal{B}_0$ along $\mathcal{V}_i$ the number ν_i of §19.9 takes the value zero, because all sides of $\mathcal{G}_i$ are equivalent with $\mathcal{V}_i$ with respect to $\mathfrak{G}_i$. Hence the process of step by step building up the region for $\mathfrak{C}$ called $\mathcal{Y}$ in the notation of §19.10, described in §19.9, comes in our present case to an end already after the first step: $\mathcal{Y}$ consists of $\mathcal{B}_0$ and all the $\tilde{\mathcal{G}}_i$. This is the essential simplification.

Formula (24) of §19.10 becomes in our present notation, $\mathfrak{B}_m$ denoting the subgroup of $\mathfrak{F}$ belonging to $\mathcal{V}_m$,

$$\mathfrak{C} = \prod_m^* (\mathfrak{B}, \mathfrak{G}_m) \text{ am } \mathfrak{B}_m. \tag{13}$$

Because $\mathfrak{G}_m$ is generated by $\mathfrak{Y}_m$ and $\mathfrak{B}_m$, and $\mathfrak{B}_m$ is already in $\mathfrak{B}$, the group $\mathfrak{C}$ is generated by $\mathfrak{B}$ and the $\mathfrak{Y}_m$, thus by $\mathfrak{B}$ and the reflections in the sides of $\mathcal{B}$ in $\mathcal{D}$. It is, of course, sufficient to use one representative of each equivalence class with respect to $\mathfrak{B}$ of sides of $\mathcal{B}$ in $\mathcal{D}$.

Finally one gets from (25), §19.10,

$$\mathfrak{F} = \prod_{m,r}^* (\mathfrak{B}, \mathfrak{G}_m, \mathfrak{W}_r) \text{ am } \mathfrak{B}_m, \mathfrak{W}_r^*. \tag{14}$$

Here the $\mathfrak{W}_r$ belonging to the reflection lines $\mathcal{W}_r$ are known. They are either $\mathfrak{A}(-)$ or $\mathfrak{A}(- \to)$, and accordingly $\mathfrak{W}_r^*$ is either 1 or $\mathfrak{A}(\to)$. For the $\mathfrak{W}_r$ we may introduce as generators the reflection $\mathfrak{t}_r$ in $\mathcal{W}_r$ and an element $\mathfrak{h}_r$, this being either 1 or a primary translation along $\mathcal{W}_r$ in the two cases, respectively. For the $\mathfrak{G}_m$ an abstract presentation has been given in the Sections 11 and 12; the subgroup $\mathfrak{B}_m$ of $\mathfrak{G}_m$ is generated by one element $\mathfrak{f}_m$ which is either 1 or a primary translation along $\mathcal{V}_m$. *Thus* (14) *reduces the abstract presentation of an* $\mathfrak{F}$*-group containing reflections to the presentation of a subgroup* $\mathfrak{F}$ *which contains no reflections.*

In the special case $\mathfrak{B} = 1$ one gets $\mathfrak{G}_m = \mathfrak{Y}_m$, and $\mathfrak{W}_r$ consists only of 1 and the reflection $\mathfrak{t}_r$ in $\mathcal{W}_r$. Thus $\mathfrak{B}_m = \mathfrak{W}_r^* = 1$. Hence, in this case, $\mathfrak{F}$ is the ordinary free product of the $\mathfrak{Y}_m$ and the $\mathfrak{W}_r$.

Assume an abstract presentation of $\mathfrak{B}$ to be known. $\mathfrak{B}$ contains the $\mathfrak{h}_r$ and the $\mathfrak{f}_m$. They are expressions in the generators of $\mathfrak{B}$ and may, in particular, be included among the generators of $\mathfrak{B}$. Then a generation of $\mathfrak{F}$ is obtained by adding as new generators the reflections $\mathfrak{t}_r$ in the $\mathcal{W}_r$ for all values of r, and for all values of m the reflections $\mathfrak{s}_{mt}$ in the sides, according to their order, in a period of b_m belonging to $\mathfrak{f}_m$, thus in all sides of b_m, if $\mathfrak{f}_m = 1$. A sufficient set of defining relations to be added to those of $\mathfrak{B}$ is the following:

$$\mathfrak{t}_r^2 = 1 \tag{15}$$

$$\mathfrak{t}_r \mathfrak{h}_r = \mathfrak{h}_r \mathfrak{t}_r \tag{16}$$

for all values of r. Here (16) is of course superfluous, if $\mathfrak{h}_r = 1$. Moreover for all values of m

$$\mathfrak{s}_{mt}^2 = 1 \tag{17}$$

$$(\mathfrak{s}_{m,t+1}\mathfrak{s}_{mt})^{\nu_{mt}} = 1 \tag{18}$$

with the meaning of the ν_{mt} as indicated in Section 5. Relation (18) holds for all values of t, if b_m is aperiodic with infinitely many sides. Otherwise t is restricted by an inequality $1 \leqq t \leqq s$, where $s = s_m$ depends on m. Relation (18) then holds for $1 \leqq t \leqq s - 1$, and one more relation has to be added:

$$(\mathfrak{f}_m \mathfrak{s}_{m1} \mathfrak{f}_m^{-1} \mathfrak{s}_{ms})^{\nu_{ms}} = 1. \tag{19}$$

This relation even holds in the case of a finite, aperiodic b_m. For, then $\mathfrak{f}_m = 1$. If $s = 2$, then $\mathfrak{s}_{m1}\mathfrak{s}_{m2}$ is the inverse of $\mathfrak{s}_{m2}\mathfrak{s}_{m1}$, thus $\nu_{m2} = \nu_{m1}$. If $s > 2$, then $\mathfrak{s}_{m1}\mathfrak{s}_{ms}$ is a translation, and thus $\nu_{ms} = \infty$.

The surface $\mathcal{D}$ mod $\mathfrak{F}$ is obtained in the following way: The surface $\tilde{\mathcal{B}}_0$ mod $\mathfrak{B}$ has as boundaries one open or closed geodesic for each value of r and one open or closed geodesic for each value of m. For each value of r these boundaries of $\tilde{\mathcal{B}}_0$ mod $\mathfrak{B}$ become reflection edges of $\mathcal{D}$ mod $\mathfrak{F}$. For each value of m an incomplete reflection strip or a crown is added to $\tilde{\mathcal{B}}_0$ mod $\mathfrak{B}$ along that boundary according as it is open or closed.

If $\mathfrak{F}$ satisfies the condition of quasi-compactness, both r and m range over a finite set of values, and no reflection strips occur, because only reflection rings arise on the surface $\mathcal{D}$ mod $\mathfrak{F}$ (§16.4).

§21 Elementary groups and elementary surfaces

21.1 Two lemmas. We shall have to apply the following simple result:

If two open, straight segments ab and cd of equal length η intersect, then in each pair of opposite sides of the quadrangle acbd at least one has its length smaller than η.

Proof. If the point of intersection is called x, then

$$[a, x] + [x, b] = \eta, \quad [c, x] + [x, d] = \eta,$$

hence

$$([a, x] + [x, c]) + ([b, x] + [x, d]) = 2\eta.$$

The two left hand terms are broken lines, and at least one of them has the sum of its lengths $\leqq \eta$. Hence, owing to the triangle inequality, either $[a, c] < \eta$ or $[b, d] < \eta$. The same applies to the other pair of opposite sides ad and bc. □

In an $\mathfrak{F}$-group $\mathfrak{F}$ let $\mathcal{P}$ denote

1) either a point in $\mathcal{D}$,

2) or a horocycle belonging to a limit-centre of $\mathfrak{F}$, the horocycle being simple modulo $\mathfrak{F}$,

3) or an axis of $\mathfrak{F}$ which is simple modulo $\mathfrak{F}$.

Let $\mathfrak{A}_{\mathcal{P}}$ denote the subgroup of $\mathfrak{F}$ whole elements leave $\mathcal{P}$ invariant, and $\mathfrak{F} \setminus \mathfrak{A}_{\mathcal{P}}$ the set of all elements of $\mathfrak{F}$ outside $\mathfrak{A}_{\mathcal{P}}$.

It should be noted that the individuals of the equivalence class $\mathfrak{F}\mathcal{P}$ do not accumulate in $\mathcal{D}$. This is evident in the cases 1) and 3). In the case 2), since the horocycle $\mathcal{P}$ is simple modulo $\mathfrak{F}$, all the horocycles of $\mathfrak{F}\mathcal{P}$ are outside each other. Therefore a circle $\mathfrak{C}$ with centre in $x = 0$ and euclidean radius $\rho < 1$ can only be cut by finitely many of the $\mathfrak{F}\mathcal{P}$. Thus $\mathfrak{C}$ contains no accumulation point of $\mathfrak{F}\mathcal{P}$. As $\rho \to 1$ the statement results.

Let also $\mathcal{Q} \neq \mathcal{P}$ denote an object 1) or 2) or 3). It may, or may not, be equivalent with $\mathcal{P}$ with respect to $\mathfrak{F}$. If not, it is assumed that $\mathcal{P}$ and $\mathcal{Q}$ are disjoint modulo $\mathfrak{F}$. In this case, let $\mathcal{U}$ denote the collection $\mathfrak{F}\mathcal{Q}$. If $\mathcal{P}$ and $\mathcal{Q}$ are equivalent with respect to $\mathfrak{F}$, let $\mathcal{U}$ denote the collection $(\mathfrak{F} \setminus \mathfrak{A}_{\mathcal{P}})\mathcal{P}$. In both cases $\mathcal{Q}$ is contained in $\mathcal{U}$.

Choose arbitrarily a point on $\mathcal{P}$ (this means $\mathcal{P}$ itself, if $\mathcal{P}$ is a point) and a point on $\mathcal{Q}$, and let δ denote their distance. If $\mathcal{P}$ is a point, let $\mathcal{K}$ denote the circular disc with centre $\mathcal{P}$ and radius δ. If $\mathcal{P}$ is a horocycle or an axis, draw two normals to $\mathcal{P}$ such that the region Ω between them contains a fundamental domain for $\mathfrak{A}_{\mathcal{P}}$. Let Ψ denote a belt containing $\mathcal{P}$ bounded by two horocycles at distance δ from the horocycle $\mathcal{P}$, or by two hypercycles at distance δ from the axis $\mathcal{P}$, and put $\mathcal{K} = \Omega \cap \Psi$. Then $\mathfrak{A}_{\mathcal{P}}\mathcal{K}$ covers Ψ.

In all three cases $\mathcal{K}$ is a bounded region, and it has points in common with the collection $\mathcal{U}$. On the other hand, only a finite number of individuals of the collection $\mathcal{U}$ enter into $\mathcal{K}$, because $\mathcal{U}$ does not accumulate in $\mathcal{D}$; none of these has points in common with $\mathcal{P}$. We can therefore select one individual out of this finite number such that its distance η from $\mathcal{P}$ is minimum. Then $\eta \leqq \delta$. It follows from this construction that η is the distance of the point sets $\mathcal{P}$ and $\mathcal{U}$. If $\mathcal{P}$ and $\mathcal{Q}$ are not equivalent, we may denote it by $\eta(\mathcal{P}, \mathcal{Q}) = \eta(\mathcal{Q}, \mathcal{P})$. It is then the distance of the point sets $\mathfrak{F}\mathcal{P}$ and $\mathfrak{F}\mathcal{Q}$. If $\mathcal{P}$ and $\mathcal{Q}$ are equivalent, we denote it by $\eta(\mathcal{P})$ and call it the *internal distance* of the collection $\mathfrak{F}\mathcal{P}$. If $\mathcal{P}$ is a point, then $\eta(\mathcal{P})$ is the distance $\eta(\mathcal{P})$ introduced in §14.4.

Let a be a point of $\mathcal{P}$ and b a point of $\mathcal{U}$ such that $[a, b] = \eta$. Since the definition of η only depends on the equivalence classes in question, we may say, by a change of notation if necessary, that b is on $\mathcal{Q}$. If both $\mathcal{P}$ and $\mathcal{Q}$ are points, then $ab = \mathcal{P}\mathcal{Q}$. If one is a point and the other a horocycle or an axis, then ab is the perpendicular from the point on the horocycle or axis. If none is a point, then ab is on the common normal of $\mathcal{P}$ and $\mathcal{Q}$. In all cases the open segment $\mathcal{T} = ab$ contains no point of $\mathfrak{F}\mathcal{P}$ nor of $\mathfrak{F}\mathcal{Q}$. Its closure $\overline{\mathcal{T}}$ is called the *join* of $\mathcal{P}$ and $\mathcal{Q}$.

We now show that the open segment $\mathcal{T}$ is simple modulo $\mathfrak{F}$: Assume that $\mathfrak{g}$ were such an element of $\mathfrak{F}$ that $\mathfrak{g}\mathcal{T}$ intersects $\mathcal{T}$, say in the point x. Then $\mathcal{P}$, $\mathcal{Q}$, $\mathfrak{g}\mathcal{P}$, $\mathfrak{g}\mathcal{Q}$ are all different, because two segments issuing from the same point, or perpendicular to the same line or horocycle, do not intersect. Hence, owing to the above result $[a, \mathfrak{g}b] < \eta$ or $[\mathfrak{g}a, b] < \eta$, which contradicts the definition of η. In all we thus get

Lemma 1. *If the open segment $\mathcal{T}$ joins $\mathcal{P}$ and $\mathfrak{f}\mathcal{P}$, where $\mathfrak{f}$ belongs to $\mathfrak{F}$ but not to $\mathfrak{A}_{\mathcal{P}}$, and the length of $\mathcal{T}$ is equal to the internal distance of $\mathfrak{F}\mathcal{P}$, then $\mathcal{T}$ contains no point of $\mathfrak{F}\mathcal{P}$ and is simple modulo $\mathfrak{F}$.*

Lemma 2. *If $\mathcal{P}$ and $\mathcal{Q}$ are not equivalent with respect to $\mathfrak{F}$ and the open segment $\mathcal{T}$ joins $\mathcal{P}$ and $\mathcal{Q}$, and its length is equal to the distance of $\mathfrak{F}\mathcal{P}$ and $\mathfrak{F}\mathcal{Q}$, then $\mathcal{T}$ contains no point of $\mathfrak{F}\mathcal{P}$ nor of $\mathfrak{F}\mathcal{Q}$ and is simple modulo $\mathfrak{F}$.*

Remark. If $\mathcal{T}$ and a reflection line $\mathcal{L}$ of $\mathfrak{F}$ have a common point, then $\mathcal{T}$ cannot meet $\mathcal{L}$ at an angle α between 0 and $\frac{\pi}{2}$. For, if $\mathfrak{s}$ denotes the reflection in $\mathcal{L}$, then $\mathcal{T}$ and $\mathfrak{s}\mathcal{T}$ would intersect. In the case of both lemmas $\mathcal{T}$ may be situated on $\mathcal{L}$, i.e. $\alpha = 0$. The case $\alpha = \frac{\pi}{2}$ cannot occur in the case of Lemma 2, because if $\mathcal{L}$ did not bisect $\mathcal{T}$, then $\mathcal{T}$ would contain either $\mathfrak{s}\mathcal{P}$ or $\mathfrak{s}\mathcal{Q}$, and if $\mathcal{L}$ did bisect $\mathcal{T}$, then one would get $\mathcal{Q} = \mathfrak{s}\mathcal{P}$. Both possibilities are excluded in Lemma 2. In case of Lemma 1, the line $\mathcal{L}$ can bisect $\mathcal{T}$.

21.2 Decomposition of $\mathcal{D}$ by $\mathfrak{F}\mathcal{S}$. From now on we consider in this §21, an $\mathfrak{F}$-group $\mathfrak{F}$ which *contains no reflections*. Consider a truncated domain $\mathcal{K}^*(\mathfrak{F})$ for $\mathfrak{F}$. Let $\mathcal{P}$ denote

1) either a centre of $\mathfrak{F}$ (thus in $\mathcal{K}^*(\mathfrak{F})$),

2) or a bounding horocycle of $\mathcal{K}^*(\mathfrak{F})$,

3) or a boundary axis of $\mathcal{K}(\mathfrak{F})$ (and thus also of $\mathcal{K}^*(\mathfrak{F})$).

Then $\mathcal{P}$ satisfies the conditions of Section 1, for $\mathcal{P}$ is simple modulo $\mathfrak{F}$. Let $\mathfrak{P}$ denote the subgroup of $\mathfrak{F}$ leaving $\mathcal{P}$ invariant, and $\mathfrak{p}$ a primary element of $\mathfrak{P}$. This is in the three cases a primary rotation, or limit-rotation, or translation, respectively.

We denote by s a straight, open segment joining $\mathcal{P}$ with one of its equivalents, in the sequel denoted by $\mathfrak{k}\mathcal{P}$ and chosen such that s contains no point of $\mathfrak{F}\mathcal{P}$ and that s in the cases 2) and 3) is situated on the common normal of $\mathcal{P}$ and $\mathfrak{k}\mathcal{P}$, and furthermore such that s is simple modulo $\mathfrak{F}$. In virtue of Lemma 1 of Section 1, such an s exists. But we need not prescribe that the length δ of s is equal to the internal distance of $\mathfrak{F}\mathcal{P}$; thus $\delta \geqq \eta(\mathcal{P})$. Since s is simple modulo $\mathfrak{F}$, it contains no centre of order > 2. Let $\bar{s}$ denote the closure of s, obtained by adding its end-points.

By $\mathcal{S}$ we denote in the case 1) the closure $\bar{s}$ of s, and in the cases 2) and 3) the complete straight line on which s is situated. Then even $\mathcal{S}$ is simple modulo $\mathfrak{F}$ in the cases 2) and 3), for the parts of $\mathcal{S}$ other than $\bar{s}$ are outside $\mathcal{K}^*(\mathfrak{F})$ in the case 2), and outside $\mathcal{K}(\mathfrak{F})$ in the case 3), and these parts are in both cases orthogonal to $\mathcal{P}$ and $\mathfrak{t}\mathcal{P}$. They cannot, therefore, be cut by any of their equivalents, nor can any of their equivalents cut $\bar{s}$, and $\bar{s}$ is simple modulo $\mathfrak{F}$ in the cases 2) and 3).

The collection $\mathfrak{F}\mathcal{S}$ does not accumulate in $\mathcal{D}$: The part of $\mathcal{D}$ outside $\mathcal{K}^*(\mathfrak{F})$ which in the cases 2) and 3) is cut off by some $\mathfrak{f}\mathcal{P}$, $\mathfrak{f} \in \mathfrak{F}$, contains no accumulation point of $\mathfrak{F}\mathcal{S}$, because the parts of $\mathfrak{F}\mathcal{S}$ contained therein, consist of a series of normals of $\mathfrak{f}\mathcal{P}$ in points of $\mathfrak{f}\mathcal{P}$ which belong to at most two equivalence classes with respect to $\mathfrak{F}$. An accumulation point of $\mathfrak{F}\mathcal{S}$ in $\mathcal{D}$ would thus in all three cases be an accumulation point of $\mathfrak{F}\bar{s}$. It cannot exist, because all segments $\mathfrak{F}\bar{s}$ have equal length δ and their end-points belong to at most two equivalence classes with respect to $\mathfrak{F}$ (§13).

Since $\mathfrak{F}\mathcal{S}$ does not accumulate in $\mathcal{D}$, we may decompose $\mathcal{D}$ by $\mathfrak{F}\mathcal{S}$. The regions of this decomposition are spoken of as $\mathcal{B}$-regions. They are convex. This is evident in the cases 2) and 3), because $\mathcal{D}$ then is decomposed by complete straight lines. It also holds in case 1), because the angle at any vertex of a $\mathcal{B}$-region is $\leqq \pi$.

Let w denote the limit-centre or the interval of discontinuity on $\mathcal{E}$ pertaining to $\mathcal{P}$ in the cases 2) and 3), respectively. We denote by $\mathcal{Y}$ the point set obtained by adding to the point set consisting of $\mathfrak{F}\mathcal{S}$ the point set $\mathfrak{F}w$. In the case 1) we define $\mathcal{Y}$ as $\mathfrak{F}\mathcal{S}$ itself. In all cases $\mathcal{Y}$ contains no fundamental point of $\mathfrak{F}$.

21.3 Boundary chains. Consider a region of decomposition $\mathcal{B}$, denote by b its boundary, by $b_{\mathcal{D}} = b \cap \mathcal{D}$ its boundary on $\mathcal{D}$, by $\overline{\mathcal{B}} = \mathcal{B} \cup b$ its closure on $\overline{\mathcal{D}}$, and by $\tilde{\mathcal{B}} = \mathcal{B} \cup b_{\mathcal{D}}$ its closure on $\mathcal{D}$. The intersections of b with the different components of $\mathcal{Y}$ are called *boundary chains* of $\mathcal{B}$. Let b_1 denote the intersection $b \cap \mathcal{Y}_1$ of b with the component $\mathcal{Y}_1$ of $\mathcal{Y}$, and put $b_{1\mathcal{D}} = b_1 \cap \mathcal{D}$. We denote by $\mathfrak{B}$ the subgroup of $\mathfrak{F}$ carrying $\mathcal{B}$ into itself, and by $\mathfrak{B}_1$ the subgroup of $\mathfrak{B}$ carrying the particular chain b_1 of $\mathcal{B}$ into itself.

If $\mathcal{P}$ is a centre of $\mathfrak{F}$, then b_1 is a polygonal line in $\mathcal{D}$, the angles of $\mathcal{B}$ at b_1 being $\leqq \pi$. If $\mathcal{P}$ is a horocycle bounding $\mathcal{K}^*(\mathfrak{F})$, then the links of $b_{1\mathcal{D}}$ are connected by limit-centres belonging to $\mathcal{Y}_1$; it is convenient from now on *to change the original notation of Section* 2, *denoting by $\mathcal{P}$ the limit-centre* of the horocycle considered in Section 2, instead of that horocycle itself; with this new notation the vertices of b_1 are a subset of $\mathfrak{F}\mathcal{P}$, and this subset is reproduced by $\mathfrak{B}_1$. If $\mathcal{P}$ is a boundary axis, the sides of $b_{1\mathcal{D}}$ are linked by arcs of $\mathcal{E}$ belonging to $\mathcal{Y}_1$.

Definition. If $\mathcal{X}'$ and $\mathcal{X}''$ are any two disjoint points or lines in $\overline{\mathcal{D}}$, we denote in the following by their *join*

a) the segment, half-line, or line, if both are points,

b) the perpendicular from the point on the line, if one is a point, the other a line,

c) the joining segment on their common normal, if both are lines.

21.4 Reversibility and non-reversibility. The segment or straight line $\mathscr{S}$ introduced in Section 2 is called *reversible*, if there exists an element of $\mathfrak{F}$ which maps $\mathscr{S}$ into itself with its end-points interchanged. Since there are no reflections in $\mathfrak{F}$, this is only possible by a half-turn about a centre c on $\mathscr{S}$, and c must be the mid-point of s. Moreover c must be of order 2, since s is simple modulo $\mathfrak{F}$. There can be no centre of order 2 other than c on s, because an element of $\mathfrak{F}$ leaving a point of s fixed must map s onto itself.

In this case of reversibility of $\mathscr{S}$, let $\mathfrak{c}$ denote the half-turn about c. Then $\mathfrak{k}\mathscr{P}$ is also $\mathfrak{c}\mathscr{P}$. Let $\mathfrak{f}$ be such an element of $\mathfrak{F}$ that $\mathscr{S}_1 = \mathfrak{f}\mathscr{S}$ is issuing from the point $\mathscr{P}$ (i.e. centre or limit-centre), or cutting the axis $\mathscr{P}$. Then $\mathfrak{f}^{-1}$ carries $\mathscr{S}_1$ into $\mathscr{S}$, thus $\mathscr{P}$ into either $\mathscr{P}$ or $\mathfrak{c}\mathscr{P}$. If $\mathfrak{f}^{-1}\mathscr{P} = \mathfrak{c}\mathscr{P}$, then $\mathfrak{c}\mathfrak{f}^{-1}\mathscr{P} = \mathscr{P}$. Both $\mathfrak{f}^{-1}$ and $\mathfrak{c}\mathfrak{f}^{-1}$ carry $\mathscr{S}_1$ into $\mathscr{S}$, and one of them leaves $\mathscr{P}$ invariant. Hence $\mathscr{S}_1$ is the image of $\mathscr{S}$ by some element of $\mathfrak{P}$. The individuals of $\mathfrak{F}\mathscr{S}$ issuing from $\mathscr{P}$, or cutting $\mathscr{P}$ if $\mathscr{P}$ is an axis, are thus equivalent with respect to $\mathfrak{P}$.

Conversely, assume that all individuals of $\mathfrak{F}\mathscr{S}$ issuing from $\mathscr{P}$ or cutting $\mathscr{P}$ are equivalent with respect to $\mathfrak{P}$. Then $\mathfrak{k}^{-1}\mathscr{S}$ is $\mathfrak{p}^{\mu}\mathscr{S}$, thus $\mathfrak{p}^{-\mu}\mathfrak{k}^{-1}\mathscr{S} = \mathscr{S}$ and $\mathfrak{p}^{-\mu}\mathfrak{k}^{-1}(\mathfrak{k}\mathscr{P}) = \mathscr{P}$. The element $\mathfrak{p}^{-\mu}\mathfrak{k}^{-1}$ then reverses $\mathscr{S}$. It is thus necessary and sufficient for the reversibility of $\mathscr{S}$ that the number of individuals of $\mathfrak{F}\mathscr{S}$ issuing from $\mathscr{P}$ or cutting $\mathscr{P}$ reduces modulo $\mathfrak{P}$ to 1.

If no element of $\mathfrak{F}$ reverses $\mathscr{S}$, then $\mathscr{S}$ is called *non–reversible*. In this case the number of individuals of $\mathfrak{F}\mathscr{S}$ issuing from $\mathscr{P}$ or cutting $\mathscr{P}$ reduces modulo $\mathfrak{P}$ to 2, the two equivalence classes with respect to $\mathfrak{P}$ being represented by $\mathscr{S}$ and by $\mathfrak{k}^{-1}\mathscr{S}$.

21.5 Rotation twins. In this section we assume that the $\mathscr{P}$ of Section 2 is a centre of order 2, and that the corresponding $\mathscr{S}$ is reversible. We denote by $\mathscr{P}'$ the mid-point of $\mathscr{S}$, which is also a centre of order 2 of $\mathfrak{F}$, by $\mathfrak{p}'$ the half-turn about $\mathscr{P}'$, and by $\mathfrak{P}'$ the group generated by $\mathfrak{p}'$. Then $\mathscr{P}$ and $\mathscr{P}'$ are not equivalent with respect to $\mathfrak{F}$, because no point of $\mathscr{S}$ other than its end-points belongs to $\mathfrak{F}\mathscr{P}$. The line on which $\mathscr{S}$ is situated is then an axis of $\mathfrak{F}$, and the corresponding subgroup of $\mathfrak{F}$ is $\mathfrak{A}(\cdot\cdot)$; it is generated by $\mathfrak{p}$ and $\mathfrak{p}'$. The element

$$\mathfrak{p}'' = \mathfrak{p}'\mathfrak{p} \tag{1}$$

is a primary translation along the axis. We denote this axis by $\mathscr{P}''$, and the group generated by $\mathfrak{p}''$ is called $\mathfrak{P}''$. Since only $\mathscr{S}$ and $\mathfrak{p}\mathscr{S}$ are issuing from $\mathscr{P}$ in virtue of the reversibility of $\mathscr{S}$ (Section 4), and both $\mathscr{S}$ and $\mathfrak{p}\mathscr{S}$ are on $\mathscr{P}''$, it is inferred that $\mathscr{P}''$ is simple modulo $\mathfrak{F}$.

The decomposition of $\mathscr{D}$ by $\mathfrak{F}\mathscr{S}$ coincides thus with the decomposition of $\mathscr{D}$ by $\mathfrak{F}\mathscr{P}''$. If $\mathscr{B}$ is one of the regions of decomposition adjacent to $\mathscr{P}''$ and $\mathfrak{B}$ the corresponding subgroup of $\mathfrak{F}$, then $\mathcal{P}''$ is a complete boundary chain b_1 of $\mathcal{B}$. The subgroup of $\mathfrak{A}_{\mathcal{P}''}$ leaving $\mathcal{B}$ invariant is generated by the primary translation $\mathfrak{p}''$, hence $\mathfrak{B}_1 = \mathfrak{P}''$. Since $\mathfrak{F}$ is not quasi-abelian, it has not $\mathcal{P}''$ as an invariant line. Hence $\mathfrak{B}_1$ does not exhaust $\mathfrak{B}$, and since $\mathcal{B}$ then contains more than one axis, $\mathfrak{B}$ is an $\mathfrak{F}$-group. Equation (6), §19, then yields

$$\mathfrak{F} = \mathfrak{B} * \mathfrak{A}_{\mathcal{P}''} \operatorname{am} \mathfrak{P}'', \tag{2}$$

thus reducing the determination of $\mathfrak{F}$ to the determination of $\mathfrak{B}$.

It should be noticed that, if $\mathcal{P}'$ is chosen instead of $\mathcal{P}$ in the construction of Section 2, then the segment s' between $\mathcal{P}'$ and $\mathfrak{p}\mathcal{P}' = \mathfrak{p}''^{-1}\mathcal{P}'$ is simple modulo $\mathfrak{F}$ and contains no point of $\mathfrak{F}\mathcal{P}'$, and this segment is reversed by $\mathfrak{p}$. This would thus lead to the same $\mathcal{P}''$, hence the same $\mathfrak{B}$ and the same determination of $\mathfrak{F}$. We call such a pair of half-turns of *rotation twins* with respect to $\mathfrak{F}$.

For comparison with cases to be considered later, it is convenient to say that $\mathfrak{A}_{\mathcal{P}''}$ is generated by the three generators $\mathfrak{p}$, $\mathfrak{p}'$, $\mathfrak{p}''$ with defining relations

$$\mathfrak{p}\mathfrak{p}'\mathfrak{p}'' = 1, \tag{3}$$

$$\mathfrak{p}^2 = 1, \quad \mathfrak{p}'^2 = 1, \quad \mathfrak{p}''^{\infty} = 1. \tag{4}$$

21.6 Non-reversibility of $\mathcal{S}$. We now revert to the general case of the Sections 2 and 3 and assume that $\mathcal{S}$ is non-reversible (Section 4). There is no centre in the interior of $\mathcal{S}$. Let $\mathcal{S}$ be provided with a direction, say from $\mathcal{P}$ towards $\mathfrak{k}\mathcal{P}$, and let this direction be carried by all elements of $\mathfrak{F}$; we then call $\mathfrak{F}\mathcal{S}$ *coherently directed.*

There is exactly one element of $\mathfrak{F}$ carrying any individual of the collection $\mathfrak{F}\mathcal{S}$ into any other; this is equivalent to stating that no element of $\mathfrak{F}$ except the identity carries $\mathcal{S}$ into itself: Assume $\mathfrak{f}\mathcal{S} = \mathcal{S}$. Then $\mathfrak{f}$ is not a translation or reversed translation, because $\mathcal{S}$ is either a finite segment or, if it is a complete line, then it is not an axis of $\mathfrak{F}$, since its end-points belong to $\mathcal{Y}$. Since there is no centre on s, the element $\mathfrak{f}$ cannot be a rotation. It cannot be a limit-rotation, because a limit-rotation leaves no line invariant. Finally there is no reflection in $\mathfrak{F}$. Thus $\mathfrak{f} = 1$.

In the case 1) we have to deal with a primary rotation $\mathfrak{p}$ with centre $\mathcal{P}$. Let ν be its order. Then

$$\mathfrak{p}^{\rho}\mathcal{S}, \quad 0 \leqq \rho \leqq \nu - 1, \tag{5}$$

are ν segments with common end-point $\mathcal{P}$ and directed from $\mathcal{P}$. Also $\mathfrak{k}\mathfrak{p}^{\rho}\mathfrak{k}^{-1}\mathcal{S}$ are ν segments with common end-point $\mathfrak{k}\mathcal{P}$ and directed towards $\mathfrak{k}\mathcal{P}$. The image of the last ν segments by $\mathfrak{k}^{-1}$, thus the segments $\mathfrak{k}^{\rho}(\mathfrak{k}^{-1}\mathcal{S})$ are ν segments with common end-point $\mathcal{P}$ and directed towards $\mathcal{P}$. They cannot coincide with the segments (5) owing to the non-reversibility of $\mathcal{S}$. There are thus 2ν segments of the collection $\mathfrak{F}\mathcal{S}$ issuing from $\mathcal{P}$, and these are, in their cyclical order around $\mathcal{P}$, alternately directed from $\mathcal{P}$ and towards $\mathcal{P}$. The two collections of ν segments each constitute the two equivalence classes with respect to $\mathfrak{P}$ of segments issuing from $\mathcal{P}$ (Section 4).

In the case 2) we have to deal with a primary limit-rotation $\mathfrak{p}$ with limit-centre $\mathcal{P}$. The order of $\mathfrak{p}$ is $\nu = \infty$. All $\mathfrak{p}^\rho \mathcal{S}$, $-\infty < \rho < \infty$, are directed from $\mathcal{P}$, and all $\mathfrak{p}^\rho(\mathfrak{k}^{-1}\mathcal{S})$ towards $\mathcal{P}$, and they alternate around $\mathcal{P}$.

In the case 3) we have to deal with a primary boundary translation $\mathfrak{p}$ with axis $\mathcal{P}$. The order of $\mathfrak{p}$ is $\nu = \infty$. All $\mathfrak{p}^\rho \mathcal{S}$, $-\infty < \rho < \infty$, are normals of $\mathcal{P}$ which in $\mathcal{K}(\mathfrak{F})$ are directed from $\mathcal{P}$, and all $\mathfrak{p}^\rho(\mathfrak{k}^{-1}\mathcal{P})$ are normals of $\mathcal{P}$ which in $\mathcal{K}(\mathfrak{F})$ are directed towards $\mathcal{P}$, and they alternate along $\mathcal{P}$.

In the decomposition of $\mathcal{D}$ by $\mathfrak{F}\mathcal{S}$, let $\mathcal{B}'$ denote the region of decomposition adjacent to $\mathcal{S}$ which is bounded positively by $\mathcal{S}$. Let b_1' be that boundary chain of $\mathcal{B}'$ to which $\mathcal{S}$ belongs. It then follows from the above description that all sides of $b'_{1\mathcal{D}}$ bound $\mathcal{B}'$ positively. If we traverse $b'_{1\mathcal{D}}$ in the direction in which it bounds $\mathcal{B}'$ positively, let $\mathcal{S}'_{-1}$ be the side which precedes $\mathcal{S}$ and $\mathcal{S}'_1$ the side which follows $\mathcal{S}$. Let $\mathcal{B}''$ be the region of decomposition bounded negatively by $\mathcal{S}$ and b''_1 its boundary chain containing $\mathcal{S}$. All sides of $b''_{1\mathcal{D}}$ bound $\mathcal{B}''$ negatively. If we traverse $b''_{1\mathcal{D}}$ in the direction in which it bounds $\mathcal{B}''$ negatively, let $\mathcal{S}''_{-1}$ be the side which precedes $\mathcal{S}$ and $\mathcal{S}''_1$ the side which follows $\mathcal{S}$. Fig. 21.1 illustrates separately the three cases 1), 2), 3), corresponding to the character of $\mathcal{P}$.

The uniquely determined element of $\mathfrak{F}$ which carries $\mathcal{S}'_{-1}$ into $\mathcal{S}$ may either be a motion or a reversion. In the latter case it is a reversed translation, because $\mathfrak{F}$ contains no reflection. We deal with the two cases separately. The Sections 7–15 deal with the case of a motion, the Section 16 with the case of a reversion.

21.7 The case of a motion. We first deal with the case in which $\mathcal{S}'_{-1}$ is carried into $\mathcal{S}$ by a motion. Let $\mathcal{R}$ denote the bisecting normal of s, $\mathcal{R}'$ the symmetry line of $\mathcal{S}'_{-1}$ and $\mathcal{S}$ in $\mathcal{B}'$, and $\mathcal{R}''$ the symmetry line of $\mathcal{S}''_{-1}$ and $\mathcal{S}$ in $\mathcal{B}''$. Let $\mathfrak{s}, \mathfrak{s}', \mathfrak{s}''$ denote reflections in $\mathcal{R}, \mathcal{R}', \mathcal{R}''$, respectively. These reflections do not belong to $\mathfrak{F}$. Then $\mathfrak{s}''\mathfrak{s}'$ carries $\mathcal{S}'_{-1}$ into $\mathcal{S}''_{-1}$ with direction preserved, and $\mathfrak{s}''\mathfrak{s}'$ is a primary element of $\mathfrak{P}$, because $\mathcal{S}'_{-1}$ and $\mathcal{S}''_{-1}$ are only separated by one line of $\mathfrak{F}\mathcal{S}$, namely $\mathcal{S}$. We put

$$\mathfrak{p} = \mathfrak{s}''\mathfrak{s}', \quad \mathfrak{p}' = \mathfrak{s}'\mathfrak{s}, \quad \mathfrak{p}'' = \mathfrak{s}\mathfrak{s}'', \tag{6}$$

and denote by $\mathfrak{P}, \mathfrak{P}', \mathfrak{P}''$ the groups generated by $\mathfrak{p}, \mathfrak{p}', \mathfrak{p}''$, respectively. It is then seen that $\mathfrak{p}'$ carries $\mathcal{S}$ into $\mathcal{S}'_{-1}$ with direction preserved, and that $\mathfrak{p}''$ carries $\mathcal{S}''_{-1}$ into $\mathcal{S}$ with direction preserved. Thus all three elements (6) are motions belonging to $\mathfrak{F}$. In virtue of the relations

$$\mathfrak{s}^2 = 1, \quad \mathfrak{s}'^2 = 1, \quad \mathfrak{s}''^2 = 1, \tag{7}$$

they satisfy the relation (3). Denoting their orders by ν, ν', ν'', respectively, each of which may become infinite, we get

$$(\mathfrak{s}''\mathfrak{s}')^\nu = 1, \quad (\mathfrak{s}'\mathfrak{s})^{\nu'} = 1, \quad (\mathfrak{s}\mathfrak{s}'')^{\nu''} = 1, \tag{8}$$

or written in $\mathfrak{p}, \mathfrak{p}', \mathfrak{p}''$

$$\mathfrak{p}^\nu = 1, \quad \mathfrak{p}'^{\nu'} = 1, \quad \mathfrak{p}''^{\nu''} = 1. \tag{9}$$

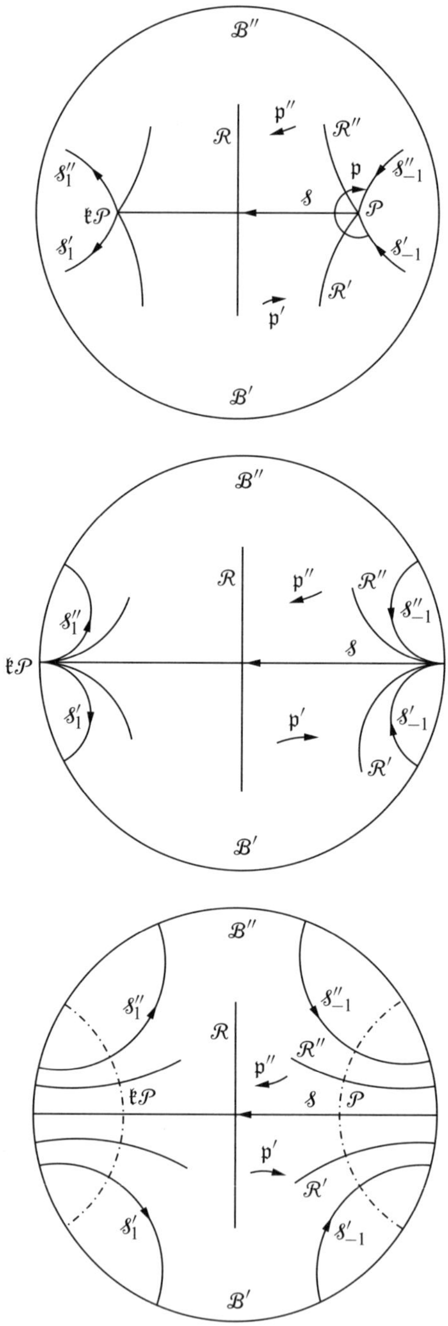

$\mathcal{B}''$
$\mathfrak{p}''$
$\mathcal{R}$
$\mathcal{R}''$
$\mathcal{S}_1''$
$\mathfrak{p}$
$\mathcal{S}_{-1}''$
$\mathcal{S}$
$\mathfrak{k}\mathcal{P}$
$\mathcal{P}$
$\mathcal{S}_1'$
$\mathcal{S}_{-1}'$
$\mathcal{R}'$
$\mathfrak{p}'$
$\mathcal{B}'$

Figure 21.1

Relation (9) takes here the place of relation (4).

We denote by $\mathfrak{R}$ the reflection group generated by $\mathfrak{s}$, $\mathfrak{s}'$, $\mathfrak{s}''$. If $\mathcal{R}$ and $\mathcal{R}'$ intersect, it cannot be at right angles, because in the cases 1) and 2) the triangle formed by $\mathcal{R}'$, $\mathcal{S}$, and $\mathcal{R}$ has a right angle between $\mathcal{S}$ and $\mathcal{R}$, and in the case 3) the quadrangle formed by $\mathcal{R}'$, $\mathcal{P}$, $\mathcal{S}$, and $\mathcal{R}$ has right angles at the three other vertices. Thus $\nu' > 2$. For corresponding reasons also $\nu'' > 2$. It follows that $\mathfrak{R}$ cannot be quasi-abelian, because none of the lines $\mathcal{R}$, $\mathcal{R}'$, $\mathcal{R}''$ is the common normal of the two others. The exponent ν is the only one for which the value 2 is possible. If $\nu = 2$, then we have case 1) with the special condition that $\mathcal{S}'_{-1}$ and $\mathcal{S}''_{-1}$ are on the same straight line. This line need not be orthogonal to $\mathcal{S}$.

Let $\mathcal{Q}$ denote the region bounded in $\mathcal{D}$ by $\mathcal{R}$, $\mathcal{R}'$, and $\mathcal{R}''$, and $\tilde{\mathcal{Q}}$ its closure on $\mathcal{D}$. Then $\tilde{\mathcal{Q}}$ is a fundamental domain for $\mathfrak{R}$, and $\mathcal{D}$ mod $\mathfrak{R}$ is a finite reflection disc (§20.5) bordered by a reflection ring (§20.4) with three links. Thus $\mathfrak{R}$ is presented by the relations (7) and (8) in the above generators; they correspond to relations (5) and (6) of §20.5. Also the area of $\mathcal{K}(\mathfrak{R})$ mod $\mathfrak{R}$ is found from (7′), §20.5,

$$\Phi(\mathfrak{R}) = 1 - \frac{1}{\nu} - \frac{1}{\nu'} - \frac{1}{\nu''} \tag{10}$$

with the usual convention that terms corresponding to infinite exponents are zero.

We denote by $\mathcal{P}'$ the join of $\mathcal{R}$ and $\mathcal{R}'$, thus a centre of $\mathfrak{F}$, if $\mathcal{R}$ and $\mathcal{R}'$ intersect, a limit-centre of $\mathfrak{F}$, if they are parallel, and an axis of $\mathfrak{F}$, if they are divergent, the axis being their common normal. Similarly $\mathcal{P}''$ denotes the join of $\mathcal{R}$ and $\mathcal{R}''$. Then $\mathcal{P}'$ is left invariant by the group $\mathfrak{P}'$ generated by $\mathfrak{p}'$, and $\mathcal{P}''$ is left invariant by the group $\mathfrak{P}''$ generated by $\mathfrak{p}''$. It should however be borne in mind that, if $\mathcal{P}'$ is an axis, the translation-group $\mathfrak{P}'$ need not exhaust the full subgroup $\mathfrak{A}_{\mathcal{P}'}$ of $\mathfrak{F}$ which carries $\mathcal{P}'$ into itself.

If none of $\mathfrak{p}$, $\mathfrak{p}'$, $\mathfrak{p}''$ is a translation, then $\overline{\mathcal{Q}}$ is a triangle $\mathcal{P}\mathcal{P}'\mathcal{P}''$, possibly with one, or two, or three, vertices on $\mathcal{E}$. This is the case of the *triangle groups*, well-known in function theory.

If, in particular, $\nu = 2$, $\nu' = 3$, $\nu'' = 7$ for a triangle group, then (10) yields $\Phi(\mathfrak{R}) = \frac{1}{42}$. A triangle with angles $\frac{\pi}{2}, \frac{\pi}{3}, \frac{\pi}{7}$ exists. This shows that the lower bound established at the end of §17, for the area $\Phi(\mathfrak{F})$ of an $\mathfrak{F}$-group containing reversions can be reached.

Whatever the character of $\mathcal{P}$ and of $\mathcal{P}'$, the element $\mathfrak{p}'$ of $\mathfrak{F}$ carries the boundary chain b'_1 of $\mathcal{B}'$ into itself, displacing $b'_{1\mathcal{D}}$ one step backward, i.e. $\mathcal{S}$ into $\mathcal{S}'_{-1}$. If $\mathfrak{p}'$ is a rotation, then b'_1 is the whole boundary of $\mathcal{B}'$. In the case 1) ($\mathcal{P}$ is a centre), b'_1 is a regular polygon with ν vertices in $\mathcal{D}$, and in the case 2) ($\mathcal{P}$ is a limit-centre) b'_1 is a polygon with ν vertices on $\mathcal{E}$, and in the case 3) ($\mathcal{P}$ an axis) b'_1 consists of ν disjoint lines connected by ν arcs of $\mathcal{E}$. In all cases the centre $\mathcal{P}'$ is inside $\mathcal{B}'$. If $\mathfrak{p}'$ is a limit-rotation, then b'_1 has infinitely many sides, the ends of b'_1 converging to $\mathcal{P}'$, and $\mathcal{P}'$ does not belong to $\mathcal{Y}$. If $\mathfrak{p}'$ is a translation, then b'_1 has infinitely many sides, the ends of b'_1 converging to the end-points u' and v' of $\mathcal{P}'$, and these end-points do not belong to $\mathcal{Y}$. The axis $\mathcal{P}'$ is inside $\mathcal{B}'$.

In order to comprise the three possibilities for $\mathcal{P}'$ under one notation, *let* $\mathcal{B}_1'$ *denote the interior of the convex hull of* b_1'. Then $\mathcal{B}_1' = \mathcal{B}'$, if $\mathfrak{p}'$ is a rotation or limit-rotation. If $\mathfrak{p}'$ is a translation, then the boundary of $\mathcal{B}_1'$ consists of b_1' and the axis $\mathcal{P}'$ together with its end-points u' and v'. The subgroup of $\mathfrak{F}$ carrying $\mathcal{B}_1'$ into itself, must carry b_1' into itself and is thus the group $\mathfrak{B}_1' = \mathfrak{P}'$, whatever the character of $\mathcal{P}'$. Then $\mathfrak{P}'$ and $\mathfrak{F}$ have equal effect in $\mathcal{B}_1'$.

Consider the case that $\mathcal{P}'$ is an axis, and let $\mathfrak{f}$ be any element of $\mathfrak{F}$. Then $\mathfrak{f}b_1'$ is a continuous curve in $\overline{\mathcal{D}}$ joining $\mathfrak{f}u'$ and $\mathfrak{f}v'$. The parts of $\mathfrak{f}b_1'$ belonging to $\mathcal{D}$ cannot enter into $\mathcal{B}_1'$, because $\mathcal{B}_1'$ is part of $\mathcal{B}'$, and $\mathcal{B}'$ does not contain any point of $\mathfrak{F}\mathcal{S}$. The parts of b_1' belonging to $\mathcal{E}$, if any, belong to $\mathcal{Y}$ and thus contain neither u' nor v', because $\mathcal{Y}$ contains no fundamental point. It is inferred that the pair $\mathfrak{f}u'$, $\mathfrak{f}v'$ cannot separate the pair u', v' on $\mathcal{E}$. Hence the axis $\mathcal{P}'$ is simple modulo $\mathfrak{F}$.

Correspondingly we may consider the region $\mathcal{B}_1''$, the interior of the convex hull of b_1'', and the corresponding group $\mathfrak{B}_1'' = \mathfrak{P}''$. If $\mathcal{P}''$ is an axis, then this axis is simple modulo $\mathfrak{F}$.

It is seen that $\mathcal{P}$ cannot be equivalent with $\mathcal{P}'$ or $\mathcal{P}''$ with respect to $\mathfrak{F}$, for $\mathcal{P}'$ and $\mathcal{P}''$ have an empty intersection with $\mathcal{Y}$, while $\mathcal{P}$ has not, and $\mathcal{Y}$ is reproduced by $\mathfrak{F}$.

21.8 Equivalence or non-equivalence of $\mathcal{P}'$ and $\mathcal{P}''$. Assume that $\mathfrak{f}$ were an element of $\mathfrak{F}$ carrying b_1'' into b_1'. Then $\mathfrak{f}$ maps $\mathcal{B}''$ onto $\mathcal{B}'$. Since b_1'' bounds $\mathcal{B}''$ negatively and b_1' bounds $\mathcal{B}'$ positively, $\mathfrak{f}$ must be a reversion. Since $\mathcal{S}$ is a side of b_1'', $\mathfrak{f}\mathcal{S}$ is a side of b_1'. Since $\mathcal{S}$ is also a side of b_1', it is carried into $\mathfrak{f}\mathcal{S}$ by an element of $\mathfrak{P}'$, thus by a motion. This contradicts the fact that there is only one element of $\mathfrak{F}$ carrying $\mathcal{S}$ into any of its equivalents.

From this follows that $\mathcal{B}_1''$ and $\mathcal{B}_1'$ are disjoint modulo $\mathfrak{F}$. Therefore, an element of $\mathfrak{F}$ carrying $\mathcal{B}''$ into $\mathcal{B}'$, if any, cannot carry $\mathcal{B}_1''$ into $\mathcal{B}_1'$, and $\mathcal{B}_1'$ and $\mathcal{B}_1''$ cannot be the whole of $\mathcal{B}'$ and $\mathcal{B}''$, respectively. Hence, if $\mathcal{P}''$ and $\mathcal{P}'$ are equivalent with respect to $\mathfrak{F}$, say $\mathfrak{h}\mathcal{P}'' = \mathcal{P}'$, $\mathfrak{h} \in \mathfrak{F}$, then both must be axes of $\mathfrak{F}$, and $\mathfrak{h}\mathcal{B}_1''$ must be adjacent to $\mathcal{B}_1'$ along $\mathcal{P}'$. In this case, $\mathcal{B}'$ consists of $\mathcal{B}_1'$, $\mathcal{P}'$ and $\mathfrak{h}\mathcal{B}_1''$, and the complete boundary of $\mathcal{B}'$ consists of b_1', $\mathfrak{h}b_1''$, u' and v'. Since $\mathfrak{h}$ maps the half-plane determined by $\mathcal{P}''$ which contains $\mathcal{P}'$ onto the half-plane determined by $\mathcal{P}'$ which does not contain $\mathcal{P}''$, it leaves two points of $\mathcal{E}$ invariant, and its axis cuts $\mathcal{P}''$ and $\mathcal{P}'$. According as $\mathfrak{h}b_1''$ bounds $\mathcal{B}'$ negatively or positively, $\mathfrak{h}$ is a translation or reversed translation. Since one gets $\mathfrak{h}\mathcal{B}'' = \mathcal{B}'$, all regions of the decomposition of $\mathcal{D}$ by $\mathfrak{F}\mathcal{S}$ are equivalent with respect to $\mathfrak{F}$. This case will be treated further in Section 16.

Apart from this, case $\mathcal{P}'$ and $\mathcal{P}''$ are not equivalent and $\mathcal{P}$, $\mathcal{P}'$, $\mathcal{P}''$ then belong to three different equivalence classes with respect to $\mathfrak{F}$.

21.9 Elementary groups. $\mathfrak{F}$-groups which are generated by two of their primary elements, each of them being either a rotation, or a limit-rotation, or a translation along a boundary axis of $\mathfrak{F}$, are called *elementary groups*. The symbol $\mathfrak{E}$ is reserved for them. For reasons which will appear later, it is often convenient to say that they are generated by three elements of this special kind with the product 1. An elementary group consists of motions only. It is emphasized that the notation "elementary group" does not comprise quasi-abelian groups, thus not the group of Section 5, which is generated by the half-turns $\mathfrak{p}$ and $\mathfrak{p}'$.

It is inferred from the defining relations (7) and (8) of the $\mathfrak{F}$-group $\mathfrak{N}$ in the generators $\mathfrak{s}$, $\mathfrak{s}'$, $\mathfrak{s}''$ that any two products of these generators representing the same element of $\mathfrak{N}$ both have an even, or both an odd, number of factors $\mathfrak{s}$, $\mathfrak{s}'$, $\mathfrak{s}''$, because application of the defining relations does not change the parity. The elements of $\mathfrak{N}$ with an even number of factors $\mathfrak{s}$, $\mathfrak{s}'$, $\mathfrak{s}''$ thus form a normal subgroup of $\mathfrak{N}$, and this subgroup is an $\mathfrak{F}$-group having the same convex domain as $\mathfrak{N}$ (§14.1). Any product of two different factors is one of the elements (6) or their inverses. The subgroup in question is thus generated by the motions (6) satisfying the relation (3) and is thus an elementary group $\mathfrak{E}$.

We want to show that (3) and (9) form a set of defining relations for $\mathfrak{E}$. Consider a product of $\mathfrak{p}$, $\mathfrak{p}'$, $\mathfrak{p}''$ which is 1. If these generators are replaced by their expressions (6) in $\mathfrak{s}$, $\mathfrak{s}'$, $\mathfrak{s}''$, then the relation follows from (7) and (8). The application of (7) for a reduction implies that the product contains a partial product $(xy)(yz)$, where x, y, z stand for some of the symbols $\mathfrak{s}$, $\mathfrak{s}'$, $\mathfrak{s}''$. If $x = z$, then the partial product is 1, and this follows from a trivial relation in the $\mathfrak{p}$'s, such as $\mathfrak{p}\mathfrak{p}^{-1} = 1$. If $x \neq z$, then the replacement of the partial product by xz means applying (3). Further the application of (8) results in an application of (9); for, even if in a transform of one of the relations (8) the transformer has an odd number of factors, it can immediately be replaced by a transformer with an even number of factors, as is indicated by

$$\begin{aligned} \mathfrak{s}''(\mathfrak{s}''\mathfrak{s}')^{\nu}\mathfrak{s}'' &= (\mathfrak{s}'\mathfrak{s})^{\nu}, \\ \mathfrak{s}'(\mathfrak{s}''\mathfrak{s}')^{\nu}\mathfrak{s}' &= (\mathfrak{s}'\mathfrak{s}'')^{\nu}, \\ \mathfrak{s}(\mathfrak{s}''\mathfrak{s}')^{\nu}\mathfrak{s} &= \mathfrak{s}\mathfrak{s}''(\mathfrak{s}'\mathfrak{s}'')^{\nu}\mathfrak{s}''\mathfrak{s}, \end{aligned}$$

and the corresponding ones arrived at by permutations.

Since the index of $\mathfrak{E}$ in $\mathfrak{N}$ is 2, the fundamental domain $\tilde{\mathcal{Q}}$ for $\mathfrak{N}$ in $\mathcal{D}$ together with its image by one of the reflections $\mathfrak{s}$, $\mathfrak{s}'$, $\mathfrak{s}''$, say by $\mathfrak{s}$, is a fundamental polygon $\mathcal{Z}$ for $\mathfrak{E}$ (§10.5). This polygon $\mathcal{Z}$ has four sides in $\mathcal{D}$, two of which correspond by $\mathfrak{p}'$, the other by $\mathfrak{p}''$. If $\mathcal{P}'$ is a centre or limit-centre, it is a complete vertex cycle; if $\mathcal{P}'$ is an axis, the arc of the polygon $\mathcal{Z}$ on $\mathcal{E}$ is fundamental for $\mathfrak{P}'$. The same holds for $\mathcal{P}''$ and $\mathfrak{P}''$. If $\mathcal{P}$ is a centre or limit-centre, then $\mathcal{P}$ together with $\mathfrak{s}\mathcal{P}$ forms a complete vertex cycle, and if $\mathcal{P}$ is an axis, two of the arcs of $\mathcal{Z}$ on $\mathcal{E}$ taken together are fundamental for $\mathfrak{P}$. The above abstract presentation of $\mathfrak{E}$ may then also be read from §10.3.

The area $\Phi(\mathfrak{E})$ of $\mathcal{K}(\mathfrak{E})$ is twice the value (10), thus

$$\Phi(\mathfrak{E}) = 2\left(1 - \frac{1}{\nu} - \frac{1}{\nu'} - \frac{1}{\nu''}\right). \tag{11}$$

Remark. For reasons to appear later, note that the fundamental polygon $\mathcal{Z}$ for $\mathfrak{E}$ in $\mathcal{D}$ is divided into two parts by the segment or full line $\mathcal{S}$. Let $\mathcal{Z}'$ denote the part of

$\mathcal{Z} \cap \mathcal{K}(\mathfrak{E})$ which has $\mathcal{S}$ and $\mathcal{R}'$ on its boundary, and $\mathcal{Z}''$ the part of $\mathcal{Z} \cap \mathcal{K}(\mathfrak{E})$ which has $\mathcal{S}$ and $\mathcal{R}''$ on its boundary. We consider in particular the case where $\mathcal{P}'$ and $\mathcal{P}''$ are axes. If $\mathcal{P}$ is a centre, and α' and α'' are the angles at $\mathcal{P}$ of $\mathcal{Z}'$ and $\mathcal{Z}''$, respectively, then the area of $\mathcal{Z}'$ is calculated from a quadrangle with two angles equal to α', the two others equal to $\frac{\pi}{2}$, thus $\Phi(\mathcal{Z}') = 1 - \frac{2\alpha'}{\pi}$. Correspondingly, $\Phi(\mathcal{Z}'') = 1 - \frac{2\alpha''}{\pi}$. Since $\alpha' + \alpha'' = \frac{\pi}{\nu}$, one gets $\Phi(\mathcal{Z}') + \Phi(\mathcal{Z}'') = 2\left(1 - \frac{1}{\nu}\right)$ in accordance with (11), since $\nu' = \nu'' = \infty$. If $\mathcal{P}$ is a limit-centre or an axis, one gets $\Phi(\mathcal{Z}') = \Phi(\mathcal{Z}'')$ which again corresponds to (11), since all three exponents are ∞.

21.10 Equal effect of $\mathfrak{F}$ and $\mathfrak{E}$ in the interior of $\mathcal{K}(\mathfrak{E})$. Since $\mathfrak{F}$ and $\mathfrak{P}'$ have equal effect in $\mathcal{B}_1'$, and $\mathfrak{F}$ and $\mathfrak{P}''$ in $\mathcal{B}_1''$, and since $\mathcal{B}_1'$ and $\mathcal{B}_1''$ are disjoint modulo $\mathfrak{F}$ (Section 8), the intersection $\mathcal{Z}_0$ of the fundamental polygon $\mathcal{Z}$ for $\mathfrak{E}$ with the union $\mathcal{B}_1' \cup \mathcal{S} \cup \mathcal{B}_1''$ contains in its interior no two points equivalent with respect to $\mathfrak{F}$. Those sides of $\mathcal{Z}_0$ which are inside $\mathcal{K}(\mathfrak{E})$ correspond in pairs, the two of them by $\mathfrak{p}'$, the other two by $\mathfrak{p}''$. From this follows that $\mathfrak{F}$ and $\mathfrak{E}$ have equal effect in the interior of $\mathcal{K}(\mathfrak{E})$, since $\mathfrak{p}'$ and $\mathfrak{p}''$ generate $\mathfrak{E}$. If $\mathcal{P}$ is an axis, then it is a boundary axis both of $\mathfrak{E}$ and $\mathfrak{F}$, and they then have equal effect also on $\mathcal{P}$ and in the corresponding half-plane. Equal effect of $\mathfrak{F}$ and $\mathfrak{E}$ on $\mathcal{P}'$ and $\mathcal{P}''$ is obvious, if they are centres or limit-centres, and also if they are boundary axes of $\mathfrak{F}$, but need not occur, if they are inner axes of $\mathfrak{F}$, as is seen from the special case considered in Section 8.

From this follows that $\mathfrak{E} = \mathfrak{F}$, if $\mathcal{P}'$ is a centre, or limit-centre, or boundary axis of $\mathfrak{F}$, and if the same holds for $\mathcal{P}''$.

In the particular case $\nu = 2$, $\nu' = 3$, $\nu'' = 7$ one gets $\mathfrak{E} = \mathfrak{F}$ and $\Phi(\mathfrak{F}) = \frac{1}{21}$. This shows that the lower bound established at the end of §17, for the area $\Phi(\mathfrak{F})$ of an $\mathfrak{F}$-group consisting of motions can be reached.

21.11 Reversibility of $\mathcal{S}$. Before we proceed further, the assumption of Section 6 has to be complemented by the contrary assumption. We thus revert to the general case of Sections 2 and 3 and assume that $\mathcal{S}$ is reversible (Section 4). The mid-point $\mathcal{P}''$ of s is then a centre of order $\nu'' = 2$. The half-turn around $\mathcal{P}''$ is denoted by $\mathfrak{p}''$. If one of the regions of decomposition of $\mathcal{D}$ by $\mathfrak{F}\mathcal{S}$ adjacent to $\mathcal{S}$ is denoted by $\mathcal{B}$, the other is $\mathfrak{p}''\mathcal{B}$, they are thus equivalent with respect to $\mathfrak{F}$. There is no way of directing $\mathfrak{F}\mathcal{S}$ coherently. $\mathcal{S}$ joins $\mathcal{P}$ and $\mathfrak{p}''\mathcal{P}$ in the cases 1) and 2), and $\mathcal{S}$ is the common normal of $\mathcal{P}$ and $\mathfrak{p}''\mathcal{P}$ in the case 3). Fig. 21.2 illustrates the three cases. The number of lines of $\mathfrak{F}\mathcal{S}$ issuing from $\mathcal{P}$ in the cases 1) and 2), and the number of lines of $\mathfrak{F}\mathcal{S}$ cutting $\mathcal{P}$ in the case 3), reduce modulo $\mathfrak{P}$ to one, instead of two in the case of non-reversibility (Section 4). Any individual of the collection $\mathfrak{F}\mathcal{S}$ is carried into any other by exactly two elements of $\mathfrak{F}$.

If $\nu = 2$, thus $\mathfrak{p}$ is a half-turn, then $\mathcal{P}$ and $\mathcal{P}''$ are a pair of rotation twins. This case has been dealt with in Section 5. We therefore assume $\nu > 2$.

Let b_1 be the boundary chain of $\mathcal{B}$ to which $\mathcal{S}$ belongs. We choose the primary element $\mathfrak{p}$ of $\mathfrak{P}$ such that $\mathcal{S}_{-1} = \mathfrak{p}^{-1}\mathcal{S}$ is the other link of b_1 ending in $\mathcal{P}$ or intersect-

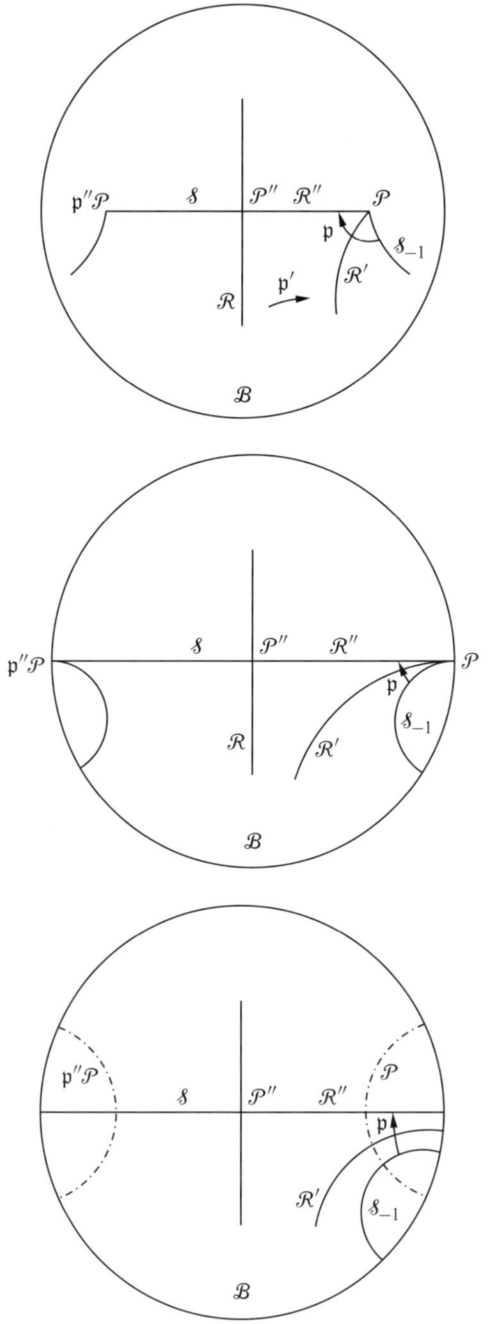
$\mathfrak{p}''\mathcal{P}$
$\mathcal{S}$
$\mathcal{P}''$
$\mathcal{R}''$
$\mathcal{P}$
$\mathfrak{p}$
$\mathcal{S}_{-1}$
$\mathcal{R}'$
$\mathfrak{p}'$
$\mathcal{R}$
$\mathcal{B}$
$\mathfrak{p}''\mathcal{P}$
$\mathcal{S}$
$\mathcal{P}''$
$\mathcal{R}''$
$\mathcal{P}$
$\mathfrak{p}$
$\mathcal{S}_{-1}$
$\mathcal{R}$
$\mathcal{R}'$
$\mathcal{B}$
$\mathfrak{p}''\mathcal{P}$
$\mathcal{S}$
$\mathcal{P}''$
$\mathcal{R}''$
$\mathcal{P}$
$\mathfrak{p}$
$\mathcal{R}'$
$\mathcal{S}_{-1}$
$\mathcal{B}$

Figure 21.2

ing $\mathcal{P}$. If there were a reversion $\mathfrak{h}$ of $\mathfrak{F}$ carrying $\mathcal{S}_{-1}$ into $\mathcal{S}$, then $\mathfrak{h}\mathfrak{p}^{-1}\mathcal{S} = \mathfrak{h}\mathcal{S}_{-1} = \mathcal{S}$, and $\mathfrak{h}\mathfrak{p}^{-1}$ would be a reversion carrying $\mathcal{S}$ into itself. It cannot be a reversed translation, because a reversed translation carries no finite segment into itself (case 1)), and $\mathcal{S}$ is in the cases 2) and 3) not an axis of $\mathfrak{F}$. It cannot be a reflection in $\mathcal{S}$, because $\mathfrak{F}$ contains no reflections. Hence no such reversion exists, and only the case of a motion occurs, as in Section 7. The motion $\mathfrak{p}''\mathfrak{p}$ carries b_1 into itself, advancing it one step.

Denote by $\mathcal{R}$ the normal of $\mathcal{S}$ in $\mathcal{P}''$, by $\mathcal{R}'$ the symmetry line of $\mathcal{S}$ and $\mathcal{S}_{-1}$ in $\mathcal{B}$, and by $\mathcal{R}''$ the line on which $\mathcal{S}$ is situated. Again, $\mathfrak{s}$, $\mathfrak{s}'$, $\mathfrak{s}''$ mean the reflections in $\mathcal{R}$, $\mathcal{R}'$, $\mathcal{R}''$, respectively, and putting $\mathfrak{p}' = \mathfrak{s}'\mathfrak{s}$ we get the relations (6), (7), (8), (9). The reflection group $\mathfrak{R}$ generated by $\mathfrak{s}$, $\mathfrak{s}'$, $\mathfrak{s}''$ is an $\mathfrak{F}$-group. Further it is seen as in Section 7, that $\nu' > 2$. One gets the fundamental domain $\tilde{\mathcal{Q}}$ for $\mathfrak{R}$ bounded on $\mathcal{D}$ by $\mathcal{R}$, $\mathcal{R}'$, $\mathcal{R}''$, where now $\mathcal{R}''$ contains $\mathcal{S}$. The area of $\mathcal{K}(\mathfrak{R})$ mod $\mathfrak{R}$ is given by (10), which in this case may be written

$$\Phi(\mathfrak{R}) = \frac{1}{2} - \frac{1}{\nu} - \frac{1}{\nu'}. \tag{11}$$

This implies that $\frac{1}{\nu} + \frac{1}{\nu'} < \frac{1}{2}$. The further discussion of $\mathfrak{R}$ goes as in Section 7. If the join $\mathcal{P}'$ of $\mathcal{R}$ and $\mathcal{R}'$ is a centre or an axis, it is inside $\mathcal{B}$. If it is a limit-centre, then it is the common point of convergence for the ends of b_1. Also $\mathcal{P}'$ cannot be equivalent with $\mathcal{P}$ with respect to $\mathfrak{F}$ for the same reasons as in Section 7. And $\mathcal{P}''$ cannot be equivalent with $\mathcal{P}$ or $\mathcal{P}'$, because it is the only centre of order 2. Thus $\mathcal{P}$, $\mathcal{P}'$, $\mathcal{P}''$ belong to three different equivalence classes with respect to $\mathfrak{F}$.

21.12 Coincidence of the cases of reversibility and non-reversibility. We revert to the case of non-reversibility considered in the Sections 6–10. In order to get a fundamental polygon for $\mathfrak{E}$ we took the fundamental polygon $\tilde{\mathcal{Q}}$ for $\mathfrak{R}$ together with $\mathfrak{s}\tilde{\mathcal{Q}}$. We might as well take $\tilde{\mathcal{Q}}$ together with $\mathfrak{s}'\tilde{\mathcal{Q}}$ in order to get a fundamental polygon for $\mathfrak{E}$. This we now do.

First, assume that the order ν of $\mathfrak{p}$ is > 2. Then, if $\mathcal{P}$ is a centre, the angle of $\tilde{\mathcal{Q}}$ at $\mathcal{P}$ is acute. Since $\nu' > 2$, also the angle of $\tilde{\mathcal{Q}}$ at $\mathcal{P}'$ is acute, if $\mathcal{P}'$ is centre. Thus in all cases the perpendicular from $\mathcal{P}''$ on $\mathcal{R}'$, or the common normal of $\mathcal{P}''$ and $\mathcal{R}'$, is inside $\mathcal{Q}$. Its image by the reflection in $\mathcal{R}'$ is then inside $\mathfrak{s}'\mathcal{Q}$. Hence the join of $\mathcal{P}''$ and $\mathfrak{s}'\mathcal{P}''$ is contained in a fundamental polygon for $\mathfrak{E}$, and its open part s^* between $\mathcal{P}''$ and $\mathfrak{s}'\mathcal{P}''$ is simple modulo $\mathfrak{E}$, and thus modulo $\mathfrak{F}$, since $\mathfrak{E}$ and $\mathfrak{F}$ have equal effect in the open part of $\mathcal{K}(\mathfrak{E})$ (Section 10). Hence we might start the construction in Section 2 with $\mathcal{P}''$ instead of $\mathcal{P}$ and s^* instead of s, if $\mathcal{P}''$ is a centre or limit-centre or a boundary axis of $\mathfrak{F}$, and s^* would be non-reversible, because there is no centre inside a fundamental polygon for $\mathfrak{E}$. Then $\mathcal{R}$ and $\mathcal{R}''$ would play the rôle previously played by $\mathcal{R}'$ and $\mathcal{R}''$, and $\mathcal{R}'$ the rôle previously played by $\mathcal{R}$; hence only an exchange of the rôles of $\mathcal{R}$ and $\mathcal{R}'$ would result.

Secondly, assume $\nu = 2$. Then $\mathcal{R}'$ and $\mathcal{R}''$ are orthogonal. Thus the perpendicular from $\mathcal{P}''$ on $\mathcal{P}'$, or the common normal of $\mathcal{P}''$ and $\mathcal{R}'$, if $\mathcal{P}''$ is an axis, is on $\mathcal{R}''$. The open part s^* of the join of $\mathcal{P}''$ and $\mathfrak{s}'\mathcal{P}''$ is still simple modulo $\mathfrak{F}$, but it is now reversible, because it passes through the centre $\mathcal{P}$ of order 2. Thus, if the segment s^* joining $\mathcal{P}''$ and $\mathfrak{s}'\mathcal{P}''$ is the starting segment of the construction, we get the case of Section 11, likewise with the interchange of the rôles of $\mathcal{R}$ and $\mathcal{R}'$, while $\mathcal{R}''$ now contains s^*, thus the new $\mathcal{S}^*$, as in Section 11.

Finally, since $\mathcal{P}'$ and $\mathcal{P}''$ play equal rôles in the Sections 6–10, the same applies, if $\mathcal{P}'$ is taken as the starting point or axis. We then have to take $\tilde{\mathcal{Q}}$ together with $\mathfrak{s}''\tilde{\mathcal{Q}}$, and we get an exchange of the rôles of $\mathcal{R}$ and $\mathcal{R}''$.

Thus whichever of $\mathcal{P}$, $\mathcal{P}'$, $\mathcal{P}''$ is taken as the start, one can choose the segment s such that one gets the same group $\mathfrak{R}$ and hence also the same group $\mathfrak{E}$.

This also shows that the distinction between the cases of reversibility and non-reversibility is unessential for $\mathfrak{E}$, because the segment s with which the start of the construction is made, can always be so chosen that it is non-reversible.

21.13 Elementary surfaces. The surface $\mathcal{K}(\mathfrak{E})$ mod $\mathfrak{E}$, where $\mathfrak{E}$ is an elementary group, will be called an *elementary surface* and denoted by the symbol ES($\mathfrak{E}$). The surface $\mathcal{D}$ mod $\mathfrak{E}$ will be called a *complete elementary surface*. The two concepts differ only if $\mathcal{K}(\mathfrak{E})$ has boundaries. Such boundaries are boundary axes, because $\mathfrak{E}$ is finitely generated and thus satisfies the condition of quasi-compactness.

Keeping the notation of the preceding sections, $\mathcal{P}$, $\mathcal{P}'$, $\mathcal{P}''$ now play equal rôles in consequence of Section 12, as far as $\mathfrak{E}$ is concerned, though possibly not as far as $\mathfrak{F}$ is concerned. We may call them the *determinators for* $\mathfrak{E}$. The surface $\mathcal{K}(\mathfrak{E})$ mod $\mathfrak{E}$ = ES($\mathfrak{E}$) has a conical point corresponding to each centre, a mast corresponding to each limit-centre, and a bounding closed geodesic corresponding to each axis among $\mathcal{P}$, $\mathcal{P}'$, $\mathcal{P}''$. These conical points, masts and closed bounding geodesics may be called the *protrusions* of ES($\mathfrak{E}$). We thus get a complete enumeration in the following table, the order in which the three symbols are put in the parentheses being arbitrary:

Table 4

Case	Surface	Case	Surface
a)	ES($\circlearrowleft_\nu$ $\circlearrowleft_{\nu'}$ $\circlearrowleft_{\nu''}$)	f)	ES($\circlearrowleft_\nu$ $\hookrightarrow$ $\rightarrow$)
b)	ES($\circlearrowleft_\nu$ $\circlearrowleft_{\nu'}$ $\hookrightarrow$)	g)	ES($\hookrightarrow$ $\hookrightarrow$ $\rightarrow$)
c)	ES($\circlearrowleft_\nu$ $\hookrightarrow$ $\hookrightarrow$)	h)	ES($\circlearrowleft_\nu$ $\rightarrow$ $\rightarrow$)
d)	ES($\hookrightarrow$ $\hookrightarrow$ $\hookrightarrow$)	i)	ES($\hookrightarrow$ $\rightarrow$ $\rightarrow$)
e)	ES($\circlearrowleft_\nu$ $\circlearrowleft_{\nu'}$ $\rightarrow$)	j)	ES($\rightarrow$ $\rightarrow$ $\rightarrow$)

In virtue of (11), the area $\Phi(\mathfrak{E})$ is $2\left(1 - \frac{1}{\nu} - \frac{1}{\nu'} - \frac{1}{\nu''}\right)$ for the type a), $2\left(1 - \frac{1}{\nu} - \frac{1}{\nu'}\right)$ for the types b), e), $2\left(1 - \frac{1}{\nu}\right)$ for the types c), f), h), and 2 for the types d), g), i), j). Since $\mathfrak{E}$ consists of motions only, one gets $\vartheta(\rightleftharpoons) = 0$. Formula (12), §17, yields $p = 0$ for all ten types.

It is seen from the remark at the end of Section 9 that an elementary surface may be divided into two parts by the simple geodetic arc σ on $\mathcal{K}(\mathfrak{E})$ mod $\mathfrak{E}$ derived from $\mathcal{S}$. Then σ either joins a conical point of $\mathcal{K}(\mathfrak{E})$ mod $\mathfrak{E}$ with itself, the ends of the arc meeting at the conical point at a certain angle, or σ is a full geodesic starting out of a mast and returning into that mast, or σ joins a closed bounding geodesic of $\mathcal{K}(\mathfrak{E})$ mod $\mathfrak{E}$ with itself, both ends of σ being orthogonal to that geodesic. In the cases d), g), i), j) of Table 4, each of the two surface parts has the area 1, half the area of the elementary surface.

21.14 The rôle of $\mathfrak{E}$ in the determination of $\mathfrak{F}$. If $\mathcal{K}(\mathfrak{E})$ has no boundary axes, it has no boundaries at all and is thus the whole of $\mathcal{D}$. Since $\mathfrak{F}$ and $\mathfrak{E}$ have equal effect in $\mathcal{K}(\mathfrak{E})$, one gets $\mathfrak{F} = \mathfrak{E}$ in these cases. This covers the types a)–d) in Table 4. For the remaining types e)–j), one also gets $\mathfrak{F} = \mathfrak{E}$ provided the boundary axes of $\mathfrak{E}$ are also boundary axes of $\mathfrak{F}$ (§14.1). Hence the case in which $\mathfrak{E}$ does not exhaust $\mathfrak{F}$ requires that some of the boundary axes of $\mathfrak{E}$ are inner axes of $\mathfrak{F}$. In this respect it is recalled that, if the determinator $\mathcal{P}$ of $\mathfrak{E}$ with which the construction of $\mathfrak{E}$ started is an axis, it is a boundary axis of $\mathfrak{F}$. This is at least the case for the type j) in Table 4. The cases which we have to consider are thus those in which either $\mathcal{P}'$ or $\mathcal{P}''$, or both, are inner axes of $\mathfrak{F}$.

In this paragraph we consider some special cases of this kind. The general investigation of $\mathfrak{F}$ is taken up in the next paragraph.

21.15 The handle and the Klein bottle. Assume that both $\mathcal{P}'$ and $\mathcal{P}''$ are inner axes of $\mathfrak{F}$ and that they are equivalent with respect to $\mathfrak{F}$ (Section 8). Let $\mathfrak{h}\mathcal{P}'' = \mathcal{P}'$, $\mathfrak{h} \in \mathfrak{F}$. As shown in Section 8, $\mathfrak{h}$ is a translation or reversed translation. It is also known from Section 7 that $\mathcal{P}'$ is simple modulo $\mathfrak{F}$. Since $\mathfrak{F}$ and $\mathfrak{E}$ have equal effect in the interior of $\mathcal{K}(\mathfrak{E})$, no axis of the collection $\mathfrak{F}\mathcal{P}'$ enters into $\mathcal{K}(\mathfrak{E})$. Consider in the decomposition of $\mathcal{D}$ by $\mathfrak{F}\mathcal{P}'$ the region $\mathcal{C}$ to which $\mathcal{K}(\mathfrak{E})$ belongs. Both $\mathcal{P}'$ and $\mathcal{P}''$ are boundaries of $\mathcal{C}$. If $\mathcal{P}$ is a centre or limit-centre, then $\mathcal{C} = \mathcal{K}(\mathfrak{E})$. If $\mathcal{P}$ is an axis, and thus a boundary axis of $\mathfrak{F}$, then $\mathcal{C}$ consists of $\mathcal{K}(\mathfrak{E})$ and the half-planes adjacent to $\mathcal{K}(\mathfrak{E})$ along the collection $\mathfrak{E}\mathcal{P}$.

The boundaries of $\mathcal{C}$ thus consist of the two collections $\mathfrak{E}\mathcal{P}'$ and $\mathfrak{E}\mathcal{P}''$, and $\mathfrak{E}$ is the subgroup of $\mathfrak{F}$ carrying $\mathcal{C}$ into itself. The two collections of boundaries of $\mathcal{C}$ are non-equivalent with respect to $\mathfrak{E}$, but are equivalent with respect to $\mathfrak{F}$. We are thus in the case I of §19.2. The subgroup of $\mathfrak{E}$ belonging to $\mathcal{P}'$ is the translation group $\mathfrak{P}'$ generated by $\mathfrak{p}'$, and this is also the subgroup of $\mathfrak{F}$ belonging to $\mathcal{P}'$, because in case I of §19.2 we got $\mathfrak{A}_{\mathcal{P}'} = \mathfrak{A}^*_{\mathcal{P}}$, with the notation there used. Thus our present $\mathfrak{A}_{\mathcal{P}'}$ corresponds to (3) in §19.2.

Fig. 21.1 shows that $\mathfrak{p}'$ and $\mathfrak{p}''$ were so defined that $\mathcal{P}'$ and $\mathcal{P}''$, when provided with the direction derived from $\mathfrak{p}'$ and $\mathfrak{p}''$, respectively, bound a fundamental polygon for $\mathfrak{E}$ in the same sense (positively in the figure). Thus, if $\mathfrak{h}$ is a translation, then $\mathfrak{h}\mathcal{P}''$ coincides with $\mathcal{P}'$ with direction reversed, whereas it coincides with direction preserved if $\mathfrak{h}$ is a reversed translation. It then follows from §19.2 that $\mathfrak{h}$ has a presentation by

$$\begin{cases} \text{generators } \mathfrak{p}, \mathfrak{p}', \mathfrak{p}'', \mathfrak{h}, \\ \text{defining relations: (3), (9) with } \nu' \text{ and } \nu'' \text{ infinite, and } \mathfrak{p}'^{\mp 1} = \mathfrak{h}\mathfrak{p}''\mathfrak{h}^{-1}, \end{cases} \tag{12}$$

the upper or lower exponent of $\mathfrak{p}'$ corresponding to a translation or reversed translation $\mathfrak{h}$, respectively. Thus $\mathfrak{F}$ is completely determined and satisfies the condition of quasi-compactness. The number of generators may be reduced to three by eliminating $\mathfrak{p}$ by applying the relation (3). One then gets for $\mathfrak{F}$ the generators $\mathfrak{p}'$, $\mathfrak{p}''$, $\mathfrak{h}$ with the relation (12) and $(\mathfrak{p}'\mathfrak{p}'')^{\nu} = 1$. If (12) then is used for eliminating $\mathfrak{p}'$, one gets generators $\mathfrak{p}''$ and $\mathfrak{h}$ with the relation $(\mathfrak{h}\mathfrak{p}''^{\mp 1}\mathfrak{h}^{-1}\mathfrak{p}'')^{\nu} = 1$. If $\mathfrak{h}$ is a translation, we may put $\mathfrak{p}''^{-1} = \mathfrak{h}'$. Then $\mathfrak{F}$ is generated by the two translations $\mathfrak{h}$ and $\mathfrak{h}'$ with defining relation $(\mathfrak{h}\mathfrak{h}'\mathfrak{h}^{-1}\mathfrak{h}'^{-1})^{\nu} = 1$. If $\mathfrak{h}$ is a reversed translation, the same holds for $\mathfrak{h}^{-1}\mathfrak{p}'' = \mathfrak{h}'$. Then $\mathfrak{F}$ is generated by the two reversed translations $\mathfrak{h}$ and $\mathfrak{h}'$ with defining relation $(\mathfrak{h}^2\mathfrak{h}'^2)^{\nu} = 1$.

If $\mathfrak{h}$ is a translation, the surface $\mathcal{K}(\mathfrak{F})$ mod $\mathfrak{F}$ will be called a *handle* and denoted by the symbol $\mathcal{H}$. According to the three possibilities for the determinators $\mathcal{P}$ of the elementary surface, we write $\mathcal{H}(\circlearrowleft_{\nu})$, $\mathcal{H}(\hookrightarrow)$ or $\mathcal{H}(\rightarrow)$.

If $\mathfrak{h}$ is a reversed translation, the surface $\mathcal{K}(\mathfrak{F})$ mod $\mathfrak{F}$ will be called a *Klein bottle* and denoted by the symbol $\mathcal{K}$ with the three signatures $\mathcal{K}(\circlearrowleft_{\nu})$, $\mathcal{K}(\hookrightarrow)$, $\mathcal{K}(\rightarrow)$.

If $\mathcal{P}$ is a centre or limit-centre, then $\mathcal{H}$ or $\mathcal{K}$ is also $\mathcal{D}$ mod $\mathfrak{F}$. If $\mathcal{P}$ is an axis, then $\mathcal{D}$ mod $\mathfrak{F}$ arises by adding a funnel along the bounding geodesic of $\mathcal{H}$ or $\mathcal{K}$. Since a fundamental polygon for $\mathfrak{E}$ in $\mathcal{K}(\mathfrak{E})$ is also a fundamental polygon for $\mathfrak{F}$ in $\mathcal{K}(\mathfrak{F})$, and since the exponents ν' and ν'' are infinite, one gets

$$\begin{cases} \Phi(\mathcal{H}(\circlearrowleft_{\nu})) = \Phi(\mathcal{K}(\circlearrowleft_{\nu})) = 2\left(1 - \dfrac{1}{\nu}\right), \\ \Phi(\mathcal{H}(\hookrightarrow)) = \Phi(\mathcal{H}(\rightarrow)) = \Phi(\mathcal{K}(\hookrightarrow)) = \Phi(\mathcal{K}(\rightarrow)) = 2. \end{cases} \tag{13}$$

Let β' and β'' denote the closed bounding geodesics of $\mathcal{K}(\mathfrak{E})$ mod $\mathfrak{E}$ corresponding to $\mathcal{P}'$ and $\mathcal{P}''$, respectively, and provided with a direction such that they bound $\mathcal{K}(\mathfrak{E})$ mod $\mathfrak{E}$ in the same sense. They have equal length, since $\mathfrak{p}'^{\mp 1}$ and $\mathfrak{p}''$ are conjugate elements of $\mathfrak{F}$.

If $\mathfrak{h}$ is a translation, then the surface $\mathcal{H}$ is obtained from $\mathcal{K}(\mathfrak{E})$ mod $\mathfrak{E}$ by making β' and β'' coincide with opposite directions. Since $\mathfrak{F}$ is generated by motions, one gets $\vartheta(\rightleftharpoons) = 0$. Equation (17.12) yields the genus $p = 1$.

If $\mathfrak{h}$ is a reversed translation, then the surface $\mathcal{K}$ is obtained from $\mathcal{K}(\mathfrak{E})$ mod $\mathfrak{E}$ by making β' and β'' coincide with concordant directions. In this case $\vartheta(\rightleftharpoons) = 1$. Equation (11), §17, yields the genus $p = 2$.

If $\mathfrak{h}$ is the translation (resp. reversed translation) with axis $\mathcal{R}$ which maps $\mathcal{P}'$ to $\mathcal{P}''$, we will call the surface $\mathcal{K}(\mathfrak{F})$ mod $\mathfrak{F}$ the *standard handle* (resp. the *standard Klein bottle*). The general handle (resp. Klein bottle) arises from the standard handle (resp. Klein bottle) by an appropriate twist which comes from a uniquely defined translation with axis $\mathcal{P}'$. (See 18.11).

21.16 The case of a reversion. We finally revert to the situation of our investigation as it stood at the end of Section 6. In the decomposition of $\mathcal{D}$ by $\mathfrak{F}\mathcal{S}$ we considered the regions of decomposition $\mathcal{B}'$ and $\mathcal{B}''$ adjacent to $\mathcal{S}$ and denoted by b_1' and b_1'', those boundary chains of $\mathcal{B}'$ and $\mathcal{B}''$, respectively, which contain $\mathcal{S}$. Let $\mathcal{S}_0$ denote one of the two links neighbouring $\mathcal{S}$ in $b'_{1\mathcal{D}}$, and assume that there is a reversion $\mathfrak{h}$ of $\mathfrak{F}$ carrying $\mathcal{S}_0$ into $\mathcal{S}$. If $\mathcal{S}$ were reversible, thus, by a half-turn $\mathfrak{c}$ around the mid-point of s, then both $\mathfrak{h}$ and $\mathfrak{ch}$ would carry $\mathcal{S}_0$ into $\mathcal{S}$, and one of them would leave $\mathcal{P}$ invariant. Both $\mathfrak{h}$ and $\mathfrak{ch}$ are reversions, thus reversed translations, and a reversed translation cannot leave a centre or limit-centre or a boundary axis of $\mathfrak{F}$ fixed. It is thus inferred that $\mathcal{S}$ cannot be reversible. Hence we now assume $\mathfrak{F}\mathcal{S}$ to be coherently directed, and the consequences drawn therefrom in Section 6 still hold. The lines $\mathfrak{F}\mathcal{S}$ with end-point $\mathcal{P}$, or intersecting $\mathcal{P}$, reduce modulo $\mathfrak{P}$ to two. All sides of $b'_{1\mathcal{D}}$ bound $\mathcal{B}'$ positively, and all sides of $b''_{1\mathcal{D}}$ bound $\mathcal{B}''$ negatively. The notations $\mathcal{S}'_{-1}$, $\mathcal{S}''_{-1}$, $\mathcal{S}'_1$, $\mathcal{S}''_1$ apply as in Section 6.

From Section 7 onwards we assumed that the uniquely determined element of $\mathfrak{F}$ carrying $\mathcal{S}'_{-1}$ into $\mathcal{S}$ was a motion. It is now assumed to be a reversion, thus a reversed translation $\mathfrak{h}$. Since $\mathfrak{h}$ carries $\mathcal{S}'_{-1}$ into $\mathcal{S}$ with direction preserved, its axis $\mathcal{A}(\mathfrak{h})$ must intersect $\mathcal{S}'_{-1}$ and $\mathcal{S}$, say, in points $\mathcal{Q}'$ and $\mathcal{Q}$, respectively. Fig. 21.3 illustrates case 1), but the text covers all three cases. $[\mathcal{Q}', \mathcal{Q}] = \lambda$ is the displacement of $\mathfrak{h}$. Also

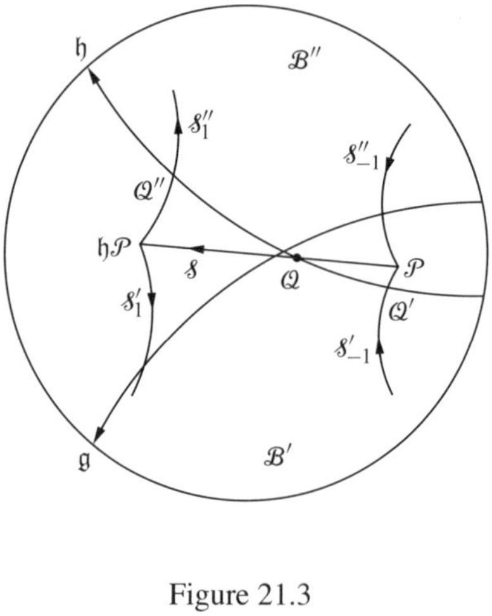

Figure 21.3

$\mathfrak{h}\mathcal{B}' = \mathcal{B}''$. The axis $\mathcal{A}(\mathfrak{h})$ cuts off from $\mathcal{B}'$ a triangle $\mathcal{Q}'\mathcal{P}\mathcal{Q}$ in the cases 1) and 2), and in the case 3) it determines together with $\mathcal{S}'_{-1}$ and $\mathcal{S}$ and the axis $\mathcal{P}$ a quadrangle in $\mathcal{B}'$. This triangle or quadrangle is by $\mathfrak{h}$ carried into a congruent triangle or quadrangle of $\mathcal{B}''$, and $\mathfrak{h}\mathcal{P}$ is the other end-point of $\mathcal{S}$ (cases 1) and 2)), or the other boundary axis cut by $\mathcal{S}$ (case 3)). Thus the $\mathfrak{k}$ of Section 2 may be taken to be $\mathfrak{h}$. (In the case 1) all angles of $\mathcal{B}'$ are equal, and equal to the angles of $\mathcal{B}''$; in the case 3) all segments of $\mathfrak{F}\mathcal{P}$ in $\mathcal{B}'$ have equal length, and equal to those in $\mathcal{B}''$.) The axis of $\mathfrak{h}$ cuts $\mathcal{S}'_1$ in a point $\mathcal{Q}''$, and $[\mathcal{Q}, \mathcal{Q}''] = \lambda$.

Consider now the uniquely determined element $\mathfrak{g}$ of $\mathfrak{F}$ which carries $\mathcal{S}$ into $\mathcal{S}_1'$. If $\mathfrak{g}$ were a motion, it would carry $\mathcal{B}'$ into itself, thus b_1' into itself advancing b_1' one step. This contradicts the fact that $\mathcal{S}_{-1}'$ is carried into $\mathcal{S}$ by a reversion. Thus $\mathfrak{g}$ is a reversed translation whose axis cuts $\mathcal{S}_{-1}''$, $\mathcal{S}$ and $\mathcal{S}_1'$. Then $\mathfrak{g}\mathfrak{h}$ is a motion carrying $\mathcal{S}_{-1}'$ into $\mathcal{S}_1'$, thus advancing b_1' two steps in itself. We denote its inverse by $\mathfrak{p}'$, and by $\mathfrak{P}'$ the group generated by $\mathfrak{p}'$. Also $\mathfrak{g}^{-1}\mathfrak{h}$ carries $\mathcal{S}_{-1}'$ into $\mathcal{S}_{-1}''$ and is thus a primary motion $\mathfrak{p}$ of $\mathfrak{P}$. If we put

$$\mathfrak{p} = \mathfrak{g}^{-1}\mathfrak{h}, \quad \mathfrak{p}' = \mathfrak{h}^{-1}\mathfrak{g}^{-1}, \quad \mathfrak{p}'' = \mathfrak{g}^2,$$

then we have three motions generating an $\mathfrak{F}$-group and satisfying (3). They thus generate an elementary group $\mathfrak{E}$.

We now consider, as in Sections 10 and 8, the interior $\mathcal{B}_1'$ of the convex hull of b_1'. One gets $\mathcal{B}_1' = \mathcal{B}'$, if $\mathfrak{p}'$ is a rotation or limit-rotation. If $\mathfrak{p}'$ is a translation, then $\mathcal{B}_1'$ is the part of $\mathcal{B}'$ cut off by the axis of $\mathfrak{p}'$, and this axis is simple modulo $\mathfrak{F}$ for the same reasons as given in Section 8. In all cases, $\mathfrak{F}$ and $\mathfrak{P}'$ have equal effect in $\mathcal{B}_1'$, and $\mathfrak{P}'$ is the group $\mathfrak{B}_1$ introduced in Section 3.

Consider the join $\mathcal{T}$ of $\mathcal{P}$ and $\mathfrak{p}'^{-1}\mathcal{P}$; this means the segment or full line joining them or their common normal according to the three cases (Fig. 21.4). $\mathcal{T}$ is contained in the region $\mathcal{B}_1'$ owing to its convexity. The collection $\mathfrak{P}'\mathcal{T}$ is a sequence inscribed in b_1'. Hence $\mathcal{T}$ is simple modulo $\mathfrak{P}'$ and then also modulo $\mathfrak{F}$. The sequence thus gives rise to a boundary chain for one of the regions of decomposition of $\mathcal{D}$ by $\mathfrak{F}\mathcal{T}$. Here $\mathcal{T}$ is reversible if and only if $\mathfrak{p}'$ is a half-turn. In this case the exponent ν for $\mathfrak{p}$ must be > 2, because otherwise $\mathcal{B}'$ would be bounded by a quadrangle with all angles equal to $\frac{\pi}{2}$. Hence the case of rotation twins considered in Section 5 does not arise.

Consider the reflection lines $\mathcal{R}$, $\mathcal{R}'$, $\mathcal{R}''$ pertaining to $\mathcal{T}$ for the reflection group $\mathfrak{R}$ corresponding to the construction in Section 7. $\mathcal{R}$ is the symmetry normal of $\mathcal{T}$, $\mathcal{R}'$ is the symmetry line of $\mathcal{T}$ and $\mathfrak{p}'\mathcal{T}$, and $\mathcal{R}''$ is the symmetry line of $\mathcal{T}$ and $\mathfrak{p}\mathfrak{p}'\mathcal{T} = \mathfrak{p}''^{-1}\mathcal{T}$. If $\mathfrak{s}$, $\mathfrak{s}'$, $\mathfrak{s}''$ denote the reflections in $\mathcal{R}$, $\mathcal{R}'$, $\mathcal{R}''$, respectively, we get the equations (6), and relations (3) and (9) for the elementary group $\mathfrak{E}$ generated by $\mathfrak{p}, \mathfrak{p}', \mathfrak{p}''$.

The axis of $\mathfrak{p}''$ is $\mathcal{P}'' = \mathcal{A}(\mathfrak{g})$, since $\mathfrak{p}'' = \mathfrak{g}^2$. We denote by $\mathfrak{P}''$ the group generated by $\mathfrak{p}''$, and by $\mathfrak{A}_{\mathcal{P}''}$ the subgroup of $\mathfrak{F}$ belonging to $\mathcal{P}''$, thus $\mathfrak{A}(\rightleftharpoons)$. We can then form the subgroup $\mathfrak{G}$ of $\mathfrak{F}$

$$\mathfrak{G} = \mathfrak{E} * \mathfrak{A}_{\mathcal{P}''} \operatorname{am} \mathfrak{P}''.$$

The result of this section may be stated as follows: One can reduce the *case of a reversion*, considered in this section, to the *case of a motion* considered in the Sections 7–15, by replacing the modulo $\mathfrak{F}$ simple segment $\mathcal{S}$ issuing from $\mathcal{P}$, or full line $\mathcal{S}$ cutting $\mathcal{P}$, by the equally modulo $\mathfrak{F}$ simple segment $\mathcal{T}$ issuing from $\mathcal{P}$, or full line $\mathcal{T}$ cutting $\mathcal{P}$, as the element of the construction whose equivalence class is used for the decomposition of $\mathcal{D}$. Therefore the case of a reversion adds nothing new. The $\mathcal{S}$ of the construction in this §21 can always be so chosen that in a boundary chain containing $\mathcal{S}$ two neighbouring links correspond by a motion.

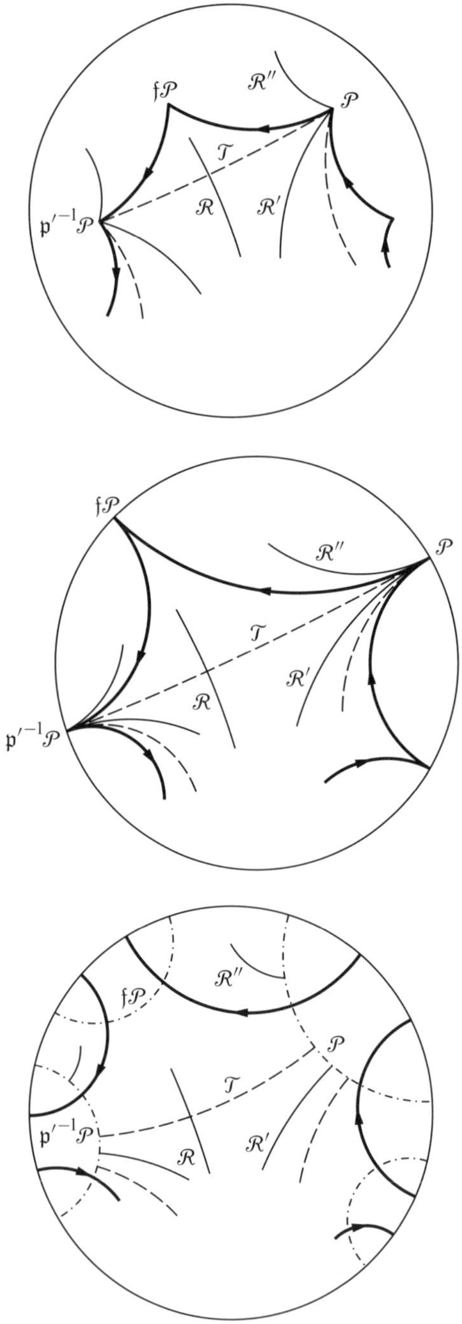

$\mathfrak{f}\mathcal{P}$
$\mathcal{R}''$
$\mathcal{P}$
$\mathcal{T}$
$\mathcal{R}$
$\mathcal{R}'$
$\mathfrak{p}'^{-1}\mathcal{P}$

Figure 21.4

21.17 Metric quantities of elementary groups. In this paragraph we have derived the 10 possible types of elementary groups and corresponding elementary surfaces. They are enumerated in the preceding Table 4 by indicating the 10 possible combinations of three determinators. In this section we are going to establish relations between certain metric quantities pertaining to elementary groups by taking as the basic set metric invariants pertaining to the determinators themselves. These three fundamental quantities will be denoted by $\alpha_1, \alpha_2, \alpha_3$.

Rotations are determined by their order ν. We denote by α half their primary angle of rotation, thus $\alpha = \frac{\pi}{\nu}$. For rotations this quantity only takes values from a discrete set, the largest possible value being $\frac{\pi}{2}$. Hence $\sin\alpha$ and $\cos\alpha$ are positive except for $\cos\frac{\pi}{2} = 0$.

Translations are determined by their displacements. We denote by α half their displacement. For translations this parameter takes values from a continuous set ranging from zero to infinity.

Limit-rotations are limiting cases both of rotations and of translations, the corresponding α tending to zero. Thus $\cos\alpha$ and $\cosh\alpha$ meet in the common value 1, and $\sin\alpha$ and $\sinh\alpha$ in the common value 0. In the deduction of the fundamental formulae the case of limit-rotations is excluded, because it can afterwards be taken into account by introducing the corresponding limit values of the α_ν. The number of possible cases then reduces to four. According as 3, 2, 1 or no rotations occur among the three determinators, the cases are denoted by the symbols (III), (IV), (V), (VI), because the main figure, which is the fundamental domain of the corresponding reflection group $\mathfrak{R}$ in $\mathcal{K}(\mathfrak{C})$, becomes a polygon with 3, 4, 5, 6 sides in these cases, respectively.

Denoting by $\mathcal{P}_\nu$, $\nu = 1, 2, 3$, the three determinators which formerly were called $\mathcal{P}$, $\mathcal{P}'$, $\mathcal{P}''$, we introduce the following metric quantities. δ_ν denotes the distance between $\mathcal{P}_{\nu-1}$ and $\mathcal{P}_{\nu+1}$, ν mod 3. These distances are the lengths of segments on the reflection lines $\mathcal{R}$, $\mathcal{R}'$, $\mathcal{R}''$ between the determinators in question. η_ν denotes the length of the perpendicular from the centre $\mathcal{P}_\nu$ on δ_ν or of the common normal of the axis $\mathcal{P}_\nu$ and δ_ν. Thus the η are half the length of the segments which were called s. η_1, η_2 and η_3 intersect in the same point (§9.6), and this point is inside the polygon except in the case of a centre of order 2, in which case it coincides with that centre.

The calculations to follow express the six quantities δ_ν, η_ν by the three fundamental quantities α_ν. They also express the α_ν by the δ_ν (*cosine relations* (IIIa) to (VIa)) and establish other relations between the α_ν and the δ_ν (*sine relations* (IIIf) to (VIf)). The common value of the three ratios in the sine relations is in all cases $\frac{\sigma}{\Delta}$.

In some of the formulae we make use of the following notations. $S\omega$ means $\sin\omega$ or $\sinh\omega$ according as ω represents an angle or a distance. Similarly $C\omega$ means $\cos\omega$ or $\cosh\omega$. All the $C\alpha_\nu$ are non-negative, and at least two of them are positive. All the $S\alpha_\nu$ are positive. We then put

$$\Delta^2 = C^2\alpha_1 + C^2\alpha_2 + C^2\alpha_3 + 2C\alpha_1 C\alpha_2 C\alpha_3 - 1.$$

This quantity is positive. That is evident, if at least one $C\alpha_\nu$ is $\cosh\alpha_\nu$, for then

$C^2\alpha_\nu > 1$. If all α_ν are angles, we may transform the expression into

$$\begin{aligned}\Delta^2 &= (C(\alpha_1+\alpha_2)+C\alpha_3)(C(\alpha_1-\alpha_2)+C\alpha_3)\\ &= 4\,C(\tfrac{1}{2}(\alpha_1+\alpha_2+\alpha_3))C(\tfrac{1}{2}(\alpha_1+\alpha_2-\alpha_3))\\ &\quad\times C(\tfrac{1}{2}(\alpha_1-\alpha_2+\alpha_3))C(\tfrac{1}{2}(-\alpha_1+\alpha_2+\alpha_3)).\end{aligned}$$

Here $C(\frac{1}{2}(\alpha_1+\alpha_2+\alpha_3)) > 0$, because $\alpha_1+\alpha_2+\alpha_3 < \pi$, and the three last factors are evidently positive.

At least two of the three α_ν are of the same kind. If α_1 and α_2 are of the same kind, thus $C\alpha_1$ and $C\alpha_2$ both cos or both cosh, then the transformation of Δ^2 into the first expression retains its meaning in the real domain. If all three α_ν are of the same kind, then the last expression for Δ^2 retains its meaning in the real domain.

By Δ we denote the positive value

$$\Delta = \sqrt{C^2\alpha_1 + C^2\alpha_2 + C^2\alpha_3 + 2C\alpha_1 C\alpha_2 C\alpha_3 - 1}.$$

Finally we introduce the positive quantity

$$\sigma = \frac{\Delta^2}{S\alpha_1\ S\alpha_2\ S\alpha_3}.$$

It is called the *sine amplitude* of the polygon. In the formulae (IIIg) to (VIg) its square is expressed as function of the δ.

Using the abbreviations Δ and σ we get the following set of formulae in the four cases (III), (IV), (V), (VI), respectively. Fig. 21.5, case III–case VI, indicate the arrangement of the pertinent quantities. In all cases the group c follows from the group b by calculating a sinh from a cosh or inversely. Also the group f follows immediately from the group c, and the group e follows from the groups c and d. Finally the group g results by expressing $C^2\alpha_\nu$ from the group a, then calculating $S^2\alpha_\nu$ from $C^2\alpha_\nu$ and using the group e.

Formulae:

$$\cosh\delta_\nu = \cosh\delta_{\nu-1}\cosh\delta_{\nu+1} - \sinh\delta_{\nu-1}\sinh\delta_{\nu+1}\cos\alpha_\nu \quad (\nu \bmod 3) \qquad \text{(IIIa)}$$

$$\cos\alpha_\nu = -\cos\alpha_{\nu-1}\cos\alpha_{\nu+1} + \sin\alpha_{\nu-1}\sin\alpha_{\nu+1}\cosh\delta_\nu \quad (\nu \bmod 3) \qquad \text{(IIIb)}$$

$$\sinh\delta_\nu = \frac{\Delta}{\sin\alpha_{\nu-1}\sin\alpha_{\nu+1}} \quad (\nu \bmod 3) \qquad \text{(IIIc)}$$

$$\sinh\eta_\nu = \frac{\Delta}{\sin\alpha_\nu} \quad (\nu \bmod 3) \qquad \text{(IIId)}$$

$$\sigma = \sinh\delta_\nu\sinh\eta_\nu = \sinh\delta_{\nu-1}\sinh\delta_{\nu+1}\sin\alpha_\nu \quad (\nu \bmod 3) \qquad \text{(IIIe)}$$

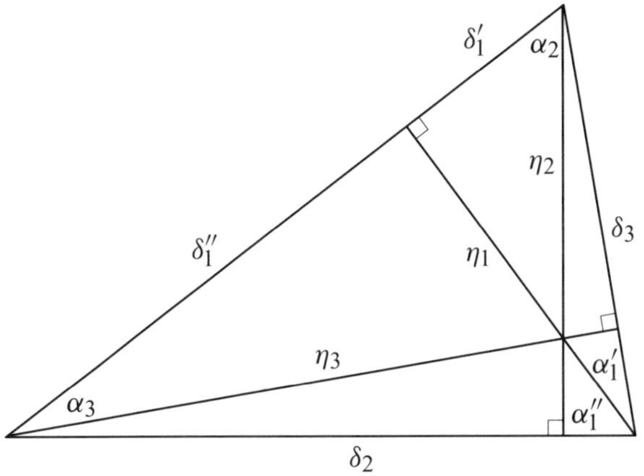
δ_1'
α_2
η_2
δ_3
δ_1''
η_1
η_3
α_1'
α_3
α_1''
δ_2

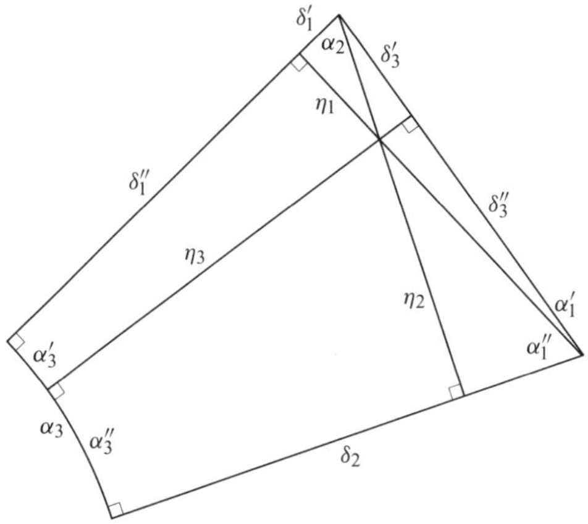
δ_1'
α_2
δ_3'
η_1
δ_1''
δ_3''
η_3
η_2
α_1'
α_1''
α_3'
α_3
α_3''
δ_2

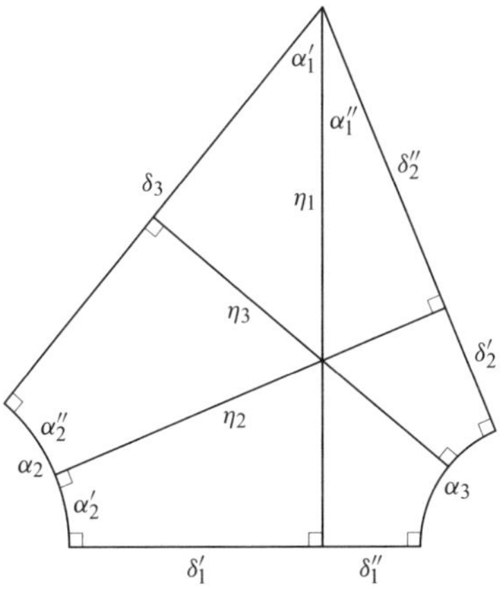
α_1'
α_1''
δ_2''
δ_3
η_1
η_3
δ_2'
α_2''
η_2
α_2
α_3
α_2'
δ_1'
δ_1''

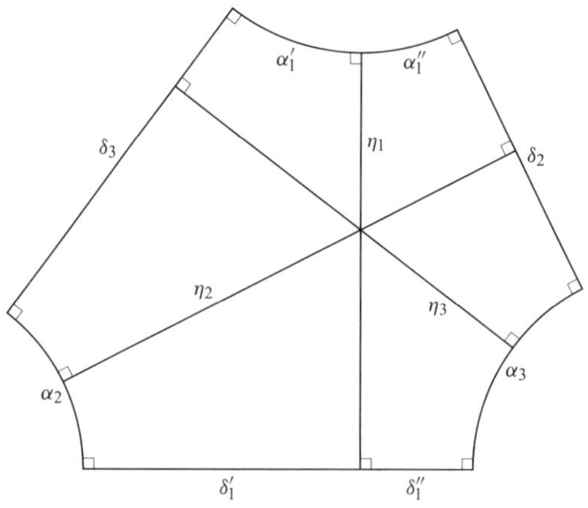

Figure 21.5

$$\frac{\sinh \delta_1}{\sin \alpha_1} = \frac{\sinh \delta_2}{\sin \alpha_2} = \frac{\sinh \delta_3}{\sin \alpha_3} \tag{IIIf}$$

$$\sigma^2 = 1 - \cosh^2 \delta_1 - \cosh^2 \delta_2 - \cosh^2 \delta_3 + 2 \cosh \delta_1 \cosh \delta_2 \cosh \delta_3. \tag{IIIg}$$

$$\begin{cases} \sinh \delta_1 & = \sinh \delta_2 \cosh \delta_3 - \cosh \delta_2 \sinh \delta_3 \cos \alpha_1 \\ \sinh \delta_2 & = \sinh \delta_1 \cosh \delta_3 - \cosh \delta_1 \sinh \delta_3 \cos \alpha_2 \\ \cosh \delta_3 & = - \sinh \delta_1 \sinh \delta_2 + \cosh \delta_1 \cosh \delta_2 \cosh \alpha_3 \end{cases} \tag{IVa}$$

$$\begin{cases} \cos \alpha_1 & = - \cos \alpha_2 \cosh \alpha_3 + \sin \alpha_2 \sinh \alpha_3 \sinh \delta_1 \\ \cos \alpha_2 & = - \cos \alpha_1 \cosh \alpha_3 + \sin \alpha_1 \sinh \alpha_3 \sinh \delta_2 \\ \cosh \alpha_3 & = - \cos \alpha_1 \cos \alpha_2 + \sin \alpha_1 \sin \alpha_2 \cosh \delta_3 \end{cases} \tag{IVb}$$

$$\begin{cases} \cosh \delta_1 & = \frac{\Delta}{\sin \alpha_2 \sinh \alpha_3} \\ \cosh \delta_2 & = \frac{\Delta}{\sin \alpha_1 \sinh \alpha_3} \\ \sinh \delta_3 & = \frac{\Delta}{\sin \alpha_1 \sin \alpha_2} \end{cases} \tag{IVc}$$

$$\begin{cases} \sinh \eta_1 & = \frac{\Delta}{\sin \alpha_1} \\ \sinh \eta_2 & = \frac{\Delta}{\sin \alpha_2} \\ \cosh \eta_3 & = \frac{\Delta}{\sinh \eta_3} \end{cases} \tag{IVd}$$

$$\begin{aligned} \sigma &= \cosh \delta_1 \sinh \eta_1 = \cosh \delta_2 \sinh \eta_2 = \sinh \delta_3 \cosh \eta_3 \\ &= \cosh \delta_1 \cosh \delta_2 \sinh \alpha_3 = \cosh \delta_1 \sinh \delta_3 \sin \alpha_2 = \cosh \delta_2 \sinh \delta_3 \sin \alpha_1 \end{aligned} \tag{IVe}$$

$$\frac{\cosh \delta_1}{\sin \alpha_1} = \frac{\cosh \delta_2}{\sin \alpha_2} = \frac{\sinh \delta_3}{\sinh \alpha_3} \tag{IVf}$$

$$\sigma^2 = -1 - \sinh^2\delta_1 - \sinh^2\delta_2 + \cosh^2\delta_3 + 2\sinh\delta_1\sinh\delta_2\cosh\delta_3. \qquad \text{(IVg)}$$

$$\begin{cases} \cosh\delta_1 &= \sinh\delta_2\sinh\delta_3 - \cosh\delta_2\cosh\delta_3\cos\alpha_1 \\ \sinh\delta_2 &= -\cosh\delta_1\sinh\delta_3 + \sinh\delta_1\cosh\delta_3\cosh\alpha_2 \\ \sinh\delta_3 &= -\cosh\delta_1\sinh\delta_2 + \sinh\delta_1\cosh\delta_2\cosh\alpha_3 \end{cases} \qquad \text{(Va)}$$

$$\begin{cases} \cos\alpha_1 &= -\cosh\alpha_2\cosh\alpha_3 + \sinh\alpha_2\sinh\alpha_3\cosh\delta_1 \\ \cosh\alpha_2 &= -\cos\alpha_1\cosh\alpha_3 + \sin\alpha_1\sinh\alpha_3\sinh\delta_2 \\ \cosh\alpha_3 &= -\cos\alpha_1\cosh\alpha_2 + \sin\alpha_1\sinh\alpha_2\sinh\delta_3 \end{cases} \qquad \text{(Vb)}$$

$$\begin{cases} \sinh\delta_1 &= \frac{\Delta}{\sinh\alpha_2\sinh\alpha_3} \\ \cosh\delta_2 &= \frac{\Delta}{\sin\alpha_1\sinh\alpha_3} \\ \cosh\delta_3 &= \frac{\Delta}{\sin\alpha_1\sinh\alpha_2} \end{cases} \qquad \text{(Vc)}$$

$$\begin{cases} \sinh\eta_1 &= \frac{\Delta}{\sin\alpha_1} \\ \cosh\eta_2 &= \frac{\Delta}{\sinh\alpha_2} \\ \cosh\eta_3 &= \frac{\Delta}{\sinh\alpha_3} \end{cases} \qquad \text{(Vd)}$$

$$\begin{aligned} \sigma &= \sinh\delta_1\sinh\eta_1 = \cosh\delta_2\cosh\eta_2 = \cosh\delta_3\cosh\eta_3 \\ &= \sinh\delta_1\cosh\delta_2\sinh\alpha_3 = \sin\delta_1\cosh\delta_3\sinh\alpha_2 = \cosh\delta_2\cosh\delta_3\sin\alpha_1 \end{aligned} \qquad \text{(Ve)}$$

$$\frac{\sinh\delta_1}{\sin\alpha_1} = \frac{\cosh\delta_2}{\sinh\alpha_2} = \frac{\cosh\delta_3}{\sinh\alpha_3} \qquad \text{(Vf)}$$

$$\sigma^2 = 1 - \cosh^2\delta_1 + \sinh^2\delta_2 + \sinh^2\delta_3 + 2\cosh\delta_1\sinh\delta_2\sinh\delta_3. \qquad \text{(Vg)}$$

$$\cosh\delta_\nu = -\cosh\delta_{\nu-1}\cosh\delta_{\nu+1} + \sinh\delta_{\nu-1}\sinh\delta_{\nu+1}\cosh\alpha_\nu \quad (\nu \bmod 3) \qquad \text{(VIa)}$$

$$\cosh\alpha_\nu = -\cosh\alpha_{\nu-1}\cosh\alpha_{\nu+1} + \sinh\alpha_{\nu-1}\sinh\alpha_{\nu+1}\cosh\delta_\nu \quad (\nu \bmod 3) \qquad \text{(VIb)}$$

$$\sinh\delta_\nu = \frac{\Delta}{\sinh\alpha_{\nu-1}\sinh\alpha_{\nu+1}} \quad (\nu \bmod 3) \qquad \text{(VIc)}$$

$$\cosh\eta_\nu = \frac{\Delta}{\sinh\alpha_\nu} \quad (\nu \bmod 3) \qquad \text{(VId)}$$

$$\sigma = \sinh\delta_\nu\cosh\eta_\nu = \sinh\delta_{\nu-1}\sinh\delta_{\nu+1}\sinh\alpha_\nu \quad (\nu \bmod 3) \qquad \text{(VIe)}$$

$$\frac{\sinh\delta_1}{\sinh\alpha_1} = \frac{\sinh\delta_2}{\sinh\alpha_2} = \frac{\sinh\delta_3}{\sinh\alpha_3} \qquad \text{(VIf)}$$

$$\sigma^2 = -1 + \cosh^2\delta_1 + \cosh^2\delta_2 + \cosh^2\delta_3 + 2\cosh\delta_1\cosh\delta_2\cosh\delta_3. \qquad \text{(VIg)}$$

Derivation of formulae:

In deriving these relations we take as our guide the construction of this paragraph. We started with the segment s joining $\mathcal{P}$ with one of its equivalents and introduced its bisecting normal $\mathcal{R}$ and the symmetry lines $\mathcal{R}'$ and $\mathcal{R}''$. If we let $\mathcal{P}$ correspond to α_1, then η_1 is the half of s, the foot of η_1 is on δ_1, which is situated on $\mathcal{R}$. The parts of α_1 and of δ_1 into which they are divided by η_1 are denoted by α_1', α_1'' and δ_1', δ_1'', respectively. Then $\eta_1, \alpha_1', \alpha_1''$ determine the polygon.

Consider first the case (III), the case of an ES$(\odot_\nu\ \odot_{\nu'}\ \odot_{\nu''})$. In Fig. 21.5, case III, none of the orders of rotation are taken to be 2, but the formulae hold even in this case. From the figure, and from the formulae for a triangle in §8, to which we refer at the end of the lines, one reads

$$\sinh\delta_1' = \sinh\delta_3 \sin\alpha_1', \qquad \sinh\delta_1'' = \sinh\delta_2 \sin\alpha_1'' \qquad \text{(III.6)}$$

$$\cosh\delta_1' = \frac{\cosh\delta_3}{\cosh\eta_1}, \qquad \cosh\delta_1'' = \frac{\cosh\delta_2}{\cosh\eta_1} \qquad \text{(III.2)}$$

and by using the formula $\frac{1}{\cosh^2} = 1 - \tanh^2$ one gets

$$\begin{aligned}
\cosh\delta_1 &= \cosh(\delta_1' + \delta_1'') = \cosh\delta_1' \cosh\delta_1'' + \sinh\delta_1' \sinh\delta_1'' \\
&= \frac{\cosh\delta_2 \cosh\delta_3}{\cosh^2\eta_1} + \sinh\delta_2 \sinh\delta_3 \sin\alpha_1' \sin\alpha_1'' \\
&= \cosh\delta_2\cosh\delta_3 - (\cosh\delta_2 \tanh\eta_1)(\cosh\delta_3 \tanh\eta_1) \\
&\quad + \sinh\delta_2 \sinh\delta_3 \sin\alpha_1' \sin\alpha_1'' \\
&= \cosh\delta_2\cosh\delta_3 - (\sinh\delta_2 \cos\alpha_1'')(\sinh\delta_3 \cos\alpha_1') \\
&\quad + \sinh\delta_2 \sinh\delta_3 \sin\alpha_1' \sin\alpha_1'' \\
&= \cosh\delta_2\cosh\delta_3 - \sinh\delta_2 \sinh\delta_3 \cos\alpha_1. \qquad \text{(III.3)}
\end{aligned}$$

Since the three determinators play equal rôles, this gives (IIIa).

$$\cos\alpha_1' = \cosh\delta_1' \sin\alpha_2, \qquad \cos\alpha_1'' = \cosh\delta_1'' \sin\alpha_3 \qquad \text{(III.4)}$$

$$\sin\alpha_1 = \frac{\cos\alpha_2}{\cosh\eta_1}, \qquad \sin\alpha_1'' = \frac{\cos\alpha_3}{\cosh\eta_1} \qquad \text{(III.4)}$$

$$\begin{aligned}
\cos\alpha_1 &= \cos(\alpha_1' + \alpha_1'') = \cos\alpha_1' \cos\alpha_1'' - \sin\alpha_1' \sin\alpha_1'' \\
&= \cosh\delta_1' \cosh\delta_1'' \sin\alpha_2 \sin\alpha_3 - \frac{\cos\alpha_2 \cos\alpha_3}{\cosh^2\eta_1} \\
&= -\cos\alpha_2\cos\alpha_3 + (\cos\alpha_2 \tanh\eta_1)(\cos\alpha_3 \tanh\eta_1) \\
&\quad + \cosh\delta_1' \cosh\delta_1'' \sin\alpha_2 \sin\alpha_3 \\
&= -\cos\alpha_2\cos\alpha_3 + (\sinh\delta_1' \sin\alpha_2)(\sinh\delta_1'' \sin\alpha_3) \\
&\quad + \cosh\delta_1' \cosh\delta_1'' \sin\alpha_2 \sin\alpha_3 \\
&= -\cos\alpha_2\cos\alpha_3 + \sin\alpha_2 \sin\alpha_3 \cosh\delta_1. \qquad \text{(III.5)}
\end{aligned}$$

This gives (IIIb). The formulae (IIIa) and (IIIb) are known as *cosine relations* of a triangle. From the last formula one calculates

$$\begin{aligned}
\sinh^2 \delta_1 &= \cosh^2 \delta_1 - 1 \\
&= \frac{(\cos\alpha_1 + \cos\alpha_2 \cos\alpha_3)^2 - (1 - \cos^2 \alpha_2)(1 - \cos^2 \alpha_3)}{(\sin\alpha_2 \sin\alpha_3)^2} \\
&= \left(\frac{\Delta}{\sin\alpha_2 \sin\alpha_3}\right)^2 .
\end{aligned}$$

Thus (IIIc) results. Also (IIId) results from (IIIc) by means of (III.6), §8. As an example, in this case (III) we indicate the method of calculation of the expression (IIIg) of the sine amplitude by the δ. From (IIIa) follows

$$\begin{aligned}
\sin^2 \alpha_1 &= 1 - \cos^2 \alpha_1 \\
&= \frac{(\cosh^2 \delta_2 - 1)(\cosh^2 \delta_3 - 1) - (\cosh\delta_2 \cosh\delta_3 - \cosh\delta_1)^2}{(\sinh\delta_2 \sinh\delta_3)^2} \\
&= \frac{1 - \cosh^2 \delta_1 - \cosh^2 \delta_2 - \cosh^2 \delta_3 + 2\cosh\delta_1 \cosh\delta_2 \cosh\delta_3}{\sinh^2 \delta_2 \sinh^2 \delta_3} .
\end{aligned}$$

Thus one gets from (IIIe) the relation (IIIg).

The case (VI) of an ES($\to\to\to$) corresponds closely to the case (III). Using Fig. 21.5, case VI, one gets from (V.2), §8:

$$\cosh\delta_1' = \sinh\delta_3 \sinh\alpha_1', \qquad \cosh\delta_1'' = \sinh\delta_2 \sinh\alpha_1''$$

$$\sinh\delta_1' = \frac{\cosh\delta_3}{\sinh\eta_1}, \qquad \sinh\delta_1'' = \frac{\cosh\delta_2}{\sinh\eta_1} .$$

Since by (V.1), §8,

$$\begin{aligned}
\cosh\delta_2 \coth\eta_1 &= \sinh\delta_2 \cosh\alpha_1'' \\
\cosh\delta_3 \coth\eta_1 &= \sinh\delta_3 \cosh\alpha_1'
\end{aligned}$$

a calculation corresponding to that of case (III) yields (VIa).

Here (VIb) expresses the same as (VIa) with the α and the δ interchanged, because geometrically the α and the δ play equal rôles. Then (VIc) follows as in the case (III). Also (V.2), §8, yields $\cosh\eta_\nu = \sinh\delta_{\nu+1} \sinh\alpha_{\nu-1}$ and thus (VId) from (VIc).

In the case (IV) of an ES($\circlearrowleft_\nu\ \circlearrowleft_{\nu'}\ \to$) only the indices 1 and 2 play equal rôles. Using Fig. 21.5, case IV, one gets

$$\sinh\delta_1' = \sinh\delta_3 \sin\alpha_1' \tag{III.5}$$

$$\cosh\delta_1' = \frac{\cosh\delta_3}{\cosh\eta_1} \tag{III.1}$$

$$\sinh\delta_1'' = \frac{\sinh\delta_2}{\cosh\eta_1} \tag{IV.6}$$

$$\cosh \delta_1'' = \cosh \delta_2 \sin \alpha_1'' \tag{IV.4}$$

In a similar way as in the case (III) this yields

$$\begin{aligned}\sinh \delta_1 &= \sinh \delta_3 \cosh \delta_2 \sin \alpha_1' \sin \alpha_1'' + \cosh \delta_3 \sinh \delta_2 \\ &\quad - (\cosh \delta_3 \tanh \eta_1)(\sinh \delta_2 \tanh \eta_1)\end{aligned}$$

and by applying (III.3) and (IV.1), §8, to the parentheses one gets

$$\sinh \delta_1 = \sinh \delta_2 \cosh \delta_3 - \cosh \delta_2 \sinh \delta_3 \cos \alpha_1,$$

thus the two first formulae of IVa. By (IV.6), §8, one gets

$$\sinh \delta_3' = \cosh \delta_1 \sinh \alpha_3', \quad \cosh \delta_3' = \frac{\sinh \delta_1}{\sinh \eta_3}$$

$$\sinh \delta_3'' = \cosh \delta_2 \sinh \alpha_3'', \quad \cosh \delta_3'' = \frac{\sinh \delta_2}{\sinh \eta_3}$$

and thus

$$\begin{aligned}\cosh \delta_3 =& (\sinh \delta_1 \coth \eta_3)(\sin \delta_2 \coth \eta_3) - \sinh \delta_1 \sinh \delta_2 \\ &+ \cosh \delta_1 \cosh \delta_2 \sinh \alpha_3' \sinh \alpha_3''.\end{aligned}$$

By applying (IV.3), §8, to the parentheses one gets the third formula of (IVa).
By the expression of the δ by the α one starts with

$$\cos \alpha_1' = \cosh \delta_1' \sin \alpha_2 \tag{III.4}$$

$$\sin \alpha_1' = \frac{\cos \alpha_2}{\cosh \eta_1} \tag{III.4}$$

$$\cos \alpha_1'' = \sinh \delta_1'' \sinh \alpha_3 \tag{IV.2}$$

$$\sin \alpha_1'' = \frac{\cosh \alpha_3}{\cosh \eta_1} \tag{IV.4}$$

and in virtue of

$$\cos \alpha_2 \tanh \eta_1 = \sin \alpha_2 \sinh \delta_1' \tag{III.5}$$

$$\cosh \alpha_3 \tanh \eta_1 = \sinh \alpha_3 \cosh \delta_1'' \tag{IV.3}$$

one gets the first, and also the second, formula of (IVb). Starting with

$$\cosh \alpha_3' = \cosh \delta_3' \sin \alpha_2, \quad \cosh \alpha_3'' = \cosh \delta_3'' \sin \alpha_1 \tag{IV.4}$$

$$\sinh \alpha_3' = \frac{\cos \alpha_2}{\sinh \eta_3}, \quad \sinh \alpha_3'' = \frac{\cos \alpha_1}{\sinh \eta_3} \tag{IV.2}$$

and applying

$$\cos \alpha_2 \coth \eta_3 = \sin \alpha_2 \sinh \delta_3', \quad \cos \alpha_1 \coth \eta_3 = \sin \alpha_1 \sinh \delta_3'' \tag{IV.5}$$

one gets the third formula of (IVb).

By calculating $\cosh \delta_1$ from the first formula of (IVb) one gets the first formula of (IVc), and similarly for the second and third formula. Also (IVd) results from (IVc) by means of (III.6) and (IV.4), §8.

Finally, the case (V) of an ES($\odot_\nu \rightarrow\rightarrow$) is similar to the case (IV). In Fig. 21.5, case V, the indices 2 and 3 play equal rôles.

$$\sinh \delta_1' = \frac{\sinh \delta_3}{\cosh \eta_1}, \quad \sinh \delta_1'' = \frac{\sinh \delta_2}{\cosh \eta_1} \qquad \text{(IV.6)}$$

$$\cosh \delta_1' = \cosh \delta_3 \sin \alpha_1', \quad \cosh \delta_1'' = \cosh \delta_2 \sin \alpha_1'' \qquad \text{(IV.4)}$$

together with

$$\sinh \delta_2 \tanh \eta_1 = \cosh \delta_2 \cos \alpha_1'', \quad \sinh \delta_3 \tanh \eta_1 = \cosh \delta_3 \cos \alpha_1' \qquad \text{(IV.1)}$$

yield the first formula of (Va).

$$\cosh \delta_2' = \sinh \delta_1 \sinh \alpha_2' \qquad \text{(V.2)}$$

$$\sinh \delta_2' = \frac{\cosh \delta_1}{\sinh \eta_2} \qquad \text{(V.2)}$$

$$\cosh \delta_2'' = \frac{\sinh \delta_3}{\sinh \eta_2} \qquad \text{(IV.6)}$$

$$\sinh \delta_2'' = \cosh \delta_3 \sinh \alpha_2'' \qquad \text{(IV.6)}$$

together with

$$\cosh \delta_1 \coth \eta_2 = \sinh \delta_1 \cosh \alpha_2' \qquad \text{(V.1)}$$

$$\sinh \delta_3 \coth \eta_2 = \cosh \delta_1 \cosh \alpha_2'' \qquad \text{(IV.3)}$$

yield the second, and thus also the third formula of (Va).

$$\cos \alpha_1' = \sinh \alpha_2 \sinh \delta_1', \quad \cos \alpha_1'' = \sinh \alpha_3 \sinh \delta_1'' \qquad \text{(IV.2)}$$

$$\sin \alpha_1' = \frac{\cosh \alpha_2}{\cosh \eta_1}, \quad \sin \alpha_1'' = \frac{\cosh \alpha_3}{\cosh \eta_1} \qquad \text{(IV.2)}$$

together with

$$\cosh \alpha_2 \tanh \eta_1 = \cosh \delta_1' \sinh \alpha_2, \quad \cosh \alpha_3 \tanh \eta_1 = \cosh \delta_1'' \sinh \alpha_3 \qquad \text{(IV.3)}$$

yield the first formula of (Vb).

$$\cosh \alpha_2' = \sinh \alpha_3 \sinh \delta_2' \qquad \text{(V.2)}$$

$$\sinh \alpha_2' = \frac{\cosh \alpha_3}{\sinh \eta_2} \qquad \text{(V.2)}$$

$$\cosh \alpha_2'' = \sin \alpha_1 \cosh \delta_2'' \tag{IV.4}$$

$$\sinh \alpha_2'' = \frac{\cos \alpha_1}{\sinh \eta_2} \tag{IV.2}$$

together with

$$\cos \alpha_1 \coth \eta_2 = \sinh \delta_2'' \sin \alpha_1 \tag{IV.5}$$

$$\cosh \alpha_3 \coth \eta_2 = \cosh \delta_2' \sinh \alpha_3 \tag{V.1}$$

yield the second, and thus also the third formula of (Vb).

Again one gets (Vc) from (Vb) by calculating the sinh from the cosh or the cosh from the sinh. One gets (Vd) from (Vc) by means of (IV.6) and (V.2), §8.

§22 Complete decomposition and normal form in the case of quasi-compactness

22.1 Decomposition based on two protrusions. Let $\mathfrak{F}$ denote an $\mathfrak{F}$-group without reflections. The conical points, masts, and closed bounding geodesics of $\mathcal{K}(\mathfrak{F})$ mod $\mathfrak{F}$ are by a common notation called the *protrusions* of $\mathcal{K}(\mathfrak{F})$ mod $\mathfrak{F}$. For an immediate visualization: conical points and masts are jutting out from the surface; the same may be said of the neighbourhood of a bounding geodesic. The number of protrusions may, in general, be finite or infinite. If $\mathfrak{F}$ satisfies the condition of quasi-compactness, then this number is known to be finite (§15.2).

Suppose that $\mathcal{K}(\mathfrak{F})$ mod $\mathfrak{F}$ possesses at least two protrusions and select arbitrarily two of them. Let the centre, or limit-centre, or boundary axis $\mathcal{P}$ of $\mathfrak{F}$ represent one of them. One may then select a representing centre, or limit-centre, or boundary axis $\mathcal{P}'$ of the other such that the open part $\mathcal{T}$ of the join $\overline{\mathcal{T}}$ of $\mathcal{P}$ and $\mathcal{P}'$ is simple modulo $\mathfrak{F}$ and contains no point of $\mathfrak{F}\mathcal{P}$ nor of $\mathfrak{F}\mathcal{P}'$. This follows from Lemma 2 in §21.1; the fact that the lemma deals with a horocycle instead of the corresponding limit-centre, has in §21.2 been shown to be irrelevant. Also $\mathcal{T}$ contains no centre of order > 2, because it is simple modulo $\mathfrak{F}$, and no centre of order 2 because it contains no point of $\mathfrak{F}\mathcal{P}$ nor of $\mathfrak{F}\mathcal{P}'$, and because $\mathcal{P}$ and $\mathcal{P}'$ are not equivalent with respect to $\mathfrak{F}$. In all cases, $\mathcal{T}$ belongs to $\mathcal{K}(\mathfrak{F})$ owing to the convexity of $\mathcal{K}(\mathfrak{F})$.

We now consider the collection $\mathfrak{F}\overline{\mathcal{T}}$. It does not accumulate in $\mathcal{D}$. This is obvious if neither $\mathcal{P}$ nor $\mathcal{P}'$ is a limit-centre. For then $\overline{\mathcal{T}}$ has finite length, and a statement of §13 applies. If $\mathcal{P}$ is a limit-centre, consider the part $\mathcal{T}_0$ of $\mathcal{T}$ situated in the disc $\mathcal{Q}$ determined by the horocycle bounding $\mathcal{K}^*(\mathfrak{F})$ and belonging to $\mathcal{P}$. Then $\mathcal{Q}$ contains the part $\mathfrak{A}_{\mathcal{P}}\mathcal{T}_0$ of $\mathfrak{F}\mathcal{T}_0$, and the rest of $\mathfrak{F}\mathcal{T}_0$ is in the same way contained in other discs of the collection $\mathfrak{F}\mathcal{Q}$, and these discs do not accumulate in $\mathcal{D}$. If also $\mathcal{P}'$ is a limit-centre, the same holds for the corresponding part $\mathfrak{F}\mathcal{T}_0'$ in $\mathfrak{F}\mathcal{Q}'$. The remaining part of $\mathcal{T}$ has finite length in all cases, and therefore admits the same conclusion as above.

One can thus decompose $\mathcal{K}(\mathfrak{F})$ by the collection $\mathfrak{F}\overline{\mathcal{T}}$. Let $\mathcal{Z}$ denote the point set consisting of the union of $\mathfrak{F}\mathcal{T}$, $\mathfrak{F}\mathcal{P}$ and $\mathfrak{F}\mathcal{P}'$. The intersections of the boundary b of a region of decomposition $\mathcal{B}$ with the different components of $\mathcal{Z}$ are called *boundary chains* of $\mathcal{B}$. If $\mathcal{P}$ and $\mathcal{P}'$ are points, then all links of a boundary chain belong to $\mathfrak{F}\mathcal{T}$. If axes occur, certain links belonging to $\mathfrak{F}\mathcal{T}$ are connected by segments of axes corresponding to the primary displacement; compare Fig. 22.1.

Let $\mathfrak{P}$ and $\mathfrak{P}'$ denote the subgroups of $\mathfrak{F}$ which leave invariant $\mathcal{P}$ and $\mathcal{P}'$, respectively. Let $\mathcal{B}$ and $\mathcal{B}'$ be the two regions of decomposition of $\mathcal{K}(\mathfrak{F})$ adjacent to $\mathcal{T}$. Let primary elements $\mathfrak{p}$ and $\mathfrak{p}'$ of $\mathfrak{P}$ and $\mathfrak{P}'$, respectively, be so chosen that $\mathfrak{p}'$ carries $\mathcal{B}$ into $\mathcal{B}'$, and $\mathfrak{p}$ carries $\mathcal{B}'$ into $\mathcal{B}$. Then $\mathfrak{p}\mathfrak{p}'$ carries $\mathcal{B}$ into itself and, in particular, the boundary chain b_1 of $\mathcal{B}$ to which $\mathcal{T}$ belongs into itself. The link $\mathfrak{p}'^{-1}\mathcal{T}$ is mapped on $\mathfrak{p}\mathcal{T}$ by $\mathfrak{p}\mathfrak{p}'$ (Fig. 22.1). We put $\mathfrak{p}\mathfrak{p}' = \mathfrak{p}''^{-1}$. If b_1 is a finite polygon, then $\mathfrak{p}''$ is a rotation; let $\mathcal{P}''$ denote its centre. If b_1 is infinite, both ends converging to the same point on $\mathcal{E}$, then $\mathfrak{p}''$ is a limit-rotation; let $\mathcal{P}''$ denote its limit-centre. If b_1 is infinite, its ends converging to two different points u and v of $\mathcal{E}$, then $\mathfrak{p}''$ is a translation with u and v as fundamental points; let $\mathcal{P}''$ denote its axis. This axis then is simple modulo $\mathfrak{F}$. For, if $\mathcal{P}''$ were cut by $\mathfrak{f}P''$, $\mathfrak{f} \in \mathfrak{F}$, thus the pair $\mathfrak{f}u$, $\mathfrak{f}v$ separating the pair u, v on $\mathcal{E}$, then b_1 would be cut by $\mathfrak{f}b_1$. This is impossible, because no segment equivalent with a link of b_1 enters into $\mathcal{B}$.

Let ν, ν', ν'' denote the orders of $\mathfrak{p}$, $\mathfrak{p}'$, $\mathfrak{p}''$, respectively, including the value ∞ for limit-rotations and translations. The angles of b_1 are $\frac{\pi}{2}$ at axes, $\frac{2\pi}{n}$ at centres of order n, and zero at limit-centres.

If, in particular, $\nu = \nu' = 2$, then b_1 is a straight line. This is the case of rotation twins considered in §21.5. Then $\mathfrak{p}''$ is a primary translation along that line.

If ν and ν' do not both take the value 2, then b_1 is a broken, convex, polygonal line. The interior $\mathcal{B}_1$ of the convex hull of b_1 then is a region contained in $\mathcal{B}$. It coincides with $\mathcal{B}$ if $\mathfrak{p}''$ is a rotation or limit-rotation. Let $\mathfrak{P}''$ denote the group generated by $\mathfrak{p}''$. In $\mathcal{B}_1$ the groups $\mathfrak{P}''$ and $\mathfrak{F}$ have equal effect.

Let s denote the open part of the join of $\mathcal{P}$ and $\mathfrak{p}'^{-1}\mathcal{P}$. One gets $\mathfrak{p}''\mathcal{P} = \mathfrak{p}'^{-1}\mathfrak{p}^{-1}\mathcal{P} = \mathfrak{p}'^{-1}\mathcal{P}$. Hence $\mathfrak{P}''\overline{s}$ is inscribed in b_1 and belongs to the convex hull of b_1. From this follows that s is simple modulo $\mathfrak{F}$, since it is contained in $\mathcal{B}_1$ and is not cut by any of the segments $\mathfrak{P}''s$.

The segment s is reversible if and only if, $\nu'' = 2$: If $\nu'' = 2$, then $\mathfrak{p}''(\mathfrak{p}'^{-1}\mathcal{P}) = \mathfrak{p}''^{-1}(\mathfrak{p}'^{-1}\mathcal{P}) = \mathfrak{p}\mathfrak{p}'\mathfrak{p}'^{-1}\mathcal{P} = \mathcal{P}$. Thus $\mathcal{P}$ and $\mathfrak{p}'^{-1}\mathcal{P}$ are interchanged by $\mathfrak{p}''$, and s is thus reversed by $\mathfrak{p}''$. Conversely, if s is reversible, thus by some half-turn $\mathfrak{g}$, then $\mathfrak{g}$ must carry $\mathcal{B}_1$ into itself. It must carry the two links adjacent to $\mathcal{P}$ in b_1, thus $\mathcal{T}$ and $\mathfrak{p}\mathcal{T}$, into the two links adjacent to $\mathfrak{p}'^{-1}\mathcal{P}$, thus $\mathfrak{p}'^{-1}\mathfrak{p}^{-1}\mathcal{T}$ and $\mathfrak{p}'^{-1}\mathcal{T}$, respectively. Since no element of $\mathfrak{F}$ except the identity leaves $\mathcal{T}$ invariant, this yields $\mathfrak{g} = \mathfrak{p}'^{-1}\mathfrak{p}^{-1}$, thus $\mathfrak{g} = \mathfrak{p}''$, thus $\nu'' = 2$.

It should be noticed that in the case $\nu'' = 2$, neither ν nor ν' can take the value 2. For, if $\nu' = 2$, then $\mathfrak{p} = \mathfrak{p}''^{-1}\mathfrak{p}'^{-1}$ would be the product of two half-turns, thus a translation whose axis is interior to $\mathcal{K}(\mathfrak{F})$, and $\mathcal{P}$ would not represent a protrusion of $\mathcal{K}(\mathfrak{F})$ mod $\mathfrak{F}$. Accordingly for the assumption $\nu = 2$. Hence in the case where the two chosen protrusions are not both conical points of order 2, the value 2 can occur for at most one among the three values ν, ν', ν''.

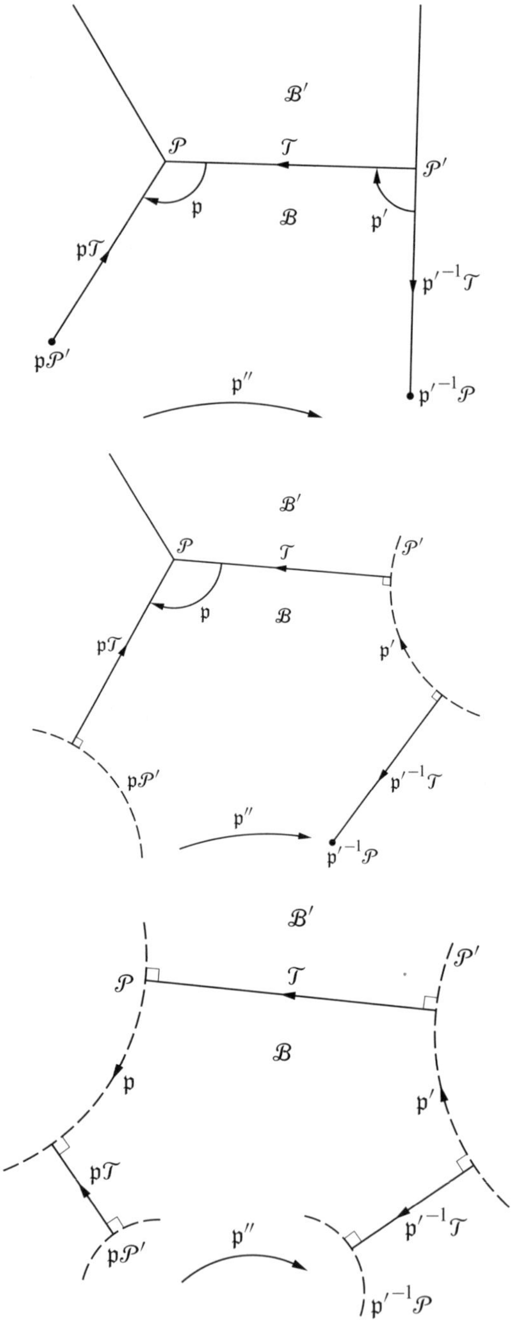
$\mathcal{B}'$
$\mathcal{P}$
$\mathcal{T}$
$\mathcal{P}'$
$\mathfrak{p}$
$\mathcal{B}$
$\mathfrak{p}'$
$\mathfrak{p}\mathcal{T}$
$\mathfrak{p}'^{-1}\mathcal{T}$
$\mathfrak{p}\mathcal{P}'$
$\mathfrak{p}''$
$\mathfrak{p}'^{-1}\mathcal{P}$

Figure 22.1

If the construction of §21 is carried out with $\mathcal{P}$ as the starting centre, or limit-centre, or boundary axis, and $\bar{s}$ as the join of $\mathcal{P}$ and its equivalent $\mathfrak{p}'^{-1}\mathcal{P}$ (called $\mathfrak{k}\mathcal{P}$ in §21), then one would get the elementary group generated by $\mathfrak{p}$, $\mathfrak{p}'$, $\mathfrak{p}''$ of the present section, these elements satisfying the relations (3) and (9) of §21. In particular, $\mathcal{T}$ is situated on the line called $\mathcal{R}'$ in §21, and $\mathfrak{p}'^{-1}\mathcal{T}$ on the line there obtained by the reflection of $\mathcal{R}'$ in $\mathcal{R}$; compare Fig. 21.1. One is thus automatically led to $\mathcal{P}'$. The difference of the present procedure as compared with that of §21 is, after the choice of $\mathcal{P}$, to arrange the choice of the segment s among the different possibilities such that a prescribed other protrusion of $\mathcal{K}(\mathfrak{F})$ mod $\mathfrak{F}$ comes into play. The third protrusion of the elementary surface obtained is then determined by the two others.

22.2 Existence of simple axes of reversed translations. Let $\mathfrak{F}$ be an $\mathfrak{F}$-group without reflections but containing reversions, thus reversed translations. Let q be a regular point in the interior of $\mathcal{K}(\mathfrak{F})$. Any two points of the equivalence class $\mathfrak{F}q$ are equivalent by exactly one element contained in $\mathfrak{F}$, because q is regular. Let $\mathfrak{F}'$ denote the subgroup of motions contained in $\mathfrak{F}$. Then $(\mathfrak{F} \setminus \mathfrak{F}')q$ is the totality of points of $\mathfrak{F}q$ which are equivalent with q by reversed translations. Let η denote the distance of q from $(\mathfrak{F} \setminus \mathfrak{F}')q$, and $\mathfrak{h}$ a reversed translation such that $[q, \mathfrak{h}q] = \eta$. Then the open segment $\mathcal{T}$ joining q and $\mathfrak{h}q$ is simple modulo $\mathfrak{F}$, as is immediately seen by applying the result at the beginning of §21.1. For, if $\mathcal{T}$ were cut by an equivalent segment, then in one of the pairs of opposite sides of the quadrangle the end-points belong to different equivalence classes with respect to $\mathfrak{F}'$. Also $\mathcal{T}$ contains no point of $\mathfrak{F}q$.

The collection $\mathfrak{h}^n\mathcal{T}$, $-\infty < n < \infty$, is a chain of segments joining the fundamental points of $\mathfrak{h}$. Every point $\mathfrak{h}^n q$ is the common end-point of exactly two segments of the collection $\mathfrak{F}\mathcal{T}$. It is inferred that the chain is not cut by any of its equivalents, and hence that the axis $\mathcal{A}(\mathfrak{h})$ is simple modulo $\mathfrak{F}$.

22.3 Reduction of the genus in the case of non-orientability. Let $\mathcal{A}_1$ denote an arbitrary axis of reversed translations which is simple modulo $\mathfrak{F}$. The preceding section has shown that such an axis exists, if $\mathfrak{F}$ contains reversed translations but no reflections. The closed geodesic a_1 corresponding to $\mathcal{A}_1$ on the surface $\mathcal{K}(\mathfrak{F})$ mod $\mathfrak{F}$ has no point in common with any protrusion of $\mathcal{K}(\mathfrak{F})$ mod $\mathfrak{F}$: It does not meet boundaries of $\mathcal{K}(\mathfrak{F})$ mod $\mathfrak{F}$, because $\mathcal{A}_1$ is an inner axis of $\mathfrak{F}$. It does not end in a mast, because the axis $\mathcal{A}_1$ does not end in a limit-centre. It also passes through no conical point; for, $\mathcal{A}_1$ passes through no centre of order > 2 because it is simple modulo $\mathfrak{F}$, and through no centre of order 2, because the product of a reversed translation and a half-turn around a point of its axis is a reflection, and $\mathfrak{F}$ is assumed without reflections.

Consider the decomposition of $\mathcal{K}(\mathfrak{F})$ by the collection $\mathfrak{F}\mathcal{A}_1$. All regions of that decomposition are equivalent with respect to $\mathfrak{F}$. Select one of them, and let $\mathfrak{F}_1$ be the subgroup of $\mathfrak{F}$ carrying that region into itself. Then $\mathfrak{F}_1$ is an $\mathfrak{F}$-group, and $\mathcal{K}(\mathfrak{F}_1)$ is the closure of the region on $\mathcal{K}(\mathfrak{F})$. All protrusions of $\mathcal{K}(\mathfrak{F})$ mod $\mathfrak{F}$ are also protrusions of $\mathcal{K}(\mathfrak{F}_1)$ mod $\mathfrak{F}_1$. Those sides of $\mathcal{K}(\mathfrak{F}_1)$ which belong to $\mathfrak{F}\mathcal{A}_1$ constitute one equivalence class with respect to $\mathfrak{F}_1$ (case II of §19). Hence $\mathcal{K}(\mathfrak{F}_1)$ mod $\mathfrak{F}_1$ possesses exactly one protrusion more than those of $\mathcal{K}(\mathfrak{F})$ mod $\mathfrak{F}$. It is obtained by the cutting up of $\mathcal{K}(\mathfrak{F})$ mod $\mathfrak{F}$ along the simple geodesic a_1, this cutting yielding only one boundary. We denote $\mathfrak{F}_1$ as a *restgroup* of the process.

Let it now be assumed that $\mathfrak{F}$ satisfies the condition of quasi-compactness.

Since $\mathfrak{F}_1$ and $\mathfrak{F}$ have equal effect in the interior of $\mathcal{K}(\mathfrak{F}_1)$, and all regions of decomposition are equivalent, one gets for the area of the surfaces $\Phi(\mathfrak{F}_1) = \Phi(\mathfrak{F})$. Therefore $\mathfrak{F}_1$ also satisfies the condition of quasi-compactness.

Assume that also $\mathfrak{F}_1$ contains reversed translations. Then (11), §17, applies to both $\mathfrak{F}$ and $\mathfrak{F}_1$. Since $\vartheta_{\mathfrak{F}_1}(\rightarrow) = \vartheta_{\mathfrak{F}}(\rightarrow) + 1$ and the type numbers $\vartheta(\hookrightarrow)$ and $\vartheta(\circlearrowleft_\nu)$ remain unchanged, it is referred from (11), §17, that $p_{\mathfrak{F}_1} = p_{\mathfrak{F}} - 1$. The process has thus decreased the genus by one.

Under the present assumption that $\mathfrak{F}_1$ contains reversed translations, one can repeat the process by introducing an axis $\mathcal{A}_2$ of reversed translations of $\mathfrak{F}_1$ which is simple modulo $\mathfrak{F}_1$. It is contained in $\mathcal{K}(\mathfrak{F}_1)$, and it is also simple modulo $\mathfrak{F}$, because $\mathfrak{F}$ and $\mathfrak{F}_1$ have equal effect in the interior of $\mathcal{K}(\mathfrak{F}_1)$. One thus arrives at a restgroup $\mathfrak{F}_2$ by decomposing $\mathcal{K}(\mathfrak{F}_1)$ by the collection $\mathfrak{F}_1\mathcal{A}_2$. This process can be repeated as long as the restgroup contains reversed translations. The area of the resulting surface remains unchanged. Since each step decreases the genus by one and since the genus is a finite, non-negative number, the process can only be repeated a finite number of times.

Assume that after m steps one arrives at a restgroup $\mathfrak{F}_m$ which contains no reversed translations and thus consists of motions only. Let $\mathcal{A}_m$ denote the axis of reversed translations in $\mathfrak{F}_{m-1}$ introduced in step number m. Let $\mathfrak{h}_m$ denote a primary reversed translation along $\mathcal{A}_m$, and let $\mathcal{A}_m$ be directed by $\mathfrak{h}_m$. In the decomposition of $\mathcal{K}(\mathfrak{F}_{m-1})$ by $\mathfrak{F}_{m-1}\mathcal{A}_m$, choose one of the regions adjacent to $\mathcal{A}_m$, say the one bounded positively by $\mathcal{A}_m$, and let $\mathfrak{F}_m$ be the subgroup of $\mathfrak{F}_{m-1}$ carrying that region into itself. The region closed on $\mathcal{K}(\mathfrak{F}_{m-1})$ is then $\mathcal{K}(\mathfrak{F}_m)$.

We now make the assumption that there is in $\mathfrak{F}_m$ an axis $\mathcal{A}$ with corresponding subgroup $\mathfrak{A}(\rightarrow)$ such that $\mathcal{A}$ is simple and non-dividing modulo $\mathfrak{F}_m$. Hence $\mathcal{A}$ corresponds to the case I of §19 in relation to $\mathfrak{F}_m$.

In the decomposition of $\mathcal{K}(\mathfrak{F}_m)$ by $\mathfrak{F}_m\mathcal{A}$ consider the region $\mathcal{B}$ adjacent to $\mathcal{A}_m$, and let $\mathfrak{B}$ be the subgroup of $\mathfrak{F}_m$ which reproduces that region (Fig. 22.2). It is known from §19 that those boundaries of $\mathcal{B}$ which belong to $\mathfrak{F}_m\mathcal{A}$ fall into two equivalence classes with respect to $\mathfrak{B}$. If the notation is such that $\mathcal{A}$ is a side of $\mathcal{B}$, one of the two classes is $\mathfrak{B}\mathcal{A}$. Let k denote the other class. Also $\mathfrak{B}\mathcal{A}_m$ is a class of boundary axes of $\mathcal{B}$, different from the other two, since an axis of the type $\mathfrak{A}(\rightleftharpoons)$ cannot be equivalent with an axis of the type $\mathfrak{A}(\rightarrow)$ with respect to $\mathfrak{F}$. The region $\mathcal{B}$ may have more classes of boundary axes with respect to $\mathfrak{B}$ than these three (it certainly has, if $m > 1$). In $\mathcal{B}$ the groups $\mathfrak{B}$ and $\mathfrak{F}_m$ (and finally $\mathfrak{F}$ itself) have equal effect, but not on certain boundaries of $\mathcal{B}$. The group $\mathfrak{B}$, being a subgroup of $\mathfrak{F}_m$, consists of motions only. In particular, the primary element of $\mathfrak{B}$ on $\mathcal{A}_m$ is $\mathfrak{h}_m^2$.

We now select a side of $\mathcal{B}$ belonging to the class $\mathfrak{B}\mathcal{A}$ such that its join $\mathcal{T}$ with $\mathcal{A}_m$ is simple modulo $\mathfrak{B}$. This is possible according to Lemma 2 of §21.1. Since it does

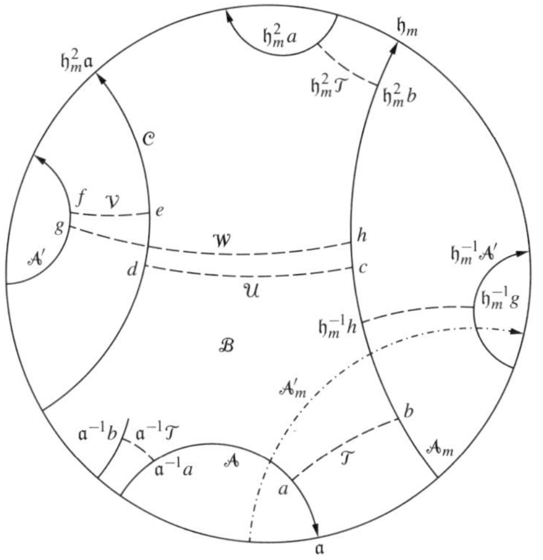

Figure 22.2

not matter which axis of the collection $\mathfrak{F}_m\mathcal{A}$ we call $\mathcal{A}$, let this axis be $\mathcal{A}$. Denote the foot of $\mathcal{T}$ on $\mathcal{A}$ by a and the foot of $\mathcal{T}$ on $\mathcal{A}_m$ by b. Let $\mathfrak{a}$ be a primary translation along $\mathcal{A}$ such that, if $\mathcal{A}$ is directed by $\mathfrak{a}$, it bounds $\mathcal{B}$ positively (Fig. 22.2). The axes of the class k then bound $\mathcal{B}$ negatively. Let $\mathcal{B}$ be composed by the collection $\mathfrak{B}\mathcal{T}$, and consider the region of decomposition bounded (in part) by a boundary chain to which $\mathfrak{a}^{-1}\mathcal{T}$, $\mathcal{T}$, and $\mathfrak{h}_m^2\mathcal{T}$ belong (Section 1). Since $\mathcal{B}$ has more than two classes with respect to $\mathfrak{B}$ of boundary axes, the primary element $\mathfrak{h}_m^2\mathfrak{a}$, which reproduces the boundary chain, cannot be a rotation or limit-rotation. Let $\mathfrak{C}$ denote the axis of the translation $\mathfrak{h}_m^2\mathfrak{a}$. Then $\mathfrak{C}$ is simple modulo $\mathfrak{B}$ (Section 1). The join $\mathcal{U}$ of $\mathfrak{C}$ and $\mathcal{A}_m$ is simple modulo $\mathfrak{B}$ and thus modulo $\mathfrak{F}_{m-1}$, because it is situated in the convex hull of the boundary chain, and in the interior of this hull the group generated by $\mathfrak{h}_m^2\mathfrak{a}$ has the same effect as $\mathfrak{F}_{m-1}$. Denote the foot of $\mathcal{U}$ on $\mathcal{A}_m$ by c and the foot of $\mathcal{U}$ on $\mathfrak{C}$ by d (Fig. 22.2).

If $\mathfrak{C}$ is a boundary axis of $\mathcal{B}$, then $\mathcal{B}$ has exactly three equivalence classes with respect to $\mathfrak{B}$ of boundary axes, and $\mathcal{B}$ mod $\mathfrak{B}$ is an ES(→→→) corresponding to the elementary group $\mathfrak{E}$ generated by $\mathfrak{a}$ and $\mathfrak{h}_m^2$. Thus $\mathfrak{C}$ must belong to the class k. – If $\mathfrak{C}$ is not a boundary axis of $\mathcal{B}$, then it is an inner axis of $\mathcal{B}$ which divides modulo $\mathfrak{B}$; for, otherwise $\mathcal{B}$ would only contain two equivalence classes with respect to $\mathfrak{B}$ of boundary axes. In the decomposition of $\mathcal{B}$ by the collection $\mathfrak{B}\mathfrak{C}$, let $\mathcal{B}'$ be the region bounded positively by $\mathfrak{C}$, thus not adjacent to $\mathcal{A}$, and $\mathfrak{B}'$ the corresponding subgroup. $\mathcal{B}'$ then contains boundary axes of the class k, and their collection in the boundary of $\mathcal{B}'$ is reproduced by the powers of $\mathfrak{h}_m^2\mathfrak{a}$. We may therefore select an axis $\mathcal{A}'$ of the class k (thus bounding $\mathcal{B}'$ negatively), such that its join $\mathcal{V}$ with $\mathfrak{C}$ is simple modulo $\mathfrak{B}'$, and thus also modulo $\mathfrak{B}$, and that the foot e of $\mathcal{V}$ on $\mathfrak{C}$ is at a distance from d not exceeding half the displacement of the primary translation $\mathfrak{h}_m^2\mathfrak{a}$ along $\mathfrak{C}$. Let f be the foot of $\mathcal{V}$ on $\mathcal{A}'$. The broken line $f\ e\ d\ c$ is then not cut by any of its equivalents with respect to $\mathfrak{B}$, because each of its three open parts is simple modulo $\mathfrak{B}$ and these three parts are disjoint modulo $\mathfrak{B}$. Hence the pair of axes $\mathcal{A}'$ and $\mathcal{A}_m$ is not separated by any equivalent pair $\mathfrak{f}\mathcal{A}'$, $\mathfrak{f}\mathcal{A}_m$, $\mathfrak{f} \in \mathfrak{B}$ and their join $\mathcal{W}$ is therefore simple modulo $\mathfrak{B}$ and thus also modulo $\mathfrak{F}_{m-1}$. Let g denote the foot of $\mathcal{W}$ on $\mathcal{A}'$ and h the foot of $\mathcal{W}$ on $\mathcal{A}_m$. (If $\mathfrak{C}$ is a boundary of $\mathcal{B}$, then $\mathcal{A}'$ is $\mathfrak{C}$ itself, and $\mathcal{W}$ is $\mathcal{U}$ itself.) Moreover, $\mathcal{W}$ and $\mathcal{T}$ are disjoint modulo $\mathfrak{B}$, and thus also modulo $\mathfrak{F}_{m-1}$, for one part of $\mathcal{W}$ is situated in $\mathcal{B}'$, which is disjoint with $\mathcal{T}$ modulo $\mathfrak{B}$, and the other part is situated in the interior of the convex hull of the above boundary chain, and is thus disjoint with $\mathcal{T}$ modulo $\mathfrak{B}$. – This fact that $\mathcal{W}$ and $\mathcal{T}$ are disjoint modulo $\mathfrak{B}$ can be stated in terms of the surface $\mathcal{B}$ mod $\mathfrak{B}$: If an orientable surface has at least three bounding geodesics, then one given boundary can be joined with each of two other given boundaries by geodetic arcs such that each of these arcs is simple and the two arcs are disjoint.

Consider now the image $\mathfrak{h}_m^{-1}\mathcal{W}$ of $\mathcal{W}$ by the reversed translation $\mathfrak{h}_m^{-1}$. It joins $\mathcal{A}_m$ and $\mathfrak{h}_m^{-1}\mathcal{A}'$, and its foot $\mathfrak{h}_m^{-1}h$ on $\mathcal{A}_m$ is at a distance smaller than the displacement of $\mathfrak{h}_m$ from b, because h is at a distance smaller than twice that displacement from b in the positive direction on $\mathcal{A}_m$. Considering the broken line $a\,b(\mathfrak{h}_m^{-1}h)(\mathfrak{h}_m^{-1}g)$ one therefore infers that the pair of axes $\mathcal{A}$ and $\mathfrak{h}_m^{-1}\mathcal{A}'$ are not separated by any pair equivalent with respect to $\mathfrak{F}_{m-1}$. Hence any segment joining these two axes is simple modulo $\mathfrak{F}_{m-1}$.

We now go back to $\mathcal{K}(\mathfrak{F}_{m-1})$ and consider its decomposition by the collection $\mathfrak{F}_{m-1}\mathcal{A}$. One of the regions of that decomposition, say $\mathcal{G}$, contains $\mathcal{A}_m$ as an inner axis, and $\mathcal{A}$ and $\mathfrak{h}_m^{-1}\mathcal{A}'$ belong to the boundaries of $\mathcal{G}$. They are equivalent with respect to $\mathfrak{F}_{m-1}$ for, $\mathcal{A}$ and $\mathcal{A}'$ though not equivalent with respect to $\mathfrak{B}$, both belong to $\mathfrak{F}_m\mathcal{A}$, and $\mathfrak{F}_m$ is a subgroup of $\mathfrak{F}_{m-1}$. Moreover, both $\mathcal{A}$ and $\mathfrak{h}_m^{-1}\mathcal{A}'$ bound $\mathcal{G}$ positively. Let $\mathfrak{f}$ be an element of $\mathfrak{F}_m$, thus a motion, for which $\mathfrak{f}\mathcal{A} = \mathcal{A}'$. Then $\mathcal{A}'$ bounds $\mathfrak{f}\mathcal{G}$ positively, because $\mathcal{A}$ bounds $\mathcal{G}$ positively. Therefore $\mathfrak{h}_m^{-1}\mathfrak{f}\mathcal{A} = \mathfrak{h}_m^{-1}\mathcal{A}'$ bounds $\mathfrak{h}_m^{-1}\mathfrak{f}\mathcal{G}$ negatively, because $\mathfrak{h}_m^{-1}$ is a reversion. Thus $\mathfrak{h}_m^{-1}\mathfrak{f}$ maps $\mathcal{G}$ on the region adjacent to $\mathcal{G}$ along $\mathfrak{h}_m^{-1}\mathcal{A}'$. It follows that the axis of the reversed translation $\mathfrak{h}_m^{-1}\mathfrak{f}$ intersects $\mathcal{A}$ and $\mathfrak{h}_m^{-1}\mathcal{A}'$, and these two axes determine the displacement of $\mathfrak{h}_m^{-1}\mathfrak{f}$ on it. We put $\mathfrak{h}_m^{-1}\mathfrak{f} = \mathfrak{h}'_m$ and call its axis $\mathcal{A}'_m$. Since $\mathcal{A}$ and $\mathfrak{h}_m^{-1}\mathcal{A}' = h'_m\mathcal{A}$ are not separated by any equivalent pair, the axis $\mathcal{A}'_m$ is simple modulo $\mathfrak{F}_{m-1}$.

The situation is now as follows. After having reached the restgroup $\mathfrak{F}_{m-1}$ we originally chose the axis $\mathcal{A}_m$ for the step number m. We now ask what happens if instead of $\mathcal{A}_m$ we choose the axis $\mathcal{A}'_m$ for the step number m. We now refer to the Fig. 22.3. The element $\mathfrak{g} = \mathfrak{a}\mathfrak{h}_m'^{-1}$ is a reversed translation. It carries $\mathfrak{h}'_m\mathcal{A}$ into $\mathcal{A}$, and $\mathfrak{h}'_m\mathcal{G}$ into $\mathfrak{a}\mathcal{G} = \mathcal{G}$. Its axis $\mathcal{A}(\mathfrak{g})$ therefore intersects $\mathfrak{h}'_m\mathcal{A}$ and $\mathcal{A}$, and these two axes cut off its displacement. Therefore $\mathcal{A}(\mathfrak{g})$ is simple modulo $\mathfrak{F}_{m-1}$. Now $\mathfrak{g}$ carries $\mathcal{A}$ into $\mathfrak{a}\mathfrak{h}_m'^{-1}\mathcal{A}$, and $\mathfrak{g}^{-1}$ carries $\mathfrak{h}'_m\mathcal{A}$ into $\mathfrak{h}'_m\mathfrak{a}^{-1}\mathfrak{h}'_m\mathcal{A} = \mathfrak{h}'_m\mathfrak{a}^{-1}\mathfrak{h}_m'^{-1}(\mathfrak{h}_m'^2\mathcal{A})$. The figure shows, taking into account the directions of $\mathfrak{a}$ and $\mathfrak{h}'_m\mathfrak{a}^{-1}\mathfrak{h}_m'^{-1}$, that the pair $\mathfrak{a}\mathfrak{h}_m'^{-1}\mathcal{A}$, $\mathfrak{h}'_m\mathfrak{a}^{-1}\mathfrak{h}'_m\mathcal{A}$ is not separated by the pair $\mathfrak{h}_m'^{-1}\mathcal{A}$, $\mathfrak{h}_m'^2\mathcal{A}$. Thus $\mathcal{A}(\mathfrak{g})$ does not intersect $\mathcal{A}'_m$. Moreover, no equivalent of $\mathcal{A}(\mathfrak{g})$ with respect to $\mathfrak{F}_{m-1}$ can cut $\mathcal{A}'_m$, since the pair $\mathcal{A}$, $\mathfrak{h}'_m\mathcal{A}$ is not separated by any equivalent pair. Also $\mathcal{A}'_m$ and $\mathcal{A}(\mathfrak{g})$ are not equivalent

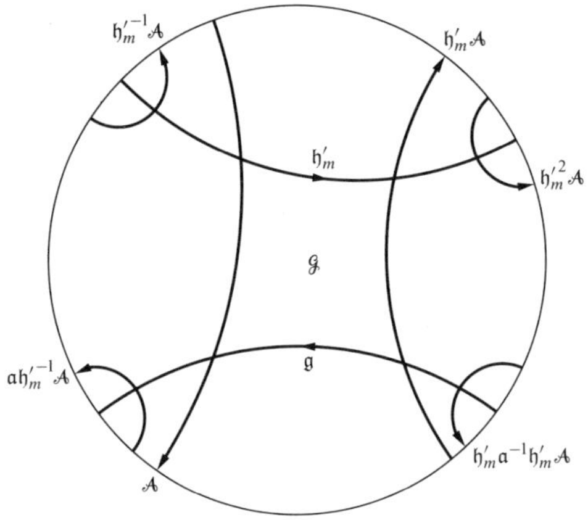

Figure 22.3

with respect to $\mathfrak{F}_{m-1}$. For, those axes of the collection $\mathfrak{F}_{m-1}\mathcal{A}_m'$ which cut $\mathcal{A}$ are derived from $\mathcal{A}_m'$ by the powers of $\mathfrak{a}$. For any power $\mathfrak{a}^n$, $n \neq 0$, the axes $\mathfrak{h}_m'\mathcal{A}$ and $\mathfrak{a}^n\mathfrak{h}_m'\mathcal{A}$ are disjoint, because $\mathcal{A}$ is simple modulo $\mathfrak{F}_{m-1}$. Thus $\mathfrak{a}^n\mathcal{A}_m'$ cannot coincide with $\mathcal{A}(\mathfrak{g})$. Therefore $\mathcal{A}_m'$ and $\mathcal{A}(\mathfrak{g})$ are disjoint modulo $\mathfrak{F}_{m-1}$.

Consider now the decomposition of $\mathcal{K}(\mathfrak{F}_{m-1})$ by the collection $\mathfrak{F}_{m-1}\mathcal{A}_m'$. Then the subgroup of $\mathfrak{F}_{m-1}$ corresponding to a region of decompositions, contains reversions, for instance, certain reversed translations conjugate to $\mathfrak{g}$. Hence, if we choose $\mathcal{A}_m'$ instead of $\mathcal{A}_m$, then the process does not stop at the step number m.

The process of successive decomposition by means of axes of reversed translations can thus be continued until, after a certain number s of steps, one arrives at a restgroup $\mathfrak{F}_s$ which consists of motions only and for which no axis of the type $\mathfrak{A}(\rightarrow)$, simple and non-dividing modulo $\mathfrak{F}_s$, exists. It will result in Section 9 that $s = p_{\mathfrak{F}}$.

Remark. In Section 21.15 we considered an elementary subgroup $\mathfrak{E}(\mathcal{P}, \mathcal{P}', \mathcal{P}'')$ of $\mathfrak{F}$, derived by starting from a centre, or limit-centre, or boundary axis $\mathcal{P}$, and we assumed $\mathcal{P}'$ and $\mathcal{P}''$ to be inner axes of $\mathfrak{F}$ equivalent with respect to $\mathfrak{F}$. Let $\mathcal{P}'$ and $\mathcal{P}''$ be equivalent by a reversed translation $\mathfrak{h}$. Then $\mathcal{K}(\mathfrak{F})$ mod $\mathfrak{F}$ is a Klein bottle. According to (14), §21, one gets $\Phi(\mathfrak{F}) = 2\big(1 - \frac{1}{\nu}\big)$, if $\nu \leqq \infty$ is the order of the primary element $\mathfrak{p}$ belonging to $\mathcal{P}$. The surface $\mathcal{K}(\mathfrak{F})$ mod $\mathfrak{F}$ has one protrusion namely the one corresponding to $\mathcal{P}$.

We may take $\mathfrak{F}$ to play the rôle of the above $\mathfrak{F}_{m-1}$. In the decomposition of $\mathcal{K}(\mathfrak{F})$ by the equivalence class of $\mathcal{P}''$ we let $\mathcal{P}''$ correspond to the above $\mathcal{A}$, also $\mathfrak{h}$ to the above $\mathfrak{h}_m'$, and $\mathcal{P}' = \mathfrak{h}\mathcal{P}''$ to the above $\mathfrak{h}_m^{-1}\mathcal{A}' = \mathfrak{h}_m'\mathcal{A}$. It then follows that $\mathfrak{g} = \mathfrak{p}''\mathfrak{h}^{-1}$ is a reversed translation carrying $\mathcal{P}'$ into $\mathcal{P}''$, and that the axes of $\mathfrak{h}$ and $\mathfrak{g}$ are disjoint modulo $\mathfrak{F}$, and each of them is simple modulo $\mathfrak{F}$. Let $\mathcal{K}(\mathfrak{F})$ be decomposed by the union of $\mathfrak{F}\mathcal{A}(\mathfrak{h})$ and $\mathfrak{F}\mathcal{A}(\mathfrak{g})$, and consider the region of decomposition $\mathcal{B}$ with corresponding subgroup $\mathfrak{B}$ which has $\mathcal{A}(\mathfrak{h})$ and $\mathcal{A}(\mathfrak{g})$ among its sides. Then $\mathcal{B}$ mod $\mathfrak{B}$ has the boundaries corresponding to $\mathcal{A}(\mathfrak{h})$ and $\mathcal{A}(\mathfrak{g})$ among its protrusions. They do not correspond to protrusions of $\mathcal{K}(\mathfrak{F})$ mod $\mathfrak{F}$, because they correspond to inner axes of $\mathfrak{F}$. Let an elementary group $\mathfrak{C}_1(\mathcal{P}_1, \mathcal{P}_1', \mathcal{P}_1'')$ of $\mathfrak{B}$ be based on them, the $\mathcal{P}_1'$ and $\mathcal{P}_1''$ being equivalent with $\mathcal{A}(\mathfrak{h})$ and $\mathcal{A}(\mathfrak{g})$, respectively. The resulting $\mathcal{P}_1$ cannot be an inner axis of $\mathcal{B}$, because if it were non-dividing modulo $\mathfrak{B}$, then no protrusion of $\mathcal{K}(\mathfrak{F})$ modulo $\mathfrak{F}$ would result, and if it were dividing modulo $\mathfrak{B}$, then $\Phi(\mathfrak{C}_1) = 2$, while $\mathcal{K}(\mathfrak{C}_1)$ mod $\mathfrak{C}_1$ would only be part of $\mathcal{K}(\mathfrak{F})$ mod $\mathfrak{F}$, and $\Phi(\mathfrak{F}) \leqq 2$. Hence $\mathcal{P}_1$ must correspond to the protrusion of $\mathcal{K}(\mathfrak{F})$ mod $\mathfrak{F}$, thus be equivalent with $\mathcal{P}$ with respect to $\mathfrak{F}$. The Klein bottle $\mathcal{K}(\mathfrak{F})$ mod $\mathfrak{F}$ is then derived from $\mathcal{K}(\mathfrak{C}_1)$ mod $\mathfrak{C}_1$ by closing the boundaries derived from $\mathcal{A}(\mathfrak{h})$ and $\mathcal{A}(\mathfrak{g})$ in the known way corresponding to a reversed translation. Hence a Klein bottle can also be thought of as a surface with two *cross caps*.

Whether the surface is arrived at in one or the other of these two ways depends on the choice of the segment, simple modulo $\mathfrak{F}$ and joining $\mathcal{P}$ with one of its equivalents, selected for the construction of §21.

22.4 Reduction of the genus in the case of orientability. In this section we assume that $\mathfrak{F}$ is a group of motions and that there exists an axis $\mathcal{A}_1$ of the type $\mathfrak{A}(\rightarrow)$ which is simple and non-dividing modulo $\mathfrak{F}$. The closed geodesic a_1 corresponding to $\mathcal{A}_1$ on $\mathcal{K}(\mathfrak{F})$ mod $\mathfrak{F}$ has no point in common with the protrusions of that surface, if any: Since $\mathcal{A}_1$ is non-dividing modulo $\mathfrak{F}$, it is an inner axis of $\mathfrak{F}$, thus a_1 does not meet the boundaries of $\mathcal{K}(\mathfrak{F})$ mod $\mathfrak{F}$, if any. Since $\mathcal{A}_1$ is an axis, it does not end in a limit-centre, thus a_1 does not end in a mast. Also $\mathcal{A}_1$ passes through no centre of order > 2, because it is simple modulo $\mathfrak{F}$, and through no centre of order 2, because its corresponding subgroup is $\mathfrak{A}(\rightarrow)$.

The decomposition of $\mathcal{K}(\mathfrak{F})$ by $\mathfrak{F}\mathcal{A}_1$ has been analysed in case I of §19. Let $\mathfrak{F}'$ be the subgroup of $\mathfrak{F}$ carrying one of the region of decomposition into itself. The closure of this region on $\mathcal{K}(\mathfrak{F})$ is then $\mathcal{K}(\mathfrak{F}')$. In the interior of $\mathcal{K}(\mathfrak{F}')$ the groups $\mathfrak{F}'$ and $\mathfrak{F}$ have equal effect, but not on the boundary. We choose one of the two regions adjacent to $\mathcal{A}_1$ as our $\mathcal{K}(\mathfrak{F}')$. Then the region corresponds to the $\mathcal{G}$ of §19, and $\mathfrak{F}'$ to the group there called $\mathfrak{G}$. The element, called $\mathfrak{h}$ in §19, which carries the region into the other region adjacent to $\mathcal{A}_1$ is here a translation; for, it cannot be a limit-rotation because $\mathcal{A}_1$ does not end in a limit-centre, and it cannot be a reversed translation, because $\mathfrak{F}$ consists of motions only.

Since a_1 passes through no protrusion of $\mathcal{K}(\mathfrak{F})$ mod $\mathfrak{F}$, since $\mathfrak{F}$ and $\mathfrak{F}'$ have equal effect in the interior of $\mathcal{K}(\mathfrak{F}')$, and since all regions of decomposition are equivalent, every protrusion of $\mathcal{K}(\mathfrak{F})$ mod $\mathfrak{F}$ is also a protrusion of $\mathcal{K}(\mathfrak{F}')$ mod $\mathfrak{F}'$. Besides these the surface $\mathcal{K}(\mathfrak{F}')$ mod $\mathfrak{F}'$ possesses two more protrusions. For, those axes of $\mathfrak{F}\mathcal{A}_1$ which contribute to the boundary of $\mathcal{K}(\mathfrak{F}')$ fall into two equivalence classes with respect to $\mathfrak{F}'$ (§19.1) and thus correspond to two boundaries of $\mathcal{K}(\mathfrak{F}')$ mod $\mathfrak{F}'$; they result from the cutting up of $\mathcal{K}(\mathfrak{F})$ mod $\mathfrak{F}$ along a_1. One of these equivalence classes is $\mathfrak{F}'\mathcal{A}_1$. Let k denote the other class.

We now base a decomposition of $\mathfrak{F}'$ on these two protrusions of $\mathcal{K}(\mathfrak{F}')$ mod $\mathfrak{F}'$ by the procedure described in Section 1. Thus we select a boundary axis $\mathcal{A}_1'$ of the class k such that the open part $\mathcal{T}$ of the join $\overline{\mathcal{T}}$ of $\mathcal{A}_1$ and $\mathcal{A}_1'$ is simple modulo $\mathfrak{F}'$, and thus also modulo $\mathfrak{F}$. Then the pair $\mathcal{A}_1, \mathcal{A}_1'$ is not separated by any equivalent pair $\mathfrak{f}\mathcal{A}_1, \mathfrak{f}\mathcal{A}_1'$. Let $\mathfrak{a}_1$ be a primary translation along $\mathcal{A}_1$ chosen such, say, that $\mathcal{A}_1$ directed by $\mathfrak{a}_1$ bounds $\mathcal{K}(\mathfrak{F}')$ positively. Let $\mathfrak{b}_1$ denote an element of $\mathfrak{F}$ carrying $\mathcal{A}_1$ into $\mathcal{A}_1'$. As already stated, it is a translation whose axis $\mathcal{B}_1$ cuts $\mathcal{A}_1$ and $\mathcal{A}_1'$. Also $\mathcal{A}_1$ and $\mathcal{A}_1'$ cut out the primary displacement of $\mathcal{B}_1$, because $\mathcal{K}(\mathfrak{F}')$ contains in its interior no axis of the collection $\mathfrak{F}\mathcal{A}_1$. The axis $\mathcal{B}_1$ is simple modulo $\mathfrak{F}$, because $\mathcal{A}_1$ and $\mathcal{A}_1'$ are not separated by any equivalent pair.

If we put $\mathfrak{b}_1\mathfrak{a}_1\mathfrak{b}_1^{-1} = \mathfrak{a}_1'$, then $\mathcal{A}_1'$ directed by $\mathfrak{a}_1'$ bounds $\mathcal{K}(\mathfrak{F}')$ negatively; see Fig. 22.4. The segment $\mathcal{T}$ is not indicated in the figure. Therefore, according to the

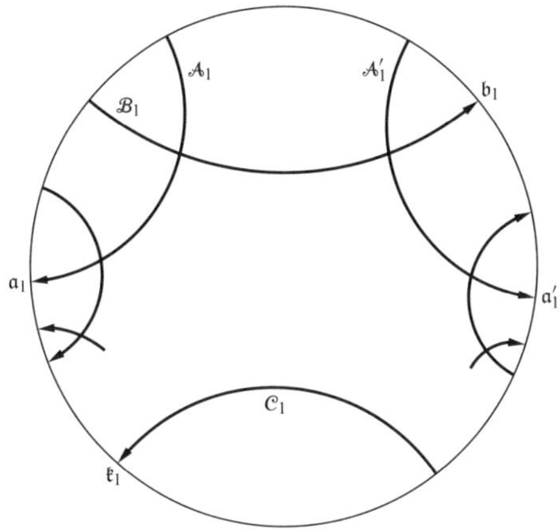

Figure 22.4

procedure of Section 1, in generating an elementary group $\mathfrak{E}_1$ we let $\mathfrak{a}_1$ and $\mathfrak{a}_1'^{-1}$ play the rôle of $\mathfrak{p}$ and $\mathfrak{p}'$ and thus get (compare Fig. 21.1)

$$\mathfrak{p}''^{-1} = \mathfrak{p}\mathfrak{p}' = \mathfrak{a}_1\mathfrak{a}_1'^{-1} = \mathfrak{a}_1\mathfrak{b}_1\mathfrak{a}_1^{-1}\mathfrak{b}_1^{-1} = \mathfrak{k}_1.$$

The commutator $\mathfrak{k}_1$ of $\mathfrak{a}_1$ and $\mathfrak{b}_1$ thus determines the third protrusion of the elementary surface ES_1 derived from $\mathfrak{E}_1$.

If $\mathfrak{k}_1$ corresponds to a protrusion of $\mathcal{K}(\mathfrak{F})$ mod $\mathfrak{F}$, then $\mathcal{K}(\mathfrak{F})$ mod $\mathfrak{F}$ is a handle (§21.15). $\mathfrak{F}$ has a presentation by

$$\begin{cases} \text{generators} & \mathfrak{a}_1, \mathfrak{b}_1 \\ \text{relation} & (\mathfrak{a}_1\mathfrak{b}_1\mathfrak{a}_1^{-1}\mathfrak{b}_1^{-1})^{\nu} = 1. \end{cases} \tag{1}$$

Here ν is finite, if $\mathfrak{k}_1$ is a rotation, thus for a handle $\mathcal{H}(\circlearrowleft_\nu)$. In the case of a $\mathcal{H}(\hookrightarrow)$ or a $\mathcal{H}(\rightarrow)$, $\mathfrak{k}$ is a limit-rotation or a translation, thus $\nu = \infty$, and the relation lapses. – In Fig. 22.4, $\mathfrak{k}_1$ is assumed to be a translation.

So let it be assumed that $\mathfrak{k}_1$ does not correspond to a protrusion of $\mathcal{K}(\mathfrak{F})$ mod $\mathfrak{F}$. Then the axis $\mathfrak{C}_1$ of $\mathfrak{k}_1$ is an inner axis of $\mathfrak{F}$, and it is known from §21.7 that $\mathfrak{C}_1$ is simple modulo $\mathfrak{F}$, and that $\mathfrak{k}_1$ is a primary translation along $\mathfrak{C}_1$. Two cases present themselves, according as $\mathfrak{C}_1$ is non-dividing or dividing modulo $\mathfrak{F}$. In both cases we consider the decomposition of $\mathcal{K}(\mathfrak{F})$ by $\mathfrak{F}\mathfrak{C}_1$.

Assume that $\mathfrak{C}_1$ is non-dividing modulo $\mathfrak{F}$. All regions of the decomposition of $\mathcal{K}(\mathfrak{F})$ by $\mathfrak{F}\mathfrak{C}_1$ are then equivalent with respect to $\mathfrak{F}$. The subgroup of $\mathfrak{F}$ corresponding to one of these regions is the *handle group* $\mathfrak{H}_1$ just established in (1) with $\nu = \infty$, and the corresponding surface is a $\mathcal{H}(\rightarrow)$. Since this surface has only one protrusion, all boundaries of the region are equivalent with respect to $\mathfrak{H}_1$. The case I of §19 does therefore not arise, and only the case II occurs. Since $\mathfrak{F}$ contains neither reflections nor reversed translations, the case II requires that $\mathfrak{C}_1$ is the axis of a subgroup $\mathfrak{A}(\cdot\cdot)$ of $\mathfrak{F}$. Then $\mathcal{K}(\mathfrak{F})$ mod $\mathfrak{F}$ arises by closing the boundary of the $\mathcal{H}(\rightarrow)$ according to §18.7–8. One gets for $\mathfrak{F}$, if $\mathfrak{c}_1$ and $\mathfrak{c}_2$ are half-turns about centres on $\mathfrak{C}_1$ whose product is $\mathfrak{k}_1^{-1}$, the presentation

$$\begin{cases} \text{generators} & \mathfrak{a}_1,\, \mathfrak{b}_1,\, \mathfrak{c}_1,\, \mathfrak{c}_2 \\ \text{relations} & \mathfrak{c}_1^2 = 1,\, \mathfrak{c}_2^2 = 1,\; \mathfrak{a}_1\mathfrak{b}_1\mathfrak{a}_1^{-1}\mathfrak{b}_1^{-1}\mathfrak{c}_1\mathfrak{c}_2 = 1. \end{cases} \tag{2}$$

Here one of the half-turns may be eliminated from the set of generators. The surface may be described as a handle with two conical points of order 2.

We are thus only left with the case in which $\mathfrak{C}_1$ is dividing modulo $\mathfrak{F}$. One of the regions adjacent to $\mathfrak{C}_1$ in the decomposition of $\mathcal{K}(\mathfrak{F})$ by $\mathfrak{F}\mathfrak{C}_1$ has as its corresponding subgroup $\mathfrak{F}$ the handle group $\mathfrak{H}_1$ with $\nu = \infty$. Let $\mathfrak{F}_1$ be the subgroup of $\mathfrak{F}$ belonging to the other region. This is an $\mathfrak{F}$-group, because $\mathfrak{C}_1$ is an inner axis of $\mathfrak{F}$. We call it the *restgroup*. It belongs to the surface obtained from $\mathcal{K}(\mathfrak{F})$ mod $\mathfrak{F}$ by cutting off the handle $\mathcal{H}(\rightarrow)$. All protrusions of $\mathcal{K}(\mathfrak{F})$ mod $\mathfrak{F}$, if any, are represented in $\mathfrak{F}_1$, and moreover $\mathcal{K}(\mathfrak{F}_1)$ mod $\mathfrak{F}_1$ has the boundary derived from $\mathfrak{C}_1$ as a protrusion. According to (14), §19, one gets

$$\mathfrak{F} = \mathfrak{F}_1 * \mathfrak{H} \text{ am } \mathfrak{K}_1 \tag{3}$$

if $\mathfrak{K}_1$ denotes the translation group generated by $\mathfrak{k}_1$. If $\mathfrak{C}_1$ is directed by $\mathfrak{k}_1$, it bounds $\mathcal{K}(\mathfrak{H}_1)$ negatively, thus $\mathcal{K}(\mathfrak{F}_1)$ positively.

We now add the further assumption that $\mathfrak{F}$ satisfies the condition of quasi-compactness.

Since $\mathfrak{C}_1$ is dividing modulo $\mathfrak{F}$, the area $\Phi(\mathfrak{F})$ of $\mathcal{K}(\mathfrak{F})$ mod $\mathfrak{F}$ is the sum of the areas of $\mathcal{K}(\mathfrak{H}_1)$ mod $\mathfrak{H}_1$ and $\mathcal{K}(\mathfrak{F}_1)$ mod $\mathfrak{F}_1$. All areas are computed from (12), §17, since $\mathfrak{F}$ contains no reversions. Thus

$$\Phi(\mathfrak{F}) = \Phi(\mathfrak{H}_1) + \Phi(\mathfrak{F}_1). \tag{4}$$

According to (14), §21, $\Phi(\mathfrak{H}_1) = 2$. Moreover, $\vartheta_{\mathfrak{F}_1}(\rightarrow) = \vartheta_{\mathfrak{F}}(\rightarrow) + 1$, while the type numbers $\vartheta(\hookrightarrow)$ and $\vartheta(\odot_\nu)$ remain unchanged. Hence, in virtue of (4), relation

(12), §17, yields

$$p_{\mathfrak{F}_1} = p_{\mathfrak{F}} - 1. \tag{5}$$

If $\mathfrak{F}_1$ possesses an axis $\mathcal{A}_2$ of the type $\mathfrak{A}(\rightarrow)$ which is simple and non-dividing modulo $\mathfrak{F}_1$, then one can repeat the process starting from $\mathcal{A}_2$. One thus gets a handle group $\mathfrak{H}_2$ with generators $\mathfrak{a}_2$ and $\mathfrak{b}_2$, their commutator being $\mathfrak{k}_2 = \mathfrak{a}_2\mathfrak{b}_2\mathfrak{a}_2^{-1}\mathfrak{b}_2^{-1}$. Now $\mathfrak{k}_2$ cannot correspond to a protrusion of $\mathcal{K}(\mathfrak{F})$ mod $\mathfrak{F}$; for, then $\mathcal{K}(\mathfrak{H}_2)$ mod $\mathfrak{H}_2$ would coincide with $\mathcal{K}(\mathfrak{F})$ mod $\mathfrak{F}$, whereas $\mathcal{K}(\mathfrak{H}_2)$ belongs to $\mathcal{K}(\mathfrak{F}_1)$ and thus has no point in common with the interior of $\mathcal{K}(\mathfrak{H}_1)$. Hence $\mathfrak{k}_2$ is a translation. Its axis $\mathfrak{C}_2$ then is simple modulo $\mathfrak{F}_1$ and thus also modulo $\mathfrak{F}$. With the same convention as to the direction of the axes $\mathcal{A}_2$ and $\mathfrak{C}_2$, the axis $\mathfrak{C}_2$ bounds $\mathcal{K}(\mathfrak{H}_2)$ negatively. Since $\mathfrak{C}_2$ does not correspond to a protrusion of $\mathcal{K}(\mathfrak{F})$ mod $\mathfrak{F}$, if it corresponds to a protrusion of $\mathcal{K}(\mathfrak{F}_1)$ mod $\mathfrak{F}_1$, it must be the one derived from $\mathfrak{C}_1$. In this case $\mathfrak{H}_2\mathfrak{C}_2$ contains $\mathfrak{C}_1$, and $\mathfrak{H}_2$ coincides with $\mathfrak{F}_1$. If $\mathfrak{C}_1 = \mathfrak{f}\mathfrak{C}_2, \mathfrak{f} \in \mathfrak{F}_1$, we may replace $\mathcal{A}_2$ by the equivalent axis $\mathfrak{f}\mathcal{A}_2$, thus $\mathfrak{a}_2$ by $\mathfrak{f}\mathfrak{a}_2\mathfrak{f}^{-1}$ and $\mathfrak{b}_2$ by $\mathfrak{f}\mathfrak{b}_2\mathfrak{f}^{-1}$, hence $\mathfrak{k}_2$ by $\mathfrak{f}\mathfrak{k}_2\mathfrak{f}^{-1}$, and $\mathfrak{C}_2$ by $\mathfrak{f}\mathfrak{C}_2 = \mathfrak{C}_1$. Hence we may from the beginning choose $\mathcal{A}_2$, such that $\mathfrak{C}_2$ and $\mathfrak{C}_1$ coincide. According to the directions chosen $\mathfrak{C}_1$ bounds $\mathcal{K}(\mathfrak{F}_1)$ positively, while $\mathfrak{C}_2$ bounds $\mathcal{K}(\mathfrak{F}_1)$ negatively. Hence $\mathfrak{k}_2 = \mathfrak{k}_1^{-1}$. Then $\mathfrak{F}$ is obtained from (3) as $\mathfrak{F} = \mathfrak{H}_1 * \mathfrak{H}_2$ am $\mathfrak{K}_1$ and has a presentation by

$$\begin{cases} \text{generators} \quad \mathfrak{a}_1, \mathfrak{b}_1, \mathfrak{a}_2, \mathfrak{b}_2 \\ \text{relation} \quad \mathfrak{a}_1\mathfrak{b}_1\mathfrak{a}_1^{-1}\mathfrak{b}_1^{-1}\mathfrak{a}_2\mathfrak{b}_2\mathfrak{a}_2^{-1}\mathfrak{b}_2^{-1} = 1. \end{cases} \tag{6}$$

In this case $\mathcal{K}(\mathfrak{F})$ mod $\mathfrak{F}$ contains no protrusions at all. $\Phi(\mathfrak{F}) = \Phi(\mathfrak{H}_1) + \Phi(\mathfrak{H}_2) = 4$, and (12), §17, yields $p = 2$.

If $\mathfrak{C}_2$ does not correspond to a protrusion of $\mathcal{K}(\mathfrak{F}_1)$ mod $\mathfrak{F}_1$, then $\mathfrak{C}_2$ is an inner axis of $\mathfrak{F}_1$. It cannot correspond to the case I of §19 for the same reasons as applied to $\mathfrak{C}_1$. It cannot correspond to the case II of §19, because $\mathcal{K}(\mathfrak{F})$ mod $\mathfrak{F}$ would then be obtained from $\mathcal{K}(\mathfrak{H}_2)$ mod $\mathfrak{H}_2$ by the closure of its boundary, whereas the interiors of $\mathcal{K}(\mathfrak{H}_1)$ and $\mathcal{K}(\mathfrak{H}_2)$ are disjoint modulo $\mathfrak{F}$. Therefore $\mathfrak{C}_2$ corresponds to case III of §19 and is thus dividing modulo $\mathfrak{F}_1$. When decomposing $\mathcal{K}(\mathfrak{F}_1)$ by $\mathfrak{F}_1\mathfrak{C}_2$ one gets a restgroup $\mathfrak{F}'$ such that $\mathcal{K}(\mathfrak{F}')$ mod $\mathfrak{F}'$ contains all protrusions of $\mathcal{K}(\mathfrak{F})$ mod $\mathfrak{F}$, if any, and moreover two boundaries derived from $\mathfrak{C}_1$ and $\mathfrak{C}_2$. These axes both bound $\mathcal{K}(\mathfrak{F}')$ positively. We now select in $\mathcal{K}(\mathfrak{F}')$ a boundary axis $\mathfrak{f}\mathfrak{C}_2$, $\mathfrak{f} \in \mathfrak{F}'$, of the class $\mathfrak{F}'\mathfrak{C}_2$ such that its join with $\mathfrak{C}_1$ is simple modulo $\mathfrak{F}'$. If $\mathfrak{f} \neq 1$, we may replace $\mathcal{A}_2$ and $\mathcal{B}_2$ by $\mathfrak{f}\mathcal{A}_2$ and $\mathfrak{f}\mathcal{B}_2$ and thus $\mathfrak{H}_2$ by $\mathfrak{f}\mathfrak{H}_2\mathfrak{f}^{-1}$. Hence we assume that $\mathcal{A}_2$ and $\mathcal{B}_2$ are from the beginning so chosen that the join of $\mathfrak{C}_1$ and $\mathfrak{C}_2$ is simple modulo $\mathfrak{F}'$. The relative position of $\mathfrak{C}_1$ and $\mathfrak{C}_2$ corresponds to the position of the axes $\mathcal{P}$ and $\mathcal{P}'$ in Fig. 21.1. We then determine the elementary group $\mathfrak{E}_2$ generated by $\mathfrak{k}_1$ and $\mathfrak{k}_2$ and put $\mathfrak{k}_1\mathfrak{k}_2 = \mathfrak{j}_2$. One then can form the group

$$\mathfrak{G}_2 = \prod\nolimits^{*}(\mathfrak{H}_1, \mathfrak{H}_2, \mathfrak{E}_2) \text{ am } \mathfrak{K}_1, \mathfrak{K}_2. \tag{7}$$

$\mathcal{K}(\mathfrak{G}_2)$ mod $\mathfrak{G}_2$ has one protrusion namely the one corresponding to the centre, or limit-centre, or boundary axis of $\mathfrak{G}_2$ belonging to $\mathfrak{j}_2$. If $\mathfrak{j}_2$ corresponds to a protrusion of $\mathcal{K}(\mathfrak{F})$ mod $\mathfrak{F}$, then $\mathfrak{F}$ coincides with $\mathfrak{G}_2$, and an abstract presentation of $\mathfrak{F}$ readily follows. Also, if $\mathfrak{j}_2$ is a translation along an inner axis of $\mathfrak{F}'$ with corresponding group $\mathfrak{A}(\cdot\cdot)$ one obtains $\mathfrak{F}$ from $\mathfrak{G}_2$ by the closure of the boundary in a way analogous to (2).

If $\mathfrak{j}_2$ is a translation along an inner axis $\mathcal{J}_2$ of $\mathfrak{F}'$ which is dividing modulo $\mathfrak{F}'$, thus with a subgroup $\mathfrak{A}(\rightarrow)$, then in the decomposition of $\mathcal{K}(\mathfrak{F})$ by $\mathfrak{F}\mathcal{J}_2$ the region bounded negatively by $\mathcal{J}_2$, when directed by $\mathfrak{j}_2$, has $\mathfrak{G}_2$ as its corresponding subgroup. Hence $\mathcal{J}_2$ is also dividing modulo $\mathfrak{F}$, because all axes $\mathfrak{G}_2\mathcal{J}_2$ bound $\mathcal{K}(\mathfrak{G}_2)$ negatively. Let $\mathfrak{F}_2$ be the subgroup of $\mathfrak{F}$ corresponding to the region bounded positively by $\mathcal{J}_2$. One then gets

$$\mathfrak{F} = \mathfrak{F}_2 * \mathfrak{G}_2 \text{ am } \mathfrak{J}_2, \tag{8}$$

where $\mathfrak{J}_2$ is the group generated by $\mathfrak{j}_2$. Hence $\Phi(\mathfrak{F}) = \Phi(\mathfrak{F}_2) + \Phi(\mathfrak{G}_2)$, and one gets $\Phi(\mathfrak{G}_2) = 6$ from (7) together with (14), §21, and Section 21.13. Since $\vartheta_{\mathfrak{F}_2}(\rightarrow) = \vartheta_{\mathfrak{F}}(\rightarrow) + 1$, one gets from (12), §17, the relation $p_{\mathfrak{F}_2} = p_{\mathfrak{F}} - 2 = p_{\mathfrak{F}_1} - 1$.

The procedure can be continued through a series of restgroups $\mathfrak{F}_1, \mathfrak{F}_2, \ldots$, as long as the resulting restgroup $\mathfrak{F}_m$ admits an inner axis $\mathcal{A}_{m+1}$ of the type $\mathfrak{A}(\rightarrow)$ which is simple and non-dividing modulo $\mathfrak{F}_m$. On the other hand, if we now assume $\mathfrak{F}$ to satisfy the condition of quasi-compactness, since the genus is a finite, non-negative number and is decreased by one in each step, the process must come to an end. Suppose it stops after the introduction of the axis $\mathcal{A}_s$ in $\mathfrak{F}_{s-1}$, thus with the step number s. There are exactly four possibilities for the process getting stopped with the step number s:

1) The element $\mathfrak{j}_s = \mathfrak{k}_1\mathfrak{k}_2\ldots\mathfrak{k}_s$ corresponds to a protrusion of $\mathcal{K}(\mathfrak{F})$ mod $\mathfrak{F}$.

One then gets as a presentation of $\mathfrak{F}$

$$\begin{cases} \text{generators} & \mathfrak{a}_1, \mathfrak{b}_1, \ldots, \mathfrak{a}_s, \mathfrak{b}_s \\ \text{relation} & (\mathfrak{k}_1\mathfrak{k}_2\ldots\mathfrak{k}_s)^{\nu} = 1, \end{cases} \tag{9}$$

where $\mathfrak{k}_m$ stands as an abbreviation for $\mathfrak{a}_m\mathfrak{b}_m\mathfrak{a}_m^{-1}\mathfrak{b}_m^{-1}$, $m = 1, 2, \ldots, s$, and ν is the order of $\mathfrak{j}_s$. Thus the relation lapses, if $\mathfrak{j}_s$ is a limit-rotation or a boundary-translation, and $\mathfrak{F}$ is then a free group. This case corresponds for $s = 1$ to (1), and $\mathcal{K}(\mathfrak{F})$ mod $\mathfrak{F}$ has exactly one protrusion.

2) $\mathfrak{j}_s$ is a translation along an axis $\mathcal{J}_s$ with corresponding subgroup $\mathfrak{A}(\cdot\cdot)$.

Then $\mathfrak{F}$ is given by

$$\begin{cases} \text{generators} & \mathfrak{a}_1, \mathfrak{b}_1, \ldots, \mathfrak{a}_s, \mathfrak{b}_s, \mathfrak{c}_1, \mathfrak{c}_2 \\ \text{relations} & \mathfrak{c}_1^2 = 1, \mathfrak{c}_2^2 = 1, \mathfrak{k}_1\mathfrak{k}_2\ldots\mathfrak{k}_s\mathfrak{c}_1\mathfrak{c}_2 = 1. \end{cases} \tag{10}$$

This case corresponds for $s = 1$ to (2), and $\mathcal{K}(\mathfrak{F})$ mod $\mathfrak{F}$ has two protrusions.

3) The commutator $\mathfrak{k}_s$ obtained in the step number s corresponds to the protrusion of $\mathcal{K}(\mathfrak{F}_{s-1})$ mod $\mathfrak{F}_{s-1}$ which is not a protrusion of $\mathcal{K}(\mathfrak{F})$ mod $\mathfrak{F}$.

Then $\mathfrak{k}_s = \mathfrak{j}_{s-1}^{-1}$, and one gets for $\mathfrak{F}$

$$\begin{cases} \text{generators} & \mathfrak{a}_1, \mathfrak{b}_1, \ldots, \mathfrak{a}_s, \mathfrak{b}_s \\ \text{relation} & \mathfrak{k}_1 \mathfrak{k}_2 \ldots \mathfrak{k}_s = 1. \end{cases} \tag{11}$$

This case corresponds for $s = 2$ to (6), and $\mathcal{K}(\mathfrak{F})$ mod $\mathfrak{F}$ has no protrusion.

4) $\mathfrak{j}_s$ is a translation along an inner axis of $\mathfrak{F}_{s-1}$ with corresponding group $\mathfrak{A}(\rightarrow)$, and this axis is dividing modulo $\mathfrak{F}_{s-1}$, but in the restgroup $\mathfrak{F}_s$ there exists no axis of the type $\mathfrak{A}(\rightarrow)$ which is simple and non-dividing modulo $\mathfrak{F}_s$.

We may then form the group

$$\mathfrak{G}_s = \prod_{m,r}^{*} (\mathfrak{H}_m, \mathfrak{E}_r) \text{ am } \mathfrak{K}_m, \mathfrak{J}_t, \quad 1 \leqq m \leqq s,\ 2 \leqq r \leqq s,\ 2 \leqq t \leqq s-1, \tag{12}$$

and thereby get

$$\mathfrak{F} = \mathfrak{F}_s * \mathfrak{G}_s \text{ am } \mathfrak{J}_s. \tag{13}$$

This corresponds for $s = 2$ to (7) and (8). The group $\mathfrak{E}_r$ is generated by $\mathfrak{k}_r$ and $\mathfrak{j}_r$, and one gets $\mathfrak{j}_{r-1} \mathfrak{k}_r \mathfrak{j}_r^{-1} = 1$ for $r = 3, \ldots, s$, while for $r = 2$ one gets $\mathfrak{k}_1 \mathfrak{k}_2 \mathfrak{j}_2^{-1} = 1$. Hence $\mathfrak{G}$ is the free group with $\mathfrak{a}_1, \mathfrak{b}_1, \ldots, \mathfrak{a}_s, \mathfrak{b}_s$ as generators, and the product of their commutators is

$$\mathfrak{j}_s = \mathfrak{k}_1 \mathfrak{k}_2 \ldots \mathfrak{k}_s. \tag{14}$$

The surface $\mathcal{K}(\mathfrak{G}_s)$ mod $\mathfrak{G}_s$ has one boundary as its only protrusion and the axis $\mathcal{J}_s$ is therefore also dividing modulo $\mathfrak{F}$.

The group $\mathfrak{G}_s$ corresponds to the case 1) for $\mathfrak{F}$ with (14) being a translation. The difference between these cases is that $\mathfrak{j}_s$ in the case 1) is a boundary axis of $\mathfrak{F}$, and the whole of $\mathfrak{F}$ thus is exhausted, whereas $\mathfrak{j}_s$ in the case 4) is an inner and dividing axis of $\mathfrak{F}$, and therefore a restgroup remains.

The relations (13) and (14) reduce the presentation of $\mathfrak{F}$ to the presentation of $\mathfrak{F}_s$. It is thus only in the case 4) that the determination of $\mathfrak{F}$ is not yet complete.

In the cases 1) and 2) the number of elementary surfaces used is $2s - 1$. Each of these has the area 2 except that in the case 1) $\Phi(\mathfrak{E}_s) = 2\left(1 - \frac{1}{\nu}\right)$, which is < 2 if ν is finite. Hence $\Phi(\mathfrak{F}) = 4s - 2 - \frac{2}{\nu}$ in the case 1) and $4s - 2$ in the case 2). By inserting this in (12), §17, and taking account of the protrusions, one gets $p_{\mathfrak{F}} = s$.

In the case 3) the number of elementary surfaces used is $2s - 2$, each of which has the area 2, and there are no protrusions of $\mathcal{K}(\mathfrak{F})$ mod $\mathfrak{F}$. By insertion in (12), §17, one again gets $p_{\mathfrak{F}} = s$.

As has already been mentioned, in the case 4) the number and type of elementary surfaces used is the same as in case 1) with $\nu = \infty$, thus $\Phi(\mathfrak{G}_s) = 4s - 2$. Since $\mathcal{K}(\mathfrak{G}_s)$ mod $\mathfrak{G}_s$ has one boundary, thus $\vartheta_{\mathfrak{G}_s}(\rightarrow) = 1$, it results from (12), §17, that $p_{\mathfrak{G}_s} = s$. It will result from Section 9 that the genus is not altered by the influence of $\mathfrak{F}_s$ in (13), and thus that $p_{\mathfrak{F}} = s$ holds even in this case.

22.5 Existence of protrusions. In this section we consider an $\mathfrak{F}$-group $\mathfrak{F}$ satisfying three conditions:

1) It consists of motions only.

2) It satisfies the condition of quasi-compactness.

3) It possesses no axis of the type $\mathfrak{A}(\rightarrow)$ which is simple and non-dividing modulo $\mathfrak{F}$.

We first want to prove the following theorem:

If an $\mathfrak{F}$-group $\mathfrak{F}$ satisfies the three conditions just mentioned, then $\mathcal{K}(\mathfrak{F})$ mod $\mathfrak{F}$ possesses protrusions.

Proof. For an indirect proof let it be assumed that $\mathcal{K}(\mathfrak{F})$ mod $\mathfrak{F}$ possesses no protrusions. All elements of $\mathfrak{F}$ are then translations with inner axes, and the subgroup pertaining to any axis is $\mathfrak{A}(\rightarrow)$ (condition 1)). Let q be an arbitrary point in $\mathcal{D}$. Any two points of $\mathfrak{F}q$ are equivalent by exactly one element of $\mathfrak{F}$, because no element $\neq 1$ leaves q fixed. Let $\mathfrak{f}$ be an element of $\mathfrak{F}$ such that the open part $\mathcal{T}$ of the join of q and $\mathfrak{f}q$ is simple modulo $\mathfrak{F}$ and contains no point of $\mathfrak{F}q$. Such an $\mathfrak{f}$ exists in virtue of Lemma 1 of §21.1. The collection $\mathfrak{F}\overline{\mathcal{T}}$ does not accumulate in $\mathcal{D}$ (§13). Consider in particular the subcollection $\mathfrak{f}^r\overline{\mathcal{T}}$, $-\infty < r < \infty$. This is a chain of segments which is reproduced by $\mathfrak{f}$. Its ends converge to the end-points u and v of the axis $\mathcal{A}$ of $\mathfrak{f}$. If the pair u, v were separated by the pair $\mathfrak{g}u$, $\mathfrak{g}v$, $\mathfrak{g} \in \mathfrak{F}$, then the chain would have points in common with its image by $\mathfrak{g}$. This is impossible, because the links of the chain are simple modulo $\mathfrak{F}$ and exactly two segments issue from every point of $\mathfrak{F}q$. Hence the axis $\mathcal{A}$ is simple modulo $\mathfrak{F}$. Now $\mathcal{A}$ is an inner axis of $\mathfrak{F}$ with corresponding group $\mathfrak{A}(\rightarrow)$, and it is seen from condition 3) that $\mathcal{A}$ is dividing modulo $\mathfrak{F}$.

In the decomposition of $\mathcal{D}$ by $\mathfrak{F}\mathcal{A}$ let $\mathfrak{F}_1$ be the subgroup of $\mathfrak{F}$ pertaining to one of the regions adjacent to $\mathcal{A}$. This region, closed on $\mathcal{D}$, is then $\mathcal{K}(\mathfrak{F}_1)$. In $\mathcal{K}(\mathfrak{F}_1)$, including its boundaries, $\mathfrak{F}$ and $\mathfrak{F}_1$ have equal effect. Therefore $\mathfrak{F}_1$ satisfies the condition of quasi-compactness, because $\mathfrak{F}$ does (condition 2)). Any inner axis of $\mathfrak{F}_1$ which is simple modulo $\mathfrak{F}_1$ is dividing modulo $\mathfrak{F}_1$, because it is also simple modulo $\mathfrak{F}$ and thus dividing modulo $\mathfrak{F}$ (condition 3)). Also $\mathfrak{F}_1$, being a subgroup of $\mathfrak{F}$, consists of motions only.

The group $\mathfrak{F}_1$ thus satisfies the three conditions imposed on $\mathfrak{F}$. It differs from $\mathfrak{F}$ in that $\mathcal{K}(\mathfrak{F}_1)$ mod $\mathfrak{F}_1$ has exactly one protrusion, namely the one derived from $\mathcal{A}$; for, since $\mathcal{A}$ is dividing modulo $\mathfrak{F}$, all sides of $\mathcal{K}(\mathfrak{F}_1)$ are equivalent with respect to $\mathfrak{F}_1$ (case III of §19.1). We now construct an elementary subgroup $\mathfrak{E}_1$ in $\mathfrak{F}_1$ by the method of §21, taking the boundary axis $\mathcal{A}$ of $\mathfrak{F}_1$ as the start $\mathcal{P}$ of the construction in §21. Then the $\mathcal{P}'$ and $\mathcal{P}''$ of the construction are axes not equivalent with $\mathcal{A}$. They are inner axes of $\mathfrak{F}_1$, because $\mathfrak{F}_1$ has only one equivalence class of boundary axes, and they are known to be simple modulo $\mathfrak{F}_1$, hence also modulo $\mathfrak{F}$, because $\mathfrak{F}$ and $\mathfrak{F}_1$ have equal effect in $\mathcal{K}(\mathfrak{F}_1)$. Each of these axes is dividing modulo $\mathfrak{F}_1$, because it is dividing modulo $\mathfrak{F}$. They are therefore not equivalent with respect to $\mathfrak{F}_1$. In the decomposition of $\mathcal{K}(\mathfrak{F}_1)$ by $\mathfrak{F}_1\mathcal{A}_1$, $\mathcal{A}_1$ being $\mathcal{P}'$, let $\mathfrak{F}_2$ be the subgroup of $\mathfrak{F}_1$ pertaining to that region adjacent to $\mathcal{A}_1$ which does not contain $\mathcal{K}(\mathfrak{E}_1)$. Then $\mathfrak{F}_2$ satisfies the three conditions imposed on $\mathfrak{F}$, and $\mathfrak{F}_2$ possesses exactly one equivalence class of boundary axes, namely the one derived from $\mathcal{A}_1$. We can therefore treat $\mathfrak{F}_2$ in the same manner as $\mathfrak{F}_1$, i.e. split off an elementary group $\mathfrak{E}_2$, and we can continue this process indefinitely. For all the $\mathfrak{E}_m, m = 1, 2, \ldots,$ one gets $\Phi(\mathfrak{E}_m) = 2$, because they are $\mathfrak{E}(\rightarrow\rightarrow\rightarrow)$. The interiors of the $\mathcal{K}(\mathfrak{E}_m)$ are disjoint modulo $\mathfrak{F}$. This contradicts the fact that $\Phi(\mathfrak{F})$ is finite (condition 2)). The assertion has thus been proved. □

The same reasoning now permits to establish the following more precise theorem:

If an $\mathfrak{F}$-group $\mathfrak{F}$ satisfies the above conditions 1), 2), 3), *then* $\mathcal{K}(\mathfrak{F})$ mod $\mathfrak{F}$ *possesses at least three protrusions.*

Proof. In virtue of the first theorem, $\mathcal{K}(\mathfrak{F})$ mod $\mathfrak{F}$ possesses at least one protrusion. Thus there exists a centre, or limit-centre, or boundary axis $\mathcal{P}_1$ of $\mathfrak{F}$. Taking $\mathcal{P}_1$ as our start, we construct an $\mathfrak{E}_1(\mathcal{P}_1, \mathcal{P}_1', \mathcal{P}_1'')$ by the method of §21. If both $\mathcal{P}_1'$ and $\mathcal{P}_1''$ represent protrusions of $\mathcal{K}(\mathfrak{F})$ mod $\mathfrak{F}$, then $\mathfrak{F} = \mathfrak{E}_1$, and $\mathcal{P}_1, \mathcal{P}_1', \mathcal{P}_1''$ represent three different protrusions, thus the theorem holds. Hence let it be assumed that at least one of them, say $\mathcal{P}_1'$, is an inner axis of $\mathfrak{F}$. It is then simple modulo $\mathfrak{F}$. If $\mathfrak{A}_{\mathcal{P}_1'} = \mathfrak{A}(\cdot\cdot)$, then we get two conical points of order 2 as additional protrusions, and the theorem holds. If $\mathfrak{A}_{\mathcal{P}_1'} = \mathfrak{A}(\rightarrow)$, then $\mathcal{P}_1'$ is dividing modulo $\mathfrak{F}$ according to condition 3). We then consider the restgroup $\mathfrak{F}_1$ for which $\mathcal{K}(\mathfrak{F}_1)$ has $\mathcal{P}_1'$ as a side and does not contain $\mathcal{K}(\mathfrak{E}_1)$. Here $\mathcal{P}_1'$ represents a protrusion of $\mathcal{K}(\mathfrak{F}_1)$ mod $\mathfrak{F}_1$. We call it $\mathcal{P}_2$ and determine in the same way an $\mathfrak{E}_2(\mathcal{P}_2, \mathcal{P}_2', \mathcal{P}_2'')$. If at least one of $\mathcal{P}_2', \mathcal{P}_2''$, say $\mathcal{P}_2'$, does not represent a protrusion of $\mathcal{K}(\mathfrak{F})$ mod $\mathfrak{F}$, then we repeat the same argument. If $\mathfrak{A}_{\mathcal{P}_2'} = \mathfrak{A}(\rightarrow)$, then we put $\mathcal{P}_2' = \mathcal{P}_3$ and go on to the construction of an $\mathfrak{E}_3$, and so forth. This process cannot go on indefinitely, because there is a positive lower bound for the $\Phi(\mathfrak{E}_m)$, and the interiors of the $\mathcal{K}(\mathfrak{E}_m)$ are disjoint modulo $\mathfrak{F}$, and $\Phi(\mathfrak{F})$ is finite. The process can only stop either by some $\mathcal{P}_m'$ corresponding to a group $\mathfrak{A}(\cdot\cdot)$, or by the $\mathcal{P}_m', \mathcal{P}_m''$ for some value m both representing protrusions of $\mathcal{K}(\mathfrak{F})$ mod $\mathfrak{F}$. Together with $\mathcal{P}_1$ one thus gets three different protrusions of $\mathcal{K}(\mathfrak{F})$ mod $\mathfrak{F}$, and the theorem is proved. □

22.6 Reduction by rotation twins. Let $\mathfrak{F}$ be an $\mathfrak{F}$-group satisfying the three conditions indicated at the beginning of the preceding section. Then, according to that section, $\mathcal{K}(\mathfrak{F})$ mod $\mathfrak{F}$ possesses at least three protrusions.

Let n be the number of conical points of order 2 of $\mathcal{K}(\mathfrak{F})$ mod $\mathfrak{F}$. If $n \geqq 2$, we select arbitrarily two of these and determine a corresponding pair of rotation twins by the method of Section 1. Let $\mathcal{A}_1$ be the corresponding axis, simple modulo $\mathfrak{F}$, of the type $\mathfrak{A}_1 = \mathfrak{A}_{\mathcal{A}_1} = \mathfrak{A}(\cdot\cdot)$. We decompose $\mathcal{K}(\mathfrak{F})$ by the collection $\mathfrak{F}\mathcal{A}_1$. All regions of decomposition are equivalent with respect to $\mathfrak{F}$, because $\mathcal{A}_1$ is non-dividing modulo $\mathfrak{F}$. Let $\mathfrak{F}_1$ be the subgroup of $\mathfrak{F}$ belonging to one of the regions adjacent to $\mathcal{A}_1$. The closure of this region of $\mathcal{K}(\mathfrak{F})$ is then $\mathcal{K}(\mathfrak{F}_1)$, and $\mathfrak{F}$ and $\mathfrak{F}_1$ have equal effect in the interior of $\mathcal{K}(\mathfrak{F}_1)$, though not on its boundary. One gets $\Phi(\mathfrak{F}_1) = \Phi(\mathfrak{F})$, and $\mathfrak{F}_1$ satisfies the three conditions imposed on $\mathfrak{F}$. The surface $\mathcal{K}(\mathfrak{F}_1)$ mod $\mathfrak{F}_1$ possesses all protrusions of $\mathcal{K}(\mathfrak{F})$ mod $\mathfrak{F}$ except the two conical points in question. Instead of these it possesses a boundary other than those of $\mathcal{K}(\mathfrak{F})$ mod $\mathfrak{F}$ (if any), namely the one corresponding to $\mathcal{A}_1$. This boundary results from the cutting up of $\mathcal{K}(\mathfrak{F})$ mod $\mathfrak{F}$ along the closed geodesic a_1 corresponding to $\mathcal{A}_1$, thus connecting the two conical points of order 2 selected, this cutting yielding one boundary of the surface $\mathcal{K}(\mathfrak{F}_1)$ mod $\mathfrak{F}_1$.

If $n \geqq 4$, and $\mathcal{K}(\mathfrak{F}_1)$ mod $\mathfrak{F}_1$ contains at least two conical points of order 2, we again select arbitrarily two of these and repeat the process by decomposing $\mathcal{K}(\mathfrak{F}_1)$ by the equivalence class of an axis of the type $\mathfrak{A}(\cdot\cdot)$ and simple modulo $\mathfrak{F}_1$. Let $\mathcal{A}_2$ denote an axis of the class bounding that region of decomposition which is adjacent to $\mathcal{A}_1$, and let $\mathfrak{F}_2$ be the subgroup pertaining to that region. If $n \geqq 6$, we repeat the process with an axis $\mathcal{A}_3$ and arrive at a restgroup $\mathfrak{F}_3$ such that $\mathcal{A}_1, \mathcal{A}_2, \mathcal{A}_3$ are among the boundary axes of $\mathcal{K}(\mathfrak{F}_3)$. Putting $m = \left[\frac{n}{2}\right]$ (the greatest integer $\leqq \frac{n}{2}$), one arrives after m steps at a restgroup $\mathfrak{F}_m$ whose corresponding surface $\mathcal{K}(\mathfrak{F}_m)$ mod $\mathfrak{F}_m$ contains none or one conical point of order 2, according as n is even or odd.

The axes $\mathcal{A}_k$, $k = 1, 2, \ldots, m$, are boundary axes of $\mathfrak{F}_m$. The groups $\mathfrak{A}_k$ are all of the type $\mathfrak{A}(\cdot\cdot)$. Let $\mathfrak{T}_k$ denote the group of translations contained in $\mathfrak{A}_k$. All regions of decomposition of $\mathcal{K}(\mathfrak{F})$ by $\bigcup_k \mathfrak{F}\mathcal{A}_k$ are equivalent with respect to $\mathfrak{F}$, one of them, closed on $\mathcal{K}(\mathfrak{F})$, being $\mathcal{K}(\mathfrak{F}_m)$. In the interior of $\mathcal{K}(\mathfrak{F}_m)$ the groups $\mathfrak{F}$ and $\mathfrak{F}_m$ have equal effect. Hence $\Phi(\mathfrak{F}_m) = \Phi(\mathfrak{F})$. Every protrusion of $\mathcal{K}(\mathfrak{F})$ mod $\mathfrak{F}$ is represented in $\mathfrak{F}_m$ by a centre, or limit-centre, or boundary axis of $\mathfrak{F}_m$ except the n or $n-1$ conical points of order 2 selected in the construction. Moreover, $\mathcal{K}(\mathfrak{F}_m)$ mod $\mathfrak{F}_m$ has m further protrusions represented by the $\mathcal{A}_k$. According to §19, one gets for $\mathfrak{F}$ the representation

$$\mathfrak{F} = \prod_k{}^{*} (\mathfrak{F}_m, \mathfrak{A}_k) \operatorname{am} \mathfrak{T}_k, \qquad k = 1, 2, \ldots, m. \tag{15}$$

22.7 Further reduction based on protrusions. Let $\mathfrak{F}$ be a finitely generated $\mathfrak{F}$-group without reflections and without modulo $\mathfrak{F}$ simple, non-dividing axes. Then $\mathfrak{F}$ consists of motions only for, otherwise it would contain a simple axis of reversed translations (Section 2), and that axis, being of the type $\mathfrak{A}(\rightleftharpoons)$, would be non-dividing modulo $\mathfrak{F}$. Moreover, $\mathfrak{F}$ contains at most one equivalence class of centres of order 2 for, otherwise, it would contain a pair of rotation twins (Section 6), and the corresponding axis of the type $\mathfrak{A}(\cdot\cdot)$ would be non-dividing modulo $\mathfrak{F}$. All axes of $\mathfrak{F}$ are then of the type $\mathfrak{A}(\rightarrow)$. Let N denote the number of protrusions of $\mathcal{K}(\mathfrak{F})$ mod $\mathfrak{F}$. Then $N \geqq 3$ according to Section 5. Let $\mathcal{Q}_1, \mathcal{Q}_2, \ldots, \mathcal{Q}_N$ denote the protrusions of $\mathcal{K}(\mathfrak{F})$ mod $\mathfrak{F}$ in an arbitrary order.

We now construct an $\mathfrak{E}_1(\mathcal{P}_1, \mathcal{P}_1', \mathcal{P}_1'')$ based on $\mathcal{Q}_1$ and $\mathcal{Q}_2$, i.e. we select arbitrarily a representative $\mathcal{P}_1$ of $\mathcal{Q}_1$ and thereafter a representative $\mathcal{P}_1'$ of $\mathcal{Q}_2$ such that the condition of Section 1 holds. The construction does not lead to a pair of rotation twins, because $\mathfrak{F}$ contains at most one equivalence class of centres of order 2. It is recalled that in the interior of $\mathcal{K}(\mathfrak{E}_1)$ the groups $\mathfrak{E}_1$ and $\mathfrak{F}$ have equal effect. Therefore the elementary surface ES_1 corresponding to $\mathfrak{E}_1$ is part of $\mathcal{K}(\mathfrak{F})$ mod $\mathfrak{F}$. The subgroups of $\mathfrak{F}$ leaving invariant $\mathcal{P}_1$, or $\mathcal{P}_1'$, or $\mathcal{P}_1''$, are called $\mathfrak{P}_1, \mathfrak{P}_1', \mathfrak{P}_1''$, respectively. The sense of rotation or translation for primary elements $\mathfrak{p}_1, \mathfrak{p}_1', \mathfrak{p}_1''$ of these three groups shall be in accordance with Fig. 21.1. The relations

$$\mathfrak{p}_r^{\nu_r} = 1, \quad \mathfrak{p}_r'^{\nu_r'} = 1, \quad \mathfrak{p}_r''^{\nu_r''} = 1, \quad \mathfrak{p}_r\mathfrak{p}_r'\mathfrak{p}_r'' = 1 \tag{16}$$

hold for $r = 1$. (In Fig. 21.1 the sense of translations among the three primary elements is so chosen that the axes correspondingly directed bound $\mathcal{K}(\mathfrak{E}_1)$ positively, and rotations of order > 2 and limit-rotations among the primary elements rotate in the negative sense.)

If $\mathcal{P}_1''$ represents a protrusion of $\mathcal{K}(\mathfrak{F})$ mod $\mathfrak{F}$, then $\mathfrak{F} = \mathfrak{E}_1$ and $N = 3$. Then $\mathfrak{F}$ is generated by $\mathfrak{p}_1, \mathfrak{p}_1', \mathfrak{p}_1''$ with relations (16).

If $\mathcal{P}_1''$ does not represent a protrusion of $\mathcal{K}(\mathfrak{F})$ mod $\mathfrak{F}$, then it is an inner axis of $\mathfrak{F}$ which is simple modulo $\mathfrak{F}$; it is dividing modulo $\mathfrak{F}$ in virtue of the assumption on $\mathfrak{F}$. We then consider the restgroup $\mathfrak{F}_1$ for which $\mathcal{K}(\mathfrak{F}_1)$ is adjacent to $\mathcal{P}_1''$ and bounded negatively by $\mathcal{P}_1''$ directed by $\mathfrak{p}_1''$, thus not containing $\mathcal{K}(\mathfrak{E}_1)$. The protrusions of $\mathcal{K}(\mathfrak{F}_1)$ mod $\mathfrak{F}_1$ are then $\mathcal{Q}^{(1)}, \mathcal{Q}_3, \dots, \mathcal{Q}_N$, where $\mathcal{Q}^{(1)}$ denotes the one derived from $\mathcal{P}_1''$. Since $\mathfrak{F}_1$ satisfies the conditions imposed on $\mathfrak{F}$, it is inferred from Section 5 that $N \geqq 4$ is a necessary and sufficient condition for $\mathcal{P}_1''$ being an inner axis of $\mathfrak{F}$. The surface $\mathcal{K}(\mathfrak{F})$ mod $\mathfrak{F}$ then consists of the ES_1 and $\mathcal{K}(\mathfrak{F}_1)$ mod $\mathfrak{F}_1$, these two parts being joined along the simple, closed geodesic corresponding to $\mathcal{P}_1''$. For $\mathfrak{F}$ one gets $\mathfrak{F} = \mathfrak{F}_1 * \mathfrak{E}_1 \operatorname{am} \mathfrak{P}_1''$.

We put $\mathfrak{p}_1''^{-1} = \mathfrak{p}_2$. Its axis $\mathcal{P}_2$ coincides with $\mathcal{P}_1''$, but when directed by $\mathfrak{p}_2$ it bounds $\mathcal{K}(\mathfrak{F}_1)$ positively, whereas $\mathcal{P}_1''$ directed by $\mathfrak{p}_1''$ bounds $\mathcal{K}(\mathfrak{E}_1)$ positively, thus $\mathcal{K}(\mathfrak{F}_1)$ negatively. We then base the construction of an $\mathfrak{E}_2(\mathcal{P}_2, \mathcal{P}_2', \mathcal{P}_2'')$ on $\mathcal{Q}^{(1)}$ and $\mathcal{Q}_3$. It is generated by $\mathfrak{p}_2, \mathfrak{p}_2', \mathfrak{p}_2''$ satisfying (16) for $r = 2$. If $N = 4$, then $\mathcal{P}_2''$ represents $\mathcal{Q}_4$. If $N > 4$, then the construction is repeated on a restgroup $\mathfrak{F}_2$ with $\mathcal{K}(\mathfrak{F}_2)$ mod $\mathfrak{F}_2$ having the protrusions $\mathcal{Q}^{(2)}, \mathcal{Q}_4, \dots, \mathcal{Q}_N$ by putting $\mathfrak{p}_2''^{-1} = \mathfrak{p}_3$, thus $\mathcal{P}_3$ representing $\mathcal{Q}^{(2)}$.

The process continues for $N - 2$ steps and then comes to an end. The surface $\mathcal{K}(\mathfrak{F})$ mod $\mathfrak{F}$ consists of $N - 2$ elementary surfaces $\mathrm{ES}_1, \mathrm{ES}_2, \dots, \mathrm{ES}_{N-2}$ arranged as a sequence by $N - 3$ simple, closed geodesics separating them. These geodesics are represented by the axes $\mathcal{P}_2, \mathcal{P}_3, \dots, \mathcal{P}_{N-2}$. One gets for $\mathfrak{F}$ the representation

$$\mathfrak{F} = \prod_r^{*} \mathfrak{E}_r \operatorname{am} \mathfrak{P}_t, \quad 1 \leqq r \leqq N-2, \ 2 \leqq t \leqq N-2, \tag{17}$$

where $\mathfrak{P}_t$ denotes the translation group generated by $\mathfrak{p}_t$.

The $\mathfrak{E}_r$ have $\mathfrak{p}_r, \mathfrak{p}_r', \mathfrak{p}_r''$ as their generators and these satisfy (16) for $r = 1, 2, \dots, N - 2$. The amalgamations are expressed by the equations

$$\mathfrak{p}_t = \mathfrak{p}_{t-1}''^{-1}, \quad 2 \leqq t \leqq N-2. \tag{18}$$

This gives a presentation of $\mathfrak{F}$. Using (18) and (16) for a reduction of the number of generators, and considering that $\nu_t = \infty$ for $t = 2, 3, \dots, N-2$, one gets the following presentation of $\mathfrak{F}$:

$$\begin{cases} \text{generators } \mathfrak{p}_1,\ \mathfrak{p}'_r,\ \mathfrak{p}''_{N-2}, & 1 \leqq r \leqq N-2, \\ \text{relations } \mathfrak{p}_1^{\nu_1} = 1,\ \mathfrak{p}'^{\nu'_r}_r = 1,\ \mathfrak{p}''^{\nu''_{N-2}}_{N-2} = 1,\ \mathfrak{p}_1\mathfrak{p}'_1\mathfrak{p}'_2 \dots \mathfrak{p}'_{N-2}\mathfrak{p}''_{N-2} = 1. \end{cases} \tag{19}$$

The N generators in (19) correspond to the N protrusions $\mathcal{Q}_1, \dots, \mathcal{Q}_N$ of $\mathcal{K}(\mathfrak{F})$ mod $\mathfrak{F}$ in succession. For, those which correspond to masts or boundaries the exponent is ∞, and the corresponding relation lapses. If there are no conical points, only the last relation remains. It can be used to eliminate one generator, and $\mathfrak{F}$ is thus, in this case, a free group with $N-1$ free generators.

The areas of the elementary surfaces corresponding to the $N-2$ elementary groups are

$$\Phi(\mathfrak{E}_1) = 2\left(1 - \frac{1}{\nu_1} - \frac{1}{\nu'_1}\right)$$

$$\Phi(\mathfrak{E}_k) = 2\left(1 - \frac{1}{\nu'_k}\right), \quad 2 \leqq k \leqq N-3$$

$$\Phi(\mathfrak{E}_{N-2}) = 2\left(1 - \frac{1}{\nu'_{N-2}} - \frac{1}{\nu''_{N-2}}\right),$$

and $\Phi(\mathfrak{F})$ is their sum. Insert this sum in (12), §17, and remember that $\vartheta(\rightarrow)+\vartheta(\hookrightarrow)$ of the terms of the form $\frac{1}{\nu}$ are zero. The sum $\Phi(\mathfrak{F})$ then reduces to $2(N-2)-2\sum_\nu \frac{1}{\nu}$, the latter sum extending over those ν which are not ∞. If the number of these ν, i.e. the number of conical points of $\mathcal{K}(\mathfrak{F})$ mod $\mathfrak{F}$, is κ, then $N = \vartheta(\rightarrow) + \vartheta(\hookrightarrow) + \kappa$. One then gets

$$\Phi(\mathfrak{F}) = -4 + 2\vartheta(\rightarrow) + 2\vartheta(\hookrightarrow) + 2\kappa - 2\sum_\nu \frac{1}{\nu}.$$

The sum of the last two terms is twice the sum of κ terms of the form $1 - \frac{1}{\nu}$, and can be split up according to the different orders of rotation occurring. One then gets $2\sum_\nu \left(1 - \frac{1}{\nu}\right)\vartheta(\circlearrowleft_\nu)$. Hence (12), §17, yields $p = 0$. One thus gets the following result:

If an $\mathfrak{F}$-group $\mathfrak{F}$ satisfies the condition of quasi-compactness and contains no reflections, if $\mathcal{K}(\mathfrak{F})$ mod *$\mathfrak{F}$ has N protrusions, and if every modulo $\mathfrak{F}$ simple, inner axis is dividing modulo $\mathfrak{F}$, then $\mathfrak{F}$ has its genus zero and is the free product with amalgamations of $N-2$ elementary groups. The surface $\mathcal{K}(\mathfrak{F})$* mod *$\mathfrak{F}$ can be decomposed into $N-2$ elementary surfaces. An abstract presentation of $\mathfrak{F}$ can be put into the form* (19). *If $\mathfrak{F}$ contains no rotations, then it is a free group.* □

22.8 A characteristic property of elementary groups. We prove the following theorem:

An elementary group $\mathfrak{E}$ contains no inner axis which is simple modulo $\mathfrak{E}$.

Proof. Consider an elementary group $\mathfrak{E}(\mathcal{P}, \mathcal{P}', \mathcal{P}'')$ and assume that $\mathcal{A}$ is an inner axis of $\mathfrak{E}$ which is simple modulo $\mathfrak{E}$. Let $\mathfrak{A}$ be the subgroup of $\mathfrak{E}$ leaving $\mathcal{A}$ invariant. Then $\mathfrak{A}$ cannot be of the type $\mathfrak{A}(\cdot\cdot)$, because $\mathfrak{E}$ contains at most one equivalence class of centres of order 2 (§21.9). Since $\mathfrak{E}$ consists of motions only, this leaves only the possibility $\mathfrak{A}(\rightarrow)$.

If $\mathcal{A}$ is non-dividing modulo $\mathfrak{E}$, we construct a handle group $\mathfrak{H}$ based on $\mathcal{A}$ by the method of Section 4. Since $\mathcal{K}(\mathfrak{H})$ mod $\mathfrak{H}$ has only one protrusion and $\mathcal{K}(\mathfrak{E})$ mod $\mathfrak{E}$ has three, the protrusion of $\mathcal{K}(\mathfrak{H})$ mod $\mathfrak{H}$ cannot be a protrusion of $\mathcal{K}(\mathfrak{E})$ mod $\mathfrak{E}$, because $\mathcal{K}(\mathfrak{E})$ mod $\mathfrak{E}$ is coherent. It thus corresponds to an inner axis of $\mathfrak{E}$. Since it does not correspond to a group $\mathfrak{A}(\cdot\cdot)$, it must be dividing modulo $\mathfrak{E}$ as shown in Section 4. Since $\mathfrak{H}$ and $\mathfrak{E}$ have equal effect in $\mathcal{K}(\mathfrak{H})$, $\mathcal{K}(\mathfrak{H})$ mod $\mathfrak{H}$ is part of $\mathcal{K}(\mathfrak{E})$ mod $\mathfrak{E}$, but not the whole of $\mathcal{K}(\mathfrak{E})$ mod $\mathfrak{E}$. This contradicts the fact that $\Phi(\mathfrak{H}) = 2$, while $\Phi(\mathfrak{E}) \leqq 2$. Thus $\mathfrak{E}$ contains no inner axis, simple and non-dividing modulo $\mathfrak{E}$, and $\mathfrak{E}$ thus satisfies the three conditions of Section 5.

So $\mathcal{A}$ is dividing modulo $\mathfrak{E}$. Let $\mathfrak{F}_1$ and $\mathfrak{F}_2$ be the two subgroups of $\mathfrak{E}$ belonging to the two regions adjacent to $\mathcal{A}$ in the decomposition of $\mathcal{K}(\mathfrak{E})$ by $\mathfrak{E}\mathcal{A}$. Both are $\mathfrak{F}$-groups, and they satisfy the conditions of Section 5. Those sides of $\mathcal{K}(\mathfrak{F}_1)$ which belong to $\mathfrak{E}\mathcal{A}$ constitute one equivalence class with respect to $\mathfrak{F}_1$, because $\mathcal{A}$ is dividing modulo $\mathfrak{E}$ (§19, case III); they therefore correspond to only one protrusion of $\mathcal{K}(\mathfrak{F}_1)$ mod $\mathfrak{F}_1$. Since $\mathcal{K}(\mathfrak{F}_1)$ mod $\mathfrak{F}_1$ has at least three protrusions (Section 5), it must have at least two protrusions which are also protrusions of $\mathcal{K}(\mathfrak{E})$ mod $\mathfrak{E}$. The same holds for $\mathfrak{F}_2$. Since $\mathcal{K}(\mathfrak{F}_1)$ mod $\mathfrak{F}_1$ and $\mathcal{K}(\mathfrak{F}_2)$ mod $\mathfrak{F}_2$ have only the closed geodesic derived from $\mathcal{A}$ in common, this contradicts the fact that $\mathcal{K}(\mathfrak{E})$ mod $\mathfrak{E}$ has only three protrusions. Thus the theorem is proved. □

Conversely, we prove:

If in an $\mathfrak{F}$-group without reflections there is no inner axis which is simple modulo $\mathfrak{F}$, then $\mathfrak{F}$ is an elementary group.

Proof. If $\mathfrak{F}$ contained reversed translations, it would according to Section 2, possess an axis of reversed translations which is simple modulo $\mathfrak{F}$, and this would be an inner axis of $\mathfrak{F}$. Hence $\mathfrak{F}$ consists of motions only. Let q be a regular point in $\mathcal{K}(\mathfrak{F})$ and $\mathfrak{f} \neq 1$ such an element of $\mathfrak{F}$ that the open part $\mathcal{T}$ of the join of q and $\mathfrak{f}q$ is simple modulo $\mathfrak{F}$ and contains no point of $\mathfrak{F}q$ (Lemma 1 of §21.1). As in Section 5, consider the chain $\mathfrak{f}^r\overline{\mathcal{T}}$, $-\infty < r < \infty$. If $\mathfrak{f}$ is a translation, then it follows as in Section 5 that its axis is simple modulo $\mathfrak{F}$, and according to the assumption on $\mathfrak{F}$ it is then a boundary axis of $\mathfrak{F}$. Hence, whatever element of $\mathfrak{F}$ the element $\mathfrak{f}$ is, $\mathcal{K}(\mathfrak{F})$ mod $\mathfrak{F}$ possesses protrusions. Let $\mathcal{P}$ represent a protrusion, and construct an elementary subgroup $\mathfrak{E}(\mathcal{P}, \mathcal{P}', \mathcal{P}'')$ of $\mathfrak{F}$ as in §21. If $\mathcal{P}'$ were not representing a protrusion of $\mathcal{K}(\mathfrak{F})$ mod $\mathfrak{F}$, then it would be an inner axis of $\mathfrak{F}$ which is simple modulo $\mathfrak{F}$ (§21) in contradiction to the assumption. Hence $\mathcal{P}'$, and for the same reason $\mathcal{P}''$, represent protrusions of $\mathcal{K}(\mathfrak{F})$ mod $\mathfrak{F}$, and therefore $\mathfrak{F} = \mathfrak{E}$, which proves the assertion. □

22.9 Normal forms. In Section 7 the N protrusions of $\mathcal{K}(\mathfrak{F})$ mod $\mathfrak{F}$ were enumerated in an arbitrary order, and this order is reflected by the succession of the generators in the last relation in (19). The partial product consisting of any two successive generators in this relations, say $\dots xy\dots$, can be replaced by

$$xy = xyx^{-1}x = y'x.$$

Since y' is conjugate to y, it represents the same protrusion, and the same exponent belong to it. By elementary exchanges of this kind every permutation of the factors in the last relation of (19) is obtainable.

For the group $\mathfrak{F}$ of Section 7, let $\mathcal{K}(\mathfrak{F})$ mod $\mathfrak{F}$ contain β boundaries, κ conical points and μ masts. Let corresponding primary elements determined by the procedure of Section 7, including the senses of translation and rotations, be

$$\mathfrak{t}_1, \mathfrak{t}_2, \dots, \mathfrak{t}_\beta, \mathfrak{r}_1, \mathfrak{r}_2, \dots, \mathfrak{r}_\kappa, \mathfrak{l}_1, \mathfrak{l}_2, \dots, \mathfrak{l}_\mu, \tag{20}$$

and let herein the $\mathfrak{r}$ be arranged according to non-decreasing orders of rotation. The relations of (19) then read

$$\mathfrak{r}_1^{\nu_1} = 1, \quad \mathfrak{r}_2^{\nu_2} = 1, \dots, \mathfrak{r}_\kappa^{\nu_\kappa} = 1, \ \ \mathfrak{t}_1\mathfrak{t}_2\dots\mathfrak{t}_\beta\mathfrak{r}_1\mathfrak{r}_2\dots\mathfrak{r}_\kappa\mathfrak{l}_1\mathfrak{l}_2\dots\mathfrak{l}_\mu = 1, \tag{21}$$

and the integers $\nu_1, \nu_2, \dots, \nu_\kappa$ form a non-decreasing sequence. Under the conditions of Section 7 at most the first, ν_1, takes the value 2.

Consider now the case of the group $\mathfrak{F}$ of Section 6. It is represented by (15). The group $\mathfrak{F}_m$ of (15) satisfies the condition of Section 7 and can thus be presented by generators (20) satisfying the relations (21). Let $\mathfrak{t}_k$, suitably directed, be one of those primary boundary translations of $\mathfrak{F}_m$ which generate the group $\mathfrak{T}_k$ of (15), this being a subgroup of a group $\mathfrak{A}_k = \mathfrak{A}(\cdot\cdot)$ in $\mathfrak{F}$. Let $\mathfrak{c}_{k_1}$, $\mathfrak{c}_{k_2}$ be half-turns of $\mathfrak{A}_k$ such that $\mathfrak{c}_{k_1}\mathfrak{c}_{k_2} = \mathfrak{t}_k$. Here k ranges over the m values of (15). We arrange the presentation of $\mathfrak{F}_m$ in the form (20), (21) such that these m translations $\mathfrak{t}_k$ are the last m in the sequence $\mathfrak{t}_1, \dots, \mathfrak{t}_\beta$ of (20). Then the formation of the free product (15) with amalgamations is simply carried out by omitting the generators $\mathfrak{t}_k$, introducing the new generators $\mathfrak{c}_{k_1}$ and $\mathfrak{c}_{k_2}$, replacing in the last relation of (21) the factor $\mathfrak{t}_k$ by $\mathfrak{c}_{k_1}\mathfrak{c}_{k_2}$ and adding the relations $\mathfrak{c}_{k_1}^2 = 1$, $\mathfrak{c}_{k_2}^2 = 1$, and this for all m values of k. If then we let β, κ, μ denote the number of boundaries, conical points, and masts of the present $\mathfrak{F}$ instead of $\mathfrak{F}_m$, then $\mathfrak{F}$ is again presented in the form (20), (21) with the understanding that now an arbitrary number of values 2 may occur at the beginning of the sequence $\nu_1, \nu_2, \dots, \nu_\kappa$.

In (12), §17, this has the effect of decreasing the term $2\vartheta(\rightarrow)$ by $2m$ and increasing in the term $2\sum_\nu \left(1 - \frac{1}{\nu}\right)\vartheta(\circlearrowleft_\nu)$ the $\vartheta(\circlearrowleft_2)$ by $2m$ corresponding to the additional $2m$

conical points of order 2, thus increasing this term by $2m$. The type number $\vartheta(\hookrightarrow)$ remains unchanged. Also $\Phi(\mathfrak{F}) = \Phi(\mathfrak{F}_m)$. For $\mathfrak{F}_m$ we found $p = 0$ in Section 7. Therefore $p = 0$ also for $\mathfrak{F}$-groups satisfying the conditions of Section 6.

Consider now a group $\mathfrak{F}$ satisfying the conditions of Section 4, and assume that it comes under the case 4) of Section 4, the only case in which $\mathfrak{F}$ was not yet fully determined. Then $\mathfrak{F}$ admits the representation (13). In (13) $\mathfrak{F}_s$ denotes a group satisfying the conditions of Section 6; it can thus be presented in the form (20), (21). It was found in Section 4 that $\mathfrak{G}_s$ was the free group with generators $\mathfrak{a}_1, \mathfrak{b}_1, \ldots, \mathfrak{a}_s, \mathfrak{b}_s$, and that $\mathfrak{G}_s$ had one equivalence class of boundary axes represented by the axis $\mathcal{J}_s$ of $\mathfrak{j}_s = \mathfrak{k}_1\mathfrak{k}_2 \ldots \mathfrak{k}_s$. It represents the only protrusion of $\mathcal{K}(\mathfrak{G}_s)$ mod $\mathfrak{G}_s$. The axis $\mathcal{J}_s$ of $\mathfrak{j}_s$ is also a boundary axis of $\mathfrak{F}_s$, and the corresponding group $\mathfrak{J}_s$ is generated by $\mathfrak{j}_s$. We may then take $\mathfrak{j}_s$ to be the $\mathfrak{t}_1$ of $\mathfrak{F}_s$ in (20), (21). The number of classes of boundary axes of $\mathfrak{F}$ is one less than those of $\mathfrak{F}_s$, because $\mathcal{J}_s$ is an inner axis of $\mathfrak{F}$. If we let β mean the number of boundaries of $\mathcal{K}(\mathfrak{F})$ mod $\mathfrak{F}$ instead of $\mathcal{K}(\mathfrak{F}_s)$ mod $\mathfrak{F}_s$ we may in (20) write $\mathfrak{j}_s, \mathfrak{t}_1, \ldots, \mathfrak{t}_\beta$ for the translations belonging to the chosen boundary axes of $\mathfrak{F}_s$. The amalgamation in (13) is then carried out by taking $\mathfrak{j}_s$ from $\mathfrak{G}_s$.

Since the interiors of $\mathcal{K}(\mathfrak{G}_s)$ and $\mathcal{K}(\mathfrak{F}_s)$ are disjoint modulo $\mathfrak{F}$, one gets

$$\Phi(\mathfrak{F}) = \Phi(\mathfrak{G}_s) + \Phi(\mathfrak{F}_s).$$

It was found in Section 4 that $\Phi(\mathfrak{G}_s) = 4s - 2$. From Section 7, with

$$\vartheta(\rightarrow) = \beta + 1, \quad \vartheta(\hookrightarrow) = \mu, \quad N = \beta + 1 + \mu + \kappa$$

one gets, since $\mathfrak{F}_s$ has the genus zero

$$\Phi(\mathfrak{F}_s) = -4 + 2\beta + 2 + 2\mu + 2\sum_{i=1}^{\kappa}\left(1 - \frac{1}{\nu_i}\right),$$

thus

$$\Phi(\mathfrak{F}) = 4s - 4 + 2\beta + 2\mu + 2\sum_{i=1}^{\kappa}\left(1 - \frac{1}{\nu_i}\right),$$

which shows that s is the genus $p_{\mathfrak{F}}$ as was announced at the end of Section 4. We thus arrive at the following presentation of $\mathfrak{F}$:

$$\text{generators } \mathfrak{a}_1, \mathfrak{b}_1, \ldots, \mathfrak{a}_p, \mathfrak{b}_p, \mathfrak{t}_1, \ldots, \mathfrak{t}_\beta, \mathfrak{r}_1, \ldots, \mathfrak{r}_\kappa, \mathfrak{l}_1, \ldots, \mathfrak{l}_\mu, \tag{22}$$

$$\text{relations } \mathfrak{r}_i^{\nu_i} = 1 \ (i = 1, \ldots, \kappa), \ \mathfrak{k}_1 \ldots \mathfrak{k}_p \mathfrak{t}_1 \ldots \mathfrak{t}_\beta \mathfrak{r}_1 \ldots \mathfrak{r}_\kappa \mathfrak{l}_1 \ldots \mathfrak{l}_\mu = 1, \tag{23}$$

where $\mathfrak{k}_j$ is an abbreviation for $\mathfrak{a}_j \mathfrak{b}_j \mathfrak{a}_j^{-1} \mathfrak{b}_j^{-1}$ $(j = 1, \ldots, p)$.

This will be called the *normal form* for the presentation of an $\mathfrak{F}$-group consisting of motions and satisfying the condition of quasi-compactness. It should be noted that it covers all cases of $\mathfrak{F}$-groups satisfying these two conditions. For, if $\mathcal{K}(\mathfrak{F})$ mod $\mathfrak{F}$ has no protrusions, the generators $\mathfrak{t}$, $\mathfrak{r}$ and $\mathfrak{l}$ disappear and one gets the presentation

(11) with $s = p$ (case 3) of Section 4). If the number of protrusions is 1 (case 1) of Section 4), then there is in the last relation in (23) only one symbol after the product $\mathfrak{k}_1 \dots \mathfrak{k}_p$. This generator can then be eliminated, and (9) results, the exponent ν being finite or infinite according to the nature of the protrusion. Also (10) in the case 2) of Section 4 is a special case of (22), (23). Finally, if the genus is zero, then the number of protrusions is at least three, and (22), (23) reduce to (20), (21). If there is at least one $\mathfrak{t}$ or $\mathfrak{l}$, but no $\mathfrak{r}$, then $\mathfrak{F}$ is a free group.

Finally, consider a group $\mathfrak{F}$ satisfying the conditions of Section 3, thus containing reversed translations, but no reflections, and satisfying the condition of quasi-compactness. It was found in Section 3 that one can introduce s equivalence classes with respect to $\mathfrak{F}$ of axes of reversed translations such that these axes are simple and disjoint modulo $\mathfrak{F}$, and that a restgroup $\mathfrak{F}_s$ pertaining to one of the regions of decomposition of $\mathcal{K}(\mathfrak{F})$ by these s classes of axes satisfies the conditions 1), 2), 3) of Section 5. Thus $\mathfrak{F}_s$ can be presented in the form (20), (21). Among the boundary axes of $\mathcal{K}(\mathfrak{F}_s)$ we select s axes $\mathcal{A}_1, \dots, \mathcal{A}_s$ such that they represent the above s classes; this is possible, because all regions of decomposition are equivalent with respect to $\mathfrak{F}$. Let $\mathfrak{h}_i$ be a primary reversed translation with axis $\mathcal{A}_i$, such that $\mathcal{A}_i$ directed by $\mathfrak{h}_i$ bounds $\mathcal{K}(\mathfrak{F}_s)$ positively ($i = 1, \dots, s$). If $\mathfrak{H}_i$ denotes the group generated by $\mathfrak{h}_i$, and $\mathfrak{T}_i$ the subgroup generated by $\mathfrak{h}_i^2$, then $\mathfrak{F}$ is the free product of $\mathfrak{F}_s$ and all the $\mathfrak{H}_i$ with amalgamation of the $\mathfrak{T}_i$ (§19). We arrange that the first s of the β translations $\mathfrak{t}$ in (20) are the $\mathfrak{h}_i^2$. The $\mathcal{A}_i$ are inner axes of $\mathfrak{F}$. Thus the number of boundaries of $\mathcal{K}(\mathfrak{F})$ mod $\mathfrak{F}$ is $\beta - s$, and we write $\vartheta_{\mathfrak{F}_s}(\rightarrow) = \vartheta_{\mathfrak{F}}(\rightarrow) + s$, while $\vartheta(\hookrightarrow)$ and also the $\vartheta(\odot_\nu)$ are the same for $\mathfrak{F}_s$ and $\mathfrak{F}$. Also $\Phi(\mathfrak{F}_s) = \Phi(\mathfrak{F})$, because the regions of decomposition of $\mathcal{K}(\mathfrak{F})$ by the collection $\bigcup_i \mathfrak{F}\mathcal{A}_i$ are all equivalent, and in the interior of $\mathcal{K}(\mathfrak{F}_s)$ the groups $\mathfrak{F}_s$ and $\mathfrak{F}$ have equal effect. Since $\mathfrak{F}$ contains no reversions, and its genus is known to be zero as seen in Section 7, one gets from (12), §17,

$$\Phi(\mathfrak{F}_s) = -4 + 2(\vartheta_{\mathfrak{F}}(\rightarrow) + s) + 2\vartheta(\hookrightarrow) + 2\sum_\nu \left(1 - \frac{1}{\nu}\right)\vartheta(\odot_\nu),$$

while $\Phi(\mathfrak{F})$ is calculated from (11), §17:

$$\Phi(\mathfrak{F}) = 2p - 4 + 2\vartheta_{\mathfrak{F}}(\rightarrow) + 2\vartheta(\hookrightarrow) + 2\sum_\nu \left(1 - \frac{1}{\nu}\right)\vartheta(\odot_\nu).$$

The two last terms are the same in both expressions. Thus equating $\Phi(\mathfrak{F})$ and $\Phi(\mathfrak{F}_s)$ one gets $p = s$, as announced in Section 3.

In order to keep the notation β for the number of boundaries of $\mathcal{K}(\mathfrak{F})$ mod $\mathfrak{F}$, we write the generators in (20)

$$\mathfrak{h}_1^2, \dots, \mathfrak{h}_p^2, \mathfrak{t}_1, \dots, \mathfrak{t}_\beta, \mathfrak{r}_1, \dots, \mathfrak{r}_\kappa, \mathfrak{l}_1, \dots, \mathfrak{l}_\mu.$$

The free product can now be obtained by introducing the additional generators $\mathfrak{h}_1, \dots, \mathfrak{h}_p$, and the amalgamation has already been expressed by the above notation.

After eliminating the redundant generators $\mathfrak{h}_1^2, \dots, \mathfrak{h}_p^2$, one gets for $\mathfrak{F}$ the presentation

$$\text{generators } \mathfrak{h}_1, \dots, \mathfrak{h}_p,\ \mathfrak{t}_1, \dots, \mathfrak{t}_\beta, \mathfrak{r}_1, \dots, \mathfrak{r}_\kappa, \mathfrak{l}_1, \dots, \mathfrak{l}_\mu, \tag{24}$$

$$\text{relations } \mathfrak{r}_i^{\nu_i} = 1\ (i = 1, \dots, \kappa), \mathfrak{h}_1^2 \dots \mathfrak{h}_p^2 \mathfrak{t}_1 \dots \mathfrak{t}_\beta \mathfrak{r}_1 \dots \mathfrak{r}_\kappa \mathfrak{l}_1 \dots \mathfrak{l}_\mu = 1. \tag{25}$$

This is called the *normal form* of an $\mathfrak{F}$-group containing reversed translations, but no reflections, and satisfying the condition of quasi-compactness.

If the surface $\mathcal{K}(\mathfrak{F})$ mod $\mathfrak{F}$ contains no protrusions, then $\mathcal{K}(\mathfrak{F}_p)$ mod $\mathfrak{F}_p$ contains the p protrusions derived from $\mathcal{A}_1, \dots, \mathcal{A}_p$, but no more than these. $\mathfrak{F}$ is then presented by

$$\mathfrak{h}_1, \mathfrak{h}_2, \dots, \mathfrak{h}_p; \quad \mathfrak{h}_1^2 \mathfrak{h}_2^2 \dots \mathfrak{h}_p^2 = 1. \tag{26}$$

Surfaces $\mathcal{K}(\mathfrak{F})$ mod $\mathfrak{F}$, where $\mathfrak{F}$ is of the type (11) or (26), are *closed surfaces* without singular points, orientable or non-orientable, respectively.

22.10 Groups containing reflections. Let $\mathfrak{F}$ be an $\mathfrak{F}$-group containing reflections and satisfying the condition of quasi-compactness. According to §16.4, all reflection chains of the surface $\mathcal{K}(\mathfrak{F})$ mod $\mathfrak{F}$ are closed, i.e. they are reflection rings. Also the number of reflections rings is finite; it is denoted by $\vartheta(-)$.

In §20 the structure of $\mathfrak{F}$-groups containing reflections was analysed in detail. In certain cases the structure of $\mathfrak{F}$ was completely determined. For groups satisfying the condition of quasi-compactness these cases are treated in the Sections 5, 6, 8, 14, 16, 17, 20 of §20. Also the special case of Section 12 in which $\mathcal{A}$ is a boundary axis of $\mathcal{K}(\mathfrak{F})$, has to be included. In all these cases, and also in the general case to be considered in the sequel, let α denote the number of closed reflection edges, i.e. the number of equivalence classes of modulo $\mathfrak{F}$ simple and disjoint axes with corresponding group $\mathfrak{A}(- \to)$, and $\gamma = \vartheta(-) - \alpha$ the number of other reflection rings, i.e. the number of equivalence classes of other boundary chains of the region called $\mathcal{B}$ in §20. Here $\vartheta(-) = \alpha + \gamma > 0$, because $\mathfrak{F}$ contains reflections. Hence at least one of the integers α and γ is positive.

We now consider the general situation of Sections 18 and 21 of §20, thus excluding for the moment the special cases just mentioned. With the notation used in Sections 18 and 21 of §20, both the $\mathcal{W}_r$ and $\mathcal{V}_s$ are axes and are finite in number, owing to the condition of quasi-compactness. Hence the groups $\mathfrak{W}_r$ are of the type $\mathfrak{A}(- \to)$, and the $\mathfrak{W}_r^*$ as well as the $\mathfrak{V}_s$ are of the type $\mathfrak{A}(\to)$. The index r ranges over the values $1 \leqq r \leqq \alpha$, if $\alpha > 0$, and s over the values $1 \leqq s \leqq \gamma$, if $\gamma > 0$.

Consider the decomposition of $\mathcal{D}$ by the collection $\bigcup_{r,s}(\mathfrak{F}\mathcal{W}_r, \mathfrak{F}\mathcal{V}_s)$. One of the regions of decomposition is the $\mathcal{B}_0$ of §20.21. The corresponding subgroup was there called $\mathfrak{B}$ and was an $\mathfrak{F}$-group. Let it here be called $\mathfrak{F}_0$. In $\mathcal{B}_0$ the groups $\mathfrak{F}$ and $\mathfrak{F}_0$ have equal effect, and $\mathcal{B}_0$ contains $\mathcal{K}(\mathfrak{F}_0)$. Thus $\mathfrak{F}_0$ satisfies the condition of quasi-compactness. If $\mathfrak{F}$ has no boundary axes of the type $\mathfrak{A}(\to)$, then $\mathcal{B}_0$ is $\mathcal{K}(\mathfrak{F}_0)$. If $\mathfrak{F}$ has boundary axes of the type $\mathfrak{A}(\to)$, then $\mathcal{B}_0$ arises from $\mathcal{K}(\mathfrak{F}_0)$ by adding the half-planes along those boundaries of $\mathcal{K}(\mathfrak{F}_0)$ which are also boundaries of $\mathcal{K}(\mathfrak{F})$. The protrusions of $\mathcal{K}(\mathfrak{F}_0)$ mod $\mathfrak{F}_0$ are the protrusions of $\mathcal{K}(\mathfrak{F})$ mod $\mathfrak{F}$, and in addition the $\alpha+\gamma$ boundaries of $\mathcal{K}(\mathfrak{F}_0)$ mod $\mathfrak{F}_0$ corresponding to the $\mathcal{W}_r$ and the $\mathcal{V}_s$.

The group $\mathfrak{F}_0$ contains no reflections. It therefore admits a presentation in one of the normal forms (24), (25) or (22), (23), according as $\mathfrak{F}_0$ contains a reversed translation or not. In both cases boundary translations $\mathfrak{t}$ occur, namely at least $\vartheta(-)$ of them. Since those corresponding to the $\mathcal{W}_r$ and the $\mathcal{V}_s$ are not boundary translations of $\mathfrak{F}$, and as we want to keep the notation β to mean the number of funnels of $\mathcal{D}$ mod $\mathfrak{F}$, i.e. the number of equivalence classes of boundary axes of the type $\mathfrak{A}(\rightarrow)$ in $\mathfrak{F}$, we write the boundary translations entering into the normal presentation of $\mathfrak{F}_0$ in the succession of the normal form

$$\mathfrak{f}_1,\dots,\mathfrak{f}_\alpha,\mathfrak{g}_1,\dots,\mathfrak{g}_\gamma,\mathfrak{t}_1,\dots,\mathfrak{t}_\beta, \tag{27}$$

the $\mathfrak{f}_1,\dots,\mathfrak{f}_\alpha$ having the $\mathcal{W}_r$ as their axes, the $\mathfrak{g}_1,\dots,\mathfrak{g}_\gamma$ the $\mathcal{V}_s$, and the $\mathfrak{t}_1,\dots,\mathfrak{t}_\beta$ belonging to boundary axes of the type $\mathfrak{A}(\rightarrow)$ in $\mathfrak{F}$, if any. The last relation of (23) then reads

$$\mathfrak{k}_1\dots\mathfrak{k}_p\mathfrak{f}_1\dots\mathfrak{f}_\alpha\mathfrak{g}_1\dots\mathfrak{g}_\gamma\mathfrak{t}_1\dots\mathfrak{t}_\beta\mathfrak{r}_1\dots\mathfrak{r}_\kappa\mathfrak{l}_1\dots\mathfrak{l}_\mu=1 \tag{28}$$

and the last one of (25)

$$\mathfrak{h}_1^2\dots\mathfrak{h}_p^2\mathfrak{f}_1\dots\mathfrak{f}_\alpha\mathfrak{g}_1\dots\mathfrak{g}_\gamma\mathfrak{t}_1\dots\mathfrak{t}_\beta\mathfrak{r}_1\dots\mathfrak{r}_\kappa\mathfrak{l}_1\dots\mathfrak{l}_\mu=1. \tag{29}$$

Now $\mathfrak{F}$ is obtained by §20 (12):

$$\mathfrak{F}=\prod_{r,s}{}^{*}(\mathfrak{F}_0,\mathfrak{G}_s,\mathfrak{W}_r)\text{ am }\mathfrak{V}_s,\mathfrak{W}_r^*.$$

Let $\mathfrak{s}_r$ denote the reflection in $\mathcal{W}_r$. Then $\mathfrak{s}_r$ and $\mathfrak{f}_r$ generate $\mathfrak{W}_r$, and $\mathfrak{f}_r$ generates $\mathfrak{W}_r^*$. The group $\mathfrak{G}_s$ is generated by $\mathfrak{g}_s$ together with the reflections $\mathfrak{s}_{s\ell}$, $1\leqq\ell\leqq\omega_s$, in the ω_s reflection lines contained in the period of the corresponding boundary chain b_s. The relations for the $\mathfrak{W}_r$, thus between $\mathfrak{s}_r$ and $\mathfrak{f}_r$, correspond to (13) and (14) of §20, and the relations for the $\mathfrak{G}_s$, thus between $\mathfrak{g}_s$ and the $\mathfrak{s}_{s\ell}$, to (15), (16) and (17) of §20, taking account of the changes in notation.

Since the $\mathfrak{f}_r$ and the $\mathfrak{g}_s$ are elements of $\mathfrak{F}_0$, the amalgamations in (12), §20, are expressed by only including each of them once among the generators of $\mathfrak{F}$. Hence a presentation of $\mathfrak{F}$ is obtained by adding to the system (24), (25) (with the change by (29)), or the system (22), (23) (with the change by (28)), respectively, those systems now corresponding to the notation (27), namely

$$\begin{aligned}&\text{generators}\quad \mathfrak{s}_r,\ \mathfrak{s}_{s\ell}\\&\text{relations}\quad \mathfrak{s}_r^2=1,\ \ \mathfrak{f}_r\mathfrak{s}_r\mathfrak{f}_r^{-1}\mathfrak{s}=1,\\&\qquad\qquad \mathfrak{s}_{s\ell}^2=1,\ \ (\mathfrak{s}_{s,t+1}\mathfrak{s}_{st})^{\nu_{st}}=1,\ \ (\mathfrak{g}_s\mathfrak{s}_{s,1}\mathfrak{g}_s^{-1}\mathfrak{s}_{s,\omega_s})^{\nu_{s,\omega_s}}=1,\end{aligned} \tag{30}$$

where $1\leqq r\leqq\alpha$, if $\alpha>0$, $1\leqq s\leqq\gamma$, if $\gamma>0$, $\alpha+\gamma=\vartheta(-)>0$, $\omega_s\geqq1$, $1\leqq\ell\leqq\omega_s$ and $1\leqq t\leqq\omega_s-1$, if $\omega_s\geqq2$.

The surface $\mathcal{K}(\mathfrak{F})$ mod $\mathfrak{F}$ is obtained from the surface $\mathcal{K}(\mathfrak{F}_0)$ mod $\mathfrak{F}_0$ by turning the α boundaries corresponding to the $\mathcal{W}_r$ into closed reflection edges (if $\alpha > 0$), and adding along each of the γ boundaries corresponding to the $\mathcal{V}_s$ a crown (if $\gamma > 0$). The first of these two operations does not affect the area $\Phi(\mathfrak{F}_0)$. It decreases the type number $\vartheta(\rightarrow)$ by α and increases $\vartheta(-)$ by α on the right hand side of the general formula (10), §17. The second operation decreases $\vartheta(\rightarrow)$ by γ and increases $\vartheta(-)$ by γ and also adds on the right hand side the term

$$\varepsilon = \sum_{\nu} \left(1 - \frac{1}{\nu}\right) \vartheta(\times_\nu) + \vartheta(\wedge) + \vartheta(\|). \tag{31}$$

This sum may be written (compare (8), §20.5)

$$\varepsilon = \sum_{s,\ell} \left(1 - \frac{1}{\nu_{s,\ell}}\right) \tag{31$'$}$$

the sum extended over all exponents $\nu_{s,\ell}$, $1 \leqq s \leqq \gamma$, $1 \leqq \ell \leqq \omega_s$, thus including those which are ∞. In §20.12 it was found that the sum (31′), when extended over all exponents belonging to a crown, thus for a fixed value of s, was equal to the area ε_s of the crown. Hence

$$\varepsilon = \sum_{s=1}^{\gamma} \varepsilon_s.$$

One thus gets

$$\Phi(\mathfrak{F}) = \Phi(\mathfrak{F}_0) + \varepsilon. \tag{32}$$

If $\mathfrak{F}_0$ contains a reversed translation, then this is also a reversed translation in $\mathfrak{F}$, and its axis does not coincide with, or meet, any reflection line of $\mathfrak{F}$, because it is situated in the interior of $\mathcal{K}(\mathfrak{F}_0)$. Thus $\vartheta_{\mathfrak{F}_0}(\rightleftharpoons) = 1$ implies $\vartheta_{\mathfrak{F}}(\rightleftharpoons) = 1$. Conversely, let $\mathcal{R}$ be an axis of reversed translations in $\mathfrak{F}$ which does not coincide with, or meet, any reflection line of $\mathfrak{F}$. Then $\mathcal{R}$ is disjoint modulo $\mathfrak{F}$ with the set $\bigcup_r \mathfrak{F}\mathcal{W}_r$, because the $\mathcal{W}_r$ are reflection lines. Assume that an axis of the collection $\mathfrak{F}\mathcal{R}$ had points in common with the interior of $\mathcal{K}(\mathfrak{G}_s)$ for some value of s. Then an axis of this collection – we may call it $\mathcal{R}$ – would have points in common with the region $\mathcal{B}_s$ adjacent to $\mathcal{V}_s$ in the construction of §20.12, the $\mathcal{B}_1$ of that section being our present $\mathcal{B}_s$ and the $\mathcal{A}$ of that section being our $\mathcal{V}_s$. No half-line of $\mathcal{R}$ is wholly contained in $\mathcal{B}_s$ for the reason indicated in §20.12. On the other hand, the boundary of $\mathcal{B}_s$ in $\mathcal{D}$ consists of $\mathcal{V}_s$ and the reflection lines of a reflection chain. Since $\mathcal{R}$ intersects no reflection line, it cannot even intersect $\mathcal{V}_s$. Hence $\mathcal{R}$ is disjoint modulo $\mathfrak{F}$ with $\bigcup_s \mathfrak{F}\mathcal{V}_s$, and it is therefore contained in the interior of $\mathcal{K}(\mathfrak{F}_0)$ or in some equivalent region. Hence $\vartheta_{\mathfrak{F}}(\rightleftharpoons) = 1$ implies $\vartheta_{\mathfrak{F}_0}(\rightleftharpoons) = 1$. Therefore the formation of the free product in (12), §20, leaves the property of orientability or non-orientability unchanged.

If now both $\Phi(\mathfrak{F})$ and $\Phi(\mathfrak{F}_0)$ are calculated from (10), §17, and inserted in (32), and if it is borne in mind that the factor $(2 - \vartheta(\rightleftharpoons))$ is the same in both, that the terms $2\vartheta(\hookrightarrow)$ and $2\sum_\nu \left(1-\frac{1}{\nu}\right)\vartheta(\odot_\nu)$, and also $2(\vartheta(\rightarrow)+\vartheta(-))$ are the same in both, and that $\Phi(\mathfrak{F})$ contains the term (31), but $\Phi(\mathfrak{F}_0)$ does not, then it results that $p_{\mathfrak{F}} = p_{\mathfrak{F}_0}$. Thus the genus remains unchanged.

Finally we want to verify that the *normal form* established in this section for $\mathfrak{F}$-groups which satisfy the condition of quasi-compactness and contain reflections also covers those special cases of §20 in which $\mathfrak{F}$ was fully determined and satisfied the condition of quasi-compactness. This is done by inserting in the following table the values of α, γ, $\vartheta_1 = \vartheta(-)$, p, β, κ, μ, $\vartheta_2 = \vartheta(\rightleftharpoons)$, corresponding to these cases. In the last column is indicated the special form to which (28) or (29) reduces, according as the surface is orientable or not. The accordance with the presentation earlier established for these groups is then easily verified.

Table 5

Surface	§20.#	α	γ	ϑ_1	p	β	κ	μ	ϑ_2	(28) or (29)
1) polygonal disc	5	0	1	1	0	0	0	0	0	$\mathfrak{g}_1 = 1$
2) polygonal cone	6	0	1	1	0	0	1	0	0	$\mathfrak{g}_1\mathfrak{r}_1 = 1$
3) polygonal mast	8	0	1	1	0	0	0	1	0	$\mathfrak{g}_1\mathfrak{l}_1 = 1$
4) crown	12	0	1	1	0	1	0	0	0	$\mathfrak{g}_1\mathfrak{t}_1 = 1$
5) reflection crown	14	1	1	2	0	0	0	0	0	$\mathfrak{f}_1\mathfrak{g}_1 = 1$
6) conical crown	16	0	1	1	0	0	2	0	0	$\mathfrak{g}_1\mathfrak{r}_1\mathfrak{r}_2 = 1$
7) cross cap crown	17	0	1	1	1	0	0	0	1	$\mathfrak{h}_1^2\mathfrak{g}_1 = 1$
8) double crown	20	0	2	2	0	0	0	0	0	$\mathfrak{g}_1\mathfrak{g}_2 = 1$

22.11 Normal form embracing all finitely generated $\mathfrak{F}$-groups.

The relation $\mathfrak{f}_r\mathfrak{s}_r\mathfrak{f}_r^{-1}\mathfrak{s}_r = 1$ of (30) has the same form as the last relation in (30), if the exponent in that last relation is replaced by 1. One may therefore unify (30) in such a way that closed reflection edges and other reflection rings appear formally alike, and that thus instead of the two numbers α and γ only their sum $\rho = \vartheta(-)$ appears. One then gets the following presentation for a finitely generated $\mathfrak{F}$-group:

Generators:

$$\left\{\begin{matrix}\mathfrak{a}_1, \mathfrak{b}_1, \dots, \mathfrak{a}_p, \mathfrak{b}_p, \\ \mathfrak{h}_1, \dots, \mathfrak{h}_p,\end{matrix}\right\} \mathfrak{g}_1, \dots, \mathfrak{g}_\rho, \mathfrak{t}_1, \dots, \mathfrak{t}_\beta, \mathfrak{r}_1, \dots, \mathfrak{r}_\kappa, \mathfrak{l}_1, \dots, \mathfrak{l}_\mu, \tag{33}$$

$$\mathfrak{s}_{11}, \dots, \mathfrak{s}_{1\omega_1}, \dots, \mathfrak{s}_{\rho 1}, \dots, \mathfrak{s}_{\rho\omega_\rho}$$

where ρ is short for $\vartheta(-)$.

Defining relations (using the abbreviation $\mathfrak{k}_i = \mathfrak{a}_i\mathfrak{b}_i\mathfrak{a}_i^{-1}\mathfrak{b}_i^{-1}$):

$$\left\{\begin{matrix}\mathfrak{k}_1 \dots \mathfrak{k}_p \\ \mathfrak{h}_1^2 \dots \mathfrak{h}_p^2\end{matrix}\right\} \mathfrak{g}_1 \dots \mathfrak{g}_\rho \mathfrak{t}_1 \dots \mathfrak{t}_\beta \mathfrak{r}_1 \dots \mathfrak{r}_\kappa \mathfrak{l}_1 \dots \mathfrak{l}_\mu = 1, \tag{34}$$

$$\mathfrak{r}_i^{\nu_i} = 1, \tag{35}$$

$$\mathfrak{s}_{s\ell}^2 = 1, \tag{36}$$

$$(\mathfrak{s}_{s,t+1}\mathfrak{s}_{st})^{\nu_{st}} = 1, \tag{37}$$

$$(\mathfrak{g}_s\mathfrak{s}_{s1}\mathfrak{g}_s^{-1}\mathfrak{s}_{s\omega_s})^{\nu_{s\omega_s}} = 1. \tag{38}$$

In the parentheses of (33) and (34) the upper or lower part is applied according as the surface $\mathcal{D}$ mod $\mathfrak{F}$ is orientable or not. These parts disappear if the genus p is zero. The numbers $\vartheta(-), \beta, \kappa, \mu$ are $\geqq 0$. The generators $\mathfrak{t}_i, \mathfrak{r}_i, \mathfrak{l}_i, \mathfrak{s}_{s\ell}$ correspond to the different funnels, conical points, masts, and reflection edges of the surface $\mathcal{D}$ mod $\mathfrak{F}$. The index s enumerates the different reflection rings of the surface, and ω_s denotes the number of reflection edges in the ring with number s.

In (35) $1 \leqq i \leqq \kappa$, and $2 \leqq \nu_i < \infty$. These relations disappear if $\kappa = 0$.

The generators $\mathfrak{g}_s$ and $\mathfrak{s}_{s\ell}$ only occur if $\vartheta(-) > 0$. They are connected by the relations (36), (37), (38).

In (36) $1 \leqq \ell \leqq \omega_s$, and $1 \leqq s \leqq \vartheta(-)$.

In (37) $1 \leqq t \leqq \omega_s - 1$, and $2 \leqq \nu_{st} \leqq \infty$. These relations disappear for those values of s for which $\omega_s = 1$.

In (38) $2 \leqq \nu_{s\omega_s} \leqq \infty$ if $\omega_s > 1$, and $1 \leqq \nu_{s\omega_s} \leqq \infty$ if $\omega_s = 1$. The case $\nu_{s\omega_s} = 1$ occurs for those, and only those values of s which correspond to a closed reflection edge.

In (37) and (38) relations with exponent ∞ cancel.

The formula (31′) for the sum of the areas of the crowns still holds if $1 \leqq \ell \leqq \omega_s$ and $1 \leqq s \leqq \vartheta(-)$, because the contribution to the sum vanishes for those $\nu_{s\ell}$ which are 1.

If $\vartheta(-) = 0$, $\kappa = 0$, and $\beta + \mu > 0$, then $\mathfrak{F}$ is a *free group*, and the surface $\mathcal{D}$ mod $\mathfrak{F}$ has no singular points.

If $\beta = 0$, $\mu = 0$, and none of the exponents $\nu_{s\ell}$ are infinite, then $\mathcal{D}$ mod $\mathfrak{F}$ is a *closed surface* in the sense that $\mathcal{D}$ is compact modulo $\mathfrak{F}$; however the surface may contain singular points.

In §17 the genus p of an $\mathfrak{F}$-group $\mathfrak{F}$ satisfying the condition of quasi-compactness was defined by equation (5), §17, and it enters into the general formula (10), §17, for the area $\Phi(\mathfrak{F})$. This quantity p, which is also called the genus of the surface $\mathcal{K}(\mathfrak{F})$ mod $\mathfrak{F}$, depends on $\mathfrak{F}$ only, because all the other quantities entering into the equations (5) or (10) of §17 obviously depend on $\mathfrak{F}$ only. It now results from the present paragraph that p can be characterized in the following way: In the case of non-orientability p is the maximum number of equivalence classes of axes of $\mathfrak{F}$ of the type $\mathfrak{A}(\rightleftharpoons)$ which are simple and mutually disjoint modulo $\mathfrak{F}$, and which have no point in common with any reflection line of $\mathfrak{F}$. In the case of orientability p is the maximum number of equivalence classes of axes of $\mathfrak{F}$ of the type $\mathfrak{A}(\rightarrow)$ which are simple and mutually disjoint modulo $\mathfrak{F}$, which have no point in common with any reflection line of $\mathfrak{F}$, and such that their totality is non-dividing modulo $\mathfrak{F}$; i.e. in the decomposition of

$\mathcal{D}$ by these p equivalence classes of axes all regions of decomposition are equivalent with respect to $\mathfrak{F}$.

In both cases p is a non-negative integer. The fact that $p \geqq 0$ was already proved in §17. The fact that p is an integer was already evident from (5), §17, in the case $\vartheta(\rightleftharpoons) = 1$. But in the case $\vartheta(\rightleftharpoons) = 0$ it was on the basis of §17 only immediately evident that $2p$ is an integer, whereas it has now been proved that p is an integer.

§23 Exhaustion in the case of non-quasi-compactness

23.1 Decomposition by a subgroup. Let $\mathfrak{F}'$ be a finitely generated $\mathfrak{F}$-subgroup of a group $\mathfrak{F}$ such that $\mathfrak{F}'$ is not the whole of $\mathfrak{F}$. Let $\mathfrak{F}'$ and $\mathfrak{F}$ have equal effect on the interior of $\mathcal{K}(\mathfrak{F}')$; this is equivalent to saying that the interior of $\mathcal{K}(\mathfrak{F}')$ is simple modulo $\mathfrak{F}$. Since $\mathfrak{F}'$ is finitely generated, all boundaries of $\mathcal{K}(\mathfrak{F}')$ are axes of $\mathfrak{F}'$ and thus of $\mathfrak{F}$; they fall into a finite number of equivalence classes with respect to $\mathfrak{F}'$ and, all the more, with respect to $\mathfrak{F}$. Some of these equivalence classes with respect to $\mathfrak{F}$ may consist of boundary axes of $\mathfrak{F}$, if any. Since $\mathfrak{F}'$ is not the whole of $\mathfrak{F}$ a certain number $\omega \geqq 1$ of these equivalence classes with respect to $\mathfrak{F}$ consist of inner axes of $\mathfrak{F}$. The collection $\mathcal{U}$ of inner axes of $\mathfrak{F}$ which make up these ω equivalence classes does not accumulate in $\mathcal{K}(\mathfrak{F})$. The fact that the interior of $\mathcal{K}(\mathfrak{F}')$ is simple modulo $\mathfrak{F}$ implies that each of these axes is simple modulo $\mathfrak{F}$, and that any two different equivalence classes with respect to $\mathfrak{F}$ of these axes are disjoint. The collection $\mathcal{U}$ produces a decomposition of $\mathcal{K}(\mathfrak{F})$. This will, in short, be called the *decomposition of* $\mathcal{K}(\mathfrak{F})$ *by* $\mathfrak{F}\mathcal{K}(\mathfrak{F}')$.

Any two regions of the decomposition of $\mathcal{K}(\mathfrak{F})$ by $\mathfrak{F}\mathcal{K}(\mathfrak{F}')$ are either equivalent with respect to $\mathfrak{F}$ or disjoint modulo $\mathfrak{F}$. One of these regions is the interior of $\mathcal{K}(\mathfrak{F}')$. It may happen that all regions of the decomposition are equivalent with the interior of $\mathcal{K}(\mathfrak{F}')$ with respect to $\mathfrak{F}$. If a region of the decomposition is not equivalent with the interior of $\mathcal{K}(\mathfrak{F}')$, then the subgroup of $\mathfrak{F}$ pertaining to it will be called *a restgroup of* $\mathfrak{F}'$ *in* $\mathfrak{F}$. Choose in any region of the decomposition a side belonging to $\mathcal{U}$. This side can be carried into a side of $\mathcal{K}(\mathfrak{F}')$ by an element of $\mathfrak{F}$, the region thus being carried either into the interior of $\mathcal{K}(\mathfrak{F}')$ or into the region adjacent to $\mathcal{K}(\mathfrak{F}')$ along that side of $\mathcal{K}(\mathfrak{F}')$. Since the sides of $\mathcal{K}(\mathfrak{F}')$ belonging to $\mathcal{U}$ fall into ω equivalence classes with respect to $\mathfrak{F}$, the number n of equivalence classes of regions satisfies the condition $1 \leqq n \leqq \omega + 1$. The number of classes of conjugate restgroups is $n - 1$.

If the number of equivalence classes with respect to $\mathfrak{F}'$ into which the boundaries or $\mathcal{K}(\mathfrak{F}')$ fall is the same as with respect to $\mathfrak{F}$, and if moreover the subgroup of $\mathfrak{F}$ pertaining to any line of $\mathcal{U}$ is not of the type $\mathfrak{A}(\cdot\cdot)$ or $\mathfrak{A}(\rightleftharpoons)$, then $\mathfrak{F}'$ and $\mathfrak{F}$ have equal effect in $\mathcal{K}(\mathfrak{F}')$, i.e. not only in the interior of $\mathcal{K}(\mathfrak{F}')$ but also on the boundary. – Conversely, if $\mathfrak{F}'$ and $\mathfrak{F}$ have equal effect in $\mathcal{K}(\mathfrak{F}')$, then these two conditions hold.

23.2 The extended hull of a subgroup. We consider in an arbitrary group $\mathfrak{F}$ an $\mathfrak{F}$-subgroup $\mathfrak{G}$ which can be generated by a finite number of its elements. The hull

$\mathfrak{H}'(\mathfrak{G})$ of $\mathfrak{G}$ in $\mathfrak{F}$, defined in §15.6, contains $\mathfrak{G}$, and $\mathfrak{H}'(\mathfrak{G})$ and $\mathfrak{F}$ have equal effect in the interior of $\mathcal{K}(\mathfrak{H}')$.

We assume that $\mathfrak{H}'(\mathfrak{G})$ is not the whole of $\mathfrak{F}$ and consider the decomposition of $\mathcal{K}(\mathfrak{F})$ by $\mathfrak{F}\mathcal{K}(\mathfrak{H}')$ (Section 1). Let the boundary axes $\mathcal{A}_r$, $1 \leqq r \leqq \omega$, of $\mathcal{K}(\mathfrak{H}')$ represent the ω different equivalence classes with respect to $\mathfrak{F}$ of boundaries of $\mathcal{K}(\mathfrak{H}')$ in the interior of $\mathcal{K}(\mathfrak{F})$. The composition in question is then effected by the collection $\mathcal{U} = \bigcup_{r=1}^{\omega} \mathfrak{F}\mathcal{A}_r$. The number n of equivalence classes of regions with respect to $\mathfrak{F}$ then satisfies the condition $1 \leqq n \leqq \omega + 1$, and there are $n - 1$ non-conjugate restgroups for $\mathfrak{H}'$ in $\mathfrak{F}$.

We now arrange the numbering of the $\mathcal{A}_r$ in the following way. For every value of r let $\mathfrak{B}_r$ denote the subgroup of $\mathfrak{F}$ which belongs to the region $\mathcal{B}_r$ adjacent to $\mathcal{K}(\mathfrak{H}')$ along $\mathcal{A}_r$. We then take $\mathcal{A}_s$, $1 \leqq s \leqq m$, such that the $\mathfrak{B}_s$ cannot be finitely generated, if that case occurs at all; otherwise $m = 0$, and there are no $\mathcal{A}_s$. For $\mathcal{A}_t$, $m + 1 \leqq t \leqq \omega$, the corresponding $\mathfrak{B}_t$ admits a finite generation, if at all $m < \omega$; if $m = \omega$, then there are no $\mathcal{A}_t$. Obviously none of the $\mathcal{B}_s$ is equivalent with one of the $\mathcal{B}_t$ or with the interior of $\mathcal{K}(\mathfrak{H}')$. Therefore the $\mathcal{B}_s$ are disjoint modulo $\mathfrak{F}$ with the $\mathcal{B}_t$ and with the interior of $\mathcal{K}(\mathfrak{H}')$. Among the $\mathcal{B}_t$ some may be equivalent with the interior of $\mathcal{K}(\mathfrak{H}')$.

Consider the decomposition of $\mathcal{K}(\mathfrak{F})$ by the collection $\mathcal{V} = \bigcup_{s=1}^{m} \mathfrak{F}\mathcal{A}_s$. The collection $\mathcal{V}$ is a subcollection of $\mathcal{U}$. Let $\mathcal{C}$ denote that region of the decomposition of $\mathcal{K}(\mathfrak{F})$ by $\mathcal{V}$ which contains the interior of $\mathcal{K}(\mathfrak{H}')$, and let $\mathfrak{C}$ denote the subgroup of $\mathfrak{F}$ pertaining to $\mathcal{C}$. The closure of $\mathcal{C}$ on $\mathcal{K}(\mathfrak{F})$ is then $\mathcal{K}(\mathfrak{C})$. (If $m = 0$, then $\mathcal{V}$ is empty, and there is no decomposition at all of $\mathcal{K}(\mathfrak{F})$ by $\mathcal{V}$, thus $\mathcal{K}(\mathfrak{C})$ means $\mathcal{K}(\mathfrak{F})$ itself, and $\mathfrak{C}$ means $\mathfrak{F}$ itself.) The decomposition of $\mathcal{K}(\mathfrak{F})$ by the collection $\mathcal{U}$ first considered is a subdivision of the one now under consideration, namely the subdivision effected by the collection $\bigcup_{t=m+1}^{\omega} \mathfrak{F}\mathcal{A}_t$, if $m < \omega$. (If $m = \omega$, then $\mathcal{U} = \mathcal{V}$, and the two decompositions coincide.) In particular, the subdivision of $\mathcal{C}$ is effected by the subcollection $\mathcal{W} = \bigcup_{t=m+1}^{\omega} \mathfrak{C}\mathcal{A}_t$, because the $\mathcal{A}_t$ are situated in $\mathcal{C}$, and in $\mathcal{C}$ the groups $\mathfrak{C}$ and $\mathfrak{F}$ have equal effect. If $m < \omega$ then any region of the decomposition of $\mathcal{C}$ by $\mathcal{W}$ has on its boundary an axis equivalent with some $\mathcal{A}_t$ with respect to $\mathfrak{C}$, and thus also with the respect to $\mathfrak{F}$. The region is therefore equivalent either with the interior of $\mathcal{K}(\mathfrak{H}')$ or with some $\mathcal{B}_t$. Thus the number of equivalence classes with respect to $\mathfrak{C}$ of regions in $\mathcal{C}$ is finite, namely at most $\omega - m + 1$. Moreover, $\mathcal{K}(\mathfrak{H}')$ mod $\mathfrak{H}'$ as well as the $\mathcal{B}_t$ mod $\mathfrak{B}_t$ have a finite area, since the groups $\mathfrak{H}'$ and the $\mathfrak{B}_t$ are finitely generated and thus satisfy the condition of quasi-compactness. Hence $\mathcal{K}(\mathfrak{C})$ mod $\mathfrak{C}$ has a finite area, and $\mathfrak{C}$ thus admits a generation by a finite number of its elements.

We now assume $m > 0$. Then $\mathfrak{F}$ cannot be finitely generated, because in $\mathcal{B}_1$ the groups $\mathfrak{F}$ and $\mathfrak{B}_1$ have equal effect, thus $\mathcal{B}_1$ mod $\mathfrak{B}_1$ is part of $\mathcal{K}(\mathfrak{F})$ mod $\mathfrak{F}$ and has an infinite area.

Consider the region $\mathcal{B}_1$ and let $\mathcal{A}'$ be one of its sides in the interior of $\mathcal{K}(\mathfrak{F})$. Let $\mathfrak{f} \in \mathfrak{F}$ be such that $\mathfrak{f}\mathcal{A}' = \mathcal{A}_r$. Then $\mathfrak{f}\mathcal{B}_1$ cannot be the interior of $\mathcal{K}(\mathfrak{H}')$, since $\mathfrak{H}'$ is infinitely generated, and $\mathfrak{B}_1$ is not. Thus $\mathfrak{f}\mathcal{B}_1 = \mathcal{B}_r$, and $\mathfrak{B}_r$ cannot be finitely generated. Hence $1 \leqq r \leqq m$, i.e. $\mathcal{A}_r$ is one of the $\mathcal{A}_s$. Therefore $\mathcal{A}'$, and thus all sides of $\mathcal{B}_1$ inside $\mathcal{K}(\mathfrak{F})$, belong to $\mathcal{V}$. Originally $\mathcal{B}_1$ was introduced as one of the regions of the decomposition of $\mathcal{K}(\mathfrak{F})$ by the collection $\mathcal{U}$. Now it is seen that it is also one of the regions of the decomposition of $\mathcal{K}(\mathfrak{F})$ by the collection $\mathcal{V}$, i.e. it contains no line of the collection $\bigcup_{t=m+1}^{\omega} \mathfrak{F}\mathcal{A}_t$ on its boundary. In other words, all lines of this latter collection belong to $\mathfrak{F}\mathcal{C}$.

We are now only concerned with the decomposition of $\mathcal{K}(\mathfrak{F})$ by the collection $\mathcal{V}$. One of the regions of that decomposition is $\mathcal{C}$, others are the regions $\mathcal{B}_s$, $1 \leqq s \leqq m$, and the latter are not equivalent with $\mathcal{C}$, since $\mathfrak{C}$ is finitely generated and the $\mathfrak{B}_s$ are not. Hence the subgroup of $\mathfrak{F}$ belonging to the inner axis $\mathcal{A}_s$ of $\mathfrak{F}$ cannot be of the type $\mathfrak{A}(\cdot\cdot)$, nor $\mathfrak{A}(\rightleftharpoons)$ nor $\mathfrak{A}(-\rightarrow)$ nor $\mathfrak{A}(-\,\|)$. It is therefore either $\mathfrak{A}(\rightarrow)$ or $\mathfrak{A}(\|)$.

Whatever the value of m, we introduce for $\mathfrak{C}$ the denotation $\mathfrak{H}(\mathfrak{G})$ and call it *the extended hull of* $\mathfrak{G}$. The group $\mathfrak{H}(\mathfrak{G})$ can play the rôle of the group $\mathfrak{F}'$ in Section 1, and the decomposition of $\mathcal{K}(\mathfrak{F})$ by the collection $\mathcal{V}$ is the decomposition of $\mathcal{K}(\mathfrak{F})$ by $\mathfrak{F}\mathcal{K}(\mathfrak{H})$. We recapitulate the essential properties of the inclusion $\mathfrak{G} \subset \mathfrak{H}' \subset \mathfrak{H} \subset \mathfrak{F}$:

If we also admit the case in which $\mathfrak{H}' = \mathfrak{F}$, then $\mathfrak{H} = \mathfrak{H}'$, of course. We assumed above that $\mathfrak{H}'$ is not the whole of $\mathfrak{F}$.

If, and only if, all regions of the decomposition of $\mathcal{K}(\mathfrak{F})$ by $\mathfrak{F}\mathcal{K}(\mathfrak{H}')$ adjacent to $\mathcal{K}(\mathfrak{H}')$ belong to subgroups of $\mathfrak{F}$ which admit no finite generation, one gets $\mathfrak{H} = \mathfrak{H}'$ (the case $m = \omega$). Thus in this case no extension of $\mathfrak{H}'$ results.

If, and only if, all regions of the decomposition of $\mathcal{K}(\mathfrak{F})$ by $\mathfrak{F}\mathcal{K}(\mathfrak{H}')$ belong to subgroups of $\mathfrak{F}$ which admit a finite generation, one gets $\mathfrak{H} = \mathfrak{F}$ (the case $m = 0$). This case occurs if and only if, $\mathfrak{F}$ admits a finite generation.

If none of these extreme cases occur, then $\mathcal{K}(\mathfrak{H}')$ is not the whole of $\mathcal{K}(\mathfrak{H})$, and $\mathcal{K}(\mathfrak{H})$ is not the whole of $\mathcal{K}(\mathfrak{F})$.

In the decomposition of $\mathcal{K}(\mathfrak{F})$ by $\mathfrak{F}\mathcal{K}(\mathfrak{H})$ none of the restgroups of $\mathfrak{H}$ in $\mathfrak{F}$ are finitely generated. This is the essential property of $\mathfrak{H}$ added to those of $\mathfrak{H}'$ by the extension. Thus in the decomposition of $\mathcal{K}(\mathfrak{F})$ by $\mathfrak{F}\mathcal{K}(\mathfrak{H})$ only the regions of the class $\mathfrak{F}\mathcal{K}(\mathfrak{H})$ belong to finitely generated subgroups of $\mathfrak{F}$. However, it is pointed out that the other regions of decomposition adjacent to $\mathcal{K}(\mathfrak{H})$ along the $\mathcal{A}_s$ need not be inequivalent with respect to $\mathfrak{F}$.

The group $\mathfrak{H}$ is its own hull as well as extended hull. If two sides of $\mathcal{K}(\mathfrak{H})$ are equivalent by an element $\mathfrak{f}$ of $\mathfrak{F}$, then $\mathfrak{f}$ obviously must carry $\mathcal{K}(\mathfrak{H})$ into itself and thus belongs to $\mathfrak{H}$. Hence $\mathfrak{H}$ and $\mathfrak{F}$ have equal effect in the whole of $\mathcal{K}(\mathfrak{H})$ (including its boundaries).

In case $\mathfrak{F}$ is not quasi-compact, let $\mathfrak{F}_0$ denote a subgroup of $\mathfrak{F}$ satisfying the following conditions:

1) $\mathfrak{F}_0$ is finitely generated.

2) $\mathfrak{F}_0$ contains $\mathfrak{G}$.

3) $\mathfrak{F}_0$ and $\mathfrak{F}$ have equal effect in $\mathcal{K}(\mathfrak{F}_0)$.

4) All subgroups of $\mathfrak{F}$ belonging to sides of $\mathcal{K}(\mathfrak{F}_0)$ are of the types $\mathfrak{A}(\rightarrow)$ or $\mathfrak{A}(\|)$.

5) No restgroup of $\mathfrak{F}_0$ in $\mathfrak{F}$ is finitely generated.

In virtue of §15.6 the group $\mathfrak{F}_0$ contains $\mathfrak{H}'(\mathfrak{G})$. Hence $\mathcal{K}(\mathfrak{H}')$ is contained in $\mathcal{K}(\mathfrak{F}_0)$. In the interior of $\mathcal{K}(\mathfrak{H}')$ the group $\mathfrak{H}'$ has the same effect as $\mathfrak{F}$, and thus also the same effect as $\mathfrak{F}_0$, because in the whole of $\mathcal{K}(\mathfrak{F}_0)$ the groups $\mathfrak{F}$ and $\mathfrak{F}_0$ have equal effect. Any element $\mathfrak{f}$ of $\mathfrak{F}$ carries $\mathcal{K}(\mathfrak{H}')$ into $\mathfrak{f}\mathcal{K}(\mathfrak{H}')$ situated in $\mathfrak{f}\mathcal{K}(\mathfrak{F}_0)$. Hence the whole collection $\mathcal{U}$ considered above belongs to $\mathfrak{F}\mathcal{K}(\mathfrak{F}_0)$, and thus also the subcollection $\mathcal{V}$ of $\mathcal{U}$.

Consider the decomposition of $\mathcal{K}(\mathfrak{F})$ by $\mathfrak{F}\mathcal{K}(\mathfrak{F}_0)$. In virtue of conditions 3) and 4), any region $\mathcal{T}$ of that decomposition adjacent to $\mathcal{K}(\mathfrak{F}_0)$ is disjoint with $\mathfrak{F}\mathcal{K}(\mathfrak{F}_0)$ and thus also with $\mathcal{V}$, and belongs to a non-finitely generated subgroup of $\mathfrak{F}_0$. Hence in the decomposition of $\mathcal{K}(\mathfrak{F})$ by $\mathcal{V}$, i.e. the decomposition of $\mathcal{K}(\mathfrak{F})$ by $\mathfrak{F}\mathcal{K}(\mathfrak{H})$, a region of decomposition which has points in common with $\mathcal{T}$ must comprise the whole of $\mathcal{T}$ and, since $\mathcal{T}$ is simple modulo $\mathfrak{F}$, belongs to a subgroup of $\mathfrak{F}$ which admits no finite generation. Now $\mathfrak{H}(\mathfrak{G})$ is finitely generated, and $\mathcal{K}(\mathfrak{H})$ contains $\mathcal{K}(\mathfrak{H}')$, which is part of $\mathcal{K}(\mathfrak{F}_0)$. It follows that $\mathcal{K}(\mathfrak{H})$ is wholly contained in $\mathcal{K}(\mathfrak{F}_0)$, and thus that $\mathfrak{H}$ is contained in $\mathfrak{F}_0$, since both of them have the same effect as $\mathfrak{F}$ in their convex domain. $\mathfrak{H}$ satisfies itself the five conditions imposed on $\mathfrak{F}_0$ and is thus the *smallest* group satisfying these conditions. We can therefore conclude this section by a statement showing that $\mathfrak{H}$ does not depend on the procedure applied in the above construction, but is uniquely determined by $\mathfrak{G}$:

The extended hull $\mathfrak{H}(\mathfrak{G})$ of a finitely generated $\mathfrak{F}$-subgroup $\mathfrak{G}$ of $\mathfrak{F}$ is the intersection of all those finitely generated subgroups of $\mathfrak{F}$ which contain $\mathfrak{G}$, have equal effect with $\mathfrak{F}$ in their convex domain, are bounded by axes with corresponding groups $\mathfrak{A}(\rightarrow)$ or $\mathfrak{A}(\|)$, and for which none of their restgroups in $\mathfrak{F}$ admit a finite generation. □

This characterization of the extended hull refers, of course, essentially to groups $\mathfrak{F}$ not satisfying the condition of quasi-compactness, because otherwise $\mathfrak{H}(\mathfrak{G})$ as well as the finitely generated subgroups in question coincide with $\mathfrak{F}$.

23.3 Exhaustion by extended hulls. Let $\mathfrak{F}$ denote an $\mathfrak{F}$-group which does not satisfy the condition of quasi-compactness, and let the infinite sequence

$$\mathfrak{f}_1, \mathfrak{f}_2, \mathfrak{f}_3, \ldots \tag{1}$$

constitute an enumeration of the elements of $\mathfrak{F}$. Let $\mathfrak{F}_n$ denote the subgroup of $\mathfrak{F}$ generated by $\mathfrak{f}_1, \mathfrak{f}_2, \ldots, \mathfrak{f}_n$.

We select an integer n_1 such that $\mathfrak{F}_{n_1}$ is an $\mathfrak{F}$-group and consider the extended hull $\mathfrak{H}_1 = \mathfrak{H}(\mathfrak{F}_{n_1})$ of $\mathfrak{F}_{n_1}$. Let $\mathcal{A}_{1s}$, $1 \leqq s \leqq m_1$, be sides of $\mathcal{K}_1 = \mathcal{K}(\mathfrak{H}_1)$ representing those equivalence classes with respect to $\mathfrak{F}$ of boundary axes of $\mathfrak{H}_1$ which are inner axes of $\mathfrak{F}$. Here $m_1 \geqq 1$, because $\mathcal{K}_1$ is not the whole of $\mathcal{K}(\mathfrak{F})$ (Section 2). The $\mathcal{A}_{1s}$ also represent all equivalence classes with respect to $\mathfrak{H}_1$ of inner axes of $\mathfrak{F}$ bounding $\mathcal{K}_1$, because $\mathfrak{F}$ and $\mathfrak{H}_1$ have equal effect not only in the interior of $\mathcal{K}_1$, but also on its boundary (Section 2).

In the decomposition of $\mathcal{K}(\mathfrak{F})$ by $\mathfrak{F}\mathcal{K}_1$ let $\mathcal{B}_{1s}$ be the region adjacent to $\mathcal{K}_1$ along $\mathcal{A}_{1s}$, and $\mathfrak{B}_{1s}$ its corresponding subgroup of $\mathfrak{F}$. None of the $\mathfrak{B}_{1s}$ admits a finite generation.

In each $\mathfrak{B}_{1s}$ we select arbitrarily a translation $\mathfrak{t}_{1s}$ whose axis $\mathcal{T}_{1s}$ is an inner axis of $\mathcal{B}_{1s}$. Then $\mathfrak{t}_{1s}$ does not belong to $\mathfrak{H}_1$, and thus, in particular, not to $\mathfrak{F}_{n_1}$. Hence in the enumeration (1) the $\mathfrak{t}_{1s}$, $1 \leqq s \leqq m_1$, appear later than $\mathfrak{f}_{n_1}$. We now select a finite set of generators of $\mathfrak{H}_1$ and determine an integer n_2 such that the section $\mathfrak{f}_1, \dots, \mathfrak{f}_{n_2}$ of the sequence (1) contains this set of generators as well as the $\mathfrak{t}_{1s}$.

Consider now the extended hull $\mathfrak{H}_2 = \mathfrak{H}(\mathfrak{F}_{n_2})$ of $\mathfrak{F}_{n_2}$. It contains $\mathfrak{H}_1$, because $\mathfrak{F}_{n_2}$ contains $\mathfrak{H}_1$. Let $\mathfrak{t}$ denote a translation in $\mathfrak{H}_1$ whose axis $\mathcal{T}$ is an inner axis of $\mathfrak{H}_1$. Then both $\mathfrak{t}$ and $\mathfrak{t}_{1s}$ belong to $\mathfrak{H}_2$, and since $\mathcal{T}$ and $\mathcal{T}_{1s}$ are separated by the boundary axis $\mathcal{A}_{1s}$ of $\mathfrak{H}_1$, this axis $\mathcal{A}_{1s}$ is an inner axis of $\mathfrak{H}_2$. This holds for all values $1 \leqq s \leqq m_1$.

Since in $\mathcal{K}_1$ the group $\mathfrak{H}_1$ has the same effect as $\mathfrak{F}$, it also has the same effect as $\mathfrak{H}_2$, because $\mathfrak{H}_2$ and $\mathfrak{F}$ have equal effect in $\mathcal{K}_2$, which contains $\mathcal{K}_1$. We can therefore decompose $\mathcal{K}_2$ by $\mathfrak{H}_2\mathcal{K}_1$, thus by the collection $\mathcal{U} = \bigcup_{s=1}^{m_1} \mathfrak{H}_2\mathcal{A}_{1s}$. One of the regions of this decomposition is the interior of $\mathcal{K}_1$. Let $\mathcal{C}_{1s}$ be the other region adjacent to $\mathcal{A}_{1s}$. and $\mathfrak{C}_{1s}$ its corresponding group. Then $\mathcal{C}_{1s}$ is contained in $\mathcal{B}_{1s}$, and $\mathfrak{C}_{1s}$ in $\mathfrak{B}_{1s}$. The $\mathcal{C}_{1s}$ need not be inequivalent with respect to $\mathfrak{F}$ and thus with respect to $\mathfrak{H}_2$. They are, however, disjoint modulo $\mathfrak{F}$ with $\mathcal{K}_1$, because the $\mathcal{B}_{1s}$ are. Let the numbering s be such that the $\mathcal{C}_{1s}$ for $1 \leqq s \leqq m_1'$ represent a complete collection of non-equivalents among the $\mathcal{C}_{1s}$. The number of equivalence classes of regions of decomposition then is $m_1' + 1$, and one gets

$$\Phi(\mathfrak{H}_2) = \Phi(\mathfrak{H}_1) + \sum_{s=1}^{m_1'} \Phi(\mathfrak{C}_{1s}). \tag{2}$$

The $\Phi(\mathfrak{C}_{1s})$ are thus finite, and the $\mathfrak{C}_{1s}$ are finitely generated.

The surface $\mathcal{K}_2$ mod $\mathfrak{H}_2$ is decomposed into $m_1' + 1$ surface parts, namely $\mathcal{K}_1$ mod $\mathfrak{H}_1$ and the $\mathcal{K}(\mathfrak{C}_{1s})$ mod $\mathfrak{C}_{1s}$, $1 \leqq s \leqq m_1'$. The m_1 geodesics a_{1s}, $1 \leqq s \leqq m_1$, corresponding to the $\mathcal{A}_{1s}$ effect this decomposition of the surface.

We now proceed from $\mathfrak{H}_2$ in the same way as from $\mathfrak{H}_1$. In detail that means: We select a set $\mathcal{A}_{2s}$, $1 \leqq s \leqq m_2$, of sides of $\mathcal{K}_2$ representing those equivalence classes with respect to $\mathfrak{H}_2$ and to $\mathfrak{F}$ of sides of $\mathcal{K}_2$ which are inner axes of $\mathfrak{F}$. In the decomposition of $\mathcal{K}(\mathfrak{F})$ by $\mathfrak{F}\mathcal{K}_2$ let $\mathcal{B}_{2s}$ be the region adjacent to $\mathcal{K}_2$ along $\mathcal{A}_{2s}$, and $\mathfrak{B}_{2s}$ its corresponding subgroup. In each $\mathfrak{B}_{2s}$ we select a translation $\mathfrak{A}_{2s}$ whose axis is an inner axis of $\mathcal{B}_{2s}$ and then determine an integer n_3 such that the section $\mathfrak{f}_1, \dots, \mathfrak{f}_{n_3}$ of (1) contains the $\mathfrak{t}_{2s}$ as well as a set of generators of $\mathfrak{H}_2$. We then treat the extended hull $\mathfrak{H}_3 = \mathfrak{H}(\mathfrak{F}_{n_3})$ in the same way as $\mathfrak{H}_2 = \mathfrak{H}(\mathfrak{F}_{n_2})$.

This process can be continued indefinitely through a sequence of groups $\mathfrak{H}_k = \mathfrak{H}(\mathfrak{F}_{n_k})$, $k = 1, 2, 3, \dots$, because the $\mathfrak{H}_k$ are all finitely generated and no $\mathfrak{H}_k$ therefore exhausts $\mathfrak{F}$. The $\mathcal{A}_{ks}$, $1 \leqq s \leqq m_k$, represent the equivalence classes of sides of $\mathcal{K}_k = \mathcal{K}(\mathfrak{H}_k)$ which are inner axes of $\mathfrak{F}$, and every restgroup of $\mathfrak{H}_k$ in $\mathfrak{F}$ is conjugate to a $\mathfrak{B}_{ks}$. Corresponding to (2) one gets an equation

$$\Phi(\mathfrak{H}_k) = \Phi(\mathfrak{H}_{k-1}) + \sum_{s=1}^{m'_{k-1}} \Phi(\mathfrak{C}_{k-1,s}). \tag{3}$$

Here $\Phi(\mathfrak{H}_{k-1})$ may be split in the same way, and by recursion one gets

$$\Phi(\mathfrak{H}_k) = \Phi(\mathfrak{H}_1) + \sum_{\ell=1}^{k-1} \sum_{s=1}^{m'_\ell} \Phi(\mathfrak{C}_{\ell s}). \tag{4}$$

The surface $\mathcal{K}_k$ mod $\mathfrak{H}_k$ consists of $\mathcal{K}_1$ mod $\mathfrak{H}_1$ and the $\mathcal{K}(\mathfrak{C}_{\ell,s})$ mod $\mathfrak{C}_{\ell s}$ for the range of values of ℓ and s as in (4). These surface parts combine along the geodesics $a_{\ell s}$ corresponding to the $\mathcal{A}_{\ell s}$ for the same range of values of ℓ and for $1 \leqq s \leqq m_\ell$.

Every $\mathcal{K}_k$ is contained in $\mathcal{K}_{k+1}$, those sides of $\mathcal{K}_k$ which are not boundary axes of $\mathfrak{F}$ being inner axes of $\mathcal{K}_{k+1}$. The integers n_k constitute an increasing sequence. An arbitrary element $\mathfrak{f}_j$ of $\mathfrak{F}$ is contained in some $\mathfrak{F}_{n_k}$ (and then also in all the following), namely at least for $j \leqq n_k$. It is then contained in $\mathfrak{H}_k$. Every finitely generated subgroup of $\mathfrak{F}$ is contained in all the $\mathfrak{H}_k$ from a certain value of k onwards. This implies that the consequences to be drawn from the above construction, especially in the next two sections, do not depend on the free choices made during this construction, especially the choice of the sequence n_k. For if $\mathfrak{H}_j^*$, $j = 1, 2, \ldots$, is the sequence of groups obtained correspondingly by other choices, then every $\mathfrak{H}_j^*$ is contained in some $\mathfrak{H}_k$, and every $\mathfrak{H}_k$ is contained in some $\mathfrak{H}_j^*$.

23.4 Coverage of points of $\mathcal{K}(\mathfrak{F})$. Let p be an arbitrary point in the interior of $\mathcal{K}(\mathfrak{F})$ and $\mathcal{S}$ a straight line through p. Each of the two half-lines $\mathcal{S}_1$ and $\mathcal{S}_2$ on $\mathcal{S}$ issuing from p meets axes of $\mathfrak{F}$ (§14.2). Let $\mathfrak{t}_1$ be a translation in $\mathfrak{F}$ whose axis cuts $\mathcal{S}_1$, and similarly $\mathfrak{t}_2$ for $\mathcal{S}_2$. Determine a value k such that $\mathfrak{t}_1$ and $\mathfrak{t}_2$ belong to $\mathfrak{H}_k$. The axes of $\mathfrak{t}_1$ and $\mathfrak{t}_2$ then belong to $\mathcal{K}_k$; owing to the convexity of $\mathcal{K}_k$ then p belongs to $\mathcal{K}_k$ (and thus also to all $\mathcal{K}_{k+\ell}$, $\ell \geqq 0$).

Let $\mathcal{A}$ be a boundary axis of $\mathfrak{F}$ and $\mathfrak{a}$ a translation along $\mathcal{A}$. If k is such that $\mathfrak{a}$ belongs to $\mathfrak{H}_k$, then $\mathcal{A}$ belongs to $\mathcal{K}_k$. It is then a boundary axis of all $\mathcal{K}_{k+\ell}$, $\ell \geqq 0$.

Let $\mathcal{L}$ be a limit side of $\mathcal{K}(\mathfrak{F})$. No axis of $\mathfrak{F}$ has points in common with $\mathcal{L}$. Since $\mathcal{K}_k$ is contained in $\mathcal{K}(\mathfrak{F})$, and all sides of $\mathcal{K}_k$ are axes of $\mathfrak{F}$, it follows that $\mathcal{L}$ and $\mathcal{K}_k$ are disjoint for all values of k. One thus gets this result:

To every point p of $\mathcal{K}(\mathfrak{F})$, except the points on limit sides of $\mathcal{K}(\mathfrak{F})$, if any, there is a smallest integer k_p such that $p \in \mathcal{K}_k$ for all $k \geqq k_p$. The limit sides of $\mathcal{K}(\mathfrak{F})$ are disjoint with $\mathcal{K}_k$ for all k. □

23.5 Coverage of points of $\mathcal{E}$. Ends of $\mathcal{K}(\mathfrak{F})$ and of $\mathcal{K}(\mathfrak{F})$ mod $\mathfrak{F}$. To every limit-centre or fundamental point u of $\mathfrak{F}$ there is a smallest integer k_u such that u

is a limit-centre, or fundamental point respectively, of $\mathfrak{H}_k$ for all $k \geqq k_u$, because a limit-rotation or translation of $\mathfrak{F}$ belonging to u appears somewhere in (1).

Let p be a point of $\mathcal{E}$ which does not belong to the closure $\overline{\mathcal{K}}_k$ of $\mathcal{K}_k$ on $\overline{\mathcal{D}}$ for any value k. Thus p is not a limit point of any $\mathfrak{H}_k$ and, in particular, not a limit-centre nor fundamental point of $\mathfrak{F}$. For every value of k let $\mathcal{I}(p; k)$ be the interval of discontinuity of $\mathcal{K}_k$ which contains p. The end-points of $\mathcal{I}(p; k)$ are fundamental points of $\mathfrak{H}_k$. Let $\mathcal{A}(p; k)$ denote the boundary axis of $\mathcal{K}_k$ which bounds $\mathcal{I}(p; k)$. The interval $\mathcal{I}(p; k+1)$ is contained in $\mathcal{I}(p; k)$, because $\mathcal{K}_{k+1}$ contains $\mathcal{K}_k$. If $\mathcal{I}(p; k+1)$ coincides with $\mathcal{I}(p; k)$, then $\mathcal{A}(p; k)$ is a boundary axis of $\mathfrak{F}$; for a boundary axis of $\mathcal{K}_k$ which is not a boundary axis of $\mathfrak{F}$ is an inner axis of $\mathcal{K}_{k+1}$ according to the construction in Section 3. Thus if $\mathcal{I}(p; k) = \mathcal{I}(p; k+1)$, then $\mathcal{I}(p; k) = \mathcal{I}(p; k+\ell)$ for all $\ell \geqq 0$. The point p then represents a *point at infinity* on a funnel or half-funnel of $\mathcal{D}$ mod $\mathfrak{F}$. – Conversely, if p belongs to a periodic interval of discontinuity of $\mathfrak{F}$, then $\mathcal{I}(p; k)$ becomes constant from a certain value of k onwards.

If $\mathcal{I}(p; k)$ does not coincide with $\mathcal{I}(p; k+1)$ for any value of k, then the intersection of the $\mathcal{I}(p; k)$ is either a point, and thus the point p itself, or a closed interval of $\mathcal{E}$. In the latter case let $\mathcal{I}$ denote the open part of that closed interval. The end-points of the $\mathcal{I}(p; k)$ converge to the end-points of $\mathcal{I}$ for $k \to \infty$, and the end-points of $\mathcal{I}$ are thus limit points of $\mathfrak{F}$. On the other hand, $\mathcal{I}$ cannot contain limit points of $\mathfrak{F}$, for then it would contain fundamental points of $\mathfrak{F}$, and a fundamental point of $\mathfrak{F}$ does not belong to an interval of discontinuity of $\mathfrak{H}_k$ for all values of k. Hence $\mathcal{I}$ is an aperiodic interval of discontinuity of $\mathfrak{F}$, and the end-points of $\mathcal{I}$ bound a limit side of $\mathcal{K}(\mathfrak{F})$. – If the intersection of the $\mathcal{I}(p; k)$ is the point p only, then p belongs to the limit-set $\overline{\mathcal{G}}(\mathfrak{F})$ of $\mathfrak{F}$.

Definition. The components of the set of those points of $\overline{\mathcal{K}(\mathfrak{F})}$ which do not belong to the closed convex domain $\overline{\mathcal{K}}(\mathfrak{F}')$ of any finitely generated subgroup $\mathfrak{F}'$ of $\mathfrak{F}$ are called *ends of* $\mathcal{K}(\mathfrak{F})$.

If $\mathcal{C}$ is such a component, and p one of its points, then the above construction by the intervals $\mathcal{I}(p; k)$ shows that $\mathcal{C}$ is either a limit side of $\mathcal{K}(\mathfrak{F})$ including its end-points on $\mathcal{E}$, or a single point of $\overline{\mathcal{G}}(\mathfrak{F})$. – Conversely, if $\mathcal{I}(k)$ is an interval of discontinuity of $\mathfrak{H}_k$ for $k = 1, 2, \dots$, and if for all values of k the interval $\mathcal{I}(k+1)$ is contained in $\mathcal{I}(k)$, but is not the whole of $\mathcal{I}(k)$, then the intersection of this sequence of intervals determines an end of $\mathcal{K}(\mathfrak{F})$; for every finitely generated subgroup $\mathfrak{F}'$ of $\mathfrak{F}$ is contained in all the $\mathfrak{H}_k$ from a certain value of k onwards. – The ends of $\mathcal{K}(\mathfrak{F})$ can thus be determined by the sequence of subgroups $\mathfrak{H}_k$, but they do not depend on the special choice of the groups $\mathfrak{H}_k$ used in the above construction.

Ends of $\mathcal{K}(\mathfrak{F})$ occur if and only if, $\mathfrak{F}$ admits no finite generation. The totality of ends of $\mathcal{K}(\mathfrak{F})$ is reproduced by $\mathfrak{F}$, as is seen almost directly from the definition. They fall into equivalence classes with respect to $\mathfrak{F}$. We say that such an equivalence class represents an *end of the surface* $\mathcal{K}(\mathfrak{F})$ mod $\mathfrak{F}$. A limit side of $\mathcal{K}(\mathfrak{F})$ corresponds to a half-line of $\mathcal{K}(\mathfrak{F})$ mod $\mathfrak{F}$ cutting off a quarter-plane or a half-plane from $\mathcal{D}$ mod $\mathfrak{F}$, according as the limit side of $\mathcal{K}(\mathfrak{F})$ is cut by a reflection line or not.

23.6 The kernel of $\mathfrak{F}$ containing a given subgroup. Let $\mathfrak{F}$ be an $\mathfrak{F}$-group which admits no finite generation and contains no reflection. Given a finitely generated $\mathfrak{F}$-subgroup $\mathfrak{G}$ of $\mathfrak{F}$, we consider the extended hull $\mathfrak{H}(\mathfrak{G})$ of $\mathfrak{G}$ (Section 2). Let $\mathcal{A}_s$, $1 \leqq s \leqq m$, represent the equivalence classes, both with respect to $\mathfrak{F}$ and with respect to $\mathfrak{H}$, of those sides of $\mathcal{K}(\mathfrak{H})$ which are inner axes of $\mathfrak{F}$. In the decomposition of $\mathcal{K}(\mathfrak{F})$ by $\mathfrak{F}\mathcal{K}(\mathfrak{H})$ let $\mathcal{B}_s$ be the region of decomposition adjacent to $\mathcal{K}(\mathfrak{H})$ along $\mathcal{A}_s$, and $\mathfrak{B}_s$ the group pertaining to $\mathcal{B}_s$. The $\mathfrak{B}_s$ admit no finite generation, and in the closure $\tilde{\mathcal{B}}_s$ of $\mathcal{B}_s$ on $\mathcal{K}(\mathfrak{F})$ the groups $\mathfrak{F}$ and $\mathfrak{B}_s$ have equal effect. Since $\mathfrak{F}$ contains no reflection, the subgroup of $\mathfrak{F}$ pertaining to $\mathcal{A}_s$ is here only of the type $\mathfrak{A}(\rightarrow)$, whereas in Section 2 we also admitted the type $\mathfrak{A}(\|)$.

As pointed out in Section 2, the regions $\mathcal{B}_s$ need not be inequivalent. We arrange the numbering s so that equivalent regions are grouped together in the beginning of the enumeration. Assume that the $\mathcal{B}_r$, $1 \leqq r \leqq q$, $1 < q \leqq m$, are equivalent with respect to $\mathfrak{F}$, but not equivalent with the $\mathcal{B}_t$, $q < t \leqq m$, if at all $q < m$. Let $\mathfrak{f}_r$ be such elements of $\mathfrak{F}$ that $\mathfrak{f}_r\mathcal{B}_r = \mathcal{B}_1$, in particular we choose $\mathfrak{f}_1 = 1$. Then the $\mathfrak{f}_r\mathcal{A}_r$ are sides of $\mathcal{B}_1$ in the interior of $\mathcal{K}(\mathfrak{F})$, and they are inequivalent with respect to $\mathfrak{F}$, and thus also with respect to $\mathfrak{B}_1$. On the other hand, since $\mathcal{B}_1$ is one of the regions of the decomposition of $\mathcal{K}(\mathfrak{F})$ by $\mathfrak{F}\mathcal{K}(\mathfrak{H})$, thus by the collection $\bigcup_s \mathfrak{F}\mathcal{A}_s$, and since $\mathcal{B}_1$ is not equivalent with the $\mathcal{B}_t$ nor with $\mathcal{K}(\mathfrak{H})$, it follows that the $\mathfrak{f}_r\mathcal{A}_r$ represent all equivalence classes with respect to $\mathfrak{B}_1$ of boundaries of $\mathcal{B}_1$ in the interior of $\mathcal{K}(\mathfrak{F})$. In other words, the surface $\tilde{\mathcal{B}}_1$ mod $\mathfrak{B}_1$ has exactly q boundaries which are not boundaries of $\mathcal{K}(\mathfrak{F})$ mod $\mathfrak{F}$; they are simple closed geodesics.

We now use the method of §22 for splitting off elementary surfaces from $\tilde{\mathcal{B}}_1$ mod $\mathfrak{B}_1$. Let $\mathfrak{p}_1$ be a primary translation along $\mathcal{A}_1$, and $\mathfrak{p}_1'$ a primary translation along a side $\mathcal{A}_2'$ of $\mathcal{B}_1$ equivalent with $\mathfrak{f}_2\mathcal{A}_2$, such that the join of $\mathcal{A}_1$ and $\mathcal{A}_2'$ is simple modulo $\mathfrak{B}_1$. We then get an elementary group $\mathfrak{E}_1$ generated by $\mathfrak{p}_1$, $\mathfrak{p}_1'$ and $\mathfrak{p}_1''$, satisfying $\mathfrak{p}_1\mathfrak{p}_1'\mathfrak{p}_1'' = 1$. Here $\mathfrak{p}_1''$ is a translation, because $\mathfrak{E}_1$ does not exhaust $\mathfrak{B}_1$. The axis of $\mathfrak{p}_1''$ is simple modulo $\mathfrak{B}_1$, and its corresponding group is of the type $\mathfrak{A}(\rightarrow)$ and generated by $\mathfrak{p}_1''$. – If $q > 2$, we put $\mathfrak{p}_1''^{-1} = \mathfrak{p}_2$ and let in the same way $\mathfrak{p}_2$ together with a primary translation $\mathfrak{p}_2'$ along a suitable axis equivalent with $\mathfrak{f}_3\mathcal{A}_3$ generate a group $\mathfrak{E}_2$, and so on. This process can be performed $q-1$ times. As is evident from §22.7 the free product $\mathfrak{P}_1$ of the $q-1$ elementary groups with amalgamations according to the axes used has a presentation by

$$\begin{cases} \text{generators} & \mathfrak{p}_1, \mathfrak{p}_1', \dots, \mathfrak{p}_{q-1}', \mathfrak{p}_{q-1}'' \\ \text{relation} & \mathfrak{p}_1\mathfrak{p}_1' \dots \mathfrak{p}_{q-1}'\mathfrak{p}_{q-1}'' = 1. \end{cases} \tag{5}$$

Compare (19), §22. Here one gets only one relation, because all the generators are translations; this holds also for $\mathfrak{p}_{q-1}''$, because $\mathfrak{P}_1$ does not exhaust $\mathfrak{B}_1$, and for the same reason the group pertaining to the axis of $\mathfrak{p}_{q-1}''$ is of the type $\mathfrak{A}(\rightarrow)$ and generated by $\mathfrak{p}_{q-1}''$. The group $\mathfrak{P}_1$ is a free group with q free generators and consists of motions only.

The axes of $\mathfrak{p}_1, \mathfrak{p}'_1, \ldots, \mathfrak{p}'_{q-1}$ represent the q equivalence classes of boundary axes of $\mathcal{B}_1$ which were represented by the $\mathfrak{f}_r \mathcal{A}_r$. We put $\mathfrak{p}''^{-1}_{q-1} = \mathfrak{p}_1 \mathfrak{p}'_1 \ldots \mathfrak{p}'_{q-1} = \mathfrak{c}_1$. The axis $\mathcal{A}(\mathfrak{c}_1)$ is an inner axis of $\mathcal{B}_1$ and is simple modulo $\mathfrak{B}_1$. In $\mathcal{K}(\mathfrak{P}_1)$ the groups $\mathfrak{P}_1$ and $\mathfrak{B}_1$ have equal effect (§22). One can decompose $\mathcal{K}(\mathfrak{B}_1) = \tilde{\mathcal{B}}_1$ by $\mathfrak{B}_1 \mathcal{K}(\mathfrak{P}_1)$, and this is the same as the decomposition of $\mathcal{K}(\mathfrak{B}_1)$ by the collection $\mathfrak{B}_1 \mathcal{A}(\mathfrak{c}_1)$, because all other equivalence classes of sides of $\mathcal{K}(\mathfrak{P}_1)$ consist of sides of $\mathcal{K}(\mathfrak{B}_1)$. One thus gets two, and only two, equivalence classes of regions by this decomposition, one represented by the interior of $\mathcal{K}(\mathfrak{P}_1)$, the other by the region $\mathcal{C}_1$ adjacent to $\mathcal{K}(\mathfrak{P}_1)$ along $\mathcal{A}(\mathfrak{c}_1)$. The group $\mathfrak{C}_1$ corresponding to $\mathcal{C}_1$ admits no finite generation, because $\mathfrak{B}_1$ is the free product of $\mathfrak{P}_1$ and $\mathfrak{C}_1$ with amalgamation of the translation group generated by $\mathfrak{c}_1$.

Any boundary of $\mathcal{B}_1$ which is an inner axis of $\mathfrak{F}$ is equivalent with a side of $\mathcal{K}(\mathfrak{P}_1)$, and not equivalent with $\mathcal{A}(\mathfrak{c}_1)$. Hence in the boundary of $\mathcal{C}_1$ only the class $\mathfrak{C}_1 \mathcal{A}(\mathfrak{c}_1)$ consists of inner axes of $\mathfrak{F}$. Consider now the decomposition of $\mathcal{K}(\mathfrak{F})$ by $\mathfrak{F}\mathcal{A}(\mathfrak{c}_1)$. The part of $\mathfrak{F}\mathcal{A}(\mathfrak{c}_1)$ in $\mathcal{B}_1$ is the collection $\mathfrak{B}_1 \mathcal{A}(\mathfrak{c}_1)$ considered above, because $\mathfrak{B}_1$ and $\mathfrak{F}$ have equal effect in $\mathcal{B}_1$. Therefore $\mathcal{C}_1$ is one of the regions of that decomposition. Since the part of $\mathfrak{F}\mathcal{A}(\mathfrak{c}_1)$ bounding $\mathcal{C}_1$ is $\mathfrak{C}_1 \mathcal{A}(\mathfrak{c}_1)$ and since the group belonging to these axes is of the type $\mathfrak{A}(\rightarrow)$, it follows from §19 that *$\mathcal{A}(\mathfrak{c}_1)$ is dividing modulo $\mathfrak{F}$.*

If among the $\mathcal{B}_t, q < t \leqq m$, two or more are equivalent, we repeat the construction for a further section $\mathcal{B}_{q+1}, \ldots, \mathcal{B}_{q'}$, and thus get a finitely generated group $\mathfrak{P}_2$ in $\mathfrak{B}_{q+1}$ and a restgroup $\mathfrak{C}_2$ belonging to a region $\mathcal{C}_2$ of $\mathcal{B}_{q+1}$ cut off by axes of the equivalence class of $\mathcal{A}(\mathfrak{c}_2)$, this axis being dividing modulo $\mathfrak{F}$.

This process is continued as long as subcollections of equivalents among the $\mathcal{B}_s$ can be found. Let this cover the values of s up to $m_1 \leqq m$. If $m_1 < m$, consider a $\mathcal{B}_\ell$, $m_1 < \ell \leqq m$, this $\mathcal{B}_\ell$ is not equivalent with any other of the $\mathcal{B}_s$, thus for that region $\mathcal{B}_\ell$ there is no corresponding group $\mathfrak{P}$ to be constructed. Then $\mathcal{B}_\ell$ itself is called the corresponding $\mathcal{C}$-region, $\mathfrak{B}_\ell$ is its $\mathfrak{C}$-group, and $\mathcal{A}_\ell$ is the corresponding $\mathcal{A}(\mathfrak{c})$ and is dividing modulo $\mathfrak{F}$.

Let N be the number of equivalence classes with respect to $\mathfrak{F}$ into which the regions $\mathcal{B}_s$, $1 \leqq s \leqq m$, fall. We then get N translations $\mathfrak{c}_1, \ldots, \mathfrak{c}_N$ and N restgroups $\mathfrak{C}_1, \ldots, \mathfrak{C}_N$ (some of which coincide with the original $\mathfrak{B}_s$, if $m_1 < m$). Each of the $\mathcal{K}(\mathfrak{C}_t)$, $1 \leqq t \leqq N$, has only one equivalence class of boundaries in the interior of $\mathcal{K}(\mathfrak{F})$, represented by the axis of the translation $\mathfrak{c}_t$. The axis $\mathcal{A}(\mathfrak{c}_t)$ is simple and dividing modulo $\mathfrak{F}$. For those values of t for which $\mathfrak{P}_t$ exists, the convex domains $\mathcal{K}(\mathfrak{C}_t)$ and $\mathcal{K}(\mathfrak{P}_t)$ are adjacent along $\mathcal{A}(\mathfrak{c}_t)$. Let this be the case for $t = 1, 2, \ldots, N_1$, $N_1 \leqq N$.

Consider now the decomposition of $\mathcal{K}(\mathfrak{F})$ by the collection

$$\mathcal{V} = \bigcup_{t=1}^{N} \mathfrak{F}\mathcal{A}(\mathfrak{c}_t).$$

One of the regions of that decomposition contains the interior of $\mathcal{K}(\mathfrak{H})$. Let it be denoted by $\mathcal{R}$, and let $\mathfrak{R}$ be the subgroup of $\mathfrak{F}$ belonging to it. The groups $\mathfrak{R}$ and $\mathfrak{F}$

have equal effect not only in $\mathcal{R}$, but also on its boundary, because each of the $\mathcal{A}(\mathfrak{c}_t)$ is dividing modulo $\mathfrak{F}$. Any point of $\mathcal{R}$ belongs either to $\mathfrak{K}\mathcal{K}(\mathfrak{H})$, or to some $\mathfrak{K}\mathcal{K}(\mathfrak{P}_t)$, $t \leqq N_1$. Since both $\Phi(\mathfrak{H})$ and $\Phi(\mathfrak{P}_t)$ are finite, also $\Phi(\mathfrak{K})$ is finite, thus $\mathfrak{K}$ can be finitely generated.

The lines bounding $\mathcal{K}(\mathfrak{C}_t)$ in the interior of $\mathcal{B}_s$ were the lines $\mathfrak{B}_s\mathcal{A}(\mathfrak{c}_t)$, if $\mathcal{B}_s$ is the $\mathcal{B}$-region containing $\mathcal{K}(\mathfrak{C}_t)$. Since $\mathfrak{B}_s$ and $\mathfrak{F}$ have equal effect in $\mathcal{K}(\mathfrak{B}_s)$ no line of $\mathcal{V}$ is in the interior of $\mathcal{K}(\mathfrak{C}_t)$, thus $\mathcal{K}(\mathfrak{C}_t)$ is the closure of one of the regions of decomposition of $\mathcal{K}(\mathfrak{F})$ by $\mathcal{V}$. We thus get the following result:

$\mathfrak{F}\mathcal{K}(\mathfrak{K})$, and thus the collection $\mathcal{V}$, decomposes $\mathcal{K}(\mathfrak{F})$ into $N+1$ equivalence classes of regions. One of these regions, closed on $\mathcal{K}(\mathfrak{F})$, is $\mathcal{K}(\mathfrak{K})$, and $\mathcal{K}(\mathfrak{K})$ is quasi-compact modulo $\mathfrak{K}$. None of the other N restgroups $\mathfrak{C}_t$ of $\mathfrak{K}$ in $\mathfrak{F}$ admits a finite generation. Each of the boundaries of $\mathcal{K}(\mathfrak{K})$ in the interior of $\mathcal{K}(\mathfrak{F})$ is dividing modulo $\mathfrak{F}$. □

Now let $\mathfrak{T}$ be a subgroup of $\mathfrak{F}$ satisfying the following five conditions:

1) $\mathfrak{T}$ is finitely generated.

2) $\mathfrak{T}$ contains $\mathfrak{G}$.

3) In $\mathcal{K}(\mathfrak{T})$ the groups $\mathfrak{T}$ and $\mathfrak{F}$ have equal effect.

4) Each side of $\mathcal{K}(\mathfrak{T})$ is dividing modulo $\mathfrak{F}$.

5) No restgroup of $\mathfrak{T}$ in $\mathfrak{F}$ admits a finite generation.

From these assumptions the following conclusions can be drawn. Since $\mathfrak{T}$ is finitely generated and $\mathfrak{F}$ is not, there are sides of $\mathcal{K}(\mathfrak{T})$ in the interior of $\mathcal{K}(\mathfrak{F})$. They fall into a finite number M of equivalence classes with respect to $\mathfrak{T}$ (condition 1)), and thus also with respect to $\mathfrak{F}$ (condition 3)). Let $\mathcal{T}_k$, $1 \leqq k \leqq M$, be sides of $\mathcal{K}(\mathfrak{T})$ representing these classes. In consequence of condition 3) we can decompose $\mathcal{K}(\mathfrak{F})$ by $\mathfrak{F}\mathcal{K}(\mathfrak{T})$. This decomposition is effected by the collection $\bigcup_{k=1}^{M} \mathfrak{F}\mathcal{T}_k$. The number of equivalence classes with respect to $\mathfrak{F}$ of regions of decomposition is then $\leqq M+1$. One of these regions, closed on $\mathcal{K}(\mathfrak{F})$, is $\mathcal{K}(\mathfrak{T})$. Let $\mathcal{L}_k$ denote the closure of $\mathcal{K}(\mathfrak{F})$ of the other region adjacent to $\mathcal{T}_k$, and $\mathfrak{L}_k$ its corresponding group. Then $\mathcal{L}_k$ is not equivalent with $\mathcal{K}(\mathfrak{T})$, because $\mathcal{T}_k$ is dividing modulo $\mathfrak{F}$ (condition 4)). Also $\mathfrak{L}_k$ admits no finite generation (condition 5)). Any two of the $\mathcal{L}_k$ are inequivalent with respect to $\mathfrak{F}$, for since $\mathcal{T}_k$ is dividing modulo $\mathfrak{F}$ all sides of $\mathcal{L}_k$ in the interior of $\mathcal{K}(\mathfrak{F})$ belong to one equivalence class, the one containing $\mathcal{T}_k$ (condition 4) and §19). Thus the number of equivalence classes of regions is equal to $M+1$.

$\mathfrak{T}$ satisfies four of the five conditions imposed on the group $\mathfrak{F}_0$ of Section 2 (conditions 1),2),3) and 5)). Hence $\mathfrak{T}$ contains the extended hull $\mathfrak{H}(\mathfrak{G})$ (Section 2). Consider now the part $\mathcal{A}_t$, $1 \leqq t \leqq m_1$, of the sides $\mathcal{A}_s$, $1 \leqq s \leqq m$, of $\mathfrak{H}(\mathfrak{G})$ used above. While all the $\mathcal{A}_s$ belong to $\mathcal{K}(\mathfrak{H})$, and thus to $\mathcal{K}(\mathfrak{T})$, the $\mathcal{A}_t$ in particular are in the interior of $\mathcal{K}(\mathfrak{T})$, because they are non-dividing modulo $\mathfrak{F}$ (all the $\mathcal{B}_t$ have more than one equivalence class of sides in the interior of $\mathcal{K}(\mathfrak{F})$), whereas all sides of $\mathcal{K}(\mathfrak{T})$ are dividing modulo $\mathfrak{F}$. The free generators $\mathfrak{p}_1, \mathfrak{p}_1', \dots, \mathfrak{p}_{q-1}'$ of $\mathfrak{P}_1$ in (5) therefore are translations along inner axes of $\mathcal{K}(\mathfrak{T})$. Hence the group $\mathfrak{P}_1$ determined by (5) is contained in $\mathfrak{T}$, thus $\mathcal{K}(\mathfrak{P}_1)$ is contained in $\mathcal{K}(\mathfrak{T})$. In particular, the translation $\mathfrak{c}_1$ then belongs to $\mathfrak{T}$. Its axis $\mathcal{A}(\mathfrak{c}_1)$ therefore is contained in $\mathcal{K}(\mathfrak{T})$, either as an inner axis or as a boundary axis. This holds for all the $\mathfrak{c}_1, \mathfrak{c}_2, \dots, \mathfrak{c}_{N_1}$ considered above. It is thus seen that all the boundary axes of the above group $\mathfrak{K}$ belong to $\mathcal{K}(\mathfrak{T})$. Hence $\mathcal{K}(\mathfrak{K})$ is contained in $\mathcal{K}(\mathfrak{T})$, and $\mathfrak{K}$ is contained in $\mathfrak{T}$, because they both have equal effect with $\mathfrak{F}$ in their respective convex domains.

On the other hand, $\mathfrak{K}$ itself satisfies the five conditions imposed on $\mathfrak{T}$. We call $\mathfrak{K} = \mathfrak{K}(\mathfrak{G})$ *the kernel of* $\mathfrak{F}$ *containing* $\mathfrak{G}$, and thus get the result:

The kernel of $\mathfrak{F}$ *containing a given, finitely generated* $\mathfrak{F}$*-subgroup* $\mathfrak{G}$ *of* $\mathfrak{F}$ *is the intersection of all subgroups* $\mathfrak{T}$ *of* $\mathfrak{F}$ *satisfying the five conditions enumerated above.*

□

This shows that the kernel $\mathfrak{K}(\mathfrak{G})$ of $\mathfrak{F}$ is uniquely determined by $\mathfrak{G}$. The kernel $\mathfrak{K}(\mathfrak{G})$ contains the extended hull $\mathfrak{H}(\mathfrak{G})$; for particular groups $\mathfrak{G}$ the kernel may coincide with the extended hull.

For a finitely generated group $\mathfrak{F}$ the only kernel is $\mathfrak{F}$ itself, independently of the group $\mathfrak{G}$ in question. The concept of kernel therefore essentially concerns groups not satisfying the condition of quasi-compactness.

23.7 Exhaustion by kernels. If $\mathfrak{F}$ does not satisfy the condition of quasi-compactness and contains no reflections, then the exhaustion of $\mathfrak{F}$ can be effected by kernels instead of extended hulls. We follow the pattern of Section 3 and use similar notations.

Corresponding to an $\mathfrak{F}$-subgroup $\mathfrak{F}_{n_1}$ of $\mathfrak{F}$ generated by the section $\mathfrak{f}_1, \dots, \mathfrak{f}_{n_1}$ of the sequence (1) we determine the kernel $\mathfrak{K}_1 = \mathfrak{K}(\mathfrak{F}_{n_1})$ of $\mathfrak{F}$ with convex domain $\mathcal{K}_1 = \mathcal{K}(\mathfrak{K}_1)$ and decompose $\mathcal{K}(\mathfrak{F})$ by $\mathfrak{F}\mathcal{K}_1$. The $\mathcal{A}_{1s}$, $1 \leqq s \leqq m_1$, are now sides of $\mathcal{K}_1$ representing those equivalence classes of sides of $\mathcal{K}_1$ which are inner axes of $\mathfrak{F}$. The regions $\mathcal{B}_{1s}$, $1 \leqq s \leqq m_1$, adjacent to $\mathcal{K}_1$ along $\mathcal{A}_{1s}$ are now inequivalent with respect to $\mathfrak{F}$. Their corresponding subgroups $\mathfrak{B}_{1s}$ of $\mathfrak{F}$ admit no finite generation.

Again, using inner translations $\mathfrak{t}_{1s}$ in each of the $\mathcal{B}_{1s}$ we determine an $\mathfrak{F}_{n_2}$ containing the $\mathfrak{t}_{1s}$ as well as a set of generators of $\mathfrak{K}_1$ and consider the kernel $\mathfrak{K}_2 = \mathfrak{K}(\mathfrak{F}_{n_2})$ of $\mathfrak{F}$ with convex domain $\mathcal{K}_2 = \mathcal{K}(\mathfrak{K}_2)$. The $\mathfrak{A}_{1s}$ are inner axes of $\mathcal{K}_2$, and they are now dividing modulo $\mathfrak{K}_2$, because they are dividing modulo $\mathfrak{F}$. The regions $\mathfrak{C}_{1s}$ adjacent to $\mathcal{K}_1$ along $\mathcal{A}_{1s}$ in the decomposition of $\mathcal{K}_2$ by $\mathfrak{K}_2\mathcal{K}_1$ are now inequivalent, because $\mathfrak{C}_{1s}$ is situated in $\mathcal{B}_{1s}$, and the $\mathcal{B}_{1s}$ are inequivalent. Therefore $m_1' = m_1$, and with this understanding and the substitution of $\mathfrak{K}_1$ and $\mathfrak{K}_2$ for $\mathfrak{H}_1$ and $\mathfrak{H}_2$ the equation (2) holds. Any side of $\mathcal{K}_2$ inside $\mathcal{K}(\mathfrak{F})$ belongs to the boundary of a region equivalent with a certain $\mathfrak{C}_{1s'}$. We select the representatives $\mathcal{A}_{2s}$ of the equivalence classes of these sides of $\mathcal{K}_2$ such that every $\mathcal{A}_{2s}$ is a side of the $\mathfrak{C}_{1s'}$ in question.

If this process is continued indefinitely through a sequence of kernels $\mathfrak{K}_1, \mathfrak{K}_2, \mathfrak{K}_3, \dots$, relating to groups $\mathfrak{F}_{n_1}, \mathfrak{F}_{n_2}, \mathfrak{F}_{n_3}, \dots$, then (3) and (4) hold with $m'_k = m_k$ and the groups $\mathfrak{K}$ substituted for the groups $\mathfrak{H}$. The sequence of kernels $\mathfrak{K}_k$ exhausts $\mathfrak{F}$ in the same sense as does the sequence of extended hulls $\mathfrak{H}_k$ in Section 3, and the consequences drawn therefrom in the Sections 3,4,5 apply.

The additional property of kernels (as compared with extended hulls) that their boundary axes are dividing modulo $\mathfrak{F}$ and correspond to subgroups of the type $\mathfrak{A}(\rightarrow)$ of $\mathfrak{F}$ implies that $\mathfrak{F}$ can, from this exhaustion, be characterized by a free product with amalgamations. The $\mathcal{A}_{ks}$ of Section 3 obviously satisfy the conditions 1), 2) and 4) imposed on the system $\mathcal{S}_\nu$ in §19.6. They also satisfy the condition 3); for any point in the interior of $\mathcal{K}(\mathfrak{F})$ is contained in a certain $\mathcal{K}_{k_0}$, and this $\mathcal{K}_{k_0}$ contains only those $\mathfrak{H}_{k_0}\mathcal{A}_{ks}$ for which $k \leqq k_0$, thus a finite number of equivalence classes. In our present case, where we use kernels, not only these four conditions hold, but also condition 5) of §19.6, because the group of an $\mathcal{A}_{ks}$ is of type $\mathfrak{A}(\rightarrow)$, and condition 6), because $\mathcal{A}_{ks}$ is dividing modulo $\mathfrak{F}$.

Consider the decomposition of $\mathcal{K}(\mathfrak{F})$ by the collection $\mathcal{U} = \bigcup_{k,s} \mathfrak{F}\mathcal{A}_{ks}$, $1 \leqq s \leqq m_k$, $k = 1, 2, \dots$. All lines of $\mathcal{U}$ are inner axes of $\mathfrak{F}$, they are simple and dividing modulo $\mathfrak{F}$, and any two of them are disjoint modulo $\mathfrak{F}$. Let $\mathfrak{A}_{ks}$ denote the subgroup of $\mathfrak{F}$ belonging to $\mathcal{A}_{ks}$, thus of the type $\mathfrak{A}(\rightarrow)$, and let $\mathfrak{a}_{ks}$ be a primary translation of $\mathfrak{A}_{ks}$. One of the regions of decomposition, closed on $\mathcal{K}(\mathfrak{F})$, is $\mathcal{K}_1$. Inside $\mathcal{K}(\mathfrak{F})$ it is bounded by the collection $\bigcup_{s=1}^{m_1} \mathfrak{K}_1\mathcal{A}_{1s}$, and this collection is also its boundary inside $\mathcal{K}_2$.

Consider now a $\mathcal{K}_k$, $k \geqq 2$. Its boundary inside $\mathcal{K}(\mathfrak{F})$ is $\bigcup_{s=1}^{m_k} \mathfrak{K}_k\mathcal{A}_{ks}$. The decomposition of $\mathcal{K}_k$ by $\mathfrak{K}_k\mathcal{K}_{k-1}$ is effected by the subcollection $\bigcup_{s=1}^{m_{k-1}} \mathfrak{K}_k\mathcal{A}_{k-1,s}$ of $\mathcal{U}$. Those regions of this decomposition, closed on $\mathcal{K}(\mathfrak{F})$, which are not equivalent with $\mathcal{K}_{k-1}$ are equivalent with the convex domain of one of the groups $\mathfrak{C}_{k-1,s}$, $1 \leqq s \leqq m_{k-1}$, used in (3). It follows that in the full decomposition of $\mathcal{K}(\mathfrak{F})$ by $\mathcal{U}$ every region of decomposition, closed on $\mathcal{K}(\mathfrak{F})$, is either equivalent with $\mathcal{K}_1$ or with the convex domain of one of the $\mathfrak{C}_{ks}$, $1 \leqq s \leqq m_k$, $k = 1, 2, \dots$. The totality of the convex domain of $\mathfrak{K}_1$ and of the $\mathfrak{C}_{ks}$ is the closure of $\mathcal{K}(\mathfrak{F})$ of the region called $\mathcal{K}$ in §19.6. The group $\mathfrak{C}$ of §19.6 here means $\mathfrak{F}$ itself, because no non-dividing line is used in the decomposition, thus the subcollection $\mathcal{U}_1$ defined in §19.6 is empty. Corresponding to (23) of §19 we get with our present notation

$$\mathfrak{F} = \prod_{k,s}^{*} (\mathfrak{K}_1, \mathfrak{C}_{ks}) \operatorname{am} \mathfrak{A}_{ks}, \quad 1 \leqq s \leqq m_k, \quad 1 \leqq k < \infty. \tag{6}$$

All factors of this free product are finitely generated. They can therefore be presented in the normal form derived in preceding paragraph. One then gets a presentation of $\mathfrak{F}$ using as generators the totality of the generators used in the normal forms for all these factors, which is an infinite set of generators for $\mathfrak{F}$, and as defining relations all relations appearing in the normal forms together with those expressing the amalgamations of the $\mathfrak{A}_{ks}$. If $\mathfrak{a}_{1s}$ is used as generator both in $\mathfrak{K}_1$ and in $\mathfrak{C}_{1s}$, and $\mathfrak{a}_{ks}$ for $k > 1$ is used as generator both in the $\mathfrak{C}_{k-1,s}$ to which it belongs and in $\mathfrak{C}_{ks}$, then the amalgamations are expressed by only listing $\mathfrak{a}_{ks}$ once among the generators of $\mathfrak{F}$.

Remark. The result of the present section renders possible an alternative procedure for the exhaustion of an $\mathfrak{F}$-group containing reflections. The general structure of such a group is given in (12), §20, where $\mathfrak{B}$ denotes a group without reflections. The simple cases where $\mathfrak{B}$ is the identity or quasi-abelian have already been characterized in §20. In the general case, if $\mathfrak{B}$ admits a finite generation, it can be put into the normal form dealt with in §22, and if not, it can be presented as a free product with amalgamations given here in the form (6). If this is inserted in (12), §20, a presentation of $\mathfrak{F}$ by generators and defining relations is obtained.

Chapter V
Isomorphism and homeomorphism

§24 Topological and geometrical isomorphism

24.1 Topological and geometrical isomorphism. In an $\mathfrak{F}$-group $\mathfrak{F}$ a translation whose axis is an inner axis of $\mathfrak{F}$, i.e. is situated in the interior of $\mathcal{K}(\mathfrak{F})$, will be called an *inner translation*, and the end-points of its axis will be called *inner fundamental points of* $\mathfrak{F}$.

We consider an $\mathfrak{F}$-group $\mathfrak{F}$ operating on a disc $\mathcal{D}$ bounded by a unit circle $\mathcal{E}$ and an $\mathfrak{F}$-group $\mathfrak{F}'$ operating on a disc $\mathcal{D}'$ bounded by a unit circle $\mathcal{E}'$. The groups $\mathfrak{F}$ and $\mathfrak{F}'$ are called *topologically isomorphic*, abbreviated t-isomorphic, and we write $\mathfrak{F} \sim \mathfrak{F}'$, if a one-to-one correspondence exists between their elements satisfying three conditions enumerated in the sequel.

Condition 1): The correspondence is an isomorphism.
Condition 2): Inner translations correspond to inner translations.

Before introducing condition 3), we first draw some conclusions from these two conditions. The isomorphic mapping of $\mathfrak{F}$ onto $\mathfrak{F}'$ is denoted by I, and we put $I(\mathfrak{f}) = \mathfrak{f}'$ for any element $\mathfrak{f} \in \mathfrak{F}$. For the inverse mapping then $I^{-1}(\mathfrak{f}') = \mathfrak{f}$.

Let v denote an arbitrary inner fundamental point of $\mathfrak{F}$. Let $\mathfrak{t}$ denote a translation of $\mathfrak{F}$ with positive fundamental point v, and such that $\mathfrak{t}$ is not a power with exponent > 1 of a translation of $\mathfrak{F}$ (it may be the square of a reversed translation of $\mathfrak{F}$). Then v is the positive fundamental point for all $\mathfrak{t}^{\rho}$, $\rho > 0$, and for no other translation of $\mathfrak{F}$. Owing to conditions 1) and 2), $\mathfrak{t}' = I(\mathfrak{t})$ is an inner translation of $\mathfrak{F}'$ and is not a power with exponent > 1 of a translation of $\mathfrak{F}'$. The positive fundamental point v' of $\mathfrak{t}'$ is the positive fundamental point for all $\mathfrak{t}'^{\rho}$, $\rho > 0$, and for no other translation of $\mathfrak{F}'$. Since $\mathfrak{t}' = I(\mathfrak{t})$ implies $\mathfrak{t}'^{\rho} = I(\mathfrak{t}^{\rho})$ (condition 1)), and since the negative fundamental point of $\mathfrak{t}$ is the positive fundamental point of $\mathfrak{t}^{-1}$, the t-isomorphism I establishes a one-to-one correspondence between the sets of inner fundamental points of the two groups.

Let κ denote the mapping of the set $\mathcal{V}$ of inner fundamental points of $\mathfrak{F}$ onto the set $\mathcal{V}'$ of inner fundamental points of $\mathfrak{F}'$.

If the inner translation $\mathfrak{t}$ of $\mathfrak{F}$ has v as its positive fundamental point, and $\mathfrak{f}$ is any element of $\mathfrak{F}$, then $\mathfrak{f}\mathfrak{t}\mathfrak{f}^{-1}$ is an inner translation of $\mathfrak{F}$, and $\mathfrak{f}v$ is its positive fundamental point. Then $I(\mathfrak{t}) = \mathfrak{t}'$ has $\kappa v = v'$ as its positive fundamental point, and the positive fundamental point of $I(\mathfrak{f}\mathfrak{t}\mathfrak{f}^{-1}) = \mathfrak{f}'\mathfrak{t}'\mathfrak{f}'^{-1}$ is $\mathfrak{f}'v' = \mathfrak{f}'\kappa v$. This is thus $\kappa\mathfrak{f}v$. Hence the mapping function $\kappa\colon \mathcal{V} \to \mathcal{V}'$ satisfies the functional equation

$$\kappa\mathfrak{f} = I(\mathfrak{f})\kappa = \mathfrak{f}'\kappa \tag{1}$$

in the whole point set $\mathcal{V}$.

If we let the inner axis $\mathcal{A}(\mathfrak{t})$ correspond to the inner axis $\mathcal{A}(\mathfrak{t}')$, a one-to-one correspondence between the inner axes of the two groups results. Bearing in mind that a boundary axis is not cut by any other axis we now introduce

Condition 3): Any two concurrent axes correspond to two concurrent axes.

The definition of t-isomorphisms given by these three conditions is symmetric between the two groups. If also $\mathfrak{F}$ and $\mathfrak{F}''$ are t-isomorphic, then $\mathfrak{F}'$ and $\mathfrak{F}''$ are. Hence all $\mathfrak{F}$-groups fall into classes of t-isomorphic groups.

Let any four different inner fundamental points of $\mathfrak{F}$ be given. We denote them by v_1, v_2, v_3, v_4 in such a way that the pair denoted by v_1, v_2 separates the pair denoted by v_3, v_4 on $\mathcal{E}$. Let v_1', v_2', v_3', v_4' be their respective images by κ. It is our intension to prove that the pair v_1', v_2' separates the pair v_3', v_4' on $\mathcal{E}'$. Let $\mathcal{I}_1, \mathcal{I}_2, \mathcal{I}_3, \mathcal{I}_4$ be disjoint intervals of $\mathcal{E}$ containing v_1, v_2, v_3, v_4 respectively, and $\mathcal{I}_1', \mathcal{I}_2', \mathcal{I}_3', \mathcal{I}_4'$ disjoint intervals of $\mathcal{E}'$ containing v_1', v_2', v_3', v_4' respectively. Let $\mathfrak{t}_1, \mathfrak{t}_2$ be translations with v_1, v_2 as positive fundamental points respectively, and $\mathfrak{t}_1', \mathfrak{t}_2'$ their images by I; like $\mathfrak{t}_1, \mathfrak{t}_2$ they are inner translations (condition 2)). For all sufficiently large m the element $\mathfrak{g} = \mathfrak{t}_1^m \mathfrak{t}_2^{-m}$ is a translation with fundamental points $v(\mathfrak{g})$ and $u(\mathfrak{g})$ in $\mathcal{I}_1$ and $\mathcal{I}_2$ respectively (§13.3). This translation $\mathfrak{g}$ is inner, because one of the two intervals of $\mathcal{E}$ joining $v(\mathfrak{g})$ and $u(\mathfrak{g})$ contains the fundamental point v_3, and the other contains v_4. Therefore also $\mathfrak{g}' = I(\mathfrak{g}) = \mathfrak{t}_1'^m \mathfrak{t}_2'^{-m}$ is an inner translation (condition 2)), and if m is chosen sufficiently large, $v(\mathfrak{g}')$ is in $\mathcal{I}_1'$ and $u(\mathfrak{g}')$ in $\mathcal{I}_2'$. In the same way we construct an inner translation $\mathfrak{h}$ of $\mathfrak{F}$ having its fundamental points in $\mathcal{I}_3$ and $\mathcal{I}_4$ and such that the corresponding $\mathfrak{h}'$ has its fundamental points in $\mathcal{I}_3'$ and $\mathcal{I}_4'$. The axes of $\mathfrak{g}$ and $\mathfrak{h}$ intersect, since the pair of intervals $\mathcal{I}_1, \mathcal{I}_2$ separates the pair $\mathcal{I}_3, \mathcal{I}_4$ on $\mathcal{E}$. Hence the axes of $\mathfrak{g}'$ and $\mathfrak{h}'$ intersect (condition 3)). Hence $\mathcal{I}_1', \mathcal{I}_2'$ separate $\mathcal{I}_3', \mathcal{I}_4'$ on $\mathcal{E}'$, and thus also v_1', v_2' separate v_3', v_4' on $\mathcal{E}'$. We have thus proved:

κ maps any four different inner fundamental points of $\mathfrak{F}$ onto four different inner fundamental points of $\mathfrak{F}'$ preserving their cyclical order on $\mathcal{E}$ and $\mathcal{E}'$. □

The set $\mathcal{V}$ of inner fundamental points of $\mathfrak{F}$ is enumerable, say v_r, $r = 1, 2, \dots$, and then $\kappa v_r = v_r'$ is an enumeration of $\mathcal{V}'$. The points v_1, v_2, v_3 in this order determine a sense of circulation on $\mathcal{E}$, and v_1', v_2', v_3' a sense on $\mathcal{E}'$. If then v_4 is inserted in the proper of the three intervals determined on $\mathcal{E}$ by v_1, v_2, v_3, it has been shown that v_4' lies in the corresponding intervals on $\mathcal{E}'$. Inserting in turn $v_5, v_6, \dots$, it follows:

The mapping κ of $\mathcal{V}$ onto $\mathcal{V}'$ preserves the cyclical order of these point sets on their respective circles. □

If $\mathfrak{f}$ is any element of $\mathfrak{F}$, the points $\mathfrak{f}v_1, \mathfrak{f}v_2, \mathfrak{f}v_3$ are inner fundamental points and determine the same sense on $\mathcal{E}$ as v_1, v_2, v_3, if $\mathfrak{f}$ is a motion, whereas the sense is reversed, if $\mathfrak{f}$ is a reversion. In virtue of (1) the points $\mathfrak{f}v_1, \mathfrak{f}v_2, \mathfrak{f}v_3$ are by κ mapped into $\mathfrak{f}'v_1', \mathfrak{f}'v_2', \mathfrak{f}'v_3'$. From the order preserving property of κ it then follows that a motion of $\mathfrak{F}$ corresponds to a motion of $\mathfrak{F}'$, and a reversion to a reversion. Reflections, being reversions of order 2, then correspond to reflections and reversed translations, being of infinite order, to reversed translations. Since the square of a reversed translation is an inner translation, the fundamental points of reversed translations correspond by κ.

Let $\mathfrak{c}$ be a primary rotation of order ν, thus with an angle of rotation equal to $\frac{2\pi}{\nu}$. Then $v_1, \mathfrak{c}v_1, \ldots, \mathfrak{c}^{\nu-1}v_1$ lie in this cyclical order on $\mathcal{E}$. Hence $v_1', \mathfrak{c}'v_1', \ldots, \mathfrak{c}'^{\nu-1}v_1'$ lie in this cyclical order on $\mathcal{E}'$. The element $\mathfrak{c}'$ has its order equal to ν (condition 1)) and is a motion. It is not the power of an element of order higher than ν, since $\mathfrak{c}$ is not. It is thus a rotation of order ν, and it is primary, because its angle of rotation is $\frac{2\pi}{\nu}$.

Let $\mathfrak{t}$ be an inner translation of $\mathfrak{F}$ and $\mathfrak{s}$ an element of order 2 for which $\mathfrak{s}\mathfrak{t}\mathfrak{s} = \mathfrak{t}^{-1}$. Then $\mathfrak{s}'$ is an element of order 2 in $\mathfrak{F}'$, and $\mathfrak{s}'\mathfrak{t}'\mathfrak{s}' = \mathfrak{t}'^{-1}$. The element $\mathfrak{s}$ is either a half-turn with centre on the axis of $\mathfrak{t}$, or the reflection in a reflection line orthogonal to that axis. Since rotations correspond to rotations and reflections to reflections, it follows that $\mathfrak{s}'$ is accordingly a half-turn with centre on the axis of $\mathfrak{t}'$, or the reflection in a reflection line orthogonal to that axis.

If $\mathfrak{s}$ is a reflection in the axis of $\mathfrak{t}$, and thus $\mathfrak{s}\mathfrak{t}\mathfrak{s} = \mathfrak{t}$, then $\mathfrak{s}'$ is a reflection, and $\mathfrak{s}'\mathfrak{t}'\mathfrak{s}' = \mathfrak{t}'$. Hence the reflection line of $\mathfrak{s}'$ coincides with the axis of $\mathfrak{t}'$.

The subgroup $\mathfrak{A}_{\mathcal{A}}$ of $\mathfrak{F}$ belonging to an inner axis $\mathcal{A}$ of $\mathfrak{F}$ takes one of the seven forms

$$\mathfrak{A}(\rightarrow), \quad \mathfrak{A}(\|), \quad \mathfrak{A}(-\rightarrow), \quad \mathfrak{A}(-\|), \quad \mathfrak{A}(\cdot\cdot), \quad \mathfrak{A}(\cdot|), \quad \mathfrak{A}(\rightleftharpoons),$$

and it is inferred from the above considerations that the subgroup $\mathfrak{A}_{\mathcal{A}'}$ of $\mathfrak{F}'$ belonging to the corresponding inner axis $\mathcal{A}'$ of $\mathfrak{F}'$ is of the same type $\mathfrak{A}_{\mathcal{A}}$. If the inner axis $\mathcal{A}$ of $\mathfrak{F}$ with end-points u and v is simple modulo $\mathfrak{F}$ then for no element $\mathfrak{f} \in \mathfrak{F}$ does the pair $\mathfrak{f}u$, $\mathfrak{f}v$ separate the pair u, v on $\mathcal{E}$. Hence by applying κ the pair $\mathfrak{f}'u'$, $\mathfrak{f}'v'$ does not separate the pair u', v' on $\mathcal{E}'$. It follows that the corresponding inner axis $\mathcal{A}'$ of $\mathfrak{F}'$ is simple modulo $\mathfrak{F}'$. If the inner axes $\mathcal{A}_1$ and $\mathcal{A}_2$ of $\mathfrak{F}$ are disjoint modulo $\mathfrak{F}$, it follows in the same way that $\mathcal{A}_1'$ and $\mathcal{A}_2'$ are disjoint modulo $\mathfrak{F}'$.

So far the correspondence by I has been determined for all elements of $\mathfrak{F}$ except boundary translations and limit-rotations. These two kinds of elements cannot be distinguished under the conditions of t-isomorphism. The totality of boundary translations and limit-rotations of one group corresponds by the t-isomorphism to the totality of boundary translations and limit-rotations of the other group. Moreover, a primary element of this kind (boundary translation or limit-rotation) corresponds to a primary element, because it is an element which is not a power with exponent > 1, and that property is preserved under an isomorphism. Also, since reflections correspond to reflections, and since a reflection of the group $\mathfrak{A}(\|)$ or $\mathfrak{A}(\wedge)$ transforms the boundary translation or limit-rotation into its inverse, the totality of boundary translations in subgroups of the type $\mathfrak{A}(\|)$ and of limit-rotations in subgroups of the type $\mathfrak{A}(\wedge)$ in one group correspond to the same totality in the other group. Then also the totality of boundary translations in subgroups of the type $\mathfrak{A}(\rightarrow)$ and of limit-rotations in subgroups of the type $\mathfrak{A}(\hookrightarrow)$ in one group correspond to the same totality in the other group. If $\mathfrak{a}$ is a boundary translation in $\mathfrak{F}$ and $\mathfrak{a}'$ is the corresponding boundary translation or limit-rotation in $\mathfrak{F}'$, then $\mathfrak{f}\mathfrak{a}\mathfrak{f}^{-1}$ corresponds to $\mathfrak{f}'\mathfrak{a}'\mathfrak{f}'^{-1}$ for all $\mathfrak{f} \in \mathfrak{F}$ and $I(\mathfrak{f}) = \mathfrak{f}'$. Hence if a boundary translation in one group corresponds to a boundary translation in the other, then the same holds for all its conjugates, and likewise if it corresponds to a limit-rotation in the other group.

A topological isomorphism in which every boundary translation in one group, if any, corresponds to a boundary translation in the other group, and thus also every limit-rotation in one group, if any, corresponds to a limit-rotation in the other, is called a *geometrical isomorphism*, abbreviated g-isomorphism, and we write $\mathfrak{F} \simeq \mathfrak{F}'$. Thus for a g-isomorphism every element of one group corresponds to an element of the same type (translation, reversed translation, rotation, limit-rotation, reflection) in the other group. The mapping κ, which for a t-isomorphism is defined in the set $\mathcal{V}$ of inner fundamental points, extends for a g-isomorphism to the whole of the set $\mathcal{G}(\mathfrak{F})$ of fundamental points.

24.2 Correspondence on $\mathcal{E}$ and $\mathcal{E}'$. If $\mathfrak{F} \sim \mathfrak{F}'$, $\mathcal{K}(\mathfrak{F}) = \mathcal{D}$ and $\mathcal{K}(\mathfrak{F}') = \mathcal{D}'$, then all fundamental points are inner fundamental points in both groups, and $\mathcal{V}$ and $\mathcal{V}'$ are dense on $\mathcal{E}$ and $\mathcal{E}'$ respectively. It then follows from the order preserving character of the mapping κ defined in $\mathcal{V}$ that κ can be extended to a topological mapping $\kappa\mathcal{E}$ of $\mathcal{E}$ onto $\mathcal{E}'$ with the inverse $\kappa^{-1}\mathcal{E}' = \mathcal{E}$. For every point of $\mathcal{E}$ which is not a fundamental point can be determined as a cut in the cyclical set $\mathcal{V}$ on $\mathcal{E}$, and the corresponding cut in the cyclical set $\mathcal{V}'$ on $\mathcal{E}'$ determines a point on $\mathcal{E}'$ which is the image of the point on $\mathcal{E}$ by the extended mapping κ. – In this case the isomorphism is a g-isomorphism.

If $\mathcal{K}(\mathfrak{F})$ is not the whole of $\mathcal{D}$, then every interval of discontinuity of $\mathfrak{F}$ closed on $\mathcal{E}$ determines a cut in $\mathcal{V}$, and this cut in $\mathcal{V}$ corresponds by κ to a cut in $\mathcal{V}'$. This determines either a single point of $\mathcal{E}'$ or an interval of discontinuity on $\mathcal{E}'$ closed on $\mathcal{E}'$, the latter of course only if also $\mathcal{K}'(\mathfrak{F}')$ is not the whole of $\mathcal{D}'$. In particular, for a cut effected by a periodic interval of discontinuity on $\mathcal{E}$ (closed on $\mathcal{E}$) the mapping κ determines a cut in $\mathcal{V}'$ effected either by a periodic interval of discontinuity (closed on $\mathcal{E}'$) or by a limit-centre on $\mathcal{E}'$. The same holds the other way for the mapping κ^{-1}.

If $\mathfrak{F}$ is finitely generated, then also $\mathfrak{F}'$ is (condition 1)). In this case all intervals of discontinuity in both groups are periodic. Hence only the correspondence of periodic intervals with either periodic intervals or limit-centres comes into play. Except for these any two corresponding cuts determine in a unique way corresponding points on $\mathcal{E}$ and $\mathcal{E}'$.

24.3 Correspondence of extended hulls and of ends of the convex domain. Assume that $\mathfrak{F}$ and thus also $\mathfrak{F}'$, admits no finite generation. Let $\mathfrak{G}$ be a finitely generated $\mathfrak{F}$-subgroup of $\mathfrak{F}$, and $\mathfrak{G}' = I(\mathfrak{G})$ its image in $\mathfrak{F}'$ by I. Since an $\mathfrak{F}$-group is not isomorphic with a quasi-abelian group, also $\mathfrak{G}'$ is an $\mathfrak{F}$-group. Let $\mathfrak{F}_0$ be a subgroup of $\mathfrak{F}$ satisfying the five conditions imposed on the $\mathfrak{F}_0$ of §23.2. Consider the image $\mathfrak{F}_0'$ of $\mathfrak{F}_0$ by I. It is finitely generated and contains $\mathfrak{G}'$. Those sides of $\mathcal{K}(\mathfrak{F}_0)$ which are not sides of $\mathcal{K}(\mathfrak{F})$ are inner axes of $\mathfrak{F}$ and thus correspond to inner axes of $\mathfrak{F}'$. Let $\mathcal{A}$ be such a side of $\mathcal{K}(\mathfrak{F}_0)$ and $\mathcal{A}'$ the corresponding axis of $\mathfrak{F}'$. The group $\mathfrak{A}_{\mathcal{A}}$ of $\mathfrak{F}$, which is either $\mathfrak{A}(\rightarrow)$ or $\mathfrak{A}(\|)$, is then mapped onto the group $\mathfrak{A}_{\mathcal{A}'}$ of $\mathfrak{F}'$ by I, and $\mathfrak{A}_{\mathcal{A}'}$ is $\mathfrak{A}(\rightarrow)$ or $\mathfrak{A}(\|)$ respectively (Section 1). Such sides $\mathcal{A}$ of $\mathcal{K}(\mathfrak{F}_0)$ are simple modulo $\mathfrak{F}$ since $\mathfrak{F}$ and $\mathfrak{F}_0$ have equal effect in $\mathcal{K}(\mathfrak{F}_0)$. Therefore also the corresponding $\mathcal{A}'$ are simple modulo $\mathfrak{F}'$ (Section 1). If $\mathcal{A}_1$ and $\mathcal{A}_2$ are non-equivalent sides of $\mathcal{K}(\mathfrak{F}_0)$ and inner axes of $\mathfrak{F}$, then they are disjoint modulo $\mathfrak{F}$, and it follows from Section 1 that $\mathcal{A}_1'$ and $\mathcal{A}_2'$ are disjoint modulo $\mathfrak{F}'$. Thus if $\mathcal{A}_r$, $1 \leqq r \leqq \omega$, represent the different equivalence classes of sides of $\mathcal{K}(\mathfrak{F}_0)$ which are inner axes of $\mathfrak{F}$, then the corresponding $\mathcal{A}_r'$ in $\mathfrak{F}'$ are all simple modulo $\mathfrak{F}'$ and mutually disjoint modulo $\mathfrak{F}'$, and they represent the equivalence classes of those boundary axes of $\mathfrak{F}_0'$ which are not boundary axes of $\mathfrak{F}'$. We now decompose $\mathcal{D}$ by the collection $\mathcal{U} = \bigcup_{r=1}^{\omega} \mathfrak{F}\mathcal{A}_r$ and $\mathcal{D}'$ by the corresponding collection $\mathcal{U}' = \bigcup_{r=1}^{\omega} \mathfrak{F}'\mathcal{A}_r'$. The axes of these two collections are in a one-to-one correspondence based on I, and all are inner axes of $\mathfrak{F}$ and $\mathfrak{F}'$ respectively. Let Ω denote an arbitrary region of the decomposition of $\mathcal{D}$ by $\mathcal{U}$, and $\mathcal{U}_\Omega$ the subcollection of $\mathcal{U}$ bounding Ω. Then two arbitrary lines of $\mathcal{U}_\Omega$ are not separated by any line of $\mathcal{U}$. Owing to the preservation of the cyclical order of $\mathcal{V}$ by the mapping κ it is inferred that two arbitrary of the lines of the subcollection $\mathcal{U}_\Omega'$ of $\mathcal{U}'$ corresponding to $\mathcal{U}_\Omega$ are not separated by any line of $\mathcal{U}'$. Hence $\mathcal{U}_\Omega'$ is part of the boundary of a certain Ω' of the decomposition of $\mathcal{D}'$ by $\mathcal{U}'$. An arbitrary side $\mathcal{A}'$ of Ω' is not separated from any line of $\mathcal{U}_\Omega'$ by a line of $\mathcal{U}'$, and it is inferred in the same way that the line $\mathcal{A}$ corresponding to it by I^{-1} is a side of Ω. One therefore gets a one-to-one correspondence of the regions Ω of the decomposition of $\mathcal{D}$ by $\mathcal{U}$ and the regions Ω' of the decomposition of $\mathcal{D}'$ by $\mathcal{U}'$, this correspondence of regions being determined by the correspondence of the totality of their boundaries derived from the isomorphism. This implies that, if $\mathfrak{F}_\Omega$ is the subgroup of $\mathfrak{F}$ carrying Ω into itself, then $I(\mathfrak{F}_\Omega)$ is the subgroup of $\mathfrak{F}'$ carrying Ω' into itself. Let Ω_0 denote the region in $\mathcal{D}$ which, if closed on $\mathcal{D}$, contains $\mathcal{K}(\mathfrak{F}_0)$. If $\mathcal{K}(\mathfrak{F}_0)$ contains no side of $\mathcal{K}(\mathfrak{F})$, then $\tilde{\Omega}_0$ is $\mathcal{K}(\mathfrak{F}_0)$ itself; otherwise $\tilde{\Omega}_0$ contains certain half-planes outside $\mathcal{K}(\mathfrak{F}_0)$. The subgroup of $\mathfrak{F}$ belonging to Ω_0 is $\mathfrak{F}_0$. The corresponding Ω_0' is then carried into itself by $I(\mathfrak{F}_0) = \mathfrak{F}_0'$ and by no other element of $\mathfrak{F}'$. The closure $\tilde{\Omega}_0'$ of Ω_0' on $\mathcal{D}'$ thus contains $\mathcal{K}(\mathfrak{F}_0')$, and all sides of Ω_0' are such boundary axes of $\mathcal{K}(\mathfrak{F}_0')$ which are inner axes of $\mathfrak{F}'$. In consequence of Section 1 the subgroup of $\mathfrak{F}'$ belonging to any side of Ω_0' is either of the type $\mathfrak{A}(\rightarrow)$ or of the type $\mathfrak{A}(\|)$, because that is the case for Ω_0. Let Ω_r denote the region of decomposition adjacent to Ω_0 along $\mathcal{A}_r$, and $\mathfrak{B}_r$ its corresponding subgroup of $\mathfrak{F}$. Then Ω_r' is adjacent to Ω_0' along $\mathcal{A}_r'$, and its corresponding subgroup of $\mathfrak{F}'$ is $\mathfrak{B}_r' = I(\mathfrak{B}_r)$. Since $\mathfrak{B}_r$ admits no finite generation, the same holds for $\mathfrak{B}_r'$. Thus no restgroup of $\mathfrak{F}_0'$ in $\mathfrak{F}'$ admits a finite generation. Hence Ω_0' is not equivalent with any of its adjacent regions. $\tilde{\Omega}_0'$ consists of $\mathcal{K}(\mathfrak{F}_0')$ and such half-planes as are adjacent to $\mathcal{K}(\mathfrak{F}_0')$ along boundary axes of $\mathcal{K}(\mathfrak{F}_0')$ which are also boundary axes of $\mathfrak{F}'$, if any. All sides of $\mathcal{K}(\mathfrak{F}_0')$ belong to subgroups of the type $\mathfrak{A}(\rightarrow)$ of $\mathfrak{A}(\|)$. This fact together with the fact that $\mathcal{K}(\mathfrak{F}_0')$ is not equivalent with any of its adjacent regions implies that $\mathfrak{F}_0'$ and $\mathfrak{F}'$ have equal effect in the whole of $\mathcal{K}(\mathfrak{F}_0')$ (including its boundaries). In all, it has thus been shown that corresponding to any subgroup $\mathfrak{F}_0$ of $\mathfrak{F}$ satisfying the five conditions enumerated in §23.2 the subgroup $\mathfrak{F}_0' = I(\mathfrak{F}_0)$ of $\mathfrak{F}'$ satisfies the same five conditions. Hence the totality of all subgroups satisfying these five conditions in the two groups correspond by the isomorphism, and therefore also their intersection corresponds. One thus gets the relation $I(\mathfrak{H}(\mathfrak{G})) = \mathfrak{H}(\mathfrak{G}') = \mathfrak{H}(I(\mathfrak{G}))$, thus the statement:

The extended hull of $\mathfrak{G}$ in $\mathfrak{F}$ corresponds to the extended hull of $I(\mathfrak{G})$ in $\mathfrak{F}'$. □

We now exhaust $\mathfrak{F}$ by extended hulls in accordance with §23.3. According to (1), §23, and putting $I(\mathfrak{f}) = \mathfrak{f}'$, the sequence $\mathfrak{f}_1', \mathfrak{f}_2', \mathfrak{f}_3', \dots$ is an enumeration of all elements of $\mathfrak{F}'$. Also $I(\mathfrak{F}_n) = \mathfrak{F}_n'$. It is then inferred that the sequence $\mathfrak{H}(\mathfrak{F}_{n_k}')$ exhausts $\mathfrak{F}'$.

It was seen that $I(\mathfrak{H}(\mathfrak{F}_{n_k})) = \mathfrak{H}(\mathfrak{F}_{n_k}')$. If we compare the convex domains $\mathcal{K}_k$ and $\mathcal{K}_k'$ of $\mathfrak{H}(\mathfrak{F}_{n_k})$ and $\mathfrak{H}(\mathfrak{F}_{n_k}')$, those sides of $\mathcal{K}_k$ which are inner axes of $\mathfrak{F}$ correspond to sides of $\mathcal{K}_k'$ which are inner axes of $\mathfrak{F}'$, and conversely. A side of $\mathcal{K}_k$ which is also a side of $\mathcal{K}(\mathfrak{F})$ corresponds either to a side of $\mathcal{K}_k'$ which is also a side of $\mathcal{K}(\mathfrak{F}')$, or to a limit-centre of $\mathfrak{F}'$, and conversely. Any such correspondence between a boundary axis or limit-centre of $\mathfrak{F}$ with either a boundary axis or a limit-centre of $\mathfrak{F}'$ becomes apparent for some value k_0 and then remains in force in the relation between $\mathcal{K}_k$ and $\mathcal{K}_k'$ for all $k \geqq k_0$.

Consider now the intervals $\mathcal{I}(k)$ dealt with in §23.5. An interval of discontinuity $\mathcal{I}(k)$ of $\mathcal{K}_k$ which is not an interval of discontinuity of $\mathfrak{F}$ is bounded by two inner fundamental points of $\mathfrak{F}$. It corresponds to an interval of discontinuity $\mathcal{I}'(k)$ of $\mathcal{K}_k'$ bounded by the two corresponding inner fundamental points of $\mathfrak{F}'$. If the interval $\mathcal{I}(k+1)$ of $\mathcal{K}_{k+1}$ is contained in $\mathcal{I}(k)$, its pair of end-points does not separate the pair of end-points of $\mathcal{I}(k)$. They are mapped by κ on the end-points of $\mathcal{I}'(k+1)$ and these do not separate the pair of end-points of $\mathcal{I}'(k)$. They are contained in $\mathcal{I}'(k)$, because the arc between them is outside $\mathcal{K}_k'$. Hence to any sequence of intervals $\mathcal{I}(k)$, $k = 1, 2, \dots$, defining an end of $\mathcal{K}(\mathfrak{F})$ the correspondence κ determines a sequence $\mathcal{I}'(k)$ defining an end of $\mathcal{K}(\mathfrak{F}')$, and vice versa. One thus gets a uniquely determined correspondence between the ends of $\mathcal{K}(\mathfrak{F})$ and those of $\mathcal{K}(\mathfrak{F}')$, and since conjugate elements of $\mathfrak{F}$ correspond to conjugate elements of $\mathfrak{F}'$, there is also a uniquely determined correspondence between the ends of $\mathcal{K}(\mathfrak{F})$ mod $\mathfrak{F}$ and those of $\mathcal{K}(\mathfrak{F}')$ mod $\mathfrak{F}'$. But is has to be observed that an end of $\mathcal{K}(\mathfrak{F})$ consisting of a closed limit side of $\mathcal{K}(\mathfrak{F})$ may well correspond to an end of $\mathcal{K}(\mathfrak{F}')$ consisting of a single limit point, and vice versa.

24.4 Correspondence of reflection lines. Let $\mathfrak{s}$ be a reflection in $\mathfrak{F}$ and $\mathcal{L}$ its reflection line, $\mathcal{D}_1$ and $\mathcal{D}_2$ the half-planes determined in $\mathcal{D}$ by $\mathcal{L}$, and $\mathcal{I}_1$ and $\mathcal{I}_2$ the open intervals on $\mathcal{E}$ on the boundary of $\mathcal{D}_1$ and $\mathcal{D}_2$ respectively. Since $\mathcal{L}$ intersects $\mathcal{K}(\mathfrak{F})$, there are inner fundamental points both in $\mathcal{I}_1$ and in $\mathcal{I}_2$. Let $\mathcal{V}_1$ and $\mathcal{V}_2$ denote the subsets of $\mathcal{V}$ situated in $\mathcal{I}_1$ and $\mathcal{I}_2$ respectively. They are interchanged by $\mathfrak{s}$.

$I(\mathfrak{s}) = \mathfrak{s}'$ is a reflection in $\mathfrak{F}'$. Let $\mathcal{L}'$ denote its reflection line. If v_1 denotes a point of $\mathcal{V}_1$, then $\mathfrak{s}v_1$ belongs to $\mathcal{V}_2$, and $\kappa\mathfrak{s}v_1 = \mathfrak{s}'\kappa v_1$ in virtue of (1). Hence κv_1 and $\kappa(\mathfrak{s}v_1)$ are interchanged by $\mathfrak{s}'$. If v_2 is another point of $\mathcal{V}_1$, then the pair v_1, v_2 is not separated by the pair $\mathfrak{s}v_1, \mathfrak{s}v_2$. Hence the pair $\kappa v_1, \kappa v_2$ is not separated by the pair

$\kappa\mathfrak{s}v_1, \kappa\mathfrak{s}v_2$, which is the pair $\mathfrak{s}'\kappa v_1, \mathfrak{s}'\kappa v_2$. Hence if v_1 and v_2 are on the same side of $\mathcal{L}$, then κv_1 and κv_2 are on the same side of $\mathcal{L}'$. It follows that $\kappa\mathcal{V}_1$ is contained in one of the intervals determined on $\mathcal{E}'$ by the end-points of $\mathcal{L}'$. Let this interval be called $\mathcal{I}_1'$ and the half-plane which it bounds in $\mathcal{D}'$ be called $\mathcal{D}_1'$. We then call $\mathcal{D}_1$ and $\mathcal{D}_1'$ *corresponding sides* of $\mathcal{L}$ and $\mathcal{L}'$, and likewise for the other two half-planes $\mathcal{D}_2$ and $\mathcal{D}_2'$ bounded on $\mathcal{E}$ and $\mathcal{E}'$ by $\mathcal{I}_2$ and $\mathcal{I}_2'$ respectively. Then $\kappa\mathcal{V}_2$ is contained in $\mathcal{I}_2'$, and $\kappa\mathcal{V}_1$ and $\kappa\mathcal{V}_2$ are interchanged by $\mathfrak{s}'$.

If ℓ_1 and ℓ_2 denote the end-points of $\mathcal{L}$ on $\mathcal{E}$, and $\mathcal{I}_1$ is traversed from ℓ_1 to ℓ_2, one passes the set $\mathcal{V}_1$ in a certain order. Then κ determines a corresponding order of $\kappa\mathcal{V}_1$, and if the end-points of $\mathcal{L}'$ on $\mathcal{E}'$ are called ℓ_1' and ℓ_2' in such a way, that this order of $\kappa\mathcal{V}_1$ corresponds to traversing $\mathcal{I}_1'$ from ℓ_1' to ℓ_2', then we may say that ℓ_1 corresponds to ℓ_1' and ℓ_2 to ℓ_2'. In the particular case where $\mathcal{L}$ is an axis, say of the translation $\mathfrak{t}$, then $\mathfrak{s}$ and $\mathfrak{t}$ are permutable, hence $\mathfrak{s}'$ and $\mathfrak{t}'$ are permutable, and $\mathcal{L}'$ is the axis of $\mathfrak{t}'$. In this case all four points are inner fundamental points of their respective groups, and the correspondence now envisaged is the correspondence $\ell_1' = \kappa\ell_1, \ell_2' = \kappa\ell_2$ already established in $\mathcal{V}$. If $\mathcal{L}$ is not an axis, then neither ℓ_1 nor ℓ_2 is a fundamental point (§11.4).

We thus get a one-to-one correspondence between the *lines* of reflection in one group and those in the other, and we have established a correspondence between the *end-points* of a reflection line in one group with the end-points of the corresponding reflection line in the other. In the case of reflection lines which are axes this is the correspondence κ defined in $\mathcal{V}$ and is therefore one-to-one. For reflection lines which are not axes a closer examination is required.

Consider two different reflection lines $\mathcal{L}$ and $\mathcal{L}^*$ of $\mathfrak{F}$ both of which are not axes of $\mathfrak{F}$, and assume that the end-point ℓ of $\mathcal{L}$ and the end-point ℓ^* of $\mathcal{L}^*$ determine the same cut in $\mathcal{V}$. This occurs either if ℓ and ℓ^* coincide, thus in a limit-centre of $\mathfrak{F}$ with corresponding group $\mathfrak{A}(\wedge)$, or if $\ell \neq \ell^*$ and one of the arcs $\ell\ell^*$ of $\mathcal{E}$ contains no point of $\mathcal{V}$, thus if $\mathcal{L}$ and $\mathcal{L}^*$ are normals of a boundary axis of $\mathfrak{F}$ with corresponding group $\mathfrak{A}(\|)$. If $\mathfrak{s}$ and $\mathfrak{s}^*$ denote reflections in $\mathcal{L}$ and $\mathcal{L}^*$ respectively, then $\mathfrak{s}^*\mathfrak{s}$ is a limit-rotation or a boundary translation in the two cases respectively. As pointed out in Section 1 then $\mathfrak{s}^{*\prime}\mathfrak{s}'$ is either a limit-rotation or a boundary translation, but not necessarily the same kind as $\mathfrak{s}^*\mathfrak{s}$. The end-points ℓ' of $\mathcal{L}'$ and $\ell^{*\prime}$ of $\mathcal{L}^{*\prime}$ which correspond to ℓ and ℓ^* in the way already defined determine the same cut in $\mathcal{V}'$, namely the one corresponding to the cut in $\mathcal{V}$ determined by ℓ and ℓ^*.

If $\mathcal{L}$ intersects a boundary axis of $\mathfrak{F}$, we may select $\mathcal{L}^*$ such that $\mathfrak{s}^*\mathfrak{s}$ is a primary translation $\mathfrak{a}$ belonging to that axis. This is the case if $\mathcal{L}$ and $\mathcal{L}^*$ are consecutive reflection lines cutting the axis. Then also $\mathfrak{s}^{*\prime}\mathfrak{s}' = \mathfrak{a}'$ is primary, namely either a primary boundary translation or a primary limit-rotation. Therefore $\mathcal{L}'$ and $\mathcal{L}^{*\prime}$ are consecutive reflection lines cutting the axis of $\mathfrak{a}'$ or ending in the limit-centre of $\mathfrak{a}'$, as the case may be. In the first case it follows that the order of the end-points of reflection lines contained in a periodic interval of discontinuity of $\mathfrak{F}$ is preserved in their correspondence with the end-points of reflection lines in the corresponding interval of discontinuity of $\mathfrak{F}'$.

Hence, in all, the mapping κ originally defined as a one-to-one and order preserving mapping of $\mathcal{V}$ onto $\mathcal{V}'$ may be extended to the set $\mathcal{U}$ of end-points of reflection lines of $\mathfrak{F}$, and $\kappa\colon \mathcal{V} \cup \mathcal{U} \to \mathcal{V}' \cup \mathcal{U}'$ is still order preserving, bearing in mind however that the infinite subset of points of $\mathcal{U}$ in a periodic interval of discontinuity of $\mathfrak{F}$ with corresponding axis of the type $\mathfrak{A}(\|)$ may be mapped into a single limit-centre of $\mathfrak{F}'$, and similarly for the inverse mapping κ^{-1}.

If $\mathfrak{f}$ is any element of $\mathfrak{F}$, then $\mathfrak{f}\mathcal{L}$ is the reflection line of the reflection $\mathfrak{f}\mathfrak{s}\mathfrak{f}^{-1}$. This element of $\mathfrak{F}$ corresponds to the element $\mathfrak{f}'\mathfrak{s}'\mathfrak{f}'^{-1}$ of $\mathfrak{F}'$, and this is the reflection in $\mathfrak{f}'\mathcal{L}'$. Let ℓ be one of the end-points of $\mathcal{L}$ and $\kappa\ell = \ell'$ the corresponding end-point of $\mathcal{L}'$. Thus ℓ and ℓ' are corresponding points of $\mathcal{V}$ and $\mathcal{V}'$, if $\mathcal{L}$ is an axis, and corresponding cuts in $\mathcal{V}$ and $\mathcal{V}'$, if $\mathcal{L}$ is not an axis. In the first case one gets from (1) that $\kappa\mathfrak{f}\ell = \mathfrak{f}'\kappa\ell = \mathfrak{f}'\ell'$, thus $\mathfrak{f}\ell$ and $\mathfrak{f}'\ell'$ are corresponding points of $\mathcal{V}$ and $\mathcal{V}'$. In the second case, since ℓ and ℓ' determine corresponding cuts in $\mathcal{V}$ and $\mathcal{V}'$ and the functional equation $\kappa\mathfrak{f} = \mathfrak{f}'\kappa$ holds in $\mathcal{V}$, it is inferred that $\mathfrak{f}\ell$ and $\mathfrak{f}'\ell'$ determine corresponding cuts in $\mathcal{V}$ and $\mathcal{V}'$. In particular, if ℓ and $\mathfrak{f}\ell$ determine the same cut in $\mathcal{V}$, thus if ℓ and $\mathfrak{f}\ell$ coincide or belong to the same interval of discontinuity, $\mathfrak{f}$ being either a limit-rotation or a boundary translation, then also ℓ' and $\mathfrak{f}'\ell'$ determine the same cut in $\mathcal{V}'$. Hence (1) holds in the extended set $\mathcal{V} \cup \mathcal{U}$, if suitably interpreted for t-isomorphisms which are not g-isomorphisms. – In the case of a g-isomorphism the mapping function κ is one-to-one in the set $\mathcal{V} \cup \mathcal{U}$.

Since the mapping κ of $\mathcal{U}$ onto $\mathcal{U}'$ is order preserving in the sense just described, it follows that, if $\mathcal{L}$ intersects a line of reflection $\mathcal{L}^*$ of $\mathfrak{F}$, then $\mathcal{L}'$ intersects $\mathcal{L}^{*\prime}$. The points of intersection are centres of the type $\mathfrak{A}(\times)$ of $\mathfrak{F}$ and $\mathfrak{F}'$ respectively, and they correspond, since $I(\mathfrak{s}^*\mathfrak{s}) = \mathfrak{s}^{*\prime}\mathfrak{s}'$. These two rotations belong to equal angles of rotation.

24.5 Relative location of corresponding centres and corresponding lines of reflection. It has been shown in Section 1 that a rotation of $\mathfrak{F}$ corresponds to a rotation of $\mathfrak{F}'$ under the t-isomorphism between $\mathfrak{F}$ and $\mathfrak{F}'$ and that, in particular, a primary rotation $\mathfrak{c}$ corresponds to a primary rotation $I(\mathfrak{c}) = \mathfrak{c}'$. The rotation group generated by $\mathfrak{c}$ and with centre c then corresponds to the rotation group generated by $\mathfrak{c}'$. If c' is the centre of $\mathfrak{c}'$, we may then call c and c' corresponding centres. The mapping function κ may thus be extended to all centres of $\mathfrak{F}$. It is one-to-one in the centres of $\mathfrak{F}$ and $\mathfrak{F}'$, and it is seen in the same way as in Section 1 that (1) holds even in centres.

We now prove the theorem:

If $\mathcal{L}$ and $\mathcal{L}'$ are corresponding lines of reflection and c and c' corresponding centres of two t-isomorphic groups $\mathfrak{F}$ and $\mathfrak{F}'$, then c and c' are situated on corresponding sides of $\mathcal{L}$ and $\mathcal{L}'$ or both on these reflection lines.

Proof. Denote by $\mathfrak{s}$ the reflection in $\mathcal{L}$. Then $\mathfrak{s}'$ is the reflection in $\mathcal{L}'$. Let $\mathcal{D}_1$, $\mathcal{D}_2$, $\mathcal{I}_1$, $\mathcal{I}_2$, $\mathcal{V}_1$, $\mathcal{V}_2$, and the corresponding notations for $\mathfrak{F}'$, be defined as in the preceding section.

The latter part of the theorem is immediate: If c is situated on $\mathcal{L}$, then $\mathfrak{A}_c$ is of the type $\mathfrak{A}(\times)$, and $\mathfrak{s}$ is an element of $\mathfrak{A}_c$. Hence $\mathfrak{s}$ transforms a rotation $\mathfrak{c}$ belonging to c into its inverse. Then $\mathfrak{s}'$ transforms $\mathfrak{c}'$ into its inverse. c' is then situated on $\mathcal{L}'$, and $I(\mathfrak{A}_c) = \mathfrak{A}_{c'}$. If c is not on $\mathcal{L}$, let it be situated in $\mathcal{D}_1$. Let $\mathcal{M}$ denote the normal of $\mathcal{L}$ through c, and $\mathfrak{s}_1$ the reflection in $\mathcal{M}$ (Fig. 24.1). Let $\mathcal{N}$ be another line through c

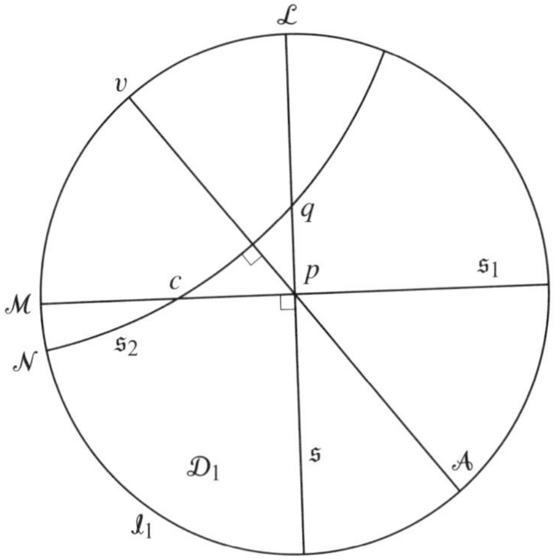

Figure 24.1

and $\mathfrak{s}_2$ the reflection in $\mathcal{N}$. We choose $\mathcal{N}$ such that $\mathfrak{s}_2\mathfrak{s}_1$ is a rotation $\mathfrak{c}$ of $\mathfrak{F}$ about c. Of course, $\mathfrak{s}_1$ and $\mathfrak{s}_2$ do not, in general, belong to $\mathfrak{F}$. Consider the product $\mathfrak{c}\mathfrak{s} = \mathfrak{s}_2\mathfrak{s}_1\mathfrak{s}$. This is a reversion belonging to $\mathfrak{F}$. Since $\mathfrak{s}_1\mathfrak{s}$ is a half-turn around the intersection p of $\mathcal{M}$ and $\mathcal{L}$, and p is not on $\mathcal{N}$, the product $\mathfrak{s}_2(\mathfrak{s}_1\mathfrak{s})$ is a reversed translation $\mathfrak{h}$ in $\mathfrak{F}$. The axis $\mathcal{A}$ of $\mathfrak{h}$, which is an inner axis of $\mathfrak{F}$, is the normal of $\mathcal{N}$ through p, and the direction determined by $\mathfrak{h}$ on $\mathcal{A}$ is from p towards $\mathcal{N}$. Now the positive fundamental point v of $\mathfrak{h} = \mathfrak{c}\mathfrak{s}$ belongs to ℓ_1 and thus to the subset $\mathcal{V}_1$ of $\mathcal{V}$: This is evident if $\mathcal{N}$ does not intersect $\mathcal{L}$; and if $\mathcal{N}$ intersects $\mathcal{L}$, say in q, then the triangle cpq has a right angle at p, hence the perpendicular $\mathcal{A}$ from the right angle is inside the triangle. Thus c and v are on the same side of $\mathcal{L}$. Equally c' and the positive fundamental point v' of $\mathfrak{c}'\mathfrak{s}'$ are on the same side of $\mathcal{L}'$. Since $v' = \kappa v$, the points v and v' are on corresponding sides of $\mathcal{L}$ and $\mathcal{L}'$. The same therefore holds for c and c'. This completes the proof of the theorem. □

We also prove the theorem:

If a reflection line contains more than one centre, then the linear order in which centres are situated on the reflection line is preserved under a t-isomorphism.

Proof. Let the reflection line $\mathcal{L} = \ell_1\ell_2$ of $\mathfrak{F}$ contain the centres c_1 and c_2 (Fig. 24.2). There are other reflection lines passing through any centre on $\mathcal{L}$. Let $\mathcal{A} = a_1a_2$ and

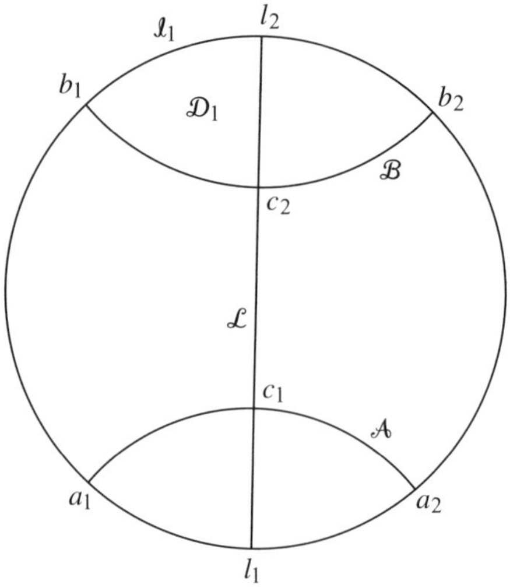

Figure 24.2

$\mathcal{B} = b_1 b_2$ be reflection lines of $\mathfrak{F}$ cutting $\mathcal{L}$ in c_1 and c_2 respectively, a_1 and b_1 being situated in $\mathcal{I}_1$. Let $\mathcal{L}' = \ell_1' \ell_2'$, $\mathcal{A}' = a_1' a_2'$, $\mathcal{B}' = b_1' b_2'$ be the corresponding reflection lines of $\mathfrak{F}'$ with corresponding end-points equally marked, thus a_1' and b_1' situated in $\mathcal{I}_1'$. Then $\mathcal{A}'$ passes through the centre $c_1' = \kappa c_1$ on $\mathcal{L}'$, and $\mathcal{B}'$ through $c_2' = \kappa c_2$. We assert that, if $\ell_1 c_1 c_2 \ell_2$ is the linear order on $\mathcal{L}$, then $\ell_1' c_1' c_2' \ell_2'$ is the linear order on $\mathcal{L}'$. If this assertion is proved, the theorem readily follows.

First, let there be no coincidence of the six points marked on $\mathcal{E}$, nor of the corresponding six points on $\mathcal{E}'$. They are then in a one-to-one correspondence by κ, and their cyclical order on the circles $\mathcal{E}$ and $\mathcal{E}'$ is preserved. If $\mathcal{A}$ and $\mathcal{B}$ are divergent (Fig. 24.2), then also $\mathcal{A}'$ and $\mathcal{B}'$ are divergent, and the assertion holds.

If $\mathcal{A}$ and $\mathcal{B}$ intersect, say in c (Fig. 24.3), then c is a centre of $\mathfrak{F}$, and $\mathcal{A}'$ and $\mathcal{B}'$ then intersect in the corresponding centre $c' = \kappa c$. If c is in $\mathcal{D}_1$, say, then it results from the first theorem, that c' is in $\mathcal{D}_1'$, and again the assertion obviously holds, since the cyclical order of the six points on $\mathcal{E}$ is preserved.

Secondly, coincidences may occur. If $\mathcal{A}$ and $\mathcal{B}$ are parallel (Fig. 24.4), say a_1 and b_1 coincide, thus in a limit-centre on $\mathcal{I}_1$, then a_2 and b_2 do not belong to one and the same interval of discontinuity of $\mathfrak{F}$; for two reflection lines ending in the same interval of discontinuity are divergent. Therefore a_2' and b_2' cannot coincide. a_1' may coincide with b_1', or they may be different points in an interval of discontinuity of $\mathfrak{F}'$. Accordingly $\mathcal{A}'$ and $\mathcal{B}'$ are either parallel or divergent, and not concurrent since $\mathcal{A}$ and $\mathcal{B}$ are not concurrent. In either case the assertion holds. This completes the proof of the theorem. □

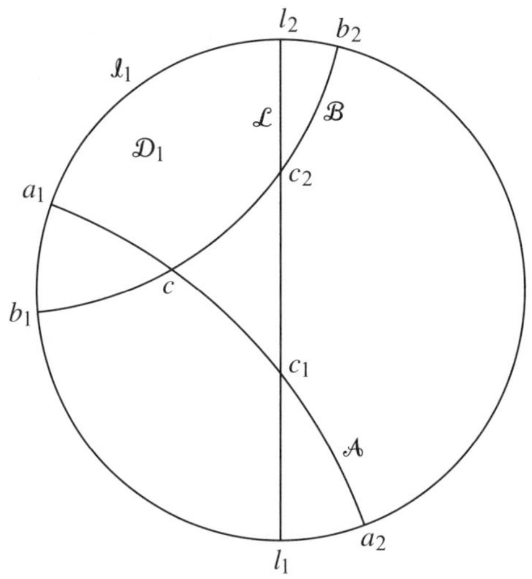

Figure 24.3

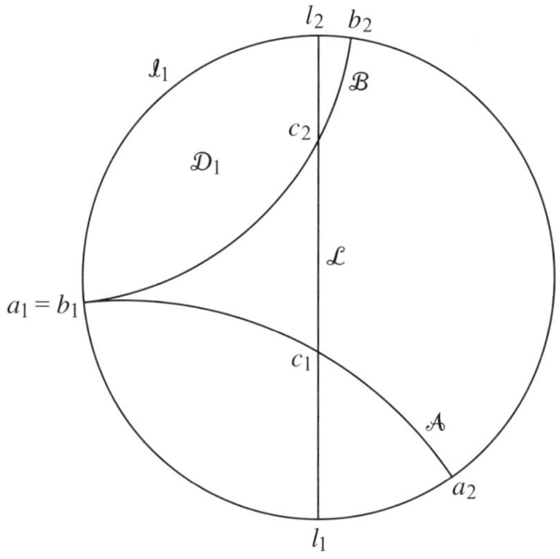

Figure 24.4

24.6 Correspondence of reflection chains and reflection rings. Let $\mathcal{R}$ be a reflection chain of $\mathfrak{F}$. If it consists of more than one link, then it has been shown above that any two consecutive links of $\mathcal{R}$ correspond to two consecutive links of a reflection chain $\mathcal{R}'$ of $\mathfrak{F}'$, whether the two links are connected by a centre, or by a limit-centre, or by an arc of an interval of discontinuity. Thus the chain $\mathcal{R}$ corresponds to the chain $\mathcal{R}'$. In particular, a reflection ring of the surface $\mathcal{D}$ mod $\mathfrak{F}$ corresponds to a reflection ring of the surface $\mathcal{D}'$ mod $\mathfrak{F}'$, since it is derived either from a closed chain, and this then corresponds to a closed chain, or it is derived from a periodic chain, and (1) holds.

If $\mathcal{R}$ consists of a single line of reflection, then also $\mathcal{R}'$ does. The end-points of $\mathcal{R}$ correspond to the end-points of $\mathcal{R}'$, as has already been stated, these end-points determining either corresponding points, or corresponding cuts in $\mathcal{V}$ and $\mathcal{V}'$. If $\mathcal{R}$ cuts a limit side of $\mathcal{K}(\mathfrak{F})$, then the cut in $\mathcal{V}$ is effected by a closed aperiodic interval of discontinuity of $\mathfrak{F}$. The corresponding cut in $\mathcal{V}'$ may well determine a single point of $\mathcal{E}'$, but this point is then also an end of $\mathcal{K}(\mathfrak{F}')$, because the ends of $\mathcal{K}(\mathfrak{F})$ correspond to the ends of $\mathcal{K}(\mathfrak{F}')$ (Section 2).

24.7 Relative location of corresponding centres and corresponding inner axes. Let $\mathcal{A}$ and $\mathcal{A}'$ be two corresponding inner axes. $\mathcal{A}$ determines in $\mathcal{D}$ two open half-planes $\mathcal{D}_1$ and $\mathcal{D}_2$. The end-points u and v of $\mathcal{A}$ determine on $\mathcal{E}$ two open intervals. Let $\mathcal{I}_1$ be the one bounding $\mathcal{D}_1$, and $\mathcal{I}_2$ the one bounding $\mathcal{D}_2$. Both $\mathcal{I}_1$ and $\mathcal{I}_2$ contain points of $\mathcal{V}$, because $\mathcal{A}$ is an inner axis of $\mathfrak{F}$. The subset $\mathcal{V}_1$ of $\mathcal{V}$ contained in $\mathcal{I}_1$ is mapped by κ onto a subset $\mathcal{V}_1'$ of $\mathcal{V}'$ situated in one of the two open intervals of $\mathcal{E}'$ determined by the end-points $u' = \kappa u$ and $v' = \kappa v$ of $\mathcal{A}'$. Let this interval be called $\mathcal{I}_1'$, and the adjacent half-plane determined by $\mathcal{A}'$ be called $\mathcal{D}_1'$. We then say that *the sides* $\mathcal{D}_1$ and $\mathcal{D}_1'$ of $\mathcal{A}$ and $\mathcal{A}'$ *correspond*, and likewise the sides $\mathcal{D}_2$ and $\mathcal{D}_2'$. – Let $\mathfrak{t}$ be a translation along $\mathcal{A}$ which is not a power with exponent > 1 of a translation. The same holds for $\mathfrak{t}'$ on $\mathcal{A}'$.

Let c be a centre of even order $\nu = 2m$, $m \geqq 1$. A primary rotation $\mathfrak{c}$ about c yields an angle of rotation $\frac{2\pi}{\nu} = \frac{\pi}{m}$.

If c is situated on $\mathcal{A}$, then $\mathfrak{c}^m$ reverses $\mathcal{A}$ and $\mathfrak{c}^m\mathfrak{t}\mathfrak{c}^{-m} = \mathfrak{t}^{-1}$. If I is applied it is inferred that $\mathfrak{c}'^m$ reverses $\mathcal{A}'$, and thus that c' is on $\mathcal{A}'$. Also $\mathfrak{t}\mathfrak{c}^m = \mathfrak{c}_1$ is a half-turn with centre on $\mathcal{A}$, and the centre c_1 of $\mathfrak{c}_1$ is of an even order $2m_1$, $m_1 \geqq 1$. All centres of even order on $\mathcal{A}$ are $\mathfrak{t}^r c$ and $\mathfrak{t}^r c_1$, $-\infty < r < \infty$. The corresponding centres of $\mathfrak{F}'$ are the centres $\mathfrak{t}'^r c'$ and $\mathfrak{t}'^r c_1'$ on $\mathcal{A}'$, c_1' being the centre of the half-turn $\mathfrak{t}'\mathfrak{c}'^m$, and the correspondence obviously preserves the linear order of these centres on the axes.

If c is not situated on $\mathcal{A}$, let it be situated in $\mathcal{D}_1$. Then $\mathfrak{c}^m\mathcal{A}$ and $\mathcal{A}$ are divergent. $\mathfrak{c}^m\mathcal{A}$ is situated in $\mathcal{D}_1$. Its end-points belong to $\mathcal{V}_1$. Thus the end-points of $\mathfrak{c}'^m\mathcal{A}'$ belong to $\mathcal{V}_1'$, the axis $\mathfrak{c}'^m\mathcal{A}'$ is situated in $\mathcal{D}_1'$, and the centre c' is in $\mathcal{D}_1'$. Hence the result:

Corresponding centres c and c′ of even order are situated on corresponding sides of corresponding inner axes $\mathcal{A}$ and $\mathcal{A}'$, or both on these axes. □

Now let c be of an odd order $\nu = 2m+1$, $m \geqq 1$. If one of the axes $\mathfrak{c}\mathcal{A}$, $\mathfrak{c}^2\mathcal{A}, \ldots,$ $\mathfrak{c}^m\mathcal{A}$ does not intersect $\mathcal{A}$, then the same argument yields that c and c' are on corresponding sides of $\mathcal{A}$ and $\mathcal{A}'$. If, in particular, $\mathcal{A}$ is simple modulo $\mathfrak{F}$, then the condition holds for all powers of $\mathfrak{c}$, and moreover $\mathcal{A}$ contains no centre of order > 2. We note this important case:

If $\mathcal{A}$ is simple modulo $\mathfrak{F}$, then corresponding centres c and c' of arbitrary order are situated on corresponding sides of the corresponding axes $\mathcal{A}$ and $\mathcal{A}'$. Only if the order of the centres is 2, there is also the possibility of c being on $\mathcal{A}$, and then also c' on $\mathcal{A}'$. □

Let again c be of an odd order $\nu = 2m + 1$, and suppose that $\mathfrak{c}^m \mathcal{A}$ intersects $\mathcal{A}$. The centre c may or may not be situated on $\mathcal{A}$. Then $\mathfrak{c}^r \mathcal{A}$ has points in common with $\mathcal{A}$ for all r. Draw a parallel to $\mathcal{A}$ through c (Fig. 24.5.).

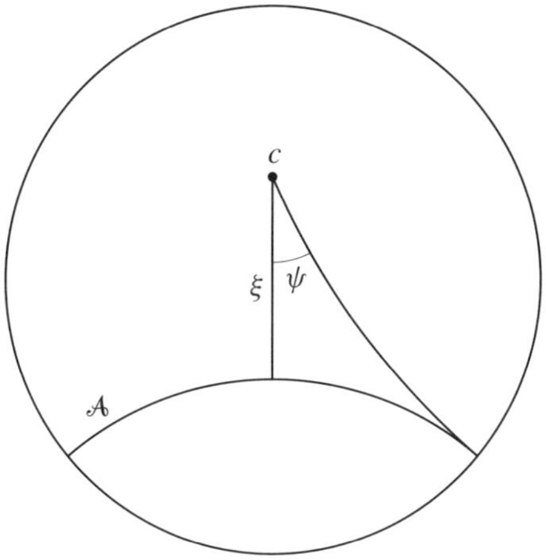

Figure 24.5

Let $\psi \leqq \frac{\pi}{2}$ denote the angle subtended by the parallel and the normal of $\mathcal{A}$ through c, and let $\xi \geqq 0$ denote the distance of c from $\mathcal{A}$. From the formula (III.4) in §8.1 one gets

$$\cosh \xi = \frac{1}{\sin \psi},$$

$\xi = 0$ corresponding to $\psi = \frac{\pi}{2}$. Since $\mathfrak{c}^m$ yields $\frac{2\pi m}{2m+1}$ as angle of rotation, the necessary and sufficient condition for $\mathfrak{c}^m \mathcal{A}$ having points in common with $\mathcal{A}$ is $\frac{\pi m}{2m+1} < \psi$ and thus

$$\cosh \xi = \frac{1}{\sin \psi} < \frac{1}{\sin \frac{\pi m}{2m+1}},$$

hence solving with respect to ξ

$$\xi < \log\ \cot \frac{\pi m}{4m+2},$$

thus in all

$$0 \leqq \xi < \log\,\cot\left(\frac{\nu-1}{\nu}\cdot\frac{\pi}{4}\right).$$

Assume that $\mathfrak{A}_c$ is of the type $\mathfrak{A}(\times_\nu)$ and that one of the reflection lines passing through c, say $\mathcal{L}$, has no point in common with $\mathcal{A}$. Then c is not on $\mathcal{A}$. Let c be situated in $\mathcal{D}_1$. If $\mathfrak{s}$ is the reflection in $\mathcal{L}$ then $\mathfrak{s}\mathcal{A}$ is in $\mathcal{D}_1$. The end-points of $\mathfrak{s}\mathcal{A}$ then belong to $\mathcal{V}_1$. The end-point of $\mathfrak{s}'\mathcal{A}'$ then belong to $\mathcal{V}_1'$. Hence the reflection line $\mathcal{L}'$ corresponding to $\mathcal{L}$ is in $\mathcal{D}_1'$, and c' is in $\mathcal{D}_1'$, since it is on $\mathcal{L}'$. Thus in this case c and c' are on corresponding sides of $\mathcal{A}$ and $\mathcal{A}'$. Hence if c and c' are on non-corresponding sides of $\mathcal{A}$ and $\mathcal{A}'$, or if one of these centres is on the pertaining axis and the other is not, then every reflection line through c has points in common with $\mathcal{A}$, and every reflection line through c' has points in common with $\mathcal{A}'$.

Hence the following theorem results:

If corresponding centres c and c' of order ν are situated on non-corresponding sides of corresponding axes $\mathcal{A}$ and $\mathcal{A}'$, or if one centre is on the pertaining axis and the other is not, then ν is odd, and the distance of c and $\mathcal{A}$, and likewise of c' and $\mathcal{A}'$, is smaller than $\log\,\cot(\frac{\nu-1}{\nu}\cdot\frac{\pi}{4})$. *In this case any two of the collection of axes $\mathfrak{A}_c\mathcal{A}$ have common points, and likewise for $\mathfrak{A}_{c'}\mathcal{A}'$.* □

The upper bound for the distance decreases to zero as ν increases to infinity. The largest value of the upper bound occurs for $\nu = 3$. Thus $\frac{1}{2}\log 3$ is an upper bound for the distance, referred to in the theorem, applicable uniformly for all orders of rotation.

24.8 Another characterization of t-isomorphisms, applicable in the case of quasi-compactness. It has been shown in Section 1 that for any two t-isomorphic groups the totality of boundary translations and limit-rotations in one group corresponds to the same totality in the other group. In this section we want to prove that this property characterizes t-isomorphism in the case of finitely generated groups, thus the theorem:

Any isomorphism between two finitely generated $\mathfrak{F}$-groups for which the totality of boundary translations and limit-rotations in one group corresponds to the same totality in the other group is a t-isomorphism.

Let I denote an isomorphism of the finitely generated $\mathfrak{F}$-group $\mathfrak{F}$ with $\mathfrak{F}'$ satisfying the condition of the theorem. For every $\mathfrak{f} \in \mathfrak{F}$ we put $I(\mathfrak{f}) = \mathfrak{f}'$. Let $\mathfrak{t}$ denote a primary boundary translation of $\mathfrak{F}$, if any. Then $\mathfrak{t}'$ is either a boundary translation or limit-rotation of $\mathfrak{F}'$, and it is primary, since $\mathfrak{t}$ is. Assume that the axis of $\mathfrak{t}$ is cut by a reflection line of $\mathfrak{F}$, say by the line of the reflection $\mathfrak{s}$. Then $\mathfrak{s}^2 = 1$ and $\mathfrak{s}\mathfrak{t}\mathfrak{s} = \mathfrak{t}^{-1}$, hence $\mathfrak{s}'^2 = 1$ and $\mathfrak{s}'\mathfrak{t}'\mathfrak{s}' = \mathfrak{t}'^{-1}$. Since no half-turn reverses a boundary axis or leaves a limit-centre fixed, it follows that $\mathfrak{s}'$ is a reflection whose line cuts the axis of the boundary translation $\mathfrak{t}'$ or terminates in the limit-centre of the limit-rotation $\mathfrak{t}'$. The reflection $\mathfrak{t}\mathfrak{s}$ of $\mathfrak{F}$ then corresponds to the reflection $\mathfrak{t}'\mathfrak{s}'$ of $\mathfrak{F}'$. It is inferred that all reflection lines of $\mathfrak{F}$ cutting the axis of $\mathfrak{t}$ are in a one-to-one and order preserving correspondence with all reflection lines of $\mathfrak{F}'$ cutting the axis of $\mathfrak{t}'$ or terminating in the limit-centre of $\mathfrak{t}'$, respectively. – The same conclusions apply if we start with a primary limit-rotation of $\mathfrak{F}$, if any. Hence: *The totality of limit-centre groups* $\mathfrak{A}(\wedge)$ *and of boundary axis groups* $\mathfrak{A}(\|)$ *of one group corresponds to the same totality of the other group. The same then holds for the totality of limit-centre groups* $\mathfrak{A}(\hookrightarrow)$ *and boundary axis groups* $\mathfrak{A}(\rightarrow)$.

Consider an inner axis $\mathcal{A}$ in $\mathfrak{F}$. Let $\mathfrak{t}$ be a primary element of $\mathfrak{F}$ belonging to $\mathcal{A}$, thus a translation or reversed translation. Then $\mathfrak{t}'$ is, like $\mathfrak{t}$, a primary element of infinite order, and $\mathfrak{t}'$ is not a boundary translation or limit-rotation, because $\mathfrak{t}$ is not. Thus $\mathfrak{t}'$ is either an inner translation or a reversed translation of $\mathfrak{F}'$. The positive fundamental point $v(\mathfrak{t})$ of $\mathfrak{t}$ is the positive fundamental point of $\mathfrak{t}^\rho$ for all $\rho > 0$, and for no other translation or reversed translation of $\mathfrak{F}$. As in Section 1 we conclude that, if we let $v(\mathfrak{t})$ correspond to the positive fundamental point $v'(\mathfrak{t}')$ of $\mathfrak{t}'$, a one-to-one correspondence between the sets of inner fundamental points of the two groups results, and that we can speak of the correspondence between inner axes. Let κ denote the mapping of the set $\mathcal{V}$ of inner fundamental points of $\mathfrak{F}$ onto the set $\mathcal{V}'$ of inner fundamental points of $\mathfrak{F}'$. It also follows as in Section 1 that κ satisfies the functional equation (1).

Let $\mathfrak{s}$ be a reflection in $\mathfrak{F}$ whose line $\mathcal{L}$ does not have both its end-points in limit-centres or intervals of discontinuity of $\mathfrak{F}$. Then it has been proved in §15.2 that, since $\mathfrak{F}$ is finitely generated, $\mathcal{L}$ is wholly inside a truncated domain $\mathcal{K}^*(\mathfrak{F})$ and that $\mathcal{L}$ is an axis of $\mathfrak{F}$. This is an inner axis, because $\mathfrak{F}$ contains the reflection in it. Let $\mathfrak{t}$ be a translation of $\mathfrak{F}$ belonging to $\mathcal{L}$. Then $\mathfrak{s}^2 = 1$ and $\mathfrak{s}\mathfrak{t}\mathfrak{s} = \mathfrak{t}$. Hence $\mathfrak{s}'^2 = 1$ and $\mathfrak{s}'\mathfrak{t}'\mathfrak{s}' = \mathfrak{t}'$. Since $\mathfrak{t}$ is an inner translation of $\mathfrak{F}$, $\mathfrak{t}'$ is either an inner translation or a reversed translation of $\mathfrak{F}'$. Since a translation or reversed translation cannot be permutable with a half-turn, $\mathfrak{s}'$ is a reflection.

It has thus been seen that reflections in one group correspond to reflections in the other by the isomorphism, if the groups at all contain reflections. We can therefore speak of a one-to-one correspondence between reflection lines of the two groups.

Let c be a centre of $\mathfrak{F}$ of order ν and $\mathfrak{c}$ a rotation of order ν with c as its centre. Then $\mathfrak{c}'$ is an element of order ν, thus a rotation, since even for $\nu = 2$ it is not a reflection. Let its centre be c'. Since c has the order ν, the element $\mathfrak{c}$ is not power of an element of order $> \nu$. The same then holds for $\mathfrak{c}'$, and c' has the order ν. If $\mathfrak{A}_c = \mathfrak{A}(\times_\nu)$, there is in $\mathfrak{F}$ a reflection $\mathfrak{s}$ such that $\mathfrak{s}\mathfrak{c}\mathfrak{s} = \mathfrak{c}^{-1}$. Hence $\mathfrak{s}'\mathfrak{c}'\mathfrak{s}' = \mathfrak{c}'^{-1}$, and $\mathfrak{A}_{c'} = \mathfrak{A}(\times_\nu)$, the reflection lines passing through c and c' being corresponding. From this follows that there is a one-to-one correspondence between the centres of the two groups, a subgroup with the symbol $\mathfrak{A}(\times_\nu)$ in one group corresponding to a subgroup with the same symbol in the other, and likewise for subgroups with the symbol $\mathfrak{A}(\odot_\nu)$.

We now set out to prove that the correspondence established above for inner axes of the two groups satisfies condition 3) of Section 1. Since two different axes cannot be parallel, this is equivalent to proving

Any two divergent inner axes correspond to two divergent inner axes.

The proof of this assertion is the main difficulty involved in proving the theorem of this section. In preparation for the proof a certain construction has to be carried out in detail.

Let x_0 be a regular point for $\mathfrak{F}$ in $\mathcal{D}$, chosen arbitrarily. The set $\mathcal{N}$ of points x of $\mathcal{D}$ for which $[x, x_0] < [x, \mathfrak{f}x_0]$ for all $\mathfrak{f} \neq 1$ in $\mathfrak{F}$ is the interior of the normal domain with x_0 as central point (§10.2). $\mathcal{N}$ is an open subset of $\mathcal{D}$ and consists of regular points only. The boundary of $\mathcal{N}$ is denoted by n; it is a polygon with a finite number of sides, since $\mathfrak{F}$ is finitely generated (§17.2). Among these sides are arcs of $\mathcal{E}$, if $\mathfrak{F}$ admits intervals of discontinuity on $\mathcal{E}$. It is recalled that centres of order 2 are included among the vertices of n even if the corresponding angle is π. We put $\mathcal{N} \cup n = \overline{\mathcal{N}}$.

Vertices of n in $\mathcal{D}$ are either regular points, or points on reflection lines of $\mathfrak{F}$ which are not centres of $\mathfrak{F}$, or centres of $\mathfrak{F}$. They comprise representatives of every equivalence class of centres of $\mathfrak{F}$. Vertices of n in $\mathcal{D}$ fall into cycles of equivalent vertices (§10). Sides of n in $\mathcal{D}$ are either made up of regular points, such sides being equivalent in pairs, or they are parts of reflection lines of $\mathfrak{F}$. To any point of any reflection line of $\mathfrak{F}$ there is an equivalent on n.

Consider vertices of n on $\mathcal{E}$. If $\mathfrak{F}$ admits limit-rotations, every equivalence class of limit-centres is represented by vertices of n on $\mathcal{E}$ constituting a vertex cycle. If $\mathfrak{F}$ admits boundary translations, every equivalence class of intervals of discontinuity contains points of n. If the interval of discontinuity $\mathcal{I}$ has points in common with n, then $\mathcal{I} \cap n$ is either a single point, the common end-point of two sides of $n \cap \mathcal{D}$, or an arc of $\mathcal{I}$ determined by the end-points of two sides of $n \cap \mathcal{D}$. The intersection of n with an equivalence class $\mathfrak{F}\mathcal{I}$ will be called a *boundary cycle*. It is the logical counterpart of vertex cycles consisting of vertices of n in $\mathcal{D}$ or consisting of limit-centres, since such vertex cycles are the intersection of n with the equivalence classes of certain points. The boundary cycle of n in $\mathfrak{F}\mathcal{I}$ contains an equivalent to every point of $\mathcal{I}$, and thus of $\mathfrak{F}\mathcal{I}$. The number of components of the intersection of n with an equivalence class of intervals of discontinuity, or of limit-centres, or of vertices in $\mathcal{D}$, will be denoted by m in the sequel. Then in all cases $m \geqq 1$, and m may, of course, vary from cycle to cycle.

$\mathfrak{F}n$ is a net in $\overline{\mathcal{D}}$. The interior of every mesh of this net is equivalent with $\mathcal{N}$. Thus $\mathfrak{F}\mathcal{N}$ is a tesselation of $\mathcal{D}$. Let $\mathcal{D}^*$ denote the point set obtained by adding to $\mathcal{D}$ all limit-centres, if any, and all intervals of discontinuity, if any, of $\mathfrak{F}$, and let $\mathcal{D}'^*$ be defined in the same way for $\mathfrak{F}'$. Then $\mathfrak{F}n$ belongs to $\mathcal{D}^*$.

In order to get a treatment common to the different cases which may present themselves, let z be a common denotation for either an arbitrary vertex of $\mathfrak{F}n$ in $\mathcal{D}$, or for an arbitrary limit-centre of $\mathfrak{F}$, or for an arbitrary interval of discontinuity of $\mathfrak{F}$. Let $\mathfrak{A}_z$ denote the subgroup of $\mathfrak{F}$ leaving z invariant. Those meshes of the tesselation $\mathfrak{F}\mathcal{N}$ which have z on their boundary fall into m equivalence classes with respect to $\mathfrak{A}_z$. A set of representatives of these classes is denoted by $\mathfrak{f}_t\mathcal{N}$, $1 \leqq t \leqq m$. Those sides of $\mathfrak{f}_t\mathcal{N}$ which end in z are denoted by $\mathcal{S}_t$ and $\mathcal{S}_{t+1}$, and their end-points in z by p_t and p_{t+1} *irrespective of whether these two end-points coincide or not.*

There are 8 different cases which can present themselves according to the 8 possibilities for $\mathfrak{A}_z$. These cases are:

$$a)\ \mathfrak{A}_z = 1,\ z = a,\ \text{a regular point in } \mathcal{D},$$
$$b)\ \mathfrak{A}_z = \mathfrak{A}(-),\ z = b,\ \text{a point on a reflection line},$$
$$\left.\begin{array}{l} c_1)\ \mathfrak{A}_z = \mathfrak{A}(\times_\nu) \\ c_2)\ \mathfrak{A}_z = \mathfrak{A}(\odot_\nu) \end{array}\right\} \quad z = c,\ \text{a centre},$$
$$\left.\begin{array}{l} u_1)\ \mathfrak{A}_z = \mathfrak{A}(\wedge) \\ u_2)\ \mathfrak{A}_z = \mathfrak{A}(\hookrightarrow) \end{array}\right\} \quad z = u,\ \text{a limit-centre},$$
$$\left.\begin{array}{l} \mathcal{I}_1)\ \mathfrak{A}_z = \mathfrak{A}(\|) \\ \mathcal{I}_2)\ \mathfrak{A}_z = \mathfrak{A}(\rightarrow) \end{array}\right\} \quad z = \mathcal{I},\ \text{an interval of discontinuity}.$$

In the first 6 cases all p_t coincide with z, in the last two cases the p_t are points in $\mathcal{I}$, some of which may coincide, but not all of them. In the cases $b), c_1), u_1), \mathcal{I}_1)$ the sides $\mathcal{S}_1$ and $\mathcal{S}_{m+1}$ are situated on reflection lines. For each of the 8 types n may give rise to more than one cycle of that type, but only to a finite number; we then choose a representing z for each of them. All the z belong to $\mathcal{D}^*$.

To each z a z' in $\mathcal{D}'^*$ is now assigned in the following way: In the case $a)$ we choose for z' an arbitrary regular point a' in $\mathcal{D}'$. If there are more than one a, then these are inequivalent with respect to $\mathfrak{F}$; we then choose the a' to be inequivalent with respect to $\mathfrak{F}'$. In the case $b)$ $z = b$ is situated on a reflection line $\mathcal{L}$ of $\mathfrak{F}$. We then choose for z' an arbitrary point b' of the corresponding reflection line $\mathcal{L}'$ of $\mathfrak{F}'$, provided that b' shall not be a centre of $\mathfrak{F}'$. If there are more than one b on $\mathcal{L}$, then they are inequivalent with respect to $\mathfrak{F}$; we then choose the b' to be points on $\mathcal{L}'$ inequivalent with respect to $\mathfrak{F}'$. In the cases $z = c$ the z' shall be the corresponding centre c' of $\mathfrak{F}'$. In the cases $z = u$, z' shall be the corresponding limit-centre u' or the corresponding interval of discontinuity $\mathcal{I}'$, as the case may be. In the cases $z = \mathcal{I}$, z' shall be the corresponding interval $\mathcal{I}'$ or the corresponding limit-centre u', as the case may be. In all cases, for inequivalent z the corresponding z' are inequivalent with respect to $\mathfrak{F}'$. In all cases the subgroup $\mathfrak{A}_{z'}$ of $\mathfrak{F}'$ corresponds to the subgroup $\mathfrak{A}_z$ of $\mathfrak{F}$ by the isomorphism.

It is now intended to establish a mapping μ of the end-points of sides of $n \cap \mathcal{D}$ into $\mathcal{D}'^*$; it will be seen in the sequel why we say *end-points of sides of* $n \cap \mathcal{D}$ rather than *vertices* of n. If z' is a point in $\mathcal{D}'$ or a limit-centre of $\mathfrak{F}'$, then we put $p'_t = z'$ for $1 \leqq t \leqq m$. If z' is an interval of discontinuity of $\mathfrak{F}'$ with corresponding group $\mathfrak{A}(\|)$, and thus z either an interval of discontinuity with corresponding group $\mathfrak{A}(\|)$ or a limit-centre with corresponding group $\mathfrak{A}(\wedge)$, then p'_1 and p'_{m+1} shall be the end-points in z' of the reflection lines of $\mathfrak{F}'$ corresponding to the reflection lines with end-points p_1 and p_{m+1} in z. If z' is an interval of discontinuity with corresponding group $\mathfrak{A}(\rightarrow)$, thus z either an interval with corresponding group $\mathfrak{A}(\rightarrow)$ or a limit-centre with corresponding group $\mathfrak{A}(\hookrightarrow)$, and $\mathcal{S}_{m+1} = \mathfrak{t}\mathcal{S}_1$, $\mathfrak{t}$ being a primary boundary translation or limit-rotation respectively, then we choose for p'_1 an arbitrary point in z' and put

$p'_{m+1} = \mathfrak{t}' p'_1$. In both cases $p'_2, \ldots, p'_m$ (if $m > 1$) shall be $m - 1$ points of z' inserted arbitrarily between p'_1 and p'_{m+1} in the order of the numbering and such that all the $m + 1$ points are different.

Consider now the mesh $\mathfrak{f}_t \mathcal{N}$ having z on its boundary. $\mathcal{S}_t$ and $\mathcal{S}_{t+1}$ are sides of $\mathfrak{f}_t \mathcal{N}$ with end-points p_t and p_{t+1}, respectively, in z. If we apply $\mathfrak{f}_t^{-1}$, then the mesh $\mathfrak{f}_t \mathcal{N}$ is carried into $\mathcal{N}$, and $\mathfrak{f}_t^{-1} \mathcal{S}_t$ and $\mathfrak{f}_t^{-1} \mathcal{S}_{t+1}$ are sides of $\mathcal{N}$. Thus $\mathfrak{f}_t^{-1} p_t$ and $\mathfrak{f}_t^{-1} p_{t+1}$ either coincide in a vertex $\mathfrak{f}_t^{-1} z$ of n in $\mathcal{D}$ or in a limit-centre of $\mathfrak{F}$, or they are both points (not necessarily different) of the interval of discontinuity $\mathfrak{f}_t^{-1} z$ of $\mathfrak{F}$. We then put

$$\mu \mathfrak{f}_t^{-1} p_t = \mathfrak{f}_t'^{-1} p'_t, \quad \mu \mathfrak{f}_t^{-1} p_{t+1} = \mathfrak{f}_t'^{-1} p'_{t+1}. \tag{2}$$

It should be noticed that if p_t and p_{t+1} coincide either in a limit-centre z or interval of discontinuity z of $\mathfrak{F}$, and z' is an interval of discontinuity of $\mathfrak{F}$, then $\mathfrak{f}_t^{-1} p_t$ and $\mathfrak{f}_t^{-1} p_{t+1}$ get different images by μ. It is therefore emphasized that *the mapping μ will only be used in relation to definite sides of n in $\mathcal{D}$.*

We now want to extend the mapping μ to the sides of $n \cap \mathcal{D}$. Let $\mathcal{S}$ be such a side and q_1 and q_2 its end-points. $\mathcal{S}$ is a straight segment, or half-line, or line, according as both q_1 and q_2 are in $\mathcal{D}$, or one or both on $\mathcal{E}$. In all cases q_1 and q_2 belong to $\mathcal{D}^*$. The images μq_1 and μq_2 of the end-points of $\mathcal{S}$ have already been fixed above; they belong to $\mathcal{D}'^*$. Also $\mu q_1 \in \mathcal{D}'$ if, and only if, $q_1 \in \mathcal{D}$, and likewise for μq_2. Let $\mathcal{S}'$ be the straight segment, or half-line, or line, joining μq_1 and μq_2. We then define μ as an arbitrary topological mapping of $\mathcal{S}$ on $\mathcal{S}'$ compatible with the given mapping of their end-points. If $\mathcal{S}$ is situated on a reflection line $\mathcal{L}$ of $\mathfrak{F}$, then $\mathcal{S}'$ is situated on the corresponding reflection line $\mathcal{L}'$ of $\mathfrak{F}'$, because that has been so arranged for the end-points by the way in which the p'_t were chosen. If $\mathcal{S}$ is not situated on a reflection line, then it consists of regular points in $\mathcal{D}$ except possibly for q_1 and q_2. In this case there is one, and only one, element $\mathfrak{f} \neq 1$ of $\mathfrak{F}$ such that $\mathfrak{f}\mathcal{S}$ is also a side of $n \cap \mathcal{D}$. Let $\mathcal{S}$ be equivalent with $\mathcal{S}_t$, independently of which of the 8 cases occurs, and $\mathfrak{f}_t^{-1} \mathcal{S}_t = \mathcal{S}$. If $t > 1$, then $\mathfrak{f}_{t-1}^{-1} \mathcal{S}_t$ is the side of n equivalent with $\mathcal{S}$, hence $\mathfrak{f} = \mathfrak{f}_{t-1}^{-1} \mathfrak{f}_t$. If $t = 1$, and $\mathfrak{f}_1^{-1} \mathcal{S}_1 = \mathcal{S}$, then $\mathfrak{f}_m^{-1} \mathcal{S}_{m+1}$ is the side of n equivalent with $\mathcal{S}$ (cases a), c_2), u_2), $\mathcal{I}_2$)), and $\mathfrak{f}_m^{-1} \mathcal{S}_{m+1} = \mathfrak{f}_m^{-1} \mathfrak{p} \mathcal{S}_1$, where $\mathfrak{p} = 1, \mathfrak{c}, \mathfrak{u}, \mathfrak{t}$, respectively, in the four cases in question. Hence $\mathfrak{f} = \mathfrak{f}_m^{-1} \mathfrak{p} \mathfrak{f}_1$. In virtue of (2) the end-point $\mathfrak{f}_t^{-1} p_t$ of $\mathcal{S}$ is mapped into $\mathfrak{f}_t'^{-1} p'_t$ by μ and the equivalent end-point $\mathfrak{f}_{t-1}^{-1} p_t$ of $\mathfrak{f}_{t-1}^{-1} \mathcal{S}_t = \mathfrak{f}\mathcal{S}$, where $t > 1$, is mapped into $\mathfrak{f}_{t-1}'^{-1} p'_t$ by μ. The point $\mathfrak{f}_t'^{-1} p'_t$ is carried into $\mathfrak{f}_{t-1}'^{-1} p'_t$ by the element $\mathfrak{f}_{t-1}'^{-1} \mathfrak{f}'_t$ of $\mathfrak{F}'$, and this is $\mathfrak{f}'$ in the case $t > 1$. In the case $t = 1$, $p_{m+1} = \mathfrak{p} p_1$ is the end-point of $\mathcal{S}_{m+1}$. In virtue of (2) one gets $\mu \mathfrak{f}_m^{-1} p_{m+1} = \mathfrak{f}_m'^{-1} p'_{m+1} = \mathfrak{f}_m'^{-1} \mathfrak{p}' p'_1$. Now $\mathfrak{f}_1'^{-1} p'_1$ is carried into $\mathfrak{f}_m'^{-1} \mathfrak{p}' p'_1$ by the element $\mathfrak{f}_m'^{-1} \mathfrak{p}' \mathfrak{f}'_1$, which is $\mathfrak{f}'$ in the case $t = 1$. It is thus seen that the mapping μ has been so fixed for the end-points of the sides $\mathcal{S}$ and $\mathfrak{f}\mathcal{S}$, that $\mu \mathfrak{f} q_1 = \mathfrak{f}' \mu q_1$ and $\mu \mathfrak{f} q_2 = \mathfrak{f}' \mu q_2$. After $\mu \mathcal{S}$ has been fixed we therefore now fix the mapping $\mu \mathfrak{f} \mathcal{S}$ by the equation

$$\mu \mathfrak{f} x = \mathfrak{f}' \mu x \tag{3}$$

where $\mathfrak{f}x$ ranges over $\mathfrak{f}\mathcal{S}$ as x ranges over $\mathcal{S}$; this is possible, because (3) has been shown to hold for the end-points of $\mathcal{S}$ and $\mathfrak{f}\mathcal{S}$. Thus μ is chosen as an arbitrary topological mapping only for one side of each pair of equivalent sides, and is then transferred to the other side of that pair by (3).

This finishes the establishment of the mapping μ of $n \cap \mathcal{D}$ into $\mathcal{D}'$. If $x \in \mathcal{D}$ and $\mathfrak{f}x$ both belong to $\overline{\mathcal{N}}$, and $\mathfrak{f} \neq 1$, then both belong to n, and x is either a vertex of n in $\mathcal{D}$, and $\mathfrak{f}$ carries that vertex either into itself or into another vertex of the same cycle, or x is a point of a side of n situated on a reflection line of $\mathfrak{F}$, and $\mathfrak{f}$ the reflection in that line, or x is a point on a side belonging to an equivalent pair, and $\mathfrak{f}$ the element carrying that side into the other side of that pair. In all cases (3) is known to hold. One can therefore immediately extend μ to a mapping of $\mathfrak{F}(n \cap \mathcal{D}) = (\mathfrak{F}n) \cap \mathcal{D}$ into $\mathcal{D}'$ by letting x in (3) range over $n \cap \mathcal{D}$ and $\mathfrak{f}$ over $\mathfrak{F}$. Then (3) holds for any two equivalent points x and $\mathfrak{f}x$ of $\mathfrak{F}n \cap \mathcal{D}$: If ξ is a point of n in the same equivalence class, then there are elements $\mathfrak{f}_1$ and $\mathfrak{f}_2$ of $\mathfrak{F}$ such that $x = \mathfrak{f}_1\xi$ and $\mathfrak{f}x = \mathfrak{f}_2\xi = \mathfrak{f}_2\mathfrak{f}_1^{-1}x$, thus $\mathfrak{f} = \mathfrak{f}_2\mathfrak{f}_1^{-1}$. Then using (3) for $\xi \in n$,

$$\mu x = \mu\mathfrak{f}_1\xi = \mathfrak{f}_1'\mu\xi$$
$$\mu\mathfrak{f}x = \mu\mathfrak{f}_2\xi = \mathfrak{f}_2'\mu\xi$$

thus

$$\mu\mathfrak{f}x = \mathfrak{f}_2'\mathfrak{f}_1'^{-1}\mu x = \mathfrak{f}'\mu x.$$

These elements $\mathfrak{f}$ and $\mathfrak{f}'$ are uniquely determined, if x, and thus also $\mathfrak{f}x$, are regular. If they are singular, the considerations are easily seen to hold for each of the finite number of possible elements $\mathfrak{f}$.

The point set $\mu(\mathfrak{F}n \cap \mathcal{D})$ is carried into itself by $\mathfrak{F}'$, because the point set $\mathfrak{F}n \cap \mathcal{D}$ is carried into itself by $\mathfrak{F}$ and (3) holds. One gets

$$\mu(\mathfrak{F}n \cap \mathcal{D}) = \mathfrak{F}'(\mu(n \cap \mathcal{D})). \tag{4}$$

By a *path* w on $\mathfrak{F}n \cap \mathcal{D}$ we understand a sequence of sides of $\mathfrak{F}n \cap \mathcal{D}$ for which the terminal point of any side of the sequence belongs to $\mathcal{D}$ and coincides with the initial point of the next side of the sequence. Since the mapping function μ is uniquely determined in $\mathfrak{F}n \cap \mathcal{D}$, the path w is mapped into μw, which is a path on $\mu(\mathfrak{F}n \cap \mathcal{D})$, thus belonging to $\mathcal{D}'$. We want to define paths on $\mathfrak{F}n$, *thus contained in* $\mathcal{D}^*$. If a path w_1 on $\mathfrak{F}n \cap \mathcal{D}$ runs into $\mathcal{E}$, let its terminal point q_1 belong to the component u of $\mathcal{D}^* \cap \mathcal{E}$, thus u denoting either a limit-centre or interval of discontinuity of $\mathfrak{F}$. It can then only return to $\mathcal{D}$ by leaving $\mathcal{E}$ in a point q_2 of the same u, this being the initial point of another side of $\mathfrak{F}n \cap \mathcal{D}$, and thus the initial point of another partial path w_2 on $\mathfrak{F}n \cap \mathcal{D}$. If q_1 and q_2 are different, the arc q_1q_2 of u is part of the total path w on $\mathfrak{F}n$. Now μq_1 and μq_2 are uniquely determined, if q_1 and q_2 are thought of as end-points in u of the last side of w_1 and the first side of w_2 respectively, and μq_1 and μq_2 belong to the component u' of $\mathcal{D}'^* \cap \mathcal{E}'$, this u' being either a limit-centre or an interval of discontinuity of $\mathfrak{F}'$, corresponding to u, but not necessarily of the same kind as u. If

μq_1 and μq_2 are different, the arc $(\mu q_1)(\mu q_2)$ of u' is taken to be a part of μw. So we can speak of the images by μ of arbitrary paths on $\mathfrak{F}n$.

A special path on $\mathfrak{F}n$ is n itself. We may denote its image by μn, bearing in mind that the mapping function μ is to be understood in the above sense in $\mathcal{D}^* \cap \mathcal{E}$. Then also $\mu(\mathfrak{F}n) = \mathfrak{F}'(\mu n)$, and (4) can be written

$$\mu(\mathfrak{F}n \cap \mathcal{D}) = (\mu(\mathfrak{F}n)) \cap \mathcal{D}' = \mathfrak{F}'\mu n \cap \mathcal{D}', \tag{5}$$

the last expression in (5) being at will interpreted as $(\mathfrak{F}'(\mu n)) \cap \mathcal{D}'$ or as $\mathfrak{F}'((\mu n) \cap \mathcal{D}')$. It should be borne in mind that μ is one-valued in $\mathfrak{F}n \cap \mathcal{D}$, but that it may map different points of $\mathfrak{F}n \cap \mathcal{D}$ into the same point of $\mathcal{D}'$. Therefore μn is a closed polygon situated in $\mathcal{D}'^*$, but it is, in general, not simple. The images by μ of the vertices of $\mathfrak{F}n \cap \mathcal{D}$ will be called *vertices of* $\mathfrak{F}'\mu n \cap \mathcal{D}'$. Other common points of any two sides of $\mathfrak{F}'\mu n \cap \mathcal{D}'$ will be called *intersection points* of $\mathfrak{F}'\mu n \cap \mathcal{D}'$. It is evident that the vertices of $\mathfrak{F}'\mu n \cap \mathcal{D}'$ belong to a finite number of equivalence classes with respect to $\mathfrak{F}'$, since that is the case for the vertices of $\mathfrak{F}n \cap \mathcal{D}$ with respect to $\mathfrak{F}$ and (3) holds. We want to prove the following lemma:

Lemma. *The intersection points of* $\mathfrak{F}'\mu n \cap \mathcal{D}'$ *belong to a finite number of equivalence classes with respect to* $\mathfrak{F}'$.

Proof. Let a truncated domain $\mathcal{K}^*(\mathfrak{F}')$ of $\mathfrak{F}'$ be chosen. For convenience of notation let u' denote either a limit-centre or an interval of discontinuity of $\mathfrak{F}'$, let $\mathcal{A}'$ denote the bounding horocycle or boundary axis respectively belonging to u', let $\mathfrak{A}_{\mathcal{A}'}$ denote the subgroup of $\mathfrak{F}'$ carrying $\mathcal{A}'$ into itself, and let $\mathcal{H}'$ denote the interior of the horocycle $\mathcal{A}'$ or the half-plane bounded by $\mathcal{A}'$ and u' respectively. Then $\mathfrak{F}'$ and $\mathfrak{A}_{\mathcal{A}'}$ have equal effect in $\mathcal{H}'$.

Let $\mathcal{S}'$ be any side of $\mathfrak{F}'\mu n \cap \mathcal{D}'$. If an end-point of $\mathcal{S}'$ belongs to u', we call $\mathcal{S}' \cap \mathcal{H}'$ the *infinite part* of $\mathcal{S}'$ belonging to u'. If $\mathcal{S}'$ is a full line, then also the other end-point of $\mathcal{S}'$ is on $\mathcal{E}'$, say in u''. Then u'' is different from u'. In this case $\mathcal{S}'$ has two infinite parts; the rest of $\mathcal{S}'$ is called the *finite part* $\mathcal{S}'_0$ of $\mathcal{S}'$. If $\mathcal{S}'$ is a half-line, then it has only one infinite part. Then, if $\mathcal{S}'$ is not wholly contained in $\mathcal{H}'$, the part of $\mathcal{S}'$ outside $\mathcal{H}'$ is called the *finite part* $\mathcal{S}'_0$ of $\mathcal{S}'$. If $\mathcal{S}'$ is a segment, it consists only of its finite part $\mathcal{S}'_0 = \mathcal{S}'$.

Consider any two finite parts $\mathcal{S}'_0$ and $\mathcal{S}''_0$, equivalent with respect to $\mathfrak{F}'$ or not. Then $\mathcal{S}'_0$ is intersected only by a finite number, if any, of segments of the equivalence class $\mathfrak{F}'\mathcal{S}''_0$, because $\mathcal{S}''_0$ is bounded and therefore the collection of point sets $\mathfrak{F}'\mathcal{S}''_0$ does not accumulate in $\mathcal{D}'$ (§13.2). Since the sides of $\mathfrak{F}'\mu n \cap \mathcal{D}'$, and thus also their finite parts, belong to a finite number of equivalence classes with respect to $\mathfrak{F}'$, it is inferred that the same holds for the set of intersection points of finite parts with finite parts.

Then consider an infinite part $\mathcal{S}' \cap \mathcal{H}'$ and a finite part $\mathcal{S}''_0$ and suppose that $\mathcal{S}''_0$ intersects $\mathcal{S}' \cap \mathcal{H}'$. The intersection point belongs to $\mathcal{S}'' \cap \mathcal{H}'$. Now

$$(\mathfrak{F}'\mathcal{S}''_0) \cap \mathcal{H}' = (\mathfrak{F}'(\mathcal{S}''_0 \cap \mathcal{H}')) \cap \mathcal{H}' = \mathfrak{A}_{\mathcal{A}'}(\mathcal{S}''_0 \cap \mathcal{H}'),$$

because $\mathfrak{F}'$ reproduces $\mathcal{K}^*(\mathfrak{F}')$ and $\mathfrak{F}'$ and $\mathfrak{A}_{\mathcal{A}'}$ have equal effect in $\mathcal{H}'$. Since $\mathcal{S}_0''$ is bounded, the part $\mathcal{S}_0'' \cap \mathcal{H}'$ is bounded, and only a finite number of its equivalents with respect to $\mathfrak{A}_{\mathcal{A}'}$ can have points in common with $\mathcal{S}'$, since, if u' is an interval of discontinuity, $\mathcal{S}'$ does not terminate in an end-point of $\mathcal{A}'$, which is outside $\mathcal{D}'^*$. It is inferred that the intersection points of finite parts with infinite parts belong to a finite number of equivalence classes with respect to $\mathfrak{F}'$.

Finally, if two infinite parts intersect, they must belong to the same $\mathcal{H}'$, and both terminate in u'. Two sides terminating in the same limit-centre are on parallel lines and thus do not intersect. Let $\mathcal{S}' \cap \mathcal{H}'$ and $\mathcal{S}'' \cap \mathcal{H}'$ intersect, both $\mathcal{S}'$ and $\mathcal{S}''$ having an end-point in the interval of discontinuity u'. The part of the collection $\mathfrak{F}'(\mathcal{S}'' \cap \mathcal{H}')$ contained in $\mathcal{H}'$ is $\mathfrak{A}_{\mathcal{A}'}(\mathcal{S}'' \cap \mathcal{H}')$, and since neither $\mathcal{S}'$ nor $\mathcal{S}''$ contains an end-point of $\mathcal{A}'$, only a finite number of members of the collection $\mathfrak{A}_{\mathcal{A}'}(\mathcal{S}'' \cap \mathcal{H}')$ can intersect $\mathcal{S}' \cap \mathcal{H}'$. It is inferred that the intersection points of infinite parts with infinite parts belong to a finite number of equivalence classes with respect to $\mathfrak{F}$. – This finishes the proof of the lemma. □

Thus in all:

The collection of all vertices of $\mathfrak{F}'\mu n \cap \mathcal{D}'$ and of all intersection points of $\mathfrak{F}'\mu n \cap \mathcal{D}'$ belong to a finite number of equivalence classes with respect to $\mathfrak{F}'$. □

Let $\mathcal{S}'$ be a side terminating in the limit-centre u'. Then $\mathcal{S}' \cap \mathcal{H}'$ contains no two different equivalent points, because $\mathfrak{F}'$ and $\mathfrak{A}_{\mathcal{A}'}$ have equal effect in $\mathcal{H}'$. – Let $\mathcal{S}'$ terminate in a point of the interval of discontinuity u', and assume that $\mathcal{S}' \cap \mathcal{H}'$ contains two equivalent points p and $\mathfrak{f}'p$, $\mathfrak{f}' \in \mathfrak{F}'$. Then $\mathfrak{f}' \in \mathfrak{A}_{\mathcal{A}'}$, because $\mathfrak{A}_{\mathcal{A}'}$ and $\mathfrak{F}'$ have equal effect in $\mathcal{H}'$. Then $\mathcal{S}' \cap \mathcal{H}'$ and $\mathfrak{f}'^{-1}\mathcal{S}' \cap \mathcal{H}'$ intersect in p. Since only a finite number of members of the collection $\mathfrak{A}_{\mathcal{A}'}\mathcal{S}' \cap \mathcal{H}'$ intersect $\mathcal{S}' \cap \mathcal{H}'$ (the special case of the above consideration in which $\mathcal{S}'$ and $\mathcal{S}''$ are equivalent), it is inferred that to any point of $\mathcal{S}' \cap \mathcal{H}'$ there is only a finite number of equivalents on $\mathcal{S}' \cap \mathcal{H}'$, if any.

From this follows that $\mathcal{S}' \cap \mathcal{H}'$ can only contain a finite number of vertices or intersection points of $\mathfrak{F}'\mu n$, because the number of equivalence classes to which such points belong is finite, and each class can only be represented a finite number of times. Also the number of sides of $\mathfrak{F}'\mu n \cap \mathcal{D}'$ terminating in u' reduce modulo $\mathfrak{A}_{u'}$ to a finite number, and the same thus holds for the number of vertices and intersection points of $\mathfrak{F}'\mu n$ contained in $\mathcal{H}'$, because they are reproduced by $\mathfrak{A}_{\mathcal{A}'}$.

In consequence of this fact the horocycles bounding $\mathcal{K}^*(\mathfrak{F}')$ may be so chosen that they contain in their interior neither vertices nor intersection points of $\mathfrak{F}'\mu n$. Let $\mathcal{O}_{u'}$ be the notation for such bounding horocycles of $\mathcal{K}^*(\mathfrak{F}')$ belonging to the limit-centres u'. Likewise for an interval of discontinuity u' of $\mathfrak{F}'$ one can determine a hypercycle $\mathcal{O}_{u'}$ belonging to $\mathcal{A}'$ such that the part of $\mathcal{H}'$ bounded by $\mathcal{O}_{u'}$ and u' does not contain any vertex or intersection point of $\mathfrak{F}'\mu n$. For any element $\mathfrak{f}'$ of $\mathfrak{F}'$ we put $\mathcal{O}_{\mathfrak{f}'u'} = \mathfrak{f}'\mathcal{O}_{u'}$. This also holds for the bounding horocycles $\mathcal{O}_{u'}$ of $\mathcal{K}^*(\mathfrak{F}')$.

After these elaborate preparations the main part of the proof of the theorem of this section proceeds as follows.

Proof of the theorem. Let $\mathcal{K}'$ denote the part of $\mathcal{D}'$ bounded on $\mathcal{D}'$ by all the $\mathcal{O}_{u'}$. This domain $\mathcal{K}'$ is reproduced by $\mathfrak{F}'$, because the collection of all the $\mathcal{O}_{u'}$ is. Let $\mathcal{M}'$ denote the intersection $\mu n \cap \mathcal{K}'$. This is a bounded point set, because all those sides of μn in $\mathcal{D}'$ which have infinite length are cut by some $\mathcal{O}_{u'}$. All equivalence classes of vertices of $\mathfrak{F}'\mu n$ in $\mathcal{D}'$ and of intersection points of $\mathfrak{F}'\mu n \cap \mathcal{D}'$ are represented on μn, and this finite set of points of $\mu n \cap \mathcal{D}'$ belongs to $\mathcal{M}'$, since outside $\mathcal{K}'$ there are no intersection points nor vertices of $\mathfrak{F}'\mu n \cap \mathcal{D}'$. Since $\mathcal{M}'$ is a bounded point set, it has points in common with a finite number of its equivalents only, say with those derived from $\mathcal{M}'$ by the elements

$$\mathfrak{f}'_1 = 1, \mathfrak{f}'_2, \ldots, \mathfrak{f}'_m \tag{6}$$

of $\mathfrak{F}'$. Then $\mathcal{U}' = \bigcup_{s=1}^{m} \mathfrak{f}'_s \mathcal{M}'$ is a bounded point set in $\mathcal{D}'$. Now $\mathcal{M}' = \mu n \cap \mathcal{K}'$ is the image by μ of a certain bounded part n_0 of $n \cap \mathcal{D}$. Then, in virtue of (5), $\mathcal{U}'$ is the image by μ of the set $\mathcal{U} = \bigcup_{s=1}^{m} \mathfrak{f}_s n_0$. This point set $\mathcal{U}$ of $\mathcal{D}$ is bounded; let δ denote its diameter. It is inferred that *if two points of $\mathfrak{F}n \cap \mathcal{D}$ have a distance exceeding δ, then their images by μ cannot coincide.* This is evident if at least one of the two points does not belong to $\mathfrak{F}n_0$, for all points of $\mathfrak{F}n \cap \mathcal{D}$ outside $\mathfrak{F}n_0$ are in a one-to-one correspondence with points of $\mathfrak{F}'\mu n \cap \mathcal{D}'$ outside $\mathcal{K}'$. If both points belong to $\mathfrak{F}n_0$, let one of them, say p_1, belong to n_0, which is part of $\mathcal{U}$. Then the other p_2, does not belong to $\mathcal{U}$, because $[p_1, p_2] > \delta$. Now μp_1 belongs to $\mu n_0 = \mu n \cap \mathcal{K}' = \mathcal{M}'$, and μp_2 does not belong to $\mathcal{U}'$. Therefore μp_1 and μp_2 do not coincide. Then also $\mu \mathfrak{f} p_1$ and $\mu \mathfrak{f} p_2$, $\mathfrak{f} \in \mathfrak{F}$, do not coincide, for in virtue of (5) they are the points $\mathfrak{f}'(\mu p_1)$ and $\mathfrak{f}'(\mu p_2)$, and $\mathfrak{f}'$ is a motion or reversion in $\mathcal{D}'$. Since every point of $\mathfrak{F}n \cap \mathcal{D}$ has an equivalent on n, the assertion is proved.

It may be mentioned that in the particular case where $\mathfrak{F}$, and thus also $\mathfrak{F}'$, contains neither boundary translations nor limit-rotations one gets $\mathcal{K}' = \mathcal{D}'$. In this case n is a simple, closed polygon in $\mathcal{D}$, and $n_0 = n$. Also $\mathcal{M}' = \mu n$ is a closed polygon in $\mathcal{D}'$, which, in general, is not simple. All conclusions are more easily established in this much simpler case.

Let $\mathcal{A}_1$ and $\mathcal{A}_2$ be two divergent inner axes of $\mathfrak{F}$, $\mathfrak{a}_1$ and $\mathfrak{a}_2$ primary translations along $\mathcal{A}_1$ and $\mathcal{A}_2$ respectively, $\mathcal{H}_1$ the half-plane determined by $\mathcal{A}_1$ and not containing $\mathcal{A}_2$, $\mathcal{H}_2$ the half-plane determined by $\mathcal{A}_2$ and not containing $\mathcal{A}_1$, $\mathcal{J}_1$ and $\mathcal{J}_2$ the open arcs of $\mathcal{E}$ bounding $\mathcal{H}_1$ and $\mathcal{H}_2$ respectively, compare Fig. 24.6.

Let h_1 and h_2 be hypercycles in $\mathcal{H}_1$ and $\mathcal{H}_2$ belonging to $\mathcal{A}_1$ and $\mathcal{A}_2$ respectively at a distance $\geqq \frac{\delta}{2}$ from these axes. They may be chosen such as not to pass through any vertex of $\mathfrak{F}n$. Since the end-points $u(\mathfrak{a}_1)$ and $v(\mathfrak{a}_1)$ of $\mathcal{A}_1$ are outside $\mathcal{D}^*$ and thus are not boundary points of any of the domains $\mathfrak{F}\mathcal{N}$, h_1 intersects $\mathfrak{F}n$. Let p be a point of $\mathfrak{F}n$ on h_1, and consider the arc of h_1 from p to $\mathfrak{a}_1 p$. It intersects a finite number of domains of the collection $\mathfrak{F}\mathcal{N}$, say $\mathfrak{f}_s \mathcal{N}$, $1 \leqq s \leqq m$. Then h_1 intersects all domains derived from these m domains by all powers of $\mathfrak{a}_1$ and no other domain of the collection $\mathfrak{F}\mathcal{N}$, because h_1 is carried into itself by $\mathfrak{a}_1$. Let q_1 denote the region

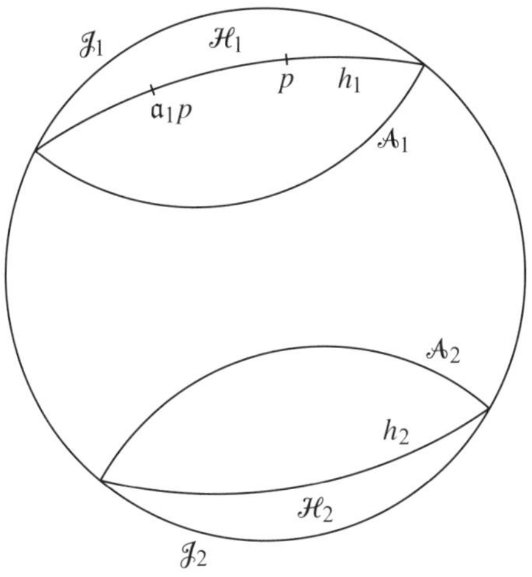

Figure 24.6

bounded by h_1, $\mathcal{J}_1$, $u(\mathfrak{a}_1)$, $v(\mathfrak{a}_1)$ and put $\mathcal{Q}_1 = q_1 \cup \mathcal{J}_1$. Consider the union

$$\mathcal{W} = \bigcup_{r=-\infty}^{\infty} \bigcup_{s=1}^{m} \mathfrak{a}_1^r \mathfrak{f}_s \overline{\mathcal{N}},$$

and put $\mathcal{R} = \mathcal{W} \cap \mathcal{Q}_1$. The boundary of $\mathcal{R}$ in $\mathcal{D}$ apart from h_1 consists of sides of $\mathfrak{F}n$ and is reproduced by the powers of $\mathfrak{a}_1$, since both $\mathcal{W}$ and $\mathcal{Q}_1$ are. The boundary of $\mathcal{R}$ may contain limit-centres of $\mathfrak{F}$ situated in $\mathcal{J}_1$, or points or arcs of intervals of discontinuity of $\mathfrak{F}$ belonging to $\mathcal{J}_1$, namely if such is the case of some of the domains $\mathfrak{f}_s \overline{\mathcal{N}}$. Thus the intersection of the boundary of $\mathcal{R}$ with $\mathcal{Q}_1$ is a continuous path $\mathfrak{C}_1$ on $\mathfrak{F}n$, reproduced by the powers of $\mathfrak{a}_1$, and thus having its ends converging to $u(\mathfrak{a}_1)$ and $v(\mathfrak{a}_1)$.

The path $\mathfrak{C}_1$ on $\mathfrak{F}n$ is now mapped into the path $\mu\mathfrak{C}_1$ on $\mathfrak{F}'\mu n$. $\mu\mathfrak{C}_1$ is reproduced by $\mathfrak{a}_1'$, because $\mathfrak{C}_1$ is reproduced by $\mathfrak{a}_1$, and (5) holds for μ. Hence $\mu\mathfrak{C}_1$ has its ends converging to the end-points of the axis $\mathcal{A}_1'$ of $\mathfrak{a}_1'$.

This process is repeated for $\mathcal{A}_2$, the set $\mathcal{Q}_2$ playing the rôle of $\mathcal{Q}_1$, and the path $\mathfrak{C}_2$ being formed in the same way as $\mathfrak{C}_1$. Then $\mu\mathfrak{C}_2$ is a continuous path on $\mathfrak{F}'\mu n$ connecting $u(\mathfrak{a}_2')$ and $v(\mathfrak{a}_2')$. The distance of the point sets $\mathcal{Q}_1$ and $\mathcal{Q}_2$ exceeds δ. Hence the distance of any point of $\mathfrak{C}_1 \cap \mathcal{D}$ from any point of $\mathfrak{C}_2 \cap \mathcal{D}$ exceeds δ, and their images by μ therefore do not coincide. Let p_1 be a point of $\mathfrak{C}_1$ in the interval of discontinuity or limit-centre u_1, and p_2 a point of $\mathfrak{C}_2$ in the interval of discontinuity or limit-centre u_2. Then μp_1 is in the u_1' corresponding to u_1, and μp_2 in the u_2' corresponding to u_2; and u_1' and u_2' are different, because u_1 and u_2 are different. Hence no point of $\mu\mathfrak{C}_1$ on $\mathcal{E}'$ coincides with a point of $\mu\mathfrak{C}_2$ on $\mathcal{E}'$. Thus, in all, $\mu\mathfrak{C}_1$ and $\mu\mathfrak{C}_2$ are disjoint. It follows that the end-points of $\mathcal{A}_1'$ are not separated on $\mathcal{E}'$ by the end-points of $\mathcal{A}_2'$. Hence $\mathcal{A}_1'$ and $\mathcal{A}_2'$ are divergent. Condition 3) in Section 1 therefore holds for $\mathfrak{F}$ and $\mathfrak{F}'$.

Until now we have not been able to exclude the possibility of an inner translation of $\mathfrak{F}$ corresponding to a reversed translation of $\mathfrak{F}'$, or vice versa. If this should happen, their squares are corresponding inner translations. After this remark it follows from condition 3) exactly in the same way as in Section 1 that the one-to-one correspondence between the sets $\mathcal{V}$ and $\mathcal{V}'$ of inner fundamental points of $\mathfrak{F}$ and $\mathfrak{F}'$ respectively is order preserving, and from this that motions correspond to motions and reversions to reversions. Then condition 2) in Section 1 is seen to hold. – This finishes the proof of the theorem of this section. □

Some special cases of the theorem may be mentioned.

Any isomorphism between two finitely generated $\mathfrak{F}$-groups for which boundary translations correspond to boundary translations and limit-rotations to limit-rotations is a geometrical isomorphism.

In particular, if both groups contain no boundary translations, thus $\mathcal{K}(\mathfrak{F}) = \mathcal{D}$, $\mathcal{K}(\mathfrak{F}') = \mathcal{D}'$, then every isomorphism for which limit-rotations correspond to limit-rotations is a geometrical isomorphism.

Any isomorphism between two finitely generated $\mathfrak{F}$-groups which contain neither boundary translations nor limit-rotations is a geometrical isomorphism.

24.9 Remarks concerning the preceding section. In view of the last statement of the preceding section it is emphasized that the concept of t-isomorphism is, in general, more restrictive than the concept of isomorphism. We mention a simple example: Let translations $\mathfrak{p}_1, \mathfrak{p}_2, \mathfrak{p}_3$ with the relation $\mathfrak{p}_1\mathfrak{p}_2\mathfrak{p}_3 = 1$ generate an $\mathfrak{E}(\to\to\to)$. This then is a free group with $\mathfrak{p}_1$ and $\mathfrak{p}_2$ as free generators. Hence $\mathfrak{p}_1' = \mathfrak{p}_1\mathfrak{p}_2^{-1}$, $\mathfrak{p}_2' = \mathfrak{p}_2$ defines an automorphism of $\mathfrak{E}$. Since $\mathfrak{p}_1$ is a boundary translation and $\mathfrak{p}_1'$ an inner translation of $\mathfrak{E}$, this isomorphism of $\mathfrak{E}$ with itself is not a t-isomorphism.

It will be seen from the proof of the theorem of the preceding section that the decisive fact is the following: The mapping μ of the simple network $\mathfrak{F}n$ of $\mathcal{D}$ into $\mathcal{D}'$ distorts, in general, that network in such a way that the images of the different meshes overlap to some degree, but that distortion is "finite"in the sense that the image of a mesh has points in common only with a finite number of its equivalents, and the distortion does not affect the neighbourhood of points of the mesh on $\mathcal{E}$. This idea of "finite distortion" is due to *Max Dehn*, who used it in the case of fundamental groups of closed surfaces without singularities, where all elements of the group are inner translations, and who for such groups proved a theorem equivalent to the statement that divergent axes correspond to divergent axes. Dehn set forth these results in some lectures in 1920–21, but no complete publication seems to exist. Some indications may, however, be found in a paper of Dehn in Mathematische Annalen, vol. 71, pg. 132–133.

The condition of the theorem of the preceding section that the isomorphic groups are finitely generated is necessary. This may be shown by a simple example: Let $\mathfrak{F}$ denote a group not satisfying the condition of quasi-compactness and such that $\mathcal{K}(\mathfrak{F})$ possesses a limit side which is not cut by a reflection line. As such one may for instance take the group $\mathfrak{F}$ of an incomplete reflection strip with an infinite number of sides (§20.11). The line $\mathcal{A}$ of §20.11 is a limit side of $\mathcal{K}(\mathfrak{F})$, and $\mathfrak{A}_{\mathcal{A}} = 1$. Let $\mathfrak{s}$ denote the reflection in $\mathcal{A}$ and $\mathfrak{c}$ the half-turn around a point of $\mathcal{A}$, and let $\mathfrak{S}$ and $\mathfrak{C}$ denote the groups of order 2 generated by $\mathfrak{s}$ and $\mathfrak{c}$ respectively. One may then form the free products

$$\mathfrak{F}_1 = \mathfrak{F} * \mathfrak{S}, \quad \mathfrak{F}_2 = \mathfrak{F} * \mathfrak{C}$$

with no amalgamation, since the factors have only the identity in common. $\mathcal{K}(\mathfrak{F}_1)$ mod $\mathfrak{F}_1$ is a complete reflection strip (§20.13), and $\mathcal{K}(\mathfrak{F}_2)$ mod $\mathfrak{F}_2$ is a conical reflection strip (§20.15). The two groups are generated by a set of generators of $\mathfrak{F}$ together with $\mathfrak{s}$ and $\mathfrak{c}$ respectively, and a complete set of defining relations is obtained by adding to a set of defining relations between the generators of $\mathfrak{F}$ the relations $\mathfrak{s}^2 = 1$ and $\mathfrak{c}^2 = 1$ respectively. Thus if one lets each generator of $\mathfrak{F}$ correspond to itself, and $\mathfrak{s}$ correspond to $\mathfrak{c}$, then an isomorphism I between $\mathfrak{F}_1$ and $\mathfrak{F}_2$ arises. This isomorphism makes boundary translations of one group correspond to boundary translations of the other, because any interval of discontinuity $\mathcal{I}_1$ of $\mathfrak{F}_1$ is derived from an interval of discontinuity $\mathcal{I}$ of $\mathfrak{F}$ by an element $\mathfrak{f}_1$ of $\mathfrak{F}_1$, and $\mathcal{I}_1$ corresponds to the interval of discontinuity $\mathcal{I}_2$ of $\mathfrak{F}_2$ derived from $\mathcal{I}$ by the element $\mathfrak{f}_2$ of $\mathfrak{F}_2$ corresponding to $\mathfrak{f}_1$; the same for limit-rotations. Even simpler: One may choose the reflection strip such that all its vertices are in $\mathcal{D}$; then neither $\mathfrak{F}_1$ nor $\mathfrak{F}_2$ contains boundary translations or limit-rotations. Hence I satisfies the conditions of the theorem of the preceding section except that the groups are not finitely generated, and I is not a t-isomorphism, because the reflection $\mathfrak{s}$ corresponds to the half-turn $\mathfrak{c}$.

24.10 Invariance of type numbers. Since in two t-isomorphic groups $\mathfrak{F}$ and $\mathfrak{F}'$ centres of the type $\mathfrak{A}(\times_\nu)$ are in one-to-one correspondence, and likewise centres of the type $\mathfrak{A}(\odot_\nu)$, the type numbers $\vartheta(\times_\nu)$ (§17.4) are the same in both groups, and likewise the type numbers $\vartheta(\odot_\nu)$. – A reflection chain has been seen to correspond to a reflection chain, and, in particular, a reflection ring to a reflection ring (Section 5). Hence the type number $\vartheta(-)$ is the same in both groups. Among the reflection rings are the simple closed reflection edges; they correspond to axes of the type $\mathfrak{A}(- \rightarrow)$. These are inner axes which are simple modulo $\mathfrak{F}$ and at the same time reflection lines, and they do not pass through any centre of $\mathfrak{F}$. They are thus not cut by any reflection line. From the preservation of the cyclical order of all inner fundamental points and end-points of reflection lines follows that such axes are in a one-to-one correspondence. Hence the type number $\vartheta(- \rightarrow)$ is also the same in both groups. This number is included in $\vartheta(-)$. – The axis of a reversed translation is an inner axis and corresponds to the axis of a reversed translation. If such axis in one group intersects no reflection line, then it again follows from the preservation of the cyclical order that the corresponding axis of the other group intersects no reflection line. Hence the type number $\vartheta(\rightleftharpoons)$ is the same in both groups.

As to the remaining type numbers $\vartheta(\rightarrow)$, $\vartheta(\hookrightarrow)$, $\vartheta(\|)$, $\vartheta(\wedge)$ introduced in §17.4, they are obviously the same in both groups, if the isomorphism is a g-isomorphism, whereas for a t-isomorphism one can only assert that the number $\vartheta(\rightarrow) + \vartheta(\hookrightarrow)$ is the same in both groups, and likewise the number $\vartheta(\|) + \vartheta(\wedge)$.

If these groups are finitely generated, then all these type numbers are finite.

It will result later that also $\Phi(\mathfrak{F}) = \Phi(\mathfrak{F}')$ and $p(\mathfrak{F}) = p(\mathfrak{F}')$.

§25 Topological and geometrical homeomorphism

25.1 ***t*-mappings.** As in the preceding paragraph, we consider an $\mathfrak{F}$-group $\mathfrak{F}$ operating on a disc $\mathcal{D}$ bounded by a unit circle $\mathcal{E}$ and an $\mathfrak{F}$-group $\mathfrak{F}'$ operating on a disc $\mathcal{D}'$ bounded by a unit circle $\mathcal{E}'$. Let t denote a topological mapping of $\mathcal{D}$ onto $\mathcal{D}'$ such that t maps every pair of points of $\mathcal{D}$ which are equivalent with respect to $\mathfrak{F}$ into a pair of points of $\mathcal{D}'$ which are equivalent with respect to $\mathfrak{F}'$, and conversely for the mapping t^{-1} of $\mathcal{D}'$ onto $\mathcal{D}$. Such mappings will be called *t-mappings*. The two groups play equal rôles in this definition.

Let x be a point of $\mathcal{D}$ which is regular with respect to $\mathfrak{F}$. Let ω be a neighbourhood of x which contains no two equivalent points with respect to $\mathfrak{F}$; for instance, any neighbourhood of x within the characteristic neighbourhood $\mathfrak{C}(x)$ of x (§14.4) has this property. Then $t\omega$ is a neighbourhood of the point tx of $\mathcal{D}'$, and the assumptions on t imply that $t\omega$ contains no two points equivalent with respect to $\mathfrak{F}'$. Hence tx is regular. Thus t carries the set $\mathcal{R}$ of regular points of $\mathfrak{F}$ in $\mathcal{D}$ into regular points of $\mathfrak{F}'$ in $\mathcal{D}'$, and conversely for t^{-1}. Thus the sets $\mathcal{R}$ and $\mathcal{R}'$ of regular points for the two groups correspond by the t-mapping and the same then holds for the sets of singular points, if any.

A centre c of $\mathfrak{F}$ with corresponding group $\mathfrak{A}(\circlearrowleft_\nu)$ is an isolated boundary point of $\mathcal{R}$ and thus corresponds to an isolated boundary point tc of $\mathcal{R}'$. Hence tc is a centre with corresponding group $\mathfrak{A}(\circlearrowleft_\nu)$. Let ω be a circular neighbourhood of c within $\mathfrak{C}(c)$; then any point of ω different from c belongs to a set of exactly ν equivalent points in ω. The same then holds for the neighbourhood $t\omega$ of tc. Hence the order ν is the same for c and tc.

Assume that $\mathfrak{F}$ contains reflections. Then $\mathcal{R}$ falls into different components, the same then holds for $\mathcal{R}'$, and such components correspond in pairs. Assume c to be a centre of $\mathfrak{F}$ with corresponding group $\mathfrak{A}(\times_\nu)$. Then c is a boundary point of 2ν components of $\mathcal{R}$. Thus tc is a boundary point of exactly 2ν components of $\mathcal{R}'$. Hence tc is a centre of $\mathfrak{F}'$ with corresponding group $\mathfrak{A}(\times_\nu)$, ν being the same for c and tc.

Points of reflection lines of $\mathfrak{F}$ not centres of $\mathfrak{F}$ then correspond to points of reflection lines of $\mathfrak{F}'$ not centres of $\mathfrak{F}'$. The boundary segments of $\mathcal{R}$ emanating from a centre c

of $\mathfrak{F}$ with corresponding group $\mathfrak{A}(\times_\nu)$ are by t mapped onto the boundary segments of $\mathcal{R}'$ emanating from tc preserving their cyclical order around the centres. This implies that a complete reflection line of $\mathfrak{F}$ is in all cases mapped upon a complete reflection line of $\mathfrak{F}'$.

Let $\mathcal{R}_1$ be a component of $\mathcal{R}$, x_0 a point of $\mathcal{R}_1$, and $\mathfrak{f}$ an element of $\mathfrak{F}$. Since $\mathfrak{f}$ reproduces $\mathcal{R}$, the point $\mathfrak{f}x_0$ is a point of the component $\mathfrak{f}\mathcal{R}_1$ of $\mathcal{R}$; this component $\mathfrak{f}\mathcal{R}_1$ either coincides with $\mathcal{R}_1$, or $\mathcal{R}_1$ and $\mathfrak{f}\mathcal{R}_1$ are disjoint. tx_0 and $t\mathfrak{f}x_0$ are equivalent with respect to $\mathfrak{F}'$. There is thus an element $\mathfrak{f}'$ of $\mathfrak{F}'$ such that $t\mathfrak{f}x_0 = \mathfrak{f}'tx_0$. This element $\mathfrak{f}'$ is uniquely determined, because tx_0 and $t\mathfrak{f}x_0$ are regular points for $\mathfrak{F}'$. Consider now the mapping function $T = t^{-1}\mathfrak{f}'^{-1}t\mathfrak{f}$. This T is a topological mapping of $\mathcal{D}$ onto itself, and T as well as its inverse $T^{-1} = \mathfrak{f}^{-1}t^{-1}\mathfrak{f}'t$ maps any point of $\mathcal{D}$ into an equivalent point of $\mathcal{D}$. Moreover $Tx_0 = x_0$, and thus $T\mathcal{R}_1 = \mathcal{R}_1$. If ξ is any point of $\mathcal{R}$ for which $T\xi = \xi$, and ω is a neighbourhood of ξ such that both ω and $T\omega$ are contained in the characteristic neighbourhood $\mathfrak{C}(\xi)$ of ξ, then for any point x of ω one gets $Tx = x$, because T maps x on an equivalent point in $\mathfrak{C}(\xi)$, and x is the only representative of its equivalence class in $\mathfrak{C}(\xi)$. Thus those points of $\mathcal{R}$ in which T is the identical mapping form an open subset of $\mathcal{R}$. In particular, let $\mathcal{R}_1^*$ denote the open subset of $\mathcal{R}_1$ in which T is the identity. Then $\mathcal{R}_1^*$ is not empty, since it contains x_0. If y is an accumulation point of $\mathcal{R}_1^*$, then $Ty = y$ in virtue of the continuity of T. Hence $\mathcal{R}_1^*$ is an open subset of $\mathcal{R}_1$ which is closed on $\mathcal{R}_1$, and thus $\mathcal{R}_1^* = \mathcal{R}_1$. – Let y be a point on a reflection line $\mathcal{L}$ bounding $\mathcal{R}_1$ and not a centre, and let $\mathfrak{s}$ denote the reflection in $\mathcal{L}$. Then $Ty = y$. If ω is a neighbourhood of y such that both ω and $T\omega$ are contained in $\mathfrak{C}(y)$, and if x is a point of $\omega \cap \mathcal{R}_1$ such that also $\mathfrak{s}x$ belongs to ω, then x and $\mathfrak{s}x$ are the only two representatives of their equivalence class in $\mathfrak{C}(y)$. Since $Tx = x$, one gets also $T\mathfrak{s}x = \mathfrak{s}x$. Since $\mathfrak{s}x$ belongs to the component $\mathfrak{s}\mathcal{R}_1$ of $\mathcal{R}$, T is the identity in $\mathfrak{s}\mathcal{R}_1$. This implies that T is the identity in the whole of $\mathcal{R}$, and thus, by continuity, in the whole of $\mathcal{D}$. One thus gets the functional equation

$$t\mathfrak{f}x = \mathfrak{f}'tx \tag{1}$$

where x ranges over the whole of $\mathcal{D}$ and $\mathfrak{f}$ over $\mathfrak{F}$, and $\mathfrak{f}'$ is uniquely determined by $\mathfrak{f}$; for its determination one can use an arbitrary regular point for $\mathfrak{F}$ in $\mathcal{D}$.

Since t is a one-to-one mapping $\mathcal{D}$ onto $\mathcal{D}'$, and the assumption on t implies that a complete equivalence class $\mathfrak{F}x$ is mapped onto a complete equivalence class $\mathfrak{F}'tx$, the element $\mathfrak{f}'$ in (1) ranges over the whole of $\mathfrak{F}'$, as $\mathfrak{f}$ ranges over $\mathfrak{F}$. Hence (1) implies

$$t^{-1}\mathfrak{f}'x' = \mathfrak{f}t^{-1}x' \tag{2}$$

where x' ranges over $\mathcal{D}'$ and $\mathfrak{f}'$ over $\mathfrak{F}'$. Thus a one-to-one correspondence between the elements of $\mathfrak{F}$ and $\mathfrak{F}'$ results. This correspondence is an isomorphism, because (1) implies

$$t\mathfrak{f}_1\mathfrak{f}_2x = \mathfrak{f}_1't\mathfrak{f}_2x = \mathfrak{f}_1'\mathfrak{f}_2'tx.$$

We denote it by I, putting $I(\mathfrak{f}) = \mathfrak{f}'$, $I^{-1}(\mathfrak{f}') = \mathfrak{f}$.

If x is an arbitrary point of $\mathcal{D}$, $tx = x'$, $\mathfrak{A}_x$ the subgroup of $\mathfrak{F}$ leaving x invariant, and $\mathfrak{f}$ an arbitrary element of $\mathfrak{A}_x$, then by (1)

$$\mathfrak{f}'x' = \mathfrak{f}'tx = t\mathfrak{f}x = tx = x',$$

and $\mathfrak{f}'$ is thus an element of the subgroup $\mathfrak{A}_{x'}$ of $\mathfrak{F}'$ leaving x' invariant; and conversely, if one starts from $\mathfrak{F}'$. Hence $I(\mathfrak{A}_x) = \mathfrak{A}_{x'}$. If x is chosen on the line $\mathcal{L}$ of the reflection $\mathfrak{s}$ of $\mathfrak{F}$ and not in a centre, it follows that $\mathfrak{s}'$ is the reflection in the line $\mathcal{L}' = t\mathcal{L}$. Thus reflections correspond to reflections. By choosing x in a centre it then follows that rotations correspond to rotations; their centres correspond by the t-mapping. Two corresponding rotations are of the same order, since the correspondence is an isomorphism. If $\mathfrak{c}$ is a primary rotation of order ν about the centre c of $\mathfrak{F}$, and if h denotes a half-line issuing from c, then the half-lines $\mathfrak{c}^r h$, $0 \leqq r \leqq \nu - 1$, are arranged around c in the order of their numbering, and the same then holds by the topological mapping t for the curves $t\mathfrak{c}^r h = \mathfrak{c}'^r th$. Hence $\mathfrak{c}' = I(\mathfrak{c})$ is also primary.

Let γ denote a simple, closed, oriented curve in $\mathcal{D}$. Then $t\gamma$ is a simple, closed curve in $\mathcal{D}'$, and an orientation of $t\gamma$ is derived from the orientation of γ by t. If $\mathfrak{f} \in \mathfrak{F}$ is a motion, then the orientations of γ and $\mathfrak{f}\gamma$ are in accordance in $\mathcal{D}$. Hence motions correspond to motions, and then also reversions to reversions. Since reflections correspond to reflections, reversed translations correspond to reversed translations. It then follows that the collection of all translations and limit-rotations in one group correspond to the same collection in the other group.

Let $\mathfrak{F}_0$ be a subgroup of $\mathfrak{F}$, and put $I(\mathfrak{F}_0) = \mathfrak{F}'_0$. Let Ω be a subset of $\mathcal{D}$ in which $\mathfrak{F}_0$ and $\mathfrak{F}$ have equal effect, and put $t\Omega = \Omega'$. Let x' and $\mathfrak{f}'x'$, $\mathfrak{f}' \in \mathfrak{F}'$, both belong to Ω'. Then $t^{-1}x'$ and $t^{-1}\mathfrak{f}'x'$ both belong to Ω, and in virtue of (2) one gets $t^{-1}\mathfrak{f}'x' = \mathfrak{f}t^{-1}x'$. Hence $\mathfrak{f}$ belongs to $\mathfrak{F}_0$, and thus $\mathfrak{f}'$ to $\mathfrak{F}'_0$. Thus $\mathfrak{F}'_0$ and $\mathfrak{F}'$ have equal effect in Ω'.

Let $\mathcal{A}$ denote either an axis of $\mathfrak{F}$, or a horocycle belonging to a limit-centre u of $\mathfrak{F}$ such that $\mathfrak{A}_u$ and $\mathfrak{F}$ have equal effect in the interior of $\mathcal{A}$. In order to treat the two cases together we write $\mathfrak{A}_{\mathcal{A}}$ for $\mathfrak{A}_u$ in the latter case. If $\mathfrak{A}_{\mathcal{A}}$ is of the type $\mathfrak{A}(\|)$ or $\mathfrak{A}(\wedge)$, let p be the intersection of $\mathcal{A}$ with the line $\mathcal{L}$ of the reflection $\mathfrak{s}$ contained in $\mathfrak{A}_{\mathcal{A}}$, and q the intersection of $\mathcal{A}$ with the line $\mathcal{L}_1$ of the reflection $\mathfrak{s}_1$, $\mathcal{L}$ and $\mathcal{L}_1$ being consecutive reflection lines cutting $\mathcal{A}$. Then $\mathfrak{s}\mathfrak{s}_1 = \mathfrak{a}$ is a primary translation or limit-rotation carrying $\mathcal{A}$ into itself. Now, tp is on the line $\mathcal{L}' = t\mathcal{L}$ of the reflection $\mathfrak{s}' = I(\mathfrak{s})$, and tq is on the line $\mathcal{L}'_1 = t\mathcal{L}_1$ of the reflection $\mathfrak{s}'_1 = I(\mathfrak{s}_1)$. The part $\mathcal{S}_0 = pq$ of $\mathcal{A}$ is by t mapped into a Jordan arc $t\mathcal{S}_0$ connecting tp and tq. The part $\mathcal{S}_0$ together with $\mathfrak{s}\mathcal{S}_0$ constitutes a part $\mathcal{S}$ of $\mathcal{A}$ joining q with $\mathfrak{s}q = \mathfrak{s}\mathfrak{s}_1 q = \mathfrak{a}q$, and $t\mathcal{S}$ is a Jordan arc joining tq and $\mathfrak{s}'tq = t\mathfrak{s}q = t\mathfrak{a}q = \mathfrak{a}'tq$. Here $\mathfrak{a}' = \mathfrak{s}'\mathfrak{s}'_1$. Fig. 25.1 illustrates the case where both $\mathfrak{a}$ and $\mathfrak{a}'$ are translations.

If $\mathfrak{A}_{\mathcal{A}}$ is of the type $\mathfrak{A}(\rightarrow)$ or $\mathfrak{A}(\hookrightarrow)$, then q shall denote an arbitrary point of $\mathcal{A}$, and $\mathfrak{a}$ a primary translation or limit-rotation carrying $\mathcal{A}$ into itself; again $\mathcal{S}$ shall denote the part of $\mathcal{A}$ from q to $\mathfrak{a}q$. Again $t\mathcal{S}$ is a Jordan arc in $\mathcal{D}'$ joining tq and $\mathfrak{a}'tq$. – The two cases concerning the existence or non-existence of reflections in $\mathfrak{A}_{\mathcal{A}}$ are now treated together.

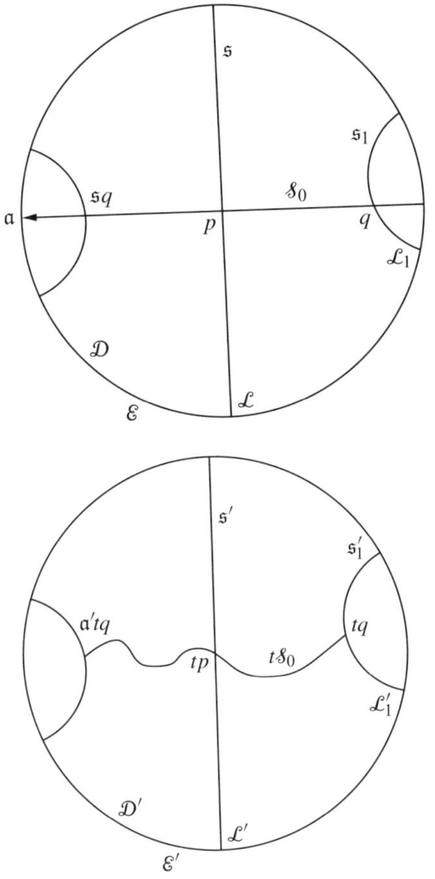

Figure 25.1

The union of the parts $\mathfrak{a}^r \mathcal{S}$ of $\mathcal{A}$, $-\infty < r < \infty$, is the whole of $\mathcal{A}$. Its image by t is, in virtue of (1),

$$t\mathcal{A} = t \bigcup_r \mathfrak{a}^r \mathcal{S} = \bigcup_r t\mathfrak{a}^r \mathcal{S} = \bigcup_r \mathfrak{a}'^r t\mathcal{S}.$$

$\mathcal{A}$ determines two open regions $\mathcal{D}_1$ and $\mathcal{D}_2$ of $\mathcal{D}$. Hence $t\mathcal{A}$ determines in $\mathcal{D}'$ two open regions $t\mathcal{D}_1$ and $t\mathcal{D}_2$ such that $\mathcal{D}' = t\mathcal{D}_1 \cup t\mathcal{D}_2 \cup t\mathcal{A}$. The Jordan arc $t\mathcal{A}$ is carried into itself by the powers of $\mathfrak{a}'$. Now $\mathfrak{a}'$ is either a translation or limit-rotation, as seen above, not necessarily of the same kind as $\mathfrak{a}$. In the first case the ends of $t\mathcal{A}$ converge to the fundamental points of $\mathfrak{a}'$, in the second they both converge to the limit-centre of $\mathfrak{a}'$. Since $\mathcal{A}$ is carried into itself by the elements of $\mathfrak{A}_{\mathcal{A}}$ and by no other element of $\mathfrak{F}$, its image $t\mathcal{A}$ is carried into itself by the elements of $I(\mathfrak{A}_{\mathcal{A}})$ and by no other element of $\mathfrak{F}'$. The Jordan arc $t\mathcal{A}$ is contained between two hypercycles ending in the fundamental points of $\mathfrak{a}'$, if $\mathfrak{a}'$ is a translation, or between two horocycles belonging to the limit-centre of $\mathfrak{a}'$, if $\mathfrak{a}'$ is a limit-rotation.

If $\mathfrak{a}$ is, in particular, not an inner translation of $\mathfrak{F}$, thus either a boundary translation or limit-rotation, then $\mathfrak{A}_{\mathcal{A}}$ and $\mathfrak{F}$ have equal effect in one of the two regions determined in $\mathcal{D}$ by $\mathcal{A}$. Let this be the region called $\mathcal{D}_1$. Hence $I(\mathfrak{A}_{\mathcal{A}})$ and $\mathfrak{F}'$ have equal effect in $t\mathcal{D}_1$. This excludes the possibility of $\mathfrak{a}'$ being an inner translation of $\mathfrak{F}'$. For, in that case there would be fundamental points on the arc of $\mathcal{E}'$ bounding $t\mathcal{D}_1$, and there would be points of $t\mathcal{D}_1$ equivalent by a translation with such fundamental point, thus not leaving the end-points of $t\mathcal{A}$ fixed, hence not belonging to $I(\mathfrak{A}_{\mathcal{A}})$.

Since the groups $\mathfrak{F}$ and $\mathfrak{F}'$ play equal rôles in the definition of t-mappings, it has thus been shown that the collection of boundary translations and limit-rotations in one group corresponds to the same collection in the other group. Therefore, finally, inner translations must correspond to inner translations.

If $\mathcal{A}_1$ and $\mathcal{A}_2$ are concurrent axes of $\mathfrak{F}$, then the Jordan arcs $t\mathcal{A}_1$ and $t\mathcal{A}_2$ intersect in one point of $\mathcal{D}'$. Hence the end-points of $t\mathcal{A}_1$ separate the end-points of $t\mathcal{A}_2$ on $\mathcal{E}'$. These are the end-points of the axes $\mathcal{A}_1'$ and $\mathcal{A}_2'$ corresponding to $\mathcal{A}_1$ and $\mathcal{A}_2$ on account of the isomorphism, and these are thus concurrent.

In all, it has thus been shown that I satisfies the conditions 1), 2), 3) of §24.1, hence the theorem:

If $\mathfrak{F}$ and $\mathfrak{F}'$ are $\mathfrak{F}$-groups in $\mathcal{D}$ and $\mathcal{D}'$, respectively, and if t maps $\mathcal{D}$ topologically onto $\mathcal{D}'$ in such a way that $\mathfrak{F}$-equivalent points of $\mathcal{D}$ are mapped into $\mathfrak{F}'$-equivalent points of $\mathcal{D}'$, and conversely by t^{-1}, then $\mathfrak{F}$ and $\mathfrak{F}'$ are t-isomorphic. This t-isomorphism is determined by (1), *$\mathfrak{f}$ and $\mathfrak{f}'$ denoting corresponding elements.* □

25.2 Extension of a t-mapping to the boundary circles. The t-isomorphism I induced by the t-mapping t as expressed in (1) and (2) determines a one-to-one and order preserving correspondence between the set $\mathcal{V}$ of inner fundamental points of $\mathfrak{F}$ and the set $\mathcal{V}'$ of inner fundamental points of $\mathfrak{F}$ (§24.1), and thereby also a one-to-one correspondence between the cuts in $\mathcal{V}$ and the cuts in $\mathcal{V}'$ (§24.2). A cut in $\mathcal{V}$ is either a single point of $\mathcal{E}$ belonging to $\overline{G}(\mathfrak{F})$, but not to $\mathcal{V}$, or an interval of discontinuity of $\mathfrak{F}$, closed on $\mathcal{E}$; and likewise for a cut in $\mathcal{V}'$.

In this section we denote by w either a single *inner* point of $\overline{G}(\mathfrak{F})$, i.e. a point of $\overline{G}(\mathfrak{F})$ which is not the end-point of an interval of discontinuity of $\mathfrak{F}$, or an interval of discontinuity of $\mathfrak{F}$, closed on $\mathcal{E}$, and we call w a *boundary unit* of $\mathfrak{F}$. Any boundary unit w of $\mathfrak{F}$ then corresponds to a boundary unit $w' = \kappa w$ of $\mathfrak{F}'$, and this mapping κ is an extension of the mapping $\kappa\colon \mathcal{V} \to \mathcal{V}'$ introduced in §24.1. Recalling that the two groups play equal rôles, we may enumerate the possible cases by starting from $\mathfrak{F}$, say.

I) Both boundary units are points. The following subcases arise:

a) If $w \in \mathcal{V}$, then $w' \in \mathcal{V}'$.

b) If w is a limit-centre, then also w' is.

c) If w is an end of $\mathcal{K}(\mathfrak{F})$, then w' is an end of $\mathcal{K}(\mathfrak{F}')$ (§24.3).

d) None of the cases a), b), c) occurs.

II) One boundary unit (say w) is a closed interval of discontinuity, the other a point. Subcases:

a) If w is periodic, then w' is a limit-centre.

b) If w is aperiodic, then w' is a limit point of $\mathfrak{F}'$ which is an end of $\mathcal{K}(\mathfrak{F}')$.

III) Both boundary units are closed intervals of discontinuity. Subcases:

a) Both are periodic.

b) Both are aperiodic.

In the case I d), w is an inner limit point of a finitely generated subgroup of $\mathfrak{F}$ and is neither a limit-centre nor an inner fundamental point of $\mathfrak{F}$. The same then holds for w' in relation to $\mathfrak{F}'$.

The t-mapping can be continuously extended to the boundary circles in the sense expressed by the following theorem:

If Ω is an unbounded subset of $\mathcal{D}$ such that all accumulation points of Ω on $\mathcal{E}$ belong to the boundary unit w of $\mathfrak{F}$, then all accumulation points on $\mathcal{E}'$ of the subset $t\Omega$ of $\mathcal{D}'$ belong to the boundary unit $w' = \kappa w$ of $\mathfrak{F}'$.

Proof. Let y' denote an inner fundamental point of $\mathfrak{F}'$ disjoint with w'. Let $\mathcal{I}_1$ and $\mathcal{I}_2$ be open interval on $\mathcal{E}'$ disjoint with y' and with w' and such that one end-point of $\mathcal{I}_1$ and one end-point of $\mathcal{I}_2$ belong to w'; they coincide, if w' is a single point. If w' is an interval of discontinuity closed on $\mathcal{E}'$, then they are the end-points of w'. Fig. 25.2 illustrates this latter case.

Both $\mathcal{I}_1$ and $\mathcal{I}_2$ contain limit points of $\mathfrak{F}'$, because they terminate either in an inner limit point w', or in the end-points of an interval of discontinuity w' of $\mathfrak{F}'$. Let $\mathcal{A}'$ denote an axis of $\mathfrak{F}'$ having its end-points u' and v' in $\mathcal{I}_1$ and $\mathcal{I}_2$, respectively (§13.3). This is an inner axis of $\mathfrak{F}'$, because both intervals determined on $\mathcal{E}'$ by u' and v' contain limit points of $\mathfrak{F}'$. Let $\mathcal{H}_1'$ denote the half-plane of $\mathcal{D}'$ determined by $\mathcal{A}'$ and containing w' on its boundary and $\mathcal{H}_2'$ the other half-plane determined by $\mathcal{A}'$, thus containing y' on its boundary. As seen in the preceding section, $t^{-1}\mathcal{A}'$ is a Jordan arc in $\mathcal{D}$ connecting $u = \kappa^{-1}u'$ and $v = \kappa^{-1}v'$, and $t^{-1}\mathcal{H}_1'$ is one of the regions determined in $\mathcal{D}$ by $t^{-1}\mathcal{A}'$. Consider the inner axis $\mathcal{B}'$ of $\mathfrak{F}'$ which has y' as an fundamental point. By the consideration of the preceding section $t^{-1}(\mathcal{B}' \cap \mathcal{H}_2')$ has $y = \kappa^{-1}y'$ as its end-point on $\mathcal{E}$, and this is thus a boundary point of $t^{-1}\mathcal{H}_2'$. Since u' and v' separate y' and w' on $\mathcal{E}'$, u and v separate y and w on $\mathcal{E}$. Hence w is on the boundary of $t^{-1}\mathcal{H}_1'$. Since all accumulation points of Ω on $\mathcal{E}$ belong to w, the intersection $\Omega \cap t^{-1}\mathcal{H}_2'$ is bounded, hence $t(\Omega \cap t^{-1}\mathcal{H}_2') = t\Omega \cap \mathcal{H}_2'$ is bounded.

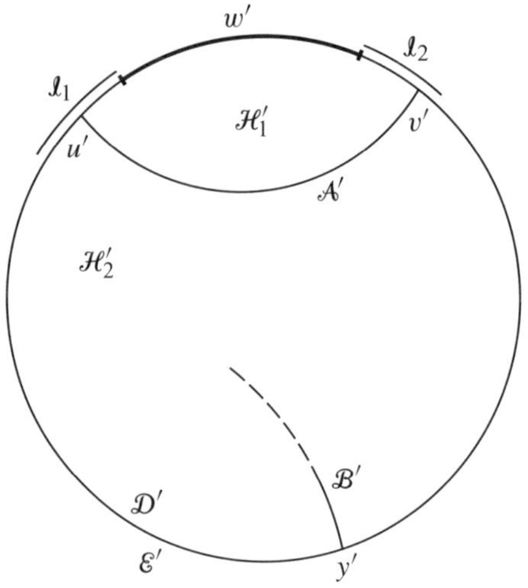

Figure 25.2

There are thus no accumulation points of $t\Omega$ on that arc $u'v'$ of $\mathscr{E}'$ which contains y'. Since $\mathscr{I}_1$ and $\mathscr{I}_2$ can be chosen arbitrarily small, this proves the theorem. □

If one of the groups, say $\mathfrak{F}'$, admits no intervals of discontinuity, then κ is a one-valued mapping function which maps $\mathscr{E}$ onto $\mathscr{E}'$ in such a way that each interval of discontinuity of $\mathfrak{F}$ (if any) is mapped into a single point of $\mathscr{E}'$, and κ is one-to-one in the rest of $\mathscr{E}$. This mapping $\kappa\colon \mathscr{E} \to \mathscr{E}'$ is the continuous extension of the topological mapping $t\colon \mathscr{D} \to \mathscr{D}'$ in this case.

If none of the groups admits intervals of discontinuity, then $\kappa\colon \mathscr{E} \to \mathscr{E}'$ is a topological mapping and is a continuous extension of $t\colon \mathscr{D} \to \mathscr{D}'$. Hence the theorem:

If $\mathscr{K}(\mathfrak{F}) = \mathscr{D}$, $\mathscr{K}(\mathfrak{F}') = \mathscr{D}'$, then any t-mapping of $\mathscr{D}$ onto $\mathscr{D}'$ can be extended to a topological mapping of $\overline{\mathscr{D}}$ onto $\overline{\mathscr{D}}'$. □

25.3 *g*-mappings. Let g denote a topological mapping of $\mathscr{K}(\mathfrak{F})$ onto $\mathscr{K}(\mathfrak{F}')$ such that g maps every pair of points of $\mathscr{K}(\mathfrak{F})$ which are equivalent with respect to $\mathfrak{F}$ into a pair of points of $\mathscr{K}(\mathfrak{F}')$ which are equivalent with respect to $\mathfrak{F}'$, and conversely for the mapping g^{-1} of $\mathscr{K}(\mathfrak{F}')$ onto $\mathscr{K}(\mathfrak{F})$. Such mappings will be called *g-mappings*. The two groups play equal rôles in this definition.

If $\mathscr{K}(\mathfrak{F}) = \mathscr{D}$, then $\mathscr{K}(\mathfrak{F})$ is an open point set. Hence $g\mathscr{K}(\mathfrak{F}) = \mathscr{K}(\mathfrak{F}')$ is an open point set, hence $\mathscr{K}(\mathfrak{F}') = \mathscr{D}'$, and the g-mapping is thus a t-mapping.

We investigate the case in which $\mathscr{K}(\mathfrak{F})$ is not the whole of $\mathscr{D}$. Let $\mathscr{A}$ be a side of $\mathscr{K}(\mathfrak{F})$ and $\mathfrak{A}_{\mathscr{A}}$ the subgroup of $\mathfrak{F}$ carrying $\mathscr{A}$ into itself. Then $\mathfrak{A}_{\mathscr{A}}$ takes one of the four forms

1) $\mathfrak{A}_{\mathcal{A}} = 1$,

2) $\mathfrak{A}_{\mathcal{A}} = \mathfrak{A}(|)$,

3) $\mathfrak{A}_{\mathcal{A}} = \mathfrak{A}(\|)$,

4) $\mathfrak{A}_{\mathcal{A}} = \mathfrak{A}(\to)$.

In the two last cases, $\mathcal{A}$ is an axis of $\mathfrak{F}$, in the two first it is not. Consider the mapping $g\mathcal{A} = \mathcal{A}'$ of $\mathcal{A}$ onto a side $\mathcal{A}'$ of $\mathcal{K}(\mathfrak{F}')$. A regular point x_0 of $\mathcal{A}$ has a neighbourhood ω on $\mathcal{A}$ containing no two equivalent points. The same therefore holds for $g\omega$, and gx_0 is regular. In the case 1), one can take $\omega = \mathcal{A}$, hence $g\mathcal{A} = \mathcal{A}'$ contains no two equivalent points, and $\mathfrak{A}_{\mathcal{A}'} = 1$. In the cases 2) and 3), a component of the regular set on $\mathcal{A}$ is carried into a component of the regular set on $\mathcal{A}'$, and since singular points on a side of the convex domain can only be points on reflection lines, it follows that $\mathfrak{A}_{\mathcal{A}'} = \mathfrak{A}(|)$ in the case 2), and $\mathfrak{A}_{\mathcal{A}'} = \mathfrak{A}(\|)$ in the case 3). Then also $\mathfrak{A}_{\mathcal{A}'} = \mathfrak{A}(\to)$ in the case 4). In the latter case a primary displacement segment of $\mathcal{A}$ is carried into a primary displacement segment of $\mathcal{A}'$, since it contains no two equivalent points except its end-points.

Let $\mathfrak{f}$ by any element of $\mathfrak{F}$ and x_0 a regular point on $\mathcal{A}$. Then $g\mathfrak{f}x_0$ is equivalent with gx_0 with respect to $\mathfrak{F}'$ and, since gx_0 is regular, there is a uniquely determined element $\mathfrak{f}'$ of $\mathfrak{F}'$ such that $g\mathfrak{f}x_0 = \mathfrak{f}'gx_0$. As in Section 1 it follows that, for any given $\mathfrak{f}$, the element $\mathfrak{f}'$ remains the same as long as x_0 ranges over the component of regular points of $\mathcal{A}$ to which it belongs, and further that $\mathfrak{f}'$ is the same for any two adjacent components. By continuity, the equation

$$g\mathfrak{f}x = \mathfrak{f}'gx \tag{3}$$

then also holds for a point x of $\mathcal{A}$ situated on a reflection line of $\mathfrak{F}$. Hence (3) holds for x ranging over a side of $\mathcal{K}(\mathfrak{f})$ and $\mathfrak{f}$ over $\mathfrak{F}$, the element $\mathfrak{f}'$ of $\mathfrak{F}'$ being uniquely determined by the element $\mathfrak{f}$ of $\mathfrak{F}$.

Let $\mathcal{H} = \mathcal{H}(\mathcal{A})$ denote the half-plane outside $\mathcal{K}(\mathfrak{F})$ determined by any side $\mathcal{A}$ of $\mathcal{K}(\mathfrak{F})$, and $\mathcal{H}' = \mathcal{H}'(\mathcal{A}')$ the half-plane outside $\mathcal{K}(\mathfrak{F}')$ determined by $\mathcal{A}' = g\mathcal{A}$. We want to extend g to a mapping of $\mathcal{H}$ onto $\mathcal{H}'$. Let a be any point of $\mathcal{A}$, $h(a)$ the normal of $\mathcal{A}$ in $\mathcal{H}$ erected in a (Fig. 25.3), $a' = ga$ the image of a on $\mathcal{A}'$, $h(a')$ the normal of $\mathcal{A}'$ in $\mathcal{H}'$ erected in a'. For any x on $h(a)$ we define gx as the point x' on $h(a')$ for which $[x', a'] = [x, a]$. Thus a point of $\mathcal{D}$ outside $\mathcal{K}(\mathfrak{F})$ and a point of $\mathcal{D}'$ outside $\mathcal{K}(\mathfrak{F}')$ shall correspond if and only if their distances from $\mathcal{K}(\mathfrak{F})$ and $\mathcal{K}(\mathfrak{F}')$, respectively, are equal and if their nearest points on $\mathcal{K}(\mathfrak{F})$ and $\mathcal{K}(\mathfrak{F}')$, respectively, correspond under the given g-mapping. The extended mapping function g is then a topological mapping of $\mathcal{D}$ onto $\mathcal{D}'$. We want to prove that it preserves equivalence. This has only to be proved for points outside $\mathcal{K}(\mathfrak{F})$. If $\mathfrak{f}$ is any element of $\mathfrak{F}$, then $\mathfrak{f}x$ is situated on $\mathfrak{f}h(a)$, which is the normal of $\mathfrak{f}\mathcal{A}$ in $\mathfrak{f}\mathcal{H}$ erected in $\mathfrak{f}a$. Since a is on $\mathcal{A}$,

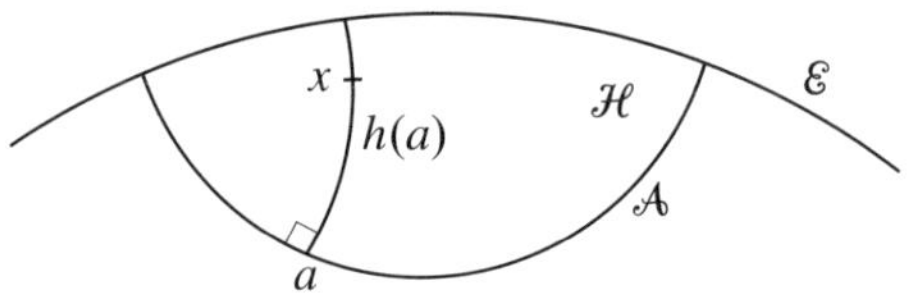

Figure 25.3

one gets by (3)

$$g\mathfrak{f}a = \mathfrak{f}'ga = \mathfrak{f}'a',$$

hence also

$$g\mathfrak{f}h(a) = gh(\mathfrak{f}a) = h(g\mathfrak{f}a) = h(\mathfrak{f}'a') = \mathfrak{f}'h(a').$$

Since

$$[\mathfrak{f}x, \mathfrak{f}a] = [x, a] = [x', a'] = [\mathfrak{f}'x', \mathfrak{f}'a'],$$

one gets

$$g\mathfrak{f}x = \mathfrak{f}'x' = \mathfrak{f}'gx.$$

This proves the assertion.

It is evident from the above construction that the inverse mapping g^{-1} maps $\mathfrak{F}'$-equivalent points of $\mathcal{H}'$ into $\mathfrak{F}$-equivalent points of $\mathcal{H}$. The extended mapping g is thus a t-mapping. It is inferred that (3) holds, if x ranges over $\mathcal{D}$ and $\mathfrak{f}$ over $\mathfrak{F}$. The correspondence $\mathfrak{f} \leftrightarrow \mathfrak{f}'$ is thus a t-isomorphism and, since boundary translations correspond to boundary translations, it is a g-isomorphism. The same is evidently true in the case where $\mathcal{K}(\mathfrak{F}) = \mathcal{D}$ and thus also $\mathcal{K}(\mathfrak{F}') = \mathcal{D}'$. Hence the theorem:

A g-mapping of $\mathcal{K}(\mathfrak{F})$ onto $\mathcal{K}(\mathfrak{F}')$ induces a g-isomorphism of $\mathfrak{F}$ and $\mathfrak{F}'$, namely the one resulting from (3), *where x ranges over $\mathcal{K}(\mathfrak{F})$ and $\mathfrak{f}$ over $\mathfrak{F}$.* □

It should be noted that g-mappings between $\mathcal{K}(\mathfrak{F})$ and $\mathcal{K}(\mathfrak{F}')$ make limit-rotations correspond to limit-rotations. In the case of non-infinitely generated groups corresponding ends of $\mathcal{K}(\mathfrak{F})$ and $\mathcal{K}(\mathfrak{F}')$ are both limit sides or both single points.

The particular topological mapping function g of $\mathcal{D}$ onto $\mathcal{D}'$ constructed above can be extended to a topological mapping of $\overline{\mathcal{D}}$ onto $\overline{\mathcal{D}}'$. In order to do this one has to map the end-point of $h(a)$ on $\mathcal{E}$ into the end-points of the corresponding $gh(a) = h(ga)$ on $\mathcal{E}'$, thus obtaining a topological mapping of the interval of discontinuity $\mathcal{I}(\mathcal{A})$ on $\mathcal{E}$ belonging to $\mathcal{A}$ onto the interval of discontinuity $\mathcal{I}(\mathcal{A}')$ on $\mathcal{E}'$ belonging to $\mathcal{A}' = g\mathcal{A}$. Then a neighbourhood of a point of $\mathcal{I}(\mathcal{A})$ in $\overline{\mathcal{D}}$ is mapped by g onto a neighbourhood of the corresponding point of $\mathcal{I}(\mathcal{A}')$ in $\overline{\mathcal{D}}'$. If this mapping is extended to the end-points of the two corresponding intervals of discontinuity, then the continuous extension of g to $\mathcal{E}$ is effected for all those boundary units of $\mathfrak{F}$ and $\mathfrak{F}'$ which consist of closed intervals of discontinuity. All other boundary units in both groups consist of single points, and the continuous extension of g to them is known from the preceding section.

25.4 t-homeomorphism and g-homeomorphism. A t-mapping $t\colon \mathcal{D} \to \mathcal{D}'$ maps a complete equivalence class of points of $\mathcal{D}$ onto a complete equivalence class of points of $\mathcal{D}'$, and conversely for t^{-1}. It thus defines a mapping τ of the surface $\mathcal{D}$ mod $\mathfrak{F}$ onto the surface $\mathcal{D}'$ mod $\mathfrak{F}'$, and this mapping is one-to-one and continuous. The two surfaces are thus homeomorphic. Since τ and τ^{-1} have the property of mapping conical and angular points of one surface into conical and angular points, respectively, of the other surface, they express a homeomorphism of a particular character; we call it a *t-homeomorphism.* Reflection edges and reflection chains of one surface correspond to reflection edges and reflection chains of the other surface, thus in particular reflection rings to reflection rings.

Similarly, a g-mapping $g\colon \mathcal{K}(\mathfrak{F}) \to \mathcal{K}(\mathfrak{F}')$ maps a complete equivalence class of points of $\mathcal{K}(\mathfrak{F})$ onto a complete equivalence class of points of $\mathcal{K}(\mathfrak{F}')$. It thus defines a mapping γ of the surface $\mathcal{K}(\mathfrak{F})$ mod $\mathfrak{F}$ onto the surface $\mathcal{K}(\mathfrak{F}')$ mod $\mathfrak{F}'$, and this mapping is one-to-one and continuous. In this case we call the homeomorphism of the two surfaces a *g-homeomorphism.* Here conical points, angular points, reflection edges, and boundaries of the two surfaces correspond in pairs.

A g-homeomorphism of $\mathcal{K}(\mathfrak{F})$ mod $\mathfrak{F}$ and $\mathcal{K}(\mathfrak{F}')$ mod $\mathfrak{F}'$ can be extended to a t-homeomorphism of $\mathcal{D}$ mod $\mathfrak{F}$ and $\mathcal{D}'$ mod $\mathfrak{F}'$ by the extension of the mapping function g described above. This means extending the topological mapping γ of $\mathcal{K}(\mathfrak{F})$ mod $\mathfrak{F}$ onto $\mathcal{K}(\mathfrak{F}')$ mod $\mathfrak{F}'$ in pairs to those funnels, half-funnels, half-planes, or quarter-planes which are added to $\mathcal{K}(\mathfrak{F})$ mod $\mathfrak{F}$ and $\mathcal{K}(\mathfrak{F}')$ mod $\mathfrak{F}'$ in order to get $\mathcal{D}$ mod $\mathfrak{F}$ and $\mathcal{D}'$ mod $\mathfrak{F}'$.

Consider a topological mapping T of the truncated domain $\mathcal{K}^*(\mathfrak{F})$ onto the truncated domain $\mathcal{K}^*(\mathfrak{F}')$ which preserves equivalence. T may well map a boundary axis $\mathcal{A}$ of $\mathcal{K}^*(\mathfrak{F})$ onto a bounding horocycle $\mathcal{A}'$ of $\mathcal{K}^*(\mathfrak{F}')$, or conversely, if the two corresponding groups are $\mathfrak{A}(\to)$ and $\mathfrak{A}(\hookrightarrow)$, or if they are $\mathfrak{A}(\|)$ and $\mathfrak{A}(\wedge)$. One may then extend T by a topological mapping to the half-plane $\mathcal{H}$ belonging to $\mathcal{A}$ onto the interior $\mathcal{H}'$ of the horocycle $\mathcal{A}'$ such that equivalent points of $\mathcal{H}$ are mapped on equivalent points of $\mathcal{H}'$, and conversely by the inverse mapping. This can be done by a construction analogous to the one described in the preceding section. Thus $T\colon \mathcal{K}^*(\mathfrak{F}) \to \mathcal{K}^*(\mathfrak{F}')$ can be extended to a topological mapping $t\colon \mathcal{D} \to \mathcal{D}'$, preserving equivalence, and t then expresses a t-homeomorphism between $\mathcal{D}$ mod $\mathfrak{F}$ and $\mathcal{D}'$ mod $\mathfrak{F}'$. Here $\mathcal{H}$ mod $\mathfrak{A}_{\mathcal{A}}$ is a funnel or half-funnel, and $\mathcal{H}'$ mod $\mathfrak{A}_{\mathcal{A}'}$ is a mast or a half-mast. The possibility of a boundary translation corresponding to a limit-rotation in the isomorphism induced by a t-mapping is thus the expression of the fact that a funnel and a mast are homeomorphic surface parts, and likewise a half-funnel and a half-mast.

§26 Construction of g-mappings. Metric parameters. Congruent groups

26.1 A lemma. We shall have to use the following

Lemma. *Let $\mathfrak{D}$ and $\mathfrak{D}'$ denote circular discs bounded by unit circles $\mathcal{E}$ and $\mathcal{E}'$, respectively, and put $\overline{\mathfrak{D}} = \mathfrak{D} \cup \mathcal{E}$ and $\overline{\mathfrak{D}}' = \mathfrak{D}' \cup \mathcal{E}'$. Let Π and Π' denote polygons in $\overline{\mathfrak{D}}$ and $\overline{\mathfrak{D}}'$, respectively, the sides of which are finite in number and straight in the sense of hyperbolic metric, and let the polygons be convex in the sense of that metric. Some or all of the vertices may be on $\mathcal{E}$ and $\mathcal{E}'$. Let g be a given topological mapping of Π onto Π' such that every vertex of Π on $\mathcal{E}$ (if any) is mapped into a vertex of Π' on $\mathcal{E}'$, and conversely for g^{-1}. Then g can be extended to a topological mapping of the region in $\mathfrak{D}$ bounded by Π onto the region in $\mathfrak{D}'$ bounded by Π'.*

Proof. If a and b are any two points of $\overline{\mathfrak{D}}$, let (a, b) denote the *euclidean* length of the arc ab of the non-euclidean straight line determined by a and b, and likewise for $\overline{\mathfrak{D}}'$. Let q and q' be points in the interior of Π and Π', respectively, chosen arbitrarily. If x is a generic point of Π, then (q, x) depends continuously on x, and $gx = x'$ is on Π'. If y is any point of the circular arc qx orthogonal to $\mathcal{E}$, we define its image as the point y' of the circular arc $q'x'$ orthogonal to $\mathcal{E}'$ for which

$$(q', y')(q, x) = (q', x')(q, y),$$

choosing $y' = q'$ if $y = q$. This yields a topological mapping of the closed region of $\overline{\mathfrak{D}}$ bounded by Π onto the closed region of $\overline{\mathfrak{D}}'$ bounded by Π' such that on Π it coincides with the given mapping g. □

26.2 g-mappings of elementary groups. Let $\mathfrak{E}$ denote an elementary group in $\mathfrak{D}$ (§21.9), thus a non-quasi-abelian group generated by three primary motions $\mathfrak{p}_1, \mathfrak{p}_2, \mathfrak{p}_3$ satisfying $\mathfrak{p}_1\mathfrak{p}_2\mathfrak{p}_3 = 1$, each of them being either a rotation, or a limit-rotation, or a boundary translation. The centre, or limit-centre, or axis of $\mathfrak{p}_i$ is called $\mathcal{P}_i$, and the subgroup of $\mathfrak{E}$ generated by $\mathfrak{p}_i$ is called $\mathfrak{P}_i$ $(i = 1, 2, 3)$.

Let $\mathfrak{E}$ be t-isomorphic with a group in $\mathfrak{D}'$, say $I\colon \mathfrak{E} \to \mathfrak{E}'$, where I is a t-isomorphism. Then $\mathfrak{E}'$ is not quasi-abelian, thus an $\mathfrak{F}$-group, on account of the isomorphism. $\mathfrak{E}'$ is generated by the elements $\mathfrak{p}_i' = I(\mathfrak{p}_i)$, $i = 1, 2, 3$, and these satisfy the relation $\mathfrak{p}_1'\mathfrak{p}_2'\mathfrak{p}_3' = 1$. Since I is, in particular, a t-isomorphism, one gets: If $\mathfrak{p}_i$ is a rotation of order ν, then also $\mathfrak{p}_i'$ is; if $\mathfrak{p}_i$ is a limit-rotation, then $\mathfrak{p}_i'$ is either a limit-rotation or a boundary translation; likewise if $\mathfrak{p}_i$ is a boundary translation. Also primary elements in one group correspond to primary elements in the other. Hence $\mathfrak{E}'$ is an elementary group. Let $\mathcal{P}_i'$ and $\mathfrak{P}_i'$ be defined as for $\mathfrak{E}$. In (12), §21.13, ten different types of elementary groups are identified and characterized by a symbol for the corresponding surface $\mathcal{K}(\mathfrak{E})$ mod $\mathfrak{E}$. In these symbols a sign $\odot_\nu$ is the same for $\mathfrak{E}$ and $\mathfrak{E}'$ with the same value for ν, whereas signs $\hookrightarrow$ and $\to$ may correspond mutually.

We now specify I to be a g-isomorphism. Then $\mathfrak{p}_i$ and $\mathfrak{p}'_i$ are elements of the same kind, and the symbols (12), §21.13, are the same for $\mathfrak{E}$ and $\mathfrak{E}'$.

A fundamental polygon for $\mathfrak{E}$ in $\mathcal{K}(\mathfrak{E})$ has been constructed in §21.9. We visualize it by Fig. 26.1, which should be compared with Fig. 21.1. In Fig. 26.1 it is assumed

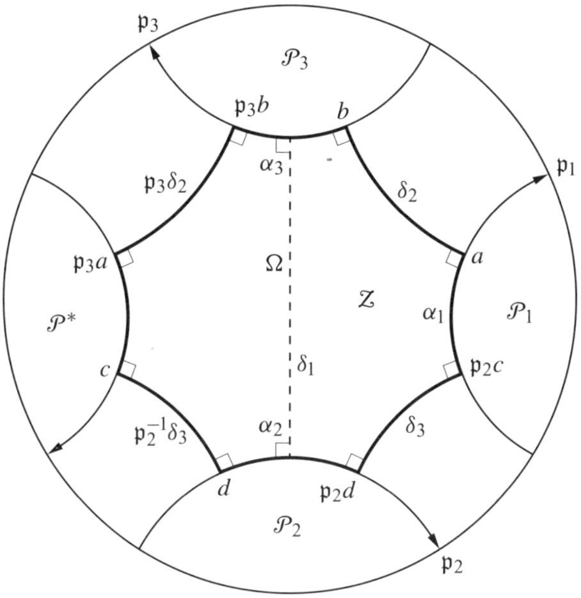

Figure 26.1

that $\mathfrak{p}_1$, $\mathfrak{p}_2$, $\mathfrak{p}_3$ are boundary translation; their axes $\mathcal{P}_1$, $\mathcal{P}_2$, $\mathcal{P}_3$ correspond to the $\mathcal{P}$, $\mathcal{P}'$, $\mathcal{P}''$ of §21. This is the case of principal interest in the sequel. It will later be easy to introduce the simplifications resulting from some, or all, of the $\mathcal{P}_1$, $\mathcal{P}_2$, $\mathcal{P}_3$ being single points (centres or limit-centres).

The right-angled octogon of Fig. 26.1 is called Π. Its vertices are in turn $a, b, \mathfrak{p}_3 b, \mathfrak{p}_3 a, c, d, \mathfrak{p}_2 d, \mathfrak{p}_2 c$. The side ab is situated on the straight line called $\mathcal{R}''$ in §21; we denote by δ_2 the closed segment ab as well as its length. Likewise the side $\mathfrak{p}_2 d$, $\mathfrak{p}_2 c$ is on the straight line called $\mathcal{R}'$ in §21; we denote this closed segment and its length by δ_3. Then $\mathfrak{p}_3\delta_2$ and $\mathfrak{p}_2^{-1}\delta_3$ are two other sides of Π. If these two sides are omitted, then the rest of Π together with the interior Ω of Π is a fundamental domain for $\mathfrak{E}$ in $\mathcal{K}(\mathfrak{E})$. The lengths of the sides of Π on $\mathcal{P}_2$ and $\mathcal{P}_3$ are equal to the displacements of $\mathfrak{p}_2$ and $\mathfrak{p}_3$, respectively. The boundary axis denoted by $\mathcal{P}^*$ in the figure is $\mathfrak{p}_3\mathcal{P}_1$ and is also $\mathfrak{p}_2^{-1}\mathcal{P}_1$. If $\mathfrak{p}_2$ is applied to the side of Π on $\mathcal{P}^*$, one gets the segment from $\mathfrak{p}_2 c$ to $\mathfrak{p}_2\mathfrak{p}_3 a = \mathfrak{p}_1^{-1} a$ on $\mathcal{P}_1$. This segment together with the side of Π on $\mathcal{P}_1$ yields a displacement segment of $\mathfrak{p}_1$, both these parts being half the displacement of $\mathfrak{p}_1$. The displacements of $\mathfrak{p}_1, \mathfrak{p}_2, \mathfrak{p}_3$ are called $2\alpha_1, 2\alpha_2, 2\alpha_3$, respectively and the distance of

$\mathcal{P}_2$ and $\mathcal{P}_3$ is called δ_1. Notations thus correspond to those in §21.21.

Let a topological mapping g of $\mathcal{P}_1$ onto $\mathcal{P}_1'$ be prescribed satisfying the equation

$$g\mathfrak{p}_1 x = \mathfrak{p}_1' g x \tag{1}$$

for x ranging over $\mathcal{P}_1$. Through repeated application of (1) one gets

$$g\mathfrak{p}_1^r x = \mathfrak{p}_1'^r g x, \qquad -\infty < r < \infty. \tag{2}$$

We extend this mapping to a mapping of the axis $\mathcal{P}^*$ onto the corresponding axis $\mathcal{P}^{*'}$ of $\mathfrak{E}'$ by letting x range over $\mathcal{P}_1$ and putting

$$g\mathfrak{p}_3 x = \mathfrak{p}_3' g x. \tag{3}$$

Then also

$$g\mathfrak{p}_3^{-1} x = \mathfrak{p}_3'^{-1} g x \tag{3'}$$

holds for x ranging over $\mathcal{P}^*$. From (1) and (3) follows, since $\mathfrak{p}_1 x$ is on $\mathcal{P}_1$, if x is on $\mathcal{P}_1$,

$$g\mathfrak{p}_3\mathfrak{p}_1 x = \mathfrak{p}_3'\mathfrak{p}_1' g x, \tag{4}$$

which may also be written

$$g\mathfrak{p}_2^{-1} x = \mathfrak{p}_2'^{-1} g x \tag{4'}$$

where x ranges over $\mathcal{P}_1$. Equations (3) and (4′) thus determine the same mapping of $\mathcal{P}^*$ onto $\mathcal{P}^{*'}$. This mapping satisfies the equation

$$g\mathfrak{p}_3\mathfrak{p}_1\mathfrak{p}_3^{-1} x = \mathfrak{p}_3'\mathfrak{p}_1'\mathfrak{p}_3'^{-1} g x \tag{5}$$

where x ranges over $\mathcal{P}^*$; this follows immediately from (4) and (3′), since $\mathfrak{p}_3^{-1} x$ belongs to $\mathcal{P}_1$. Here $\mathfrak{p}_3\mathfrak{p}_1\mathfrak{p}_3^{-1}$ is a primary translation along $\mathcal{P}^*$.

Let also a topological mapping g of $\mathcal{P}_2$ onto $\mathcal{P}_2'$ and of $\mathcal{P}_3$ onto $\mathcal{P}_3'$ be prescribed satisfying

$$g\mathfrak{p}_2 x = \mathfrak{p}_2' g x \tag{6}$$

$$g\mathfrak{p}_3 x = \mathfrak{p}_3' g x \tag{7}$$

respectively; in (6), x ranges over $\mathcal{P}_2$ and in (7) over $\mathcal{P}_3$. – The mapping now defined on $\mathcal{P}_i$ $(i = 1, 2, 3)$ determines a topological mapping of $\mathcal{P}_i$ mod $\mathfrak{P}_i$ onto $\mathcal{P}_i'$ mod $\mathfrak{P}_i'$.

So far a topological mapping g of each of the four axes of Fig. 26.1 onto the corresponding axis of $\mathfrak{E}'$ is given, satisfying (1), (6), (7), (5), when x ranges over $\mathcal{P}_1, \mathcal{P}_2, \mathcal{P}_3, \mathcal{P}^*$, respectively. This yields a mapping of four sides of Π onto four segments of the corresponding boundary axes $\mathcal{P}_1', \mathcal{P}_2', \mathcal{P}_3', \mathcal{P}^{*'}$ of $\mathfrak{E}'$, thereby fixing the images by g of all eight vertices of Π.

Since I is a g-isomorphism, it defines a mapping κ of $\mathfrak{g}(\mathfrak{E})$ onto $\mathfrak{g}(\mathfrak{E}')$ which preserves the cyclical order (§24.1). Hence the directed axes $\mathcal{P}_1', \mathcal{P}_3', \mathcal{P}^{*'}, \mathcal{P}_2'$ are arranged in this cyclical order in $\mathcal{D}'$. They are the boundaries on $\mathcal{D}'$ of a convex subset of $\mathcal{D}'$.

We now extend g by an arbitrary topological mapping of δ_2 onto the straight segment ga, gb of $\mathcal{D}'$, and of δ_3 onto the straight segment $g\mathfrak{p}_2 c$, $g\mathfrak{p}_2 d$ of $\mathcal{D}'$, and finally to a mapping of the side $\mathfrak{p}_3\delta_2$ by (3), letting x range over δ_2, and to a mapping of $\mathfrak{p}_2^{-1}\delta_3$ by (4′), letting x range over δ_3. Thus a mapping g of Π onto a closed polygon Π' in $\mathcal{D}'$ is obtained, and Π' is a convex octogon in $\mathcal{D}'$. In general, Π' is not right-angled. If the sides $g\mathfrak{p}_3\delta_2$ and $g\mathfrak{p}_2^{-1}\delta_3$ are omitted from Π', then the rest of Π' together with the interior Ω' of Π' is a fundamental domain for $\mathfrak{E}'$ in $\mathcal{K}(\mathfrak{E}')$.

In virtue of the lemma of Section 1, the mapping g now given on Π can be extended to a topological mapping g of $\overline{\Omega}$ onto $\overline{\Omega}'$ such that on Π it coincides with the mapping $g\colon \Pi \to \Pi'$ already obtained.

We now consider the general case. If $\mathcal{P}_3$ is a point (centre or limit-centre), then also $\mathcal{P}_3'$ is a point, namely the corresponding centre or limit-centre determined by I. The points b and $\mathfrak{p}_3 b$ coincide, and they are mapped in the corresponding centre or limit-centre of $\mathfrak{E}'$. Similarly for $\mathcal{P}_2$, if $\mathcal{P}_2$ is a point. If $\mathcal{P}_1$ is a point, then the same holds for $\mathcal{P}^*$. According to the possible cases of axes being replaced by points Π reduces to a heptagon, hexagon, pentagon, or quadrangle, and the same holds accordingly for $g\Pi = \Pi'$. The application of the lemma is still valid, because vertices of Π on $\mathcal{E}$ (in limit-centres) are by g mapped into vertices of Π' on $\mathcal{E}'$ (in limit-centres). In all cases one gets a topological mapping $g\colon \overline{\Omega} \to \overline{\Omega}'$. Any two points of $\overline{\Omega}$ equivalent with respect to $\mathfrak{E}$, say x and $\mathfrak{f}x$, $\mathfrak{f} \in \mathfrak{E}$, belong to Π, and they are mapped into points gx and

$$g\mathfrak{f}x = \mathfrak{f}'gx, \quad \mathfrak{f}' = I(\mathfrak{f}), \tag{8}$$

of Π'. One can therefore immediately extend g to a topological mapping of $\mathcal{K}(\mathfrak{E})$ onto $\mathcal{K}(\mathfrak{E}')$ by (8), letting x range over the whole of $\overline{\Omega}$ and $\mathfrak{f}$ over $\mathfrak{E}$. This yields a g-mapping of $\mathcal{K}(\mathfrak{E})$ onto $\mathcal{K}(\mathfrak{E}')$ which induces I, i.e. which satisfies (8) when x ranges over $\mathcal{K}(\mathfrak{E})$ and $\mathfrak{f}$ over $\mathfrak{E}$. Hence the result:

For any given g-isomorphism I between two elementary groups $\mathfrak{E}$ and $\mathfrak{E}'$ there exists a g-mapping of $\mathcal{K}(\mathfrak{E})$ onto $\mathcal{K}(\mathfrak{E}')$ which induces I. □

26.3 Metric parameters of elementary groups. If $\mathcal{P}_i$ is a centre, then α_i, as in §21.21, shall mean half the angle of rotation of the primary rotation $\mathfrak{p}_i$, thus $\frac{\pi}{\nu}$, if ν is the order of $\mathfrak{p}_i$. Hence $\alpha_i \leqq \frac{\pi}{2}$, the equality sign occurring at most for one value of i. Thus for a given symbol (12), §21.13, such α_i are fixed numbers. If $\mathcal{P}_i$ is a limit-centre, then $\alpha_i = 0$. For those $\mathcal{P}_i$ which are axes α_i shall denote half the displacement of the primary translation $\mathfrak{p}_i$. We call the displacements $2\alpha_i$ of primary boundary translations *metric parameters of* $\mathfrak{E}$. Thus the number of metric parameters of $\mathfrak{E}$ is 3, 2, 1, or 0. We set out to prove the following theorem:

Theorem. *There exists an elementary group of any of the types* (12), §21.13, *and with any prescribed values of its metric parameters.*

Proof. First assume that there are at least two metric parameters. We take α_2 and α_3 to be any prescribed lengths. These lengths are laid off on normals $\mathcal{P}_2$ and $\mathcal{P}_3$ erected at the end-points of a straight segment, the length of which is called δ_1. Then draw a normal δ_3 to $\mathcal{P}_2$ in the end-point of α_2 and a normal δ_2 to $\mathcal{P}_3$ in the end-point of α_3; compare Fig. 26.1. If δ_1 is chosen as the length δ_1^* calculated from

$$\cosh \delta_1^* = \frac{1 + \cosh \alpha_2 \cosh \alpha_3}{\sinh \alpha_2 \sinh \alpha_3}, \tag{9}$$

which is possible since the right hand term in (9) is > 1, then the formulae (VIb) and (Vb) of §21.21 show that the lines δ_2 and δ_3 become parallel. Hence an $\mathfrak{E}(\hookrightarrow\rightarrow\rightarrow)$ with prescribed values of the two metric parameters exists.

If δ_1 is increased beyond the value δ_1^*, then the lines δ_2 and δ_3 become divergent; their distance α_1 is then obtained from (VIb) $(\nu = 1)$. The length α_1 increases continuously from 0 to infinity, as δ_1 increases from δ_1^* to infinity. Hence an $\mathfrak{E}(\rightarrow\rightarrow\rightarrow)$ with all three metric parameters arbitrarily prescribed exists.

If δ_1 is decreased below the value δ_1^*, then the lines δ_2 and δ_3 intersect; their angle of intersection α_1 is obtained from the first formula (Vb) of §21.21. Here $\cos \alpha_1$ decreases continuously from 1 to 0 as $\cosh \delta_1$ decreases from the value (9) to the value $\coth \alpha_2 \coth \alpha_3$. Hence α_1 assumes the values $\frac{\pi}{\nu}$ for all $\nu \geqq 2$, and $\mathfrak{E}(\odot_\nu \rightarrow\rightarrow)$ exists for all integers $\nu \geqq 2$ and the two metric parameters arbitrarily prescribed.

Secondly, assume that there are less than two metric parameters. Let $\alpha_1 = \frac{\pi}{\nu_1}$, $\alpha_2 = \frac{\pi}{\nu_2}$, at most one of the integers ν_1 and ν_2 being 2, and let these angles be laid off from a straight segment δ_3 from its end-points and to the same side of δ_3. If δ_3 is chosen as the length δ_3^* calculated from

$$\cosh \delta_3^* = \frac{1 + \cos \alpha_1 \cos \alpha_2}{\sin \alpha_1 \sin \alpha_2}, \tag{10}$$

which is possible since the right hand term is > 1, then formula (IIIb) $(\nu = 1)$ and the first formula (IVb) of §21.21 show that the lines δ_1 and δ_2 become parallel. Hence an $\mathfrak{E}(\odot_{\nu_1} \odot_{\nu_2} \hookrightarrow)$ exists.

If δ_3 is increased beyond the value δ_3^*, then the lines δ_1 and δ_2 become divergent; their distance α_3 is obtained from formula (IVb). The length α_3 increases continuously from 0 to infinity as δ_3 increases from δ_3^* to infinity. Hence an $\mathfrak{E}(\odot_{\nu_1} \odot_{\nu_2} \rightarrow)$ with the metric parameter arbitrarily prescribed exists.

If δ_3 is decreased below the value δ_3^*, then the lines δ_1 and δ_2 intersect; their angle α_3 is obtained from formula (IIIb). Thus α_3 increases continuously from 0 to $\frac{\pi}{2}$ as $\cosh \delta_3$ decreases from the value (10) to the value $\cot \alpha_1 \cot \alpha_2$, provided both ν_1 and ν_2 are > 2. If one of them say ν_1, takes the value 2, then $\cos \alpha_3 = \sin \alpha_2 \cosh \delta_3$, thus α_3 tends to $\frac{\pi}{2} - \alpha_2$ as δ_3 tends to 0. Hence, in all cases, an $\mathfrak{E}(\odot_{\nu_1} \odot_{\nu_2} \odot_{\nu_3})$ exists, provided $\frac{1}{\nu_1} + \frac{1}{\nu_2} + \frac{1}{\nu_3} < 1$.

If $\mathfrak{p}_1$ is a rotation and $\mathfrak{p}_2$ a limit-rotation, then the line δ_1 can be chosen at will. This shows that an $\mathfrak{E}(\odot_{\nu_1} \hookrightarrow\rightarrow)$ exists for every prescribed length of α_3.

If both $\mathfrak{p}_1$ and $\mathfrak{p}_2$ are limit-rotations, then both δ_1 and δ_2 can be chosen at will. This shows that an $\mathfrak{E}(\hookrightarrow\hookrightarrow\rightarrow)$ exists for every prescribed length of α_3, that an $\mathfrak{E}(\hookrightarrow\hookrightarrow\hookrightarrow)$ exists, and that an $\mathfrak{E}(\hookrightarrow\hookrightarrow \circlearrowleft_{\nu_3})$ exists for every value $\nu_3 \geqq 2$.

This covers all cases of (12), §21.13, and the theorem has been proved. □

26.4 Congruence of elementary groups. Let $\mathfrak{F}$ and $\mathfrak{F}'$ be $\mathfrak{F}$-groups such that a g-mapping of $\mathcal{K}(\mathfrak{F})$ onto $\mathcal{K}(\mathfrak{F}')$ exists. If g can be chosen as a Möbius transformation

$$x \mapsto x' = \frac{ax+b}{\bar{b}x+\bar{a}}, \quad a\bar{a} - b\bar{b} > 0, \tag{11}$$

where x ranges over $\mathcal{K}(\mathfrak{F})$ and x' over $\mathcal{K}(\mathfrak{F}')$, then $\mathfrak{F}$ and $\mathfrak{F}'$ are called *congruent*. The mapping function (11) then maps $\overline{\mathcal{D}}$ onto $\overline{\mathcal{D}}'$. We then also speak of $\mathcal{K}(\mathfrak{F})$ mod $\mathfrak{F}$ and $\mathcal{K}(\mathfrak{F}')$ mod $\mathfrak{F}'$, and likewise of $\mathcal{D}$ mod $\mathfrak{F}$ and $\mathcal{D}'$ mod $\mathfrak{F}'$ as *congruent surfaces.*

If $\mathfrak{F}$ is an $\mathfrak{F}$-group in $\overline{\mathcal{D}}$ and M denotes an arbitrary mapping function of the form (11) of $\overline{\mathcal{D}}$ onto $\overline{\mathcal{D}}'$, then $M\mathfrak{F}M^{-1}$ is an $\mathfrak{F}$-group $\mathfrak{F}'$ in $\overline{\mathcal{D}}'$. Points of $\overline{\mathcal{D}}$ which are equivalent with respect to $\mathfrak{F}$ are by M mapped into points of $\overline{\mathcal{D}}'$ which are equivalent with respect to $\mathfrak{F}'$, and conversely by M^{-1}. Corresponding elements of $\mathfrak{F}$ and $\mathfrak{F}'$ are of the same kind. Hence M is a g-mapping and induces a g-isomorphism between $\mathfrak{F}$ and $\mathfrak{F}'$. – We now prove:

Two elementary groups $\mathfrak{E}$ and $\mathfrak{E}'$ are congruent if and only if they are of the same type (12), §21.13, *and corresponding metric parameters are equal.*

Proof. If $\mathfrak{E}$ and $\mathfrak{E}'$ are congruent, then there exists a g-mapping (11) which carries each of the generators $\mathfrak{p}_1, \mathfrak{p}_2, \mathfrak{p}_3$ of $\mathfrak{E}$ into an element of the same kind in $\mathfrak{E}'$, preserving the order of rotations. Hence $\mathfrak{E}$ and $\mathfrak{E}'$ belong to the same type (12), §21.13. Also corresponding metric parameters, if any, are equal, since (11) preserves distance. The conditions of the theorem are thus necessary.

In order to show the sufficiency of the conditions we consider the closed region $\mathcal{Z}$ of $\overline{\mathcal{D}}$ bounded by $\mathcal{P}_1, \mathcal{P}_2, \mathcal{P}_3, \delta_1, \delta_2, \delta_3$ in the notation of the two preceding sections. In the case of Fig. 26.1 it is a hexagon inside $\mathcal{D}$. If one or more of the $\mathcal{P}_i$ is a point, it becomes a pentagon, quadrangle, or triangle, which may have vertices in limit-centres on $\mathcal{E}$. In all cases the polygon $\mathcal{Z}$ is completely determined by $\alpha_1, \alpha_2, \alpha_3$, as shown by the formulae of §21.18. Hence $\mathcal{Z}$ is congruent with $\mathcal{Z}'$, because $\alpha_1, \alpha_2, \alpha_3$ are in turn equal to $\alpha_1', \alpha_2', \alpha_3'$. Moreover $\mathcal{Z}$ together with its image by the reflection in one of the sides δ is a fundamental polygon Ω for $\mathfrak{F}$, and the same for $\mathfrak{F}'$. There is thus a fractional linear mapping (11), say M, which maps $\overline{\Omega}$ onto $\overline{\Omega}'$ such that points on the boundary Π of Ω which correspond by an element $\mathfrak{f}$ of $\mathfrak{E}$ are mapped into points of Π' which correspond by the element $\mathfrak{f}' = M\mathfrak{f}M^{-1}$. The mapping M thereby extends to the whole of the convex domains, and indeed to $\overline{\mathcal{D}}$ and $\overline{\mathcal{D}}'$. □

26.5 g-mappings of handle groups. In §21.15 a handle group was derived from an elementary group, and in §22.4 a handle group was determined as a subgroup of a given $\mathfrak{F}$-group $\mathfrak{F}$, if $\mathfrak{F}$ possesses a modulo $\mathfrak{F}$ simple and non-dividing axis with corresponding subgroup $\mathfrak{A}(\rightarrow)$. We now show:

Any modulo $\mathfrak{H}$ simple, inner axis of a handle group $\mathfrak{H}$ is non-dividing modulo $\mathfrak{H}$, and its corresponding subgroup is of the type $\mathfrak{A}(\rightarrow)$.

Proof. Let $\mathcal{A}$ be a modulo $\mathfrak{H}$ simple, inner axis of $\mathfrak{H}$. Assume that it were dividing modulo $\mathfrak{H}$. Let $\mathcal{K}(\mathfrak{H})$ be decomposed by $\mathfrak{H}\mathcal{A}$. Let $\mathfrak{F}_1$ and $\mathfrak{F}_2$ denote the subgroups carrying the regions of decomposition adjacent to $\mathcal{A}$ into themselves respectively. They are $\mathfrak{F}$-groups, and the regions, closed on $\mathcal{K}(\mathfrak{H})$, are $\mathcal{K}(\mathfrak{F}_1)$ and $\mathcal{K}(\mathfrak{F}_2)$, respectively. Then one, and only one, of the corresponding surfaces, say $\mathcal{K}(\mathfrak{F}_1)$ mod $\mathfrak{F}_1$, contains the protrusion of $\mathcal{K}(\mathfrak{H})$ mod $\mathfrak{H}$. All boundaries of $\mathcal{K}(\mathfrak{F}_1)$ belong to one equivalence class with respect to $\mathfrak{F}_1$, because $\mathcal{A}$ is dividing modulo $\mathfrak{H}$ (case III of §17.3). Hence $\mathcal{K}(\mathfrak{F}_1)$ mod $\mathfrak{F}_1$ has only one protrusion. Therefore $\mathcal{K}(\mathfrak{F}_1)$ contains a modulo $\mathfrak{F}_1$ simple and non-dividing axis with corresponding group $\mathfrak{A}(\rightarrow)$ (second theorem of §22.5). Hence $\mathfrak{F}_1$ contains a handle group $\mathfrak{H}_1$, and $\Phi(\mathfrak{H}_1) = 2$, because $\mathfrak{F}_1$ contains no rotation. This contradicts the fact that $\Phi(\mathfrak{H}) \leqq 2$, and that $\mathcal{K}(\mathfrak{H}_1)$ mod $\mathfrak{H}_1$ is not the whole of $\mathcal{K}(\mathfrak{H})$ mod $\mathfrak{H}$. Hence $\mathcal{A}$ is non-dividing modulo $\mathfrak{H}$. Since $\mathfrak{H}$ contains no reversions and at most one equivalence class of rotations of order 2, the only possibility for $\mathfrak{A}_{\mathcal{A}}$ is $\mathfrak{A}(\rightarrow)$. This completes the proof of the theorem. □

Let $\mathfrak{H}$ be a handle group in $\mathcal{D}$, $\mathfrak{H}'$ a group in $\mathcal{D}'$, and $I: \mathfrak{H} \rightarrow \mathfrak{H}'$, a t-isomorphism. Then $\mathfrak{H}'$ is an $\mathfrak{F}$-group, because $\mathfrak{H}$ is. Since $\mathfrak{H}$ is not an elementary group, it possesses a modulo $\mathfrak{H}$ simple, inner axis (§22.8), say $\mathcal{A}$. In virtue of the preceding theorem $\mathcal{A}$ is non-dividing modulo $\mathfrak{H}$, and $\mathfrak{A}_{\mathcal{A}}$ is $\mathfrak{A}(\rightarrow)$. Let $\mathfrak{a}$ denote a primary translation along $\mathcal{A}$. Then $\mathfrak{a}' = I(\mathfrak{a})$ is a primary translation in $\mathfrak{H}'$ belonging to an inner axis of $\mathfrak{H}'$, since $\mathcal{A}$ is an inner axis of $\mathfrak{H}$, and I is a t-isomorphism. Moreover $\mathcal{A}'$ is simple modulo $\mathfrak{H}'$, because $\mathcal{A}$ is simple modulo $\mathfrak{H}$, and $\mathfrak{A}_{\mathcal{A}'} = \mathfrak{A}(\rightarrow)$, because $\mathfrak{A}_{\mathcal{A}} = \mathfrak{A}(\rightarrow)$ (§24.1). Let $\mathcal{K}(\mathfrak{H})$ be decomposed by $\mathfrak{H}\mathcal{A}$ and $\mathcal{K}(\mathfrak{H}')$ by $\mathfrak{H}'\mathcal{A}'$. The axes of these two collections are in a one-one-correspondence derived from I. The region of decomposition, closed on $\mathcal{K}(\mathfrak{H})$, adjacent to $\mathcal{A}$, and bounded positively by $\mathcal{A}$, directed by $\mathfrak{a}$, is the convex domain $\mathcal{K}(\mathfrak{E})$ of an elementary group $\mathfrak{E}$, a subgroup of $\mathfrak{H}$. Then $I(\mathfrak{E}) = \mathfrak{E}'$ is a subgroup of $\mathfrak{H}'$ which is also elementary (Section 2). In the subcollection $\mathcal{S}$ of $\mathfrak{H}\mathcal{A}$ which contributes to the boundary of $\mathcal{K}(\mathfrak{E})$ any two axes are not separated by any other. Since the mapping $\kappa: \mathcal{V} \rightarrow \mathcal{V}'$ derived from I (§24.1) preserves the cyclical order, the same holds for the corresponding subcollection $\mathcal{S}'$ of $\mathfrak{H}'\mathcal{A}'$. Since $\mathcal{S}$ is reproduced by $\mathfrak{E}$ and by no other elements of $\mathfrak{H}$, it follows from the isomorphism that $\mathcal{S}'$ is reproduced by $\mathfrak{E}'$ and by no other element of $\mathfrak{H}'$. Thus $\mathcal{S}'$ is the boundary of $\mathcal{K}(\mathfrak{E}')$ inside $\mathcal{K}(\mathfrak{H}')$. The axes of $\mathcal{S}$ fall into two equivalence classes with respect to $\mathfrak{E}$; it then follows from the isomorphism that the axes of $\mathcal{S}'$ fall into two equivalence classes with respect to $\mathfrak{E}'$. It is known from §21.15 or §22.4 that $\mathfrak{H}$ contains a translation $\mathfrak{b}$

such that the axis $\mathcal{A}_1$ of $\mathfrak{bab}^{-1}$ is a side of $\mathcal{K}(\mathfrak{E})$ not equivalent with $\mathcal{A}$ with respect to $\mathfrak{E}$, that the axis $\mathcal{B}$ of $\mathfrak{b}$ is simple modulo $\mathfrak{H}$, and that $\mathcal{B}$ cuts $\mathcal{A}$ and $\mathcal{A}_1$ (note: $\mathcal{B}$ cuts $\mathcal{A}$ and $\mathcal{A}_1$ under equal angles). It then follows from the t-isomorphism and from the preservation of the cyclical order of corresponding inner fundamental points, that $\mathfrak{b}' = I(\mathfrak{b})$ has the corresponding properties in relation to $\mathfrak{H}'$; the axis $\mathcal{A}_1$ bounds $\mathcal{K}(\mathfrak{E})$ positively, if it is directed by $\mathfrak{ba}^{-1}\mathfrak{b}^{-1}$. Hence the product of these two translations,

$$\mathfrak{aba}^{-1}\mathfrak{b}^{-1} \tag{12}$$

is a rotation, or limit-rotation, or boundary translation of $\mathfrak{H}$. If (12) is a rotation of order ν, then the same holds for the element

$$\mathfrak{a}'\mathfrak{b}'\mathfrak{a}'^{-1}\mathfrak{b}'^{-1} \tag{13}$$

of $\mathfrak{H}'$. If (12) is a limit-rotation, then (13) is either a limit-rotation or a boundary translation of $\mathfrak{H}$ on account of the t-isomorphism, and likewise if (12) is a boundary translation of $\mathfrak{H}$. Hence $\mathfrak{H}'$ is a handle group.

We now assume that the t-isomorphism is, in particular, a g-isomorphism. Then $\mathfrak{E}$ and $\mathfrak{E}'$ belong to the same type (12), §21.13, and $\mathfrak{H}$ and $\mathfrak{H}'$ determine a handle with the same symbol $\mathcal{H}(\odot_\nu)$, or $\mathcal{H}(\hookrightarrow)$, or $\mathcal{H}(\rightarrow)$, (§21.16). We want to show:

For any g-isomorphism I between two handle groups $\mathfrak{H}$ and $\mathfrak{H}'$ there exists a g-mapping of $\mathcal{K}(\mathfrak{H})$ onto $\mathcal{K}(\mathfrak{H}')$ inducing I.

Proof. The construction of the mapping g is based on Section 2 in the following way: We let the above $\mathfrak{a}$ correspond to the $\mathfrak{p}_2$ of Section 2 and $\mathfrak{ba}^{-1}\mathfrak{b}^{-1}$ to $\mathfrak{p}_3$. Then $\mathfrak{p}_2\mathfrak{p}_3$ is the element (12), and this then corresponds to $\mathfrak{p}_1^{-1}$. It is a rotation, or limit-rotation, or boundary translation of $\mathfrak{H}$ according as $\mathcal{K}(\mathfrak{H})$ mod $\mathfrak{H}$ is a $\mathcal{H}(\odot_\nu)$, or a $\mathcal{H}(\hookrightarrow)$, or a $\mathcal{H}(\rightarrow)$ (§21.15). Thus there are either two or three equivalence classes with respect to $\mathfrak{E}$ of boundary axes of $\mathfrak{E}$, and this number is the same for $\mathfrak{E}'$. We first establish the mapping g of $\mathcal{A}$ onto $\mathcal{A}'$ as an arbitrary topological mapping satisfying

$$g\mathfrak{a}x = \mathfrak{a}'gx \tag{14}$$

for x ranging over $\mathcal{A}$. We then define the mapping g of $\mathcal{A}_1 = \mathfrak{b}\mathcal{A}$ onto $\mathcal{A}_1' = \mathfrak{b}'\mathcal{A}'$ by

$$g\mathfrak{b}x = \mathfrak{b}'gx, \tag{15}$$

where x ranges over $\mathcal{A}$. In an analogous way as for (5) in Section 2, it then follows from (14) and (15) that this mapping satisfies the equation

$$g\mathfrak{bab}^{-1}x = \mathfrak{b}'\mathfrak{a}'\mathfrak{b}'^{-1}gx \tag{16}$$

for x ranging over $\mathcal{A}_1$. Finally, if (12) and thus also (13), is a translation, the mapping g of the axis of (12) onto the axis of (13) may be any prescribed topological mapping satisfying

$$g\mathfrak{aba}^{-1}\mathfrak{b}^{-1}x = \mathfrak{a}'\mathfrak{b}'\mathfrak{a}'^{-1}\mathfrak{b}'^{-1}gx, \tag{17}$$

where x ranges over the axis of (12).

The situation is now the same as in Section 2. Thus g can be extended to a mapping of the closed fundamental domain $\overline{\Omega}$ of $\mathfrak{E}$ onto the closed fundamental domain $\overline{\Omega}'$ of $\mathfrak{E}'$. The difference between our present case and that of Section 2 has been that the mapping of $\mathcal{A}$ and $\mathcal{A}_1$ onto the corresponding axes $\mathcal{A}'$ and $\mathcal{A}'_1$, respectively, cannot be prescribed independently of each other; they are linked by (15), and therefore only one of them is arbitrary. Any two points of $\overline{\Omega}$ equivalent with respect to $\mathfrak{H}$ (not only with respect to $\mathfrak{E}$!), say x and $\mathfrak{f}x$, $\mathfrak{f} \in \mathfrak{H}$, belong to the boundary Π of Ω, and they are mapped into points gx and

$$g\mathfrak{f}x = \mathfrak{f}'gx, \quad \mathfrak{f}' = I(\mathfrak{f}), \tag{18}$$

of the boundary Π' of Ω'; here $\mathfrak{f}$ may also be the element $\mathfrak{b}$, which is outside $\mathfrak{E}$. After the omission of certain boundary points $\overline{\Omega}$ is a fundamental domain for $\mathfrak{H}$ in $\mathcal{K}(\mathfrak{H})$. One can now extend g to a mapping of $\mathcal{K}(\mathfrak{H})$ onto $\mathcal{K}(\mathfrak{H}')$ by (18), letting x range over the whole of $\overline{\Omega}$ and $\mathfrak{f}$ over $\mathfrak{H}$. This yields a g-mapping of $\mathcal{K}(\mathfrak{H})$ onto $\mathcal{K}(\mathfrak{H}')$ satisfying (18) for x ranging over $\mathcal{K}(\mathfrak{H})$ and $\mathfrak{f}$ over $\mathfrak{H}$, thus inducing the given g-isomorphism I. This completes the proof of the theorem. □

We also show the following theorem:

Every isomorphism between two handle groups is a t-isomorphism.

Proof. Let $I : \mathfrak{H} \to \mathfrak{H}'$ be any given isomorphism between the handle groups $\mathfrak{H}$ and $\mathfrak{H}'$.

If $\mathcal{K}(\mathfrak{H})$ mod $\mathfrak{H}$ is a $\mathcal{H}(\odot_\nu)$, and $\mathfrak{H}$ thus contains rotations, then also $\mathfrak{H}'$ does. Hence $\mathcal{K}(\mathfrak{H}')$ mod $\mathfrak{H}'$ is a $\mathcal{H}(\odot_\nu)$ with the same ν. In this case the theorem is an immediate consequence of the last sentence in §24.8, because $\mathfrak{H}$ and $\mathfrak{H}'$ are finitely generated groups which contain no boundary elements (limit-rotations or boundary translations). I is, in particular, a g-isomorphism.

If $\mathcal{K}(\mathfrak{H})$ mod $\mathfrak{H}$ is a $\mathcal{H}(\hookrightarrow)$ or a $\mathcal{H}(\rightarrow)$, then also $\mathcal{K}(\mathfrak{H}')$ mod $\mathfrak{H}'$ is, but the two are not necessarily of the same kind. Let $\mathfrak{a}$ and $\mathfrak{b}$ be two elements of $\mathfrak{H}$ chosen as above; then (12) is a primary boundary element of $\mathfrak{H}$. In the same way we select in $\mathfrak{H}'$ two elements $\mathfrak{a}'$ and $\mathfrak{b}'$, and (13) is then a primary boundary element in $\mathfrak{H}'$. Now $\mathfrak{H}$ is a free group, and $\mathfrak{a}, \mathfrak{b}$ is a set of free generators of $\mathfrak{H}$; likewise $\mathfrak{a}', \mathfrak{b}'$ for $\mathfrak{H}'$. Therefore the mapping $\mathfrak{a} \mapsto \mathfrak{a}'$, $\mathfrak{b} \mapsto \mathfrak{b}'$ defines an isomorphism I_1 between $\mathfrak{H}$ and $\mathfrak{H}'$. Since it maps (12) on (13), it makes all the conjugates of (12) correspond to all the conjugates of (13). It thus maps boundary elements on boundary elements, and I_1 is thus a t-isomorphism. Consider now the automorphism $I_2 = I_1^{-1} I$ of $\mathfrak{H}$. The elements $I_2(\mathfrak{a}) = \mathfrak{a}^*$ and $I_2(\mathfrak{b}) = \mathfrak{b}^*$ are two free generators of $\mathfrak{H}$. A theorem of M. Dehn, proved in the paper "Die Isomorphismen der allgemeinen unendlichen Gruppe mit zwei Erzeugenden" by J. Nielsen in Mathematische Annalen vol. 78, states that a necessary and sufficient condition for any correspondence $\mathfrak{a} \mapsto \mathfrak{a}^*$, $\mathfrak{b} \mapsto \mathfrak{b}^*$ determining an automorphism of the free group generated by $\mathfrak{a}, \mathfrak{b}$ is that $\mathfrak{a}^*\mathfrak{b}^*\mathfrak{a}^{*^{-1}}\mathfrak{b}^{*^{-1}}$ is a conjugate of $(\mathfrak{a}\mathfrak{b}\mathfrak{a}^{-1}\mathfrak{b}^{-1})^{\pm 1}$. Hence in our present case I_2 maps boundary elements of $\mathfrak{H}$ on boundary elements of $\mathfrak{H}$ and is thus a t-automorphism and, in particular, a g-automorphism. Hence $I = I_1 I_2$ is a t-isomorphism. This completes the proof of the theorem. □

Consider now an arbitrary handle group $\mathfrak{H}$; thus $\mathcal{K}(\mathfrak{H})$ mod $\mathfrak{H}$ is either 1) a $\mathcal{H}(\odot_\nu)$, or 2) a $\mathcal{H}(\hookrightarrow)$, or 3) a $\mathcal{H}(\rightarrow)$. Let $\mathfrak{a}$, $\mathfrak{b}$ be two generators of $\mathfrak{H}$ chosen as in the beginning of this section. An element $\mathfrak{a}^*$ of $\mathfrak{H}$ will be called a *prime element* of $\mathfrak{H}$, if there exists in $\mathfrak{H}$ an element $\mathfrak{b}^*$ such that the mapping $\mathfrak{a} \mapsto \mathfrak{a}^*$, $\mathfrak{b} \mapsto \mathfrak{b}^*$ determines an automorphism I of $\mathfrak{H}$. If $\mathfrak{H}$ is a free group (cases 2) and 3)), this simply requires that $\mathfrak{a}^*$ and $\mathfrak{b}^*$ generate $\mathfrak{H}$. According to the preceding theorem I is a t-automorphism, and thus in this particular case, a g-automorphism. Since $\mathfrak{a}$ is a primary, inner translation of $\mathfrak{H}$ with modulo $\mathfrak{H}$ simple axis, it follows from §24.1 that also $\mathfrak{a}'$ is. On the other hand, if $\mathfrak{a}^*$ is a primary inner translation with modulo $\mathfrak{H}$ simple axis, then it has been seen that there exists a primary inner translation $\mathfrak{b}^*$ such that $\mathfrak{a}^*$ and $\mathfrak{b}^*$ generate $\mathfrak{H}$. Thus in the cases 2) and 3), the element $\mathfrak{a}^*$ is a prime element of $\mathfrak{H}$. In the case 1) the commutator $\mathfrak{a}^*\mathfrak{b}^*\mathfrak{a}^{*^{-1}}\mathfrak{b}^{*^{-1}}$ is known to be a primary rotation, and since $\mathfrak{H}$ only contains one equivalence class of centres, it follows that $\mathfrak{a}^*\mathfrak{b}^*\mathfrak{a}^{*^{-1}}\mathfrak{b}^{*^{-1}}$ is a conjugate of $(\mathfrak{a}\mathfrak{b}\mathfrak{a}^{-1}\mathfrak{b}^{-1})^{\pm 1}$, hence that $\mathfrak{a} \mapsto \mathfrak{a}^*$, $\mathfrak{b} \mapsto \mathfrak{b}^*$ determines an automorphism of $\mathfrak{H}$. Hence $\mathfrak{a}^*$ is a prime element of $\mathfrak{H}$ even in the case 1).

This yields the theorem:

Theorem. *The totality of prime elements of $\mathfrak{H}$ coincides with the totality of primary inner translations of $\mathfrak{H}$ with modulo $\mathfrak{H}$ simple axes.* □

It thus turns out that even in the case 1) the above definition of prime elements does not depend on the choice of the pair $\mathfrak{a}$, $\mathfrak{b}$.

In the case of a free handle group $\mathfrak{H}$, if a prime element $\mathfrak{a}^*$ is given as an expression $\mathfrak{a}^*(\mathfrak{a}, \mathfrak{b})$ in $\mathfrak{a}$ and $\mathfrak{b}$, then the paper quoted above analyses the structure of that expression. This yields a method for deciding whether a given word $w(\mathfrak{a}, \mathfrak{b})$ in $\mathfrak{a}$ and $\mathfrak{b}$ determines a primary inner translation with modulo $\mathfrak{H}$ simple axis, or not.

Remark. If similar considerations are applied to a free elementary group, for instance to an $\mathfrak{E}(\rightarrow\rightarrow\rightarrow)$, it results that the commutators of all pairs of prime elements of $\mathfrak{E}$ belong to one equivalence class of axes, thus to one and the same closed geodesic of $\mathcal{K}(\mathfrak{E})$ mod $\mathfrak{E}$. This geodesic has three double points; compare §9, Section 5 and following. These points, the "Löbell points" of the surface are uniquely determined by the metric parameters of $\mathfrak{E}$.

26.6 Metric parameters and congruence of handle groups. The number of metric parameters of $\mathfrak{E}$ in the preceding section is three or two according as (12) is a translation or not. Two of them are equal, namely the displacements of $\mathfrak{a}$ and $\mathfrak{b}\mathfrak{a}\mathfrak{b}^{-1}$. This provides together with the twist-parameter which measures the deviation from the standard-handle (§21.15) three metric parameters, if the handle in question is a $\mathcal{H}(\rightarrow)$, and two, if it is a $\mathcal{H}(\odot_\nu)$ or $\mathcal{H}(\hookrightarrow)$.

Moreover, if the handle is given, i.e. if the group $\mathfrak{H}$ is given, then there is more than one axis satisfying the conditions imposed on $\mathcal{A}$. In fact, the axis of every prime element (such as for instance $\mathcal{B}$) satisfies the same conditions. In order to arrive at a reasonable definition we use the fact proved in §15.2, that the displacements of the translations in a finitely generated $\mathfrak{F}$-group have no accumulation value. Therefore among the displacements of primary translations of $\mathfrak{H}$ with modulo $\mathfrak{H}$ simple inner axes there is a smallest one. Let $\mathfrak{a}$ be a translation of that kind and $\mathcal{A}$ the axis of $\mathfrak{a}$. If $\mathfrak{E}$ is, as in the last section, the according elementary group, we consider a boundary axis $\mathcal{A}_1$ of $\mathfrak{E}$, not equivalent with $\mathcal{A}$ modulo $\mathfrak{E}$ but equivalent with $\mathcal{A}$ modulo $\mathfrak{H}$ and with minimal distance from $\mathcal{A}$ (§21.1). If $\mathfrak{b}_0$ is the translation along the common normal of $\mathcal{A}$ and $\mathcal{A}_1$ that maps $\mathcal{A}$ onto $\mathcal{A}_1$ we choose a translation $\mathfrak{t}$ with axis $\mathcal{A}$, directed as $\mathfrak{a}$ and with minimal displacement such that $\mathfrak{b} = \mathfrak{t}\mathfrak{b}_0$ is an element of $\mathfrak{H}$ (which maps $\mathcal{A}$ onto $\mathcal{A}_1$). The displacements $\lambda(\mathfrak{a})$ and $\lambda(\mathfrak{t})$ are called the *inner metric parameters* of $\mathfrak{H}$. In the case of a handle $\mathcal{H}(\rightarrow)$ the other metric parameter of $\mathfrak{E}$ may be called the *boundary metric parameter* of $\mathfrak{H}$.

We now prove the theorem:

Theorem. *If two handle groups $\mathfrak{H}$ and $\mathfrak{H}'$ with the same symbol $\mathcal{H}(\odot_\nu)$, or $\mathcal{H}(\hookrightarrow)$, or $\mathcal{H}(\rightarrow)$ have their metric parameters equal, then they are congruent.*

Proof. If in the case of a $\mathcal{H}(\odot_\nu)$ or $\mathcal{H}(\hookrightarrow)$ the inner metric parameters are equal, then the pentagons called $\mathcal{Z}$ and $\mathcal{Z}'$ in Section 4 together with the transversals $\mathcal{B}$ and $\mathcal{B}'$ which result from the common normals of $\mathcal{A}$ and $\mathcal{A}_1$, respectively $\mathcal{A}'$ and $\mathcal{A}'_1$, by twisting are congruent figures, and the g-mapping g can then be chosen as a congruence (i.e. as a direct Möbius transformation). The same holds for the hexagons $\mathcal{Z}$ and $\mathcal{Z}'$ in the case of two groups $\mathfrak{H}$ and $\mathfrak{H}'$ with the same symbol $\mathcal{H}(\rightarrow)$, if their inner metric parameters are equal, and likewise their boundary metric parameters. In both cases $\mathfrak{H}$ and $\mathfrak{H}'$ are congruent. This implies that any translation in $\mathfrak{H}$ corresponds to a translation in $\mathfrak{H}'$ with the same displacement. □

Remark. If both handle groups $\mathfrak{H}$ and $\mathfrak{H}'$ have only one conjugacy class of primary translations with simple and non-dividing inner axes and minimal displacement then, obviously, their metric parameters are equal if they are congruent. If, however, there is more than one such class of primary translations there are different sets of inner metric parameters depending on the first choice of a primary translation with simple and non-dividing inner axis and minimal displacement.

26.7 Finitely generated groups of motions with $p > 0$. Let $\mathfrak{F}$ and $\mathfrak{F}'$ be two finitely generated $\mathfrak{F}$-groups consisting of motions, and let a g-isomorphism $I: \mathfrak{F} \rightarrow \mathfrak{F}'$ be given. Our aim is to construct a g-mapping of $\mathcal{K}(\mathfrak{F})$ onto $\mathcal{K}(\mathfrak{F}')$ inducing I. In this section the case with genus $p > 0$ is treated.

In the sequel we follow closely the pattern of §22.4 and use the notations there introduced. Assuming $p > 0$, we determine the handle group $\mathfrak{H}_1$ given by (1), §22.4. Since the case of a single handle group has already been dealt with, we assume that

$\mathfrak{k}_1 = \mathfrak{a}_1\mathfrak{b}_1\mathfrak{a}_1^{-1}\mathfrak{b}_1^{-1}$ is an inner translation of $\mathfrak{F}$. Let g denote a g-mapping of $\mathcal{K}(\mathfrak{H}_1)$ onto $\mathcal{K}(\mathfrak{H}_1')$, $\mathfrak{H}_1' = I(\mathfrak{H}_1)$, inducing the restriction of I to $\mathfrak{H}_1$, i.e. the g-isomorphism between $\mathfrak{H}_1$ and $\mathfrak{H}_1'$ derived from I. It has been shown in Section 5 that such a g exists. In constructing g we choose g arbitrarily on the axis $\mathcal{A}_1$ of $\mathfrak{a}_1$, subject of course to the functional equation

$$g\mathfrak{a}_1 x = \mathfrak{a}_1' g x \tag{19}$$

for x ranging over $\mathcal{A}_1$; compare (14).

If the axis $\mathfrak{C}_1$ of $\mathfrak{k}_1$ is of the type $\mathfrak{A}(\rightarrow)$, then the same holds for the axis $\mathfrak{C}_1'$ of $\mathfrak{k}_1' = I(\mathfrak{k}_1)$. We then choose g arbitrarily on $\mathfrak{C}_1$ subject to the functional equation

$$g\mathfrak{k}_1 x = \mathfrak{k}_1' g x \tag{20}$$

for x ranging over $\mathfrak{C}_1$. For simple axes of the type $\mathfrak{A}(\rightarrow)$ it is always understood that such functional equation holds for g in relation to a primary translation along the axis, even if it is not always mentioned in the sequel; it is a necessary and sufficient condition for g preserving equivalence, as far as points of such axes are concerned.

If $\mathfrak{C}_1$ is of the type $\mathfrak{A}(\cdot\cdot)$, then $\mathfrak{k}_1^{-1} = \mathfrak{c}_1\mathfrak{c}_2$, the centres c_1 and c_2 of the half-turns $\mathfrak{c}_1$ and $\mathfrak{c}_2$ being neighbouring centres on $\mathfrak{C}_1$. We then map the segment c_1c_2 arbitrarily onto the segment $c_1'c_2'$ of $\mathfrak{C}_1'$, where c_1' and c_2' are the centres of $I(\mathfrak{c}_1) = \mathfrak{c}_1'$ and $I(\mathfrak{c}_2) = \mathfrak{c}_2'$, respectively. This mapping g is extended to $\mathfrak{C}_1$ by putting

$$g\mathfrak{c}_i x = \mathfrak{c}_i' g x, \quad i = 1, 2, \tag{21}$$

for x ranging over $\mathfrak{C}_1$. Then also (20) holds. In this latter case, $\mathfrak{F}$ is the group (2), §22.4. The mapping g of $\mathcal{K}(\mathfrak{F})$ is obtained from the mapping g of $\mathcal{K}(\mathfrak{H}_1)$ by (8) for x ranging over $\mathcal{K}(\mathfrak{H})$ and $\mathfrak{f}$ over $\mathfrak{F}$. Since there is an amalgamation parameter for $\mathfrak{F}$ (see §18.7), the number of metric parameters for $\mathfrak{F}$ is one more as for $\mathfrak{H}_1$, namely 4.

If $\mathfrak{C}_1$ is of the type $\mathfrak{A}(\rightarrow)$, and $p > 1$, then we split off a handle group $\mathfrak{H}_2$ from the restgroup $\mathfrak{F}_1$ as described in §22.4. We first consider the case in which the axis $\mathfrak{C}_2$ of $\mathfrak{k}_2 = \mathfrak{a}_2\mathfrak{b}_2\mathfrak{a}_2^{-1}\mathfrak{b}_2^{-1}$ is equivalent with $\mathfrak{C}_1$ with respect to $\mathfrak{F}$. As described in §22.4 we can then choose $\mathfrak{H}_2$ such that $\mathfrak{C}_2 = \mathfrak{C}_1$ and $\mathfrak{k}_2 = \mathfrak{k}_1^{-1}$, thus obtaining for $\mathfrak{F}$ the group (6) of §22.4, $\mathcal{K}(\mathfrak{F})$ mod $\mathfrak{F}$ being a closed surface of genus $p = 2$. In this case g is chosen arbitrarily on $\mathcal{A}_2$ (subject to the relevant functional equation $g\mathfrak{a}_2 x = \mathfrak{a}' g x$), but on $\mathfrak{C}_2$ the mapping g shall be up to a translation with axis $\mathfrak{C}_1$, the one already existing on $\mathfrak{C}_1$ and satisfying (20), thus also $g\mathfrak{k}_2 x = \mathfrak{k}_2' g x$. The mapping g thus obtained on the union of fundamental domains of $\mathfrak{H}_1$ and $\mathfrak{H}_2$ is then extended to $\mathcal{K}(\mathfrak{F})$ by (8), x ranging over that union and $\mathfrak{f}$ over $\mathfrak{F}$. The number of metric parameters of $\mathfrak{F}$ is 6, because the primary displacement of $\mathfrak{k}_1$ and $\mathfrak{k}_2$ is the same, but a twist parameter arises.

If $\mathfrak{C}_1$ and $\mathfrak{C}_2$ are not equivalent with respect to $\mathfrak{F}$, we split off from the restgroup $\mathfrak{F}_1$ an elementary group $\mathfrak{E}_2$ generated by $\mathfrak{k}_1$ and $\mathfrak{k}_2$ as described in §22.4, and construct a g-mapping for $\mathfrak{E}_2$ as in Section 2. In this construction $g\mathfrak{C}_1$ and $g\mathfrak{C}_2$ shall be up to translations the mappings already existing on $\mathfrak{C}_1$ and $\mathfrak{C}_2$. If $\mathfrak{j}_2 = \mathfrak{k}_1\mathfrak{k}_2$ is a translation and its axis $\mathcal{J}_2$ is of the type $\mathfrak{A}(\rightarrow)$, then g is chosen arbitrarily on $\mathcal{J}_2$. If $\mathcal{J}_2$ is of the type $\mathfrak{A}(\cdot\cdot)$, then g is chosen in the way corresponding to (21); there is a one parameter family of possible choices. If $\mathfrak{j}_2$ is a rotation or limit-rotation, then also $\mathfrak{j}_2' = I(\mathfrak{j}_2)$ is. In this latter case the group $\mathfrak{G}_2$ given by (7), §22.4, is the whole of $\mathfrak{F}$. In all cases the mapping is extended to $\mathcal{K}(\mathfrak{G}_2)$ by (8), where x ranges over the union of fundamental domains of $\mathfrak{H}_1$, $\mathfrak{H}_2$, and $\mathfrak{E}_2$, and $\mathfrak{f}$ over $\mathfrak{G}_2$. The number of metric parameters for $\mathfrak{G}_2$ is 10, 9 or 8, according as $\mathcal{J}_2$ is of type $\mathfrak{A}(\cdot\cdot)$ or $\mathfrak{A}(\rightarrow)$ in the first two cases, or $\mathfrak{j}_2$ is a rotation or limit-rotation in the last case.

If $\mathcal{J}_2$ is an inner axis of $\mathfrak{F}$ of the type $\mathfrak{A}(\rightarrow)$ and $p > 2$, then we proceed in the same way with the restgroup $\mathfrak{F}_2$ on the pattern of §22.4 by splitting off a $\mathfrak{H}_3$. The process comes to an end after p steps. We examine the four cases denoted by 1), 2), 3), 4) in §22.4. In the cases 1), 2), 3), $\mathfrak{F}$ is fully determined and the construction of a g-mapping of $\mathcal{K}(\mathfrak{F})$ onto $\mathcal{K}(\mathfrak{F}')$ inducing I has been achieved by the above construction. We want to calculate the number of metric parameters in these cases.

In the case 1), $\mathfrak{F}$ is the group (9), §22.4. In all, p handles and $p-1$ elementary surfaces are involved. There are $2p$ inner metric parameters, p boundary metric parameters and p twist parameters for the boundary axis corresponding to the p handles. For each of the $p-1$ elementary groups two metric parameters are the same as those intervening in the previous step, but a twist parameter arises for each pair. So it introduces two new metric parameters, except for the last one, $\mathfrak{E}_p$. Here only one arises if $\mathfrak{j}_p = \mathfrak{k}_1\mathfrak{k}_2\dots\mathfrak{k}_p$ is a translation and none, if it is a rotation or limit-rotation. Hence the number of metric parameters is $6p-3$, if $\mathfrak{j}_p$ is a translation, thus if $\mathcal{K}(\mathfrak{F})$ mod $\mathfrak{F}$ has one boundary, and $6p-4$, if $\mathfrak{j}_p$ is a rotation or limit-rotation. This also holds for $p=1$, in which case $\mathfrak{F}$ is a handle group.

In the case 2), $\mathfrak{F}$ is the group (10), §22.4. In this case one has a rotation twin and $\mathfrak{j}_p$ is a translation. Hence there are $6p-2$ metric parameters. This also holds for $p=1$, as seen above.

In the case 3), $\mathfrak{F}$ is the group (11), §22.4. After $p-1$ steps one has introduced $p-1$ handles and $p-2$ elementary groups, and $\mathfrak{j}_{p-1}$ is a translation. This gives $4(p-1)+2(p-3)+1 = 6p-9$ metric parameters. In the step number p one more handle is introduced, thus one has two more inner metric parameters. The boundary metric parameter of that handle is equal to the displacement of $\mathfrak{j}_{p-1}$, and thus there results only one more twist parameter. Hence the number of metric parameters for the group (11), §22.4, is $6p-6$. This number is known in function theory as the number of modules of a closed Riemann surface of genus $p > 1$.

In the case 4), $\mathfrak{F}$ has not been fully exhausted. $\mathcal{J}_p$ is an inner axis of $\mathfrak{F}$ of the type $\mathfrak{A}(\rightarrow)$, and a restgroup $\mathfrak{F}_p$ of $\mathfrak{G}_p$ in $\mathfrak{F}$ remains. $\mathcal{K}(\mathfrak{G}_p)$ mod $\mathfrak{G}_p$ has one boundary and corresponds to the case 1). Thus for $\mathfrak{G}_p$ the number of metric parameters is $6p-3$. A mapping function g has been constructed for $\mathfrak{G}_p$. The restgroup $\mathfrak{F}_p$ has its genus equal to zero. The case of $\mathfrak{F}_p$ is studied in the next section, and the case of $\mathfrak{F}$ in Section 9. The full group $\mathfrak{F}$ is given by (13), §22.4, with $s = p$.

26.8 Finitely generated groups $\mathfrak{F}$ without reflections and without modulo $\mathfrak{F}$ simple, non-dividing axes. $\mathfrak{F}$ satisfies the conditions of §22.7. We follow the pattern of that section in order to construct a g-mapping inducing a given g-isomorphism

$I : \mathfrak{F} \to \mathfrak{F}'$. By (17), §22.7, $\mathfrak{F}$ is the free product with certain amalgamations of $N-2$ elementary groups $\mathfrak{E}_r$, the convex domains of which are contiguous along $N-3$ inner, modulo $\mathfrak{F}$ simple and dividing axes $\mathcal{P}_t$, $2 \leqq t \leqq N-2$, of $\mathfrak{F}$. The amalgamated subgroups are the groups $\mathfrak{P}_t$ of translations belonging to these $N-3$ axes $\mathcal{P}_t$. We construct g for the fundamental domains of $\mathfrak{E}_1, \mathfrak{E}_2, \dots, \mathfrak{E}_{N-2}$ in succession as in Section 2. For all those axes which are boundary axes of $\mathfrak{F}$ the mapping g is chosen arbitrarily subject to the relevant functional equation, and likewise on $\mathcal{P}_t$ for the construction of g in $\mathfrak{E}_{t-1}$. This mapping of $\mathcal{P}_t$ then remains up to a translation with axis $\mathcal{P}_t$ the same for the construction of g in $\mathfrak{E}_t$. Finally g is extended to the whole of $\mathcal{K}(\mathfrak{F})$ by (8), where x ranges over the union of fundamental domains of the $N-2$ elementary groups $\mathfrak{E}_r$ and $\mathfrak{f}$ over $\mathfrak{F}$. It then satisfies (8) in the whole of $\mathcal{K}(\mathfrak{F})$.

If $\beta = \vartheta(\rightarrow)$ denotes the number of equivalence classes of boundary axes of $\mathfrak{F}$, then the number of metric parameters of $\mathfrak{F}$ is $2(N-3)+\beta$, namely two for each of the $N-3$ inner axes $\mathcal{P}_t$ and one for each of those boundary axes of $\mathfrak{F}$ which are involved in the successive construction of g. Since

$$N = \vartheta(\rightarrow) + \vartheta(\hookrightarrow) + \sum_{\nu} \vartheta(\circlearrowleft_{\nu}),$$

the number of metric parameters of $\mathfrak{F}$ may be written

$$3\vartheta(\rightarrow) + 2\vartheta(\hookrightarrow) + 2\sum_{\nu} \vartheta(\circlearrowleft_{\nu}) - 6. \tag{22}$$

26.9 Finitely generated $\mathfrak{F}$-groups without reflections. Consider now the general case of a finitely generated group $\mathfrak{F}$ without reflections. The type numbers $\beta = \vartheta(\rightarrow)$, $\mu = \vartheta(\hookrightarrow)$, $\kappa = \sum_{\nu} \vartheta(\circlearrowleft_{\nu})$ denote the number of boundaries, masts, and conical points of $\mathcal{K}(\mathfrak{F})$ mod $\mathfrak{F}$, respectively. We are in the case of Section 9 of §22 and follow the pattern of that section by gradually loosening the restrictions imposed on $\mathfrak{F}$. Again any g-isomorphism $I : \mathfrak{F} \to \mathfrak{F}'$ is prescribed.

a) First, let $\mathfrak{F}$ be of genus $p = 0$, but possess an arbitrary number n of equivalence classes of centres of order 2. As in §22.6, we put $m = \left[\frac{n}{2}\right]$ and represent $\mathfrak{F}$ in the form (15), §22.6. The subgroup $\mathfrak{F}_m$ of $\mathfrak{F}$ satisfies the conditions of the preceding section. One gets

$$\beta(\mathfrak{F}_m) = \beta + m, \quad \mu(\mathfrak{F}_m) = \mu, \quad \kappa(\mathfrak{F}_m) = \kappa - 2m. \tag{23}$$

For the g-isomorphism $I : \mathfrak{F}_m \to \mathfrak{F}'_m$ derived from I as its restriction to $\mathfrak{F}_m$ a g-mapping can be constructed as in the preceding section. The m axes $\mathcal{A}_k$, $1 \leqq k \leqq m$, of §22.6 are inner axes of $\mathfrak{F}$ with corresponding group $\mathfrak{A}(\cdot\cdot)$. As far as the group $\mathfrak{F}_m$ is concerned, the mapping g on these axes was arbitrary subject to the functional equation

$$g\mathfrak{t}_k x = \mathfrak{t}'_k g x, \tag{24}$$

where $\mathfrak{t}_k$ denotes, as in §22.6, a primary translation along $\mathcal{A}_k$. In $\mathfrak{F}$ this is $\mathfrak{t}_k = \mathfrak{c}_{k1}\mathfrak{c}_{k2}$, the product of two half-turns. Thus g is chosen on $\mathcal{A}_k$ so as to satisfy

$$g\mathfrak{c}_{ki} x = \mathfrak{c}'_{ki} g x, \quad i = 1, 2, \tag{25}$$

and then (24) also holds; compare equations (20), (21). The mapping g as now defined in a fundamental domain for $\mathfrak{F}_m$ is then extended to $\mathcal{K}(\mathfrak{F})$ by (8) in the usual way.

The number of metric parameters for $\mathfrak{F}_m$ is given by (22), thus

$$3\beta(\mathfrak{F}_m) + 2\mu(\mathfrak{F}_m) + 2\kappa(\mathfrak{F}_m) - 6. \tag{26}$$

This then is also the number of metric parameters for $\mathfrak{F}$. If (23) is inserted in (26) and if one remembers that for every rotation twin there is an amalgamation parameter, it is seen that (22) gives the number for $\mathfrak{F}$.

b) Next, let $\mathcal{K}(\mathfrak{F})$ mod $\mathfrak{F}$ be non-orientable. Then $p > 0$ for $\mathfrak{F}$. As in §22.3 we introduce p modulo $\mathfrak{F}$ simple and disjoint axes $\mathcal{A}_m$, $1 \leqq m \leqq p$, with corresponding subgroups $\mathfrak{A}(\rightleftharpoons)$ generated by reversed translations $\mathfrak{h}_m$, thereby reducing $\mathfrak{F}$ to a subgroup $\mathfrak{F}_p$. Compare the main theorem of §22.3, the equation $s = p$ resulting from §22.9. Here

$$\beta(\mathfrak{F}_p) = \beta + p, \quad \mu(\mathfrak{F}_p) = \mu, \quad \kappa(\mathfrak{F}_p) = \kappa, \tag{27}$$

and $\mathfrak{F}_p$ is a group of motions with $p = 0$, thus coming under the case a). In constructing a g-mapping g for $\mathfrak{F}_p$ inducing the restriction $I : \mathfrak{F}_p \to \mathfrak{F}'_p$ of I to $\mathfrak{F}_p$ we choose g on $\mathcal{A}_m$ such that

$$g\mathfrak{h}_m x = \mathfrak{h}'_m g x. \tag{28}$$

In virtue of (28) it then satisfies

$$g\mathfrak{h}^2_m x = \mathfrak{h}'^2_m g x,$$

this being the equation needed for the choice of g on $\mathcal{A}_m$ in relation to $\mathfrak{F}_p$. Again g is extended by (8) from a fundamental domain for $\mathfrak{F}_p$ to $\mathcal{K}(\mathfrak{F})$.

The number of metric parameters for $\mathfrak{F}_p$ is determined by (22), thus given by (26), reading $\mathfrak{F}_p$ instead of $\mathfrak{F}_m$. The number is the same for $\mathfrak{F}$. By inserting (27) in (26) one gets

$$3p - 6 + 3\beta + 2\mu + 2\kappa \tag{29}$$

as the number of metric parameters for $\mathfrak{F}$.

c) Finally, let $\mathcal{K}(\mathfrak{F})$ mod $\mathfrak{F}$ be orientable and $p > 0$. Thus $\mathfrak{F}$ consists of motions. As in §22.4, we split off from $\mathfrak{F}$ a group $\mathfrak{G}_p$ by (13), §22.4, this $\mathfrak{G}_p$ being given by (12), §22.4, the equation $s = p$ resulting from Section 9 of §22. The case in which $\mathfrak{G}_p$ is the whole of $\mathfrak{F}$ has been completely covered in Section 7. We are thus concerned with the case 4) of Section 7, where a restgroup $\mathfrak{F}_p$ exists, thus where the $\mathcal{J}_p$ of (13), §22.4, is an inner axis of $\mathfrak{F}$ with corresponding group $\mathfrak{A}(\rightarrow)$ generated by $\mathfrak{j}_p = \mathfrak{k}_1\mathfrak{k}_2\dots\mathfrak{k}_p$; see (14) of §22.4.

Here a mapping g for $\mathcal{K}(\mathfrak{G}_p)$ inducing the restriction $I : \mathfrak{G}_p \to \mathfrak{G}'_p$ of I to $\mathfrak{G}_p$ has been constructed in Section 7. The restgroup $\mathfrak{F}_p$ is a group of motions with $p = 0$, thus coming under the case a). In constructing g for $\mathfrak{F}_p$ the only additional condition to be observed is that the choice of g on $\mathcal{J}_p$ is not arbitrary, but is up to a translation with axis $\mathcal{J}_p$ the same as the g on $\mathcal{J}_p$ already used in $\mathfrak{G}_p$. With this specification g

may be constructed in $\mathfrak{F}_p$ as above in the case a). The mapping g existing in the union of fundamental domains for $\mathfrak{G}_p$ and $\mathfrak{F}_p$ is then extended to $\mathcal{K}(\mathfrak{F})$ by (8).

It was seen in Section 7 that $\mathfrak{G}_p$ has $6p-3$ metric parameters, and in the case a) that $\mathfrak{F}_p$ has its number of metric parameters given by (22), thus by the expression (26) reading $\mathfrak{F}_p$ instead of $\mathfrak{F}_m$. The number for $\mathfrak{F}$ is equal to the sum of these two numbers. Now

$$\beta(\mathfrak{F}_p)=\beta+1, \quad \mu(\mathfrak{F}_p)=\mu, \quad \kappa(\mathfrak{F}_p)=\kappa.$$

Hence the number of metric parameters of $\mathfrak{F}$ is

$$6p-6+3\beta+2\mu+2\kappa. \tag{30}$$

If (30) is compared with the cases 1), 2), 3) of Section 7, it is seen that the numbers of metric parameters in these cases are special cases of (30).

If all is expressed by type numbers of $\mathfrak{F}$, §17.3-4, one gets from (29) and (30) the

Theorem. *The number of metric parameters for finitely generated $\mathfrak{F}$-groups without reflections and with $p \geqq 0$ is*

$$3(2-\vartheta(\rightleftharpoons))p-6+3\vartheta(\rightarrow)+2\vartheta(\hookrightarrow)+2\sum_{\nu}\vartheta(\circledcirc_{\nu}). \tag{31}$$

26.10 Finitely generated $\mathfrak{F}$-groups containing reflections. At the outset we consider the special cases of such a group $\mathfrak{F}$ enumerated in Table 5 of §22.10.

Consider the first four cases of the table. There is one reflection ring, thus $\vartheta(-)=1$. Let ω denote the number of reflection lines in a period of the reflection chain from which this ring is derived. Let $\mathcal{L}_\ell$, $1 \leqq \ell \leqq \omega$, denote reflection lines in the order of the chain, after $\mathcal{L}_1$ has been chosen as an arbitrary link of the chain. Let $\mathfrak{s}_\ell$ denote the reflection in $\mathcal{L}_\ell$. Let $\mathfrak{g}$ denote a primary element belonging to the period of the chain, thus a rotation, or limit-rotation, or translation in the cases 2), 3), 4), respectively, while in the case 1) we put $\mathfrak{g}=1$. We then put

$$\mathcal{L}_{\omega+1}=\mathfrak{g}\mathcal{L}_1, \quad \mathfrak{s}_{\omega+1}=\mathfrak{g}\mathfrak{s}_1\mathfrak{g}^{-1}.$$

In the cases 2), 3), 4), let $\mathcal{V}$ denote the centre, or limit-centre, or axis of $\mathfrak{g}$, respectively, and α half the primary angle of rotation, or zero, or half the displacement for $\mathfrak{g}$. The element

$$\mathfrak{f}_\ell=\mathfrak{s}_{\ell+1}\mathfrak{s}_\ell, \quad 1 \leqq \ell \leqq \omega, \tag{32}$$

of $\mathfrak{F}$ is a rotation, or limit-rotation, or translation, according as $\mathcal{L}_\ell$ and $\mathcal{L}_{\ell+1}$ are concurrent, or parallel, or divergent. Let $\mathcal{A}_\ell$, correspondingly, denote the centre, or limit-centre, or axis of $\mathfrak{f}_\ell$, and α_ℓ the angle $\frac{\pi}{\nu_\ell}$, or zero, or half the displacement of $\mathfrak{f}_\ell$. In the two last cases the notation $\nu_\ell=\infty$ was agreed upon (§20.5).

Let a g-isomorphism $I:\mathfrak{F}\to\mathfrak{F}'$ be given. By I the elements $\mathfrak{g}$, $\mathfrak{s}$, $\mathfrak{f}$ correspond to elements $\mathfrak{g}'$, $\mathfrak{s}'$, $\mathfrak{f}'$ of the same kind in $\mathfrak{F}'$. In particular, $\mathfrak{s}'_{\omega+1}=\mathfrak{g}'\mathfrak{s}'_1\mathfrak{g}'^{-1}$. The notations $\mathcal{L}'_\ell$, $\mathcal{V}'$ are understood correspondingly. If α_ℓ is the angle $\frac{\pi}{\nu_\ell}$, then it is equal to α'_ℓ, since $\nu_\ell = \nu'_\ell$; likewise if $\alpha_\ell = 0$. But if α_ℓ denotes half a displacement, then it need not be equal to α'_ℓ. Likewise, if $\mathfrak{g}$ is a translation, half the displacements α and α' of $\mathfrak{g}$ and $\mathfrak{g}'$ need not be equal.

In the case 1), the case of a finite polygonal disc, let Π denote the polygon determined by vertices in or sides on the $\mathcal{A}_\ell$ and sides on the $\mathcal{L}_\ell$ for $1 \leqq \ell \leqq \omega$ (Fig. 26.2). In the cases 2), 3), 4), let $\mathcal{S}$ denote the join of $\mathcal{V}$ and $\mathcal{L}_1$. Then $\mathfrak{g}\mathcal{S}$ is the join of $\mathcal{V}$

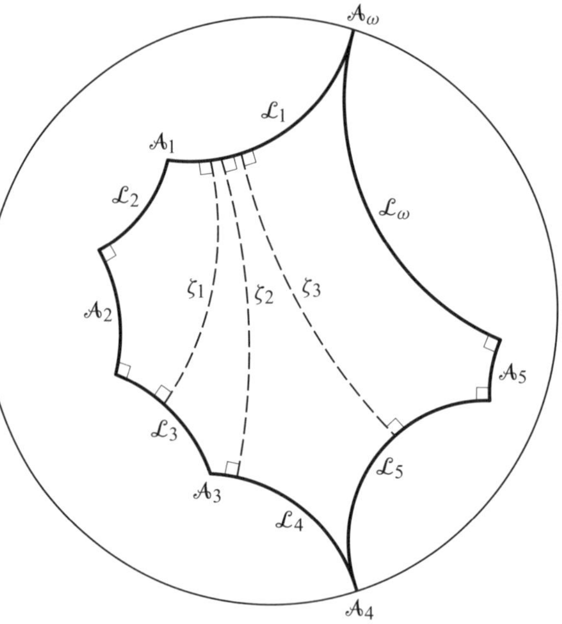

Figure 26.2

and $\mathfrak{g}\mathcal{L}_1 = \mathcal{L}_{\omega+1}$. In these cases let Π denote the polygon determined by the sides $\mathcal{S}$ and $\mathfrak{g}\mathcal{S}$, by vertices in or sides on $\mathcal{V}$ and the $\mathcal{A}_\ell$, $1 \leqq \ell \leqq \omega$, and sides on the $\mathcal{L}_\ell$, $1 \leqq \ell \leqq \omega + 1$ (Fig. 26.3–26.5).

In all four cases there is in $\overline{\mathcal{D}}'$, a polygon Π' with sides corresponding in a uniquely determined way to those of Π; here $\mathcal{S}'$ in the cases 2), 3), 4), means the join of $\mathcal{V}'$ and $\mathcal{L}'_1$, and $\mathfrak{g}'\mathcal{S}'$ is then the join of $\mathcal{V}'$ and $\mathcal{L}'_{\omega+1} = \mathfrak{g}'_1\mathcal{L}'_1$. One can thus establish a topological mapping g of Π onto Π' which maps each side of Π onto the corresponding side of Π'. In the cases 2), 3), 4), we let this g satisfy an additional condition: After a mapping g of $\mathcal{S}$ onto $\mathcal{S}'$ has been chosen, the mapping of $\mathfrak{g}\mathcal{S}$ onto $\mathfrak{g}'\mathcal{S}'$ shall be determined by

$$g\mathfrak{g}x = \mathfrak{g}'gx \tag{33}$$

for x ranging over $\mathcal{S}$. This mapping $g\colon \Pi \to \Pi'$ is extended to the interiors Ω and Ω'

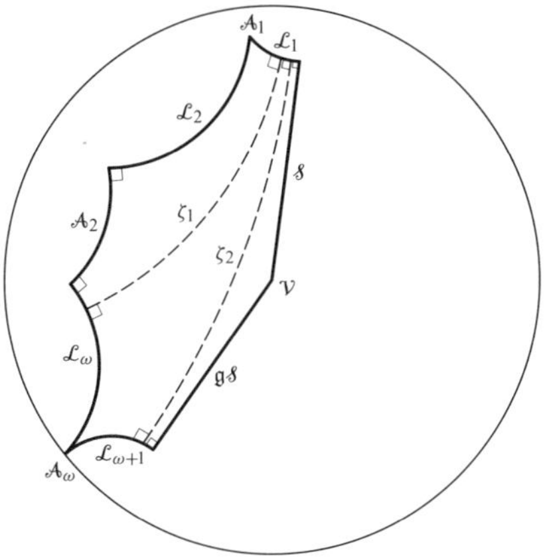

Figure 26.3

of Π and Π' in virtue of the lemma of Section 1. In the case 1) the region Ω closed on $\mathcal{K}(\mathfrak{F})$ is a fundamental domain for $\mathfrak{F}$ in $\mathcal{K}(\mathfrak{F})$. In the cases 2), 3), 4), the same holds after the omission of $\mathfrak{g}\mathcal{S}$. Hence in all four cases, and considering (33) in the last three cases, g can be extended to $\mathcal{K}(\mathfrak{F})$ by (8), letting x range over this fundamental domain and $\mathfrak{f}$ over $\mathfrak{F}$. Then (8) holds throughout $\mathcal{K}(\mathfrak{F})$. *This yields a g-mapping of $\mathcal{K}(\mathfrak{F})$ onto $\mathcal{K}(\mathfrak{F}')$ inducing the given g-isomorphism I.*

As metric parameters for $\mathfrak{F}$ we choose

a) those $\vartheta(\|)$ values among the $2\alpha_\ell$ for which the $\mathfrak{f}_\ell$ of (32) is a translation,

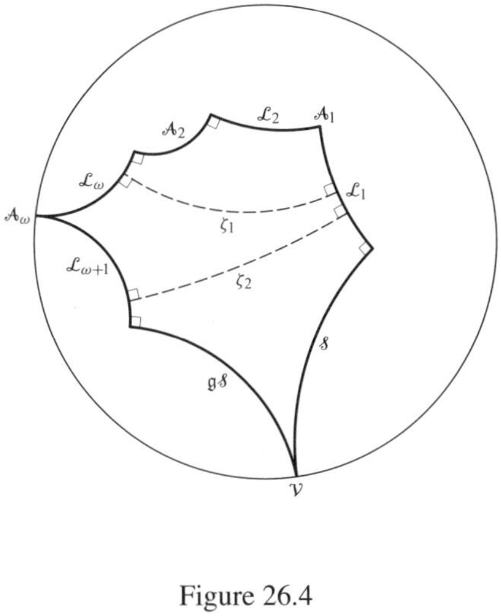

Figure 26.4

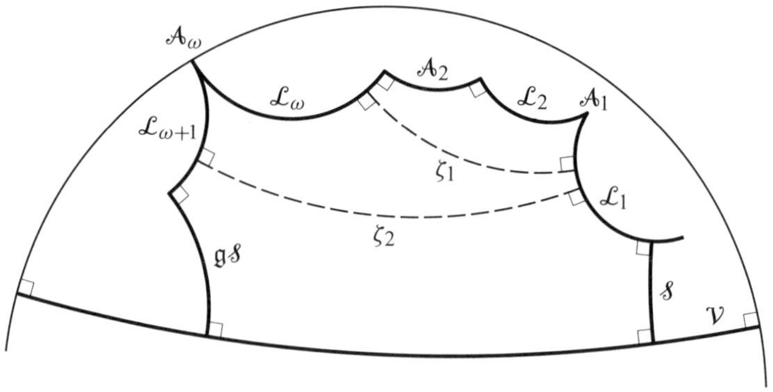

Figure 26.5

b) the displacement 2α of $\mathfrak{g}$, if $\mathfrak{g}$ is a translation,

c) the length ζ_t of the join of $\mathcal{L}_1$ and $\mathcal{L}_{t+2}$ for those values of t for which this join belongs to Ω. These ζ_t are half the displacements of the translations $\mathfrak{s}_{t+2}\mathfrak{s}_1$.

In the case 1) one gets $1 \leqq t \leqq \omega - 3$; see Fig. 26.2, which is drawn for $\omega = 6$. Thus for the minimum value of ω in this case, which is $\omega = 3$, there is no ζ_t.

In the cases 2), 3), 4), one gets $1 \leqq t \leqq \omega - 1$, $\zeta_{\omega-1}$ being the join of $\mathcal{L}_1$ and $\mathcal{L}_{\omega+1}$; see the Fig. 26.3–26.5, which are drawn for $\omega = 3$.

This system a), b), c), of metric parameters is uniquely determined by $\mathfrak{F}$ after the link $\mathcal{L}_1$ of the reflection chain has been chosen. On the other hand, for any given values of these parameters there is in each of the four cases a corresponding group $\mathfrak{F}$. To see this, let any system of ω values α_ℓ be given, each of these α_ℓ being either a length, or an angle $\frac{\pi}{\nu}$, or zero.

In the case 1), if $\omega = 3$, the three values α_ℓ determine a polygon Ω, provided $1/\nu_1 + 1/\nu_2 + 1/\nu_3 < 1$, of course. This results from §21.21. If $\omega > 3$, then the polygon Ω of Fig. 26.2 is divided into $\omega - 2$ parts by the $\omega - 3$ lines of the ζ_t. Each of these parts is a polygon in which either one of the ζ_t and two of the α_ℓ, or two of the ζ_t and one of the α_ℓ is known. This polygonal part is determined by these three arbitrarily given values, as seen from the formulae of §21.21. Then a polygon Ω with the prescribed values of the metric parameters results.

In the cases 2), 3), 4), one first gets in the same way $\omega - 1$ fully determined polygonal parts belonging to the prescribed values of the metric parameters a) and c). Consider the lines $\mathcal{L}_1$ and $\mathcal{L}_{\omega+1}$ obtained in this construction. They are divergent, $\zeta_{\omega-1}$ being their distance. Let $\mathcal{L}_1$ be directed towards $\mathcal{A}_1$ and $\mathcal{L}_{\omega+1}$ be directed away from $\mathcal{A}_\omega$. Then there is a rotation of any prescribed order ν, and a limit-rotation, and a translation with any prescribed displacement 2α, each of which carries $\mathcal{L}_1$ into $\mathcal{L}_{\omega+1}$ with coincident direction and for which $\mathcal{V}$ is situated on the concave side of the reflection chain. This is illustrated by Fig. 26.6. In each case $\mathcal{V}$ is uniquely determined

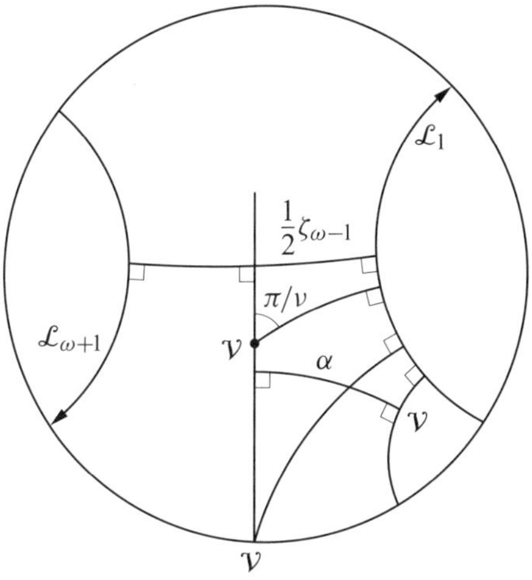

Figure 26.6

by the corresponding value α belonging to $\mathfrak{g}$. This yields the theorem:

Theorem. *Two g-isomorphic groups determining polygonal discs, or polygonal cones, or polygonal masts, or crowns, are congruent if corresponding metric parameters are equal.* □

We are going to express the number of metric parameters by the type numbers of $\mathfrak{F}$ (§17.3–17.4). In all four cases one gets

$$\omega = \vartheta(\|) + \vartheta(\wedge) + \sum_{\nu} \vartheta(\times_\nu). \tag{34}$$

The number of metric parameters introduced under c) is then $\omega - 3$ in the case 1) and $\omega - 1$ in the three other cases with (34) inserted for ω. The number of metric parameters introduced under a) is $\vartheta(\|)$. The type numbers $\vartheta(\rightarrow)$, $\vartheta(\hookrightarrow)$, $\sum_\nu \vartheta(\circlearrowright_\nu)$ are all zero in the case 1), and exactly one of them is 1, the other two being zero, in the other three cases. Thus their sum is 1 in these three cases. Hence the number of metric parameters introduced under b) is $\vartheta(\rightarrow)$.

The number of metric parameters of a finite polygonal disc is

$$2\vartheta(\|) + \vartheta(\wedge) + \sum_{\nu} \vartheta(\times_\nu) - 3, \tag{35a}$$

whereas for the cases 2), 3), 4), one gets

$$\vartheta(\rightarrow) + 2\vartheta(\|) + \vartheta(\wedge) + \sum_{\nu} \vartheta(\times_\nu) - 1. \tag{35b}$$

One easily verifies that both formulas are special cases of the general formula given in the following

Theorem. *For any finitely generated $\mathfrak{F}$-group, the number of metric parameters expressed by the type numbers of the group is given by*

$$3(2-\vartheta(\rightleftharpoons))p-6+3\vartheta(\rightarrow)+2\vartheta(\hookrightarrow) \\ +2\sum_{\nu}\vartheta(\circlearrowleft_{\nu})+2\vartheta(\|)+\vartheta(\wedge)+\sum_{\nu}\vartheta(\times_{\nu})+3\vartheta(\rightarrow -)+3\vartheta(-). \tag{36}$$

Proof. The formula (31) is the special case of (36) in which the last five terms cancel, because the group contains no reflections. Thus the proof of the above theorem will be complete, if (36) is seen to hold for the cases still to be covered in the sequel.

We first treat the last four cases of Table 5 of §22.10.

The *reflection crown* is derived from a crown by letting $\mathfrak{A}_{\mathcal{V}}$ be of the type $\mathfrak{A}(\rightarrow -)$. The fundamental domain and its g-mapping remains the same as for the crown. Again g is extended to $\mathcal{K}(\mathfrak{F})$ by (8). The number of metric parameters is the same as for the crown. Since $\vartheta(\rightarrow)$ is decreased by 1, while $\vartheta(-)$ is increased by 1, and all other type numbers remain unchanged, (36) holds.

The *conical crown* is derived from a crown by letting $\mathfrak{A}_{\mathcal{V}}$ be of the type $\mathfrak{A}(\cdot\cdot)$. On the side of Ω on $\mathcal{V}$ there are two centres c_1 and c_2 belonging to half-turns $\mathfrak{c}_1$ and $\mathfrak{c}_2$, and $\mathfrak{c}_1\mathfrak{c}_2=\mathfrak{g}$. The mapping g of this side, which was arbitrary for a crown, has to be so chosen as to satisfy the equations $g\mathfrak{c}_i x=\mathfrak{c}_i' gx$, $i=1,2$. It then also satisfies

$$g\mathfrak{g}x=\mathfrak{g}'gx \tag{37}$$

for x ranging over $\mathcal{V}$; compare (25) and (24). Then again a g-mapping of $\mathcal{K}(\mathfrak{F})$ onto $\mathcal{K}(\mathfrak{F}')$ is derived from the g-mapping of Ω by (8). There is one metric parameter more than for the crown fixing the position of the centres on $\mathcal{V}$. In particular, $[c_1,c_2]=\alpha$, half the displacement of $\mathfrak{g}$. All type numbers remain the same except that $\vartheta(\rightarrow)$ is decreased by 1 and $\vartheta(\circlearrowleft_2)$ is increased by 2. Hence (36) holds.

The *cross cap crown* is derived from a crown by letting $\mathfrak{A}_{\mathcal{V}}$ be of the type $\mathfrak{A}(\rightleftharpoons)$. If $\mathfrak{h}$ is a primary reversed translation along $\mathcal{V}$, thus with the displacement α, the mapping g on $\mathcal{V}$ has to be so chosen as to satisfy the equation $g\mathfrak{h}x=\mathfrak{h}'gx$. It then also satisfies (37), since $\mathfrak{g}=\mathfrak{h}^2$. Then again the g-mapping of $\mathcal{K}(\mathfrak{F})$ onto $\mathcal{K}(\mathfrak{F}')$ is derived from the mapping g of Ω by (8). The number of metric parameters remains the same as for the crown. All type numbers remain the same except that $\vartheta(\rightarrow)$ is decreased by 1 and $\vartheta(\rightleftharpoons)$ is increased by 1, while p is increased from 0 to 1. Hence (36) holds.

The group $\mathfrak{F}$ of the *double crown* is derived as the free product of two crown groups $\mathfrak{F}_1$ and $\mathfrak{F}_2$ with amalgamation of the translation group belonging to the common boundary axis $\mathcal{V}_1=\mathcal{V}_2$ of $\mathfrak{F}_1$ and $\mathfrak{F}_2$. After the mapping g has been chosen for the Ω_1 of $\mathfrak{F}_1$, closed on $\mathcal{K}(\mathfrak{F}_1)$, the mapping of $\mathcal{V}_1$ onto $\mathcal{V}_1'$ has been fixed. In constructing g for the Ω_2 of $\mathfrak{F}_2$, closed on $\mathcal{K}(\mathfrak{F}_2)$, up to a translation with axis $\mathcal{V}_1$ this same g has to be used on $\mathcal{V}_2=\mathcal{V}_1$. Then a g-mapping of $\mathcal{K}(\mathfrak{F})$ onto $\mathcal{K}(\mathfrak{F}')$ is derived from the mapping g established in the union of the fundamental domains of the two groups by (8). The number of metric parameters for $\mathfrak{F}$ is equal to the sum of those for $\mathfrak{F}_1$ and $\mathfrak{F}_2$ since $\mathcal{V}_1$ and $\mathcal{V}_2$ yield the same α and an amalgamation parameter arises. Any type number for $\mathfrak{F}$ is the sum of the type numbers of the same kind for $\mathfrak{F}_1$ and $\mathfrak{F}_2$, except that $\vartheta(\rightarrow)$ is 1 for both $\mathfrak{F}_1$ and $\mathfrak{F}_2$, and 0 for $\mathfrak{F}$. This shows that (36) holds.

We are thus left with the general case of a finitely generated $\mathfrak{F}$-group $\mathfrak{F}$ containing reflections treated in Section 10 of §22 and not belonging to the eight special cases of Table 5. Keeping the notations there used, we reduce $\mathfrak{F}$ to the finitely generated subgroup $\mathfrak{F}_0$, the boundaries of which inside $\mathcal{K}(\mathfrak{F})$ are the equivalence classes of the axes $\mathcal{W}_r$, $1 \leqq r \leqq \alpha$, and of the axes $\mathcal{V}_s$, $1 \leqq s \leqq \gamma, \alpha + \gamma = \vartheta(-)$ being the number of reflection rings of $\mathcal{K}(\mathfrak{F})$ mod $\mathfrak{F}$. The subgroup $\mathfrak{V}_s$ of $\mathfrak{F}$ belonging to $\mathcal{V}_s$ is of the type $\mathfrak{A}(\rightarrow)$. It is generated by the translations $\mathfrak{g}_s$, this being the primary boundary translation of a crown group $\mathfrak{G}_s$. It is recalled that the subgroup $\mathfrak{W}_r$ of $\mathfrak{F}$ belonging to $\mathcal{W}_r$ is of the type $\mathfrak{A}(\rightarrow\ -)$, that $\mathfrak{W}_r^*$ means the translation group belonging to $\mathcal{W}_r$, and that $\mathfrak{F}$ is given by the free product with amalgamations

$$\mathfrak{F} = \prod_{s,r}^{*} (\mathfrak{F}_0, \mathfrak{G}_s, \mathfrak{W}_r) \operatorname{am} \mathfrak{V}_s, \mathfrak{W}_r^*. \tag{38}$$

We now construct a g-mapping for the Ω_s, closed on $\mathcal{K}(\mathfrak{G}_s)$, $1 \leqq s \leqq \gamma$, in the way described above. From this results a topological mapping of $\mathcal{V}_s$ onto $\mathcal{V}_s'$ satisfying $g\mathfrak{g}_s x = \mathfrak{g}_s' g x$. In constructing a mapping g for a fundamental domain of $\mathfrak{F}_0$ by the method of the preceding section g can be prescribed arbitrarily on the $\mathcal{W}_r$ and $\mathcal{V}_s$, because they are boundary axes of $\mathfrak{F}_0$. We now fix that the mapping g of $\mathcal{V}_s$ shall be up to a translation with axis $\mathcal{V}_s$, the one resulting from the mapping of Ω_s. Then the mapping g given in the union of fundamental domains for $\mathfrak{F}_0$ and the $\mathfrak{G}_s$ extends to the whole of $\mathcal{K}(\mathfrak{F})$ by (8). *Thus there exists a g-mapping of $\mathcal{K}(\mathfrak{F})$ onto $\mathcal{K}(\mathfrak{F}')$ which induces a given g-isomorphism.*

The number of metric parameters of $\mathfrak{F}$ is equal to the sum of the numbers of metric parameters for $\mathfrak{F}_0$ and the $\mathfrak{G}_s$. For $\mathfrak{F}_0$ this number is given by (31) provided the type numbers refer to $\mathfrak{F}_0$. We here refer to type numbers belonging to $\mathfrak{F}_0$ and the $\mathfrak{G}_s$ by adding a subscript 0 or s, respectively, while type numbers without subscript refer to $\mathfrak{F}$. Since the $\mathcal{W}_r$ and the $\mathcal{V}_s$ are inner axes of $\mathfrak{F}$, but boundary axes of $\mathfrak{F}_0$, one gets

$$\vartheta_0(\rightarrow) = \vartheta(\rightarrow) + \vartheta(-) \tag{39}$$

while the numbers $\vartheta(\hookrightarrow)$, $\vartheta(\odot_\nu)$, $\vartheta(\rightleftharpoons)$, and p are the same for $\mathfrak{F}_0$ and $\mathfrak{F}$. All numbers referring to reflections are zero for $\mathfrak{F}_0$. Hence the number of metric parameters for $\mathfrak{F}_0$ is by (31)

$$3(2 - \vartheta(\rightleftharpoons))p - 6 + 3\vartheta(\rightarrow) + 3\vartheta(-) + 2\vartheta(\hookrightarrow) + 2\sum_\nu \vartheta(\odot_\nu). \tag{40}$$

For $\mathfrak{G}_s$ the number of metric parameters is (35*b*) which reduces to

$$2\vartheta_s(\|) + \vartheta_s(\wedge) + \sum_\nu \vartheta_s(\times_\nu), \tag{41}$$

since $\vartheta_s(\rightarrow) = 1$. Hence the number of metric parameters of $\mathfrak{F}$ results by adding (40) and the expressions (41) for all values of s. This yields (36). The proof for the formula (36) is thus complete. □

26.11 g-mappings of infinitely generated groups. Let $\mathfrak{F}$ be an $\mathfrak{F}$-group not satisfying the condition of quasi-compactness, and $I: \mathfrak{F} \rightarrow \mathfrak{F}'$ a g-isomorphism. A g-isomorphism being a special case of a t-isomorphism, the extended hull $\mathfrak{H}(\mathfrak{G})$ of a subgroup $\mathfrak{G}$ of $\mathfrak{F}$ corresponds by I to the extended hull of $I(\mathfrak{G})$ in $\mathfrak{F}'$ (§24.3).

Consider the exhaustion of $\mathfrak{F}$ by the sequence of extended hulls described in §23.3. We use the notations there introduced. The subgroup $\mathfrak{F}_{n_k}$ of $\mathfrak{F}$ corresponds by I to the subgroup $I(\mathfrak{F}_{n_k}) = \mathfrak{F}'_{n_k}$ of $\mathfrak{F}'$. Then also the extended hulls $\mathfrak{H}_k = \mathfrak{H}(\mathfrak{F}_{n_k})$ and $\mathfrak{H}'_k = \mathfrak{H}(\mathfrak{F}'_{n_k})$ correspond by the g-isomorphism derived for them from $I: \mathfrak{F} \rightarrow \mathfrak{F}'$, i.e. the restriction of I to $\mathfrak{F}_{n_k}$. Hence a g-mapping g of $\mathcal{K}_k = \mathcal{K}(\mathfrak{H}_k)$ onto $\mathcal{K}'_k = \mathcal{K}(\mathfrak{H}'_k)$ inducing this isomorphism exists in virtue of the preceding sections.

We now consider $\mathfrak{H}_{k+1}$ and its image $\mathfrak{H}'_{k+1}$ by I. The convex domain $\mathcal{K}_{k+1}$ of $\mathfrak{H}_{k+1}$ is decomposed by $\mathfrak{H}_{k+1}\mathcal{K}_k$, thus by the equivalence classes with respect to $\mathfrak{H}_{k+1}$ of those boundary axes $\mathcal{A}_{kr}$, $1 \leqq r \leqq m_k$, of $\mathcal{K}_k$ which are inner axes of $\mathfrak{F}$. The number of equivalence classes of regions of decomposition is $m'_k + 1$, where $m'_k \leqq m_k$. Those m'_k classes of regions which are not equivalent with $\mathcal{K}_k$ are represented by the regions $\mathfrak{C}_{ks}$, $1 \leqq s \leqq m'_k$, with corresponding groups $\mathfrak{C}_{ks}$, the $\mathfrak{C}_{ks}$, closed on $\mathcal{K}(\mathfrak{F})$, being the convex domains of the $\mathfrak{C}_{ks}$. Since the group $\mathfrak{C}_{ks}$ is finitely generated, a g-mapping of its convex domain $\mathfrak{C}_{ks}$ onto the convex domain $\mathfrak{C}'_{ks}$ of $\mathfrak{C}'_{ks} = I(\mathfrak{C}_{ks})$ exists such that it induces the restriction of I to $\mathfrak{C}_{ks}$. Some sides of $\mathfrak{C}_{ks}$ are equivalent with respect to $\mathfrak{H}_{k+1}$ with sides $\mathcal{A}_{kr}$ of $\mathcal{K}_k$, for at least one value of r. If $\mathcal{B}$ denotes a boundary axis of $\mathfrak{C}_{ks}$ such that $\mathcal{B} = \mathfrak{f}\mathcal{A}_{kr}$, $\mathfrak{f} \in \mathfrak{H}_{k+1}$, then on $\mathcal{B}$ the mapping g will be defined by

$$g\mathfrak{f}x = \mathfrak{f}'gx, \quad \mathfrak{f}' = I(\mathfrak{f}), \tag{42}$$

where x ranges over $\mathcal{A}_{kr}$, where the mapping g is already defined. Thus g is prescribed on one or more classes of boundary axes of $\mathfrak{C}_{ks}$. That is the only additional condition to be imposed on the construction of g for $\mathfrak{C}_{ks}$. A g-mapping for K_{k+1} is then obtained by (42), where x now ranges over the union of fundamental domains for $\mathfrak{H}_k$ and the $\mathfrak{C}_{ks}$, $1 \leqq s \leqq m'_k$, and $\mathfrak{f}$ over $\mathfrak{H}_{k+1}$. It is recalled that the subgroup of $\mathfrak{F}$ belonging to $\mathcal{A}_{kr}$ is either $\mathfrak{A}(\rightarrow)$ or $\mathfrak{A}(\|)$; in both cases $g\mathcal{A}_{kr}$ is known from $g\mathcal{K}_k$.

If a g-mapping is chosen for $\mathcal{K}_1$ and this rule for the extension of g from $\mathcal{K}_k$ to $\mathcal{K}_{k+1}$ is observed, one gets a g-mapping for the infinite sequence of the $\mathcal{K}_k$, $1 \leqq k < \infty$, such that the mapping of $\mathcal{K}_k$ is not altered by the extension to $\mathcal{K}_{k+\ell}$, $1 \leqq \ell < \infty$.

Since any point of the interior of $\mathcal{K}(\mathfrak{F})$ and any boundary axis of $\mathcal{K}(\mathfrak{F})$ belongs to $\mathcal{K}_k$ for all sufficiently large k (§23.4), a mapping g inducing I is obtained for the whole of $\mathcal{K}(\mathfrak{F})$ with the exception of all limit sides of $\mathcal{K}(\mathfrak{F})$, if any. The extension of g to limit sides of $\mathcal{K}(\mathfrak{F})$ is not always possible; to see this, it is sufficient to remark that, even for a g-isomorphism, a limit side of $\mathcal{K}(\mathfrak{F})$ may well correspond to a single limit point of $\mathcal{K}(\mathfrak{F}')$, or conversely.

Remark. Consider any $\mathfrak{F}$-group $\mathfrak{F}$, whether finitely generated or not. Let $\mathcal{K}^\bullet(\mathfrak{F})$ denote the point set obtained from $\mathcal{K}(\mathfrak{F})$ by the omission of limit sides of $\mathcal{K}(\mathfrak{F})$, if any; and likewise $\mathcal{K}^\bullet(\mathfrak{F}')$ for $\mathfrak{F}'$. Let $I : \mathfrak{F} \to \mathfrak{F}'$ denote a g-isomorphism and $g : \mathcal{K}^\bullet(\mathfrak{F}) \to \mathcal{K}^\bullet(\mathfrak{F}')$ a g-mapping inducing I, thus

$$g\mathfrak{f}x = I(\mathfrak{f})gx \tag{43}$$

for x ranging over $\mathcal{K}^\bullet(\mathfrak{F})$ and $\mathfrak{f}$ over $\mathfrak{F}$. Let g_1 be a g-mapping of $\mathcal{K}^\bullet(\mathfrak{F})$ onto $\mathcal{K}^\bullet(\mathfrak{F}')$ which effects the same correspondence between equivalence classes of points in $\mathcal{K}^\bullet(\mathfrak{F})$ and $\mathcal{K}^\bullet(\mathfrak{F}')$ as g, thus the same mapping of $\mathcal{K}^\bullet(\mathfrak{F})$ mod $\mathfrak{F}$ onto $\mathcal{K}^\bullet(\mathfrak{F}')$ mod $\mathfrak{F}'$. If x is a point of $\mathcal{K}^\bullet(\mathfrak{F})$, regular with respect to $\mathfrak{F}$, then there is a uniquely determined element $\mathfrak{x}'$ of $\mathfrak{F}'$ such that

$$g_1x = \mathfrak{x}'gx. \tag{44}$$

It then follows as in §25.1 that $\mathfrak{x}'$ remains the same as long as x ranges over the set of regular points in $\mathcal{K}^\bullet(\mathfrak{F})$, and that (44) then holds for x ranging over $\mathcal{K}^\bullet(\mathfrak{F})$. By (43) and (44) one then gets

$$g_1\mathfrak{f}x = \mathfrak{x}'g\mathfrak{f}x = \mathfrak{x}'I(\mathfrak{f})gx = \mathfrak{x}'I(\mathfrak{f})\mathfrak{x}'^{-1}\ g_1x.$$

Hence g_1 induces the g-isomorphism $I_1 : \mathfrak{F} \to \mathfrak{F}'$ given by

$$I_1(\mathfrak{f}) = \mathfrak{x}'I(\mathfrak{f})\mathfrak{x}'^{-1}.$$

If $g_1 \neq g$, thus $\mathfrak{x}' \neq 1$, then $\mathfrak{x}'$ is not permutable with all elements of $\mathfrak{F}'$, hence $I_1 \neq I$. Hence: If only the correspondence of equivalence classes of points is given for the two groups, then the g-isomorphism is determined only up to an inner automorphism for one of the two groups.

26.12 Congruence of $\mathfrak{F}$-groups. Alignment lengths. Two $\mathfrak{F}$-groups $\mathfrak{F}$ and $\mathfrak{F}'$ are congruent (Section 4), if there exists a g-mapping of the form (11) of $\overline{\mathcal{D}}$ onto $\overline{\mathcal{D}}'$. Then $\mathfrak{F}$ and $\mathfrak{F}'$ are g-isomorphic (§25.3), and by g a system of metric parameters for $\mathfrak{F}'$ is derived from a system of metric parameters for $\mathfrak{F}$, corresponding metric parameters of the two groups being equal, since g preserves distances.

If $\mathfrak{F}$ is either an elementary group, or a handle group, or a crown group, and $I : \mathfrak{F} \to \mathfrak{F}'$ a g-isomorphism, then also $\mathfrak{F}'$ is an elementary group, or a handle group, or a crown group, respectively. In these cases it has been seen in Sections 4, 6, 10, respectively, that the groups are congruent, if corresponding metric parameters are equal, and that the mapping function of the form (11) inducing I is uniquely determined. The same holds for the groups of a finite polygonal disc, (§20.5), of a polygonal cone (§20.6), and of a polygonal mast (§20.8), as seen in Section 10.

If g is the congruent mapping preserving equivalence in relation to groups $\mathfrak{F}$ and $\mathfrak{F}'$, then also $g_1 = \mathfrak{x}'g$ is, if $\mathfrak{x}'$ denotes an arbitrary element of $\mathfrak{F}'$, because $\mathfrak{x}'$ is also a fractional linear transformation. Besides the congruent mapping functions arrived at in this way by letting $\mathfrak{x}'$ range over $\mathfrak{F}'$, there may still be more congruent mappings preserving equivalence in relation to $\mathfrak{F}$ and $\mathfrak{F}'$, and these then determine other isometric mappings of $\mathcal{D}$ mod $\mathfrak{F}$ onto $\mathcal{D}'$ mod $\mathfrak{F}'$. However, *if a particular, g-isomorphism* $I: \mathfrak{F} \to \mathfrak{F}'$ *is prescribed*, then there cannot be more than one congruent mapping inducing it. This has already been mentioned for the groups of a finite polygonal disc, or a polygonal cone, or a polygonal mast. Every other $\mathfrak{F}$-group is known to contain either an elementary group, or a handle group, or a crown group. The mapping function g of the form (11) for a fundamental domain of such a subgroup is then uniquely determined, if I is given, and g is thereby determined in the whole of $\overline{\mathcal{D}}$.

Let a g-isomorphism $I: \mathfrak{F} \to \mathfrak{F}'$ be given for two $\mathfrak{F}$-groups $\mathfrak{F}$ and $\mathfrak{F}'$. Let $\mathcal{A}$ denote a modulo $\mathfrak{F}$ simple, straight line in the interior of $\mathcal{K}(\mathfrak{F})$. The subgroup $\mathfrak{A}_{\mathcal{A}}$ of $\mathfrak{F}$ leaving $\mathcal{A}$ invariant takes one of the eleven forms:

1) $\mathfrak{A}(\vert)$,	2) $\mathfrak{A}(\Vert)$,	3) $\mathfrak{A}(\cdot\vert)$,	4) $\mathfrak{A}(-\ \Vert)$,
5) $\mathfrak{A}(-)$,	6) $\mathfrak{A}(\to\ -)$,	7) $\mathfrak{A}(\rightleftharpoons)$,	8) identity,
9) $\mathfrak{A}(\cdot)$,	10) $\mathfrak{A}(\cdot\cdot)$,	11) $\mathfrak{A}(\to)$.	

In the cases 2), 3), 4), 6), 7), 10), 11), the line $\mathcal{A}$ is an inner axis of $\mathfrak{F}$. Thus its end-points are inner fundamental points of $\mathfrak{F}$, and $\mathfrak{F}\mathcal{A}$ does not accumulate in $\mathcal{D}$. The same holds in $\mathfrak{F}'$ for the corresponding axis $\mathcal{A}'$.

In the case 5) there is in $\mathfrak{F}'$ a uniquely determined reflection line $\mathcal{A}'$ corresponding to $\mathcal{A}$, and $\mathfrak{F}\mathcal{A}$ and $\mathfrak{F}'\mathcal{A}'$ do not accumulate in $\mathcal{D}$ and $\mathcal{D}'$, respectively.

In the remaining cases 1), 8), 9), we *assume* that $\mathfrak{F}\mathcal{A}$ does not accumulate in $\mathcal{D}$ and that the cuts determined by the end-points of $\mathcal{A}$ in the set of fundamental points of $\mathfrak{F}$ determine corresponding cuts in the set of fundamental points of $\mathfrak{F}'$ which are not effected by closed aperiodic intervals of discontinuity; these cuts then determine single points on $\mathcal{E}'$. There is then a uniquely determined line $\mathcal{A}'$ in $\mathcal{K}(\mathfrak{F}')$ corresponding to $\mathcal{A}$, and it is assumed that $\mathfrak{F}'\mathcal{A}'$ does not accumulate in $\mathcal{D}'$. Only lines satisfying these conditions will be needed in the sequel.

In the cases 1) and 9), the end-points of $\mathcal{A}$ are interchanged by one and only one element of $\mathfrak{F}$. The same then holds for $\mathcal{A}'$, and $\mathfrak{A}_{\mathcal{A}'} = \mathfrak{A}_{\mathcal{A}}$, because every element of $\mathfrak{F}$ corresponds to an element of the same kind in $\mathfrak{F}'$. In the case 5), there is one and only one element of $\mathfrak{F}$ leaving the end-points of $\mathcal{A}$ fixed; the same then holds in $\mathfrak{F}'$. In the case 8) no element of $\mathfrak{F}$ interchanges the end-points of $\mathcal{A}$ or leave them fixed; the same then holds in $\mathfrak{F}'$. Thus in all eleven cases the subgroup of $\mathfrak{F}'$ belonging to $\mathcal{A}'$ is of the same type as $\mathfrak{A}_{\mathcal{A}}$, since this is known for axes (§24). From the preservation of the cyclical order of the union of all fundamental points and all end-points of reflection lines follows that $\mathcal{A}'$ is simple modulo $\mathfrak{F}'$, since $\mathcal{A}$ is simple modulo $\mathfrak{F}$.

Let $\mathcal{K}(\mathfrak{F})$ be decomposed by $\mathfrak{F}\mathcal{A}$ and $\mathcal{K}(\mathfrak{F}')$ by $\mathfrak{F}'\mathcal{A}'$. Each region of decomposition in $\mathcal{K}(\mathfrak{F})$ corresponds to a definite region of decomposition in $\mathcal{K}(\mathfrak{F}')$, namely the one bounded by the corresponding lines; compare §24.3. Let Ω_1 and Ω_2 be the regions adjacent to $\mathcal{A}$, and Ω_1' and Ω_2' their corresponding regions in $\mathcal{K}(\mathfrak{F}')$, adjacent to $\mathcal{A}'$. Let $\mathfrak{F}_1$ and $\mathfrak{F}_2$ be the subgroups of $\mathfrak{F}$ carrying Ω_1 and Ω_2 into themselves, respectively. Then the subgroups $\mathfrak{F}_1' = I_1(\mathfrak{F}_1)$, $\mathfrak{F}_2' = I_2(\mathfrak{F}_2)$ derived from I belong in the same way to Ω_1' and Ω_2', respectively. Here I_1 and I_2 denote the restriction of I to $\mathfrak{F}_1$ and $\mathfrak{F}_2$, respectively. They are g-isomorphisms.

Let it now be assumed that there exists a congruent g-mapping $g_1 : \Omega_1 \to \Omega_1'$ thus of the form (11), which induces I_1, and likewise a congruent g-mapping $g_2 : \Omega_2 \to \Omega_2'$ which induces I_2. As seen above, the mapping functions g_1 and g_2 are then uniquely determined. Thus g_1 and g_2 induce isometric mappings of $\mathcal{A}$ onto $\mathcal{A}'$. There is thus a translation $\mathfrak{t}$ along $\mathcal{A}$, in general not belonging to $\mathfrak{F}$, such that

$$g_1 \mathfrak{t} x = g_2 x \tag{45}$$

for x ranging over $\mathcal{A}$. This may also be written $g_1 x = \mathfrak{t}' g_2 x$, where x ranges over $\mathcal{A}$ and $\mathfrak{t}_1' = g_1 \mathfrak{t}^{-1} g_1^{-1}$ is a translation along $\mathcal{A}'$. The displacement of $\mathfrak{t}$, which is also the displacement of $\mathfrak{t}'$, will be called the *alignment length for* $\mathcal{A}$ in relation to g_1 and g_2.

For any element $\mathfrak{f}$ of $\mathfrak{F}$ the line $\mathfrak{f}\mathcal{A}$ is the common boundary of $\mathfrak{f}\Omega_1$ and $\mathfrak{f}\Omega_2$. Then $I(\mathfrak{f})g_1\mathfrak{f}^{-1}$ is a congruent mapping of $\mathfrak{f}\Omega_1$ onto $I(\mathfrak{f})\Omega_1'$. If x is any point of $\mathfrak{f}\Omega_1$ and $\mathfrak{h}$ is any element of the group $\mathfrak{f}\mathfrak{F}_1\mathfrak{f}^{-1}$, which is the subgroup of $\mathfrak{F}$ carrying $\mathfrak{f}\Omega_1$ into itself, then

$$(I(\mathfrak{f})g_1\mathfrak{f}^{-1})\mathfrak{h}x = I(\mathfrak{f})g_1(\mathfrak{f}^{-1}\mathfrak{h}\mathfrak{f})\mathfrak{f}^{-1}x = I(\mathfrak{f})I(\mathfrak{f}^{-1}\mathfrak{h}\mathfrak{f})g_1\mathfrak{f}^{-1}x = I(\mathfrak{h})(I(\mathfrak{f})g_1\mathfrak{f}^{-1})x,$$

since $\mathfrak{f}^{-1}x$ belongs to Ω_1, and $\mathfrak{f}^{-1}\mathfrak{h}\mathfrak{f}$ is an element of $\mathfrak{F}_1$, and g_1 induces the restriction of I to $\mathfrak{F}_1$. Hence $I(\mathfrak{f})g_1\mathfrak{f}^{-1}$ induces for the group $\mathfrak{f}\mathfrak{F}_1\mathfrak{f}^{-1}$ the restriction of I to that subgroup of $\mathfrak{F}$. The same for $I(\mathfrak{f})g_2\mathfrak{f}^{-1}$ in relation to the group $\mathfrak{f}\mathfrak{F}_2\mathfrak{f}^{-1}$ belonging to $\mathfrak{f}\Omega_2$. Both $I(\mathfrak{f})g_1\mathfrak{f}^{-1}$ and $I(\mathfrak{f})g_2\mathfrak{f}^{-1}$ map $\mathfrak{f}\mathcal{A}$ isometrically onto $I(\mathfrak{f})\mathcal{A}'$. Since both $I(\mathfrak{f})$ and $\mathfrak{f}^{-1}$ preserve distances, the alignment length for $\mathfrak{f}\mathcal{A}$ is the same as for $\mathcal{A}$. It thus belongs to the equivalence class of $\mathcal{A}$.

If I determines a congruence for $\mathfrak{F}$ and $\mathfrak{F}'$, thus if there exists a congruent mapping g of $\mathcal{K}(\mathfrak{F})$ onto $\mathcal{K}(\mathfrak{F}')$ inducing I, then g coincides with g_1 in Ω_1 and with g_2 in Ω_2, thus, in particular, g_1 and g_2 induce the same mapping of $\mathcal{A}$ onto $\mathcal{A}'$. Hence, in this case, the alignment length for $\mathcal{A}$ is zero.

Conversely, if the alignment length for $\mathcal{A}$ is zero, then it has been seen that it is zero for all lines of the collection $\mathfrak{F}\mathcal{A}$. The congruent mappings determined in each of the regions of decomposition then combine into a congruent mapping g of the whole of $\mathcal{K}(\mathfrak{F})$ onto $\mathcal{K}(\mathfrak{F}')$. Hence the theorem:

Theorem. *If $\mathfrak{F}_1$ and $\mathfrak{F}_1'$, and likewise $\mathfrak{F}_2$ and $\mathfrak{F}_2'$ are congruent in relation to the restriction of I to $\mathfrak{F}_1$ and $\mathfrak{F}_2$, respectively, then a necessary and sufficient condition for $\mathfrak{F}$ and $\mathfrak{F}'$ being congruent in relation to I is that the alignment length for $\mathcal{A}$ is zero.*

□

We now examine the different types enumerated above for the subgroup $\mathfrak{A}_{\mathcal{A}}$ of $\mathfrak{F}$ belonging to the line $\mathcal{A}$.

In the cases 1), 2), 3), 4), there is a reflection line $\mathcal{L}$ of $\mathfrak{F}$ intersecting $\mathcal{A}$ at right angles. Let $\mathfrak{s}$ denote the reflection in $\mathcal{L}$. Then $I(\mathfrak{s})$ is the reflection in a line $\mathcal{L}'$ intersecting $\mathcal{A}'$, and this intersection of $\mathcal{L}'$ and $\mathcal{A}'$ is at right angles; for $I(\mathfrak{s})$ interchanges the end-points of $\mathcal{A}'$, since $\mathfrak{s}$ interchanges those of $\mathcal{A}$. Now $\mathcal{L}$ enters into both Ω_1 and Ω_2, hence $\mathfrak{s}$ belongs to both $\mathfrak{F}_1$ and $\mathfrak{F}_2$. Therefore both g_1 and g_2 map $\mathcal{L}$ on $\mathcal{L}'$, and thus the intersection point of $\mathcal{L}$ and $\mathcal{A}$ on the intersection point of $\mathcal{L}'$ and $\mathcal{A}'$. It follows that the alignment length for $\mathcal{A}$ is zero.

In the cases 5) and 6) the reflection in $\mathcal{A}$ belongs to $\mathfrak{F}$. If it is denoted by $\mathfrak{f}$, then $I(\mathfrak{f})$ is the reflection in $\mathcal{A}'$. One then gets

$$\mathfrak{f}\Omega_1 = \Omega_2 \tag{46}$$
$$I(\mathfrak{f})\Omega_1' = \Omega_2' \tag{47}$$
$$g_2 = I(\mathfrak{f})g_1\mathfrak{f}^{-1}. \tag{48}$$

Here (48) expresses a special case of the above statement about the mapping function belonging to $\mathfrak{f}\Omega_1$. Let x be any point of $\mathcal{A}$ and g_1x its image on $\mathcal{A}'$ by g_1. Then by (48) one gets

$$g_2x = I(\mathfrak{f})g_1\mathfrak{f}^{-1}x = I(\mathfrak{f})g_1x = g_1x, \tag{49}$$

because $\mathfrak{f}^{-1}$ leaves x fixed, and $I(\mathfrak{f})$ leaves g_1x fixed. (49) shows that the alignment length for $\mathcal{A}$ is zero.

In the case 7) let $\mathfrak{f}$ denote a primary reversed translation along $\mathcal{A}$. Then $I(\mathfrak{f})$ is a primary reversed translation along $\mathcal{A}'$. Both g_1 and g_2 satisfy on $\mathcal{A}$ the equation

$$g_i\mathfrak{f}^2x = I(\mathfrak{f})^2g_ix, \quad i = 1, 2,$$

because $\mathfrak{f}^2$ is an element of both $\mathfrak{F}_1$ and $\mathfrak{F}_2$. They therefore also satisfy on $\mathcal{A}$ the equation

$$g_i\mathfrak{f}x = I(\mathfrak{f})g_ix, \quad i = 1, 2, \tag{50}$$

because they are isometric mappings and the displacements of $\mathfrak{f}^2$ and $I(\mathfrak{f}^2)$, and thus of $\mathfrak{f}$ and $I(\mathfrak{f})$, are equal. Also (46), (47), (48) hold. One then gets from (50)

$$g_2x = I(\mathfrak{f})g_1\mathfrak{f}^{-1}x = I(\mathfrak{f})I(\mathfrak{f}^{-1})g_1x = g_1x.$$

This shows that the alignment length for $\mathcal{A}$ is zero.

In the cases 9) and 10), let $\mathfrak{f}$ denote the half-turn around a centre c on $\mathcal{A}$. Then $I(\mathfrak{f})$ is a half-turn in $\mathfrak{F}'$, and its centre c' is on $\mathcal{A}'$, because $I(\mathfrak{f})$ interchanges the end-points of $\mathcal{A}'$. Again (46), (47), (48) hold. Since $\mathfrak{f}^{-1}c = c$, one gets

$$g_2c = I(\mathfrak{f})g_1\mathfrak{f}^{-1}c = I(\mathfrak{f})g_1c.$$

Hence $g_2c = g_1c$ if and only if $g_1c = c'$. This need not be the case, for g_1 is determined only as a congruent mapping function for Ω_1 in relation to the restricted g-isomorphism $I_1 : \mathfrak{F}_1 \to \mathfrak{F}_1'$, and $\mathfrak{f}$ does not belong to $\mathfrak{F}_1$. Hence, in general, the alignment length for $\mathcal{A}$ is equal to the distance of g_1c and g_2c, this being twice the distance of c' both from g_1c and from g_2c.

Of course, the concept of alignment length only plays a useful rôle if there is the possibility of its being different from zero. One thus gets this result:

The concept of alignment length refers to lines $\mathcal{A}$ for which the corresponding subgroup $\mathfrak{A}_{\mathcal{A}}$ is either $\mathfrak{A}(\cdot)$, or $\mathfrak{A}(\cdot\cdot)$, or $\mathfrak{A}(\rightarrow)$, or identity. □

From the preceding considerations one gets the theorem:

Theorem. *If the groups $\mathfrak{F}$ and $\mathfrak{F}'$ are derived from congruent groups $\mathfrak{F}_1$ and $\mathfrak{F}_1'$ by the adjunction of the reflection in corresponding boundaries of $\mathcal{K}(\mathfrak{F}_1)$ and $\mathcal{K}(\mathfrak{F}_1')$, then $\mathfrak{F}$ and $\mathfrak{F}'$ are congruent. The same holds for the adjunction of reversed translations belonging to corresponding boundary axes of $\mathcal{K}(\mathfrak{F})$ and $\mathcal{K}(\mathfrak{F}')$.* □

This covers, for instance, the case of the *complete reflection strip* (§20.13), since the assumptions made on $\mathcal{A}$ are fulfilled by the boundary line of the incomplete reflection strip from which the complete reflection strip is derived.

If $\mathfrak{F}_1$ is an elementary group, or a handle group, or a crown group, then $\mathfrak{F}$ and $\mathfrak{F}'$ are congruent if the corresponding metric parameters are equal. This covers, for instance, the case of a *reflection crown* (§20.14), a *cross cap crown* (§20.17), and a *Klein bottle* (§21.16).

If the groups $\mathfrak{F}$ and $\mathfrak{F}'$ are derived from congruent groups $\mathfrak{F}_1$ and $\mathfrak{F}_1'$ by the adjunction of half-turns with centres on corresponding boundaries $\mathcal{A}$ and $\mathcal{A}'$ of $\mathfrak{F}_1$ and $\mathfrak{F}_1'$, then $\mathfrak{F}$ and $\mathfrak{F}'$ are congruent if and only if the alignment length for $\mathcal{A}$ is zero.

This applies, for instance, to the *conical reflection strip* (§20.15), the assumptions on $\mathcal{A}$ being evidently fulfilled, and to the *conical crown* (§20.16).

Let $\mathfrak{F}$ be the free product of two $\mathfrak{F}$-groups $\mathfrak{F}_1$ and $\mathfrak{F}_2$ whose convex domains are adjacent along a common boundary $\mathcal{A}$, the subgroup $\mathfrak{A}_{\mathcal{A}}$ of $\mathfrak{F}$ being either the identity or $\mathfrak{A}(\rightarrow)$, with amalgamation of $\mathfrak{A}_{\mathcal{A}}$ in the latter case. Let a g-isomorphism $I : \mathfrak{F} \rightarrow \mathfrak{F}'$ be given such that $\mathfrak{F}_1$ and $I(\mathfrak{F}_1)$ are congruent in relation to the restriction of I to $\mathfrak{F}_1$, and the same for $\mathfrak{F}_2$ and $I(\mathfrak{F}_2)$. Then $\mathfrak{F}$ and $\mathfrak{F}'$ are congruent in relation to I if and only if the alignment length for $\mathcal{A}$ is zero.

This applies, for instance, to be the case of a *double reflection strip* (§20.19), the assumptions on $\mathcal{A}$ being evidently fulfilled, and to a *double crown* (§20.20). It is the basis of the considerations to follow later.

We first illustrate the notion of alignment length by two simple examples, both referring to axes of the type $\mathfrak{A}(\rightarrow)$:

First, let $\mathfrak{G}$ be a crown group, $\mathcal{V}$ a boundary axis of $\mathfrak{G}$, and $\mathfrak{A}_{\mathcal{V}}$ the subgroup of $\mathfrak{G}$ belonging to $\mathcal{V}$; it consists of translations. Let Ω be the fundamental polygon for $\mathfrak{G}$ illustrated by Fig. 26.5. It contains a side on $\mathcal{V}$ equal in length to the displacement of the primary translation along $\mathcal{V}$. Let $\mathfrak{s}$ denote the reflection in $\mathcal{V}$. Then $\mathfrak{s}\Omega$ is a fundamental polygon for the crown group $\mathfrak{s}\mathfrak{G}\mathfrak{s}^{-1} = \mathfrak{G}_1$. Then

$$\mathfrak{F}_1 = \mathfrak{G} * \mathfrak{G}_1 \text{ am } \mathfrak{A}_{\mathcal{V}}$$

is the group of a double crown. (The reflection $\mathfrak{s}$ does not belong to $\mathfrak{F}_1$!). Let $\mathfrak{h}$ denote a reversed translation with axis $\mathcal{V}$. Then $\mathfrak{h}\Omega$ is a fundamental polygon for the crown group $\mathfrak{h}\mathfrak{G}\mathfrak{h}^{-1} = \mathfrak{G}_2$. Then

$$\mathfrak{F}_2 = \mathfrak{G} * \mathfrak{G}_2 \text{ am } \mathfrak{A}_{\mathcal{V}}$$

is the group of a double crown. (The reversed translation $\mathfrak{h}$ does not belong to $\mathfrak{F}_2$!). If one lets every element $\mathfrak{f}$ of $\mathfrak{G}$ correspond to itself and the element $\mathfrak{s}\mathfrak{f}\mathfrak{s}^{-1}$ of $\mathfrak{G}_1$ correspond to the element $\mathfrak{h}\mathfrak{f}\mathfrak{h}^{-1}$ of $\mathfrak{G}_2$, then every element of $\mathfrak{A}_{\mathcal{V}}$ corresponds to itself, and a g-isomorphism $I\colon \mathfrak{F}_1 \to \mathfrak{F}_2$ arises. The identical mapping of $\mathcal{K}(\mathfrak{G})$ onto itself induces $I\colon \mathfrak{G} \to \mathfrak{G}$, and the congruent mapping $\mathfrak{h}\mathfrak{s}^{-1}$ of $\mathcal{K}(\mathfrak{G}_1)$ onto $\mathcal{K}(\mathfrak{G}_2)$ induces $I\colon \mathfrak{G}_1 \to \mathfrak{G}_2$. The alignment length for $\mathcal{V}$ is equal to the displacement of $\mathfrak{h}$. The two groups $\mathfrak{F}_1$ and $\mathfrak{F}_2$ admit no congruent mapping inducing I.

Secondly, consider the group $\mathfrak{F}$, with $p = 2$, given by (6), §22. The mapping

$$\mathfrak{a}_1 \mapsto \mathfrak{a}_1, \quad \mathfrak{b}_1 \mapsto \mathfrak{b}_1, \quad \mathfrak{a}_2 \mapsto \mathfrak{k}^{\rho}\mathfrak{a}_2\mathfrak{k}^{-\rho}, \quad \mathfrak{b}_2 \mapsto \mathfrak{k}^{\rho}\mathfrak{b}_2\mathfrak{k}^{-\rho}$$

with the abbreviation

$$\mathfrak{k} = \mathfrak{a}_1\mathfrak{b}_1\mathfrak{a}_1^{-1}\mathfrak{b}_1^{-1} = \left(\mathfrak{a}_2\mathfrak{b}_2\mathfrak{a}_2^{-1}\mathfrak{b}_2^{-1}\right)^{-1}$$

determines an automorphism I of $\mathfrak{F}$. This is a g-automorphism, because $\mathfrak{F}$ contains no boundary elements (§24.8). The axis $\mathcal{A}$ of $\mathfrak{k}$ determines in the decomposition of $\mathcal{D}$ by $\mathfrak{F}\mathcal{A}$ two subgroups $\mathfrak{F}_1$ and $\mathfrak{F}_2$ of $\mathfrak{F}$ belonging to its adjacent regions. $\mathfrak{F}_i$ is generated by $\mathfrak{a}_i$ and $\mathfrak{b}_i$, $i = 1, 2$. The identical mapping of $\mathcal{K}(\mathfrak{F}_1)$ onto itself induces $I\colon \mathfrak{F}_1 \to \mathfrak{F}_1$, and the mapping $\mathfrak{k}^{\rho}$ of $\mathcal{K}(\mathfrak{F}_2)$ onto itself induces $I\colon \mathfrak{F}_2 \to \mathfrak{F}_2$ and is also a congruent mapping. The alignment length for $\mathcal{A}$ is ρ times the displacement for $\mathfrak{k}$. There is no congruent mapping function inducing the g-automorphism I of $\mathfrak{F}$, if the integer ρ is different from zero.

Consider now the case of a g-isomorphism $I\colon \mathfrak{F} \to \mathfrak{F}'$ of $\mathfrak{F}$-groups $\mathfrak{F}$ and $\mathfrak{F}'$ which are finitely generated, and assume that $\mathfrak{F}$ is not an elementary group, nor a handle group, nor a crown group, nor derived from any of these groups by the closing of boundaries or by turning boundaries into reflection lines. By the method of §22, $\mathcal{K}(\mathfrak{F})$ is decomposed into regions each of which, closed on $\mathcal{K}(\mathfrak{F})$, is the convex domain of an elementary group with at least one equivalence class of boundary axes, or of a handle group of the type $\mathcal{H}(\to)$, or of a crown group, and these regions belong to a finite number M of equivalence classes with respect to $\mathfrak{F}$. The collection $\mathcal{U}$ of lines which effects the decomposition of $\mathcal{K}(\mathfrak{F})$ consists of inner, modulo $\mathfrak{F}$ simple axes of $\mathfrak{F}$ with corresponding subgroups of $\mathfrak{F}$ of the type $\mathfrak{A}(\to)$, or $\mathfrak{A}(\cdot\cdot)$, or $\mathfrak{A}(\rightleftharpoons)$, or $\mathfrak{A}(\to\, -)$, and these axes belong to a finite number N of equivalence classes.

Consider an arbitrary among the regions of decomposition and let $\mathfrak{F}_0$ denote the subgroup $\mathfrak{F}$ carrying that region into itself. This region, closed on $\mathcal{K}(\mathfrak{F})$, is $\mathcal{K}(\mathfrak{F}_0)$.

The subgroup $I(\mathfrak{F}_0) = \mathfrak{F}'_0$ of $\mathfrak{F}'$ has as its convex domain $\mathcal{K}(\mathfrak{F}'_0)$ one of the regions of decomposition of $\mathcal{K}(\mathfrak{F}')$ by the collection $\mathcal{U}'$ of axes which correspond to $\mathcal{U}$ by I. If it is assumed that any metric parameter of $\mathfrak{F}_0$ and the metric parameter of $\mathfrak{F}'_0$ which corresponds to it by I are equal, then, in virtue of the fact that $\mathfrak{F}_0$ is either elementary, or a handle group, or a crown group, a congruent mapping g of $\mathcal{K}(\mathfrak{F}_0)$ onto $\mathcal{K}(\mathfrak{F}'_0)$ exists such that

$$g\mathfrak{f}_0 x = I(\mathfrak{f}_0) g x$$

holds for x ranging over $\mathcal{K}(\mathfrak{F}_0)$ and $\mathfrak{f}_0$ over $\mathfrak{F}_0$. If $\mathfrak{f}$ is any element of $\mathfrak{F}$, then $I(\mathfrak{f})g\mathfrak{f}^{-1}$ is a congruent mapping of $\mathfrak{f}\mathcal{K}(\mathfrak{F}_0)$ onto $I(\mathfrak{f})\mathcal{K}(\mathfrak{F}'_0)$ inducing the restriction of I to $\mathfrak{f}\mathfrak{F}_0\mathfrak{f}^{-1}$. If $\mathfrak{F}_0$ ranges over a set of representatives of the M classes of conjugate subgroups belonging to the regions of decomposition, and if the assumption concerning the metric parameters is valid for these M groups $\mathfrak{F}_0$, then a uniquely determined congruent mapping is obtained for each of the regions. Thereby an alignment length is determined on each line of $\mathcal{U}$, and this is the same for two equivalence lines of $\mathcal{U}$, so that N alignment lengths result. These are known to be zero for the lines of $\mathcal{U}$ of the types $\mathfrak{A}(\rightleftharpoons)$ or $\mathfrak{A}(\rightarrow\ -)$. If and only if they all are zero for the lines of $\mathcal{U}$ of the types $\mathfrak{A}(\cdot\cdot)$ or $\mathfrak{A}(\rightarrow)$, the congruent mappings defined in all the regions of decomposition closed on $\mathcal{K}(\mathfrak{F})$ combine into one congruent g-mapping of $\mathcal{K}(\mathfrak{F})$ onto $\mathcal{K}(\mathfrak{F}')$ inducing I. *This is thus the necessary and sufficient condition for the finitely generated $\mathfrak{F}$-groups $\mathfrak{F}$ and $\mathfrak{F}'$ being congruent in relation to I.*

Finally, let $I\colon \mathfrak{F} \to \mathfrak{F}'$ be a g-isomorphism of groups which do not satisfy the condition of quasi-compactness. Then $\mathfrak{F}$ can be exhausted by an infinite sequence of extended hulls $\mathfrak{H}_k$, $1 \leqq k < \infty$, (§23.3). In Section 11 this fact was used in order to prove the existence of a g-mapping inducing I. We follow the pattern of Section 11. It is recalled that in the decomposition of $\mathcal{K}(\mathfrak{F})$ by the collection

$$\mathcal{U} = \bigcup_{k=1}^{\infty} \bigcup_{r=1}^{m_k} \mathfrak{F}\mathcal{A}_{kr},$$

the regions of decomposition, closed on $\mathcal{K}(\mathfrak{F})$, are the equivalence classes with respect to $\mathfrak{F}$ of the convex domain $\mathcal{K}_1$ of $\mathfrak{H}_1$ and the convex domains $\mathcal{C}_{ks}$, $1 \leqq s \leqq m'_k$, $1 \leqq k < \infty$, of the groups $\mathfrak{C}_{ks}$. Let I_1 and I_{ks} denote the restriction of I to $\mathfrak{H}_1$ and $\mathfrak{C}_{ks}$, respectively. The groups $\mathfrak{H}_1$ and $\mathfrak{C}_{ks}$ are finitely generated.

We now assume that $\mathfrak{H}_1$ is congruent with $I_1(\mathfrak{H}_1)$ and $\mathfrak{C}_{ks}$ with $I_{ks}(\mathfrak{C}_{ks})$. Necessary and sufficient conditions for that being the case are those just stated for finitely generated groups, since $\mathfrak{H}_1$ and the $\mathfrak{C}_{ks}$ are finitely generated. Let g_1 and g_{ks} denote those uniquely determined congruent mappings of the convex domains of $\mathfrak{H}_1$ onto that of $I_1(\mathfrak{H}_1)$ inducing I_1, and of $\mathfrak{C}_{ks}$ onto that of $I_{ks}(\mathfrak{C}_{ks})$ inducing I_{ks}. The subgroups of $\mathfrak{F}$ belonging to the lines of $\mathcal{U}$ are known to be of the type $\mathfrak{A}(\|)$ or $\mathfrak{A}(\rightarrow)$. One thus gets the result:

If and only if the alignment lengths for those $\mathcal{A}_{kr}$ which belong to the type $\mathfrak{A}(\rightarrow)$ are all zero, do the congruent mappings derived from g_1 and the g_{ks} in all regions of *decomposition combine to a congruent mapping* g *of* $\mathcal{K}(\mathfrak{F})$ *onto* $\mathcal{K}(\mathfrak{F}')$ *inducing* I. □

In this case g is, of course, a mapping of $\overline{\mathfrak{D}}$ onto $\overline{\mathfrak{D}}'$. Thus g exists also on limit sides of $\mathcal{K}(\mathfrak{F})$, if any.

In a wording which gets its precision by the above detailed considerations of all cases one may thus formulate the result of the present section in the main theorem:

Theorem. *Two g-isomorphic* $\mathfrak{F}$*-groups are congruent in relation to a given g-isomorphism if and only if corresponding metric parameters are equal and all alignment lengths are zero.* □

Symbols and definitions

$(x_1 y_1 x_2 y_2)$: cross ratio 2.1
$\mathcal{D}$: open unit disc 6.1
$\mathcal{E}$: unit circle = boundary of $\mathcal{D}$ 6.1
$\overline{\mathcal{D}}$: closed unit disc = $\mathcal{D} \cup \mathcal{E}$ 6.1
$\mathcal{D}^*$: exterior of unit disc = complement of $\overline{\mathcal{D}}$ 6.1
$[x, y]$, $[x, \mathcal{G}]$, $[\mathcal{G}, \mathcal{H}]$: non-euclidean distance 6.2
$\mathcal{A}(\mathfrak{f})$: axis of translation or reversed translation $\mathfrak{f}$ 7.3
$\lambda(\mathfrak{f})$: translation length of translation or reversed translation $\mathfrak{f}$ 7.3
$c(\mathfrak{f})$: centre of rotation $\mathfrak{f}$ 7.3
$\varphi(\mathfrak{f})$: angle of rotation $\mathfrak{f}$ 7.3
$u(\mathfrak{f})$: (limit-) centre of limit-rotation $\mathfrak{f}$ 7.3
$\mathcal{L}(\mathfrak{s})$: line of reflection $\mathfrak{s}$ 7.3
$\chi = \chi(\mathcal{P})$: characteristic of polygon $\mathcal{P}$ 8.3
$\chi' = \chi'(\mathcal{P})$: boundary-characteristic of polygon $\mathcal{P}$ 8.3
$\Phi = \Phi(\mathcal{P})$: non-euclidean area of polygon $\mathcal{P}$, normalized by a zero-angled triangle having area = 1 8.3
σ: sine amplitude of triangle 8.4
$\mathrm{I}(\rightarrow\rightarrow\rightarrow)$, $\mathrm{I}(\rightarrow\rightarrow \circlearrowleft)$, $\mathrm{I}(\rightarrow \circlearrowleft \circlearrowleft)$, $\mathrm{I}(\circlearrowleft \circlearrowleft \circlearrowleft)$ 9.3
$\mathfrak{k}_{\mathfrak{f}\mathfrak{g}} = \mathfrak{k}(\mathfrak{f}\mathfrak{g})$: commutator of $\mathfrak{f}$ and $\mathfrak{g}$ 9.5
$\mathcal{C}(a; \rho)$: non-euclidean circle (circular disc) with centre a and radius ρ 10.1
$\mathfrak{G}$: group of motions and reversions 10.2
$\varepsilon(x) = \varepsilon(\mathfrak{G}; x)$: distance function of $\mathfrak{G}$ 10.3
$\mathcal{N}(x_0) = \mathcal{N}(\mathfrak{G}; x_0)$: normal domain for $\mathfrak{G}$ with central point x_0 10.8
$\mathfrak{A}$: quasi-abelian group of motions and reversions 11.1
$\mathfrak{A}_c$: quasi-abelian group with invariant point $c \in \mathcal{D}$ 11.2
$\mathfrak{A}_c(\circlearrowleft_\nu)$: cyclic group generated by rotation with centre c and angle $2\pi/\nu$ 11.2
$\mathfrak{A}_c(\cdot)$: $\mathfrak{A}_c(\circlearrowleft_2)$ 11.2
$\mathfrak{A}_c(\times_\nu)$: dihedral group generated by reflections in two lines through c at an angle π/ν 11.2
$\mathfrak{A}_u$: quasi-abelian group with invariant point $u \in \mathcal{E}$ 11.3
$\mathfrak{A}_u(\hookrightarrow)$: infinite cyclic group generated by limit-rotation with centre u 11.3
$\mathfrak{A}_u(\wedge)$: non-abelian group generated by reflections in two lines through u 11.3
$\mathfrak{A}_{\mathcal{A}}$: quasi-abelian group with invariant line $\mathcal{A}$ 11.5
$\mathfrak{A}_{\mathcal{A}}(-)$: cyclic group generated by reflection in line $\mathcal{A}$ 11.5
$\mathfrak{A}_{\mathcal{A}}(|)$: cyclic group generated by reflection in normal to $\mathcal{A}$ 11.5
$\mathfrak{A}_{\mathcal{A}}(\rightarrow)$: infinite cyclic group generated by translation with axis $\mathcal{A}$ 11.5
$\mathfrak{A}_{\mathcal{A}}(\rightleftharpoons)$: infinite cyclic group generated by reversed translation with axis $\mathcal{A}$ 11.5
$\mathfrak{A}_{\mathcal{A}}(- \rightarrow)$: abelian group generated by translation with axis $\mathcal{A}$ and reflection in $\mathcal{A}$ 11.5

$\mathfrak{A}_{\mathcal{A}}(\|)$: non-abelian group generated by reflections in two normals to $\mathcal{A}$ 11.5
$\mathfrak{A}_{\mathcal{A}}(\cdot\cdot)$: non-abelian group generated by two half-turns with centres on $\mathcal{A}$ 11.5
$\mathfrak{A}_{\mathcal{A}}(\cdot|)$: non-abelian group generated by a half-turn with centre on $\mathcal{A}$ and a reflection in a normal to $\mathcal{A}$ 11.5
$\mathfrak{A}_{\mathcal{A}}(-\ \|) = \mathfrak{A}_{\mathcal{A}}(-\cdot\cdot)$: non-abelian group generated by reflections in two normals to $\mathcal{A}$ and reflection in $\mathcal{A}$ = non-abelian group generated by two half-turns with centres on $\mathcal{A}$ and reflection in $\mathcal{A}$ 11.5
$\mathfrak{F}$-group: discontinuous group of motions and reversions without invariant point or line 13.1
$\mathcal{G}(\mathfrak{F})$: fundamental set of $\mathfrak{F}$ = the set of endpoints (in $\mathcal{E}$) of axes belonging to translations or reversed translations in $\mathfrak{F}$ 13.3
$\overline{\mathcal{G}}(\mathfrak{F})$: limit set of $\mathfrak{F}$ 13.4
$\mathcal{K}(\mathfrak{F})$: convex domain of $\mathfrak{F}$ = $\mathcal{D} \cap$ convex cover($\mathcal{G}(\mathfrak{F})$) 14.1
$\tilde{\mathcal{K}}(\mathfrak{F})$: closure of $\mathcal{K}(\mathfrak{F})$ in $\mathcal{D}$ 14.1
$\overline{\mathcal{K}}(\mathfrak{F})$: closure of $\mathcal{K}(\mathfrak{F})$ in $\overline{\mathcal{D}}$ 14.1
$\mathfrak{A}_x$: quasi-abelian subgroup of $\mathfrak{F}$ leaving $x \in \mathcal{D}$ invariant 14.4
$\eta(x) = \eta(\mathfrak{F}; x) : \min_{\mathfrak{g} \in \mathfrak{F} \setminus \mathfrak{A}_x} [x, \mathfrak{g}x]$ 14.4
$\mathcal{D}(x) = \mathcal{D}(\mathfrak{F}; x)$: characteristic neighbourhood of x = interior of $\mathfrak{C}(x; \frac{1}{2}\eta(x))$ 14.4
$\xi(\mathfrak{F}; x, y)$: distance modulo $\mathfrak{F}$ of x and $y = \min_{\mathfrak{g},\mathfrak{h} \in \mathfrak{F}} [\mathfrak{g}x, \mathfrak{h}y]$ 14.5
$\chi(x)$: radius of $\mathcal{D}(x) = \frac{1}{2}\eta(x)$ 14.6
$\mathcal{I}(x) = \mathcal{I}(\mathfrak{F}; x)$: isometric neighbourhood of x = largest open circular disc with centre x with $\xi(\mathfrak{F}; x_1, x_2) = \xi(\mathfrak{A}_x; x_1, x_2)$ for all points x_1, x_2 in the disc 14.6
$\iota(x)$: radius of $\mathcal{I}(x)$ 14.6
$\mathfrak{A}_u$: quasi-abelian subgroup of $\mathfrak{F}$ leaving limit-centre $u \in \mathcal{E}$ invariant 14.7
$\delta(\mathcal{O}, \mathcal{O}_1)$: distance between horocycles $\mathcal{O}$ and $\mathcal{O}_1$ 14.7
$\mathcal{D}(\mathfrak{F}; u)$: characteristic neighbourhood of limit-centre u 14.7
$\mathcal{I}(\mathfrak{F}; u)$: isometric neighbourhood of limit-centre u 14.8
$\mathcal{K}^*(\mathfrak{F})$: truncated domain of $\mathfrak{F}$ 14.10
$\mathfrak{H}'(\mathfrak{G})$: hull of subgroup $\mathfrak{G}$ of finitely generated $\mathfrak{F}$-group $\mathfrak{F}$ 15.6
$[X, Y]$: distance between points on surface $\mathcal{D}$ mod $\mathfrak{G}$ 16.1
$\vartheta(\odot_\nu)$: number of classes of centres of order ν in $\mathfrak{F}$ = number of conical points of order ν of $\mathcal{D}$ mod $\mathfrak{F}$ or $\mathcal{K}(\mathfrak{F})$ mod $\mathfrak{F}$ 17.4
$\vartheta(\hookrightarrow)$: number of classes of limit-centres of $\mathfrak{F}$ = number of masts of order ν of $\mathcal{D}$ mod $\mathfrak{F}$ or $\mathcal{K}(\mathfrak{F})$ mod $\mathfrak{F}$ 17.4
$\vartheta(\rightarrow)$: number of classes of boundary axes of $\mathfrak{F}$ = number of funnels of $\mathcal{D}$ mod $\mathfrak{F}$ = number of boundaries of $\mathcal{K}(\mathfrak{F})$ mod $\mathfrak{F}$ 17.4
$\vartheta(\times_\nu)$: number of classes of centres of $\mathfrak{F}$ for subgroups of type $\mathfrak{A}(\times_\nu)$ = number of angular points of order ν of $\mathcal{D}$ mod $\mathfrak{F}$ or $\mathcal{K}(\mathfrak{F})$ mod $\mathfrak{F}$ 17.4
$\vartheta(\wedge)$: number of classes of limit-centres of $\mathfrak{F}$ for subgroups of type $\mathfrak{A}(\wedge)$ = number of half-masts of $\mathcal{D}$ mod $\mathfrak{F}$ or $\mathcal{K}(\mathfrak{F})$ mod $\mathfrak{F}$ 17.4
$\vartheta(\|)$: number of classes of boundary axes of $\mathfrak{F}$ for subgroups of type $(\ \|)$ = number of half-funnels of $\mathcal{D}$ mod $\mathfrak{F}$ = number of geodetic boundaries of $\mathcal{K}(\mathfrak{F})$ mod $\mathfrak{F}$ joining two reflection edges 17.4

$\vartheta(-\to)$: number of classes of reflection lines of $\mathfrak{F}$ for subgroups of type $(-\to)$ and containing no centre = number of closed reflection edges of $\mathcal{D}$ mod $\mathfrak{F}$ or $\mathcal{K}(\mathfrak{F})$ mod $\mathfrak{F}$ passing through no angular point 17.4

$\vartheta(-)$: number of reflection rings of $\mathcal{D}$ mod $\mathfrak{F}$ 17.4

$\vartheta(\rightleftharpoons)$: 0 if $\mathcal{D}$ mod $\mathfrak{F}$ is orientable, and otherwise 1 17.5

$\chi(\mathfrak{F})$: Euler characteristic of $\mathcal{K}^*$ mod $\mathfrak{F}$ 17.6

p: genus of $\mathcal{D}$ mod $\mathfrak{F}$ or $\mathcal{K}(\mathfrak{F})$ mod $\mathfrak{F}$ or $\mathcal{K}^*(\mathfrak{F})$ mod $\mathfrak{F}$ 17.6

$\mathfrak{G} = \prod^* \mathfrak{G}_i \mathrm{am} \mathfrak{H}_{ij}, \quad i, j \in I$: free products of $\mathfrak{G}_i$ with amalgamated subgroups $\mathfrak{H}_{ij}$ 18.1

$\mathfrak{E}$: elementary group 21.9

$\mathrm{ES}(\mathfrak{E}) = \mathcal{K}(\mathfrak{E})$ mod $\mathfrak{E}$: elementary surface, when $\mathfrak{E}$ is an elementary group 21.13

Alphabets

The Greek Alphabet:

A,α: alpha	B,β: beta	Γ, γ: gamma	Δ, δ: delta	E, ε: epsilon
Z,ζ: zeta	H, η: eta	Θ, ϑ: theta	I,ι: iota	K,κ: kappa
Λ, λ: lambda	M,μ: mu	N,ν: nu	Ξ, ξ: xi	Π, π: pi
R,ρ: rho	Σ, σ: sigma	T,τ: tau	Υ, υ: upsilon	Φ, φ: fi
X,χ: chi	Ψ, ψ: psi	Ω, ω: omega		

The Gothic Alphabet:

$\mathfrak{A}$, $\mathfrak{a}$: A, a	$\mathfrak{B}$, $\mathfrak{b}$: B, b	$\mathfrak{C}$, $\mathfrak{c}$: C, c	$\mathfrak{D}$, $\mathfrak{d}$: D, d	$\mathfrak{E}$, $\mathfrak{e}$: E, e	$\mathfrak{F}$, $\mathfrak{f}$: F, f
$\mathfrak{G}$, $\mathfrak{g}$: G, g	$\mathfrak{H}$, $\mathfrak{h}$: H, h	$\mathfrak{I}$, $\mathfrak{i}$: I, i	$\mathfrak{J}$, $\mathfrak{j}$: J, j	$\mathfrak{K}$, $\mathfrak{k}$: K, k	$\mathfrak{L}$, $\mathfrak{l}$: L, l
$\mathfrak{M}$, $\mathfrak{m}$: M, m	$\mathfrak{N}$, $\mathfrak{n}$: N, n	$\mathfrak{O}$, $\mathfrak{o}$: O, o	$\mathfrak{P}$, $\mathfrak{p}$: P, p	$\mathfrak{Q}$, $\mathfrak{q}$: Q, q	$\mathfrak{R}$, $\mathfrak{r}$: R, r
$\mathfrak{S}$, $\mathfrak{s}$: S, s	$\mathfrak{T}$, $\mathfrak{t}$: T, t	$\mathfrak{U}$, $\mathfrak{u}$: U, u	$\mathfrak{V}$, $\mathfrak{v}$: V, v	$\mathfrak{W}$, $\mathfrak{w}$: W, w	$\mathfrak{X}$, $\mathfrak{x}$: X, x
$\mathfrak{Y}$, $\mathfrak{y}$: Y, y	$\mathfrak{Z}$, $\mathfrak{z}$: Z, z				

The Calligraphic Alphabet:

$\mathcal{A}$: A	$\mathcal{B}$: B	$\mathcal{C}$: C	$\mathcal{D}$: D	$\mathcal{E}$: E	$\mathcal{F}$: F	$\mathcal{G}$: G	$\mathcal{H}$: H	$\mathcal{I}$: I	$\mathcal{J}$: J
$\mathcal{K}$: K	$\mathcal{L}$: L	$\mathcal{M}$: M	$\mathcal{N}$: N	$\mathcal{O}$: O	$\mathcal{P}$: P	$\mathcal{Q}$: Q	$\mathcal{R}$: R	$\mathcal{S}$: S	$\mathcal{T}$: T
$\mathcal{U}$: U	$\mathcal{V}$: V	$\mathcal{W}$: W	$\mathcal{X}$: X	$\mathcal{Y}$: Y	$\mathcal{Z}$: Z				

Bibliography

[1] W. Abikoff: *The Real Analytic Theory of Teichmüller Space*. Lecture Notes in Math. 820, Springer-Verlag, Berlin–Heidelberg–New York, 1980.

[2] R. D. M. Accola: *Riemann Surfaces, Theta Functions, and Abelian Automorphisms Groups*. Lecture Notes in Math. 483, Springer-Verlag, Berlin–Heidelberg–New York, 1975.

[3] R. D. M. Accola: *Topics in the Theory of Riemann Surfaces*. Lecture Notes in Math. 1595, Springer-Verlag, Berlin–Heidelberg–New York, 1994.

[4] L. V. Ahlfors: *Möbius Transformations in Several Dimensions*. School of Mathematics, University of Minnesota, 1981.

[5] L. V. Ahlfors: *Collected Mathematical Papers*, 1–2. Editor: G.-C. Rota. Birkhäuser, Boston–Basel–Stuttgart, 1982.

[6] L. V. Ahlfors, L. Sario: *Riemann Surfaces*. Princeton University Press, Princeton, NJ, 1960.

[7] L. V. Ahlfors et al. (editors): *Contribution to the Theory of Riemann Surfaces*. Centennial Celebration of Riemann's Dissertation, Ann. of Math. Stud. 30, Princeton University Press, Princeton, NJ, 1953.

[8] L. V. Ahlfors et al. (editors): *Advances in the Theory of Riemann Surfaces*. Proceedings of the 1969 Stony Brook Conference, Ann. of Math. Stud. 66, Princeton University Press and University of Tokyo Press, Princeton, NJ, 1971.

[9] L. V. Ahlfors et al. (editors): *Contributions to Analysis*. Dedicated to L. Bers, Academic Press, New York–San Francisco–London, 1974.

[10] B. N. Apanasov: *Discrete Groups in Space and Uniformization Problems*. Kluwer Academic Publishers, Dordrecht–Boston–London, 1991.

[11] B. N. Apanasov: *Conformal Geometry of Discrete Groups and Manifolds*. De Gruyter Exp. Math. 32, Walter de Gruyter, Berlin–New York, 2000.

[12] P. Appell, E. Goursat: *Théorie des fonctions algébriques d'une variable*, 2. édition. Tome II, P. Fatou: *Fonctions automorphes*, Gautier-Villars, Paris, 1930.

[13] R. Baldus, F. Löbell: *Nichteuklidische Geometrie*. Sammlung Göschen 970/970a, Walter de Gruyter, Berlin, 1964.

[14] L. Balke: *Diskontinuerliche Gruppen als Automorphismengruppen von Pflasterungen*. Bonner Mathematischen Schriften, Band 211, Universität Bonn, Mathematisches Institut, Bonn, 1990.

[15] A. F. Beardon: *The Geometry of Discrete Groups*. Graduate Texts in Mathematics 91, Springer-Verlag, Berlin–Heidelberg–New York, 1983.

[16] A. F. Beardon: *A Primer on Riemann Surfaces*. London Math. Soc. Lecture Note Ser. 78, Cambridge University Press, Cambridge, 1984.

[17] L. Bers: *Riemann Surfaces*. Notes by E. Rodlitz, R. Pollack, New York University, 1957–58.

[18] L. Bers: *On Moduli of Riemann Surfaces*. Notes by L. M. and R. J. Sibner, Eidgenössische Technische Hochschule, Zürich, 1964.

[19] L. Bers and I. Kra (editors): *A Crash Course on Kleinian Groups*. Lecture Notes in Math. 400, Springer-Verlag, Berlin-Heidelberg–New York, 1974.

[20] J. Birman: *Braids, Links, and Mapping Class Groups*. Ann. of Math. Stud. 82, Princeton University Press, Princeton, NJ, 1974.

[21] R. Bonola: *Non-Euclidean Geometry*. Dover, New York, 1955.

[22] F. E. Browder (editor): *The Mathematical Heritage of Henri Poincaré*. Proc. Sympos. Pure Math. 39, 1–2, American Mathematical Society, Providence, RI, 1984.

[23] A. Casson: *Automorphisms of Surfaces after Nielsen and Thurston*. Cambridge University Press, Cambridge, 1988.

[24] H. Cohn: *Conformal Mapping on Riemann Surfaces*. McGraw–Hill, New York, 1967; Dover, New York, 1980.

[25] J. L. Coolidge: *A Treatise on the Circle and the Sphere*. The Clarendon Press, Oxford University Press, Oxford, 1916.

[26] H. S. M. Coxeter: *Non–Euclidean Geometry*. The University of Toronto Press, Toronto, 1980; 6. edition, MAA Spectrum, Mathematical Association of America, Washington, DC, 1998.

[27] H. S. M. Coxeter, W. O. J. Moser: *Generators and Relations for Discrete Groups*. Ergeb. Math. Grenzgeb. 14, Springer-Verlag, Berlin–Heidelberg–New York, 1980.

[28] J. Elstroedt, F. Grunewald, J. Mennicke: *Groups Acting on Hyperbolic Space*. Harmonic Analysis and Number Theory, Springer-Verlag, Berlin, 1998.

[29] H. M. Farkas, I. Kra: *Riemann Surfaces*. Grad. Texts in Math. 71, Springer-Verlag, Berlin–Heidelberg–New York, 1980; 2. edition 1992.

[30] W. Fenchel: *Nielsen's Contribution to the Theory of Discontinuous Groups of Isometries of the Hyperbolic Plane*. J. Nielsen: *Collected Mathematical Papers*, 2, pp. 427–440, 1986.

[31] W. Fenchel: *Elementary Geometry in Hyperbolic Space*. De Gruyter Stud. Math. 11, Walter de Gruyter, Berlin–New York, 1989.

[32] L. R. Ford: *Automorphic Functions*. Chelsea, New York, 1951.

[33] O. Forster: *Lectures on Riemann Surfaces*. Grad. Texts in Math. 81, Springer-Verlag, Berlin–Heidelberg–New York, 1991.

[34] R. Fricke, F. Klein: *Vorlesungen über die Theorie der automorphen Funktionen* 1. Teubner, Leipzig, 1897; Johnson Reprint Corp., New York and Teubner, Stuttgart, 1965.

[35] R. Fricke, F. Klein: *Vorlesungen über die Theorie der automorphen Funktionen* 2. Teubner, Leipzig, 1912; Johnson Reprint Corp., New York and Teubner, Stuttgart, 1965.

[36] J. J. Gray: *Ideas of Space. Euclidean, Non–Euclidean, and Relativistic*. The Clarendon Press, Oxford University Press, New York, 1989.

[37] L. Greenberg (editor): *Discontinuous Groups and Riemann Surfaces*. Annals of Mathematics Studies No 79, Princeton University Press and University of Tokyo Press, Princeton, NJ, 1974.

[38] J. Hadamard: *Non–Euclidean Geometry in the Theory of Automorphic Functions*. History of Mathematics 17, American Mathematical Society, Providence, RI; London Mathematical Society, Cambridge, 1999.

[39] W. J. Harvey (editor): *Discrete Groups and Automorphic Functions*. Academic Press, London–New York–San Francisco, 1977.

[40] D. A. Hejhal: *The Selberg Trace Formula for* $PSL(2, \mathbb{R})$, Vol. 1. Lecture Notes in Math. 548, Springer-Verlag, Berlin–Heidelberg–New York, 1976;

[41] D. A. Hejhal: *The Selberg Trace Formula for* $PSL(2, \mathbb{R})$ vol. 2. Lecture Notes in Mathematics 1001, Springer-Verlag, Berlin–Heidelberg–New York, 1983;

[42] D. A. Hejhal, P. Sarnak, A. A. Terras (editors): *The Selberg Trace Formula and related Topics*. Contemp. Math. 53, American Mathematical Society, Providence, RI, 1993.

[43] A. Hurwitz: *Mathematische Werke,* 1 (*Funktionentheorie*). Editor: G. Pólya. Birkhäuser, Boston–Basel–Stuttgart, 1932; reprint 1962.

[44] A. Hurwitz, R. Courant: *Funktionentheorie*. Springer-Verlag, Berlin, 1964.

[45] R. A. Johnson: *Advanced Euclidean Geometry*. Dover, New York, 1960.

[46] F. Klein, R. Fricke : *Vorlesungen über die Theorie der elliptischen Modulfunktionen* I. Teubner, Leipzig, 1890;

[47] F. Klein, R. Fricke : *Vorlesungen über die Theorie der elliptischen Modulfunktionen* II. Teubner, Leipzig, 1892;

[48] I. Kra: *Automorhic Forms and Kleinian Groups*. Benjamin, Reading, MA, 1972.

[49] I. Kra and B. Maskit (editors): *Riemann Surfaces and related Topics*. Ann. of Math. Stud. 97, Princeton University Press, Princeton, NJ, 1981.

[50] S. Lang: *Introduction to Modular Forms*. Grundlehren Math. Wiss. 222, Springer-Verlag, Berlin–Heidelberg–New York, 1976.

[51] S. Lauritzen: *En Indledning til en gruppeteoretisk Behandling af de ikke orienterbare Flader*. Jul. Gjellerup, København, 1942.

[52] J. Lehner: *Discontinuous Groups and Automorphic Functions*. American Mathematical Society, Providence, RI, 1964.

[53] J. Lehner: *A Short Course in Automorphic Functions*. Holt, Rinehart and Winston, New York, 1966.

[54] O. Lehto: *Univalent Functions and Teichmüller spaces*. Grad. Texts in Math. 109, Springer-Verlag, Berlin–Heidelberg–New York, 1988.

[55] H. Lenz: *Nichteuklidische Geometrie*. Hochschultaschenbücher 123/123a, Bibliographisches Institut, Mannheim, 1967.

[56] H. Liebmann: *Nichteuklidische Geometrie*. G. J. Göschen, Berlin, 1912.

[57] R. C. Lyndon, P. E. Schupp: *Combinatorial Group Theory*. Ergeb. Math. Grenzgeb. 89, Springer-Verlag, Berlin, 1977.

[58] W. Magnus: *Noneuclidean Tesselations and their Groups*. Academic Press, New York–London, 1974.

[59] W. Magnus: *Discrete Groups and Automorphic Functions*. Academic Press, New York, 1977.

[60] W. Magnus, A. Karass, D. Solitar: *Combinatorial Group Theory*. John Wiley-Interscience, New York–London–Sidney, 1966.

[61] B. Maskit (editor): *Proceedings of the Conference on Quasi–conformal Mappings, Moduli, and Discontinuous Groups*. Tulane University, 1965.

[62] B. Maskit: *Kleinian Groups*. Grundlehren Math. Wiss. 287, Springer-Verlag, Berlin–Heidelberg–New York–London–Paris–Tokyo, 1988.

[63] K. Matsuzaki, M. Taniguchi: *Hyperbolic Manifolds and Kleinian Groups*. Oxford Math. Monogr. The Clarendon Press, Oxford University Press, New York, 1998.

[64] T. Miyake: *Modular Forms*. Springer, Berlin–Heidelberg–New York–London–Paris–Tokyo–Hong Kong, 1989.

[65] R. Narasimhan: *Compact Riemann Surfaces*. Lectures in Mathematics ETH Zürich, Birkhäuser, Boston–Basel–Stuttgart, 1992.

[66] J. Nielsen: *Collected Mathematical Papers*, 1–2. Editor: V. Lundsgaard Hansen. Birkhäuser, Boston–Basel–Stuttgart, 1986.

[67] O. Perron: *Nichteuklidische Elementargeometrie der Ebene*. Teubner, Stuttgart, 1962.

[68] H. Poincaré: Théorie des groupes Fuchsiens. *Acta Math.* 1 (1882), 1–62.

[69] H. Poincaré: Théorie des groupes Kleiniens. *Acta Math.* 3 (1883), 49–92.

[70] H. Poincaré: *Three Supplementary Essays on the Discovery of Fuchsian Functions* (French). Editors: J. J. Gray, S. A. Walter; Publications of the Henri Poincaré–Archives, 2. Akademie Verlag, Berlin; Albert Blanchard, Paris, 1997.

[71] R. A. Rankin: *Modular Forms and Functions*. Cambridge University Press, Cambridge, 1977.

[72] E. Reyssat: *Quelques aspects des surfaces de Riemann*. Progr. Math. 77, Birkhäuser, Boston–Basel–Stuttgart, 1989.

[73] B. Riemann: *Gesammelte Mathematische Werke*. Editor: H. Weber. Teubner, Leipzig, 1892; Dover, New York, 1953.

[74] L. Schlesinger: *Automorphe Funktionen*. Teubner, Leipzig, 1924.

[75] B. Schoeneberg: *Elliptic Modular Functions*. Grundlehren Math. Wiss. 203, Springer-Verlag, Berlin–Heidelberg–New York, 1974.

[76] J.-P. Serre: *A Course in Arithmetic*. Springer-Verlag, New York, 1996.

[77] G. Springer: *Introduction to Riemann Surfaces*. Addison–Wesley, Reading, MA, 1957.

[78] J. Stillwell: *Sources in Hyperbolic Geometry*. Hist. Math. 10, American Mathematical Society, Providence, RI; London Mathematical Society, London, 1996.

[79] K. Strebel: *Quadratic Differentials*. Ergeb. Math. Grenzgeb. (3) 5, Springer-Verlag, Berlin–Heidelberg–New York, 1984.

[80] O. Teichmüller: *Gesammelte Abhandlungen*. Editors: L. V. Ahlfors, F. W. Gehring. Springer-Verlag, Berlin–Heidelberg–New York, 1982.

[81] W. P. Thurston: *The Geometry and Topology of* 3-*Manifolds*. Princeton Univ. Xeroxed Notes, Princeton, NJ, 1977–78.

[82] W. P. Thurston: *Three-Dimensional Geometry and Topology*, Vol. 1. Princeton Math. Ser. 35, Princeton University Press, Princeton, NJ, 1997.

[83] H. Weyl: *Die Idee der Riemannschen Fläche*. Teubner, Stuttgart, 1913; 3. edition, Teubner, Stuttgart, 1955; reprint of 1913 edition, Teubner, Stuttgart, 1997.

[84] H. Zieschang: *Finite Groups of Mapping Classes of Surfaces*. Lecture Notes in Math. 875, Springer-Verlag, Berlin–Heidelberg–New York, 1981.

[85] H. Zieschang, E. Vogt and H.-D. Coldewey: *Surfaces and Planar Discontinuous Groups*. Lecture Notes in Math. 835, Springer-Verlag, Berlin, 1980.

[86] H. Zieschang, E. Vogt and H.-D. Coldewey: *Surfaces and Discontinuous Groups* (Russian). Nauka, Moscow, 1988.

Editor's note: In the main text of this book there are references in §24.9 to a paper by M. Dehn and in §26.5 to a paper by J. Nielsen (see also [66]). Special attention should be given to the article [30], which contains Werner Fenchel's own review on the development of various subjects of this book and stimulated by copies of the Fenchel–Nielsen manuscript in private circulation. This article also contains a valuable list of references.

Index